Integrated Circuit Fabrication

Master fundamental technologies for modern semiconductor integrated circuits with this definitive textbook.

Key features

- Early introduction of a state-of-the-art CMOS process flow exposes students to big-picture thinking from the outset, and encourages a practical integration mindset.
- Extensive use of process and TCAD simulation, using industry tools such as Silvaco Athena and Victory Process, provides students with deeper insight into physical principles, and prepares them for applying these tools in a real-world setting.
- Accessible framing assumes only a basic background in chemistry, physics and mathematics, providing a gentle introduction for students from a wide range of backgrounds.
- Over 450 color figures, and over 280 end-of-chapter problems, to support and cement student understanding.

Accompanied by lecture slides and solutions for instructors, this is the ideal introduction to semiconductor fabrication technologies for senior undergraduate and graduate students in electrical engineering, materials science, and physics, and for semiconductor engineering professionals seeking an authoritative introductory reference.

James D. Plummer is a Professor of Electrical Engineering and former Dean of Engineering at Stanford University. He has over thirty years of experience in teaching semiconductor fabrication and device physics, and has served on the Board of Directors of companies including Intel and Cadence. He is a Member of the National Academy of Engineering and the American Academy of Arts and Sciences, a Fellow of the Institute of Electrical and Electronics Engineers, and a co-author of *Silicon VLSI Technology – Fundamentals, Practice and Modeling* (Prentice Hall, 2000).

Peter B. Griffin is a Senior Research Engineer in Electrical Engineering at Stanford University, and a recognized expert on microfabrication. He has significant hands-on experience in building semiconductor structures, and teaching semiconductor technology courses, at the Stanford Nanofabrication Facility. He also is a co-author of *Silicon VLSI Technology – Fundamentals, Practice and Modeling* (Prentice Hall, 2000).

Integrated Circuit Fabrication
Science and Technology

James D. Plummer
Stanford University, California

Peter B. Griffin
Stanford University, California

 CAMBRIDGE
UNIVERSITY PRESS

Shaftesbury Road, Cambridge CB2 8EA, United Kingdom

One Liberty Plaza, 20th Floor, New York, NY 10006, USA

477 Williamstown Road, Port Melbourne, VIC 3207, Australia

314–321, 3rd Floor, Plot 3, Splendor Forum, Jasola District Centre, New Delhi – 110025, India

103 Penang Road, #05–06/07, Visioncrest Commercial, Singapore 238467

Cambridge University Press is part of Cambridge University Press & Assessment,
a department of the University of Cambridge.

We share the University's mission to contribute to society through the pursuit of
education, learning and research at the highest international levels of excellence.

www.cambridge.org
Information on this title: www.cambridge.org/highereducation/isbn/9781009303583

DOI: 10.1017/9781009303606

First published 2024

Printed in the United Kingdom by CPI Group Ltd, Croydon CR0 4YY

A catalogue record for this publication is available from the British Library.

Library of Congress Cataloging-in-Publication Data
Names: Plummer, James D., author. | Griffin, Peter B., author.
Title: Integrated circuit fabrication : science and technology / James D. Plummer, Stanford University, California,
Peter B. Griffin, Stanford University, California.
Description: Cambridge ; New York, NY, USA : Cambridge University Press, 2024.
Identifiers: LCCN 2023016901 | ISBN 9781009303583 (Hardback) | ISBN 9781009303606 (ebook)
Subjects: LCSH: Integrated circuits – Design and construction. | Semiconductors – Design and construction.
Classification: LCC TK7874 .P5895 2024 | DDC 621.3815–dc23/eng/20230705
LC record available at https://lccn.loc.gov/2023016901

ISBN 978-1-009-30358-3 Hardback

Additional resources for this publication at www.cambridge.org/plummer-griffin

Contents

Preface

Semiconductor devices have transformed modern life. Silicon integrated circuits provide the critical components for the computers and communications systems that power our information age, and are key components of virtually every modern system from cars to washing machines. Modern electronic systems are made possible by the ability to integrate millions or billions of individual devices on a common substrate.

Basic discoveries and inventions between 1945 and 1970 laid the foundations for these chips. In the past 50 years, silicon chip complexity has increased at an exponential rate, primarily because of the constant shrinking of device geometries, improved manufacturing practices and clever inventions enabling specific functions to be implemented more efficiently. Shrinking geometries permit more devices to be placed in a given area of silicon; improved manufacturing permits larger chips to be economically fabricated; and clever inventions permit functions to be realized in smaller areas. Silicon chips contained a few dozen components in 1970; today, the most complex chips contain more than 50 billion transistors.

For much of this period, increasingly complex chips were fabricated with a single layer of devices built in a silicon chip with up to 15 layers of metal interconnects above the devices to enable the transistors to be wired into complex circuits. Within the past decade, as lateral scaling of device dimensions has become more difficult, "three-dimensional integration" has become increasingly important as a means of continuing to increase circuit complexity without depending only on two-dimensional lateral scaling. Innovations in stacking chips, in packaging and in vertically integrating multiple layers of active devices have driven continued progress. It is widely expected that these trends will continue for at least another decade, resulting in chips that have hundreds of billions of transistors. Such chips will have extraordinary capabilities. As a point of reference, the human brain contains approximately 100 billion neurons, although the neurons are interconnected through hundreds of trillions of synapses in the brain.

Compound semiconductor devices are key elements in applications, in which their unique properties enable system performance unachievable by silicon. Many compound semiconductor materials efficiently convert electrical power to light, enabling light-emitting diodes (LEDs) and lasers, which are the core components of modern lighting systems and long-distance communication systems. Some compound semiconductors are useful in very-high-frequency applications, in which their high carrier mobilities are essential, while other compound materials are finding applications in power management because of their ability to withstand large electric fields. It is difficult to imagine how different our world would be without all these semiconductor devices.

The beginnings of the silicon chip industry 50 years ago involved simple, empirically determined processing steps and simple manually operated machines. Undergraduate teaching laboratories today are more advanced than those early factories. Fabrication steps like silicon oxidation were understood as simple chemical reactions ($Si + O_2 \rightarrow SiO_2$), and the growth of the SiO_2 was accomplished in inexpensive furnaces. Printed patterns on chips were tens of microns (μm, micrometers) in size, were easily visible in simple optical microscopes and were produced by simple contact printing methods.

Today, the most complex silicon chips are built in automated manufacturing facilities that cost more than $10 billion. These factories use machines that are among the most precise (and most expensive) tools found in any manufacturing process. The processing methods that are used in these factories have a scientific basis derived from physics and chemistry that today is understood at the level of individual atoms. The optical steppers that print microscopic patterns on wafers represent one of the most advanced applications of the principles of Fourier optics. Plasma etching involves some of the most complex chemistries used in manufacturing today. Ion implantation draws upon understanding from research in high-energy physics. Thin films on silicon wafers can be deposited today with single-atomic-layer control using complex chemical reactions that are self-limiting. Etching of those layers to pattern them can also be accomplished today with atomic-scale precision. And, of course, silicon devices themselves are approaching physical sizes at which molecular- and atomic-scale phenomena involving ideas from quantum mechanics are important.

One of the great challenges in integrated circuit manufacturing is the need to draw on scientific principles and engineering developments from such an extraordinarily wide range of disciplines. Scientists and engineers who work in this field need broad understanding and the ability to seek out, integrate and use ideas from many fields. Our goal in this book is not only to describe the processing methods used in semiconductor manufacturing but also to discuss the underlying scientific principles that make those processing methods work.

Often scientific knowledge is incorporated in a "model," which may be a mathematical equation describing a process or an atomistic picture of how a particular process works. Models codify knowledge and are an elegant way of expressing what is known. They also provide a way of exchanging ideas between researchers in a particular field. Finally, models can be tested experimentally to assess their predictive capability.

Today, models of semiconductor fabrication methods are embedded in computer simulation tools that can simulate the various technologies used in fabricating chips. Some simulation tools today use well-established scientific principles to predict experimental results. Optical lithography simulators, which are based on mathematical descriptions of Fourier optics, are a good example. Such tools can now accurately predict the image that will be printed in photoresist on a semiconductor wafer, given a particular mask design and a specific exposure system. Other simulation tools rest on less solid scientific ground. Models of dopant diffusion in silicon and compound semiconductors, for example, use ideas that are still debated in the scientific literature, and which are clearly incomplete in terms of describing all the physical phenomena involved. Nevertheless, these models are very useful today.

In most of the chapters in this book, models are discussed in the context of computer simulation programs that have incorporated those models and use them to simulate technology steps. We make extensive use of simulation examples to illustrate how technologies work and to help in visualizing features of the technologies that are not easily seen any other way. We have found these tools to be powerful teaching aids. In order to provide a consistent set of simulation examples throughout the book, we have chosen to use a suite of tools available from Silvaco, including Athena and Victory Process. These tools and others from companies such as Synopsys are widely used in the semiconductor industry today. They can result in very substantial cost savings in developing new generations of technology and are very useful in solving manufacturing problems.

This book is organized somewhat differently than other texts on this general topic, in two principal ways. The first is the extensive use of simulation examples throughout the text. These serve several purposes. The first is simply to help explain the scientific principles involved in each chapter. Simulations help to illustrate things like the time evolution of a growing oxide layer, a diffusing dopant profile or a depositing thin film. They are also very useful in illustrating the effects of specific physical phenomena in a process step because it is straightforward in simulators to add or eliminate specific physical models. Simulators provide the only real way in which complex interactions between process steps can be illustrated and understood. Finally, students who spend their careers in this industry will certainly use these tools, and understanding their capabilities and limitations will be important in their future work.

The second way in which this book is organized differently is the discussion of a complete process flow early in the book (Chapter 2). While readers new to this field may not appreciate many of the complexities of a complementary metal-oxide–semiconductor (CMOS) silicon process before studying the later chapters, we have found that an early broad exposure to a complete chip manufacturing process is very helpful in establishing the context for the specific technologies discussed in later chapters. In teaching the material in this book, we usually cover the CMOS process in the first or second lecture somewhat superficially, and then return to the same topic in the last lecture, at which point the details can be more fully discussed. Returning to the complete process flow in the last lecture also serves as a very useful way to review much of the material covered during the course.

The material in this book can be covered in a one quarter senior/graduate-level course, which is how this material is taught at Stanford, although not all of the material in each chapter can be covered in one quarter. A semester-long course would provide more time to cover the full range of material in the book.

Follow-on courses to a basic integrated circuit fabrication course can make more extensive use of the simulation tools discussed in this text. We have not used the simulation tools described in this book for homework assignments or for lab assignments in connection with a first course in integrated circuit fabrication. We believe that the simulation examples are better used simply as teaching tools in such courses, to illustrate ideas and to clarify physical principles. But, in follow-on courses, hands-on experience with these simulation tools is easily possible and highly desirable. The computer tools we use in the book are commercially available, and the vendors of these tools are generally anxious to work with college or university instructors to make the tools available for teaching purposes.

This book began as an attempt to revise an earlier text by the same authors plus Mike Deal (J. D. Plummer, M. D. Deal and P. B. Griffin, *Silicon VLSI Technology – Fundamentals, Practice and Modeling*, Prentice Hall, 2000). However, as we began work on this book, we realized that so much has changed in the more than 20 years since that book was published, that a completely new book was called for. Thus, while the general outline of this book is similar to our earlier text, the material is essentially completely new. One of the many changes we have made compared to our previous book is to include significant sections in most chapters on compound semiconductor technologies. The core technologies used to build compound semiconductor devices and integrated circuits largely derive from silicon technology because of the much larger investments in research and development that the silicon industry permits. However, the application of these technologies to compound semiconductor manufacturing often involves interesting differences because of the different materials involved. Discussing compound semiconductor fabrication provides useful insights into the extendibility and limitations of the physical models developed for silicon.

Lecture slides based on the figures in the book that are suitable for a course being taught, based on this book, are available for instructors, as is a solutions manual for the end-of-chapter problems. These can be found on the Cambridge University Press website at www .cambridge.org/plummer-griffin. Our intention is to continue to update this book in the coming years. We also ask that individuals who use material from this book in their own teaching or research, acknowledge the source of the material appropriately.

The material in this book has been used in various draft versions in classes we have taught. The inputs and suggestions of students in those classes have made this a better book. For many years we have worked with an energetic group of Ph.D. students and faculty colleagues at Stanford who have helped to develop some of the models and software tools described in this book. We particularly acknowledge Professor Bob Dutton and former Ph.D. students, now Professor Mark Law and Dr. Conor Rafferty, who developed the Stanford process simulation program SUPREM. We would like to acknowledge inputs and suggestions from Dr. Heyward Robinson, Dr. Scott Luning and Dr. Conor Rafferty in their areas of expertise. Professor Chris Chidsey provided helpful inputs on atomic layer deposition. Dr. Jin (H. J.) Cho at Silvaco provided superb support for many of the advanced simulations. We would also like to specifically acknowledge the suggestions and contributions of Dr. Jim McVittie in the etching chapter and Dr. Ted Kamins in the deposition chapter. Both contributed their course notes and provided careful reviews of the chapters. We welcome comments or suggestions on this text by email to plummer@stanford.edu and griffin@stanford.edu.

1   Historical Perspective, Moore's Law and Future Technology Prospects

1.1   Introduction

Silicon integrated circuits (ICs) are pervasive in our world, and the global semiconductor industry today exceeds $500 billion in annual sales. The devices and chips this industry produces support global industries, including consumer electronics, transportation, avionics and many others, that collectively represent a major part of global markets. Devices and chips built with other semiconductor materials such as GaAs, SiC and GaN provide critical components for specific application areas, including high-frequency communications systems, solid-state lighting and power management. It is not incorrect to say that the technical foundation of our modern world is based on semiconductors. The critical role that chips play has led to global competition to design, fabricate and build into advanced systems these remarkable components. Their importance to our world is unlikely to change in the foreseeable future.

    Silicon ICs were invented over 60 years ago. They had very humble beginnings. In the 1960s it was possible to integrate, that is, fabricate on a common substrate, only a few hundred electronic components (transistors, resistors and capacitors primarily). In 1965, Gordon Moore, in a famous paper, observed that the number of components integrated on a silicon chip seemed to be doubling roughly every year [1]. Moore actually described his now famous "law" in terms of the component count that produced the minimum cost at any point in time, and he predicted that within a decade (1975) chips would have 65,000 components. In 1975 his prediction turned out to be correct, and Moore was asked to predict the future again. He did so, but slowed his doubling rate to roughly every two years. That remarkable prediction has held true for more than another 45 years, with today's silicon chips containing in excess of 10 billion transistors in a few square centimeters of silicon. Figure 1.1 illustrates the increasing complexity of silicon chips over more than 60 years of IC history.

1.2   Integrated Circuits and the Planar Process: Key Inventions that Made It All Possible

The devices that are used in today's ICs, primarily metal-oxide–semiconductor (MOS) and bipolar transistors, were invented long before the technologies were available to manufacture

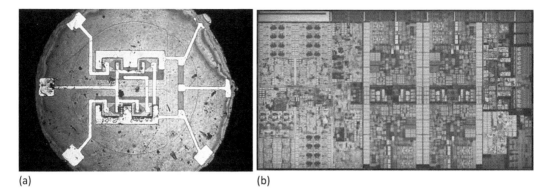

(a)   (b)

Figure 1.1 Photomicrographs (top views) of (a) the world's first integrated circuit from Fairchild Semiconductor (1959) and (b) a recent microprocessor (Intel). The chip in (a) is an RTL (resistor–transistor logic) flip-flop and contains four transistors and several resistors. It is about 1.5 mm in diameter. The chip in (b) contains several billion transistors and is about 2 × 4 cm. (a) Photo reprinted with permission from onsemi. (b) Photo copyright © Intel Corporation, reprinted with permission.

them in high volume. Many of the basic properties of semiconductors, such as rectification and photoconductivity, were discovered before 1900, although they were not understood at that time. Simple devices based on these properties were available in the form of selenium rectifiers and photodetectors by the mid-1930s. It was also during this time period when the physical principles underlying the behavior of metal–semiconductor contacts began to be understood, particularly through the work of Schottky and Mott.

The Austro-Hungarian American physicist Julius E. Lilienfeld filed a US patent in 1926 for a "Method and Apparatus for Controlling Electric Currents" in which he described a three-electrode amplifying device using copper sulfide semiconductor material. There is no evidence that he built or even tried to build a working amplifier. His patent, however, had sufficient resemblance to the later field effect transistor (FET) to deny future patent applications for that structure. The origins of the MOSFET thus pre-date the more celebrated invention of the bipolar transistor at Bell Telephone Labs. several decades later.

World War II interrupted much of the basic work on semiconductors that was ongoing at the time, particularly at Bell Telephone Labs., where an effort was underway to find a solid-state device to replace vacuum tubes for switching telephone signals. Shortly after the end of the War, this effort resumed, and it was not long until a major breakthrough was made with the demonstration of the point contact transistor in December 1947 (Figure 1.2). This and subsequent work, which resulted in the invention of the bipolar transistor, resulted in the Nobel Prize in Physics for John Bardeen, Walter Brattain and William Shockley in 1956. An interesting historical account of many of these events can be found in [2].

In the 1950s, interest in semiconductor surfaces was renewed when it became apparent that reliability problems associated with the new bipolar transistor structures were related to surface effects. In a classic experiment in 1953, Brattain and Bardeen found that the surface properties of semiconductors could be controlled by exposing them to oxygen, water or ozone ambients. Many experiments over the next few years led to the first high-quality SiO_2 layers grown on Si

(a)   (b)

Figure 1.2 The point contact transistor invented at Bell Telephone Labs. in 1947 by Bardeen, Brattain and Shockley. (a) The cover of *Electronics* magazine, September 1948. (b) The slab on the bottom right is polycrystalline Ge that forms the base region of the transistor. The triangular piece of plastic carries electrical connections down the two edges to make point contacts with the Ge. The third contact is on the backside of the Ge slab. (b) Photo copyright © Nokia Corporation and AT&T Archives, reprinted with permission.

substrates around 1960. At about this same time, the stable properties of SiO_2 layers on Si began to be understood theoretically.

Although the first point contact transistors in 1947 were built in polycrystalline germanium, within a year or two after that the device was also demonstrated in silicon and in single-crystal material. These two changes would also have a major impact on future ICs. Single crystals provide uniform and reproducible device characteristics when billions of identical components are integrated side-by-side on a chip. Silicon provides a controllable, stable and reproducible surface passivation layer, SiO_2, that has made possible modern IC technology. The properties of semiconductor/insulator interfaces are discussed in detail in Chapter 6. Much of the credit for developing single-crystal source material (Si and Ge) belongs to Gordon Teal of Bell Telephone Labs. Chapter 3 describes crystal growth and crystal properties in more detail.

By the mid-1950s, both grown junction and alloy junction bipolar transistors were commercially available. Germanium was still the dominant material used at that time. Figures 1.3 and 1.4 illustrate the methods used to fabricate these devices.

In the grown junction process, a single crystal of silicon or germanium was grown using a method such as the Czochralski technique described in Chapter 3. During the growth, the liquid silicon bath from which the crystal was pulled was doped with an N-type dopant initially. As the crystal was pulled, a P-type dopant was introduced into the liquid, followed by an N-type dopant, resulting in the NPN doping profile illustrated in

Figure 1.3 Grown junction technology of the 1950s. After [3].

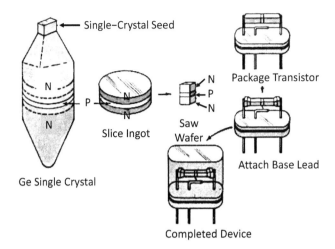

Figure 1.4 Bipolar transistor alloy junction technology of the 1950s, for (a) vertical and (b) lateral devices.

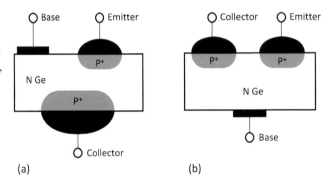

(a)   (b)

Figure 1.3. Sawing removed most of the upper and lower N-type material, leaving a thin wafer with an NPN structure suitable for bipolar transistors. Further sawing of this wafer into pieces resulted in a number of individual transistor structures to which leads were attached to provide external connections. In silicon-based devices, it was common to use Al wires to connect to the middle base (P) region since Al forms an ohmic (conducting) contact to P-type material and a rectifying (Schottky) contact to lightly doped N regions. As a result, it was not critical that the middle connection be placed exactly on the P region. In Ge-based structures of this type, PNP-type devices were usually made, with Au contacts to the middle N region.

The alloy junction technology illustrated in Figure 1.4 was even simpler. A metal such as indium was placed on the semiconductor (usually Ge). The structure was then heated, melting the In and allowing it to dissolve into the Ge. Indium is a P-type dopant, so P regions were formed, creating a PNP structure along with contacts to the P regions. As illustrated in Figure 1.4, the emitter and collector regions could be placed either on opposite sides of the Ge crystal, or on the same side.

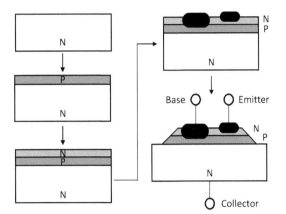

Figure 1.5 Bipolar transistor double-diffused mesa transistor technology of the late 1950s.

While such devices were very useful components, the technologies used to build them were not extendible to multi-transistor ICs. Exposed junctions were present on the semiconductor surface and no means to electrically isolate and interconnect multiple devices was provided. New inventions were needed to overcome these problems.

Part of the solution was provided by the invention of gas-phase diffusion processes, again at Bell Telephone Labs. That led to the commercial availability of diffused mesa bipolar transistors by 1957. Figure 1.5 illustrates this device structure. Beginning with an N-type crystal, the wafer was exposed in a high-temperature furnace to a gaseous source of a P-type dopant such as boron. The boron diffused into the crystal by solid-state diffusion, resulting in the P-type layer. The process was then repeated with an N-type gaseous source, producing the final NPN structure. After contacts were alloyed to the surface N and P regions, a silicon etch created the mesa structure to localize the N and P regions on the surface. In the structure shown, the base contact must make an ohmic contact to the P base region and a rectifying contact to the N emitter region, so that the N and P regions are not shorted together. This double-diffused process had the great advantage that multiple devices could be produced from a single substrate, and multiple substrates (wafers) could be produced from one grown N-type crystal. However, all of the processes described to this point still had the great disadvantage that exposed junctions were present on the wafer surface or at the wafer edges.

The next breakthrough came with the invention of the planar process, illustrated in Figure 1.6, by Jean Hoerni of Fairchild Semiconductor [4]. This process relied on the gas-phase diffusion of dopants to produce N- and P-type regions, but also on the ability of SiO_2 to block or mask these diffusions. This was a major advance, and it was largely responsible for the switch from Ge to Si. Silicon is unique among semiconductor materials in its ability to be oxidized to produce a stable insulating coating. SiO_2 has many properties that make it almost an ideal surface layer for semiconductor devices. We will discuss oxidation and SiO_2 in much more detail in Chapter 6, but one of those properties is the ability to block or mask most of the common dopants that are used in semiconductor fabrication. In other words, most dopants like As, B and P have much smaller rates of diffusion in SiO_2 than they do in silicon. The planar

Figure 1.6 The planar process invented by Jean Hoerni of Fairchild Semiconductor in the late 1950s.

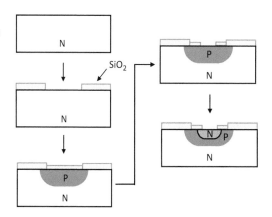

process makes use of this property by using SiO_2 to selectively block the dopants from diffusing into the silicon substrate. This produces junctions that terminate under an SiO_2 surface and which, as a result, are passivated. We will define "passivation" more carefully in Chapter 6, but we basically mean by this term that the junction electrical properties are not significantly degraded at the surface, compared to their bulk properties.

Today the doping method used to produce the P and N regions is ion implantation, a process we will discuss in detail in Chapter 8. This technology accelerates ions of the desired dopant, under vacuum, to energies of 10^3 to 10^6 eV and "shoots" them at the semiconductor wafer. The ions penetrate into the wafer and lose their energy by collisions with substrate atoms and through other processes that we will discuss in Chapter 8. The range of the ions intuitively should depend on their energy, and the concentration of the dopant ions in the substrate depends on the implanted dose. Not surprisingly, the energy transfer process from the implanted ions to the substrate can damage the semiconductor crystal. Damage repair is a critical topic that we will discuss in detail in Chapter 8. The oxide mask plays the same role in ion implantation that it played in gas-phase diffusion doping methods. The SiO_2, or another mask, must be thick enough to stop the implanted ions so they enter the substrate only in the areas where there is no mask.

A second essential part of the planar process, illustrated in Figure 1.7, is the ability to pattern the SiO_2. This is normally accomplished by a process called photolithography or often just lithography, which we will discuss in detail in Chapter 5. The essence of this process involves the use of a light-sensitive resist (or "film"), which is deposited onto the SiO_2 surface. The resist is then exposed through a mask and developed to produce a pattern in the resist. The photons that strike the resist during the exposure chemically change the resist so that, during the developing, the resist washes away in the regions where diffusions are desired, but remains in the regions where masking is desired. The resist is then used as a mask to selectively etch the SiO_2 layer. Modern lithography systems use 193 nm light to print features that are as small as approximately 25 nm. Such systems cost more than $50 million and are arguably among the most advanced optical systems in the world. As we will see in Chapter 5, the next phase of lithography tools use extreme ultraviolet (EUV) wavelengths (13.5 nm) and use

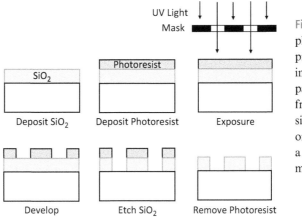

Figure 1.7 Basic photolithography process. The integrated circuit pattern is transferred from a mask to the silicon by printing it on the wafer using a light-sensitive resist material.

reflecting systems (mirrors) rather than today's refracting (lens-based) systems. These systems cost well in excess of $100 million and are beginning to be used in manufacturing the most advanced chips today.

One final key invention was needed to make modern IC technology possible. That was the ability to integrate multiple components on the same chip and to interconnect them to form a circuit. The invention of the IC occurred in 1959 and was due to Jack Kilby of Texas Instruments and Robert Noyce of Fairchild Semiconductor. An interesting historical account of these events is given in [5]. The key concept extended the ideas in Figures 1.6 and 1.7 to using masking and lithography to provide multiple devices and interconnects on the same chip. Kilby's IC is shown in Figure 1.8 and wired multiple components together to make an "integrated circuit." Noyce had the idea to use wires patterned on the chip surface above the devices built in the silicon.

Figure 1.9 illustrates these extensions for a simple circuit example. Kilby and Noyce have been widely recognized for their work leading to the IC. By combining P- and N-type diffusions and SiO_2 passivation layers, many types of devices, including transistors, resistors and capacitors, are possible in modern IC structures.

Although it is perhaps obvious from Figure 1.9, it should be noted that, when multiple diffusions and thin-film layers are used to construct an IC, the masks associated with each of the layers must be precisely aligned with respect to each other. That is, the N-type diffusion in the center NPN transistor must be placed inside the P-type region, and the contact holes that allow the Al metal to contact the device regions must be placed inside those regions. In general, the placement accuracy during photolithography must be on the order of one-fourth to one-third of the linewidth being printed. In today's modern factories, this alignment process is carried out automatically in the lithography tools that print the patterns on the wafer surface. As mentioned earlier in connection with Figure 1.7, these machines can print patterns as small as 25 nm today, which means that they must also be capable of positioning those patterns on the wafer surface to an accuracy of better than 10 nm. This is a remarkably precise capability that we will discuss in detail in Chapter 5.

Figure 1.8 First integrated circuit constructed by Jack Kilby in 1958. Photo courtesy of Texas Instruments.

Figure 1.9 Integrated circuits use photolithography and masking to fabricate multiple components in a common substrate.

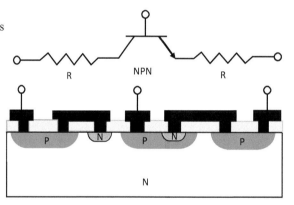

Since 1960, the principles on which IC manufacturing are based have not changed. The technologies implementing these principles have, of course, been significantly improved, and many new ways of depositing, etching, diffusing and patterning have been developed. These improvements have been more evolutionary than revolutionary, however. In spite of this fact, the cumulative impact of rapid evolutionary developments over the past 60 years in IC technology has been enormous.

Figure 1.10 shows actual cross-sections of a modern silicon IC. Figure 1.10(a) shows the wiring layers (typically Cu) that sit above the silicon wafer, insulated from each other by layers of SiO_2 or another dielectric layer. There may be as many as 15 such wiring layers in modern chips, allowing the billions of individual devices on the chip to be interconnected into specific circuit functions. The actual transistors, which are not visible in this photomicrograph, are shown at higher

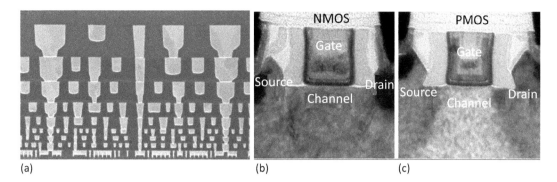

Figure 1.10 Cross-sections of Intel 32 nm technology. Image (a) shows the interconnect layers. The MOS transistors are on the bottom of this image but are not visible unless observed at much higher magnification, as shown in (b) and (c). The width of the gates in (b) and (c) and the width of the smallest metal lines (bottom of (a)) are on the order of 32 nm. (a) Photo © IEEE 2008, reprinted with permission from [6]. (b,c) Photos © IEEE 2008, reprinted with permission from [7].

magnification in Figure 1.10(b) and (c). These are modern MOSFET devices, and the technology is called complementary metal-oxide–semiconductor (CMOS) technology since both NMOS and PMOS devices are used. We discuss MOS transistors and CMOS circuits in more detail in the next section of this chapter. The technology illustrated in Figure 1.10 is similar to the technology we will describe in Chapter 2 as a means of illustrating in detail how modern silicon chips are built.

The inventions that led to modern silicon ICs also provide the foundation for devices and circuits made from other semiconductor materials. The high-frequency GaAs ICs used today in communication systems, the high-voltage SiC and GaN devices used in power control systems, the GaN light-emitting diodes (LEDs) that are widely used for solid-state lighting, and many other examples, all use lithography, etching, deposition, metal interconnects and the myriad other technologies developed for silicon chips. Of course, the specific ways in which these technologies are used may be different in different applications. Throughout this book, we will consider both silicon and other semiconductor materials in the fabrication methods we discuss.

1.3   MOS Transistors and CMOS Circuits

While the bipolar transistor was the switching element used in early digital ICs and also the gain element used in early analog ICs, by the late 1960s the bipolar junction transistor (BJT) was largely replaced by the MOS transistor, at least in digital circuits. The MOS device has a high-impedance input terminal, requiring no DC current for control, and is therefore much easier to design with in complex circuits. Just as important, once CMOS circuits became practicable in the late 1960s, it became possible to design digital circuits that consumed essentially no power unless they were actively switching from one logic state to another. Primarily for these reasons, CMOS devices have dominated digital silicon chips for the past 50 years, and it is very likely that this will continue to be the case in the coming decades. Since a major portion of this book deals with

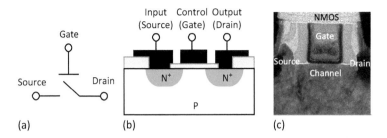

Figure 1.11 The MOS transistor. (a) Equivalent circuit, (b) cross-section and (c) actual device cross-section. (c) Photo © IEEE 2008, reprinted with permission from [7].

fabrication principles and processes needed to build CMOS devices, an overview of MOS transistors and CMOS logic is included in this section. We also include a brief discussion of bipolar transistors later in this section, because these were the earliest switching devices used in ICs and because bipolar devices still find application today in some silicon and compound semiconductor chips.

In virtually all digital and analog circuits today, three terminal switching devices are used, as shown schematically in Figure 1.11(a). Circuits are much easier to design if the control terminal is electrically isolated from the output. In digital circuits, the simplest representation of the switching element is simply a switch which is closed and opened by an isolated control terminal. The two states, closed and open, represent digital 0 and 1 states. In fact, semiconductor switches have a finite resistance in the closed or ON state, and, in analog circuits, the gate voltage is used to control the value of this resistance and hence the amount of current flowing through the switch.

As discussed, the dominant switching device used in digital silicon ICs today is the MOSFET shown in Figure 1.11(b). A photomicrograph of an actual MOSFET is shown in Figure 1.11(c) [7]. As we saw in Section 1.2, the FET was invented almost a hundred years ago (but not built then), but it was not until about 40 years later, in the 1960s, that MOSFETs were actually used in commercially available ICs. This delay was caused by manufacturing issues associated with controlling the device's electrical properties and in producing stable devices. We will discuss these issues in more detail in later chapters. The basic MOS device shown in Figure 1.11 is an NMOS or N-type device. The device is a three-terminal structure with an input (source), output (drain) and a control terminal (gate).

The basic operation of MOS devices is discussed in detail in all textbooks on semiconductor devices [8] and is illustrated in Figure 1.12. The MOSFET operates using the field effect principle, hence the name MOSFET for MOS field effect transistor. The N^+ regions at the source and drain of the device provide a means of contacting the region under the gate, which is the active part of the device. The N^+ regions also provide a source of electrons in normal device operation. Consider the central part of the device, under the gate electrode. The structure here is metal on top of an insulator (generally SiO_2) on top of silicon, and thus provides the name for the device (metal-oxide–semiconductor or MOS). The silicon is P type and hence there are very few electrons normally present. Both PN junctions in the device are normally at zero bias or are reverse-biased, which means that very little current flows across these junctions. Thus, with a negative or zero voltage on the gate, there is no connection between the input and output N^+

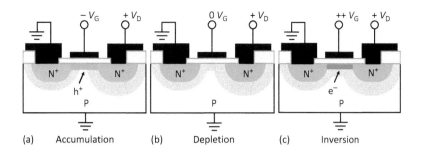

Figure 1.12 Simplified cross-section of a MOS transistor biased in its (a) OFF (accumulation), (b) inter-mediate (depletion) and (c) ON (inversion) states. The dotted areas are the depletion regions of the N^+P diodes. The gate vertical electric field controls the device and gives the device its FET "field effect transistor" name.

diffusions and the device operates as an open circuit. This condition is illustrated in Figure 1.12(a). With negative bias on the gate, majority carrier holes are attracted to the surface, which makes the surface more strongly P type. With two zero-biased or reverse-biased junctions in series, no current flows.

If we now apply zero volts or a slightly positive voltage to the gate electrode, the vertical electric field produced across the insulating gate oxide will attract electrons to the silicon surface and repel positively charged holes. Initially, this creates a depletion region in the central part of the device as the holes are driven away from the surface, as shown in Figure 1.12(b). If a sufficiently positive voltage is applied to the gate, the electrons can actually become the dominant carrier in a narrow layer at the surface, effectively making the surface N type, as shown in Figure 1.12(c). This process is called "surface inversion" and results in an inversion layer of electrons a few nanometers thick at the silicon surface. With this inversion layer present, the input and output N^+ diffusions are now effectively connected, resulting in a closed switch. The gate electrode thus allows the device to be turned on $(+V_G)$ or turned off (zero or $-V_G$). When the MOS transistor is turned on, the number of electrons in the inversion layer is proportional to the gate voltage. Thus the resistance of the inversion layer depends on the gate voltage, and therefore the current that flows in the device depends on the gate voltage.

The process of turning the MOS device on or off would actually be very slow if it were not for the presence of the N^+ regions. Because the substrate is P type, there are very few electrons present there and the only way for the substrate to provide the electrons needed for the inversion layer would be by actually generating them, which is a slow process. The only way for the substrate to get rid of the electrons in the inversion layer when the device is turned off would be by electron–hole recombination. These processes would limit the turn-on and turn-off times to milliseconds or seconds, depending on the rates of generation and recombination, and would certainly not allow the nanosecond or picosecond switching times that are common in today's ICs. The solution to this problem resides in the N^+ regions. These regions are rich in electrons and can provide a source or a sink for these carriers in the times required in modern circuits (picoseconds). The inversion layer electrons thus flow out of the source and drain when the device is switched on, and back into these regions when it is switched off.

The depletion regions indicated in Figure 1.12 (the gray areas) are the regions of the substrate in which there are essentially no mobile carriers. Beneath the gate, the holes in the P-type substrate are driven away by positive gate voltages. In the regions around the source and drain, the electric fields due to reverse bias on the junctions or due to the built-in junction fields that are present even with zero applied external bias, also drive the mobile carriers away from the immediate vicinity of the junctions.

The field effect principle described above for the MOSFET is used in many other semiconductor devices. When the gate electrode directly contacts the silicon (no SiO_2 layer in between), a Schottky diode can be formed. When the bias on this diode is changed, the diode depletion region under the gate expands or contracts, modulating the conductivity of the channel region. This device is called a metal–semiconductor field effect transistor (MESFET). When a P region is used on top of the channel, the same effect occurs through the changing reverse bias on the PN junction. This device is called a junction field effect transistor (JFET). Various other types of FET devices, such as modulation-doped field effect transistors (MODFETs) and high-electron-mobility transistors (HEMTs), are common in compound semiconductor technologies. We will discuss the technologies used to fabricate all of these devices in this book. For readers interested in more detail on the device physics issues for all these FETs, standard texts provide detailed discussions [8].

To fabricate the MOS transistor, we need techniques to dope the silicon N type (for the NMOS transistor shown here) or P type for PMOS devices. We will see that ion implantation followed by solid-state diffusion provides this capability. We also need the ability to deposit or grow very thin insulating layers to produce the gate dielectric. Thermal growth or deposition of SiO_2 provides this capability and can also provide an Si/SiO_2 interface which is almost perfect electrically. By this we mean that there are very few charges, electron or hole traps or other undesirable features of the interface that would degrade the operation of MOS devices. We also need to deposit and define thin films of polysilicon (and other materials) to build the gate electrode and thin metal films to provide a wiring capability on the chip. IC technology provides all these capabilities.

While the field effect mechanism to control semiconductor surfaces and convert them from P to N type or vice versa was known long before the invention of the IC, practical MOS devices were not manufactured until the mid-1960s. The principal reason for this was simply that attempts to build the devices earlier than this usually resulted in structures whose electrical characteristics were unstable or non-reproducible. Many laboratories investigated these problems during the 1960s, and the problem was finally traced to charges (defects) present at the Si/SiO_2 interface and to minute (parts per million) concentrations of alkali ions in the MOS gate dielectrics. Charges at the Si/SiO_2 interface are sensitive to fabrication conditions and were gradually brought under control and minimized by better fabrication methods. Ions like Na^+ and K^+ are normally charged and mobile in SiO_2 even at room temperature and can drift around in the gate dielectric as the gate voltage is switched under normal device operating conditions. Charges moving inside the gate dielectric act exactly like a changing gate voltage because they change the vertical electric field in the active part of the MOS device. The result is a MOS device whose current–voltage $(I - V)$ characteristics change as the alkali ions drift around.

Many readers may have already drawn the conclusion that, since the human body has large concentrations of Na^+ and K^+, the origin of the contamination problem may well have been handling of the IC wafers during device manufacturing. This is in fact correct, although other sources (chemicals and manufacturing equipment) were also identified. Strict attention to cleanliness in handling and processing the wafers largely solved the MOS stability problem by the end of the 1960s and made possible their large-scale manufacturing. Today's IC plants employ cleanrooms with highly controlled environments to prevent a recurrence of these MOS stability problems, but, even today, occasional MOS device instabilities arise because of improperly trained technicians or a batch of contaminated chemicals. We discuss these issues in Chapter 4.

Charges present in the insulator layer, or at the Si/SiO_2 interface, affect the underlying semiconductor surface properties simply because these charges provide termination points for the electric field lines originating on the gate in MOS devices. If no such charges are present, then applying a voltage on the gate means that the electric field lines must penetrate through the insulator and into the underlying semiconductor where they can then terminate on mobile carriers or on ionized impurity atoms (n, p, N_D^+ and N_A^-). Thus depletion, inversion and accumulation are possible. If large numbers of charges are present in the insulator or at the insulator/semiconductor interface, the field lines can terminate before they reach the underlying semiconductor and thus the field effect mechanism does not operate. Nature has provided us with only one superb insulator/semiconductor structure that we know of today, in which these charges are practically zero, the Si/SiO_2 interface. This is one of the main reasons why silicon is the material of choice for today's ICs.

Within the last decade, alternatives to the thermally grown Si/SiO_2 interface have been successfully put into manufacturing. These alternatives include "high-K" dielectrics in silicon CMOS devices (dielectrics like HfO_2) and deposited dielectrics on other semiconductors like Ge, GaN and GaAs. While none of these structures achieves the electrical perfection of the Si/SiO_2 interface, many of them have been improved to the point where they are manufacturable and are used in commercially available devices and ICs today. We will discuss many of these options in more detail later in this book, particularly in Chapter 6.

The earliest semiconductor devices that were manufactured in the 1950s were bipolar transistors, as discussed earlier. The basic operation of these devices is also discussed in detail in all semiconductor texts [8] and is illustrated in Figure 1.13.

In bipolar transistors, a direct connection is made to the P region separating the N^+ input and output terminals. Rather than using the field effect principle to reduce the potential barrier between the N^+ and P regions, the control terminal does this directly in bipolar transistors. When the potential barrier is reduced by applying a positive voltage to the P region, electrons are injected by the N^+ emitter into the P base region. Unlike the MOS device in which this injection is confined to the surface inversion layer, injection occurs across the entire N^+P junction in bipolar transistors. If the physical thickness (the base width) of the P region is small, these injected electrons can move by diffusion through the base before they recombine with the majority carrier holes. The injected electrons are then "collected" by the N-type output collector terminal, providing a current path between input and output. Unlike the MOS transistor in which no DC current needs to flow into the control terminal since an insulating

Figure 1.13 The bipolar transistor: (a) equivalent circuit and (b) cross-section.

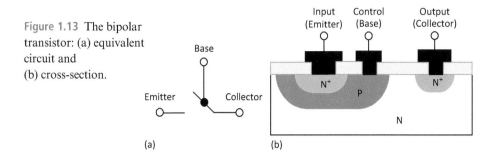

gate structure is used, current does flow into the control terminal of the bipolar transistor. This is because reducing the potential barrier between the N^+ emitter and P base also allows holes to be injected from the P base into the N^+ emitter and these holes must be supplied by the control terminal. The bipolar transistor is thus a current-controlled device in contrast to the voltage control used in MOS devices. In a well-designed bipolar transistor, the control or base current is much smaller than the main current flowing between the emitter and collector.

Because the control terminal in bipolar transistors is not electrically isolated from the main current path (input to output) in the device, bipolar transistor circuits are generally more complex to design than MOS circuits. More importantly, because there is no simple equivalent to the CMOS circuit topology (discussed next) in the bipolar world, power dissipation is a major issue in complex bipolar circuits. Primarily for these reasons, bipolar transistors have been largely replaced by MOS devices in silicon chips. However, they still find application in some compound semiconductor applications because they are capable of high speeds. In addition, the carrier injection mechanisms used in bipolar devices have been incorporated into some types of power switching devices such as insulated gate bipolar transistors (IGBTs) because they reduce the switch resistance below what is achievable with purely MOS structures.

As mentioned earlier, for 50 years, silicon ICs have used MOS devices in a CMOS circuit configuration to implement complex digital systems in silicon. Figure 1.14 illustrates the basic CMOS circuit arrangement in which stacked NMOS and PMOS devices are used to realize inverters, NAND and NOR gates and, by extension, much more complex circuits. The advantage of this arrangement can be seen in the simple inverter circuit in Figure 1.14(a). When the input is high (V_{DD}), the NMOS device is turned ON and the PMOS device is OFF. The NMOS switch connects the output to ground ("0"). When the input is low (ground), the PMOS device is ON and the NMOS device is OFF. The PMOS switch connects the output to V_{DD} ("1"). Thus an input 0 produces an output 1, and an input 1 produces an output 0 to realize the inverter function. Importantly, in both logic states, one or the other of the MOS devices is OFF while the other is ON. There is no path between V_{DD} and ground and hence no DC current flows. A slightly more complex logic circuit is shown in Figure 1.14(b). Only a 1 on In_1 and a 1 on In_2 produces a 0 at the output. All other input combinations produce a 1 output and hence this is a NAND gate. Again, very importantly, for all combinations of inputs, there is no DC current path between V_{DD} and ground and hence the circuit dissipates power only when it is switching between logic states.

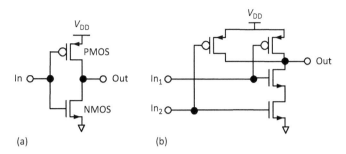

Figure 1.14 (a) A simple CMOS inverter, composed of PMOS and NMOS devices in series. (b) A CMOS NAND gate.

Figure 1.15 shows a cross-section of the basic NMOS and PMOS devices integrated in a modern CMOS process. This cross-section was actually produced by a technology computer-aided design program (TCAD simulator) in which the fabrication steps such as etching, deposition, lithography, etc. were inputs to the program. This example uses Athena and Victory Process, which are commercially available simulators from Silvaco [9] that we will use throughout this book to illustrate fabrication steps. In Chapter 2 we will go through the complete CMOS fabrication sequence that results in the structure in Figure 1.15, in order to introduce many of the topics to be covered in detail in this book. The color contours in the silicon represent the N- and P-type doping profiles. The MOS gate structures are in the centers of each transistor and the darker materials above the surface are the first level of metal wiring used to interconnect the transistors. This simulated cross-section can be compared with actual device images in Figure 1.10.

1.4 Moore's Law and MOS Device Scaling

As is apparent from Figure 1.1, the two dominant contributors to increasing chip complexity have been increasing chip size and decreasing component dimensions over time. The maximum size chip that can be economically manufactured is critically related to manufacturing defects. In most circuits, a single defect will make the circuit fail. Some chips, such as memory products, can build in some redundancy so that a small number of failures can be tolerated. Nevertheless, the manufacturing defect density is a critical parameter. We will discuss these issues in detail in Chapter 4. For much of the 60-year history of silicon ICs, maximum chip size increased at a rate of 10–20% per year, reflecting improving manufacturing practice over time. In recent years, chip size increase has slowed for a variety of reasons associated with the manufacturing technology, such as lithography (printing) limits, with the result that maximum chip size today is increasing at less than 10% per year.

The most important contributor to increasing chip complexity has been the ability to make smaller devices on a chip. This progress has historically been measured in terms of the "minimum feature size" or the smallest lateral feature built into a transistor structure. When MOSFETs were first introduced in commercially available ICs, device lateral dimensions were larger than 10 μm.

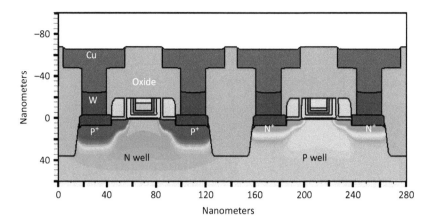

Figure 1.15 Cross-section of the basic devices in a CMOS integrated circuit, with a PMOS device on the left and an NMOS device on the right [9]. The 0 on the vertical scale corresponds to the original silicon wafer surface.

Figure 1.16 Ideal or Dennard scaling. The device on the right is scaled down in size by 2× (k = 2) [10].

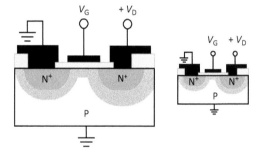

Today, minimum feature sizes are three orders of magnitude smaller (10 nm) and Moore's law is approaching 60 years old. One of the first attempts to determine how long scaling might continue was published in a seminal paper in 1974 [10]. Dennard or "ideal" scaling provided a path forward for several decades. The idea behind ideal scaling was actually quite straightforward. Figure 1.16 illustrates the basic idea, with the scaling factors being listed in Table 1.1.

The MOS transistor in the center is scaled down in size by 2×. Dennard and his colleagues reasoned that, to maintain similar device operation in the scaled device, the electric fields in the smaller device should be maintained constant. This means that all dimensions (vertical and horizontal) scale by 2×, doping concentrations increase by 2× and voltages and currents scale down by 2×. The result is that the transistor gets smaller, faster and dissipates less power. If $k = 2$, a chip fabricated in the new technology can incorporate four times as many transistors in the same area, and the total power consumed by those transistors is the same in the more complex chip as it was in the older chip. This seems like a perfect combination, and for several decades it worked.

Ideal or constant electric field scaling provided a predictable roadmap for the silicon chip industry. In 1994 the Semiconductor Industry Association (SIA) organized a workshop in

Table 1.1 **Scaling factors for Dennard scaling.**

Device or circuit parameter	Scaling factor
Device dimension, t_{ox}, L, W	$1/k$
Doping concentration	k
Voltage, V	$1/k$
Current, I	$1/k$
Capacitance, $\varepsilon A/t_{ox}$	$1/k$
Delay time, VC/I	$1/k$
Power dissipation, VI	$1/k^2$
Power density, VI/WL	1

Table 1.2 **Selected data taken from the 2003 version of the ITRS.**

Year of production	2000	2002	2004	2007	2010	2013	2016	2018
Technology node (nm)	180	130	90	65	45	32	22	14
Microprocessor gate length	100	70	53	35	25	18	13	10
DRAM bits/chip (GB)	0.512	1	4	16	32	64	128	256
MPU transistors (10^6)			550	1100	2200	4400	8800	14,000
Supply voltage (V)	1.5–1.8	1.2–1.5	0.9–1.2	0.8–1.1	0.7–1.0	0.6–0.9	0.5–0.8	0.5–0.7

Boulder, Colorado, which was charged with formalizing this roadmap. The result was the National Technology Roadmap for Semiconductors or NTRS [11]. This roadmap was updated periodically through a series of industry-wide meetings. By the end of the 1990s, the NTRS became an international roadmap, involving silicon chip companies worldwide and became known as the ITRS (International Technology Roadmap for Semiconductors) [12]. The ITRS became the most widely cited "strategic plan" for the semiconductor industry. Its fundamental premise was that industry progress, based on CMOS technology, should be described by Moore's law maintaining a 2–3 year cadence for increasing device density by 2× with each successive technology generation, that is, decreasing lateral dimensions by 0.7× in each generation. Table 1.2 shows a portion of the 2003 version of the ITRS. Note that all of the table columns beyond 2003 were future projections at the time this was published, providing a consensus roadmap for the companies in the silicon chip industry to try to achieve.

Until shortly after the year 2000, the technology node name in the ITRS was roughly synonymous with the minimum feature size lithographically printed in the technology. Thus the 180 nm node meant that the minimum printed feature size was on the order of 180 nm. Each node or generation represented a 0.7× decrease in minimum feature size. Note that, in each of these technology generations, the microprocessor gate length specified is actually smaller than the minimum feature size. This was done because this is the single most important dimension

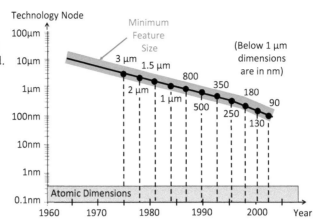

Figure 1.17 Technology nodes versus time during the "ideal" scaling period. The red-shaded region corresponds to the minimum feature size printed with lithography in each generation.

determining the performance of MOS transistors. In Chapter 5 we will discuss in depth the various lithography "tricks" that were and are used to print these very small features.

Because of the existence of the ITRS, it was very common for the semiconductor industry to plot its progress with plots like Figure 1.17. The data points in this figure show roughly a three-year cadence prior to 1995 and roughly a two-year cadence after 1995. These data points represent the period of "ideal" or Dennard scaling. The shaded area represents the minimum feature size associated with each technology generation. This closely tracks the technology node name in this time period, reflecting the fact that ideal scaling achieves density improvements through scaling lateral dimensions.

Although each new generation of technology required major investments in research and development, in retrospect this period of scaling was straightforward. Improved lithography methods allowed smaller features to be printed over time. The basic CMOS device structure remained unchanged. Density increased by 2× every 2–3 years, circuits ran faster in each new generation, and power dissipation in chips remained relatively constant.

What stopped ideal scaling 15–20 years ago can be understood from a more careful look at the numbers in Table 1.1. The transition of a MOS transistor from the OFF state to the ON state can be no steeper than the rate at which a PN diode can be turned on or off. As discussed in connection with Figure 1.12, the electrons for the inversion layer in the transistor come from the N^+ source region and there is a potential barrier between this region and the P-type substrate. This potential barrier is pushed down by the applied gate voltage, allowing electrons from the N^+ region to spill into the channel region to form the inversion layer. The rate at which this potential barrier can be reduced by the gate voltage is limited by thermionic emission and is derived in standard texts on semiconductor devices [8]. The resulting current–voltage characteristic of the transistor is shown in Figure 1.18 (discussed shortly). In the low-current portion, the slope is

$$\text{inverse slope} = n\left(\frac{kT}{q}\right)\ln(10) = n\left(\frac{60\text{ mV}}{\text{decade current }(I)}\right) \quad \text{where } n \geq 1. \tag{1.1}$$

This means that, in order to change the current flowing through a MOS device by 10×, the gate voltage must change by at least 60 mV. Practical circuits require an ON/OFF current ratio of 10^6 or larger, which means that the gate voltage must swing at least 6×(60 mV) = 360 mV. Actual MOS devices usually do not achieve the 60 mV/decade current lower limit ($n > 1$) because the field effect mechanism used in MOS devices to decrease the N^+P potential barrier acts indirectly on this barrier across the gate insulator between the barrier and the gate. Bipolar transistors do achieve the 60 mV/decade current limit ($n = 1$) because the control terminal directly contacts the P region and all the applied control voltage is effective in reducing the N^+P barrier potential.

In addition, noise margins and other practical issues imply that V_G in MOS devices needs to swing more like 0.5–0.7 V between the two states. Thus it is very difficult to reduce the circuit supply voltage in CMOS circuits below this range. Once ideal scaling reduced the supply voltage to the order of a volt, which happened around the 90 nm generation soon after the year 2000, ideal scaling was finished. The supply voltage numbers in Table 1.1 clearly show this effect. The ITRS projected dimensions continuing to decrease, but operating voltages remaining roughly constant.

1.5 Beyond Ideal Scaling: Materials and Device Innovations

The consequences of not being able to continue scaling supply voltage are profound. Since the 90 nm generation, dimensional scaling has continued but, because voltages have not scaled as much, electric fields have increased in devices, resulting in the need for device and material innovations to continue scaling. The physics is simple. The average electric field is simply $\mathcal{E} = V/L$, where V is the voltage and L is the distance over which that voltage is dropped. If L decreases and V does not, the field increases, resulting in increased power dissipation, reliability issues and many other concerns. These consequences and the device and material innovations that have been adopted to deal with them can be better understood by considering Figure 1.18.

All semiconductor device texts derive these $I-V$ characteristics and discuss them in detail (see [8], for example). The vertical axis is the current flowing through the MOS channel region (the switch current). The horizontal axis is the applied gate (control) voltage. Equation (1.1) describes the straight-line portion for V_G less than ≈ 0.45 V, the threshold voltage. In this region, the gate voltage is reducing the N^+P source–channel potential barrier and the current increases at $\leq 10\times$ per 60 mV of gate voltage. Once the potential barrier is reduced to ≈ 0, the transistor is considered turned ON. Further increases in V_G increase the current but at a much slower rate. In this region above V_{TH}, $I_{DS} \propto V_{GS}^2$ in larger MOS transistors and $I_{DS} \propto V_{GS}$ in small MOS transistors in this higher-current region. Here V_{TH} is the threshold voltage or "turn-on" voltage of the transistor.

In the example in Figure 1.18, the supply voltage (V_{DD}) is assumed to be 1 V, so in a CMOS circuit the logic "1" and logic "0" states correspond to the two black dots. I_{ON} and I_{OFF} are the

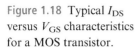

Figure 1.18 Typical I_{DS} versus V_{GS} characteristics for a MOS transistor.

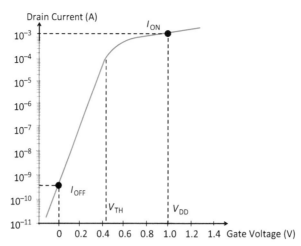

two critical parameters. I_{ON} is critical in determining circuit speed, since the charging and discharging times of circuit capacitances are proportional to I_{ON}. I_{OFF} is critical to circuit power dissipation, since the leakage current that flows through OFF transistors is simply I_{OFF}. Note that if each transistor in a 10-billion-transistor circuit had an I_{OFF} current as shown in Figure 1.18, the chip power dissipation when no switching was occurring would be (10 billion)$(3.5 \times 10^{-10}$ A$)(1$ V$) = 3.5$ watts. Continued scaling beyond the "ideal scaling" era was therefore aimed at making I_{ON}/I_{OFF} as large as possible. We will discuss some of the device and materials innovations aimed at doing this shortly, but first we consider some of the system implications of the end of ideal scaling.

Minimum feature sizes are an order of magnitude smaller today than they were when ideal scaling ended. The increased device density made possible by this scaling has been used to build more complex systems on a chip, as illustrated in Figure 1.19 for microprocessors [13].

But note in Figure 1.19 that by 2005 both clock frequency and power dissipation had saturated in microprocessor designs. Since dynamic (switching) power dissipation is given by $P = CV^2f$, where C is the capacitance being driven, V is the supply voltage and f is the frequency, the inability to continue scaling V means that power increases with frequency. A practical limit for cooling a chip is 100 W and hence both P and f have saturated. The number of transistors given by Moore's law has continued to increase, with the latest data points approaching 100 billion. The consequence of P and f saturating, however, means that microprocessor designers must now extract improved performance through parallel hardware (multiple processor cores per chip and more memory on chip) and through architectural changes (pipelining, executing multiple instructions per clock cycle, speculative execution, etc.).

Today, the system benefits obtainable through these methods are also beginning to diminish, with the result that many system architects are resorting to designing special-purpose chips with architectures optimized to solve specific problems with higher performance, rather than using general-purpose microprocessors that can solve many different problems through software programming. One consequence of this trend is that many companies which historically have

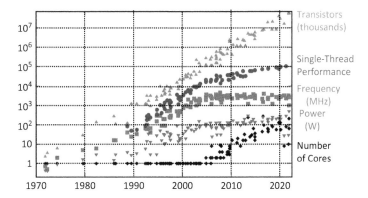

Figure 1.19 Microprocessor trends over time showing saturation of power dissipation and clock frequency but continued device density improvements. After Rupp [13].

simply purchased general-purpose microprocessors, now have in-house chip design teams that craft designs specifically for the company's products.

The scaling era beginning roughly with the 90 nm generation is best characterized as an era of materials and device innovation. Ideal scaling was not possible because supply voltages could not be scaled, so electric fields were increasing. How did device and technology designers deal with this? Figure 1.20 illustrates some of the most important innovations.

More complex doping profiles were introduced in device structures (lightly doped drain (LDD) profiles) designed to reduce peak electric fields. The year 2003 saw the introduction of compressive and tensile strain in MOS devices to improve carrier mobilities. This improves I_{ON} since MOS transistor currents are proportional to mobility. Later, 2007 saw the introduction of new dielectric materials ("high-K" to replace SiO_2) and metal gates to replace polysilicon. This allowed the effective thickness of the gate dielectric to continue to scale, helping to keep the transistor subthreshold slope close to 60 mV/decade I. The year 2011 saw the introduction of 3D device structures ("trigate" or "FinFET" devices), to mitigate the effects of shorter channel lengths in MOS devices. In essence the FinFET geometry allows the gate to better control the channel potential, which allows smaller transistors to continue to operate well even with increasing electric fields because V_{DD} has not scaled.

We will discuss these technology and device structure changes in detail in later chapters of this book, but, in many ways, the last 15 years have seen more innovation in silicon technology than occurred in the previous several decades. All of this was driven by the fact that ideal scaling was not possible and the economic factors pushing technology advances drove creative solutions to allow continued shrinking in devices.

There is one additional very interesting observation that should be made about this era of materials and device innovation. Figure 1.21 extends the data in Figure 1.17 to the present time and into the future. At the time of writing, the most advanced technology is labeled on the figure as "5 nm"; the "3 nm" technology is scheduled to be available in 2022. Note in Figure 1.21 that the minimum feature size actually printed in recent generations of technology

2003   2005   2007   2009   2011   2014
90 nm   65 nm   45 nm   32 nm   22 nm   14 nm

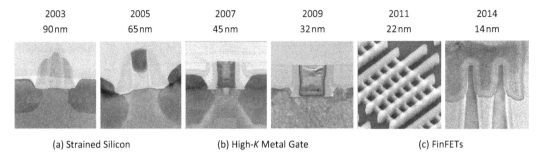

(a) Strained Silicon   (b) High-*K* Metal Gate   (c) FinFETs

Figure 1.20 Material and device structure innovations since ideal scaling ended. These examples are taken from Intel technology generations. Other manufacturers have followed a similar path. Copyright 2017 IEEE. Reprinted with permission from [14].

Figure 1.21 Technology nodes versus time through the current 5 nm generation. The red dots are expected technology generations in the next decade. The red-shaded region corresponds to the minimum feature size printed with lithography in each generation.

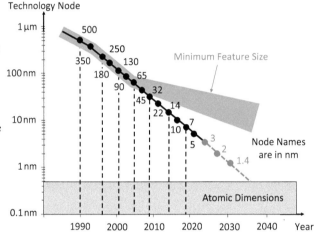

is a significantly bigger number (shaded area) than the technology node name. The red-shaded area is shown as a band because different manufacturers have taken individual approaches to new generations.

This observation also means that the current node names used in the semiconductor industry have little physical meaning, since they are not connected to physical dimensions on the chip. So node names are merely marketing tools and they generally do not even have the same meaning when two companies both advertise "5 nm" technology, for example. If "3 nm" technology actually meant that printed features were 3 nm, then it is obvious from Figure 1.21 that scaling and the density improvements it provides would have to end fairly soon, simply because we are approaching atomic dimensions ($\approx$ 0.3 nm). The devices we use today (MOS transistors) simply do not work at those dimensions.

The ambiguity about silicon technology node names has led to a number of proposals recently to change the node naming system [15]. Whether this will actually happen or not in an industry-wide standard remains to be seen, but, for now, the traditional naming system remains in common use and we will use it in this book, subject to all the limitations apparent in Figure 1.21. Interestingly, very recently Intel announced that they were changing the system they use to name their technology generations [16]. Intel's new naming system is 10 nm, I7, I4, I3, I20A, I18A, ... They are thus dropping the nm nomenclature, although the names they have chosen roughly correspond in technology generations to the 10 nm, 7 nm, 4 nm, 3 nm, 2 nm and 1.8 nm names still used by TSMC and Samsung. The A designation in Intel's naming system is meant to represent ångströms ($10 \text{ Å} = 1$ nm). In any case, the reader is cautioned to interpret these naming systems from all manufacturers as primarily marketing tools and not to infer very much about actual printed feature sizes in the technology.

One might ask, with printed features not scaling as fast as they have historically, how is it possible that transistor density is continuing to increase at historical rates as the upper curve in Figure 1.19 indicates? The answer is that printed features are still decreasing and this provides some of the density improvement. Additionally, process technology improvements such as self-aligned contacts also contribute to increased device density. Finally, the density increases from node to node are simply not 2×, as has been the case historically, and so the upper curve in Figure 1.19 is beginning to bend over somewhat.

1.6   Future Prospects for Semiconductor Technology

When plots like Figure 1.21 began to suggest even 20 years ago that dimensional scaling of MOS transistors could not continue beyond nanometer dimensions, many research efforts began to search for the "next switch," a device that could be scaled to atomic dimensions or in some other way continue Moore's law. To date, no such device has emerged, although research continues in many laboratories around the world. So, for now, we will assume that the basic MOS switch and CMOS architecture will remain dominant as we consider future prospects for semiconductor technology.

As long as there is an economic driving force to increase device density, one should not underestimate the creativity of semiconductor engineers to solve scaling challenges. Gordon Moore predicted in 1965 that the optimum (lowest-cost) chip would have a transistor density that increased by 2× every year or two. His prediction has been correct for more than 55 years. While the scale of investment needed to achieve increasing device densities has continued to increase, the semiconductor industry's growing size has provided the resources to fund the research and development needed to sustain Moore's law. Today the question is how much longer can this continue? Modern silicon chip factory costs are approaching $10 billion. Only a small number of companies (Intel, TSMC and Samsung today) can afford the investments needed to create next-generation technologies. And the costs of designing and building chips in the latest technology are now extraordinarily large, limiting this technology to applications that have very large chip markets (computers, cell phones, data centers, etc.). Of course, emerging applications, such as autonomous vehicles, artificial intelligence, virtual reality and others, all require massive amounts of computing and could provide the economic incentive to continue the historic improvements in silicon technology.

At almost every point in the more than 60-year history of silicon technology, visibility into solutions for continued progress in scaling device dimensions was limited to 5–10 years. This reality led to many papers being published on a regular basis predicting "the end of Moore's law" within 10 years. All these predictions proved to be incorrect, of course, primarily because they underestimated the creativity of semiconductor engineers in solving difficult problems, and perhaps because they underestimated the economic driving forces for finding solutions to difficult scaling problems. Today the situation is no different in the sense that the end of Moore's law is being predicted within the next decade, but today this view is being embraced by many more people than has been the case historically, driven by plots like Figure 1.21.

We have seen that lateral device scaling has become more difficult because of the costs of advanced lithography tools and because traditional device structures (planar MOSFETs and FinFETs) are reaching their scaling limits. As a result, advanced process flows using techniques like self-aligned contacts and other process innovations that are used by TSMC, Samsung and Intel have become company-specific solutions to continued scaling.

Nevertheless, a general "roadmap" for how device structures may continue to evolve has become fairly clear in recent years. Figure 1.22 illustrates a general roadmap for CMOS device evolution over the next decade or so. In all of these sketches, the cross-section shows the channel region of the device with the current flowing into the page. The source and drain are not shown; they are into and out of the page. The FinFET structure on the left has been the mainstream technology for advanced CMOS technologies since 2011. This structure provides full gate control of three sides of the silicon MOSFET channel (the top and two sides of the fin). The gate, however, does not directly control the silicon beneath the fin and, as devices have continued to scale, this has become an issue because this provides a leakage current path between source and drain that is not well controlled by the gate.

The gate-all-around (GAA) device shown beside the FinFET is the next natural evolution in device structures, and there have been many experimental demonstrations of devices of this type

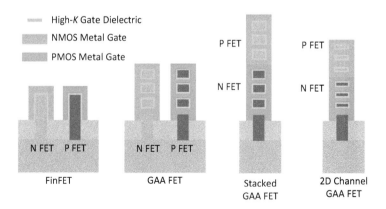

Figure 1.22 General roadmap for CMOS device structures beyond the FinFET. GAA is gate-all-around since the gate surrounds all four sides of the silicon wire or nanosheet.

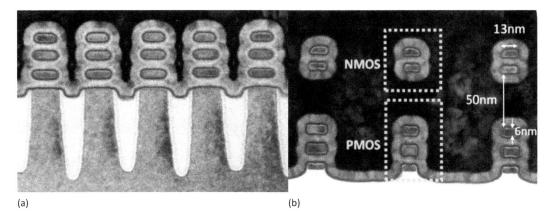

Figure 1.23 (a) GAA MOSFET technology reported by IBM at the 2017 VLSI Symposium. © 2017 Japan Society of Applied Physics. Reprinted with permission from [17]. (b) Stacked PMOS and NMOS GAA devices reported by Intel at the 2020 IEDM. © 2020 IEEE. Reprinted with permission from [18].

in universities and in semiconductor companies. The silicon fin is generally grown epitaxially as a layered structure with alternating Si and SiGe layers. Selective etchants that remove the SiGe material, but do not etch the Si, leave Si "wires" supported at the ends (into and out of the page). Conformal deposition of high-K gate dielectrics and metal gates with work functions appropriate for the P and N channel devices produce the structure shown in Figure 1.22. These devices are sometimes referred to as "nanosheet" transistors. Typical experimental devices with this architecture are shown in Figure 1.23(a).

The next natural evolution in device structure is the stacked GAA FET technology shown in Figure 1.22. In this technology, the NMOS and PMOS devices are vertically stacked, which obviously improves transistor density. These devices are generally fabricated using the same Si/SiGe stacked structure as described above for the GAA technology. However, it should be pointed out that the actual implementation of such a technology in manufacturing is enormously challenging. It is one thing to draw simple pictures like Figure 1.22, it is quite another to actually build billions of such devices at nanometer dimensions with high yield! An experimental example of a stacked GAA architecture is shown in Figure 1.23(b).

The final device architecture shown on the right in Figure 1.22 replaces the Si with a sheet semiconductor material. There is considerable research today on materials like $MoSe_2$, MoS_2 and literally hundreds of other similar materials that can be grown in thin sheets and which have semiconducting properties that might be interesting for CMOS applications. One reference that describes the range of such materials and some of their properties is [19]. These materials are often grown on separate substrates and then must be transferred to silicon wafers.

There have been many attempts over the years to replace silicon as the basic semiconductor used in ICs, none of which to date has been successful. Usually, both process complexity and marginal performance improvements have combined to allow silicon to maintain its dominance. So the nanosheet 2D GAA structure shown on the right in Figure 1.22 may not ever make its way to manufacturing, but it is a very important current research area.

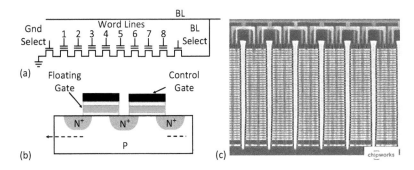

Figure 1.24 3D NAND structure showing (a) the circuit arrangement and (b) transistor floating gate structure. (c) Vertical stacking of the transistors in a Samsung 3D NAND chip cross-section. Photo from Chipworks (now TechInsights).

The device structures shown in Figure 1.22 achieve higher device densities even if lateral dimensions are not scaled because they use the third dimension. The first important commercial application of this approach has been the introduction of 3D NAND memory technologies within the past few years. Lateral scaling of NAND memories had reached a practical limit about a decade ago, as device dimensions were reduced below 20 nm, primarily because NAND device operation requires high voltages and smaller devices simply could not sustain the required electric fields.

The architecture of NAND chips basically places many MOS devices in series, as shown in Figure 1.24(a). The word line connections are numbered and BL is the bit line in the memory. Individual memory transistors have the structure shown in Figure 1.24(b). These are basically MOSFETs, but the gate structure is modified so a charge storage layer is inserted under the control gate. This charge storage layer either is usually a floating polysilicon gate or it may be a dielectric layer such as Si_3N_4 that can trap and hold charge. In operation, electrons from the substrate can quantum mechanically tunnel through the bottom SiO_2 layer and get trapped on the floating gate layer. This changes the threshold voltage of the transistor, which can be detected by the amount of current that flows through the transistor, providing a memory function.

Clever process technologies allowed such structures to be built by stacking the transistors vertically rather than horizontally, as shown in Figure 1.24(c). This fundamentally changed the density issue, since now increased device density could be achieved in the third dimension rather than in 2D. In 3D NAND technologies, dozens or hundreds of layers are vertically deposited on top of each other to form the stacked memory cells. Highly anisotropic etching is then used to etch vertical columns through the deposited layers. The etched trenches are then refilled, as shown in Figure 1.24. The control gates are contacted on the sides of the memory array (not shown, but in the dimension into the figure). These technologies provide very dense memory arrays but require very sophisticated etching and deposition technologies. Techniques for etching and depositing films with very anisotropic features will be discussed in Chapters 9 and 10.

Many companies and research laboratories today are pursuing 3D approaches for more general logic applications via approaches like wafer bonding and heterogeneous integration using

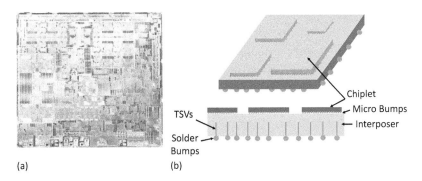

(a)   (b)

Figure 1.25 (a) System on chip (SOC); Apple M2 Max. Copyright © Apple Corporation, reprinted with permission. (b) Complex system realized by packaging multiple chips, each more specialized than the SOC, in a package. TSV = through-silicon via.

advanced packaging methods. In Figure 1.25(b), for example, the concept of chiplets and partitioning a complex system on chip (SOC) into smaller components is illustrated. This approach allows different technologies to be used for different parts of the system, which is often more economical than building the entire system on a single chip, which requires using a technology node for the whole system set by the most advanced part of the system. The chiplet approach obviously also allows different kinds of chips such as GaN or SiC power devices or optical components made from III–V semiconductors to be combined in close proximity. The SOC chip shown in Figure 1.25(a) is the Apple M2 chip, which contains about 20 billion transistors.

These concepts are beginning to be used in commercial products. In some cases chiplets can be stacked on top of each other and interconnected through dense vertical connections. This provides a pathway to continue Moore's law without the need for aggressive lateral scaling of device dimensions. Of course, continuing to increase device density using these 3D methods also creates challenging problems with power dissipation, which must be dealt with through sophisticated heat removal methods. These advanced packaging methods are described in more detail in Chapter 11.

In addition to using 3D methods to implement more complex systems, there is another view of the future of semiconductor technology that is equally interesting. This view of the future looks upon silicon technology as a "sandbox," a set of tools to build amazingly complex things. There have already been many examples of this, with silicon technology being used to build sensors of many kinds (accelerometers, pressure sensors, humidity sensors, etc.) that have found their way into many products that we use daily. In many universities today that have silicon chip research facilities, most of the faculty and students who use these facilities are working on projects that are not aimed at advancing silicon chips. Rather, the "technology sandbox" laboratory is being used to explore many other applications which benefit from the advanced fabrication tools that silicon technology provides.

Silicon technology has become the foundation for building other types of semiconductor devices. For example, wide-bandgap semiconductors like GaN and SiC show promise today for revolutionizing the power management market, with applications in electric vehicles, motor control, renewable energy systems and electric grid management. Biotechnology applications of silicon technology are equally interesting. Silicon-developed etching and deposition technologies can be used to create microfluidic systems that are useful in DNA sequencing, in blood analysis and in drug delivery systems. Silicon technologies applied to polymer materials can create wearable electronics for medical diagnostic systems. The list is really endless and in many ways is limited only by the creative energies of people looking for ways to make all of our lives better.

So the thrust of this book is only partly about Moore's law and silicon-based computing systems. It is also about exploring the new things that this amazing technology makes possible in other application areas. Today, students who study silicon technology, and who read books such as this one, come from many disciplines beyond the traditional electrical engineering and materials science disciplines that have played central roles in developing silicon technology. Engineering students, science majors, medical researchers and many others use silicon technology today because it enables breakthroughs in their disciplines.

One final point is important before we conclude this chapter. As the complexity of silicon chips has increased over time, so has the cost of designing these chips. Gordon Moore pointed out in 1979 [20] that the cost required to design an IC was increasing at about the same rate as the complexity of the technology. Transistor count was doubling in ICs every two years; design costs were not far behind. Had this continued, there would be very few ICs being designed today at the capability of the technology (billions of transistors), simply because it would be difficult to define products that could return the huge investment needed to design such chips.

Beginning in the late 1970s, however, a revolution took place, which dramatically reduced design costs. That revolution was the introduction of very powerful computer-aided design (CAD) tools, which take care of many of the design details. The industry that supports these tools today – the electronic design automation (EDA) industry – provides a powerful set of tools to design, debug, simulate and test complex chips. These tools are refreshed with each new technology generation. They have been extraordinarily successful in keeping chip design costs reasonable. Today, these tools are being extended to allow design and simulation of complex systems involving multiple chips, and chip design is being integrated with other important design elements such as thermal modeling, electromagnetic field modeling and fluid flow.

CAD tools also exist in the technology arena (TCAD tools or technology CAD tools). Such tools are very useful in designing new technologies, in creating new products based on the "sandbox" that silicon technology represents and in solving manufacturing problems in silicon chip manufacturing plants. We will make extensive use of these tools in this book because they provide insights into silicon process technology that are very difficult to obtain any other way. Figure 1.15 was an example.

1.7   The Plan for This Book

Modern silicon IC processes are built from a number of what are often called unit process steps. Examples include oxidation, ion implantation and lithography. Complete process flows are combinations of such steps, with often many hundreds of individual steps. Understanding such a complex fabrication sequence begins, of course, with understanding the individual steps.

Most books on silicon technology begin with a thorough discussion of each of the unit process steps. Only after this is done is the integration of these steps into an overall manufacturing sequence considered. Our experience is that considering the integration first has a significant advantage. That advantage is that the context for each of the unit processes is established first, i.e. the "big picture" is established. The disadvantage, of course, is that a reader unfamiliar with silicon technology may not fully appreciate a description of a complete process flow before the individual steps have been discussed in detail. We devote Chapter 2 to such a process flow (a relatively state-of-the-art silicon CMOS process) and recommend that it be read before the other chapters. Readers relatively new to silicon technology will likely not completely understand all the details of the CMOS process at first. A second reading of Chapter 2 after reading the later chapters in the book should bring together many of the ideas discussed in the later chapters.

Following the overview of CMOS technology in Chapter 2, individual chapters will then discuss in detail the unit process technologies (diffusion, ion implantation, etching, deposition, lithography, etc.) that silicon technology provides. Our goal in each of these discussions will be to try to understand the physical mechanisms involved – the chemistry, the physics and the materials science – in order to understand how these technologies work. We will consider how they were developed, what the capabilities are today and what the future may offer in terms of improved fabrication methods.

The history of scientific discovery has until quite recently usually been characterized by laboratory experiments followed by the development of physical models or theory. The invention of the IC described briefly in Section 1.2 is a good example. The basic contributions of Kilby, Noyce, Hoerni and many others were largely based on experiments done in laboratories, and on insight gained from those experiments. In fact, as we shall see throughout this book, the basic physical mechanisms controlling IC technology are in some cases still not fully understood. However, this lack of full scientific understanding has obviously not prevented widespread industrial application of the technology. The invention of the bipolar transistor described by Shockley [2] is another classic example of experimental science preceding the development of a complete theoretical understanding. In his historical description of those events in the 1940s and 1950s, Shockley wrote of the "creative failure methodology," by which he meant that experiments that produce unexpected or undesired results often lead scientists down new paths to invention. There certainly are examples in science of theoretical predictions preceding and suggesting experiments. However, the reverse has more often been the case, at least in areas of science relevant to this book.

With the remarkable developments in computer capability in the past several decades (due partly to IC technology) has come an opportunity to do experiments in a new way. In fields of science in which theories based on first principles are known, or in which models have been developed, it is quite common today to program those theories or those models in a computer. To the extent that such models are correct, it is then possible to do "computer experiments" or computer simulations. Chemicals can be mixed in a simulated beaker and the resulting reactions observed on a computer screen. Bridges can be constructed in a computer, with stresses placed on them to determine failure points and failure mechanisms.

In the design, synthesis, debugging and testing of complex ICs, computer simulation methods have been indispensable for several decades. The fact that most ICs, even if they have several billion transistors, work or mostly work the first time they are built, is testament to the power and accuracy of these CAD tools. No one today would think about designing a new chip without using an extensive suite of CAD tools to create the chip design and simulate its performance thoroughly before committing the design to fabrication.

In the IC device and technology worlds, such tools have become powerful adjuncts to real laboratory experiments in recent decades. The basic equations, which describe electron and hole transport in semiconductors due to drift and diffusion, have been incorporated into powerful simulators that can today accurately predict the electrical characteristics of proposed device structures. In a similar way, models describing the steps used in fabricating ICs have also been incorporated into process simulators. It is therefore quite possible today to "build" new semiconductor structures and predict their performance using these computer tools.

The state of the art in such simulators today is that they are very useful, but they cannot completely replace real laboratory experiments. This is simply because the models used in the simulators are not complete in some cases, or are purely empirical in other cases. As the models are improved with additional research, the simulators will become more robust and therefore more generally useful. There is great motivation to do this because real laboratory experiments are very expensive and very time-consuming, especially as chip technology continues to evolve. In an industry as competitive as the semiconductor industry, the time required to get new products to market is a key indicator of a company's success. To the extent that simulators can decrease this time and the development costs, they can improve competitiveness.

We will make extensive use of process or TCAD simulators in this book. This is not only because they are becoming increasingly important as industrial tools, but also because we believe they can greatly enhance the learning process. We believe they provide remarkable physical insight and intuition, and we will use them throughout the book to illustrate physical principles and to provide accurate illustrations of what real IC structures look like. These tools also provide the ability to "look inside" structures or devices and to observe physical phenomena in action that often cannot be easily observed in the laboratory. While a number of TCAD tools are commercially available, we have chosen to use the suite of tools provided by Silvaco in this book, primarily Athena and Victory Process [9], in order to provide a consistent set of examples throughout the book. These tools are generally available to the university and industry communities and can be used in connection with courses that are taught based on this book.

One final point should be made. Throughout this book we will assume that readers have a basic knowledge of the physics and mathematics of semiconductor materials. While this book does not require a deep understanding of band diagrams, doping in semiconductors and concepts like the Fermi level, some understanding of these concepts is required. Many readers will have studied these concepts in undergraduate courses. For readers who have not, and for those who might feel a little rusty on these concepts, Appendix 1 provides a discussion at the level necessary for this book. We encourage readers to at least skim that appendix because it will be helpful in more easily understanding concepts throughout this book. A number of the problems at the end of this chapter are based on the material in Appendix 1 for students who may wish to check their understanding of these concepts.

1.8   Summary of Key Ideas

Integrated circuits are one of the key components of today's world. They are pervasive in their application. Modern CAD tools have enabled a wide variety of people from many different disciplines to design and use ICs in many different kinds of applications. Historically "Moore's law" described the fact that every two years the complexity of ICs doubled; that is, the technology used to make them allowed twice as many components to be integrated on a single chip as was possible two years earlier. In addition, historically, the performance of the chips improved every year through scaling down of device sizes.

After more than 60 years of Moore's law, today we are seeing increasing difficulty in continuing this progress. In fact, "ideal" scaling ended 15–20 years ago. The challenges are increasingly fundamental as IC dimensions approach atomic sizes, but just as importantly they are economic. The cost of developing new generations of technology has become prohibitive except for a small number (three today) of companies. In addition, since ideal or "Dennard" scaling is no longer possible because operating voltages cannot be reduced at the same rate as dimensions, improvements in transistor performance and chip power consumption have become major concerns. Nevertheless, it is likely that the economic driving forces that have pushed silicon technology for six decades will continue to do so for at least another decade and likely longer.

Just as important as the applications of silicon technology in mainstream systems, like computers, communications, data centers, etc., are emerging applications in many areas. The enormous impact of and huge quantities of data provided today by sensors, cameras, social media and the like will drive high-performance computing needs for many decades to come. Autonomous vehicles are an interesting example, with their enormous data analysis requirements.

In addition to all these widely discussed applications, silicon technology has become a "sandbox" for engineers and scientists to use in attacking many important problems. The fabrication methods developed for silicon chips have been directly applied to many compound semiconductor devices and chips. Deposition and etching technologies have allowed complex sensors to be built. Biotechnology systems rely on microfluidic systems that are made using silicon-developed fabrication methods. New power semiconductor device technologies using wide-bandgap materials like SiC and GaN are also based on silicon-developed fabrication methods.

We will take the approach in this text that the applications of silicon technology are very broad and many are unexplored today. One of the very exciting opportunities for the future is to make the silicon "sandbox" available for many more innovative people to use. As we discuss unit process technologies, we will keep in mind that they are used for much more than producing an integrated CMOS fabrication process.

1.9 REFERENCES AND NOTES

[1] G. E. Moore, "Cramming more components onto integrated circuits," *Electronics*, vol. 38, no. 8 (April 19), 1965.

[2] W. Shockley, "The path to the conception of the junction transistor," *IEEE Transactions on Electron Devices*, vol. 31, no. 11, pp. 1523–1546, 1984.

[3] R. M. Ryder, "10 years of transistors," *Radio Electronics*, May 1958. See http://www.rfcafe.com/references/radio-electronics/ten-years-transistors-may-1958-radio-electronics.htm.

[4] J. A. Hoerni, "Method of manufacturing semiconductor devices," US Patent 3025589A, 1962.

[5] J. S. Kilby, "Invention of the integrated circuit," *IEEE Transactions on Electron Devices*, vol. 23, no. 7, pp. 648–654, 1976.

[6] S. Natarajan, M. Armstrong, M. Bost, *et al.*, "A 32nm logic technology featuring 2nd-generation high-k + metal-gate transistors, enhanced channel strain and 0.171μm^2 SRAM cell size in a 291Mb array," in *2008 IEEE International Electron Devices Meeting*, IEEE, pp. 1–3, 2008.

[7] P. Packan, S. Akbar, M. Armstrong, *et al.*, "High performance 32nm logic technology featuring 2nd generation high-k + metal gate transistors," in *2009 IEEE International Electron Devices Meeting*, IEEE, pp. 1–4, 2009.

[8] C. Hu, *Modern Semiconductor Devices for Integrated Circuits*, Prentice Hall, 2010.

[9] Athena and Victory Process are TCAD tools developed by Silvaco Inc. that provide a variety of 2D and 3D modeling capabilities.

[10] R. H. Dennard, F. H. Gaensslen, H.-N. Yu, V. L. Rideout, E. Bassous and A. R. LeBlanc, "Design of ion-implanted MOSFETs with very small physical dimensions," *IEEE Journal of Solid-State Circuits*, vol. 9, no. 5, pp. 256–268, 1974.

[11] "National Technology Roadmap for Semiconductors," (NTRS), Semiconductor Industry Association (SIA), 1994.

[12] "International Technology Roadmap for Semiconductors," (ITRS). See http://www.itrs2.net/itrs-reports.html.

[13] K. Rupp, "50 years of microprocessor trend data." Data and plot available under Creative Commons License. See https://github.com/karlrupp/microprocessor-trend-data.

[14] M. T. Bohr and I. A. Young, "CMOS scaling trends and beyond," *IEEE Micro*, vol. 37, no. 6 (November/December), pp. 20–29, 2017.

[15] H.-S. P. Wong, K. Akarvardar, D. Antoniadis, *et al.*, "A density metric for semiconductor technology," *Proceedings of the IEEE*, vol. 108, no. 4, pp. 478–482, 2020.

[16] P. Moorehead, "Intel updates IDM 2.0 strategy with new node naming and transistor and packaging technologies," *Forbes*, July 26, 2021. See https://www.forbes.com/sites/patrickmoorhead/2021/07/26/intel-updates-idm-20-strategy-with-new-node-naming-and-technologies/?sh=2d611e1e29d5.

[17] N. Loubet, T. Hook, P. Montanini, *et al.*, "Stacked nanosheet gate-all-around transistor to enable scaling beyond FinFET," in *2017 Symposium on VLSI Technology*, IEEE, pp. T230–T231, 2017,.

[18] C.-Y. Huang, G. Dewey, E. Mannebach, *et al.*, "3-D self-aligned stacked NMOS-on-PMOS nanoribbon transistors for continued Moore's law scaling," in *2020 IEEE International Electron Devices Meeting*, IEEE, pp. 20.6.1–20.6.4., 2020.

[19] F. Schwierz, J. Pezoldt and R. Granzner, "Two-dimensional materials and their prospects in transistor electronics," *Nanoscale*, vol. 7, no. 18, pp. 8261–8283, 2015.

[20] G. Moore, "Are we really ready for VLSI 2?," in *1979 IEEE International Solid-State Circuits Conference. Digest of Technical Papers*, vol. 22, IEEE, pp. 54–55., 1979.

1.10 PROBLEMS

A number of these problems rely on material covered in Appendix A.1 of this book – Basics of Semiconductor Materials in Equilibrium. The material covered in that appendix will be used through this text, so readers who are not familiar with the physics and mathematics of semiconductor materials should read that appendix.

1.1 Short-answer questions.
  (a) Is "intrinsic silicon" necessarily undoped? Why or why not?
  (b) For an N-type doped silicon, is the Fermi level position closer to the conduction band or valence band? What about P-type doped silicon? In each case, explain your answer in terms of the hole and electron population in the material.
  (c) Why does the intrinsic carrier concentration change with temperature?
  (d) Explain why, at the same temperature, $n_{i\,\text{GaAs}} < n_{i\,\text{Si}} < n_{i\,\text{Ge}}$.
  (e) Is the position of the Fermi level E_F of a semiconductor always located within the bandgap region? Why or why not?
  (f) Suppose silicon atoms are introduced as dopants in GaAs and replace Ga atoms in the lattice. In this case, are the silicon atoms donors or acceptors?
  (g) When looking at electron concentration versus energy, does the largest electron concentration occur at the edge of the conduction band? Explain.

1.2 Assuming that dopant atoms are uniformly distributed in a silicon crystal, find the distance between these dopant atoms when the doping concentration is
  (a) 10^{14} cm^{-3},
  (b) 10^{18} cm^{-3} and
  (c) 5×10^{20} cm^{-3}.
  (d) Suppose the channel region in a minimum geometry MOS transistor built in 10 nm technology is 10 nm (length) × 10 nm (width) × 10 nm (depth). If the channel region has a doping concentration of 10^{18} cm^{-3}, how many doping atoms on average are there in the channel region? Is this a problem?

1.3 In "10 nm" node technology, assume the "pitch" of minimum-size transistors is 50 nm and the area of a typical CPU chip is 1 cm × 1 cm. Roughly estimate how many minimum-area transistors there could be on such a CPU chip.

1.4 Estimate the resistivity of pure silicon in ohm cm at
  (a) room temperature,
  (b) 77 K and
  (c) 1000 °C.

  You may neglect the temperature dependence of the carrier mobility in making these estimates.

1.5 (a) Show that the minimum conductivity of a semiconductor sample occurs when

$$n = n_i \sqrt{\frac{\mu_p}{\mu_n}}.$$

  (b) What is the expression for the minimum conductivity?

(c) Is this value greatly different than the value calculated in problem 1.4 for the intrinsic conductivity?

1.6 When a gold (Au) atom sits on a lattice site in a silicon crystal, it can act as either a donor or an acceptor. E_D and E_A levels both exist for the Au and both are close to the middle of the silicon bandgap.

    (a) If a small concentration of Au is placed in an N-type silicon crystal, will the Au behave as a donor or an acceptor? Explain.

    (b) If the N region doping is 5×10^{17} cm^{-3} and the Au concentration is 10^{15} cm^{-3}, calculate the electron and hole concentrations in the material.

1.7 Show that E_F is approximately in the middle of the bandgap for intrinsic silicon.

1.8 Find the equilibrium electron and hole concentrations inside a uniformly doped sample of Si under the following conditions:

    (a) $T = 300$ K, $N_A = 10^{16}$ cm^{-3}, $N_D \ll N_A$,

    (b) $T = 300$ K, $N_A = 9 \times 10^{15}$ cm^{-3}, $N_D = 10^{16}$ cm^{-3},

    (c) $T = 450$ K, $N_A = 0$, $N_D = 10^{14}$ cm^{-3} and

    (d) $T = 650$ K, $N_A = 0$, $N_D = 10^{14}$ cm^{-3}.

1.9 Calculate the electron concentration, hole concentration and Fermi level for the following conditions:

    (a) a silicon sample doped with 10^{15} cm^{-3} boron at 300 K;

    (b) a silicon sample doped with both 10^{15} cm^{-3} of boron and 10^{14} cm^{-3} of phosphorus at 300 K; and

    (c) a silicon sample doped with 10^{17} cm^{-3} of boron at 1000 °C.

1.10 A silicon wafer has a phosphorus background doping level of 10^{15} cm^{-3}. Arsenic doping is introduced with a level of 10^{17} cm^{-3}. Calculate the Fermi level position before and after the arsenic doping, respectively, at room temperature. You may assume the Boltzmann approximation holds.

1.11 The Maxwell–Boltzmann distribution is often used to approximate the Fermi–Dirac distribution function. On the same set of axes, sketch both distributions as a function of $(E - E_F)/kT$. Consider only positive values of $E - E_F$. For what range of $(E - E_F)/kT$ is the Maxwell–Boltzmann approximation accurate to within 10%?

1.12 Construct a diagram similar to Figure A1.17 for P-type material. Explain physically, using this diagram, why the capture of the minority carrier electrons in P-type material is the rate-limiting step in recombination.

1.13 A silicon diode has doping concentrations on the N and P sides of $N_D = 1 \times 10^{19}$ cm^{-3} and of $N_A = 1 \times 10^{15}$ cm^{-3}.

    (a) Calculate the process temperature at which each of the two sides of the diode become intrinsic. (Intrinsic is defined as $n_i = N_D$ or N_A.)

    (b) If you wanted to have the diode operate in a high-temperature environment (around 150 °C), would you choose Si or Ge as the material? Why?

1.14 A silicon sample is doped with $N_A = 10^{17}$ cm^{-3} boron atoms.

(a) Find the hole and electron concentrations and the location of the Fermi level relative to the valence band edge at room temperature, $T = 300$ K, assuming full ionization of the boron atoms.

(b) Assuming that the acceptor level lies 50 meV above the valence band, check the assumption of full ionization of the acceptor states using your answer from part (a). You should find the probability that the acceptor state is not occupied by an electron.

1.15 Consider three materials with the acceptor concentration $N_A = 10^{17}$ cm^{-3} and following energies:

Material	E_A
Silicon	E_V + 50 meV
4H silicon carbide	E_V + 200 meV
Hexagonal GaN	E_V + 160 meV

(a) Assuming full ionization, calculate the electron and hole concentrations in the materials and the Fermi level positions (relative to E_C or E_V) at 300 K.

(b) Check the full ionization assumption using the calculated Fermi level and the acceptor levels in the table using (A1.15) in Appendix A1 and the Fermi levels you calculated in part (a). Calculate answers with and without degeneracy included. Which materials are fully ionized and which ones are not? Why are the numbers you calculated for percent ionization in SiC and GaN not actually correct?

(c) Repeat part (b), with and without including degeneracy, now using (A1.16)–(A1.18) in Appendix A1. Why should these answers be more accurate than the answers you calculated in part (b)?

(d) For each of the four methods you used in parts (b) and (c), plot the ionization levels from 30 K to 600 K for each material. Include any source code with your submission.

1.16 The rate of recombination in trap-assisted recombination (Shockley–Read–Hall) depends on the trap energy level. For an N-doped Si semiconductor with $N_D = 10^{15}$ cm^{-3} with excess carriers of $\Delta p = 10^9$ cm^{-3}, consider traps with density of 10^{13} cm^{-3} and capture cross-section of 10^{-15} cm^2. Take $T = 300$ K.

(a) Plot the net recombination rate for traps with energy (E_T) from the valence band (E_V) to the conduction band (E_C). Submit any code used for plotting.

(b) For what values of E_T is the value of τ ($= \Delta p / U$) within 10% of the value at $E_T = E_i$?

1.17 One of the classic semiconductor experiments is the demonstration of drift and diffusion of minority carriers, first performed by J. R. Haynes and W. Shockley in 1951 at Bell Telephone Labs. The experiment allows independent measurement of the minority carrier mobility μ and diffusion coefficient D. The basic principles of the Haynes–Shockley experiment are as follows (see diagram below).

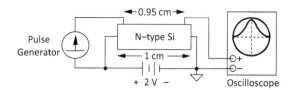

Experimental setup to measure minority carrier parameters.

A pulse of holes is created in an N-type bar (for example) that contains an electric field. As the pulse drifts in the field at a velocity v_d and spreads out by diffusion, the excess hole concentration is monitored at some point down the bar. The time required for the holes to drift a given distance in the field gives a measure of the mobility. And the spreading of the pulse during a given time is used to calculate the diffusion coefficient. If a voltage pulse is applied to the silicon sample at $t = 0$, the trace below is observed on the oscilloscope.

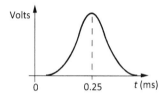

Output signal from experimental setup.

What is μ_p given that $v_d = \mu_p E$?

2 Modern Complementary Metal-Oxide–Semiconductor (CMOS) Technology

2.1 Introduction

In this chapter, we discuss the fabrication of a modern complementary metal-oxide–semiconductor (CMOS) integrated circuit using the individual process steps that are combined in a complete process flow sequence to make the chips. Such an ordered process flow from the sandbox of tools available in different combinations would be used to make any kind of device, such as a biosensor, a microfluidic device or a micro-electromechanical systems (MEMS) device. The wafer's past history and the future process steps can greatly influence how one chooses to order the individual steps. For example, high-temperature steps at the end of a process could disturb delicate doping profiles introduced early in the process. For this reason, we believe it is worth understanding the choices made in assembling a modern CMOS process flow. Seeing the "big picture" of a complete process flow should also help to put the individual process steps we discuss in subsequent chapters into perspective.

The process we have chosen is at the 28 nm node (Intel's 32 nm node), which was the most advanced planar process made and which is still widely used by circuit designers today [1]. This node entered high-volume manufacturing in about 2009. At the next 22 nm node, a FinFET process was introduced and, as the name implies, the field effect transistor was built in a silicon fin to provide better gate control of the channel [2]. The FinFET process has carried all the way to the 5 nm node today and represents the present state of the art. But its three-dimensional structure is difficult to understand at first pass, so we discuss the 28 nm planar process in detail and introduce a simplified version of the FinFET structure towards the end of this introductory chapter.

For readers new to silicon technology, this chapter will provide significant understanding of the basic logic behind the ordering of the process steps. However, full appreciation of the details will likely only come through study of the unit steps in more detail in the later chapters. We will revisit this CMOS process flow in later chapters and discuss in more detail the reasons for the particular choices made in this chapter.

2.2   Basic CMOS Technology and Device Structures

The simple CMOS inverter in Figure 2.1 consists of a PMOS and an NMOS transistor in series and became the dominant device used to fabricate circuits for a subtle but important reason, which we discussed briefly in Chapter 1 – since one of the transistors is always OFF for either a 1 (V_{DD}) or a 0 (ground) input voltage, it dissipates no static power except during the brief period when the transistors switch from one state to the other. Because there are billions of transistors in a modern chip, this seemingly small advantage turned into a huge power savings, giving the CMOS topology a dominant advantage over NMOS and bipolar circuits. The same basic CMOS topology can be combined to form a NAND gate composed of two PMOS devices in parallel and two NMOS devices in series, as also shown in Figure 2.1. A NAND gate (or its companion NOR gate) is a universal gate from which any Boolean combinatorial circuit can be designed. For these reasons, CMOS topology became the most widely used device structure in all modern silicon chips [3].

The end result of the process flow is shown in Figure 2.2, where a cross-section shows the NMOS and PMOS devices isolated by a shallow trench from each other, each fabricated in their own "wells." The first of the many wiring levels is also shown, which would link up devices in different circuit configurations. Through the rest of this chapter, we will follow the steps needed to produce a structure like that shown in Figure 2.2.

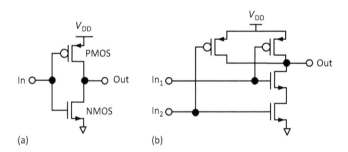

Figure 2.1 (a) A simple CMOS inverter, composed of a PMOS and NMOS device in series. (b) A CMOS NAND gate, a universal gate capable of implementing any Boolean function.

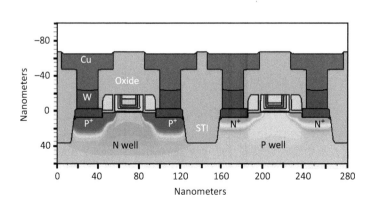

Figure 2.2 A cross-section of the final CMOS integrated circuit, with a PMOS device on the left and an NMOS device on the right.

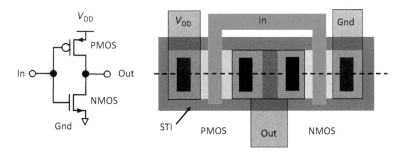

Figure 2.3 Top view of a simple layout for a CMOS inverter circuit.

Figure 2.2 is actually a Silvaco Victory Process [4] simulation of the process flow that we will discuss in this chapter. In fact, most of the figures in this chapter were generated by this program. The physically based models that the simulator uses produce much more realistic cross-sections than simple sketches of the structure. We will also extract from the simulations some quantitative information, such as specific doping profiles, as we discuss the CMOS process.

Almost all of the figures we will use in this book will be cross-sections like Figure 2.2. This is simply because this is a book about fabrication and such cross-sections show the details of the structure being built. For readers new to silicon chip fabrication and design, we include in Figure 2.3 a top view of what the layout of a simple CMOS inverter circuit might look like. The black rectangles are contact holes allowing the metal to make contact to the device source and drain regions. The blue band around and between the transistors is the shallow trench isolation (STI) region that is discussed in Section 2.4. The STI region provides lateral electrical isolation between transistors on the chip. The dashed line through the center of the layout represents the cutline that corresponds to the cross-section in Figure 2.2. Note that the connection between the NMOS and PMOS gates (in red) is made with the gate material outside the plane of Figure 2.2 and hence is not shown in Figure 2.2.

2.3 Picking a Starting Substrate

We start the process with a bare silicon wafer. Logically, the way to fabricate a functioning chip is to do the steps that form the transistors first, such as isolating them from each other, doping the silicon to make it more conductive in places and depositing the gate material to control the switching, before the structure is covered by the metal layers that interconnect the transistors together. All the doping, deposition and etching steps used to form the transistors are called the "front-end process" steps, and, by analogy, the interconnect layers on top of the devices are called the "back-end" steps.

The starting material is a bare silicon wafer. Such a silicon wafer costs approximately $1 per square inch, though the price can vary significantly depending on the size and quantity. To order such a wafer, you would be asked to specify four principal parameters:

1. the wafer diameter,
2. the crystal orientation,
3. the wafer doping type and
4. the doping concentration in the wafer.

Because both the transistor and the method of growing silicon wafers were invented in the USA, wafer diameters have historically been given in inches, until recently, when metric measurements have become more common. Common silicon wafer diameters are 4, 6, 8 or 12 inches or their metric equivalents 100, 150, 200 or 300 mm. Since many of the process steps (such as etching or deposition) operate on a wafer, larger wafers produce more chips at lower cost.

Methods of silicon crystal growth and the details of the crystal structure are discussed in more detail in Chapter 3. Semiconductor wafers are crystalline, which means that they have a periodic atomic structure with a basic unit cell repeating in three dimensions. Depending on which orientation of the unit cell is on the surface, that defines the orientation of the wafer. Because of the cubic symmetry in the silicon crystal, any of the faces of the cube end up being (100) surfaces (see Figure 3.2 in Chapter 3). The (100) crystal orientation provides the best electrical interface between the silicon substrate and the insulator silicon dioxide. We will discuss the reasons for this in more detail in Chapter 6, but the key idea is that the (100) Si/SiO$_2$ interface was found to be the best in terms of lack of defects, like unbonded atoms, charges, etc. It is almost universally used for building CMOS devices.

The doping type (P or N) in the starting wafer is somewhat flexible, since we end up building the transistors in separate N- and P-type regions or wells, so in some sense the doping type of the substrate does not appear critical. P type is more often used as a practical matter because it is somewhat easier to grow uniformly, as we will see in Chapter 3. Looking at Figure 2.2, one might wonder why the NMOS device cannot be built directly in the P-type substrate – indeed it can, and that is a viable choice. However, the twin-well process illustrated in Figure 2.2 is much more common because the doping method used to produce the P well is much better controlled in manufacturing than is the substrate doping. Also, since the P-well and N-well doping concentrations are similar, it is easier to start with a much more lightly doped substrate and then tailor the wells for the NMOS and PMOS devices individually.

The doping concentration of the substrate, expressed in atoms per cubic centimeter (atoms cm^{-3}) or alternatively in resistivity (ohm cm, Ω cm) typically corresponds to very light doping levels of parts per million of doping atoms in the silicon wafer. Numbers in the range of 10^{15} atoms cm^{-3} correspond to a substrate resistivity of 5–20 ohm cm (see Figure A1.7 in Appendix 1). From the periodic table, silicon is in column IV and anything from column III could be used as P-type dopant and anything from column V for an N-type dopant. For the P type we could use B, Al, Ga or In, but when one considers the solubility and diffusivity of these elements in silicon, it quickly becomes clear that boron is the P-type dopant of choice, as we will see in Chapter 3.

So the wafer bought from a vendor will be P type, (100) orientation with approximately 10^{15} atoms cm^{-3} doping concentration and will cost approximately \$100 for a 300 mm wafer. The cost of the wafer is remarkably small considering the quality of the material. Silicon in this form is the purest material used in any manufacturing process on Earth. Impurities are in the range of parts per trillion, except for O and C (Chapter 3) and, of course, deliberately introduced doping atoms, which are typically in the parts per million range in starting wafers. There are no mechanical defects and every atom is in the right lattice position over the entire wafer.

2.4 Preparing the Substrate for Transistors

Circuit designers assume that individual devices do not interact with each other except through deliberately placed interconnects. So our first step is to make sure the individual devices are electrically isolated from each other, and we do this by building a shallow trench between the devices. If the liner on the trench were electrically leaky, this would defeat the purpose of the isolation, so it must be carefully constructed. This region is the shallow trench isolation (STI) region labeled on Figure 2.2.

The first step is to put down three layers uniformly on the substrate: a thin silicon dioxide layer, a silicon nitride layer and a photoresist layer that is used to pattern the underlying layers, as shown in Figure 2.4(a). The thin silicon dioxide layer is put down through a chemical reaction by exposing the silicon surface to oxygen or water vapor at high temperature, typically 900–1200 °C, in a process known as thermal oxidation. We have a source of silicon atoms in the substrate, and passing oxygen or water vapor at high temperature over the surface creates a reaction to form the glassy SiO_2 layer. The silicon nitride cannot be put down in the same way using a reaction with nitrogen, because the oxide blocks the substrate silicon atoms from reacting with the nitrogen gas. Also, the bond energy of Si–N is higher than that of Si–O, so the chemical reaction to form silicon nitride occurs more slowly and tends to be self-limiting.

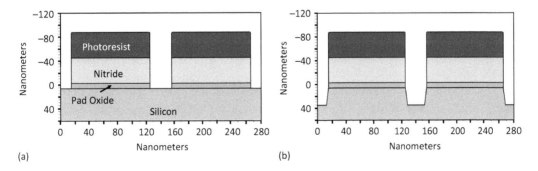

Figure 2.4 Shallow trench isolation formation. The photoresist, nitride and pad oxide layers are all uniformly deposited or grown on the silicon surface to start. (a) The structure after lithography and patterning of the nitride and pad oxide. (b) The silicon trench etch at an 85 ° tapered angle.

How, then, do we put down the Si_3N_4 when we cannot grow it using the silicon from the substrate and nitrogen gas? One way is to provide a gas source of silicon and nitrogen in a furnace, typically silane (SiH_4) or dichlorosilane (Si_2Cl_2) and ammonia (NH_3), which react in the gas phase and deposit silicon nitride on the surface. This process is called an LPCVD nitride layer, referring to a low-pressure, chemical vapor deposition (CVD) process. The low pressure is used in order to obtain uniform films over a large number of wafers, as we will see in Chapter 10.

We want the isolation layer just in places between the transistors, so we must introduce local geometric information, and we do that using our first lithographic or masking step. A light-sensitive photoresist layer is uniformly coated on the wafer by applying droplets of liquid photoresist with a honey-like consistency to a spinning wafer. A low-temperature bake around 100 °C hardens the photoresist to a plastic-like consistency. Then we shine light through a glass mask with patterned light and dark areas, which changes the properties of the photoresist locally so that a positive photoresist will develop away in the regions exposed to light. This photolithography process is one of the most complex and expensive in manufacturing chips, and several innovative tricks are used to enable features smaller than the wavelength of the light source to be built, as we will see in Chapter 5.

To give a sense of the scale, the printed pattern in Figure 2.4 in the photoresist in a 28 nm CMOS technology node would be on the order of 28–30 nm. We would make this dimension small because the trenches we are going to build are "wasted" space in the sense that they simply provide separation or isolation between active devices. This feature is printed today with lithography tools using 193 nm light. Perhaps readers might recall the guideline that optical systems cannot easily print features smaller than about half of the wavelength of the light used. Clearly, here, we are doing much better than that! How is this possible? We will study this in detail in Chapter 5.

The pattern in the photoresist layer is transferred to the underlying nitride, oxide and silicon by etching. A gas source of fluorine atoms and ions is used to perform the etching, forming volatile SiF_4. The process is called plasma etching, which we will discuss in detail in Chapter 9. It occurs in a plasma similar to that in a fluorescent light bulb. Note that three materials must be etched sequentially, Si_3N_4, SiO_2 and finally Si. By varying the gas source and the plasma parameters, the etch rates can be tailored for different materials. By using a gas source like CF_4, a competition can be set up between carbon deposition and fluorine etching to modify the etch profile, resulting in a tapered sidewall in the silicon trench. Avoiding sharp corners minimizes the electric fields and improves the isolation efficiency. At this point the structure looks like that in Figure 2.4(b).

Once the etching is completed, we are finished with the photoresist and it can be chemically wet etched in an H_2SO_4/H_2O_2 mixture or dry etched in an O_2 plasma, neither of which attacks the underlying nitride, oxide or silicon layers. At this point, after the wafers are cleaned, a thin layer of silicon dioxide is grown in the trench by thermal oxidation in a high-temperature furnace, as shown in Figure 2.5(a). The nitride film is very dense and acts as a blocking layer for oxygen or water vapor, so the oxidation only happens in the exposed silicon regions. This process is called LOCOS (local oxidation of silicon) and was invented by accident in 1970 when engineers noticed that an incompletely removed nitride layer blocked oxidation in certain

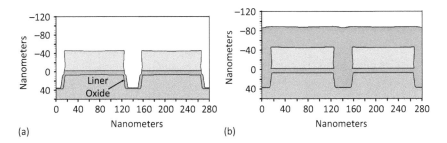

Figure 2.5 (a) A thin liner oxide is thermally grown to passivate the trench sidewalls. (b) A CVD deposited oxide fills the trench. The oxide deposition actually overfills the trench.

regions of a wafer [5]. More precisely, LOCOS refers to a thick oxide grown on the planar surface of a silicon wafer, while here, in this STI, we only grow a thin layer because of mechanical stress in the corners. A 30 min, 800 °C, O_2 oxidation would grow $\approx$ 5 nm of SiO_2 and would passivate the sides of the etched trench.

The oxidation process works by converting the Si layer-by-layer to silicon dioxide so the interface moves into the substrate, atomic layer by atomic layer. Once an oxide layer exists, the next layer forms by oxygen or water vapor diffusing through the existing oxide layer to react at the Si/SiO_2 interface. Thus it is a diffusion–reaction process that forms the thermal oxide. When this process happens, the substrate is the source of the silicon atoms and one can think of it as breaking the silicon–silicon bonds and inserting an oxygen atom between them to form silicon–oxygen bonds. There is a volume expansion by roughly a factor of 2.2 when that happens.

When one unit of silicon is oxidized, the oxide both grows into the silicon and out of the silicon, creating 2.2 units of silicon dioxide. Why is this important? First, because if you locally oxidize the planar surface of silicon, the surface is no longer flat because the oxide sticks up. More importantly, because the oxide is expanding when it grows, yet is attached to the silicon substrate, mechanical forces are generated in the structure as the oxide expands. These mechanical stresses are particularly important at the corners of the trench structure and are the reason we only grow a thin liner oxide. The details of the shape and the mechanical stresses under the nitride layer and in the corners are quite complicated, and are dealt with in detail in Chapter 6. The key idea is that thermally growing the liner oxide produces the best possible Si/SiO_2 interface and also helps to round the corners of the trenches due to the expansion and viscoelastic flow properties of the oxide at high temperatures.

The next step is to fill and indeed overfill the trench with a deposited oxide layer from a gas-phase LPCVD reactor, as shown in Figure 2.5(b). Precursor gases like SiH_4 and O_2 react to produce SiO_2. It is important that the filling process not leave gaps or voids in the trenches, especially as the trench sidewalls become more vertical to improve density. The details of these deposition systems are described in Chapter 10.

The next step in the STI process sounds incredibly crude compared to the sophisticated techniques generally used to fabricate chips. It involves pressing the silicon surface on a rotating abrasive wheel with a chemical slurry, in a process known as chemical–mechanical polishing (CMP). This polishes off the excess oxide from the top surface of the wafer, leaving a planar

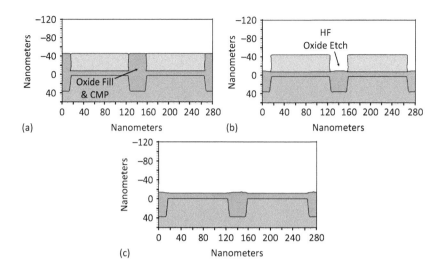

Figure 2.6 (a) The CVD deposited oxide is polished back to the nitride stop layer. (b) The nitride is exposed by etching any residual oxide, and also the height of the oxide above the trench is controlled by an HF etch. (c) The nitride is stripped, producing the final STI structure.

substrate with oxide-filled trenches, as shown in Figure 2.6(a). The hard nitride layer serves as a polishing stop. This is a typical result in CMP when materials with different hardness are being polished. While the CMP process sounds crude, it was a significant invention and its most common use now is in leveling or planarizing the many interconnect layers in the back-end processing. We will discuss this technology in detail in Chapter 9.

Once the CMP operation is complete, the nitride layer can be chemically removed by wet etching in phosphoric acid. Because this chemical etch has a very high selectivity to oxide (i.e., it etches nitride but not oxide), it is important to completely expose the nitride layer by etching any residual oxide using hydrofluoric acid (HF). This process can also control the oxide height above the trench, as shown in Figure 2.6(b). After the nitride is removed, we are left with the flat structure shown in Figure 2.6(c).

2.5   Well Formation

At this stage, the wafers are ready for device fabrication in each of the isolated regions. The first step is to put in the P- and N-type wells for the two types of transistors. We want to put in a P-type well doping, creating a local substrate for the NMOS transistors, and an N-type well doping for the PMOS transistors. We use photoresist to pattern where the doping goes, using the next two masks in the process as shown in Figure 2.7.

While the first mask for the STI had features near the minimum feature size to form a narrow trench and avoid wasting space, the well mask only needs to select an entire transistor area for doping. A cheaper lithography tool could be used for these steps, and these trade-offs are common and are called mix and match in the lithography toolset.

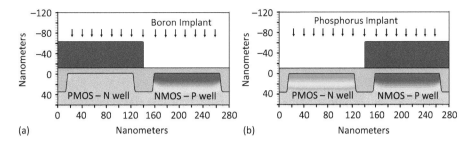

Figure 2.7 The well implants are performed to create the N well (b) and P well (a) regions, which act as local substrates for the PMOS and NMOS devices (boron, 3 keV, 10^{13} cm^{-2}; phosphorus, 10 keV, 10^{13} cm^{-2}). The photoresist masks the implants where it is present.

Now that we have defined the region for doping, we want to replace some fraction of silicon atoms in the P well with boron atoms so that we have a P-type well. The process used is called ion implantation, shown in Figure 2.7(a), because it is precise, controllable and reproducible. Note that, while the implant (arrows) occurs everywhere on the wafer, it is masked or blocked by the photoresist or another suitable mask in regions where we do not want it to penetrate into the silicon. We will discuss implant masking in detail in Chapter 8.

The ion implantation concept does not necessarily sound good at first sight – a small version of a linear accelerator is used to accelerate boron ions to high energies, shooting them at the silicon wafer [6]. The process is precise because the accelerating voltage can be dialed in with high accuracy, determining how far the ions travel into the silicon, and it is reproducible because it is easy to measure the ion current. You might imagine that you could dope the silicon by passing a boron-containing gas (diborane) over the silicon surface at high tempera-ture, and indeed this was the process used before ion implantation was invented. But it is difficult to accurately control the dose or amount of dopant that gets into the silicon and the peak doping is always at the surface. On the other hand, ion implantation can create a retrograde well with higher doping deep in the silicon and lower doping close to the surface. This can reduce the capacitance between the source/drain regions and the well and improve the switching speed of the transistor.

But ion implantation comes at a cost and this is because the ions slow down due to two processes – electronic and nuclear stopping. Electronic stopping occurs because the boron ion is charged and we can think of the silicon as nuclei surrounded by a sea of electrons. This drag force slows the ions down. The nuclear stopping process occurs because the boron ion can easily strike a silicon atom, knocking it off its lattice site, and thus creating damage in the lattice. This is easy to imagine because the incoming ion has an energy of typically 10–100 keV and can collide with a lattice silicon atom, which has a binding energy of only four Si–Si bonds (about 12 eV). The silicon atom can recoil a significant distance from its original lattice site and can itself create further damage until it too comes to rest. This damage must be repaired somehow since the devices we want to build require virtually perfect crystalline substrates. Fortunately, the damage can be repaired by a simple high-temperature anneal, otherwise the ion implant-ation process could not be used. Indeed, the main claim in Shockley's original patent [6] was not for ion implantation, but for the annealing step to repair the damage. A historical review described the process as "a sophisticated, expensive, yet brutal method for simply doping delicate semiconductors" [7].

So, it is clear what happens when the boron ion with hundreds or thousands of electron volts of energy hits the silicon – a silicon atom is knocked off its lattice site, often sent some large distance, like an elastic billiard ball collision, and the boron atom loses energy and repeats this process many times before coming to rest some distance into the silicon. This is clearly a random statistical process. So, if we were to plot the resting place of many boron atoms, we would find an approximately Gaussian distribution, where the peak occurs at the projected range, which is the average distance that a boron ion goes before it stops. But there is a distribution, because it is a statistical process, and some go further and some stop sooner. The peak concentration is determined by how much boron we implant, a parameter called the dose, measured in atoms per square centimeter (atoms cm^{-2}). This is the integral or area under the Gaussian distribution, the total number of boron ions we implant into the substrate. If we plot the doping concentration versus depth, in atoms cm^{-3}, this gives the volume concentration at any point. The parameters we can adjust are the energy of the implant by tuning the accelerator voltage and the time of the implant which is essentially the dose. The longer we leave the beam on, the more boron ions we implant and the higher the dose of boron. So we have a lot of flexibility with ion implantation to adjust the concentration or position of the elements we put in.

Ion implantation is really the only doping technique that has this flexibility. So it is great in all respects, very flexible and reproducible, but it does damage to the silicon. What is the probability of the ion striking a silicon atom? To first order, each incoming ion hits a silicon atom at each silicon plane. This is certainly true for heavy ions like arsenic and antimony, less true for light atoms like boron. It does not take a lot of dose coming in to significantly damage the substrate. There are secondary collisions as well, because if a silicon atom is knocked off its lattice site, it also knocks off other atoms before it comes to rest. The multiplier effect can be a factor of 10 or 100 where a single incoming ion ends up displacing 10 or more silicon atoms per plane that it travels through. The silicon can become completely amorphous with no crystal order and the atoms randomly distributed – which would not be usable as a semiconductor material for high-performance devices. So, we have to fix it to take advantage of the positive aspects of ion implantation and return the substrate to the perfect crystal lattice that we need.

Before we do that, we do the flip side and put in an N well as shown in Figure 2.7(b). We mask the other side and do an N-type implant. We have choices for the N-type dopant and could use phosphorus, arsenic or antimony. The reason phosphorus was picked in this particular case is that, when we finish up with this process, we want the wells to be symmetrical. We would not want one well being very deep and the other being very shallow because it might result in different capacitance values, for example, in the transistors. We would like P- and N-type dopants to diffuse at roughly the same rate, so they end up at the same depth in the silicon. It turns out that phosphorus and boron have almost identical diffusion coefficients, so after the implant we can diffuse these into the substrate and they will move at the same rate, producing roughly the same P and N well depths. Arsenic and antimony diffuse a lot slower, so we would have ended up with a deep P well and a shallower N well.

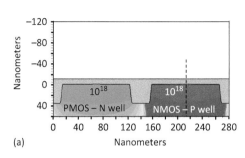
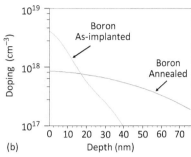

Figure 2.8 (a) The well profile after the drive-in anneal. (b) A 1D cutline (dashed line) through the center of the boron-doped well shows the boron doping profile as implanted and after the drive-in anneal (10 min at 1000 °C).

There are some subtle points we will briefly mention. There is a thin oxide on the surface of the silicon so the ion energy must be high enough to penetrate that layer and get into the silicon. It is an amorphous glassy layer that helps randomize the path of the ions so they do not go down the regular channels in the periodic silicon lattice. But the ion energy cannot be so high that it penetrates the photoresist, which acts as the blocking layer. So we need to choose the ion energies carefully for both device and physical process reasons.

After the ion implant steps, we have damaged the silicon in these regions. What we have to do now is to diffuse the dopants down to their final junction depths and get back the perfect silicon crystal that we need for the devices. Magically perhaps, it is easy to do this and to first order all this takes is an anneal at high temperature. Below the damaged silicon is perfect single-crystal silicon and what happens is that it provides a seed for regrowing the top part of the silicon that is damaged. As long as thermal energy is provided for the atoms in the top region to move around, they can find the right place to attach themselves to the substrate and layer-by-layer regrow a perfect silicon crystal. This process is called solid-phase epitaxy (SPE) and it is remarkable that it works as well as it does. We will discuss it in detail in Chapter 8. The anneal step performs three things – it provides the thermal energy for the solid-state diffusion process to drive the dopants to the correct depth, it activates the dopants by placing them on substitutional sites and it repairs the damage. Now we have created the localized wells. Figure 2.8(a) shows the structure after this implant drive-in/anneal. The doping profiles in Figure 2.8(b) were calculated by the Victory Process simulator using models we will discuss in Chapters 7 and 8.

2.6 Threshold-Adjust Implants

Now that we have created these wells, we want to put in the P and N regions near the surface. The reader might guess that we would put in the deep source/drain doping regions first, but that is not the process sequence. The regions that end up being the source/drain are designed to align to the gate edge because we want to minimize the overlap capacitance. So these regions are put in later after the polysilicon gate. So, the next steps are the implants in the channel region under the gate, so that we do not have to implant the ions all the way through the polysilicon gate, which would damage the fragile and critical thin gate dielectric. We use the same masks as the early well implant steps using a lower-resolution (less expensive) lithography tool.

The purpose of these implants in the channel region is to fine-tune the doping concentration under the gate of the transistor to set the threshold or turn-on voltage (V_{TH}) of the transistor. These V_{TH} implants are adjusted to control the threshold to whatever we want. Some transistors will have a high threshold voltage and very low leakage currents, while others will have a low threshold voltage and switch faster but have higher leakage. One can implant either P or N type to shift the threshold voltage up or down to the value that is desired. You can see in Figure 2.9 that the V_{TH} implant goes everywhere in the transistor but it will get swamped out by the heavier source/drain implants in those regions later in the process. These are called threshold-adjust implants and there is one for each type of transistor. In a modern process, there would be several threshold voltages to provide options for circuit designers. The change in V_{TH} from this implant is

$$\Delta V_{TH} = qQ_i/C_{ox} , \tag{2.1}$$

i.e., the implant dose qQ_i (ion charge q multiplied by Q_i atoms cm^{-2}) divided by C_{ox} (the capacitance of the gate oxide).

In Figures 2.7 and 2.9, it is apparent that the same dopant is implanted into each well, although at different doses and energies for the well implant and the V_{TH} adjust implant. A modern process might use a single masking step and multiple implants at different energies to form the well structure, with a shallow V_{TH} adjust implant, a deeper implant to set the bulk well concentration and a very-high-energy implant to form a "punch-through" implant layer at the base of the trench. A short anneal would repair the damage and leave the implant profiles largely unchanged (unlike the drive-in process in the process described here). This kind of retrograde well profile (the doping is higher as you go deeper in the substrate) may have advantages for avoiding an electrical problem called "latch-up" in CMOS devices [8]. Without adequate isolation or doping deep in the substrate, CMOS devices are prone to latch-up, which creates electrical shorting between the PMOS and NMOS regions. After this threshold-adjust implant, there is a short anneal to repair the damage.

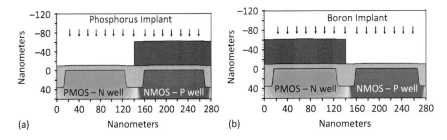

Figure 2.9 The low-dose threshold-adjust implants in the wells change the surface doping to obtain the desired threshold voltage in the devices (phosphorus, 15 keV, 6×10^{12} cm^{-2}; boron, 5 keV, 6×10^{12} cm^{-2}). These slight modifications in the surface doping tune the threshold voltages.

2.7   Gate Oxide and Polysilicon "Dummy" Gate

In Figure 2.2, there are small doped regions at the edge of the gate, called extensions, or tips or lightly doped drain (LDD) regions. These features are much smaller than anything that can be defined lithographically and are much smaller than the minimum feature size in the technology. It is very likely that the dimension of the gate is the best that can be done lithographically, since that dimension sets the performance of the transistor and is as small as the technology can make it. So we have some interesting tricks in order to get features smaller than the gate length. The next step is to put down the polysilicon gate, but first we need to strip the thin oxide and regrow the gate oxide after fixing the damage from the implant. After regrowing the gate oxide, we deposit the polysilicon immediately to avoid contamination.

Since we do not generally have deposition techniques that are selective, the usual technique is to deposit material everywhere, do lithography and etch it away in the regions where we do not want it. We deposit the polysilicon from a gas source (like silane) using LPCVD and it deposits as an amorphous or partially crystalline layer (called polysilicon) because it deposits on an amorphous insulator. The same deposition technique could deposit a perfect crystalline layer if it were deposited on a bare crystalline wafer, a process called silicon epitaxy. The polysilicon needs to be an N-type conductor on the NMOS device and a P-type conductor on the PMOS device, which could be achieved by masking the large transistor region and implanting the N- or P-type dopants, respectively. An anneal at high temperature would distribute the dopant uniformly in the polysilicon and the diffusion would occur quickly, as the polysilicon grain boundaries would help redistribute the dopant. This would set the work function for the transistor gates to the valence or conduction band edges, which is generally what is desired to get the correct device threshold voltages.

Process engineers are always looking for clever ways to combine steps, so one interesting option might be to use the source/drain implants (to be done later) to also dope the gate. That might work, but the source/drain anneals are the shortest possible to avoid deleterious device effects and may not be enough to fully redistribute the dopant in the polysilicon gate.

In the process flow we are considering, the polysilicon will act as only a temporary or "dummy" gate and will be replaced later with a high-K dielectric/metal gate stack. Thus doping of the polysilicon is not an issue. Earlier generations of CMOS technology directly used the polysilicon material as the gate electrode and so required polysilicon doping.

The next step is to etch the polysilicon gate, which is shown in Figure 2.10(a). This dimension is probably the smallest dimension on the entire integrated circuit surface that is formed by lithography and, for the 28 nm process being described here, it would likely be somewhat smaller than 28 nm. (Recall Table 1.1 in Chapter 1, which lists microprocessor gate lengths that are smaller than the technology node.) This dimension is pushed as small as possible because it determines the performance of the transistor. Very clever patterning techniques are used to enable features much smaller than the wavelength of the light source to be made, which we will discuss in Chapter 5. For the moment, we will assume that a very expensive exposure tool and a very expensive mask are used to pattern the polysilicon. After patterning the polysilicon, another very expensive etching step, requiring high selectivity, is required to etch the polysilicon and stop on the thin gate oxide.

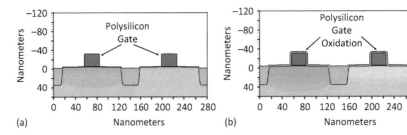

Figure 2.10 (a) Polysilicon gate regions after being patterned and etched with high selectivity to the underlying gate oxide. (b) Polysilicon gates after a short reoxidation to passivate the sidewalls (30 min at 800 °C in dry O_2).

Notice the sharp straight edges on the silicon gate, an example of a very anisotropic etch process. Taller than wide, this complicates the definition of the polysilicon gate.

To faithfully replicate the gate dimension, we need both very good anisotropy and selectivity. This is a really good example of the selectivity issue. The polysilicon thickness could be as much as 200 nm, although it is shown thinner than that in Figure 2.10. The gate oxide is less than 2 nm. Thus there could be a 100:1 difference between the thickness of the polysilicon and the gate oxide, and we have to over-etch in manufacturing to allow for tolerances and wafer-to-wafer variations, by typically 10–20%. The same plasma dry etch chemistry that etches the polysilicon would certainly etch the substrate silicon if we were to etch through the thin gate oxide. We cannot time the process exactly but we still must not etch through the thin gate oxide. This is probably the best example of why selectivity is a really key issue in pattern definition. Spectroscopic techniques exist to detect the endpoint of an etching process, but usually some amount of over-etching is necessary to make certain that all areas on a wafer are completely etched.

Anisotropy is another key issue because we are usually concerned about the shape of the edges on etched regions. Ideally, we would like the edges of the etched materials to be nearly vertical to preserve the mask dimensions in the etched layers. Anisotropic etches do this. Isotropic etches go sideways as much as they go down, so are unsuitable for trying to define small geometries. Wet etching with chemicals is typically isotropic, while plasma etching can be tailored to produce anisotropic profiles. We will discuss these issues in detail in Chapter 9.

After defining the polysilicon gates, a short oxidation is used to passivate the polysilicon. This is shown in Figure 2.10(b). Note that all oxides get thicker during this step because the silicon oxidizes everywhere.

2.8 Lightly Doped Drain Formation, Halo Implants and Source/Drain Doping

The next step is to perform the two most critical implants (aside from the V_{TH} implant) that determine the device performance. The first is a halo or angled implant that helps prevent the deep source/drain regions from punching through to each other. The second is a very shallow tip or extension region that reaches to just under the gate edge.

One might wonder why we need these complicated structures if we are simply following Dennard scaling, where all dimensions reduce by the same factor, keeping the electric fields constant. That would be true, except that, in practice, system-level pressures to maintain a constant power supply level and the fact that the turn-on or subthreshold region of MOSFETs does not scale (Chapter 1) constrained the scaling of the device voltage, and consequently the electric fields increased starting in about 2000. High fields cause problems in semiconductor devices, often called "hot electron" problems because most of them are due to the high energies that electrons (or holes) can reach in high fields. At high energies, carriers can cause impact ionization, which creates a multiplier effect, making additional electron–hole pairs by breaking Si–Si bonds. These carriers can even gain sufficient energy to surmount large electron barriers such as the 3.2 eV barrier between the silicon conduction band and the gate oxide conduction band. The result is that carriers can be injected into the gate dielectric where they may become trapped and cause device reliability problems.

The high fields occur on the drain side of the device because of the high gate voltage and drain voltage there, while the source is usually grounded. The tips are symmetrically placed for simplicity even though that may actually only be required on the drain side. The lightly doped drain (LDD) region grades the doping profile between the drain and the channel and allows the drain voltage to be dropped over a larger distance than would be the case if an abrupt junction were formed. Since many of the deleterious effects of high electric fields depend exponentially on the field strength, even modest reductions in the field can make a significant difference in device reliability.

Before completing the LDD structure, a deeper implant is often performed, called a halo or angled implant. This implant is designed to prevent subsurface punch-through between the source and drain in the transistor. The halo implants (Figure 2.11) use the same doping type as the channels and can be formed by implanting at a tilt or an angle under the gate edge to locally raise the channel doping and help avoid short channel effects. These local pocket implants would have to be put in carefully to cover all possible device orientations on the wafer. Since these complex implant profiles around the gate edges determine the device performance and cannot be seen using any physical methods, sophisticated simulation tools that model these effects are extensively used for device optimization [4].

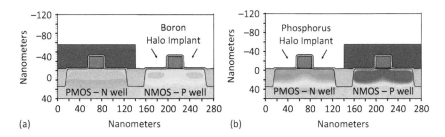

Figure 2.11 Angled halo implants introduce extra pockets of well doping under the source/drain extensions to minimize punchthrough between the source and drain (boron, 5 keV, four rotations, each $5 \times 10^{12}\,\mathrm{cm}^{-2}$; phosphorus, 15 keV, four rotations, each $5 \times 10^{12}\,\mathrm{cm}^{-2}$). The halo implants are done at a 25° implant angle. Four implants are done with 90° rotations, so transistors oriented in both the x and y directions are equally implanted.

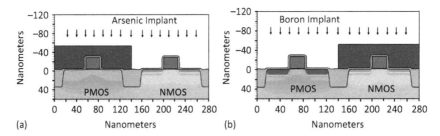

Figure 2.12 Shallow extension implants next to the gate edge form the NMOS and PMOS shallow doped regions that contact the channel (arsenic, 1.5 keV, 10^{15} cm^{-2}; boron, 300 eV, 10^{15} cm^{-2}).

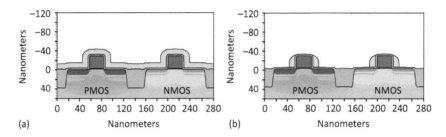

Figure 2.13 Spacer formation. (a) Conformal deposition of a nitride or oxide layer. (b) Anisotropic (directional) etch of the nitride layer leaves a sublithographic nitride spacer on the polysilicon gate edge (10 nm nitride deposition).

The sublithographic tips are formed by a two-step process. First, the very shallow tip implants are performed using the gate edge as a mask, as shown in Figure 2.12, so the implants are self-aligned to the edges of the gates and extend slightly underneath the gate edges because of the lateral scattering of the implant and the later sideways diffusion during the anneal step.

We now proceed to the second step in forming the small LDD regions. Spacers are formed by putting down a thin layer of conformal oxide or nitride that is uniform everywhere, as shown in Figure 2.13(a). Because the deposition is conformal, it is much thicker from the top to the bottom along the sidewall than it is in the flat regions. So, if we now do a highly anisotropic or directional etch of the film we just deposited, it will strip the film in the flat regions and leave a narrow spacer on the sidewalls with the thickness equal to the deposited film thickness, as shown in Figure 2.13(b). This clever trick makes the sidewall spacers, produced with no lithography, just a blanket deposition and the blanket etch to produce sublithographic features. In Chapter 5 we will see how these spacer technologies can also be used to extend the limits of conventional lithography. In some process flows, the halo implants in Figure 2.11 might be done after the sidewall spacer formation.

The final step is to put in the deep source/drain regions that are self-aligned to the edge of the spacer regions in Figure 2.14. During the spacer formation, the over-etch required to clear the film in the flat regions would etch the underlying thin oxide by varying amounts. For this reason, the oxide would be wet etched in HF down to bare silicon and a thin "screen" oxide

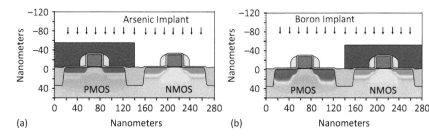

Figure 2.14 Deep source/drain implants, doping both polysilicon gate and the substrate (arsenic, 3 keV, 3×10^{15} cm^{-2}; boron, 600 eV, 3×10^{15} cm^{-2}).

would be regrown on the surface. The purpose is to randomize the incoming ion beam and prevent channeling. A mask exposes the source/drain regions on the NMOS device and an arsenic implant is performed. After stripping the photoresist, a new lithography step exposes the PMOS device and the boron or BF$_2$ implant is performed.

The only thing that remains is to repair the implant damage. A short high-temperature rapid thermal anneal (RTA) or flash anneal is used to minimize the amount of dopant diffusion in this critical step. The trick is to repair the damage in the substrate as efficiently as possible and not allow a significant amount of dopant diffusion to occur in the source/drain and tip regions. We do not want the annealing step to smear out the doping profiles because of diffusion. It might seem that a low-temperature anneal to repair the damage would minimize the diffusion, but there is an interesting anomalous effect called transient enhanced diffusion (TED) where dopants move at very high rates because of the damage. It turns out that the transient diffusion is minimized at high temperatures so there was a lot of effort to develop rapid thermal or flash annealing tools that can accomplish very-short-time high-temperature anneals. We will discuss this topic in detail in Chapter 8. The transistor structure after this anneal is shown in Figure 2.15. The reader may also note that, while the implants in Figures 2.11–2.14 also go into the polysilicon "dummy" gates, doping contours are not shown in the polysilicon material. This is because the poly gates will be removed later in the process and hence the doping profile is unimportant for device operation, but also because diffusion in polysilicon is very fast due to the grain boundaries in the material and hence the doping tends to be uniform in the poly gates. We discuss dopant diffusion in polysilicon in more detail in Chapter 7.

2.9   Adding Strain to MOSFET Channel Regions

The structure shown in Figure 2.15 was precisely the process that was used up to the 90 nm node and represents a fully functional self-aligned polysilicon gate CMOS technology. However, at the 90 nm node, the push for higher performance required that the channel be strained to improve the hole and electron mobilities. This was accomplished by replacing the deep source/drain implants in the PMOS device by etching the silicon in the

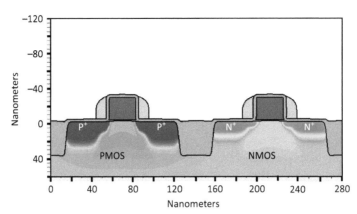

Figure 2.15 Final source/drain profiles after anneal (20 ms, 1000 °C).

source/drain regions and regrowing the region with a boron-doped SiGe layer (a process called epitaxial growth). The germanium atom is bigger than silicon, so this process introduces compressive strain in the channel, which improves the hole mobility. A similar process using phosphorus-doped SiC epitaxy introduces tensile strain in the NMOS device and improves the electron mobility. We will discuss this heteroepitaxy growth process in Chapter 10.

The epitaxial growth process takes place at temperatures around 750–850 °C, which is precisely where TED is at a maximum, so a short flash anneal is required to activate and stabilize the tip regions before the epitaxial growth. Another problem is that we cannot use photoresist to mask one of the transistors at these high temperatures. Instead, a hard mask of nitride is used, meaning that it will withstand the high temperatures involved. A nitride layer is deposited everywhere by LPCVD and a photoresist mask is used to selectively remove the nitride over the PMOS device using etching. The PMOS source/drain recess is dry etched and the SiGe epitaxial layer is grown. During growth, it is doped *in-situ* using a boron gas source such as diborane. Because the deposition occurs everywhere, HCl gas is also added to provide selectivity by introducing a competing etching reaction, so we only grow the layer in the exposed silicon regions. The Si, Ge and B precursors are typically dichlorosilane (SiH_2Cl_2), 10% germane (GeH_4) and 1% diborane (B_2H_6) in an H_2 ambient with HCl gas to obtain selectivity against the oxide and nitride layers on the wafers. The Ge content in the layers has steadily increased at each node from 20% to 60% as experience with the layer quality and defect control has improved in manufacturing.

After the compressive strain is introduced in the PMOS device, a second nitride layer is deposited and a photoresist mask is used to open the NMOS device, followed by stripping the photoresist. The NFET source/drain recess is formed and a phosphorus-doped silicon carbon layer is grown by epitaxy. The small carbon atom introduces tensile strain in the adjacent channel, which improves the electron mobility. After the nitride is stripped, this represents a high-performance strained-channel CMOS device and only requires the contacts to be fully functional.

2.10 Contact Formation

The first step in forming the contacts is to lower the sheet resistance of the source/drain regions by a process called silicidation. By depositing a reactive metal such as Ti, Ni or Co using sputtering and performing a short anneal at low temperatures, such as 350 or 450 °C, a metal silicide layer is formed, as shown in Figure 2.16.

The sputtering process occurs using a metal target and a plasma of argon ions that accelerate towards the electrically biased target and knock off the metal atoms onto the wafer. We will discuss sputtering in Chapter 10. The process is largely conformal (uniform deposition), so, after the metal silicide is formed, the unreacted metal on the dielectric regions must be etched off the wafer. This thin layer of metal silicide forms in the source/drain regions and gate region and lowers the sheet resistance that connects the contacts to the channel.

Since the anneal is done in an N_2 atmosphere, if Ti is used, the top surface reacts to form a TiN layer. The two reactions produce a $TiN/TiSi_2$ composite layer in regions where a Si supply is available and a pure TiN layer where the Ti sits on layers such as SiO_2 and no free Si is available for reaction. In the process shown in Figure 2.16, the TiN layer is etched away, leaving only the $TiSi_2$ layer. It is also possible to use the TiN layer as a local interconnect since TiN is a fairly good conductor, and this is done in some manufacturing processes. If it is desired to use the TiN as a local conductor, a photolithography step would be performed to pattern the TiN. We discuss these silicide processes in more detail in Chapter 11.

One minor point should be clarified. The process flow to this point is a viable polysilicon gate process, such as was used prior to the introduction of high-K metal gate structures at the 45 nm node by Intel. In poly gate processes, the tops of the poly gates would be silicided, as shown in Figure 2.16(b). However, in the high-K metal gate structure described next, the poly gate will be removed, and so it generally would not be silicided. This could be easily accomplished by using a hard mask in Figure 2.10 to etch the polysilicon and then by leaving that hard mask on the poly during subsequent steps. This would prevent silicidation on the poly gates.

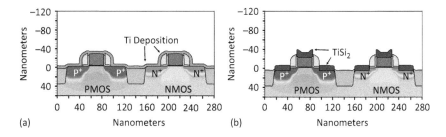

Figure 2.16 Silicide formation steps. (a) Titanium is sputter deposited. (b) A low-temperature anneal in nitrogen forms titanium silicide ($TiSi_2$) in exposed silicon or polysilicon regions and TiN on the dielectric regions. The TiN is wet etched, leaving low-resistance silicide on the source/drain and gate.

2.11   High-K, Metal Gate Formation

This strained-channel polysilicon gate process was used up to the 45 nm node, when a fundamental limit was reached. At that node, the gate oxide thickness became so thin, because of scaling, that direct quantum mechanical tunneling through the gate oxide produced unacceptable leakage currents. Since quantum mechanical tunneling increases exponentially as the dielectric thickness reduces, this was once considered to be a fundamental, hard limit on Moore's law scaling [9]. However, the invention of high-K gate dielectric materials alleviated the problem.

The idea is to use a gate insulator with a higher dielectric constant than SiO_2 and trade that off for a thicker layer which maintains the same gate capacitance and hence the same electrical control over the channel. The gate capacitance is given by

$$C = K\varepsilon_0/d\,, \tag{2.2}$$

so that an increase in dielectric constant K enables a thicker insulator (minimizing quantum mechanical tunneling) for the same capacitance. Since these high-K dielectrics are somewhat unstable, it made sense to introduce them as the very last step in the front-end process. A number of companies initially used a "gate-first" approach to high-K dielectrics in which the high-K material and the metal gate were deposited at the process stage of Figure 2.10 in place of the SiO_2 gate and polysilicon gate electrode. But this choice subjects the high-K dielectric and metal gate materials to all of the subsequent high-temperature processing, and in the end this proved to be an unworkable process flow. As a result, the industry converged on the "gate-last" process flow, which uses a polysilicon "dummy" gate as described above and then replaces it at this stage in the process with a high-K/metal gate stack. To accomplish this, a thick deposited oxide layer is polished back all the way to the gate polysilicon using CMP, shown in Figure 2.17(a). CMP is discussed in detail in Chapter 9. The polysilicon gate is removed in a very selective plasma etch process that stops on the gate oxide, shown in Figure 2.17(b).

The thin gate oxide is removed by a quick wet etch revealing the bare silicon channel surface, as shown in Figure 2.18(a). To stabilize the interface, a very thin (<1 nm) chemical oxide is formed on the silicon surface. At this stage, a sophisticated atomic layer is deposited by atomic layer deposition (ALD) of the high-K material to a thickness of 2–3 nm, performed in an ALD reactor, as shown in Figure 2.18(b).

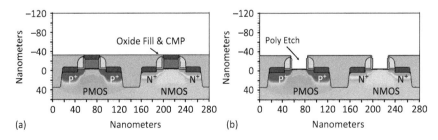

Figure 2.17 (a) A CVD deposited oxide covers the structure and is polished back by CMP. (b) Polysilicon is removed by a highly selective etch to the underlying gate oxide.

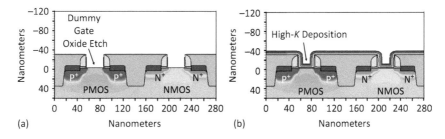

Figure 2.18 (a) The dummy gate oxide is wet etched. (b) The ALD high-K stack is deposited, consisting of interfacial oxide, high-K HfO_2, TiN and TaN barrier layer.

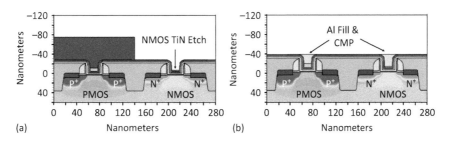

Figure 2.19 (a) A thick titanium nitride fill is thinned in the NMOS device, so that the subsequent aluminum fill can set the NMOS work function. (b) A thick aluminum deposition fills the gate and is planarized by CMP, setting the NMOS and PMOS work functions.

An ALD system uses two chemical precursors that are introduced sequentially. These precursors react with the surface atoms and completely saturate the surface reaction sites to a thickness of a single atomic layer. By iterating between the two precursors, a dielectric film such as HfO_2 can be built up one atomic layer at a time. ALD is discussed in detail in Chapter 10. This atomic-level thickness control is immensely important for thin gate dielectrics. To protect the high-K dielectric, a thin capping layer of titanium nitride (TiN) is deposited by ALD or sputtering. A thin etch stop layer of tantalum nitride (TaN) is then deposited, which enables a different metal to be formed in the NMOS and PMOS transistors for gate threshold voltage control, as shown in Figure 2.18(b).

A thick TiN layer can be deposited everywhere and this sets the PMOS work function. The PMOS device can then be masked and the TiN layer etched in the NMOS gate stack (hence the TaN etch stop), as shown in Figure 2.19(a). A deposited aluminum layer sets the work function for the NMOS device. Finally, both gate stacks are filled with a final metal such as thick aluminum or tungsten and the whole surface is planarized using CMP to the top of the gate stacks, giving the final replacement metal gate (RMG) structure shown in Figure 2.19(b). There may be one or more low-temperature (∼ 400 °C) anneals performed during the stack depositions to cause some interdiffusion of the metal layers and correctly set the work function in the NMOS and PMOS devices, respectively.

Figure 2.20 Detailed cross-section of the high-K layer stack near the gate edge in the NMOS device. For scale, the HfO_2 layer is ≈ 3 nm thick.

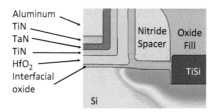

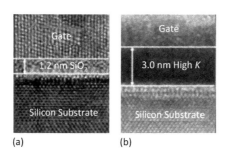

(a)   (b)

Figure 2.21 Actual cross-sections of (a) older polysilicon gate/SiO_2 gate technologies (Intel's 65 nm technology) and (b) newer metal/high-K gate technologies (Intel's 45 nm technology). (a) Photo © Intel, reprinted with permission. (b) Photo © 2007 IEEE. Reprinted with permission from [11].

Clearly, it was an enormous engineering effort to replace the canonical Si/SiO_2 gate interface with a new material stack with low defect densities based on a high-K/metal gate process at the 45 nm node, and the details of the high-K/metal gate stack formation are among the most closely held trade secrets in modern semiconductor device manufacturing [10]. We will look at some of the processing issues associated with these materials in Chapter 6. Intel was the first company to introduce this technology into manufacturing, and it has allowed Moore's law to continue for many more generations. Figure 2.20 illustrates in more detail the complicated materials combinations needed to implement this new gate stack.

Figure 2.21 shows actual cross-sections of an older polysilicon/SiO_2 gate structure and a newer metal/high-K gate structure. Figure 2.22 shows actual cross-sections of these gate technologies incorporated into MOS transistor structures. The many details of the high-K metal gate process are generally not visible in simple scanning electron micrograph (SEM) or transmission electron micrograph (TEM) images like Figure 2.22.

2.12 The Back-End-of-Line (BEOL) Process

The next steps are representative of a whole series of steps in the back end of the process. BEOL processing consists basically of a series of insulator/metal depositions that are stacked on top of one another to produce the interconnect or wiring levels. They rely heavily on CMP to planarize or flatten the deposited layers. Inter-level dielectric (ILD) layers in the back-end structure are typically deposited using plasma-enhanced CVD at low temperatures. The oxide layers are doped with

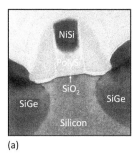
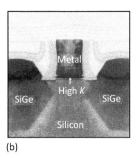

(a)   (b)

Figure 2.22 Cross-sections of (a) older polysilicon gate/SiO$_2$ gate technologies (Intel's 65 nm technology) and (b) newer metal/high-K gate technologies (Intel's 45 nm technology). (a) Photo © 2005 IEEE. Reprinted with permission from [10]. (b) Photo © 2007 IEEE. Reprinted with permission from [11].

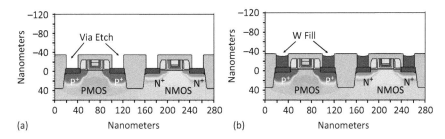

(a)   Nanometers   (b)   Nanometers

Figure 2.23 (a) Via holes are patterned in ILD1 using lithography and then etched. (b) Tungsten deposited by CVD (or sputtering) overfills the vias and is then planarized by CMP.

fluorine or carbon to reduce the dielectric constant, which is discussed in Chapter 11. This minimizes the coupling capacitance between layers and between wires. The deposition is conformal, so as to avoid the build-up of step heights with each additional layer, but ILD layers must be planarized using CMP.

In Figure 2.23(a), contact holes are etched to the source/drain silicide regions and the gate metal regions. The gate contacts are not shown since, until very recently, they were typically made outside the active gate area of the transistor because the size of the gate was too small to allow alignment of a contact hole. This is the structure illustrated in Figure 2.3. In very recent technologies, self-aligned contacts have been adopted that can be placed directly over the gate. A very thin TiN barrier layer is deposited by sputtering and a CVD tungsten (W) fill is performed. The W is overfilled to completely cover the top surface. A CMP step polishes the W, removing it from the oxide regions, leaving the W plugs as shown in Figure 2.23(b). The iterative portion of the BEOL processing begins next.

An inter-level dielectric (ILD1) is deposited and the damascene (or embedded) vias are defined in the dielectric by two levels of lithography, shown in Figure 2.24(a). These damascene vias will house the metal 1 (M1) copper interconnect. A TiN or TaN barrier/adhesion layer is deposited by

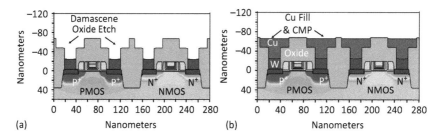

Figure 2.24 (a) The Cu interconnect pattern is formed and etched in ILD2 in preparation for copper seed layer deposition and electrodeposition of thick Cu. (b) Copper is electrodeposited in patterns etched in ILD1 and ILD2, overfilled and then planarized by CMP.

sputtering and a copper seed layer (Cu) is deposited, followed by a thicker electroplated Cu layer. These details are not shown in Figure 2.24 but are discussed in Chapter 11.

Copper is a lifetime killer in MOS devices, so the barrier layers and temperature control in the back-end process are critical to avoid copper contamination. Earlier processes used aluminum, but the push for performance meant lower-resistance copper interconnects were needed, and they were first introduced by IBM in 1997. The copper is planarized using CMP to complete the M1 interconnect level, as shown in Figure 2.24(b).

It is important to note that no etching of the Cu interconnect layer was required in this process. The Cu was "inlaid" in previously formed holes in ILD1 and then CMP was used to remove the Cu everywhere except in the ILD1 patterned regions. This process is called a damascene process, named after metalworking methods developed hundreds of years ago. It is critical in the case of Cu interconnects because Cu cannot easily be plasma etched, as we will see in Chapter 9 when we discuss etching. For plasma etching to work effectively, the byproducts of the reaction must be volatile at the etching temperature, and Cu does not form volatile byproducts with normally used etching chemistries.

Figure 2.25 illustrates some of the details that were not shown in the earlier BEOL figures. The barrier layers and Cu seed layers are now shown in the inlaid Cu patterns. Also shown is the fact that the insulating layers in the BEOL structures are actually multilayer structures. The dual lithography steps needed for each Cu layer to form the vias and then the M1, M2, ... wiring patterns need an etch stop layer to be manufacturable. These etch stop layers are the lighter blue/green-colored layers in the dielectric layers. Chapter 11 will discuss these process details.

This process of ILD/Cu/CMP is then iterated to produce a dozen or more layers of interconnect in modern chips. The metal layers get thicker and wider in the upper layers to distribute the power supply lines across the chip with low resistance. Low-dielectric-constant ILD layers (such as fluorinated oxides) are used to minimize capacitance between layers. The final layer typically uses a nitride dielectric to minimize contamination from the environment. Sophisticated automated wire routing programs would optimize the wiring distributions and minimize parasitics.

This process in Figure 2.25 represents a state-of-the-art high-K/metal gate planar CMOS device at the 28 nm node. An amazing number of inventions were required to push this process from the original transistors that Gordon Moore was familiar with in 1965, and they continue

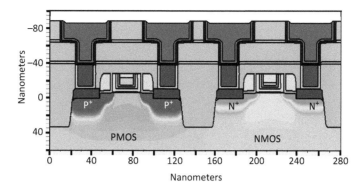

Figure 2.25 More detailed BEOL technology structure, showing some of the details discussed in Chapter 11.

to this day. This "circuit and device cleverness" was one of the reasons Moore cited for progress, along with manufacturing experience allowing larger die sizes to be made and the finer minimum dimensions enabled by lithography advances. Each contribution was (approximately) equally important to progress, but there is an outsized impact from lithography because its contribution gets squared (it happens in x and y).

2.13   The Introduction of the FinFET

At the 22 nm node in 2011, Intel introduced a FinFET device structure which is essentially a vertical silicon channel that has a gate on both sides for better gate control of the channel. It was competing with an extension of the planar process described above, which was called ultra-thin-body silicon-on-insulator (UTB-SOI). In the UTB-SOI usually fabricated with the "Smart Cut" process, an ultra-thin layer of silicon was formed on a thermal oxide on a silicon wafer. The process was complex and involved bonding a silicon wafer with a hydrogen implant to an oxide-coated base wafer. The hydrogen coalesced under heat treatment to form bubbles and led to cracking or splitting of a thin layer of silicon from the top wafer. A CMP step smoothed the split layer and thinned it so that it was fully depleted during device operation, thus minimizing leakage currents and improving gate control over the channel. We will discuss SOI structures in more detail in Chapter 8 as examples of ion implantation applications. For readers who wish to jump ahead, the "Smart Cut" process is described in Section 8.12 and illustrated in Figure 8.50.

It was not obvious initially which device structure would win out. Devices basically compete on three axes [12]:

1. the drive current or ON current I_{ON}, which determines how fast the circuit will be;
2. the leakage current or OFF current I_{OFF}, which determines how long the battery will last; and
3. the manufacturing complexity, which determines the yield and cost of the devices.

The UTB-SOI device had a low leakage current and a well-known manufacturing flow in spite of the higher cost of the initial wafer. But it had only a single channel, so its drive current was limited to approximately half that of the FinFET. In spite of the enormously complex process flow, the FinFET has become the device of choice for almost all high-performance CMOS applications.

We will describe a very simplified process for a single NMOS FinFET using a 3D simulator from Silvaco [13]. In the process flow pictures, we do not show the photolithography steps, since by this stage it should be obvious that, if a nitride stripe appears on a wafer, it was produced by depositing a blanket nitride layer, masking it with photoresist and etching the nitride in the exposed photoresist areas.

Several initial implants are used to prepare the substrate for the fin formation. Since the base of the fin is connected to the substrate and we do not want different fins interacting electrically through the substrate connection, the first step is to perform a heavily doped P-type implant at the depth of the fin's base. The implant could be performed after the fin is made, but it is perhaps simpler to do the implant at the very start of the process in the areas where the NMOS FinFETs will be built. At the same time, a second implant or series of implants can adjust the doping in the channel and thus the V_{TH} of the fins. An anneal is performed to activate the doping and repair the damage.

A nitride layer is deposited on the surface of the wafer to act as a hard mask for the fin etch. Next, an expensive lithography step is used to pattern the photoresist that will define the fin's lateral dimension and the nitride is patterned to act as the mask. The actual lithography step is more complicated than the simple flow presented here, and the techniques to produce the nitride hard mask at subwavelength dimensions are described in more detail in Chapter 5. The nitride hard mask is shown in Figure 2.26(a). This is followed by a deep anisotropic (directional) etch of the silicon to form the fin itself, stopping at the depth of the isolation implant, producing the structure shown in Figure 2.26(b). A short oxidation may be used to further narrow the fin dimension and remove etching damage on the critical channel sidewalls.

The next step forms the equivalent of the STI isolation in the planar process. An oxide layer is deposited using LPCVD fully covering the fins (Figure 2.26(c)). A CMP step planarizes the surface and stops on the fin's hard nitride mask (Figure 2.27(a)). An oxide wet etch can then be performed to reveal the portion of the fin that will act as the channel. At this stage, the FinFET structure looks like that in Figure 2.27(b).

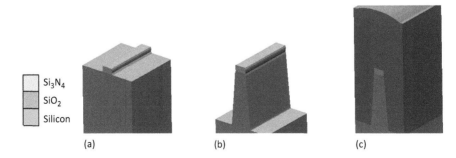

Si₃N₄
SiO₂
Silicon

(a)   (b)   (c)

Figure 2.26 (a) A nitride hard mask defines the fin dimension. (b) A deep anisotropic etch forms the silicon fin. (c) A thick oxide is deposited prior to CMP.

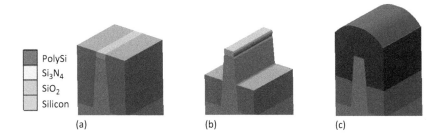

Figure 2.27 (a) CMP of the deposited oxide stops on the nitride stop layer. (b) FinFET structure after "fin reveal" step to expose the active channels surrounded by isolation oxide (equivalent to STI isolated devices in a planar process). (c) After nitride strip, a thick polysilicon layer is deposited on the fin, which will become the gate.

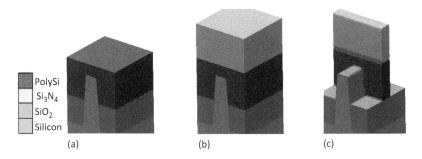

Figure 2.28 (a) Polysilicon is CMP polished to form a planar surface. (b) A thick nitride layer is deposited. (c) The nitride hard mask is patterned and etched. Then the polysilicon is anisotropically etched to the isolation oxide.

After stripping the nitride hard mask, the next steps are to form the gate oxide and gate polysilicon, but we know, because this is an advanced node, that these will eventually be replaced with a high-K/metal gate stack. A thin gate oxide is grown on the channel sidewalls and surface and a thick layer of polysilicon is deposited completely covering the fins, as shown in Figure 2.27(c).

The polysilicon surface is polished back by CMP to form a planar surface for the next high-resolution lithography step (Figure 2.28(a)). A thick CVD nitride is deposited to act as a hard mask for the polysilicon gate etch (Figure 2.28(b)). Another expensive lithography step defines the gate dimension which runs at right angles to the fin and the nitride is etched to form the hard mask, as shown in Figure 2.28(c). A highly anisotropic polysilicon etch is then performed, which stops on the isolation oxide, also shown in Figure 2.28(c).

The dummy gate crosses the fin at right angles. At this stage, the gate surrounds the channel on the sides and top and the fin sections extending beyond the gate function as the source/drain regions. Just as in the planar MOSFET, a tip or LDD implant is performed to connect to the edge of the channel. However, this implant is not only at the top surface of the fin, but also all along the sidewalls, so it must be performed at an angle to uniformly span from the top to the bottom of the FinFET channel. The damage is removed by a short, high-temperature anneal. Special care must be taken with narrow fins to ensure that there is a crystalline region from which the regrowth can occur. In later FinFET nodes, this is difficult because of the shadowing from closely spaced, deep fins. Instead of implanting the fin, the fin extending past the spacer edge is etched away and an epitaxial growth step is used to extend and dope the source/drain regions. We will discuss more advanced doping options later in the book.

Spacer formation occurs by depositing a thin nitride layer (Figure 2.29(a)), which is then anisotropically etched to the isolation layer (Figure 2.29(b)). This offsets the fin source/drain region from the gate edge with a spacer, just as in the 28 nm planar process.

At this stage, the equivalent of the "deep source/drain" is formed by doping the fin extensions. It is possible to do this by implantation, but exceptional care must be taken to avoid fully amorphizing or damaging the fin in Figure 2.29(b), so that the damage can be repaired by an anneal and SPE. To avoid the geometrical problems of implanting such a fin at steep angles, where multiple adjacent fins may also cause shadowing, an implantation technique that uses plasma doping can be used. The plasma surrounds all three-dimensional structures with the same fields, to uniformly implant a shallow layer. Plasma implantation is discussed in Section 8.13 of Chapter 8.

An alternative technique that is more common in advanced fin processes uses a strained epitaxial growth of a silicon layer with *in-situ* doping on the fin extension, forming the doped source/drain. This process is shown in Figure 2.30, where the fin source/drain extension is etched away to the edge of the channel (shown in Figure 2.30(b)) and then a doped SiGe layer is epitaxially grown to strain the channel and improve the mobility. Because of the crystal orientation dependence of the epitaxial growth, these extensions take on the final triangular shapes shown in Figure 2.30(f).

The final step is to perform the high-*K*/metal gate replacement process. A thick oxide is deposited and is etched back to the nitride stop layer using CMP (Figure 2.31(a)). After etching the nitride stop layer (Figure 2.31(a)), the polysilicon dummy gate can be removed (Figure 2.31(b)). This is followed by ALD depositions of the high-*K* dielectric stack (Figure 2.31(c)). A sputtered thin-film metal stack is then deposited and planarized to produce the final structure with the appropriate work function (shown in Figure 2.32).

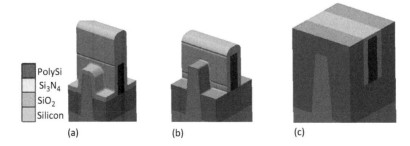

Figure 2.29 (a) Thin nitride is conformally deposited everywhere. (b) After anisotropic nitride etch, the spacer on the gate sidewall offsets the fin source/drain regions from the channel. (c) A thick oxide is deposited and planarized by CMP to the nitride etch stop as the first step in forming the RMG.

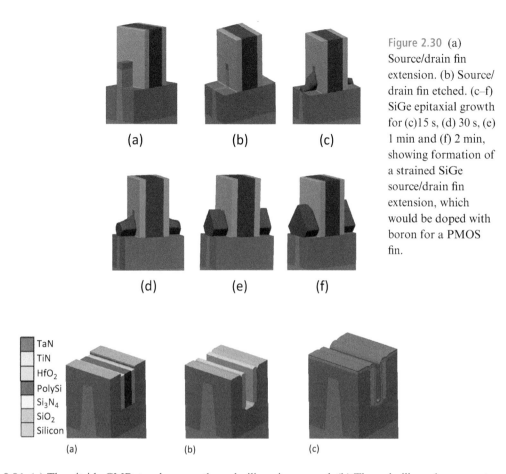

Figure 2.30 (a) Source/drain fin extension. (b) Source/drain fin etched. (c–f) SiGe epitaxial growth for (c)15 s, (d) 30 s, (e) 1 min and (f) 2 min, showing formation of a strained SiGe source/drain fin extension, which would be doped with boron for a PMOS fin.

Figure 2.31 (a) The nitride CMP stop layer on the polysilicon is removed. (b) The polysilicon dummy gate and oxide are removed. (c) The high-K gate dielectric stack is deposited.

It is not trivial to visualize the complex three-dimensional structure of the FinFET, so simulation tools like those used here are very helpful in figuring out if the fins can be doped by tilted implants or if alternative options are required. It should be clear even from this simplified description of the FinFET process that it is an enormously complex engineering operation, and there was no guarantee that FinFETs would be the structure of choice for advanced device technologies in high-volume manufacturing today. Figure 2.33 shows actual cross-sections of FinFET devices from several manufacturers of these devices.

In Section 1.6 of Chapter 1, the possible future evolution of the FinFET structure into gate-all-around (GAA) devices and perhaps into stacked GAA structures was discussed. These future possibilities are shown in Figures 1.22 and 1.23. The technologies required to build these structures, and to make them manufacturable in chips with billions of transistors, are formidable. There is active research and development today on the deposition and etching techniques that will be required. We will see over the next few years if they can be successfully implemented in products.

Figure 2.32 The final FinFET structure with the planarized RMG stack on the gate crossing at right angles to the fin channel, simulated with Victory Process [13].

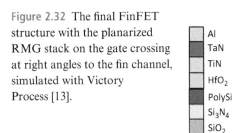

	Al
	TaN
	TiN
	HfO$_2$
	PolySi
	Si$_3$N$_4$
	SiO$_2$
	Silicon

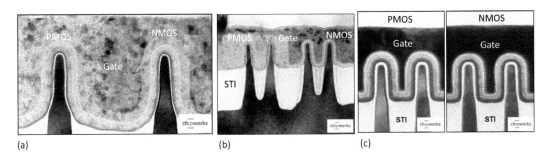

(a)   (b)   (c)

Figure 2.33 Cross-sections of 14 nm FinFET devices from (a) Samsung, (b) Intel and (c) TSMC. Photos © 2016 IEEE. Reprinted with permission from [14].

2.14 Summary of Key Ideas

The purpose of this chapter was to describe in some detail a complete modern CMOS process flow. For readers relatively new to silicon technology, many new ideas were presented, often without full explanation or justification. All of these will be described in later chapters as we deal with the individual process steps. At this point, the context in which such processes are used should be clear.

A final point which is important to make is that the process we have described is not a unique way of achieving the final result shown in Figure 2.2. Many commercial companies and research laboratories today build chips with final cross-sections similar to Figure 2.2, with, however, quite different process details. The reasons for these differences from one process to another may have to do with specific types of equipment a particular laboratory or plant has, or they may have to do with the applications targeted for the technology. Trade-offs in technology complexity and device performance may lead an individual company to a process flow quite different than the one we have described. Some of these trade-offs will become clearer as we discuss the individual process steps in later chapters.

Simulation tools are widely used today and are powerful adjuncts to experiments. Throughout this chapter, we have used Silvaco's Victory Process simulator to produce the cross-sections and other quantitative information resulting from the process flows we have described. We will continue this practice throughout this book, but as we discuss individual process steps like ion implantation, etching, deposition, etc., we will also discuss the physical models behind these technologies. These models are embedded in simulators such as Victory Process and, in reality, the simulations and predictions made by these tools are only as useful as the underlying models permit. So it will be important to understand both the power and the limitations of these models as we discuss individual process technologies.

2.15 REFERENCES AND NOTES

[1] P. Packan, S. Akbar, M. Armstrong, *et al.*, "High performance 32nm logic technology featuring 2nd generation high-k + metal gate transistors," in *2009 IEEE International Electron Devices Meeting*, IEEE, pp. 1–4, 2009.

[2] C. Auth, C. Allen, A. Blattner, *et al.*, "A 22nm high performance and low-power CMOS technology featuring fully-depleted tri-gate transistors, self-aligned contacts and high density MIM capacitors," in *2012 Symposium on VLSI Technology*, IEEE, pp. 131–132, 2012.

[3] S. Chih-Tang, "Evolution of the MOS transistor – from conception to VLSI," *Proceedings of the IEEE*, vol. 76, no. 10, pp. 1280–1326, 1988.

[4] Athena and Victory Process are Silvaco process simulation tools. See https://silvaco.com/published-papers/process-simulation.

[5] E. Kooi, *The Invention of LOCOS*, IEEE, 1991.

[6] W. Shockley, "Forming semiconductive devices by ionic bombardment," US Patent 2787564, 1957.

[7] L. Wegmann, "Historical perspective and future trends for ion implantation systems," *Nuclear Instruments and Methods in Physics Research*, vol. 189, no. 1, pp. 1–6, 1981.

[8] R. R. Troutman, *Latchup in CMOS Technology: The Problem and Its Cure*, Springer, 1986.

[9] P. A. Packan, "Pushing the limits," *Science*, vol. 285, no. 5436, pp. 2079–2081, 1999.

[10] S. Tyagi, C. Auth, P. Bai, *et al.*, "An advanced low power, high performance, strained channel 65nm technology," in *IEEE International Electron Devices Meeting. IEDM Technical Digest*, IEEE, pp. 245–247, 2005.

[11] K. Mistry, C. Allen, C. Auth, *et al.*, "A 45nm logic technology with high-k + metal gate transistors, strained silicon, 9 Cu interconnect layers, 193nm dry patterning, and 100% Pb-free packaging," in *2007 IEEE International Electron Devices Meeting*, IEEE, pp. 247–250, 2007.

[12] S. Luning, personal communication.

[13] Victory Process is a Silvaco Process simulation tool. See https://silvaco.com/tcad/victory-process-3d/.

[14] D. James, "Moore's law continues into the 1x-nm era," in *2016 21st International Conference on Ion Implantation Technology*, IEEE, pp. 1–10, 2016.

2.16 PROBLEMS

2.1 Short-answer questions.

    (a) Briefly describe two methods to electrically isolate adjacent MOS transistors.

    (b) Explain the two main reasons why single-crystal silicon is chosen as the wafer substrate material.

(c) Why was Al first used for wiring interconnects before Cu, even though Cu has higher conductivity?

(d) Ion implantation has become the dominant doping method in CMOS fabrication. What is a benefit of ion implantation? What is the biggest challenge in using it? Explain.

(e) For modern CMOS technology, explain why high-K dielectrics are used as gate dielectrics but low-K dielectrics are preferred as insulating materials between metal wires.

2.2 The sketch below represents a CMOS structure typical of state-of-the-art processes in the 1980s and 1990s. LOCOS isolation was used rather than the STI structure. Only one well (the N well) was used. Only one level of Al wiring is shown. Sketch a process flow that could result in the structure shown by drawing a series of drawings similar to those in this chapter.

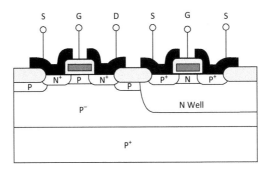

2.3 During the 1970s, the dominant logic technology was NMOS. A cross-sectional view of this technology is shown below. The depletion mode device is identical to the enhancement mode device except that a separate channel implant is done to create a negative threshold voltage. Design a plausible process flow to fabricate such a structure, following the ideas of the CMOS process flow in this chapter. You do not have to include any quantitative process parameters (times, temperatures, doses, etc.). Your answer should be given in terms of a series of sketches of the structure after each major process step. Briefly explain your reasoning for each step and the order you choose to do things.

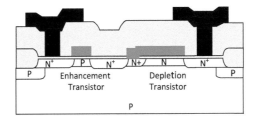

2.4 The cross-section below illustrates a simple bipolar transistor fabricated as part of a silicon IC. The structure uses one level of Al metallization. Design a plausible process flow to fabricate such a structure, following the ideas of the CMOS process flow in this chapter. You do not have to include any quantitative process parameters (times, temperatures, doses, etc.). Your answer should be given in terms of a series of sketches of the structure after each major process step. Briefly explain your reasoning for each step and the order you choose to do things.

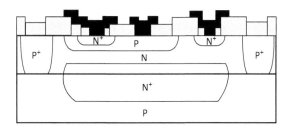

2.5 Simplified NMOS transistors are shown below. The earliest generations of these devices used Al gates as shown on the right and the Al gate was defined after the source and drain regions were doped. This order was necessary because the Al cannot withstand the high temperatures needed to diffuse the N regions (Al melts at 660 °C). When polysilicon technology was developed in the 1970s, the gate poly could be deposited before the N regions were doped. This allowed the N regions to be "self-aligned" to the edges of the poly and reduced device capacitances. Starting from P-type silicon substrates, sketch the process flows to get structures (a) and (b), respectively. No calculations are needed.

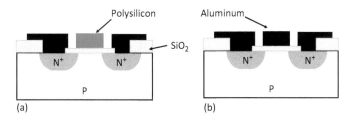

3   Semiconductor Materials: Crystal Growth, Wafer Preparation and Material Properties

3.1   Introduction

Almost from the very beginning, it was clear that silicon was the best choice for the material on which to base the integrated circuit (IC) industry. The abundance of silicon, the availability of simple techniques for refining it and growing single crystals, the essentially ideal properties of the Si/SiO_2 interface and the invention of manufacturing techniques based on the planar process, all led to the dominance of silicon-based devices by the early 1960s.

However, while silicon has dominated this $500 billion industry, other semiconductors have found markets where they outperform silicon or do things that silicon simply cannot do. The compound semiconductor market today is worth approximately $15 billion, dominated by GaAs devices that operate at higher frequencies than Si devices. SiC and GaN are opening multi-billion-dollar market opportunities in power devices. Light-emitting diodes (LEDs) for general lighting and other displays are a $15 billion market today. And, finally, even though it is silicon-based, the micro-electromechanical systems (MEMS) market today is also roughly $15 billion, producing a wide variety of sensors. All of these industries are built on the basic toolkit of fabrication technologies developed for mainstream silicon chips.

The silicon industry depends on a ready supply of inexpensive, high-quality single-crystal wafers. In this chapter, we will discuss the methods by which such wafers are prepared and some of the basic properties of these wafers. IC manufacturers typically specify physical parameters (diameter, thickness, flatness, mechanical defects like scratches, crystallographic defects like dislocation density, etc.), electrical parameters (N or P type, dopant, resistivity, etc.) and, finally, impurity levels (oxygen and carbon in particular) when purchasing wafers. All of these parameters must be tightly controlled in order to use the wafers in the high-volume manufacturing of complex chips. We saw a specific example of this type of specification in Chapter 2 in the CMOS process. In that example, we specified a starting substrate that was P type (10^{15} cm^{-3} boron-doped), (100) orientation, with a resistivity of roughly 5–20 Ω cm. Although these are the principal parameters that need to be specified, a more complete wafer description would actually include many additional parameters, as outlined above. In the case of the other important semiconductors used today, similar specifications on the starting wafers would also apply.

3.2   Crystal Structure

Although the first transistors were built in polycrystalline germanium, all modern ICs depend on single-crystal material for optimum device operation. Polycrystalline or even amorphous materials can be used to build semiconductor devices, but the defects present in these materials usually act as generation and recombination centers and have significant effects on basic device properties like mobility, carrier lifetime and, very importantly, the reproducibility of device parameters from one device to another. The result is poor performance in the devices – lower output currents for a given applied voltage, higher leakage currents in junctions and variation from one device to another. These same issues apply to compound semiconductor materials used in the other applications described above. We begin this chapter with a discussion of some of the basic properties of crystals.

In crystalline materials, the atoms are arranged spatially in a periodic fashion. A basic unit cell can be defined which repeats in all three dimensions. A standard terminology is used to describe crystal directions and planes, which we will briefly describe here. We will first consider cubic crystals, which include Si, Ge and some compound materials like GaAs. In some cases, the electrical properties of devices depend on the particular crystal orientation used for the starting wafer. (Recall that in Chapter 2 in our discussion of the CMOS process, we chose a (100) crystal.) Quite often, these orientation-dependent effects are associated with wafer surfaces or material interfaces, where the crystal terminates. Here, atoms may be incompletely bonded and the exact atomic arrangement can make a difference to the electrical properties. Bulk properties in cubic crystals like silicon are generally isotropic (independent of direction) because of the cubic symmetry of the crystal.

Figure 3.1 illustrates three simple cubic crystal unit cells. The body-centered cubic (BCC) cell has an extra atom in the center of the cube (shown in red); the face-centered cubic (FCC) cell has an extra atom in the center of each face of the cube (shown in red). Silicon and other semiconductors have more complex unit cells than these simple examples, but we can illustrate all of the basic ideas about crystal directions and planes with these simpler structures. An *xyz* coordinate system defines the directions in the crystal. The dimension *a* is called the lattice constant and is the basic distance over which the unit cell repeats in the cubic crystal. Directions in cubic crystals are expressed in terms of three integers, which are the same as the components of a vector in that direction. For example, to get from atom A in the cubic crystal to atom B, we

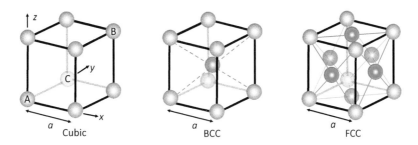

Figure 3.1 Unit cells in simple cubic crystals.

move 1 unit in each (x, y and z) direction. This is then the [111] direction in this crystal. To get from atom A to atom C, we would move in the [010] direction. By symmetry, many directions in a crystal are often equivalent, simply depending on our choice of reference. For example, the [100], [010] and [001] directions are equivalent in this example and by convention are referred to as ⟨100⟩ directions.

It is also useful to describe planes in a crystal, and a convention has been adopted for doing this. Figure 3.2 illustrates three simple planes in a cubic crystal. Such planes are described by Miller indices, which are a set of three integers calculated as follows. For a particular plane, the intercepts of that plane with the three crystal axes are determined. In the middle example in Figure 3.2, the intercepts of the (110) plane with the x, y and z axes are 1, 1 and ∞, respectively. The reciprocals of these three intercepts are then taken, which gives 1/1, 1/1 and 1/∞ or simply the (110) plane. (The reciprocals eliminate infinities in the notation).

A minor complication can arise in this procedure if we pick a plane with intercepts outside the basic unit cell. For example, if we picked a plane with x, y and z intercepts of 3, 4 and 2, respectively, the reciprocals would be 1/3, 1/4 and 1/2. In this case the smallest set of integers that have the same relative values are chosen for the Miller indices, which would result in the (436) plane in this case (by multiplying by 12). There is also no reason why we could not pick a plane with a negative intercept, for example 1, −1 and 2. In this case, the reciprocals are 1/1, 1/(−1) and 1/2, respectively. The nomenclature in this case places a bar over the corresponding Miller index, so that this plane (multiplying by 2) becomes the ($2\bar{2}1$) plane. Finally, as was the case for crystal directions, many crystal planes are also equivalent. The (100), (010) and ($0\bar{1}0$) planes are equivalent and are designated as {100} planes. Note that there are conventions that are used here for the style of bracket used to describe crystal directions and planes.

It should be noted that, in cubic lattices, the direction [hkl] is perpendicular to a plane with the identical three integers (hkl), as illustrated in Figure 3.2. In this figure the red arrows represent directions and they are perpendicular to the red planes. This is sometimes helpful in visualizing directions and planes in these crystals.

Silicon has a diamond cubic lattice structure, shown in Figure 3.3. This structure is most easily visualized as two merged FCC lattices with the origin of the second lattice offset from the first by $a/4$ in all three directions. Equivalently, one could think of the structure as an FCC lattice with two atoms at each lattice point, with the two offset from each other by $a/4$ in all three

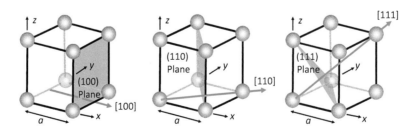

Figure 3.2 Crystal planes and major directions for a cubic lattice. The heavy arrows in each case illustrate crystal directions, designated [hkl]. The cross-hatched areas are the corresponding crystal planes, designated (hkl).

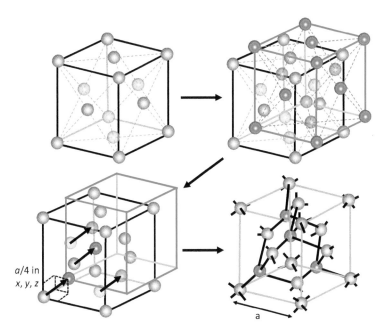

Figure 3.3 Diamond crystal structure of silicon. The unit cell may be visualized as two merged FCC cells, offset from each other by $a/4$ in all three directions. Top left shows the basic FCC unit cell. Top right shows two merged cells, offset by $a/4$. In the bottom left, only the four atoms in the second FCC cell that lie within the first FCC cell are shown. The arrows show the corresponding $a/4$ displacement of these four atoms. In the bottom right, the actual tetragonal bonding between the atoms in the diamond cell is shown. Note that only the four red atoms have all four covalent bonds within the unit cell.

directions. Each atom is individually covalently bonded to four nearest neighbors, providing the electrical properties that are described in Appendix 1. Figure 3.3 illustrates the construction of the diamond lattice from the two merged FCC lattices.

There are two principal silicon crystal orientations that are used in manufacturing integrated circuits, (111) and (100), meaning that the crystal terminates at the wafer surface on {111} or {100} planes, respectively. As mentioned earlier, the bulk properties of silicon are generally isotropic since the cubic crystal is symmetric. Thus we find, for example, that dopant diffusion coefficients are independent of the direction in which the dopant is diffusing as long as surfaces play no role in the process (Chapter 7). However, real devices are always built near surfaces, and which crystal plane the surface terminates on can make a difference in the surface electrical and physical properties.

The {111} planes in silicon have the largest number of silicon atoms per cm^2, the {100} planes the lowest. This results in a number of differences in properties. The {111} planes oxidize faster than {100} because the oxidation rate is proportional to the number of silicon atoms available for reaction (Chapter 6). The {111} surfaces have higher densities of electrical defects (interface states) because at least some of these defects are believed to be associated with dangling silicon bonds (Chapter 6). Finally, dopant diffusion coefficients and other properties in bulk silicon usually do depend on the surface orientation of the crystal because surfaces are often active interfaces. We will see in Chapter 7 that point defects (missing or extra silicon atoms in the crystal) play crucial roles in dopant diffusion, and active surfaces can perturb the concentrations of these defects. Primarily because of the superior electrical properties of the (100) Si/SiO$_2$ interface, (100) silicon is dominant in manufacturing today. Virtually all MOS-based technologies use crystals with this orientation (like the CMOS process in Chapter 2).

Other important semiconductors have the same crystal structure as Si, for example, Ge and GaAs. In the case of compound semiconductors like GaAs that have this structure, the Ga and As atoms each occupy sites on the two FCC sublattices that make up the overall unit cell. In compound materials, the structure is called the zincblende lattice rather than the diamond lattice. In Figure 3.3, in the lower right unit cell, the blue atoms would be As and the red atoms Ga. Note that each As atom has four Ga nearest neighbors, and each Ga atom has four As nearest neighbors, arranged in a tetragonal structure. Figure 3.4 shows the lattice constants versus bandgap for a number of important semiconductor materials. Most of the semiconductors shown on the right-hand side of Figure 3.4 have a diamond or zincblende lattice, as shown in Figure 3.3. Semiconductors on the far left-hand side of the figure have a hexagonal crystal structure, as discussed below, and are labeled, for example, hex-GaN in the figure. The dashed lines on the left represent data points for ternary compounds such as Ga$_x$In$_{1-x}$N or In$_x$Al$_{1-x}$N. Many ternary compounds exist for elements on the right-hand side, such as Ga$_x$Al$_{1-x}$As, but are not shown for clarity.

The colors in Figure 3.4 correspond to the color a photon would have if it were emitted by the semiconductor through direct band-to-band recombination. If all the energy released by the recombination of a hole and an electron is given to a photon, then the photon wavelength λ is

$$\lambda = hc/E_G , \tag{3.1}$$

where E_G is the bandgap of the semiconductor.

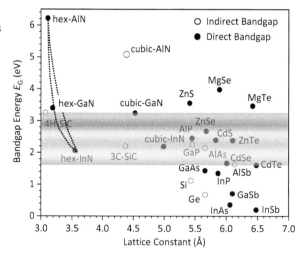

Figure 3.4 Bandgap versus lattice constant for many important semiconductor materials.

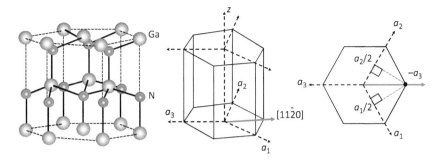

Figure 3.5 Hexagonal crystal structure of GaN (Ga blue, N red) using the Miller–Bravais notation.

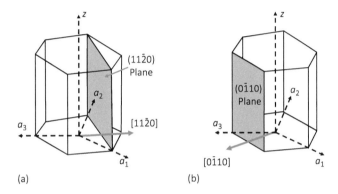

Figure 3.6 Directions and planes in the Miller–Bravais system for hexagonal crystals.

In hexagonal crystals, such as those on the far left-hand side of Figure 3.4, four axes are used to describe the crystal (the Miller–Bravais notation). Figure 3.5 illustrates this for GaN. The cell consists of alternating vertically stacked planes of the two atoms, arranged in a hexagon, with an atom in the center of each hexagon. Three of the axes are separated by equal angles and lie in the same plane (a_1, a_2 and a_3). The fourth axis is perpendicular to the other three (z axis). The algorithm used to define crystal directions is also illustrated in Figure 3.5 for the specific example of the $[11\bar{2}0]$ direction. First, the vector defining the direction is repositioned, if necessary, to pass through the origin of the coordinate system. Then the projections on the four axes are read off in terms of the cell dimensions. In this example these projections are 1/2, 1/2, −1 and 0. These projections are then rewritten in terms of the smallest integer values, and square brackets are used to indicate the direction, i.e., $[11\bar{2}0]$. The bar over the 2 indicates a negative direction. This procedure is very similar to the process described earlier for cubic crystals. Note that in the Miller–Bravais system, since four indices are used to describe a 3D structure, the four indices are not all independent. In an [$hkil$] system, it is always the case that $i = -(h + k)$.

In the Miller–Bravais system, crystal directions are perpendicular to crystal planes with the same indices. Figure 3.6 shows two examples. The $[11\bar{2}0]$ example from Figure 3.5 is shown in

Figure 3.6(a). Independently of the direction, the crystal plane can be found as follows. First, find the intercepts of the plane with the four axes. In the example in Figure 3.6(a) these are 1, 1, $-1/2$ and ∞. We then take the reciprocals of these intercepts (to avoid infinity values) and find the lowest set of integer values that describe these reciprocal intercepts, in this case $(11\bar{2}0)$. The round brackets are used to indicate a plane. In the example in Figure 3.6(b), the plane has intercepts of ∞, -1, 1 and ∞, so the plane is the $(0\bar{1}10)$ plane. The vector showing the direction has projections on the four axes of 0, -1, 1 and 0, so the direction is $[0\bar{1}10]$.

3.3   Defects in Crystals

All semiconductor crystals contain defects, and these defects can play important roles in device fabrication and in device operation. While the raw materials used to form semiconductor crystals are highly purified, they do contain impurities that may sit on lattice sites or may exist in interstitial sites between lattice atoms. Even if the concentration of impurities could be reduced to zero, there are still defects that are found in otherwise perfect crystals. It is useful to classify these defects into four categories, depending on their dimensionality. The usual categories are point defects, line defects, area defects and volume defects. Figure 3.7 illustrates some of the more common defects in a simple two-dimensional representation.

Point defects are the simplest to visualize and they play crucial roles in impurity diffusion (Chapter 7) and in ion implantation damage (Chapter 8) and lesser roles in oxidation kinetics (Chapter 6) and other phenomena. In general, anything other than a silicon atom on a lattice site constitutes a point defect. By this definition, for example, a substitutional doping atom is a point defect, and might be referred to as an impurity-related defect. In compound semiconductors, if element A sits on a site normally occupied by element B or vice versa, this point defect is called an antisite defect.

There are two principal types of point defects we shall be primarily concerned with, which are often referred to as native point defects. The first is simply a missing lattice atom or vacancy, which we shall designate simply as V. The other is an extra atom that we shall designate simply as

Figure 3.7 Simple 2D representation of some of the common defects found in crystals. V and I are point defects, the dislocation represents a typical line defect, the stacking fault is an area defect and the precipitate is a volume defect.

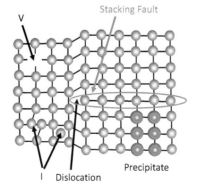

I. Figure 3.7 illustrates two possible arrangements of such an extra atom in the lattice. The first case illustrates a purely interstitial atom, that is, an atom sitting unbonded in one of the available sites between silicon atoms. The second illustrates two atoms sharing one lattice site, a defect usually referred to as an interstitialcy. Most of the literature on silicon technology does not distinguish between these two defects, designating both of them as simply a silicon interstitial or I. Since the distinction is not usually important in modeling processes in silicon, we shall adopt this strategy in this book and simply use the symbol I to refer to an excess silicon atom. Most workers in the field today believe that the interstitialcy is likely the defect that is important in silicon processes, because of the lower energies required to create this defect compared to a pure interstitial. We will discuss native point defects (I and V) in more detail in the next section of this chapter.

One-dimensional defects in crystals are known as dislocations. These are normally classified into two main types – edge and screw dislocations – but for our purposes these details are not critical. An example is illustrated in Figure 3.7. The right-hand part of the crystal contains an extra plane of atoms, which terminates at a dislocation. The dislocation itself, then, is a linear defect in the direction into the plane of the figure. Dislocations either terminate at the edge of the crystal or they form a closed loop within the crystal. Macroscopic edge dislocations are never present in silicon wafers at the start of IC manufacturing; silicon wafers are normally specified as "dislocation-free." However, such defects can be generated during the high-temperature steps normally used in wafer fabrication, particularly if thin films present on the wafer surface generate high stresses.

Another way in which such stresses can be generated is through temperature gradients. These can occur during crystal growth, discussed later in this chapter, when the seed is first inserted into the melt and the outside of the seed heats up more rapidly than the inside, or during wafer processing if a heating system is used which is non-uniform. In the crystal growth case, dislocation formation is inevitable when the seed is first inserted into the liquid silicon, as we discuss later in this chapter, but such dislocations can be grown out to the edge of the crystal during the initial stages of growth, as discussed in Section 3.5. In the wafer heating case, large temperature gradients are not normally a problem during gradual heating or cooling cycles such as those encountered in furnaces. However, they can be a problem in rapid thermal processing systems, which use high-intensity lamps to heat wafers at a rate of hundreds of degrees Celsius per second and are commonly used in semiconductor device fabrication. If the lamps in such systems are not configured to maintain a uniform temperature across the wafer, temperature gradients can result. Note that maintaining a uniform temperature is not necessarily the same thing as maintaining uniform energy input into the wafer because heat loss due to radiation or conduction may vary with position across a wafer, especially at the edges. If temperature gradients exist, the resulting thermal stresses are given approximately by

$$\sigma = \alpha Y \Delta T, \tag{3.2}$$

where α is the thermal expansion coefficient of silicon, Y is Young's modulus and ΔT is the temperature gradient. The yield strength of silicon is on the order of 0.5×10^9 dyne cm^{-2} at process temperatures. Such stress levels can be reached, and dislocations therefore generated, with ΔT values on the order of $100\,°C\,cm^{-1}$.

Microscopic dislocation loops usually are present in silicon starting wafers. At the high temperatures associated with crystal growth, point defect concentrations are quite high, as we will see shortly. As the crystal is pulled from the melt and cools, there may not be sufficient time for the populations of these defects to reduce to the much lower equilibrium populations associated with lower temperatures. (The V and I concentrations could be reduced by recombination of the two defects with each other, or diffusion of the individual defects to the crystal surface.) Small dislocation loops can form during the cooling process as a result of the agglomeration of excess V or I. If, for example, the extra half-plane of atoms on the right of Figure 3.7 extended only over a small number of lattice sites, the dislocation bounding the extra plane would form a closed loop, often a circle, since a circle has the shortest dislocation length and lowest energy. Such a defect could be formed by a group of I forming the extra half-plane on the right of Figure 3.7, or by a group of V forming the missing half-plane on the left.

Dislocations are active defects in crystals, that is, they can move when subjected to stresses or when excess point defects are present. The process of "climb" occurs when excess point defects are absorbed by the dislocation. It is easy to visualize in Figure 3.7, for example, the dislocation moving one lattice plane to the left by absorbing an I, or moving one lattice plane to the right by absorbing a V. In fact, these kinds of processes are often useful in determining whether a particular processing step increases or decreases V or I concentrations, by observing whether dislocations grow or shrink during that step.

The most common kind of 2D or area defect found in silicon is the stacking fault. Such a defect is also shown in Figure 3.7. Stacking faults always form along {111} planes, and are simply the insertion or removal of an extra {111} plane. In the silicon crystal structure shown in Figure 3.3, there are actually three parallel {111} planes in each unit cell commonly referred to as A, B and C. In a perfect crystal, the stacking order is ABCABC.... When a stacking fault is present, either an extra plane is inserted (ABCACBC...) or a plane is missing (ABCABABC...). Such faults are referred to as "extrinsic" if there is an extra plane of atoms, or "intrinsic" if a plane is missing. Stacking faults are bounded by dislocations and, when they intersect the wafer surface, are usually referred to as surface stacking faults. It is relatively easy to measure the density and the size of stacking faults in silicon using chemical etchants, which preferentially attack the highly stressed regions associated with the dislocations that bound the faults.

Most of the stacking faults observed in silicon have been determined to be of the extrinsic type, implying that they formed in an environment of excess interstitials. Since the size of the faults can be determined experimentally, this measurement can be used as a means of determining whether a particular process increases or decreases point defect concentrations. If the faults are extrinsic, then they will grow by absorbing more I, or shrink by absorbing V. Experiments like this have been used to determine that oxidation injects excess interstitials into the underlying substrate, since stacking faults are observed to grow during oxidation. Stacking faults that are used as markers in such experiments are sometimes called oxidation-induced stacking faults (OISFs). An example is shown as a top view in Figure 3.8(a). In this case, oxidation was used to produce stacking faults in a silicon wafer. The length of the faults or the rate at which they grow

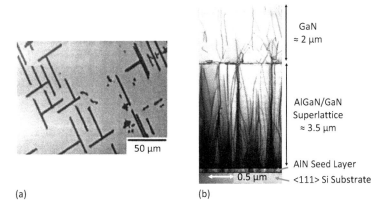

Figure 3.8 (a) Top view of oxidation-induced stacking faults in silicon. Reprinted from [1] with the permission of AIP Publishing. (b) Cross-sectional view of a GaN epitaxial structure grown on a silicon ⟨111⟩ substrate. Courtesy S. Chowdhury.

during oxidation can be used to study the injection of interstitials during oxidation. We will discuss this in more detail in Chapter 6.

Many modern semiconductor devices are fabricated using heteroepitaxy – that is, the epitaxial growth of one semiconductor material on a substrate of a different material. Often this is done so that the bandgap of the structure can be changed throughout the layers of a device. One interesting example is shown in Figure 3.8(b) in which a GaN layer is grown on a silicon ⟨111⟩ substrate. Si is a cubic crystal and GaN has a hexagonal unit cell, so one might wonder how crystals of such different types could be grown on each other. We will discuss heteroepitaxy in Chapter 10 more deeply, but in this specific case, the Si ⟨111⟩ surface actually has a hexagonal atomic layout that can seed the GaN growth. Nevertheless, the lattice mismatch is substantial and one can observe in Figure 3.8 that a complex multi-material structure is used to achieve the GaN on Si structure. It is also clear from the figure that large numbers of crystal defects ($10^9 \, \mathrm{cm}^{-2}$ dislocations and stacking faults) are present in the layers of the structure.

Volume defects in crystals can include agglomerations of point defects such as voids caused by a collection of vacancies, precipitates of dopants or impurities, which may be crystalline or amorphous or other 3D defects. One example is shown in Figure 3.7, where a cluster of doping atoms is shown, in this case as a crystalline precipitate (presumed in this example to extend into the third dimension into the plane of the figure). The term "crystalline precipitate" is used in this case because each of the precipitated atoms occupies a lattice site. Such defects may nucleate either heterogeneously, because of an existing crystal defect such as a dislocation, or homogeneously when, through random diffusion, several of the species involved come close enough together to form a critical size nucleus for the precipitate.

There are several such situations of importance in silicon technology. When doping concentrations are pushed too high (beyond the solubility of the dopant in the crystal), precipitation of dopant atoms in amorphous or crystalline clusters is often observed. The doping atoms in these clusters are usually not electrically active. Another example occurs in Czochralski (CZ) silicon because of the relatively high oxygen impurity level. The oxygen is usually incorporated at a level corresponding to its solubility at the crystal growth temperature ($\approx 10^{18}$ cm^{-3}). This is far higher than oxygen's solubility at normal device fabrication temperatures, with the result that oxygen precipitation and clustering can occur during normal device fabrication. In some situations, this can actually be a useful event, since, if the resulting dislocations and damage are kept away from active device regions, these heavily faulted regions can preferentially getter or collect unwanted impurities like heavy metals, keeping them away from device active regions. This is usually accomplished by precipitating the oxygen in the bulk of the wafer, away from the surface where the active devices are located. In CMOS processes that use an epitaxial layer (P on P$^+$, for example), the epitaxial P layer is normally much lower in oxygen concentration than the P$^+$ substrate, so oxygen precipitation is naturally kept away from the region in which devices are fabricated. We will discuss this process in more detail in Chapter 4 when we consider intrinsic gettering.

We will find many examples throughout this book in which crystalline defects play important roles in process steps. In some cases the role is very beneficial; in others the defects are minimized as much as possible to prevent unwanted effects. In both cases, it has become very important in recent years to both understand and control these defects.

3.4   Point Defects

Much of the understanding of fabrication processes such as impurity diffusion, ion implantation damage annealing, gettering and other similar process steps is based today on the role that native point defects, interstitials (I) and vacancies (V) play in these processes. As we saw in the previous section, many different types of crystal defects can exist in crystalline semiconductors. Native point defects are "fundamental" in the sense that thermodynamics predicts their existence and also gives us the mathematical tools to quantify concepts regarding these defects. The other types of defects we discussed are generally process-induced and can be minimized by appropriate choice of fabrication technologies. In this section we will develop the quantitative and qualitative tools to apply point defect concepts throughout this book.

Thermodynamics predicts from fundamental principles that I and V defects will exist in equilibrium at all temperatures above 0 K because the presence of such defects minimizes the free energy of the crystal. The concentrations of both V and I increase with temperature and are given by [2, 3]

$$C^*_{V^0}, C^*_{I^0} = N_S \exp\left(\frac{S^f}{k}\right) \exp\left(\frac{-H^f}{kT}\right), \tag{3.3}$$

where $C^*_{V^0}$ and $C^*_{I^0}$ are the equilibrium concentrations (denoted by *) of the vacancies and interstitials in their neutral charge states, N_S is the number of lattice sites, S^f is the formation entropy of the defect and H^f is the enthalpy of formation of the defect. There is no requirement from thermodynamics that these two concentrations be equal, so in general the S^f and H^f values for I and V are different.

We should consider this a little more carefully because it will be important in later chapters. V and I can be generated in the interior of a crystal by simply moving a silicon atom off a lattice site. This is known as the Frenkel process and it necessarily creates equal numbers of V and I. However, other processes are possible which can affect the two concentrations. Generation processes at crystal surfaces can create a single type of defect. For example, an I is created if a silicon atom at the surface moves into the bulk. Recombination events can also take place between V and I in the bulk (which eliminates one defect of each type), or at the surface (which eliminates only one type of defect). There may also be generation or recombination events that take place at other types of defects in crystals. Stacking faults may capture either type of defect, for example, and grow or shrink by one lattice site in doing so. The net result of all these processes is that the crystal has a number of means to achieve different, yet equilibrium, populations of V and I. Whenever a change is made in the temperature of the crystal, the equilibrium population of the native point defects will change through a combination of the processes described above.

The time required for the crystal to achieve new equilibrium populations is not specified by thermodynamics and depends on the kinetics of the processes taking place (i.e., generation and recombination rates). An important assumption which is usually made in modeling processes in silicon is that temperature changes instantaneously change $C_{V^0}^*$ and $C_{I^0}^*$. This assumption really says that the generation and recombination rates are fast compared to times of interest in process steps like oxidation and diffusion. The actual generation and recombination rates for V and I are not known with any accuracy in silicon. However, the assumption of instantaneously achieved equilibrium populations at the start of any process step generally seems to produce reasonable models of silicon processes. One clear exception to this statement occurs when ion implantation is used to introduce impurities into silicon. The resulting damage introduces large excess concentrations of V and I and other types of defects, which do not disappear immediately upon heating the crystal. Thus large, non-equilibrium populations of the defects may be present in the crystal for extended periods of time, with important implications for process phenomena like diffusion. Other situations can also arise which produce non-equilibrium point defect concentrations. We will pursue these issues further in later chapters, particularly in Chapters 7 and 8.

The equilibrium concentrations of V and I have never been measured directly in silicon at process temperatures, primarily because they are very small. Most of the estimates of $C_{V^0}^*$ and $C_{I^0}^*$ come from fitting impurity diffusion data with models that involve the native point defect concentrations (Chapter 7). Since such estimates are subject to the particular model used, and since there is still controversy over such models, exact values for the parameters in (3.3) are still in question. Values fitted to the diffusion models that we will discuss in Chapter 7 give [3]

$$C_{I^0}^* \cong 5 \times 10^{22} \exp\left(-\frac{2.36 \text{ eV}}{kT}\right), \tag{3.4}$$

$$C_{V^0}^* \cong 2.31 \times 10^{21} \exp\left(-\frac{1.08 \text{ eV}}{kT}\right). \tag{3.5}$$

These expressions, along with n_i, are plotted in Figure 3.9 for silicon as a function of inverse temperature. A more detailed discussion of these values is given in [4].

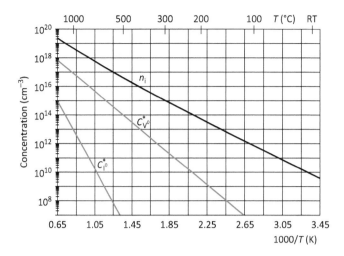

Figure 3.9 Intrinsic carrier concentration n_i, $C^*_{V^0}$ and $C^*_{I^0}$ versus inverse temperature in silicon.

As shown in Figure 3.9, $C^*_{I^0} \approx 10^{14}$ cm^{-3} and $C^*_{V^0} \approx 10^{17}$ cm^{-3} at 1000 °C. At this temperature, the intrinsic carrier concentration n_i is about 7.14×10^{18} cm^{-3}. Normal doping concentrations in silicon devices are 10^{15}–10^{20} cm^{-3}. It is easy to see then why it is so difficult to measure the point defect concentrations directly. Their concentrations are much smaller than either the impurity concentrations or the intrinsic carrier concentration at all temperatures, and, since silicon has $\approx 5 \times 10^{22}$ atoms cm^{-3}, they represent only about 1 in 10^5 to 1 in 10^{10} lattice sites, and are well below the detection limit of physical techniques.

There are a few conditions under which point defect concentrations are large enough to be measurable. At Si crystal growth temperatures (Section 3.6), equilibrium point defect concentrations are high. If the crystal is cooled rapidly as it is pulled from the melt, there may not be time for these point defects to recombine or out-diffuse, with the result that non-equilibrium, excess concentrations may be frozen into the crystal. This often results in what are referred to as swirl defects in silicon crystals, which are believed to be condensed I defects. Such macroscopic defects can be delineated in silicon by etchants and rough estimates can be made of defect populations. Annealing of the crystal for extended periods at lower temperatures can dissolve such defect precipitates by allowing time for recombination and diffusion processes to eliminate the excess point defects. A second example relates to the ion implantation process. Ion implantation damage often results in very large concentrations of point defects, large enough to completely destroy the crystalline nature of the semiconductor in many cases. Such concentrations are easily measurable by physical techniques like Rutherford backscattering. We will discuss this particular case in more detail in Chapter 8.

Point defects are extremely mobile in the silicon lattice. While their diffusivity cannot be directly measured experimentally (because of their small concentrations), it can be inferred from other experimental measurements of dopant diffusivities (Chapter 7). Point defect diffusivities can also be estimated by doing a computer simulation of an interstitial or vacancy in the silicon lattice at different lattice temperatures and following their migration path. The diffusivity can be estimated this way at different temperatures and can then be expressed in Arrhenius form. Current estimates for the diffusivities of interstitials and vacancies are [3, 4]

$$D_I \cong 600 \exp\left(-\frac{2.44 \text{ eV}}{kT}\right) \text{ cm}^2\text{s}^{-1}, \tag{3.6}$$

$$D_V \cong 13 \exp\left(-\frac{2.92 \text{ eV}}{kT}\right) \text{ cm}^2\text{s}^{-1}. \tag{3.7}$$

These equations, which are plotted in Figure 3.10, give numbers for D_I and D_V that are orders of magnitude larger than dopant diffusivities. This will have significant implications in Chapter 4 when we discuss gettering, and in Chapters 7 and 8 when we discuss dopant diffusion and ion implantation.

It is well established that native point defects in silicon can exist in charged as well as neutral states [4, 5]. Singly charged (+, −) and doubly charged (++, =) vacancies have been identified experimentally using electronic techniques after excess point defects were introduced by high-energy electron bombardment. Similar charge states for silicon I are assumed to exist, although they have not been measured experimentally, probably because of the fast diffusion of these defects in silicon.

Charged point defects have energy levels in the silicon bandgap. That is, the defects can take on various charge states depending on the location of the Fermi level E_F. Shallow donors and acceptors are usually ionized, at least at room temperature, while impurity levels lying deeper in the bandgap are ionized only when E_F is above them (acceptors) or below them (donors). (These ideas are discussed in detail in Appendix A.1 for readers who wish to review them.) The charged point defect levels behave very much like deep donors and acceptors.

The positions of the vacancy energy levels are illustrated qualitatively in Figure 3.11 at a particular temperature. The Fermi level E_F will be at the E_i position for intrinsic conditions

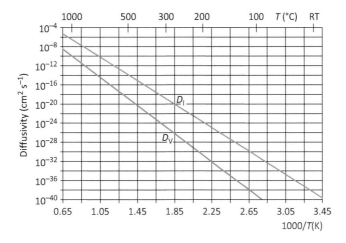

Figure 3.10 Vacancy and interstitial diffusivities in silicon inferred from diffusion experiments [3].

and would be higher in the bandgap, as illustrated, for N-type doping. We saw in Figure 3.9 that $n_i = 7.14 \times 10^{18}$ cm^{-3} at 1000 °C, so that, at this temperature, for all doping levels below this value, $E_F = E_i$. For higher doping levels, the material will be extrinsic and E_F will move up or down in the bandgap for N- or P-type doping, respectively. Consider first the intrinsic case for which $E_F = E_i$. In this situation, E_F is below the vacancy acceptor levels V$^-$ and V$^=$. It is also above the vacancy donor level V$^+$. Thus, none of these levels will be ionized and the dominant vacancy charge state in the silicon will be the neutral vacancy V^0. There is some evidence that a V^{++} vacancy charge state also exists, as mentioned earlier. However, modeling processing technologies such as dopant diffusion do not require using this charge state, so it is not shown in Figure 3.11.

If the silicon is extrinsic, the dominant vacancy charge state will be different. Consider the case illustrated in Figure 3.11 for extrinsic N-type doping. There, E_F is now above the V$^-$ level. Thus the V$^-$ level will be populated by an electron; that is, the vacancy will be acting as an acceptor and will be charged negatively. Since the background material is N type in this example, there will be many electrons available from the shallow donors to charge the vacancy. It is therefore likely that the dominant vacancy charge state will be V$^-$ and not the neutral vacancy. Similarly, in P-type material, the dominant vacancy charge state will be V$^+$ since E_F will be in the lower half of the bandgap, below the V$^+$ level.

The same statistics that we used in Appendix A.1 to describe donors and acceptors can also be used to describe the equilibrium populations of V and I in their various charge states. Shockley and Last [6] first presented these results:

$$C_{V^+}^* = C_{V^0}^* \exp\left(\frac{E_{V^+} - E_F}{kT}\right), \tag{3.8}$$

$$C_{V^-}^* = C_{V^0}^* \exp\left(\frac{E_F - E_{V^-}}{kT}\right), \tag{3.9}$$

$$C_{V^=}^* = C_{V^0}^* \exp\left(\frac{2E_F - E_{V^-} - E_{V^=}}{kT}\right). \tag{3.10}$$

We can rewrite these equations in a form that will be particularly convenient in later chapters when we discuss diffusion, ion implantation and other specific processes. For example, using (A1.13) from Appendix A.1, we have

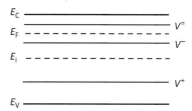

Figure 3.11 Approximate location of charged vacancy energy levels in the silicon bandgap. Here E_i is the intrinsic Fermi level and E_F is the Fermi level in N-type material.

$$C_{V^-}^* = C_{V^0}^* \exp\left(\frac{E_F - E_i}{kT}\right) \exp\left(-\frac{E_{V^-} - E_i}{kT}\right) = \frac{n}{n_i} C_{V^0}^* \exp\left(-\frac{E_{V^-} - E_i}{kT}\right). \tag{3.11}$$

In a similar fashion, we can also show that

$$C_{V^+}^* = \frac{p}{n_i} C_{V^0}^* \exp\left(-\frac{E_i - E_{V^+}}{kT}\right), \tag{3.12}$$

$$C_{V^=}^* = \left(\frac{n}{n_i}\right)^2 C_{V^0}^* \exp\left(-\frac{E_{V^-} - E_i}{kT}\right) \exp\left(-\frac{E_{V^=} - E_i}{kT}\right). \tag{3.13}$$

Similar equations can be written for I defects.

There are several important observations that need to be made regarding these equations and the populations of charged point defects in silicon. The first is that the equilibrium populations of the neutral and charged defects are always small compared to either n_i or dopant concentrations. This is important because it means that the number of electrons or holes bound to V and I is always a negligible number compared to the total number of electrons or holes present in the material (at least, in equilibrium). This greatly simplifies calculations involving the charged point defects. To find the charged defect populations, we simply do the following:

1. Calculate the position of E_F in the usual way (described in Appendix A.1), ignoring the point defects.
2. Calculate the concentrations of the charged point defects using this value of E_F and (3.8)–(3.10) or (3.11)–(3.13). The resulting numbers for the charged point defect populations will always be much smaller than either n_i or n or p (whichever is dominant). Because of this, there is no need to take the charged defects into account when calculating the position of E_F. If the numbers of charged point defects were comparable to n or p, then the charging of the defects would necessarily reduce the free electron or hole population and hence change E_F.

A second point that turns out to be very important is that, as the doping in the silicon changes and hence E_F changes, not only does the dominant charge state of the V and I change, but so also does the total number of each type of defect. This is not an obvious result at first glance but can be explained as follows. The neutral vacancy and interstitial concentrations are a function only of temperature, not of doping. Equations (3.4) and (3.5) depend only on temperature. If we change the population of charged defects by doping the material and moving E_F, then $C_{V^0}^*$ and $C_{I^0}^*$ do not change. So where do the charged vacancies and interstitials come from?

Our initial guess would likely have been that we get the charged defects simply by adding an electron or hole to a neutral defect. That cannot be correct, however, because that would decrease the population of neutral defects. The correct answer, therefore, must be that new V and I are created by the crystal to supply the need for charged defects. This, of course, can occur through the generation processes described earlier. Thus in doped material, we should expect to find not only more charged point defects than in intrinsic material, but also more total V and I. In fact, at the same temperature, the $C_{V^0}^*$ and $C_{I^0}^*$ will be identical in extrinsic and intrinsic silicon. The difference will be that the extrinsic material will have many more charged defects: V^-, $V^=$, I^- and $I^=$ in N-type material, and V^+ and I^+ in P-type material. These points are discussed in more detail by Shockley and Last [6].

Equations (3.11)–(3.13) also provide us with some very important insights. They show that the charged point defect populations are proportional to n/n_i or p/n_i. At processing temperatures, n_i is on the order of 10^{18}–10^{19} cm^{-3} (Figure 3.9), so, for any doping concentrations less than this, the material will be intrinsic. In intrinsic material, $n/n_i = p/n_i = 1$, and charged V or I concentrations are generally small. At processing temperatures, doping levels can exceed n_i by 10–100 times, resulting in huge increases in the concentrations of charged point defects. We will see in later chapters that dopant diffusivities are proportional to n/n_i or p/n_i in extrinsic material. This provides a direct connection to the underlying mechanisms – charged point defects – that we will discuss in Chapter 7.

In compound semiconductors the number of point defect types is larger, as illustrated in a simple 2D representation in Figure 3.12. Two atom types (A and B in Figure 3.12) form the lattice, so there are two types of interstitials and two types of vacancies. Both the pure interstitial and the interstitialcy are shown for each type, as was the case in Figure 3.7. Both types of native point defects can exist in several charge states, as was the case for silicon considered above. Finally, antisite defects can also exist in compound semiconductor crystals. Antisite defects such as As_{Ga} or Ga_{As} often play important roles in devices made in these materials (see Section 4.4.5 in Chapter 4).

The same mathematical description of the concentrations of these various defects that was discussed above for silicon also applies to compound crystals [7]. Of course, there are more such equations because there are more defect types. In general, quantitative modeling using point defects is not as advanced today in compound materials as it is in silicon, but the same qualitative concepts apply, as we will see in later chapters. Qualitatively, we should expect that charged point defect populations should increase with doping in extrinsic material and total point defect concentrations should also, in other semiconductors, just as they do in silicon.

Figure 3.12 Simple 2D representation of compound semiconductor point defects.

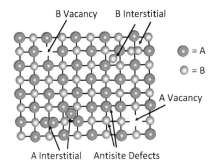

3.5   Silicon Crystal Growth Methods: Czochralski and Float Zone

Silicon is a very abundant material, representing about 25.7% of the Earth's crust. Naturally occurring minerals containing silicon are very impure, however, which means that the silicon must be refined as well as converted into the crystalline form. This is usually a multi-stage process, beginning with quartzite, a type of sand. Chemically, quartzite is SiO_2. The first step in the refining process is to convert the quartzite to metallurgical-grade silicon (MGS). This process usually takes place in a furnace in which a mixture of the quartzite and a carbon source (usually coal or coke) is heated to temperatures approaching 2000 °C. A number of reactions take place in the furnace, the overall result of which is that liquid silicon is drawn off and CO is given off as a gas.

To convert the MGS to electronic-grade silicon (EGS), several steps are required. In the first of these, the MGS reacts with gaseous HCl, usually by grinding the MGS to a fine powder and then reacting it in the presence of a catalyst at elevated temperatures. This process can form any of a number of SiHCl compounds; $SiHCl_3$ is most commonly used today. $SiHCl_3$ is a liquid at room temperature, so it can be purified by fractional distillation. In this process, the $SiHCl_3$ is boiled along with the impurities it contains and separated by its boiling point. After this process, the $SiHCl_3$ is extremely pure and is ready to be converted back to purified polysilicon. This is accomplished in a large chemical vapor deposition (CVD) reactor.

A thin silicon rod is used as the nucleation surface, on which the silicon deposits. Since the rod is not a single crystal, polysilicon is deposited. Typically the rod is heated by passing an electric current through it. The polysilicon, which is deposited, may be several meters long and several hundred millimeters in diameter. The overall deposition can take many days. Finally, the resulting polysilicon is broken up into pieces to load into crucibles for CZ crystal growth or the polysilicon rod could itself be used for float zone (FZ) crystal growth, which is described later. The overall result of these purification steps is that the final product (polysilicon) has only parts per billion (ppb) or on the order of 10^{13}–10^{14} cm^{-3} impurities. This represents about an eight-order-of-magnitude improvement from the starting quartzite and makes EGS the purest material routinely available on Earth.

Starting with EGS silicon raw material, the two methods used to grow silicon crystals today are the CZ and the FZ techniques. CZ is much more common because it is capable of easily producing large-diameter crystals, from which large-diameter wafers can be cut. The only significant drawback to the CZ method is that the silicon is contained in liquid form in a crucible during growth and, as a result, impurities from the crucible are incorporated in the growing crystal. Oxygen and carbon are the two most significant contaminants. These impurities are not always a drawback, however. Oxygen, in particular, can be very useful in mechanically strengthening the silicon crystal and in providing a means for gettering other unwanted impurities during device fabrication, as discussed in Chapter 4.

The CZ growth technique is illustrated conceptually in Figure 3.13. Pieces of EGS are placed in a silica (SiO_2) crucible along with a small amount of doped silicon and melted. The melt temperature is stabilized at just above the silicon melting point (1417 °C). A small single-crystal seed is then lowered into the melt. The crystal orientation of this seed will determine the

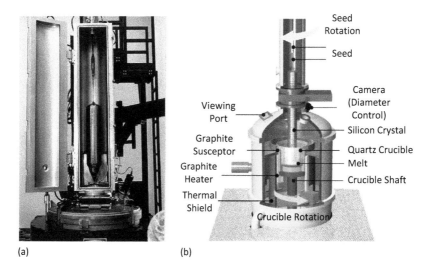

Figure 3.13 (a) Basic Czochralski (CZ) crystal growing apparatus and (b) schematic. (a) Photo courtesy of R. Carranza. (b) Schematic reproduced from [8], open access publication.

orientation of the resulting pulled crystal and wafers. The amount of dopant placed in the crucible with the silicon charge will determine the doping concentration in the resulting crystal. The seed is then slowly pulled out of the melt. Silicon atoms from the melt bond to the atoms in the seed, lattice plane by lattice plane, forming a single crystal as the seed is pulled upwards. The diameter of the resulting crystal is controlled primarily by the rate at which the crystal is pulled. Faster pulling results in a smaller-diameter crystal, an intuitive result that we will quantify later in the chapter.

When the cold seed is first inserted into the molten silicon, very large temperature gradients exist through the seed volume. This results in high stresses and the generation of dislocations in the crystal. As discussed earlier, dislocations are active defects in a crystal. They can grow or shrink during high-temperature processing and are very detrimental to device electrical characteristics. Fortunately, it is possible to eliminate them from the crystal during the growth process. Dash first demonstrated these ideas [9]. By beginning the growth at a high pull rate, a thin neck is produced just below the seed. This grows out the dislocations to the crystal (neck) surface where they terminate. The growth is then slowed down, the diameter increases, and a dislocation-free, large-diameter crystal is grown, provided, of course, that the rest of the growth is done under conditions that do not introduce large stresses or growth perturbations.

During CZ crystal growth, the seed and crucible are normally rotated in opposite directions to promote mixing in the liquid and more uniform growth. Unfortunately, this also has the effect of increasing the corrosion of the crucible by the melt. Very few options are available in terms of crucible materials that are relatively inert to molten silicon. The dominant choice today is quartz, but these crucibles are slowly dissolved by the silicon, resulting in both silicon and oxygen being incorporated into the melt. Much of the oxygen evaporates as SiO, but some is incorporated into the growing crystal, producing oxygen levels of 10^{17}–10^{18} cm^{-3}. (The solid solubility of oxygen in silicon is $\approx 10^{18}$ cm^{-3} at melt growth temperatures.) The quartz crucible requires a graphite susceptor for mechanical support. During crystal growth, carbon evaporation from this source and carbon from the EGS charge itself result in carbon incorporation in the crystal, typically at levels of 10^{15}–10^{16} cm^{-3}. Both the oxygen and carbon concentrations are higher than the equilibrium bulk solubilities at typical IC fabrication temperatures. Therefore, the Si is in non-equilibrium during processing and there is a driving force to precipitate both C and O. To avoid additional impurities from the ambient, the growth is normally performed in an argon ambient.

Since the carbon and oxygen impurities evolve from time-dependent sources, they are usually not incorporated into the growing crystal uniformly along the length of the crystal. Often the seed end of the crystal will contain higher oxygen concentrations than the tail end because, as the level of silicon in the melt reduces, there is less dissolution of the crucible by the molten silicon. It is possible to more precisely control the oxygen and carbon concentrations in the crystal using more advanced growth methods. The rotation (mixing) of the melt is an important variable. Magnetic fields can be added with external magnets to help control the mixing processes in the crucible.

Oxygen has three principal effects in the silicon crystal. The first is beneficial. In an as-grown crystal, the oxygen is believed to be incorporated primarily as dispersed single atoms designated O_I occupying interstitial positions in the silicon lattice, but covalently bonded to two silicon atoms. The oxygen atoms thus replace one of the normal Si–Si covalent bonds with a Si–O–Si structure. The oxygen atom is neutral in this configuration. Such interstitial oxygen atoms improve the yield strength of silicon by as much as 25%, making the silicon wafers more robust in a manufacturing facility. The improvement, due to solution hardening, increases with oxygen concentration, as long as the oxygen stays on interstitial sites and does not precipitate.

The second principal effect of oxygen in silicon is the formation of oxygen donors. A small amount of the oxygen in the crystal forms what are thought to be SiO_4 complexes, which act as donors. They can be detected by changes in the silicon resistivity corresponding to the free electrons donated by the oxygen complexes. As many as 10^{16} cm^{-3} donors can be formed in this way, which is sufficient to dramatically increase the resistivity of lightly doped P-type wafers under extreme conditions. The donors form most readily at temperatures around 400–500 °C at a rate proportional to $[O_I]^4$. During the CZ growth process, the crystal cools slowly through this temperature range and oxygen donors normally form. The SiO_4 complexes are unstable at temperatures above 500 °C and so they can be removed by annealing wafers above this temperature. The donors can re-form, however, during normal IC manufacturing, if a thermal step around 400–500 °C is used.

The third effect of oxygen in silicon is the tendency of the oxygen to precipitate under normal device processing conditions, forming SiO_2 regions inside the wafer. The precipitation arises because the oxygen was incorporated at the melt temperature and is therefore supersaturated in the silicon at process temperatures. The formation of SiO_2 precipitates is a critical part of a process called "intrinsic gettering" that we will study in detail in Chapter 4 because of its importance in removing unwanted impurities from silicon wafers.

The carbon present in CZ grown crystals comes from the graphite components in the crystal pulling machine (see Figure 3.13). The actual mechanism of incorporation is believed to be the following. The melt contains silicon and modest concentrations of oxygen as discussed above. This results in the formation of SiO that evaporates from the melt surface. Generally, the ambient in the crystal puller is Ar flowing at reduced pressure, and the SiO can be transported in the gas phase to the graphite crucible and other support fixtures. SiO reacts with graphite (carbon) to produce CO that again transports through the gas phase back to the melt. From the melt, the carbon is incorporated into the growing crystal.

At the low concentrations in which it is normally present in CZ crystals, carbon is mostly substitutional in the silicon lattice. Since it is a column IV element, it does not act as a donor or acceptor in silicon and one might think that at 1 ppm, it might be relatively benign. This is not necessarily the case, however. Carbon is known to affect the precipitation kinetics of oxygen in silicon even at such very small concentrations. This is likely because there is a volume expansion when oxygen precipitates and a volume contraction when carbon precipitates because of the relative sizes of O and C. There is thus a tendency for precipitates that are complexes of C and O to form to minimize stresses in the crystal. Since precipitated SiO_2 is crucial in intrinsic gettering, this can have an effect on gettering efficiency.

The basic CZ process has not changed very much since it was first applied to silicon and germanium in the early 1950s by Teal [10, 11]. It is really quite remarkable that large-diameter silicon wafers (up to 300 mm or 12 inches today) with essentially ideal properties can be routinely purchased today at a cost of about $100, which is a small fraction of the total wafer fabrication costs associated with IC manufacturing.

Simple analytic models have been developed that show the relationship between pull rate and crystal diameter. Intuitively one would expect that faster pulling would produce a smaller-diameter crystal, and this is in fact the case. These models are based on heat transfer, as illustrated in Figure 3.14. The control parameters during the process are temperature and pull rate, and modern crystal pullers use computer control in a feedback system to monitor temperature and crystal diameter and to control both the pull rate and the power into the heater. Secondary parameters that influence the growth rate are the gas flows in the system and crystal and crucible rotation. Typically, an infrared temperature sensor is focused on the melt/crystal interface and lasers may be used to monitor the crystal diameter and the level of liquid in the melt. These inputs to the control system are used to adjust the pull rate and power levels to the heater.

In Figure 3.14, process A is the transfer of the latent heat of crystallization from the liquid to the solid. Process B is the conduction of the heat away from the freezing interface up the solid crystal. Process C is the transfer of the heat by radiation away from the crystal. In steady state, all three of these must be equal. Simple models for these processes result in the growth rate of the crystal being described by [12]

$$v_{PMAX} = \frac{1}{LN} \sqrt{\left(\frac{2\sigma\varepsilon k_M T_M^5}{3r} \right)},$$

(3.14)

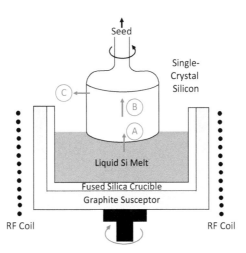

Figure 3.14 Freezing processes occurring during CZ crystal growth.

where v_{PMAX} is the maximum pull rate, L is the latent heat of fusion, N is the density of silicon, σ is the Stefan–Boltzmann constant, ε is the emissivity of the silicon, k_M is the thermal conductivity of silicon, T_M is the melting temperature of silicon and r is the radius of the crystal. This relationship confirms our intuition that faster pull rates result in smaller-diameter crystals, although the relationship is an inverse square-root relationship, not a linear relationship. Substituting in numbers produces a maximum pull rate v_{PMAX} of about 24 cm h^{-1} for a 150 mm (6 inch) diameter crystal. Today, powerful numerical modeling tools are available for simulating the growth process that can calculate not only the crystal pull rate, but also important quantities like the oxygen profile in the crystal [13, 14]. An example using commercially available software is shown in Figure 3.15.

Some crystal growing machines use a magnetic field during CZ growth, resulting in magnetic Czochralski (MCZ) ingots and wafers. The principle is simple. Surrounding the crucible containing the molten silicon with a magnetic field results in suppression of the thermal convection currents in the melt. This occurs because, according to Lenz's law, currents are induced in a conductor whenever that conductor crosses magnetic field lines. The direction of the magnetic force tends to reduce the currents flowing in the silicon conductor due to thermal convection. This produces a more uniform ingot diameter and resistivity because the temperature fluctuations are smaller, and it produces lower and more uniform oxygen concentrations in the crystal because there is less dissolution of the crucible walls.

Another modification to the basic CZ process that has been investigated to improve the uniformity of crystal properties is the double crucible method. In this process, there are actually two crucibles of molten silicon. Growth occurs in one of them; the second is a reservoir of additional molten silicon. The two crucibles are usually concentric, with the inner one used for growth. Silicon pellets can be supplied to the outer crucible during growth, feeding silicon to the inner crucible continuously during growth. This can result in longer crystals but it also allows a more nearly constant doping concentration in the growth crucible. The result is better control and uniformity of doping and oxygen concentration. This process is not widely used today but may find increased use in the future.

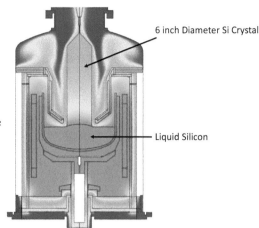

Figure 3.15 Example of numerical modeling of the CZ growth process showing the temperature distribution throughout the CZ puller. This example was taken from the FEMAG Soft website using FEMAG/CZ software (https://www.femagsoft.com/products/femag-cz.html). FEMAG Soft is a commercial supplier of software to simulate crystal growth.

6 inch Diameter Si Crystal

Liquid Silicon

The FZ process is illustrated in Figure 3.16. Since this same process can also be used to refine silicon, it is also sometimes referred to as zone refining. The principal difference between this process and the CZ process is that no crucible is used. This markedly reduces impurity levels in the resulting crystal, particularly oxygen, and makes it easier to grow high-resistivity material. FZ material is used primarily today in applications that require high resistivity and low oxygen content, or both, such as in some types of detectors and power devices.

In the FZ process, a rod of EGS polysilicon is clamped at both ends, with the bottom in contact with a single-crystal seed. An RF coil provides power, which generates large currents in the silicon and locally melts it through I^2R heating. Usually the molten zone is about 2 cm long. Surface tension and levitation due to the RF field keep the system stable. If the melting is initiated in a zone at the seed end and slowly moved up the rod, a single crystal results. As was the case in the CZ process, atoms from the liquid phase bond to the single-crystal solid material, atomic plane by atomic plane, as the zone slowly moves up the rod (or as the rod is translated downwards through a stationary coil). Doping of the crystal can be accomplished by starting with either a doped polysilicon rod or a doped seed, or by maintaining a gas ambient during the FZ process that contains a dilute concentration of the desired dopant.

It is not as straightforward to scale up the crystal diameter using the FZ process, because of stability problems associated with the liquid zone in a gravity environment. For these same reasons, FZ wafers usually have greater microscopic resistivity variations than do CZ wafers. It is thus likely that the CZ technique will remain the dominant growth method for large-diameter silicon crystals. Modern numerical simulation methods, similar to the example shown in Figure 3.15, have also been applied to the FZ method of growth [15]. An example is shown in Figure 3.17. This example involves a multi-physics simulation including heat transfer, fluid flow and electromagnetics to simulate the power transfer from the inductive heating coil to the silicon. Note the shape

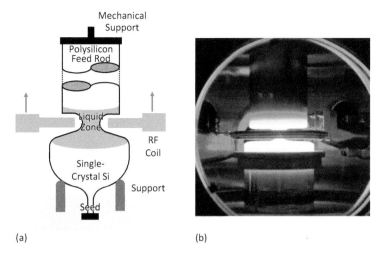

Figure 3.16 Basic float zone (FZ) crystal growth or zone refining apparatus, shown (a) conceptually and (b) in an actual machine. Photo courtesy Topsil GlobalWafers A/S. Photographer Lars Dalby Nielsen.

of the liquid/solid interface in the simulation, which is semicircular. Float zone silicon wafers are generally available with diameters up to 6 inches, with limited availability of larger-diameter wafers.

3.6 Compound Semiconductor Crystal Growth Methods

Compound semiconductor crystals can often be grown with methods that are similar to those used for silicon crystals. However, modifications are often required because one of the elements in the compound material is volatile at the melting temperature. GaAs is a good example. GaAs melts at 1240 °C. However, at this temperature, the equilibrium vapor pressure of the GaAs is $\approx$ 80 kPa (just under 1 atm) and the As preferentially evaporates compared to the Ga [16, 17]. Thus a free GaAs liquid surface is not a good approach for GaAs CZ growth.

To mitigate the high As vapor pressure issue, two crystal growth techniques are commonly used that are modifications of the silicon methods described earlier [17]. These are illustrated in Figure 3.18. In the liquid-encapsulated Czochralski (LEC) process shown in Figure 3.18(a), boron trioxide (B_2O_3) is placed in a high-purity BN or SiO_2 crucible along with the GaAs. As the crucible is heated, the B_2O_3 melts at 460 °C and forms a thick viscous liquid that covers the GaAs. This layer along with the As vapor pressure in the crystal puller prevents sublimation of the As from the melt. Continued heating results in the GaAs melting at 1240 °C. Then a single-crystal seed is inserted through the B_2O_3 layer and a crystal is pulled just as in the Si CZ process.

In the horizontal Bridgman method shown in Figure 3.18(b), a two-zone furnace is used. The left-hand zone is maintained at $\approx$ 610 °C. This provides sufficient As overpressure in the sealed system to prevent As loss from the GaAs. The right side contains a single-crystal seed on the left

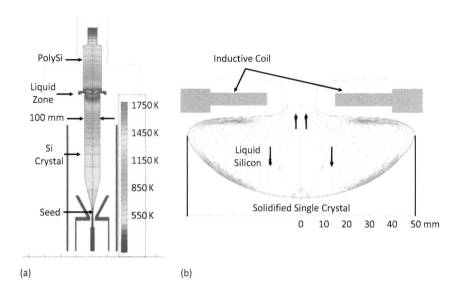

(a)   (b)

Figure 3.17 Numerical modeling of the Si FZ growth process, simulating the growth of a 100 mm Si crystal grown at a 1 mm min^{-1} growth rate. (a) The temperature distribution around the liquid zone. (b) A simulation of the flow lines in the liquid silicon region. After https://www.femagsoft.com/products/femag-fz.html. FEMAG Soft is a commercial supplier of software to simulate crystal growth.

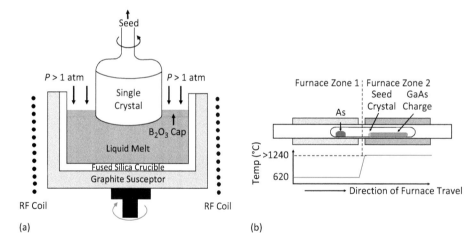

(a)   (b)

Figure 3.18 (a) Liquid-encapsulated Czochralski (LEC) crystal growth method. (b) Bridgman two-zone furnace crystal growth apparatus [18].

and a GaAs polycrystalline charge. As the furnace moves to the right, the GaAs crystallizes at the liquid/solid interface. This growth method does not produce circular-shaped boules (typically D-shaped), so either the wafer processing has to be modified to deal with the unusual wafer shape, or circular wafers are cut from the boule, wasting some of the crystalline GaAs. The

Bridgman method can also be accomplished with the furnaces oriented vertically, which produces more circular-shaped boules.

The LEC method is used today for most GaAs single crystals. Wafers up to 150 mm (6 inches) in diameter are available. The most dominant defect in melt-grown GaAs crystals is the so-called EL2 defect that is believed to be an As_{Ga} antisite defect (see Figure 3.12). This defect has a deep level in the GaAs bandgap and is often purposefully introduced during growth to create semi-insulating material. The concentration of As_{Ga} defects can be controlled during growth through the As overpressure used during the growth.

InP is another commercially important semiconductor. Boules up to 150 mm (6 inches) in diameter can be grown with the LEC technology described above for GaAs [19]. At the 1062 °C melting temperature of InP, P has a high vapor pressure so the B_2O_3 liquid performs the same function it did in GaAs LEC. Both horizontal and vertically oriented Bridgman two-zone furnaces have also been used to grow InP single crystals, but the dominant method today is LEC. Many of the other compound semiconductors clustered around InP in Figure 3.4 can also be grown in single-crystal form using either CZ or LEC growth methods, the choice depending on the volatility of the column V element.

Several wide-bandgap semiconductors, primarily SiC and GaN, are also of considerable current interest because they have much higher critical electric fields than Si and hence dramatically improve power devices that operate at high voltages and high currents. Both of these materials have much higher melting temperatures than the semiconductors we have discussed to this point and also other issues that make single-crystal growth very challenging.

SiC does not exist in the liquid phase. Rather, it sublimes at temperatures in excess of 2000 °C. This means single crystals generally must be grown either by a sublimation process or by a CVD that we will discuss in Chapter 9 in detail. Figure 3.19 illustrates the basic idea of the sublimation process, known as the Lely process [20]. A SiC powder source is heated to 2300–2400 °C that causes sublimation. The growth occurs on a single-crystal seed. Alternatively, the Si and C can be introduced by gaseous precursors such as SiH_4 and C_3H_8 that react to grow SiC on a single-crystal seed at temperatures above 1000 °C. Growth rates are slow and a seed with the full diameter of the growing boule must be used. The sublimation process is most commonly used today and SiC wafers up to 150 mm (6 inches) in diameter are commercially available.

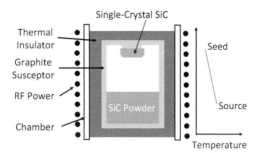

Figure 3.19 Sublimation growth of SiC single-crystal growth.

Finally, GaN, the other wide-bandgap material of interest today, is even more difficult to grow into large single crystals. GaN melts at ≈2500 °C, but it decomposes at temperatures above 1000 °C. To exist in the liquid state at its melting temperature, the N overpressure must be 4.5 GPa (45,000 atm). This rules out most of the common crystal growth methods. As a result, CVD methods must generally be used, which result in low growth rates [21, 22]. Today, up to 100 mm (4 inch) diameter GaN single-crystal wafers are commercially available, but they are very expensive, and the pathway to much lower cost is challenging. Because of this, current applications of GaN generally involve growing single-crystal GaN on other substrates that are lattice-matched to GaN as closely as possible. A very large industry exists today using GaN on SiC or sapphire (Al_2O_3) to produce GaN LEDs for lighting applications. A growing industry exists using GaN on Si (111) substrates to support the power device industry (see Figure 3.8). Both of these are examples of heteroepitaxy that we will consider in Chapter 9.

3.7   Wafer Preparation

After a crystal is grown by any of the methods described above, individual wafers must then be prepared for IC manufacturing. This involves a series of mechanical steps, which result in wafers that are almost perfectly flat, polished to a mirror finish on the top side and free of any mechanical defects from the sawing and other operations. The processes are similar for all of the semiconductor materials, but we will describe the steps for silicon wafers since that is the most common semiconductor used today.

The process of creating the individual wafers begins with shaping the grown crystal or boule to a uniform diameter. Modern crystal growers cannot maintain perfect control over the crystal diameter during growth, so the crystal is normally grown slightly oversized and then trimmed to the desired final diameter. This shaping operation involves placing the boule in a lathe-like machine and grinding it with a diamond wheel.

After the boule is ground to an appropriate diameter, one or more "flats" are normally ground along its length. These will end up as straight edges on otherwise circular wafers. They serve several purposes. The first is as a reference plane for many types of automatic equipment that handle wafers during manufacturing. The second purpose is to indicate the type and crystal orientation of the wafers. The flats are normally ground along a major crystal axis. The longest ("primary") flat is, by convention, oriented perpendicular to the ⟨110⟩ direction. In larger-diameter wafers (200 and 300 mm), the flats have been replaced by a simple notch to indicate orientation, since a flat along one edge loses too much area otherwise usable for chip fabrication. The notch normally indicates the ⟨110⟩ direction on a (100) wafer.

The next step in wafer preparation is the actual sawing of the boule into individual wafers. This is usually accomplished with a diamond-tipped blade that cuts on its inside edge. Using an inside edge cutting procedure allows the outer part of the saw blade to be thicker and hence more rigid, without wasting large amounts of the boule in the saw cut regions. The two common surface orientations for silicon wafers today are (100) and (111). Wafers with these surface orientations are usually produced using seeds of the appropriate orientation, which then allows cutting of the individual wafers perpendicular to the boule.

Two additional mechanical steps are performed on the individual wafers at this point. The first is a mechanical lapping operation that removes about 50 μm of silicon using pressure and a mixture of Al_2O_3, water and glycerin. This step serves to improve the flatness of the wafers to about ±2 μm, removing most of the taper and bow that result from the sawing operation. The lapping also removes much of the saw damage from the wafer surfaces. Successively finer grit Al_2O_3 powders are used in the lapping to produce the final uniform wafer. It is important that the wafers be flat because later in the IC manufacturing process, high-resolution printing processes with limited depth of focus will be used to produce the patterns needed for the various device structures. The final purely mechanical process normally performed on the wafers produces rounded edges. This is again performed on a tool with diamond cutting edges and grinds a radius on the wafer edges. This greatly decreases chipping and the introduction of dislocations and other defects at the wafer edges during IC manufacturing.

At this stage, the wafers are ready for the final few steps that produce a mirror finish on one surface. A two-step process is used, chemical etching, followed by chemical–mechanical polishing (CMP). The chemical etching is done as a batch process, with the wafers loaded into cassettes and immersed in a mixture of nitric, hydrofluoric and acetic acids. The acetic acid serves to dilute the mixture and control the etch rate. The actual etching takes place with the nitric acid oxidizing the silicon and the HF dissolving the resulting oxide. Overall the reaction that takes place is the following:

$$3Si + 4HNO_3 + 18HF \rightarrow 3H_2SiF_6 + 4NO + 8H_2O \,. \tag{3.15}$$

Generally about 20–25 μm of silicon is etched off each side of the wafer. The process removes the surface layers containing damage from the various mechanical operations performed earlier.

The final step in preparing the wafers is CMP. This process is illustrated in Figure 3.20. The wafers are polished under pressures on the order of 20 psi (≈ 140 kPa). The wafers are rotated in the polishing machine in a slurry consisting of a suspension of fine (10 nm) SiO_2 particles in an aqueous solution of NaOH. The rotation and pressure generate heat that drives a chemical reaction in which OH^- radicals from the NaOH oxidize the silicon. The SiO_2 particles abrade the oxide away. The overall process is thus a combination of chemical and mechanical polishing. Typically, about 25 μm of silicon is polished away, producing a surface that is defect-free and has a mirror finish suitable for IC fabrication. CMP has become a pervasive technology during wafer manufacturing. We saw a number of examples of this in the CMOS process in Chapter 2. We will discuss CMP in detail in Chapter 9.

A final point regarding wafer preparation should be mentioned. It is often important in IC manufacturing to track individual wafers through the fabrication process. Lot tracking and individual wafer tracking require that the wafers carry some identification. Generally, either an alphanumeric code or a bar code is used, which is unique to each wafer. Typically, the code is written with a laser on the wafer near the primary flat. This operation could be performed by

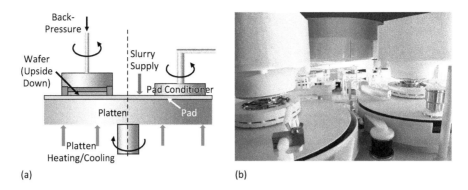

Figure 3.20 CMP process shown (a) schematically and (b) with an actual machine (the Applied Materials Reflexion CMP tool). Photo courtesy of Applied Materials, Inc.

either the wafer or the IC manufacturer. Since the laser process does damage to the wafer surface and may create particle contamination of the surface, this step is often performed by the wafer supplier and is done before some of the final etching and polishing steps described above.

3.8 Dopant and Impurity Incorporation during Crystal Growth

The liquid melt used in many crystal growing methods always contains impurities. In addition to dopants that are purposely placed in the melt, trace levels of other impurities are always present because the refining processes used to produce the raw materials are not perfect. In addition, the crystal growing process itself may introduce impurities. We saw a specific example of this with O and C incorporation during the CZ growth process, O from the SiO_2 crucible that holds the molten Si, and C from the graphite components of the crystal puller. It is important to understand the transfer of these dopants and impurities from the liquid melt to the growing crystal, since the devices later fabricated in the wafers may be affected by specific impurities.

We will consider two examples of analytic models that describe dopant and impurity incorporation in growing crystals. These models provide insight into CZ and FZ crystal growth. Modeling dopant and impurity incorporation is not as straightforward as might be first thought because of a process called segregation. All impurities segregate between the liquid and solid phases when the two phases are in intimate contact. By this we mean that, if the concentrations of the impurity are C_S in the solid and C_L in the liquid, then in general $C_S \neq C_L$. In fact, we can define a segregation coefficient k_0 as

$$k_0 = \frac{C_S}{C_L}, \tag{3.16}$$

where C_S and C_L are the concentrations just on either side of the solid/liquid interface and k_0 is usually referred to as the equilibrium segregation coefficient. If $C_S > C_L$, the impurity prefers to be in the solid phase; if $C_S < C_L$, the reverse is true. Physically, segregation occurs because

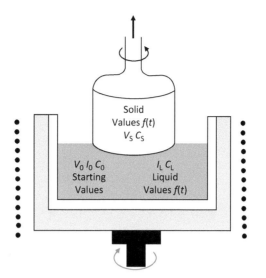

Figure 3.21 Dopant behavior during CZ crystal growth.

impurities have different solubilities in the two phases. Stated differently, the chemical potential of an impurity must be the same in the two phases across the interface, and this results in segregation. Values of k_0 are normally experimentally measured quantities for a particular impurity in a particular system at a particular temperature. If the experimental conditions are far from equilibrium, then an effective value different from k_0 may be measured. This might happen, for example, if the freezing process occurred so rapidly that the impurities did not have time to adjust their concentrations across the interface before they were frozen into the solid. Segregation will also be important in later chapters when we deal with oxidation and diffusion of impurities, since in those cases we generally have interfaces like SiO_2/Si across which dopants will also partition themselves.

We first consider CZ crystal growth with the basic model shown in Figure 3.21. For most impurities in silicon, $k_0 < 1$. This means that these impurities prefer to be in the liquid phase. If we start with a certain concentration C_L in the melt at the beginning of growth, then, as the crystal grows, C_L will increase over time as the silicon from the melt is incorporated into the growing crystal faster than the impurity is. The result will be a crystal that is doped more heavily towards the end of growth than it is at the beginning. In other words, both C_L and C_S will be functions of time during the growth and both will increase if $k_0 < 1$. Table 3.1 contains equilibrium segregation coefficients for many of the impurities of interest in Si CZ growth.

In Figure 3.21, we define V_0, I_0 and C_0 to be the initial volume, number of impurities and impurity concentration in the melt, V_L, I_L and C_L to be the volume, number and concentration of impurities in the melt during growth, and finally V_S and C_S to be the corresponding quantities in the solid crystal. Note that V_L, I_L, C_L, V_S and C_S will all be functions of time (or alternatively functions of position along the crystal) if $k_0 \neq 1$.

If during the growth process an additional volume of melt dV freezes, it will remove from the melt a number of impurities given by

Table 3.1 **Equilibrium segregation coefficients during CZ crystal growth for Si.**

Impurity	Segregation coefficient
As	0.3
Bi	7×10^{-4}
C	0.07
Li	10^{-2}
O	0.5
P	0.35
Sb	0.023
Al	2.8×10^{-3}
Ga	8×10^{-3}
B	0.8
Au	2.5×10^{-5}

$$dI = -k_0 C_L \, dV = -k_0 \frac{I_L}{V_0 - V_S} \, dV. \tag{3.17}$$

Integrating, we can write this as

$$\int_0^{I_L} \frac{dI}{I_L} = -k_0 \int_0^{V_S} \frac{dV}{V_0 - V_S}, \tag{3.18}$$

$$\log\left(\frac{I_L}{I_0}\right) = \log\left[\left(1 - \frac{V_S}{V_0}\right)^{k_0}\right], \tag{3.19}$$

giving

$$I_L = I_0 \left(1 - \frac{V_S}{V_0}\right)^{k_0}, \tag{3.20}$$

which gives the number of impurities in the melt as a function of how much of the melt has been frozen. We are really interested, of course, in C_S, the impurity concentration in the solid crystal. When an incremental volume of the liquid freezes,

$$C_S = -\frac{dI_L}{dV_S}, \tag{3.21}$$

$$C_S = C_0 k_0 (1 - f)^{k_0 - 1}. \tag{3.22}$$

This is the desired result, expressing C_S in terms of $f = V_S/V_0$, the fraction frozen. Equation (3.22) is plotted in Figure 3.22 for the common dopants used in silicon technology.

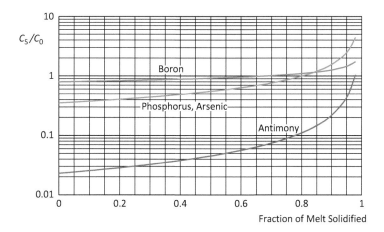

Figure 3.22 Doping concentration versus position along the grown CZ crystal for common dopants in silicon.

For dopants like antimony, where $k_0 \ll 1$, the doping concentration increases dramatically along the length of the pulled crystal. The common P-type dopant, boron, produces a much flatter profile because k_0 is closer to 1. In actual pulled crystals, (3.22) is not followed exactly because of non-idealities that we did not model in the derivation above. These include inhomogeneities in the impurity concentration in the melt, small temperature gradients in the melt, convection currents in the melt and boundary layers that form at the freezing interface. The rotation of the crystal and the crucible during growth help to minimize all these effects, but, in real crystals, there are doping variations both radially and axially on microscopic and more macroscopic scales due to these non-idealities. Radial doping variations also result from the fact that the freezing interface is actually concave into the melt, as illustrated conceptually in Figure 3.21 and in the simulation in Figure 3.17, with the result that freezing at each radial point on a wafer occurs at a different time in the crystal growth process.

Segregation effects also play an important role in the FZ process that is also used for zone refining. For most impurities in the liquid/solid silicon system, $k_0 < 1$. As we saw in the CZ case, this means that most impurities prefer to be in the liquid phase. We can now understand what is meant by zone refining. As the liquid zone sweeps up the polycrystalline rod, impurities will tend to stay in the liquid and be swept to the top of the rod. The resulting single crystal will have lower impurity levels than the starting rod. The process thus refines the material or improves its purity. Consider this process a little more carefully. Referring to the idealized geometry in Figure 3.23, the zone length is L. This is a simplified version of Figure 3.16.

Assume that the rod has an initial uniform impurity concentration of C_0. If the molten zone moves by a distance dx, the number of impurities in the liquid zone will change, since some will be dissolved into the melting liquid on the right and some will be lost to the freezing solid on the left. Thus

Figure 3.23 Float zone crystal growth process. The liquid zone moves left to right.

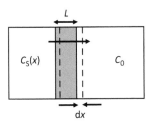

$$dI = (C_0 - k_0 C_L)\, dx, \tag{3.23}$$

where I is the number of impurities in the liquid. But $C_L = I/L$. Substituting and integrating, we find that

$$\int_0^x dx = \int_{I_0}^I \frac{dI}{C_0 - k_0 I/L}, \tag{3.24}$$

where I_0 is the number of impurities in the zone when it is first formed at the bottom. Carrying out the integration and noting that $I_0 = C_0 L$ and $C_S = k_0 I/L$, we have finally that

$$C_S(x) = C_0\left\{1 - (1 - k_0)e^{-k_0 x/L}\right\}. \tag{3.25}$$

This result is plotted in Figure 3.24. Note that the initial single-crystal material will have an impurity level of $k_0 C_0$, which is generally much less than C_0. Thus the single-crystal material is purified and the impurities are swept up towards the top of the rod. The simple solution given by (3.27) applies only for a single pass through the solid rod because C_0 is assumed constant in the derivation. However, C_S/C_0 asymptotically approaches 1 as the concentration in the liquid zone rises during refining. Note that, when k_0 is close to 1, the impurity profile is relatively flat, which is the same conclusion we drew for the CZ crystal growth method.

The refining process can be repeated using multiple passes of the zone up the rod. In this case, C_0 is not constant (after the first pass). This problem can be modeled using a simple spreadsheet analysis or with programs like MATLAB (see problem 3.11 at the end of this chapter). The resulting impurity profile is illustrated in Figure 3.25 for the particular case in which $k_0 = 0.1$.

In practice, to reduce the refining time, multiple zones can be melted and moved along a horizontal boat. This does create the problem of finding a boat material that will not contaminate the molten silicon and to which the freezing silicon will not stick. Orienting the silicon vertically eliminates the need for a boat since the liquid zone can be supported by surface tension and electromagnetic forces, but only a single liquid zone can be used in this case.

In most applications, the silicon wafers that result from the FZ process are required to be doped, usually very lightly since high-resistivity material is needed for most FZ-based devices. It is clear from the analysis above that starting with uniform doping in the

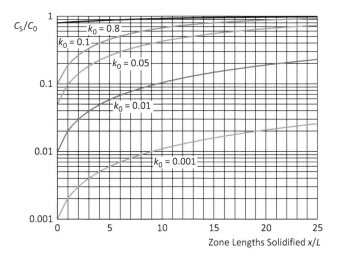

Figure 3.24 Behavior of impurities during FZ growth or zone refining. The solid is assumed to start with a uniform doping concentration C_0. The curves correspond to one pass of the molten zone through the solid.

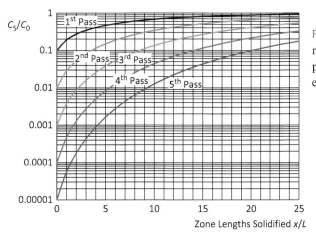

Figure 3.25 Zone refining with multiple passes. In this example, $k_0 = 0.1$.

polycrystalline rod will not produce uniform doping along the length of the single-crystal ingot. Several approaches can be used to solve this problem. In some cases, the FZ process is carried out in a gas ambient that contains the dopant species. Dilute concentrations of PH_3 or B_2H_6 could be used, for example. In this case the dopants are incorporated into the liquid zone, in which they diffuse rapidly, producing a reasonably uniform doping profile. Another alternative is to dope only the seed end of the polysilicon rod. In this case, no additional dopant is added as the liquid zone moves up the rod and a fairly uniform doping profile can be obtained.

The final possibility is to grow the FZ crystal undoped and then to use a process known as neutron transmutation doping (NTD) to uniformly dope the ingot. In this process, the FZ ingot is placed inside a nuclear reactor and exposed to thermal neutrons. Silicon naturally occurs in three isotopes, ^{28}Si, ^{29}Si and ^{30}Si, with relative abundances of approximately 91%, 6% and 3%, respectively. The neutrons in the reactor react with the ^{30}Si atoms to produce an unstable isotope ^{31}Si, which then decays to a stable phosphorus isotope ^{31}P, with a half-life of about 2.6 h. The overall reaction is

$$^{30}\text{Si} + (\text{n}, \gamma) \rightarrow {}^{31}\text{Si} \rightarrow {}^{31}\text{P} + \beta^- . \tag{3.26}$$

Since the neutron absorption depth is large compared to the ingot diameter, very uniform doping tends to result from this process. The ingot must be shielded during the radioactive decay process that produces the ^{31}P. Radiation damage does result from this technique, but it can be annealed at modest temperatures after the process is complete. NTD silicon can generally be purchased only with resistivities greater than about 5 Ω cm because more heavily doped ingots would require excessive times to become safe to handle. This doping method also works in SiC crystals since half the crystal atoms are Si.

3.9   Other Wafer Approaches: Silicon-on-Insulator and Bonded Wafers

A portion of the silicon industry today uses starting wafers that are silicon-on-insulator (SOI) structures. An example of such a starting wafer is shown in Figure 3.26. The advantages of such a structure come about because the devices are built in a thin silicon film on an insulator (SiO$_2$ in this case). This can reduce parasitic capacitances and, as a result, speed up circuits. SOI devices also easily provide lateral isolation between devices, since the silicon layer can be etched away or oxidized in areas where there are no devices.

Several approaches are being used to produce such wafers. The first involves starting with a normal CZ or FZ wafer and ion implanting a large dose ($\approx 10^{18}$ cm^{-2}) of oxygen. We will discuss this in more detail in Chapter 8 since ion implantation is involved. Subsequent annealing of the implanted oxygen at high temperature forms a buried SiO$_2$ layer [23]. The silicon above the buried SiO$_2$ can be of reasonable crystal quality if the structure is properly annealed, although the quality is generally not as good as the starting wafer. This process is known as SIMOX for "separation by implanted oxygen" and such wafers are commercially available and are used in manufacturing.

A second approach to realizing "wafers" like Figure 3.26 is called the BESOI or "bonded and etchback" technology. It starts with two oxidized conventional CZ wafers. If these wafers are placed face-to-face after proper surface treatment they will "stick" together. After annealing in a high-temperature furnace, an intimate bond is formed between the two wafers, which is about as strong as the original wafers. One of the two wafers can then be etched or lapped away to leave a thin silicon film on an insulating SiO$_2$ substrate. The advantage of the bonded wafer approach is that the resulting silicon should be of the same quality as the starting wafer. Wafer bonding in various forms is finding increasing application in manufacturing because it is

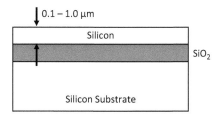

Figure 3.26 Silicon-on-insulator (SOI) wafer technology. The buried oxide layer is typically several hundred nanometers thick.

actually a more general approach than just using it for SOI wafers. The bonded wafers can be made from different semiconductor materials or they can even be fully processed wafers containing different types of devices that can be wired together through vertical connections. This latter approach is one of many being considered today for so-called "3D" integration. We will discuss some of these "3D" approaches in Section 11.8 in Chapter 11.

   Another approach to SOI substrates is the "Smart Cut" process [24], which we will also discuss in Chapter 8 since it involves ion implantation. In this approach, a high-dose hydrogen implant is done into an oxidized silicon wafer. The energy of the implant is chosen to place the peak of the hydrogen implant at a depth below the silicon surface corresponding to slightly more than the desired final silicon thickness in the SOI structure. This wafer is then "stuck" to a second oxidized silicon wafer as in the BESOI process. The two wafers are then annealed in a two-step process. During the first, low-temperature anneal (400–600 °C), the first wafer splits apart at the location of the hydrogen implant. This leaves a thin and very uniform silicon layer from the first wafer sitting on an oxide, on the second wafer. The second, high-temperature anneal (>1000 °C) strengthens the bond between the two wafers. A final polish of the thin silicon layer produces the SOI structure. This process has two significant advantages. First, it produces a thin, well-controlled and high-quality silicon layer in the SOI material. Second, it uses only one starting wafer per SOI substrate (unlike the two wafers in the BESOI process), because, after the wafers split apart, the first wafer can be reused. Smart Cut wafers are commercially available today and are used in manufacturing some products. Chapter 8 contains a more detailed discussion of these technologies.

3.10 Summary of Key Ideas

In this chapter, we have described the techniques by which silicon wafers are produced. Through a process of refining, raw materials like quartzite (SiO_2) are turned into silicon with a purity unmatched by any other routinely available material on Earth. This raw material is then used to grow single crystals, principally using the CZ method. Today, the resulting boules are up to 300 mm (12 inches) in diameter and up to a meter long. Larger 450 mm (15 inch) wafers have been demonstrated, but to this point they have not been adopted in manufacturing, presumably because the high costs of larger machines to handle these wafers have not justified the larger-diameter wafers. Silicon wafers are essentially defect-free and the only significant impurities in the crystals, aside from dopants, are usually oxygen and carbon that come from the machines used to grow the crystals. A series of sawing, mechanical lapping and CMP steps finally produce the individual wafers that are supplied to IC manufacturers. These wafers are mirror polished on one side and are defect-free. They contain dopants typically in the parts per million range with tight tolerances, controlled oxygen concentrations in the 10^{17}–10^{18} cm^{-3} range, controlled carbon concentrations in the 10^{16} cm^{-3} range and levels of all other undesirable impurities in the parts per billion range or lower. These wafers then form the starting material for IC manufacturers.

Compound semiconductor wafers are produced by modified CZ growth methods such as LEC. In this method, a liquid layer, usually B_2O_3, floats on top of the liquid semiconductor in the growth crucible, in order to mitigate the high vapor pressure of some column V elements. Other III–V semiconductor crystals are produced using sublimation (SiC) or CVD methods (SiC and GaN). The wafers produced by these methods are generally significantly more expensive than Si wafers because growth rates are much slower than with the standard CZ method. Compound semiconductor wafers are also generally available only in smaller diameters, 100–150 mm today.

Semiconductor wafers are normal-doped during the CZ or FZ growth process. We saw in this chapter that segregation of the dopant atoms between the liquid and solid phases plays a critical role in determining the doping in the finished wafers. Segregation across material interfaces will also be important in later chapters.

We also introduced in this chapter various types of defects that can be present in crystalline semiconductor wafers. Many of these are process-induced and can be minimized by appropriate choice of process conditions. Point defects are fundamental and are always present. Their concentration depends on temperature and on E_F (doping) because these defects exist in both neutral and charged states. Point defects play key roles in many fabrication processes and will be important in our discussion in later chapters.

3.11   REFERENCES AND NOTES

[1] H. Wang, D. Yang, X. Yu, *et al.*, "Effect of oxygen precipitates and induced dislocations on oxidation induced stacking faults in nitrogen-doped Czochralski silicon," *Journal of Applied Physics*, vol. 96, no. 5, p. 3031, 2004.

[2] M. Lannoo, *Point Defects in Semiconductors I: Theoretical Aspects*, Springer, 2012.

[3] H. Park and M. E. Law, "Point defect based modeling of low dose silicon implant damage and oxidation effects on phosphorus and boron diffusion in silicon," *Journal of Applied Physics*, vol. 72, no. 8, pp. 3431–3439, 1992.

[4] P. M. Fahey, P. Griffin and J. Plummer, "Point defects and dopant diffusion in silicon," *Reviews of Modern Physics*, vol. 61, no. 2, p. 289, 1989.

[5] M. Tang, L. Colombo, J. Zhu and T. D. De La Rubia, "Intrinsic point defects in crystalline silicon: tight-binding molecular dynamics studies of self-diffusion, interstitial–vacancy recombination, and formation volumes," *Physical Review B*, vol. 55, no. 21, p. 14279, 1997.

[6] W. Shockley and J. Last, "Statistics of the charge distribution for a localized flaw in a semiconductor," *Physical Review*, vol. 107, no. 2, p. 392, 1957.

[7] D. Hurle, "Point defects in compound semiconductors," in *Crystal Growth – from Fundamentals to Technology*, Elsevier, pp. 323–343, 2004.

[8] J. Zhang, H. Liu, J. Cao, W. Zhu, B. Jin and W. Li, "A deep learning based dislocation detection method for cylindrical crystal growth process," *Applied Sciences*, vol. 10, no. 21, p. 7799, 2020.

[9] W. C. Dash, "Evidence of dislocation jogs in deformed silicon," *Journal of Applied Physics*, vol. 29, no. 4, pp. 705–709, 1958.

[10] G. Fisher, M. R. Seacrist, and R. W. Standley, "Silicon crystal growth and wafer technologies," *Proceedings of the IEEE*, vol. 100, Special Centennial Issue, pp. 1454–1474, 2012.

[11] G. K. Teal, "Single crystals of germanium and silicon – basic to the transistor and integrated circuit," *IEEE Transactions on Electron Devices*, vol. 23, no. 7, pp. 621–639, 1976.

[12] J. D. Plummer, M. D. Deal and P. B. Griffin, *Silicon VLSI Technology – Fundamentals, Practice and Modeling*, Prentice Hall, chap. 3, 2000.

[13] M. Kirpo, "Global simulation of the Czochralski silicon crystal growth in ANSYS FLUENT," *Journal of Crystal Growth*, vol. 371, pp. 60–69, 2013.

[14] B. Nacke and A. Muiznieks, "Numerical modelling of the industrial silicon single crystal growth processes," *GAMM-Mitteilungen*, vol. 30, no. 1, pp. 113–124, 2007.

[15] X. F. Han, X. Liu, S. Nakano, H. Harada, Y. Miyamura and K. Kakimoto, "3D global heat transfer model on floating zone for silicon single crystal growth," *Crystal Research and Technology*, vol. 53, no. 5, p. 1700246, 2018.

[16] C. Smith and A. R. Barron, "Synthesis and purification of bulk semiconductors: electronic-grade gallium arsenide," in *Chemistry of the Main Group Elements*, ed. A. R. Barron, OpenStax CNX, p. 414, 2014. See https://cnx.org/contents/9G6Gee4A@25.9:u-A3POtF@2/Electronic-Grade-Gallium-Arsenide.

[17] S. Hegewald, K. Hein, C. Frank, M. John and E. Buhrig, "Investigation on the equilibrium vapour pressure over a GaAs melt," *Crystal Research and Technology*, vol. 29, no. 4, pp. 549–554, 1994.

[18] P. Rudolph and M. Jurisch, "Bulk growth of GaAs – an overview," *Journal of Crystal Growth*, vol. 198, pp. 325–335, 1999.

[19] I. R. Grant, "Indium phosphide crystal growth," in *Bulk Crystal Growth in Electronic, Optical and Optoelectronic Materials*, ed. P. Capper, Wiley, chap. 4, pp. 121–147, 2005.

[20] P. J. Wellmann, "Review of SiC crystal growth technology," *Semiconductor Science and Technology*, vol. 33, no. 10, p. 103001, 2018.

[21] K. Fujito, S. Kubo, H. Nagaoka, T. Mochizuki, H. Namita and S. Nagao, "Bulk GaN crystals grown by HVPE," *Journal of Crystal Growth*, vol. 311, no. 10, pp. 3011–3014, 2009.

[22] A. Yoshikawa, E. Ohshima, T. Fukuda, H. Tsuji and K. Oshima, "Crystal growth of GaN by ammonothermal method," *Journal of Crystal Growth*, vol. 260, no. 1–2, pp. 67–72, 2004.

[23] K. Izumi, "History of SIMOX material," *MRS Bulletin*, vol. 23, no. 12, pp. 20–24, 1998.

[24] M. Bruel, "Silicon on insulator material technology," *Electronics Letters*, vol. 31, no. 14, pp. 1201–1202, 1995.

3.12 PROBLEMS

3.1 Short-answer questions.

(a) The CMOS process we described could start with either a P-type or an N-type Si substrate. Usually P type is used. Why?

(b) Why do modern CMOS processes typically start with a quite lightly doped wafer (P type, 5–20 Ω cm in the process we discussed in Chapter 2), as opposed to choosing a higher doping that would provide the correct threshold voltage for one of the MOS transistor types?

    (c) Explain the two main reasons why single-crystal silicon is chosen as the wafer substrate material as opposed to polycrystalline or amorphous silicon?

3.2 Calculate the temperature difference across a (100) silicon wafer necessary for the silicon to reach its yield strength. This gradient sets an upper bound on the temperature non-uniformity that is acceptable in a rapid thermal annealing (RTA) heating system.

3.3 Using values for the atomic weight and lattice constant of Si given in Appendix A.5, calculate its mass density (g cm^{-3}). How does your answer compare to the mass density given in Appendix A.5?

3.4 A silicon wafer is doped N type at 1×10^{14} cm^{-3}. We saw in this chapter that the equilibrium neutral point defect concentrations $C_{I^0}^*$ and $C_{V^0}^*$ increase with temperature and can become quite high at processing temperatures. At what temperature does $C_{V^0}^*$ become equal to the electron concentration in this silicon sample? Explain your answer.

3.5 In general, thermodynamics does not require that interstitial and vacancy concentrations are equal. Is there a temperature in silicon at which the neutral interstitial and neutral vacancy populations actually are equal? If so, what is that temperature?

3.6 In all of our models involving point defects, we made the fundamental assumption that the concentrations of these defects (at least in equilibrium) are much less than the intrinsic carrier concentration n_i. Show quantitatively, at least for the equilibrium concentrations of the neutral point defects I^0 and V^0, that this is a good assumption at all temperatures even up to the melting point of silicon.

3.7 Assume that the source/drain region for a PMOS transistor has doping level of 5×10^{19} cm^{-3}. Find the Fermi level at 300 K (room temperature) and 1000 °C. What can you say qualitatively (no calculation required) about the dominant vacancy charge state and the ionization of the vacancy energy levels in each case?

3.8 Assume that the source/drain region for a PMOS transistor has a doping level of 5×10^{19} cm^{-3}. Calculate the equilibrium total vacancy concentration at both 300 K (room temperature) and 1000 °C. You do not need to consider $V^=$ states for this problem. You may assume $(E_C - E_{V^-}) = 0.57$ eV and $(E_{V^+} - E_V) = 0.05$ eV, both values being independent of temperature.

3.9 Where do the $+$ and $-$ charges come from to create charged point defects such as V^+ and V^-? Answer this question for both intrinsic and doped silicon.

3.10 The maximum pull rate in the CZ process can be modeled based on a heat balance model. The model predicts that the maximum pull rate for a 150 mm (6 inch) diameter wafer is about 24 cm h^{-1}. A manufacturer of silicon wafers uses CZ crystal pullers to produce 300 mm (12 inch) ingots. Estimate how long it takes this manufacturer to pull a crystal that is 3 m long.

3.11 A boron-doped crystal pulled by the CZ technique is required to have a resistivity of 10 Ω cm when half the crystal is grown. Assuming that a 100 g pure silicon charge is used, how much 0.01 Ω cm boron-doped silicon must be added to the melt? For this crystal, plot resistivity as a function of the fraction of the melt solidified. Assume $k_0 = 0.8$ and the hole mobility $\mu_p = 550$ cm^2 V^{-1} s^{-1}.

3.12 A CZ crystal is pulled from a Si melt containing 5×10^{16} cm^{-3} boron and 5×10^{16} cm^{-3} arsenic. What is the initial doping type (N type or P type) of the silicon solid and at what point does the pulled crystal change to the opposite doping type?

3.13 A CZ crystal is grown with an initial Sb concentration in the melt of 1×10^{16} cm^{-3}. After 80% of the melt has been used up in pulling the crystal, pure silicon is added to return the melt to its original volume. Growth is then resumed. What will the Sb concentration be in the crystal after 50% of the new melt has been consumed by growth? Assume $k_0 = 0.02$ for Sb.

3.14 A new dopant with an unknown segregation coefficient is added to the liquid silicon in a CZ crystal growth experiment. It is found that the concentration of the dopant in the pulled solid crystal, when the crystal is 90% grown, is twice the concentration of the dopant in the crystal at the start of the growth. What is the segregation coefficient of the unknown dopant?

3.15 During CZ crystal growth, the dopant concentration varies along the boule's length due to dopant segregation effects. A wafer manufacturer uses the CZ method to grow a phosphorus-doped crystal. When 80% of the liquid is solidified, the phosphorus doping concentration is beyond the designed tolerance, so the growth has to be stopped temporarily. Assume the initial melt phosphorus concentration is C_0, and the initial melt volume is V_0.

(a) What is the melt dopant concentration when 80% of the liquid is solidified?

(b) How much silicon (in terms of V_0) would need to be added to the liquid to recover the initial liquid concentration?

(c) If the silicon calculated in part (b) is added to the melt, what is the phosphorus concentration of the melt after 30% of the new melt has been consumed by the growth?

3.16 In the raw material (polycrystalline silicon) used for CZ crystal growth, some level of impurities is inevitable. The International Technology Roadmap for Semiconductors (ITRS) specifies maximum levels of impurities like Fe and Au in starting wafers. Suppose a wafer manufacturer is required to meet a specification of $<10^{10}$ cm^{-3} Au in starting wafers produced by the CZ method. Suppose also that the CZ wafer manufacturer wants to use 99% of the melt in a crucible to grow an ingot and produce wafers.

(a) What is the upper bound on the Au contamination level in the polycrystalline silicon raw material that the wafer manufacturer uses?

(b) Suppose that the only polycrystalline silicon available to you has 10 times more Au in it than the answer you calculated in part (a). Suggest two approaches you could use to solve this problem.

3.17 The maximum pull rate in the CZ process can be modeled based on a heat balance model. The model predicts that the maximum pull rate for a 6 inch diameter wafer is about 24 cm h^{-1}. Now assume that the silicon melt is pulled at a rate of 24 cm h^{-1} and the melt is arsenic-doped with an initial concentration of 10^{16} cm^{-3}.

(a) Calculate the rate ($cm^3 \, s^{-1}$) at which liquid Si needs to be added to the crucible to maintain a constant dopant concentration in the crystal.

(b) If the melt is arsenic-doped with an initial concentration of $5 \times 10^{17} \, cm^{-3}$, what is the pulling rate required to still maintain a constant dopant concentration?

3.18 Consider the zone refining process illustrated in Figures 3.16 and 3.23. Set up a simple spreadsheet (or use MATLAB) to analyze this problem and use it to generate a plot like Figure 3.25. Divide the crystal up into n segments as shown below. Consider a zone length dx which steps in increments dx up the crystal during refining. As the zone moves an amount dx, it incorporates an impurity concentration given by $C_S(n + 1)$, where $n + 1$ is the next zone to be melted and $C_S(n)$ is the impurity concentration in that zone from the previous pass. The liquid zone also leaves behind an impurity concentration given by $k_0 C_L$, where C_L is the concentration in the liquid during the current pass. Note that the impurity concentration in the liquid consists of C_S plus all the impurities "swept up" by the liquid during the current pass. Note that the final zone at the end simply solidifies and can be neglected.

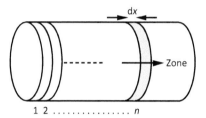

3.19 Some of the very early transistors were manufactured using the grown junction technology illustrated in Figure 1.3. A student studying semiconductor technology decides that she can make such a device using CZ crystal growth. She sets up a liquid Si melt that has three dopants in it, B, P and Al, all at $1 \times 10^{15} \, cm^{-3}$, expecting that segregation effects will produce the desired PNP structure. If there is enough Si in the melt to produce a crystal with a total length of 1 cm, calculate the width of the base region (the N region) that this crystal growth would produce.

3.20 Suppose your company was in the business of producing silicon wafers for the semiconductor industry by the CZ growth process. Suppose you had to produce the maximum number of wafers per boule that met a fairly tight resistivity specification.

(a) Would you prefer to grow N-type or P-type crystals? Why?

(b) What dopant would you use in growing N-type crystals? What dopant would you use in growing P-type crystals? Explain.

4 Semiconductor Manufacturing: Cleanrooms, Wafer Cleaning, Gettering and Chip Yield

4.1 Introduction and Basic Concepts

If workers from one of today's multi-billion-dollar integrated circuit (IC) manufacturing plants were suddenly transported to a 1960s semiconductor plant, they would likely be amazed that chips could be successfully manufactured in such a place. Such factories were "dirty" by today's standards, and wafer cleaning procedures were poorly understood. Of course, chips were manufacturable even in those days, but they were very small and contained very few components by today's standards. Since defects on a chip tend to reduce yields (fraction of good chips on a wafer) exponentially as chip size increases, small chips can be manufactured with a yield greater than zero even in quite dirty environments. However, all of the progress that has been made in the past six decades in shrinking device sizes and designing very complex chips would have been for naught if similar advances had not been made in manufacturing capability, especially in defect density.

In a typical silicon or compound semiconductor fabrication process, a single defect generally renders the chip inoperable. There are a few exceptions to this, such as in devices like memories in which extra rows and columns can be built on the chip and electrically connected during testing to replace defective cells. But, in most cases, with chip sizes today as large as many square centimeters, it is easy to see why defect densities must be much smaller than 1 defect cm^{-2} to achieve reasonable yields. It is also important to keep in mind that the manufacturing process consists of hundreds of sequential steps and a defect in almost any of these steps can be fatal. So the required defect density in each process step must be very small. Since wafers typically contain hundreds of identical chips, the yield at the end of the process can often determine the profitability (or not) of a chip manufacturing company.

In order to understand some of the issues that arise in semiconductor manufacturing, it is useful first to briefly discuss how a manufacturing process is actually developed. The silicon industry has been driven by Moore's law for decades, and this means that a new generation of technology and a new manufacturing process is introduced roughly every 2–3 years. This happens through a rigorously defined process illustrated in Figure 4.1 (the names for the various steps vary from company to company).

In order for a new technology to enter manufacturing in year 0 in this example, research on the devices and the process technologies to be used would typically start a decade earlier. Some

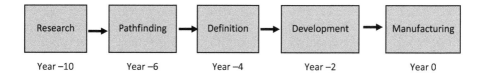

Figure 4.1 Typical industrial process for developing a new generation of silicon technology.

of this research would be done in universities, some in industrial research labs and some in shared R&D programs like imec [1]. Often this research is openly published at conferences and in technical journals because it is generally precompetitive, although specific breakthroughs may be kept as proprietary information within individual companies. Today, work in this category would include, for example, carbon nanotubes and 2D electronic materials such as $MoSe_2$.

Approximately six years before entering manufacturing (on the two year cadence illustrated in this example), pathfinding is a focus. This is essentially a process in which the many options identified in basic research are narrowed down to a much smaller number that have promise of being successful in manufacturing. This usually happens through a combination of experiments, simulation and feasibility studies. Today, some of the nanoribbon and nanowire extensions to FinFET devices such as vertically stacked gate-all-around (GAA) NMOS and PMOS devices (see Figure 1.23 in Chapter 1) would be in this stage or perhaps even in the next definition stage.

The next stage in development is the box labeled "definition", which happens roughly four years before a process enters high-volume manufacturing. This step is what its name implies. The new technology is defined explicitly with all of the process steps defined quantitatively. Ion implant doses, doping profiles, dielectric thicknesses, etc., are all specified at this point, so that wafers can be run through the complete process in a development line. Again, this involves lots of experiments, simulations and optimizations.

The final stage of "development" involves running wafer lots with carefully designed test chips on them through a complete manufacturing process and evaluating the process yield and device characteristics. This is the stage during which yields are improved so that the process will be economic, that is, so that it will produce chips that are profitable. Typically, what happens during this phase is illustrated in Figure 4.2.

To understand what is happening in Figure 4.2, we need to distinguish between systematic and random defects [2, 3]. Random defects are primarily what we will discuss in this chapter. These are particles and contaminants that randomly impact individual devices or circuits on a wafer (much more on this later). These types of defects normally dominate in high-volume manufacturing, the right-hand part of both Figures 4.1 and 4.2. During the development phase, systematic defects are dominant. These are things like an etch process being marginally too short so that contact holes do not reliably open, or a lithography step having too large a process variation so that significant numbers of devices either do not work or perform outside of design specifications. Often one of the most difficult types of systematic defects to find and fix occurs when process modules that were developed separately are placed in series in a manufacturing

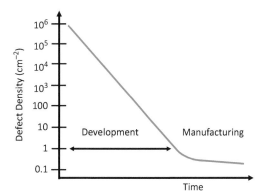

Figure 4.2 Illustration of the yield improvement that occurs during the development phase.

line and unexpected interactions between the process modules occur. The development stage, in other words, is about centering and optimizing the manufacturing process flow so that complex circuits can be built at high yield.

In the beginning stages of development, defect densities are usually huge, so large that no working chips would be obtained if the process were put into manufacturing. The goal is to eliminate the many systematic defects that are always present when a new process is developed and to get well below 1 defect cm^{-2} that is typical of successful high-volume silicon chip manufacturing lines. Often, the red curve illustrated in Figure 4.2 is called the "learning curve." While it is shown as monotonically decreasing over the roughly two-year development phase, in reality it often looks more like a stairstep curve. Test wafers are processed, text vehicles reveal a systematic defect, and that defect is fixed by "tweaking" the manufacturing process or the design rules or perhaps by a change in the process flow. New wafers are run; if the "fix" is successful, then the defect density improves.

Since there are usually many different kinds of systematic defects, they often cannot be resolved in parallel. Fixing one uncovers the next layer of problems. This has been compared to "peeling back the layers of an onion." In state-of-the-art processes, the time required for wafer lots to go through the development line is generally many weeks, which limits the number of "learning cycles" that are possible per quarter. There is an obvious premium on making this learning cycle as short as possible because the new process is not producing any revenue for the company as long as it is in development. Sometimes "short loop" experiments can be run in which wafers do not go through the entire process flow if a systematic defect is known to be in a particular part of the process. In the end, the defect density is reduced to the random defects typical of a successful high-volume manufacturing process.

Figure 4.3 illustrates the situation once we are in the high-volume manufacturing phase and random defects dominate chip yield. During the manufacturing process, semiconductor wafers often are coated with photoresist as part of the lithography process. Particles may well be present from the room air or from the process steps to which the wafers have been exposed; these must be removed. Small concentrations of undesirable elements such as metals (Fe, Au, Cu, etc.) and alkali ions (Na, K, etc.) may be present from machines, chemicals or the people to

Table 4.1 2013 ITRS specifications for Si CMOS processing in the early 2020s [4].

Critical defect size	15 nm
Particles on starting wafer	< 0.18 cm^{-2}
Front-end processing defect density	< 0.01 cm^{-2}
Mobile ions (Na$^+$, K$^+$) on surface	$< 2 \times 10^{10}$ cm^{-2}
Metal ions on surface	$< 5 \times 10^9$ cm^{-2}

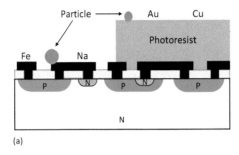

(a)   (b)

Figure 4.3 (a) Example of some of the contaminants and defect-causing particles to which semiconductor wafers are exposed during chip manufacturing. (b) The photomicrograph shows a human hair (≈ 50 μm in diameter) on a memory chip. (b) Photo courtesy of and © H. Föll (Electronic Materials – Script).

whom the wafers have been exposed; these must also be removed. The wafers themselves contain a variety of thin films, doped regions and metal films that generally must not be significantly attacked by cleaning processes.

Until 2015, the international semiconductor community met annually to produce the "International Technology Roadmap for Semiconductors" (ITRS) [4]. This roadmap laid out in great detail how the industry expected to keep Moore's law on track for the subsequent 10–15 years. The ITRS contained many detailed quantitative roadmaps for device dimensions, film thicknesses and hundreds of other parameters. The ITRS was discontinued in 2015 because individual companies by that time were following their own roadmaps, since it had become clear that there was no longer a universal roadmap. In one of the last versions of the ITRS, the numbers in Table 4.1 were set out as goals for the early 2020s.

Front-end processing defect density specifies the particle count introduced on the wafer during a typical front-end process step. The critical defect size in Table 4.1 is an indication of how large a random particle needs to be in order to create a malfunctioning chip. For example, if a particle is present on the wafer surface when photoresist is being exposed, then that particle will likely prevent exposure beneath it. If such a defect is a significant fraction of the feature size being printed and it is located in a "critical area" of the chip, it will likely lead to a malfunctioning device. It has been estimated that about 50–75% of the yield loss in a modern silicon IC manufacturing plant is due directly to defects caused by particles on the wafer. Note that the projected critical defect size is only 15 nm or about 60 atoms across. Also note that the specification of <0.01 particles cm^{-2} corresponds to less than 10 such particles on a 300 mm diameter silicon wafer.

Mobile ions and metal ions are expressed in area densities since these are commonly deposited on the surface of the wafer during process steps. These specifications (metal contamination levels on the order of 10^{10} cm^{-2}) represent only about 0.001% of a monolayer. If this metal ion area density were uniformly distributed throughout a wafer 700 μm thick, the volume density would be $\approx 10^{11}$ cm^{-3}, or less than one atom per trillion in the substrate.

The defect density numbers in Table 4.1 arise from the fact that economic manufacturing of complex chips requires that most of the ICs on a wafer work at the end of the process. The numbers are driven by the desire to achieve chip yields on the order of 90% at the end of the manufacturing process. This implies that the yield on each individual step must be far higher (> 99%), since so many steps are performed in series in a manufacturing process. Fortunately, not all steps are critical, and so optimizing the yield of chips requires identifying the critical steps, understanding the types of defects or contaminants that affect the yield in each step and then setting upper bounds on those types of defects for each step.

Two simple device examples will help to illustrate why contamination levels are such a concern in semiconductor fabrication. Mobile ions (Na$^+$ and K$^+$) are specified to be less than 2×10^{10} cm^{-2} in Table 4.1. These ions are mobile in dielectrics like SiO$_2$. If a metal-oxide–semiconductor (MOS) transistor were built with a gate dielectric 2 nm thick and that dielectric were contaminated with mobile ions, then the threshold voltage of that transistor would be given by [5]

$$V_{\mathrm{TH}} = V_{\mathrm{FB}} + 2\phi_{\mathrm{f}} + \frac{\sqrt{2\varepsilon_{\mathrm{S}} q N_{\mathrm{A}} (2\phi_{\mathrm{f}})}}{C_{\mathrm{ox}}} - \frac{q Q_{\mathrm{M}}}{C_{\mathrm{ox}}}, \tag{4.1}$$

where Q_{M} is the mobile charge density (number of charges per cm^2) associated with Na$^+$ or K$^+$ in the gate oxide and C_{ox} is the oxide capacitance. The other terms are defined in standard texts on MOS devices [5]. The last term represents the instability in V_{TH} that the mobile charges can cause. With $Q_{\mathrm{M}} = 2 \times 10^{10}$ cm^{-2} and $t_{\mathrm{ox}} = 2$ nm ($C_{\mathrm{ox}} = \varepsilon A / t_{\mathrm{ox}}$), the last term has a magnitude of about 1 mV, which would be of little concern in normal device operation. However, much larger Q_{M} values would cause instabilities large enough to be of concern. Identifying the sources of Na$^+$ or K$^+$ contamination and learning to control them dominated research on the MOS system in the 1960s and largely prevented MOS technologies from being important commercially until towards the end of that decade.

As a second example, consider the dynamic random access memory (DRAM). The basic memory cell structure is shown in Figure 4.4. Charge stored on a MOS capacitor represents the "1" or "0" state. The MOS transistor is used to write information onto the capacitor and to access the stored information. The stored charge will decay because of leakage currents that discharge the capacitor. The drain–substrate junction of the MOS transistor is electrically in parallel with the capacitor, and the junction leakage will discharge it. Periodically, circuitry on the DRAM chip must refresh the stored charge so that the data are not permanently lost. This typically happens automatically every few milliseconds (the "refresh" time in a DRAM).

In silicon devices, junction leakage currents are often dominated by Shockley–Read–Hall (SRH) recombination (see Appendix A.1). To provide a refresh time of a few milliseconds, the

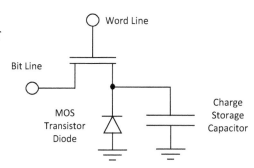

Figure 4.4 Equivalent circuit of a DRAM memory cell. Charge is stored on a MOS capacitor and the stored information is accessed through the MOS transistor switch.

minority carrier lifetime in the silicon must typically be on the order of 1 ms. The generation lifetime associated with SRH centers from Appendix A.1 is

$$\tau_G = \frac{1}{\sigma v_{th} N_t},$$ (4.2)

where τ_G is the generation lifetime, σ is the capture cross-section of the trap (on the order of the atom cross-section or about 10^{-16} cm^2), v_{th} is the minority carrier thermal velocity (about 10^7 cm s^{-1}) and N_t is the density of traps per cm^3. If $\tau_G = 1$ ms, then using typical values for σ and v_{th} gives $N_t \approx 10^{12}$ cm^{-3}. SRH traps are normally associated with deep-level impurities like Au, Cu, Fe and other elements (the metal ions in Table 4.1). This simple calculation therefore suggests that proper DRAM operation requires impurity levels of these elements in the range of the values given in Table 4.1.

Modern IC manufacturing plants employ a three-tiered approach to control unwanted impurities: cleanrooms, wafer cleaning and gettering. The first tier is implemented by building the chips in a clean environment. The air is highly filtered. Machines are designed to minimize particle production. Ultrapure chemicals and gases are used in wafer processing. The second line of defense is to chemically clean the wafers often and thoroughly. This removes particles and contaminant films from the wafer surface before they can get into the thin films on the wafer or into the semiconductor substrate. The final line of defense is called gettering and is a method used in silicon IC manufacturing in which these unwanted impurities are made to collect in non-critical parts of the wafer, typically the wafer backside or the wafer bulk, far away from the active devices on the top wafer surface. We will discuss these three levels in the next three sections.

4.2   Level 1 Contamination Reduction: Clean Factories

The factories in which ICs are manufactured today clearly must be clean facilities. Particles that might deposit on a silicon wafer and cause a defect may originate from many sources, including people, machines, chemicals and process gases. Such particles may be airborne or may be suspended in liquids or gases. The cleanliness of the air in IC manufacturing plants is described

Table 4.2 **ISO (International Standards Organization) cleanroom classification. The number in each cell is the maximum concentration limit (particles/m^3) in the air that are equal to or larger than the specified size. Prior to adoption of the ISO standards, Federal Standard 209 was used for cleanroom specification.**

ISO class number	≥ 0.1 μm	≥ 0.2 μm	≥ 0.3 μm	≥ 0.5 μm	≥ 1 μm	≥ 5 μm	Equiv. FS209 class
ISO 1	10	2					
ISO 2	100	24	10	4			
ISO 3	10^3	237	102	35	8		1
ISO 4	10^4	2370	1020	352	83		10
ISO 5	10^5	2.37×10^4	1.02×10^4	3520	832	29	100
ISO 6	10^6	2.37×10^5	1.02×10^5	3.52×10^4	8320	293	1000
ISO 7				3.52×10^5	8.32×10^4	2930	10,000
ISO 8				3.52×10^6	8.32×10^5	2.93×10^4	100,000
ISO 9				3.52×10^7	8.32×10^6	2.93×10^5	

by ISO standards that are shown in Table 4.2. Prior to the adoption of these ISO standards, cleanrooms were described by their "class," which is also shown in Table 4.2. The "class" had a simple meaning. Class X simply meant that, in each cubic foot of air in the factory, there were fewer than X total particles greater than 0.5 μm in size. A typical office building is about ISO 8 or Class 100,000, while room air in state-of-the-art Si manufacturing facilities today is typically ISO 1 in critical areas. This level of cleanliness is obtained through a combination of air filtration and circulation, cleanroom design, and through careful elimination of particulate sources.

Particles are always present in a distribution of sizes and shapes, and it has been estimated that particles are responsible for 50–75% of the yield loss in modern IC manufacturing plants [6]. Those particles that are of most concern in semiconductor plants are between about 10 nm and 10 μm in size, as illustrated in Figure 4.5. In room air, particles smaller than 10 nm tend to coagulate into larger sizes; those larger than 10 μm tend to be heavy enough to precipitate fairly quickly.

Particles between 10 nm and 10 μm can remain suspended in the air for very long periods of time. Such particles deposit on surfaces primarily through two mechanisms, Brownian motion and gravitational sedimentation. The first is the random motion of the particles in the air that occasionally brings them into contact with surfaces; the second is the gravitational force acting on the particles. Assuming a cleanroom environment in which the airflow is laminar at 50 cm s^{-1}, Hu has estimated the rate of particle deposition on horizontal surfaces [7]. He finds that Brownian motion dominates for particles smaller than about 0.5 μm and gravitational effects dominate for larger particles. The net result is that, in a class ISO 5 cleanroom, about five particles larger than 0.1 μm will deposit on each square centimeter of surface area per hour. Of course, the numbers are much smaller in ISO 1 cleanrooms, but, nevertheless, it is

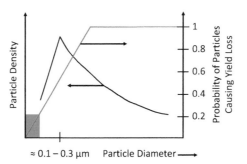

Figure 4.5 Empirical particle size distribution in room air (left axis) and probability of a particle causing a chip defect (right axis).

not a good idea to leave in-process wafers exposed to the air in a cleanroom, a topic we will revisit shortly.

The ISO specification for particle counts can be expressed mathematically as [8]

$$\text{particles of size } \geq D \text{ in } 1 \text{ m}^3 \text{ of air} = 10^N \left[\frac{0.1}{D} \right]^{2.08}, \tag{4.3}$$

where N is the ISO class. This formula must fail for very small particles and so, in practice, the distribution function, which cannot be easily measured at small sizes, is assumed also to fall off at small sizes, as illustrated in Figure 4.5 [9].

The second issue with respect to particles and yield, is how large a particle needs to be to cause a defect. Generally, particles on the order of the technology minimum feature size or larger will cause defects unless they happen to occur in a non-critical area of the chip. For example, an unintended isolated area of metal due to a photolithography defect during the metal etching may not cause a yield problem if it is well away from other metal lines. If the particle is in a critical area, then the probability it will cause a defect behaves empirically as shown in Figure 4.5. The probability is 1 for large particles and falls off as the particle or defect size decreases. Note that the probability may not be zero for very small particles, depending on the type of process involved (the red area near the origin in Figure 4.5). For example, a pinhole in a gate oxide is a catastrophic failure no matter how small the pinhole is.

Particles in the air in a manufacturing plant generally come from several main sources. These include the people who work in the plant, machines that operate in the plant and supplies that are brought into the plant. Many studies have been done to identify particle sources and the relative importance of various sources. For example, people typically emit several hundred particles per minute from each square centimeter of surface area. The actual rate is different for clothing versus skin versus hair, but the net result is that a typical person emits 5–10 million particles per minute. This rate also varies with activity level, ranging from less than 10^6 to more than 10^7 particles/min for sitting versus running.

The first step in reducing particles is to minimize these sources. People in the plant wear "bunny suits" which cover their bodies and clothing and which block particle emission from these sources. Often face masks and sometimes even individual air filters are worn to prevent exhaling particles into the room air. Air showers at the entrance to the cleanroom blow loose particles off people before they enter, and cleanroom protocols are enforced to minimize particle generation. Machines that handle the wafers in the plant are specifically designed to minimize particle generation and materials are chosen for use inside the plant that minimize particle emission.

Since the sources of particles can never be completely eliminated, constant air filtration is used to remove particles as they are generated. This is accomplished by recirculating the factory air through high-efficiency particulate air (HEPA) filters, as shown in Figure 4.6 (red arrows). Interestingly, these filters were developed during World War II for the removal of airborne fissionable particulates. HEPA filters are composed of thin porous sheets of ultrafine glass fibers (<1 μm diameter). Room air is forced through the filters with a velocity of about 50 cm s^{-1}. Large particles are trapped by the filters, small particles impact the fibers as they pass through the filter and "stick" to these fibers primarily through electrostatic forces. The airborne particles may be charged when they impact the filter. Even if they are neutral, differences in the electron work function between the particle and the filter material can result in charge transfer and electrostatic forces. The net result is that HEPA filters are 99.97% efficient at removing particles from the air. The pumps, RF power, etc. may be located in a basement below the cleanroom or in the equipment area. Makeup air accounts for less than 100% recirculated air.

Actual modern silicon IC manufacturing plants are shown in Figure 4.7. In addition to the features illustrated in Figure 4.6, the photos also show the wafer transport boxes carrying wafers from machine to machine (in the ceiling tracks). The use of these boxes minimizes the exposure of the wafers to the air in the cleanroom. Some pictures that the reader might find on the internet of modern IC manufacturing plants show yellow lighting in the rooms. This is because those areas of the facility are used for photolithography operations that are sensitive to ultraviolet light. As a result, the room lighting is normally yellow. We will discuss this in more detail in Chapter 5.

Particles can also be introduced to wafer surfaces during chemical cleaning steps. The chemicals that are used in semiconductor manufacturing are ultrapure or "semiconductor

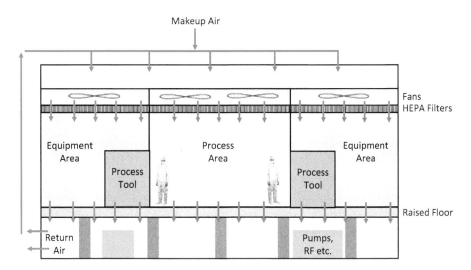

Figure 4.6 Conceptual drawing of a modern semiconductor cleanroom.

(a)   (b)

Figure 4.7 Modern silicon chip manufacturing facilities. The wafers are carried in sealed containers on the ceiling tracks from machine to machine. (a) Photo courtesy Intel. (b) Photo courtesy Taiwan Semiconductor Manufacturing Co. Ltd (TSMC).

grade," which means that particles and metal impurity concentrations are very small, typically parts per billion or less. Point-of-use ultrafiltration is used to further minimize the particles in these liquids in the manufacturing process. An interesting alternative approach to creating ultrapure chemicals is point-of-use generation using gaseous sources with deionized (DI) water [6]. In the next section when we discuss wafer cleaning procedures we will look more carefully at how particles can be "cleaned" from the wafer surface.

Most IC manufacturing facilities produce their own clean water on site, starting with water from the local water supply. This water is filtered to remove dissolved particles and organics. Dissolved ionic species are removed by ion exchange or reverse osmosis. The result is high-purity water that is used in large quantities in the plant. The most common impurities in DI water are ions. A simple measurement of water resistivity is often used to assess the "quality" of DI water. In pure water the resistivity is not infinite because the following reaction, at equilibrium, provides finite concentrations of H^+ and OH^- ions:

$$H_2O \leftrightarrow H^+ + OH^- . \tag{4.4}$$

Clean factories provide the first level of defense against contamination affecting modern IC manufacturing. While most of the examples discussed above relate to silicon technology, very similar approaches are used in all semiconductor manufacturing factories whether the product is advanced CMOS microprocessors, GaAs high-speed amplifiers or SiC or GaN power devices or even light-emitting diodes (LEDs). Figure 4.8 shows two examples of recent compound semiconductor facilities. The basic cleanroom strategy is apparent in both. Generally such facilities are much smaller than state-of-the-art silicon manufacturing plants, there is much less automation and, of course, the cost is much smaller ($ millions versus $ billions).

By itself, having clean factories is not sufficient. No matter how carefully designed the factory is, no matter how careful the people operating the factory are, no matter how purified and

Example

At room temperature in equilibrium, $[H^+] = [OH^-] \approx 6 \times 10^{13}$ cm^{-3} in water. The diffusivities of the H^+ and OH^- ions are 9.3×10^{-5} cm^2 s^{-1} and 5.3×10^{-5} cm^2 s^{-1}, respectively. Using the Nernst–Einstein relationship $\mu = zqD/kT$, where z is the charge on the ion, calculate the ion mobilities and the DI water resistivity.

Answer

Using the Nernst–Einstein relationship for each ion, we have

$$\mu_{H^+} = \frac{qD}{kT} = \frac{9.3 \times 10^{-5}\,\text{cm}^2\,\text{s}^{-1}}{25.9 \times 10^{-3}\,\text{V}} = 3.59\,\text{cm}^2\,\text{V}^{-1}\,\text{s}^{-1},$$

$$\mu_{OH^-} = \frac{qD}{kT} = \frac{5.3 \times 10^{-5}\,\text{cm}^2\,\text{s}^{-1}}{25.9 \times 10^{-3}\,\text{V}} = 2.04\,\text{cm}^2\,\text{V}^{-1}\,\text{s}^{-1}.$$

Using (A1.1) in Appendix A.1 for the resistivity, we finally have

$$\rho = \frac{1}{1.6 \times 10^{-19}[3.59(6 \times 10^{13}) + 2.04(6 \times 10^{13})]} \cong 18.5\ \text{M}\Omega\ \text{cm}.$$

Deionized water in an IC manufacturing facility is normally monitored to make certain its resistivity is ≥ 18 MΩ cm; DI water used for rinsing at the end of cleaning procedures can also have its resistivity monitored to determine when the rinsing is complete.

(a)

(b)

Figure 4.8 (a) A recent 16,000 sq. ft GaAs ISO 4 fabrication facility at Skyworks Solutions. (b) An SiC manufacturing facility at NY Power Electronics Manufacturing Consortium that processes 6 inch (150 mm) SiC wafers for power electronic devices. The yellow color in (b) is typical of the parts of manufacturing facilities that are used for photolithography because yellow light will not expose the photoresist. (b) Photo courtesy of SUNY Polytechnic Institute.

filtered the chemicals used in manufacturing are, and no matter how well designed the machines used in manufacturing are, particles and contaminants will get onto wafers. In addition, the very manufacturing process itself puts materials like photoresist on wafers as part of the manufacturing process. The second tier of attack on contamination and particles is wafer cleaning, the topic we take up in the next section.

4.3   Level 2 Contamination Reduction: Wafer Cleaning

Contamination on wafers consists of particles, which we discussed in the last section, films such as photoresist which must be removed after they serve their purpose in lithography, and trace levels of any element that has not been purposely introduced. A particular example is chemical–mechanical polishing (CMP), which was introduced in the process flow described in Chapter 2 and is widely used today. After CMP, the wafer surface may contain a variety of particles from the polishing slurry as well as residues of all of the materials that have been polished on the wafer. The second level of contamination reduction is wafer cleaning. There is probably no step that is used more often in IC manufacturing. Cleaning must remove particles, films such as photoresist and trace concentrations of any other elements present on the wafer surface.

Figure 4.9 illustrates the general strategy used in cleaning wafers. Most cleaning procedures begin with photoresist removal, since photolithography normally precedes each processing step. Photoresists are organic compounds and can be stripped in a variety of chemicals. For front-end processing (no metals on the wafers), two methods are common. The first uses an acid (usually H_2SO_4) and a strong oxidant (usually H_2O_2) to decompose the resist into CO_2 and H_2O. The second method uses an oxygen plasma to convert the resist to gaseous byproducts (again CO_2 and H_2O). The oxygen plasma method offers the advantages of reduced pollution problems and very good selectivity to almost all underlying materials. For removal of photoresists in back-end processes, the oxygen plasma approach can often be used, as can a variety of phenol-based organic strippers that do not significantly attack metals.

The most important aspects of wafer cleaning are the steps that are used following photoresist removal for front-end processes. Front-end processes often involve high temperatures

Figure 4.9 Typical cleaning strategies in a modern IC fabrication process: front-end cleaning (left) and back-end cleaning (right).

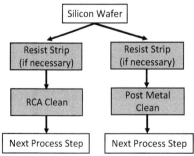

(oxidations, anneals, film deposition, etc.) and it is these steps which can allow diffusion of contaminants into thin films on the silicon or into the silicon itself. It is thus essential that the cleaning procedure remove contaminants from the wafer surface prior to these steps. Back-end processes are normally low temperature, so the wafer cleaning, while important, is not as critical. (Both front-end and back-end cleaning procedures must remove particles from the wafers, however, since these translate directly into defects.)

Prior to about 1970, a variety of front-end chemical cleaning procedures were used in the semiconductor industry. In that year, however, a classic paper was published [10] which provided both a "standard" cleaning procedure and the beginnings of a scientific basis for wafer cleaning. This paper described work done at RCA over the previous several years in which detailed measurements were made of the effectiveness of various cleaning procedures in removing contaminants from silicon wafers. The resulting cleaning procedure became known as the "RCA clean" and has been the mainstay of the industry since that time. The RCA clean uses aqueous chemical processes involving H_2O_2 mixtures. The large chemical consumption and the corresponding disposal issues of these cleaning procedures have led to the development of methods with reduced chemical use in recent years. An excellent recent general reference on wafer cleaning issues is [6].

Modern silicon and other semiconductor structures have introduced many new materials such as high-K and low-K dielectrics, metal gates, SiGe and other semiconductor materials into the device structures. This has complicated cleaning methods because chemical reactions between the cleaning chemistries and these materials must obviously be considered. In addition, modern-day structures often have high-aspect-ratio structures like contact holes, which means that cleaning approaches must be able to penetrate to the bottom of these structures.

4.3.1 Photoresist Removal

Photoresist materials are hydrocarbon polymers that consist of a resin, a photoactive compound and a solvent. We will discuss these materials in detail in Chapter 5. Removal of photoresist from wafers can be accomplished by submerging the wafers in a bath of 98 wt.% (weight percent) H_2SO_4 and 30 wt.% H_2O_2. Volume ratios of 2:1 to 4:1 are used at a temperature of 100–130 °C for 10–15 minutes. Organics like photoresist are destroyed and eliminated by this wet chemical oxidation. However, inorganic contaminants such as metals are not desorbed by this process and the silicon surface is contaminated with sulfur (S) from the H_2SO_4. Vigorous rinsing with DI H_2O is required to completely remove the liquid mixture from the wafer surface. The wet oxidation process forms a thin SiO_2 layer on the silicon surface that is contaminated with metals and S, and is therefore generally removed by submerging the wafers in a dilute (50:1) H_2O:HF solution followed by another DI rinse.

Wet chemistry removal of organics is not possible for back-end structures because the H_2SO_4/H_2O_2 solution attacks metals and other back-end materials. In addition to this, chemical disposal and cost issues associated with wet chemical removal of organics have led to increasing use of vapor-phase or plasma-enhanced approaches to organic removal. The most common method today for removing photoresist utilizes downstream plasma generation of atomic

Figure 4.10 Resist stripping details added to cleaning strategies in a modern silicon IC fabrication process: front-end cleaning (left) and back-end cleaning (right).

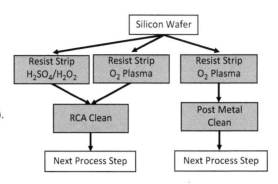

oxygen from O_2. "Downstream" means that the actual plasma environment is kept away from the wafer surface to minimize damage and to minimize the impact of charged ions on the wafer surface. The atomic oxygen diffuses to the wafer surface where it oxidizes the organic photoresist, with H_2O and CO_2 as the primary byproducts. In some cases, additional species are added to the plasma system such as fluorine-containing gases or H_2O vapor to speed up the photoresist stripping rate. We will discuss these types of systems in more detail in Chapter 9. Figure 4.10 adds the detail on the resist stripping options commonly used in silicon manufacturing today.

An alternative to the plasma-enhanced systems uses ultraviolet light and O_2. A mercury discharge lamp is used which generates UV radiation at many wavelengths. Photons of 185 nm wavelength are absorbed by O_2 to create ozone (O_3), which is highly reactive. Additional wavelengths are absorbed by the organic molecules, aiding the reactions between the O_3 and the photoresist. As in the plasma systems, H_2O and CO_2 are the primary byproducts.

4.3.2 Front-End Cleaning: the RCA Clean

The original RCA clean (Figure 4.11) [10] is based on a two-step oxidizing and complexing treatment using hydrogen peroxide solutions. The first solution (called SC-1 in the literature for "Standard Clean 1") is a high-pH solution consisting of 5:1:1 to 7:2:1 H_2O:H_2O_2:NH_4OH in which the wafers are placed at 75–85 °C for 10–20 min.

The SC-1 solution was designed to remove from Si, oxide and quartz surfaces organic contaminants that are attacked by both the solvating action of the NH_4OH and the powerful oxidizing action of the alkaline H_2O_2. The NH_4OH also serves to remove by complexing some periodic group IB and IIB metals, such as Cu, Au, Ag, Zn and Cd, and some elements from other groups, such as Ni, Co and Cr. In fact, Cu, Ni, Co and Zn are known to form amine complexes. The SC-1 solution slowly dissolves the thin native oxide layer on silicon and continuously grows a new oxide layer by oxidation. This combination of etching and reoxidation helps to dislodge particles from the wafer surface. NH_4OH etches silicon and this solution can produce micro-roughening of the silicon. In recent years, the concentration of NH_4OH in the SC-1 solution has generally been reduced to minimize these effects. This improves the quality of very thin oxides grown in modern devices on these cleaned surfaces.

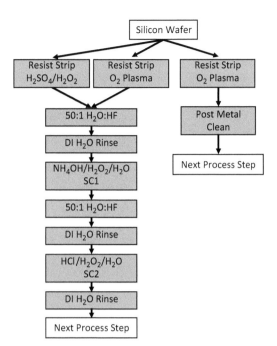

Figure 4.11 RCA clean details added to cleaning strategies in a modern silicon IC fabrication process: front-end cleaning (left) and back-end cleaning (right).

The second solution (called SC-2 in the literature for "Standard Clean 2") is a low-pH solution consisting of 6:1:1 to 8:2:1 $H_2O:H_2O_2:HCl$ in which the wafers are placed at 75–85 °C for 10–20 min. The SC-2 solution was designed to dissolve and remove from the Si surface alkali residues, any residual trace metals (such as Au and Ag) and metal hydroxides, including $Al(OH)_3$, $Fe(OH)_3$, $Mg(OH)_2$ and $Zn(OH)_2$. These metals precipitate onto the wafer surface in the SC-1 solution. Displacement replating from solution is prevented by formation of soluble metal complexes with the dissolved ions. The solution does not etch Si or SiO_2 and does not have the beneficial surfactant activity of SC-1 for removing particles. SC-2 has better thermal stability than SC-1, so that the treatment temperature need not be as closely controlled.

Several improvements to the original RCA cleaning procedure were reported by Kern and others in a number of papers following the original paper in 1970. The most influential of these changes was the introduction of ultrasonic or megasonic agitation for cleaning and rinsing of wafers [11]. Particles may sometimes adhere tenaciously to surfaces primarily due to electrostatic forces, so additional steps may be necessary to remove them. We depend on these forces in HEPA filters that clean the room air in semiconductor fabrication plants. On wafer surfaces, however, we need to fully remove such particles. The original RCA cleaning procedure was really designed for removing contaminant films and atomic elements rather than particles, so additional strategies have evolved for dealing with particles. Ultrasonic scrubbing works well and can be combined with the RCA cleaning chemicals. Ultrasonic agitation is produced usually with piezoelectric transducers operating at 20–50 kHz. The sound waves in an appropriate liquid cause the generation, expansion and violent collapse of tiny vapor bubbles under the alternating tensile and compressive stresses of the ultrasound. The bubble collapsing is known as cavitation and it literally knocks particles off the wafers or loosens them.

Ultrasonic cleaning generally becomes less efficient as the frequency increases, because at higher frequencies only very small bubbles can collapse in the compressive portion of the sound cycle. However, very high frequencies (megasonic, ≈ 1 MHz) are also quite effective for wafer cleaning, and some commercial systems use these frequencies. Megasonic treatment in an SC-1 bath is especially advantageous for physically dislodging particles from the wafer surface because of the high level of kinetic energy. It allows a reduction in solution temperature and offers a much more efficient mode of rinsing than immersion tank processing.

Other important improvements include the simplification of the composition ratios for both SC-1 and SC-2 to 5:1:1 and a reduction of treatment temperature and time to 70–75 °C for just 5–10 min. Finally, an optional process step was introduced by stripping the hydrous oxide film formed after SC-1 with high-purity, particle-free, 50:1 H_2O:HF for 10 s to re-expose the Si surface for the subsequent SC-2 step. Figure 4.11 adds the details of the RCA cleaning procedure to the overall cleaning sequence.

The original implementation of the RCA cleaning procedure utilized simple immersion tanks for the various liquid solutions (Figure 4.12(a)). Today, these units are quite elaborate (Figure 4.12(b)), featuring a series of tanks for different bath solutions, all enclosed with HEPA-filtered airflow, and a fully automated robotic processing capability. In some manufacturing facilities today, centrifugal spray machines for batch or single-wafer dynamic wet cleaning use freshly mixed and separately introduced cleaning solutions. Closed systems designed for batch or single-wafer processing retain the wafers in an enclosed module for the entire cleaning, rinsing and drying sequence. Isopropyl alcohol (IPA, propan-2-ol) is generally the purest organic solvent and is widely used for vapor drying of wafers at the end of the cleaning process.

(a)

(b)

Figure 4.12 (a) A simple cleaning station. Wet chemical cleaning baths are in the sinks. The robot is loading a cassette of wafers into a spin dryer. (b) A modern fully enclosed automated system for RCA cleaning. (a) Photo courtesy Ruth Carranza. (b) Photo courtesy SCREEN Semiconductor Solutions, reprinted with permission, https://www.screen.co.jp/spe/en.

Some of the scientific basis for the SC-1 and SC-2 solutions comes from reaction chemistry and oxidation potentials. Consider the simple experiment illustrated in Figure 4.13, in which a silicon wafer is immersed in H_2O. A metal ion is also present in the liquid. To remove metal atoms from the surface of a silicon wafer, they need to be converted into ions that are soluble in the cleaning solution. This process involves oxidizing the metal atoms. Oxidation is defined as a process that removes electrons from an atom, while reduction is the opposite process in which an atom gains electrons. Consider the following reactions, which could take place on a silicon wafer in the simple experiment illustrated in Figure 4.13:

$$Si + 2H_2O \leftrightarrow SiO_2 + 4H^+ + 4e^-, \tag{4.5}$$

$$M \leftrightarrow M^{z+} + ze^-, \tag{4.6}$$

where M represents a metal atom and z is the charge on the metal ion. Oxidation reactions go to the right, reduction reactions to the left. If electrons are available to drive reactions, the stronger oxidant will consume them.

Table 4.3 summarizes the standard oxidation potentials of a number of reactions of interest in wafer cleaning (Cu, Ni and Fe are usually the dominant impurities in silicon). These potentials are the open-circuit voltages that would be measured in an electrochemical cell operating with the given reaction, referenced to a standard hydrogen cell. In this table, stronger oxidants have more negative oxidation potentials. In the chemical reactions, oxidants are on the right side of the equations, reductants are on the left side. Standard chemistry references discuss these concepts in more detail (see for example [12]).

Suppose we now have a water solution containing a silicon wafer and metal atoms (dissolved in the liquid or on the wafer surface) as in Figure 4.13. Consider the SiO_2/Si and Fe^{3+}/Fe reactions as examples. Here the Fe^{3+}/Fe reaction has the stronger oxidation potential and hence

Table 4.3 Oxidation–reduction reactions for a number of species of interest in silicon wafer cleaning.

Oxidant/reductant	Standard oxidation potential (V)	Oxidation–reduction reaction
Mn^{2+}/Mn	1.05	$M \leftrightarrow Mn^{2+} + 2e^-$
SiO_2/Si	0.84	$Si + 2H_2O \leftrightarrow SiO_2 + 4H^+ + 4e^-$
Cr^{3+}/Cr	0.71	$Cr \leftrightarrow Cr^{3+} + 3e^-$
Ni^{2+}/Ni	0.25	$Ni \leftrightarrow Ni^{2+} + 2e^-$
Fe^{3+}/Fe	0.17	$Fe \leftrightarrow Fe^{3+} + 3e^-$
H_2SO_4/H_2SO_3	−0.20	$H_2O + H_2SO_3 \leftrightarrow H_2SO_4 + 2H^+ + 2e^-$
Cu^{2+}/Cu	−0.34	$Cu \leftrightarrow Cu^{2+} + 2e^-$
O_2/H_2O	−1.23	$2H_2O \leftrightarrow O_2 + 4H^+ + 4e^-$
Au^{3+}/Au	−1.42	$Au \leftrightarrow Au^{3+} + 3e^-$
H_2O_2/H_2O	−1.78	$2H_2O \leftrightarrow H_2O_2 + 2H^+ + 2e^-$
O_3/O_2	−2.07	$O_2 + H_2O \leftrightarrow O_3 + 2H^+ + 2e^-$

Figure 4.13 Conceptual experiment in which a Si wafer is immersed in water.

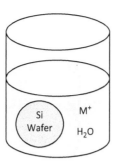

this reaction will go to the left, plating out Fe atoms on the wafer surface. The SiO_2/Si reaction will go to the right, oxidizing the silicon. This is exactly what we do not want to happen in a cleaning procedure! In this system, metal ions would be plated out on the Si wafer.

If we now consider adding H_2O_2 to the solution, the H_2O_2/H_2O reaction near the bottom of the table will dominate the Fe^{3+}/Fe reaction because of its stronger oxidation potential. Thus the H_2O_2 will take electrons from the metal atoms, creating ions that are soluble in aqueous solutions. This is the basis for the use of H_2O_2 in the RCA cleaning procedure. At the same time, the silicon will also be oxidized by the H_2O_2 since the SiO_2/Si reaction is also dominated by the H_2O_2/H_2O reaction. The general rule is that the lowest reaction in Table 4.3 dominates, going to the left and driving all reactions above it to the right. Only ozone O_3 has a stronger oxidation potential than H_2O_2. Ozone-based cleaning procedures have been investigated [6] and we will discuss these briefly below at the end of this subsection.

We can now see on a more fundamental level how the RCA cleaning procedure works. The first solution (SC-1) is a high-pH solution consisting of $H_2O/H_2O_2/NH_4OH$. This solution oxidizes organic films into water-soluble compounds (CO_2, H_2O, etc.) and complexes group IB and IIB metals as well as other metals like Au, Ag, Cu, Ni, Zn, Cd, Zn, Co and Cr. For example, Cu forms a $Cu(NH_3)_4^{2+}$ soluble amino-complex.

The second solution (SC-2) is a low-pH solution consisting of $H_2O/H_2O_2/HCl$. This solution removes alkali ions and cations like Al^{+3}, Fe^{+3} and Mg^{+2} that form NH_4OH-insoluble hydroxides in basic solutions like SC-1. The SC-2 solution also completes the removal of metallic contaminants such as Au through reactions like (4.6) that may not have been completely removed by the SC-1 step. Both SC-1 and SC-2 solutions depend on the strong oxidation potential of H_2O_2 to remove metal atoms from the wafer surface. The kinetics of the cleaning process depend, of course, on the reaction rate constants associated with reactions like (4.6). These are temperature-dependent so the necessary cleaning time is reduced at elevated temperatures.

In recent years, modifications to the basic RCA cleaning process have been adopted in the semiconductor industry [6]. The RCA cleaning method uses large quantities of chemicals and DI water at high temperatures. This generates large amounts of chemical vapors that must be exhausted from the cleanrooms and then treated. The chemicals themselves must also be treated before disposal. These environmental issues are becoming increasingly important as the scale of

IC manufacturing plants increases. One simple change that is being used by some wafer manufacturers is to simply use SC-1 and SC-2 solutions that are more dilute. These have been found to be as effective as the original concentrations and they reduce the environmental issues associated with chemical disposal. In addition, H_2O_2 is unstable at the temperatures used in the SC-1 and SC-2 solutions, especially in acidic solutions, decomposing into H_2O and O_2. As a result, the composition of the solutions can change with time, resulting in less effective cleaning.

In the SC-1 solution, if the concentration of H_2O_2 drops too much, the NH_4OH will attack silicon. Reduction of the NH_4OH concentration in the SC-1 solution by 2–10 times helps to prevent surface micro-roughening and aids particle removal. In some cases the SC-2 solution has been replaced with a very dilute, room-temperature HCl solution because Ag and Au are now no longer present in high-purity process chemicals. Other metals and their hydroxides are readily soluble in dilute HCl. Finally, in some situations, a final dip in 50:1 H_2O:HF is used to strip the chemical oxide from the cleaned wafers and to create a hydrogen-passivated surface, depending on the next processing step the wafers will see.

Beyond these relatively minor changes to the basic RCA cleaning process, more radical changes to wet chemical cleaning have also been investigated [6, 13, 14] and to some degree incorporated in modern silicon wafer cleaning. Not surprisingly, these alternative processes have focused on using ozone (O_3), given its position in Table 4.3. Ohmi's process [14] uses ozonized ultrapure water, is all at room temperature, uses far fewer chemicals than the standard RCA clean and is claimed to be as effective in cleaning wafers. The principles behind Ohmi's cleaning procedure are similar to the RCA clean. A strong oxidant is used to oxidize organics and metals. Megasonic agitation is used to loosen particles that may have adhered to the wafers. The imec process [13] also uses ozonized ultrapure water, megasonic agitation and a dilute HCl/ HF second step. Again, the principles behind this procedure are similar to those governing the effectiveness of the RCA clean. The details of these and several other alternative cleaning procedures are described in the references cited above.

There has also been considerable work on "dry" or vapor-phase cleaning procedures [3]. This work has been motivated both by the environmental issues associated with wet chemistries and by the increasing use of cluster tools in IC manufacturing. In these tools, multiple process steps, often on single wafers, are performed in a single machine by moving wafers from one chamber to another. There is great interest in accomplishing the cleaning process in one of these chambers because the wafers would then move directly to the process step under well-controlled ambient conditions, minimizing contamination between cleaning and the next step. Removing liquid cleaning solutions from the very small and often high-aspect-ratio structures being fabricated today is also an issue that vapor-phase cleaning would help to solve. However, removing particles using vapor-phase cleaning approaches is often challenging, and H_2O rinsing may still be required to completely remove reaction products. Some applications of vapor-phase cleaning have begun to be used in the semiconductor industry, and this continues to be an active area of research and development.

4.3.3   Back-End Cleaning

Cleaning strategies for back-end structures are somewhat different than for the front-end structures described in the previous section. These differences are largely driven by the fact that back-end structures contain metals (Cu or Al), low-K dielectrics that are not very chemically resistant or mechanically strong and a variety of other materials like silicides, W, TaN and many others. In addition, CMP is used frequently in back-end processing, as we saw in the CMOS example in Chapter 2. The CMP process uses a slurry to thin down films and form a planar surface. The slurry is composed of small particulates, the abrasive material, which mechanically removes the film, and a chemical etchant that assists with the chemical film removal. The process is very effective, but leaves a large quantity of small slurry particles, such as silica (SiO_x), alumina (AlO_x) or ceria (CeO_x), on the wafer.

As a result of all these issues, there generally is no standard back-end cleaning procedure analogous to the RCA clean used for front-end processing. Photoresist generally must be stripped using vapor-phase or plasma-enhanced approaches to organic removal, as described earlier, because the wet chemical options involving H_2SO_4/H_2O_2 react with many back-end materials. Wet cleaning chemicals are generally formulated by manufacturers for specific applications. These formulations contain compounds like NH_4F, H_2O, aprotic solvents (molecules that do not contain an O–H bond, e.g., acetone), amines, glycols and other components [15]. There are also a number of vapor-phase cleaning options available for back-end cleaning. A number of these are described in [16].

4.3.4   Compound Semiconductor Cleaning Procedures

The compound semiconductor industry is much smaller and much more fragmented than the silicon industry is. In addition, most of the chips that are built in materials like GaAs, InP, SiC and GaN are significantly smaller than chips like Si microprocessors and memories. Finally, all of these materials except SiC are direct-bandgap materials, which means that they are less sensitive to deep levels introduced by trace contaminants because SRH recombination does not dominate carrier lifetime in these materials. One might therefore conclude that particle (yield) and contamination issues are less critical in chips based on these materials than they are in silicon manufacturing plants.

While there is some validity to these arguments, particle and contamination control are still very important in compound semiconductor manufacturing plants. The photos in Figure 4.8 look very much like silicon manufacturing plants although the scale is obviously smaller and generally the level of cleanliness does not need to be as high as in the latest silicon manufacturing plants. The GaAs facility shown in Figure 4.8 is ISO 4, whereas state-of-the-art silicon facilities are ISO 1 in the most critical areas.

Cleaning procedures for compound semiconductors generally follow the same philosophy as in silicon manufacturing. Particles are removed using ultrasonic or megasonic agitation in liquid baths. Photoresist is stripped using the same methods as described earlier for silicon. And cleaning procedures based on the same scientific principles as the RCA clean are used. The differences that exist are a result of the fact that chemical reactions with the compound materials must be considered and the oxidation process that is part of Si cleaning is obviously different in compound materials. Another important difference is that compound semiconductors are used for a wide range of products (high-speed digital and analog circuits, optoelectronic circuits, LEDs, lasers, etc.). Differences in these manufacturing processes result in much more variety in cleaning approaches than is found in Si manufacturing. We consider two examples below to illustrate these differences.

GaAs is a material that is used for a variety of semiconductor products, including RF power amplifiers for smartphones and communication base stations, LEDs and optoelectronic circuits. Cleaning GaAs wafers generally requires that particles and contaminants are removed from the surface, surface oxide layers are removed and that the surface remains flat. GaAs can be cleaned with wet solutions very similar to the RCA SC-1 and SC-2 used in Si processing, although often with more dilute solutions than the standard SC-1 and SC-2 and often at room temperature [17]. This process leaves a native oxide on the GaAs surface that is a mixture of GaO_x, As_2O_3 and As_2O_5. Generally, a final rinse in a dilute HF or HCl or TMAH ($N(CH_3)_4^+$ OH^-, tetramethyl ammonium hydroxide) followed by spin drying is used to remove the native oxide layer [18]. (TMAH is a strong base which is often used as a photoresist developer.) The GaAs surface is highly reactive and readily forms a new native oxide after cleaning. Some recent work has been aimed at *in-situ* methods to remove this oxide prior to critical next steps such as epitaxial growth [19].

SiC is another compound semiconductor material of considerable commercial interest. SiC is a wide-bandgap material, which makes it very attractive for power applications. A variety of power MOSFETs, insulated-gate bipolar transistors (IGBTs), PIN diodes and other devices have been demonstrated and are commercially available. SiC is an indirect-bandgap material like Si. All of the particle and contamination issues that are of concern in Si manufacturing are also important in SiC device fabrication, although generally SiC chips are smaller than state-of-the-art Si chips, and so manufacturing facilities are simpler and less stringently controlled than modern Si plants (see Figure 4.8). SiC wafer cleaning is actually very similar to Si wafer cleaning. Photoresist can be stripped with any of the methods described earlier for Si, including H_2SO_4/H_2O_2 liquid mixtures. The RCA cleaning process can be used directly on SiC [20]. The only real difference with Si cleaning is in the final step, which strips the native oxide prior to the next process step. In Si, a simple dilute HF dip strips the SiO_2 oxide and leaves a hydrogen-passivated surface ready for the next process step. In SiC, the surface chemistry is different and an HF dip leaves a surface that is hydrophilic and terminated with O or OH bonds [21]. As a result, H plasma techniques are sometimes used to strip the native oxide on SiC wafers [22].

4.4   Level 3 Contamination Reduction: Gettering in Silicon

The term "gettering" has its origins in the vacuum tube industry where it referred to the use of materials like barium or cesium that react with or absorb trace gases in vacuum tubes and thus help to maintain a vacuum environment during the lifetime of the tubes. It is very difficult to reduce the concentration of undesirable elements such as metals (Fe, Au, Cu, etc.) and alkali ions (Na^+, K^+, etc.) in silicon materials to levels where they would have no effect on device performance using only clean factories and wafer cleaning strategies. The deleterious effects of both types of contaminants were illustrated in examples in the introduction to this chapter.

Gettering is the third tier defense and is a means of collecting these unwanted elements in regions of the chip where they do minimal harm. This is not as difficult as it might seem, because the active devices in most silicon chips are located in a thin region at the top of the wafer. In addition, the undesirable elements tend to have very high diffusivities (which is part of the reason why they are so "bad") and they tend to be easily captured either in regions with mechanical defects or in regions that chemically trap them. The elements we are most concerned about are highlighted on the periodic table in Figure 4.14. The transition metals in the middle of the periodic table generally cause deep levels in the Si bandgap and act as SRH recombination centers. The alkali ions are mobile in dielectrics like SiO_2 and cause device instabilities.

Gettering therefore consists of three steps. First, the elements to be gettered must be "freed" from any trapping sites they may currently occupy and made mobile. Second, they must diffuse to the gettering site. Finally, they must be trapped. Figure 4.15 illustrates some of the key points, with the three steps labeled for metal atoms being trapped either in the bulk or on the backside of the wafer.

Alkali ions can be trapped in deposited phosphosilicate glass (PSG) layers. Transition metals can be trapped in the silicon bulk or on the wafer backside. Since, in silicon ICs, the devices are usually fabricated in a thin layer on the top surface, intrinsic or bulk gettering is the most common approach, since almost the entire thickness of the wafer can be used to trap metal impurities in a location where they will not degrade device performance.

Some types of silicon devices, for example, solar cells and power devices, use most of the thickness of the wafer as part of the active device, and hence intrinsic gettering is not an option.

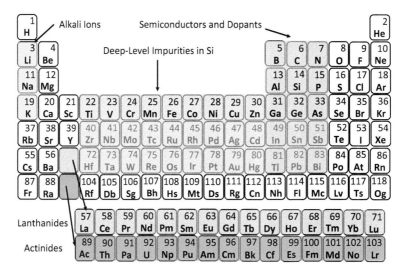

Figure 4.14 Periodic table indicating the elements that are of most concern in gettering. The alkali ions are in column I of the table; the metal atoms that form deep-level SRH centers are largely in the central portion of the table.

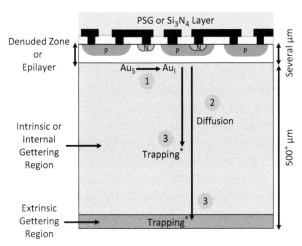

Figure 4.15 Wafer gettering strategies showing a surface denuded zone, intrinsic or bulk gettering regions and backside or extrinsic gettering. Au_S is Au on a substitutional site in the silicon lattice; Au_i is Au in an interstitial site.

In these cases, gettering sites can be introduced in places where trapped metals will not degrade device operation, often the backside of the wafer. This is known as extrinsic gettering.

4.4.1 Gettering Alkali Ions

In the early years of silicon IC manufacturing (1960–1980), instabilities especially in MOS structures were common problems, for example, the V_{TH} stability issue discussed at the beginning of this chapter. As was described in that example, alkali ions (Na^+ and K^+ in particular) were found to be the cause of these instabilities. The sources of these contaminants in manufacturing plants were traced to people handling the wafers (our bodies are full of these elements), process chemicals and the machines used in manufacturing plants. Tier 1 and 2 strategies (clean factories and wafer cleaning) helped enormously in mitigating these instabilities, but several approaches to gettering such elements were also explored [23, 24, 25, 26]. The PSG or Si_3N_4 layer on the top of the wafer in Figure 4.15 is a common example. PSG is a P_2O_5/SiO_2 glass that is normally deposited by chemical vapor deposition (CVD) or low-pressure chemical vapor deposition (LPCVD). Usually it is about 5 wt.% phosphorus. This glass has been found to be a very efficient trap for alkali ions, forming a stable complex that binds Na^+ or K^+ ions. Depositing such a glass as one of the upper thin-film layers on a chip can thus effectively keep such ions from penetrating down to sensitive gate oxides or field oxides which are adjacent to the silicon surface. PSG layers are effective getters for such ions. Na^+ and K^+ are very mobile in SiO_2 at temperatures above room temperature and they will easily diffuse to and be trapped by a PSG layer. Effective gettering of these ions from dielectric layers below the PSG can thus take place.

There are some drawbacks to PSG layers, however. These layers contain charge dipoles that can affect surface electric fields [26], they are susceptible to absorbing water vapor and they can cause Al corrosion problems. These effects can be minimized by keeping the P concentration to a few percent. An alternative or an additional strategy is to use a deposited Si_3N_4 layer as the top dielectric layer on a chip. These layers are impermeable to alkali ions and can form effective barriers to in-diffusion. Using a barrier layer like this requires that earlier processing steps be free from significant alkali ion contamination, since if Na^+ or K^+ ions were introduced earlier in the process, they would simply be locked into the IC structure by the surface nitride layer and could therefore cause instability problems.

In modern silicon structures with many levels of Cu interconnect, low-K dielectric layers to minimize capacitances and generally low-temperature processing for back-end structures, the use of PSG layers is less practicable. In early Si technologies, the PSG layer was "reflowed" at a temperature on the order of $900\,°C$ to help in planarizing back-end structures, but such temperatures are not possible today in back-end processing. And, of course, today CMP is the dominant technology used for planarization, and so PSG reflow is unnecessary. As a result, it is common today to eliminate the PSG gettering layers, and use a final Si_3N_4 layer as the top dielectric. In modern Si manufacturing plants, Na^+ and K^+ contaminants are low enough in concentration that this strategy is viable.

4.4.2 Gettering Metal Impurities in Silicon

Most of the metals of interest in gettering (those that create deep levels in silicon) exist in two states in the silicon crystal, occupying either a substitutional lattice site M_S or an interstitial site M_i. These metals are pure interstitials in the latter case, and generally do not occupy interstitialcy sites (we distinguished between these two in Figure 3.7 in Chapter 3). Metals can diffuse in silicon in either state, but their diffusivities are generally orders of magnitude higher in the interstitial form, as we will see below. We will discuss diffusion mechanisms in much more detail in Chapter 7, but it should be easy to imagine that substitutional diffusion involves hopping from lattice site to lattice site, a process that involves breaking bonds. Interstitial diffusion simply involves moving through the interstitial spaces between lattice atoms with no bond breaking. In general, the interstitial process is much faster. The data in Figure 4.16, which we will discuss shortly, illustrate this. Interstitial diffusers like Au and Cu have orders of magnitude higher diffusivities than substitutional dopants like P or even Si itself. Thus the first step in gettering these metals – making the metal atoms mobile – generally involves getting the metal atoms into their interstitial form.

The importance of this first step depends on the specific metal atom being considered. Some metals (Cu and Ni, for example) have much higher solubilities in their interstitial form and so most such atoms are in the interstitial form naturally. The release step is therefore unnecessary unless these atoms have formed precipitates due to very high concentrations. A second category of metals (Au and Pt are examples) have much higher solubilities in their substitutional form but diffuse rapidly once they become interstitials. For these elements, the release process is very important. Finally, there are some elements (Ti and Mo, for example) that are primarily substitutional, but which have relatively slow diffusion rates in either interstitial or substitutional form (see Figure 4.16). These elements are the most difficult to getter effectively because they will not diffuse over long distances (to the wafer backside, for example), unless very long thermal cycles are used.

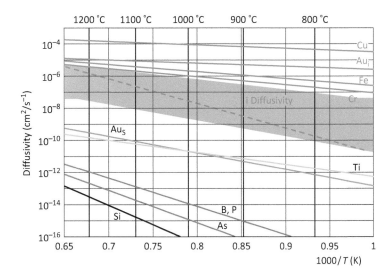

Figure 4.16 Diffusivities of various species in silicon. Metal atoms that diffuse interstitially are shown in red. Au_S refers to gold in substitutional form (on a lattice site); Au_i to gold in an interstitial site. The silicon interstitial (I) diffusivity is also shown and will be discussed later. The red-shaded region indicates the uncertainty in this parameter. After [29, 30].

Metals such as Au and Pt are the most interesting case, because they need to be converted from substitutional to interstitial sites to be gettered effectively. This can happen through one of two reactions, where we have taken Au as a specific example:

$$Au_S + I \leftrightarrow Au_i, \tag{4.7}$$

$$Au_S \leftrightarrow Au_i + V. \tag{4.8}$$

In (4.7), a silicon interstitial exchanges places with the Au atom on a lattice site, creating a Au interstitial. This is called the "kick-out" mechanism. In (4.8), the Au atom jumps off its lattice site, creating a silicon vacancy. This is known as the "dissociative" or "Frank–Turnbull" mechanism [27]. Process conditions that create excess concentrations of I should be useful in driving Au_S to Au_i and thus help the gettering process. Similarly, processes that create excess V should hinder the gettering process by driving Au_i back on to substitutional sites. It has been proposed that many of the effective gettering procedures inject excess concentrations of I and thus drive (4.7) to the right [28]. For example, we will see in Chapter 7 that high-concentration phosphorus diffusions produce very large supersaturations of I. Backside P diffusions are known to be among the most effective gettering treatments, and such diffusions are thought to inject large quantities of I. Similarly, ion implantation damage creates excess I, as we will see in Chapter 8. Such damage is also effective at backside gettering. Intrinsic gettering involves oxidation (SiO_2 formation) and this process also injects I.

Once transition metals are in their interstitial forms, most metals have very high diffusion coefficients in silicon. Figure 4.16 shows some representative data [29, 30]. There are several interesting things to note in Figure 4.16. In general, diffusion coefficients increase exponentially with temperature and are of the form

$$D = D_0 \, \exp\left(-\frac{E_A}{kT}\right), \tag{4.9}$$

where E_A is the activation energy. Plotting D on a semi-log plot versus $1/T$ therefore produces a straight line, as shown in Figure 4.16. We will discuss the physical basis for this mathematical form in Chapter 8.

Note that the diffusivities of the dopants in silicon (As, B, P) are orders of magnitude smaller than the D values for the metal contaminants. Because of this, in the same time that standard dopants are diffusing to produce a PN junction 1 μm below the silicon surface, metals such as Au can easily diffuse completely through a silicon wafer. From a positive perspective, this makes it easy to getter these elements to regions away from active devices. From a negative perspective, metal contamination anywhere on a wafer can redistribute throughout an entire wafer. Note also the large difference in D for Au_S versus Au_i that emphasizes the importance of getting impurities like Au into their interstitial forms for gettering to be effective. Finally, note the relatively small Ti diffusion coefficient compared to other transition metals. This explains the statement made above that metals like Ti are difficult to getter.

Once the metal atoms are in their interstitial (fast-diffusing) form, they can diffuse through the wafer to a backside sink, or to the trap sites in the bulk of the wafer if intrinsic gettering is being used. Obviously, the diffusion distance is a lot smaller for the intrinsic gettering case, which is one of the reasons why it is more popular today.

The final thing to note about Figure 4.16 is the region labeled "I diffusivity." The red dashed line is a plot of (3.6) from Chapter 3. There is still uncertainty about the diffusivity of silicon interstitials and the shaded region is intended to indicate the range of possible values. The I diffusivity is much larger than dopant diffusivities but considerably smaller than those of most metal contaminants in their interstitial form. We will return to this topic later in Section 4.4.4 on extrinsic gettering because effective backside trapping regions also inject interstitials, and we will see how this affects the conversion of the metal atoms from their substitutional to interstitial form and, as a result, how this affects the gettering of the metal atoms.

4.4.3   Intrinsic or Internal Gettering Trapping Sites in Silicon

Once the metals we wish to getter are mobile and can diffuse, we next need to create trapping sites for those metals that are away from the active device regions. The most common strategy today for doing this uses intrinsic or internal gettering that creates trapping sites throughout the bulk of Czochralski (CZ) grown silicon crystals. The other alternative is to create backside trapping sites, which we will discuss in the next subsection on extrinsic gettering.

The intrinsic gettering region shown in Figure 4.15 is created using an interesting byproduct of the CZ crystal growth method. In Chapter 3 we saw that the liquid silicon in CZ growth is contained in an SiO_2 crucible and that, because the liquid silicon is corrosive at its melting temperature, some SiO_2 is dissolved in the liquid silicon. As a result, oxygen is incorporated in

CZ crystals at concentrations in the range of 10^{17}–10^{18} cm^{-3}. We saw in Chapter 3 that this dissolved oxygen has several effects on the silicon wafers grown with the CZ method. It improves the yield strength of silicon by as much as 25%, making the silicon wafers more robust in a manufacturing facility. Second, SiO_4 complexes can form, which act as donors, and these can compensate the resistivity of P-type silicon. Most importantly, the third effect of oxygen in silicon is the tendency of the oxygen to precipitate under normal device processing conditions, forming SiO_2 regions inside the wafer. It is this property of oxygen that we will discuss in more detail here because it provides the foundation for intrinsic gettering.

The solubility of oxygen in silicon increases at higher temperatures and is given by [31]

$$C_0^* = 5.5 \times 10^{20} \exp\left(-\frac{0.89 \text{ eV}}{kT}\right), \tag{4.10}$$

where C_0^* is the equilibrium solubility. This gives a value of about 1.2×10^{18} cm^{-3} at the melting point of 1417 °C. At normal processing temperatures of about 1000 °C, the solubility is much lower, about 1×10^{17} cm^{-3}. There is thus a strong driving force for the grown-in oxygen to precipitate during wafer processing.

If a wafer containing oxygen in this form is heated as part of a normal IC fabrication process, precipitation of the oxygen can occur. The process can really be thought of as an internal oxidation, occurring in the bulk of the wafer, with the oxidant provided by the grown-in oxygen. Precipitates can be nucleated by either heterogeneous or homogeneous means. Heterogeneous nucleation occurs when there is a particular lattice site (perhaps a defect or impurity cluster) that provides a disturbance in the otherwise perfect lattice at which precipitation can take place. Homogeneous nucleation occurs when enough oxygen atoms come together through random diffusion to form a cluster of critical size. The distinction between the two is often not important. From a more macroscopic point of view, small SiO_2 precipitates are constantly forming and either growing or shrinking as oxygen atoms are either added or escape. The growth process is controlled by the in-diffusion of the oxygen atoms. These ideas are illustrated in Figure 4.17. Oxygen's diffusivity in silicon is well characterized and is given by [32]

$$D_{O_i} = 0.13 \exp\left(-\frac{2.53 \text{ eV}}{kT}\right) \text{ cm}^2 \text{s}^{-1}. \tag{4.11}$$

Qualitatively, what occurs is the following. At any temperature, small SiO_2 precipitates or embryos are constantly forming as diffusing oxygen atoms randomly encounter each other. Normally these embryos dissolve because, when they are small, there is a strong tendency for them to shrink. This occurs simply because in-diffusing oxygen atoms do not arrive at a rate sufficient to overcome the flow of oxygen atoms that leave the embryo due to normal thermal bond breaking and diffusion processes. If the embryo manages to reach a critical size of something like 1 nm, then it is stable. That is, it will only grow beyond that point and is considered to have nucleated. Because all of these processes are temperature-dependent, there is an optimum temperature at which SiO_2 precipitates can be nucleated. That temperature is around 700 °C in silicon. At lower temperatures, the diffusivity of oxygen is too low to provide

Figure 4.17 Illustration of point defect and diffusion mechanisms that contribute to the growth or shrinkage of SiO$_2$ precipitates in silicon. O$_i$ is interstitial oxygen. After Kennel [33].

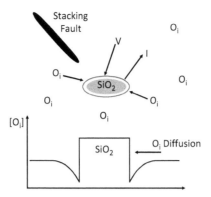

an appreciable probability of nuclei forming; at higher temperatures, the embryos tend to break up before they reach the critical size.

The growth of these SiO$_2$ precipitates can be described in a slightly different way, which is also illustrated in Figure 4.17. An important observation regarding their growth is that volume must be provided for the SiO$_2$ to form. The volume occupied by the SiO$_2$ precipitates is larger by about a factor of 2 than the volume occupied by the silicon atoms used to form the SiO$_2$. The oxygen atoms that help to form the SiO$_2$ were originally located on interstitial sites (O$_i$), not lattice sites. In the SiO$_2$ precipitates, however, the oxygen atoms take up a lattice site, requiring the volume expansion. When we discuss surface oxidation in Chapter 6, we will see that volume expansion is inherent to the oxidation process. The same thing happens here during the "internal" oxidation.

Because of the need for volume during the precipitation process, we can consider the roles that point defects present in the crystal might play. Vacancies can be thought of as microscopic empty sites in the crystal or, in other words, they could provide some of the volume or lattice sites needed for the SiO$_2$ precipitates to form. Stated another way, if a V happened to be present at a precipitate, an in-diffusing oxygen atom could occupy the site of the V. Thus the V could provide the volume required. Growing precipitates should thus consume V, as illustrated in Figure 4.17. By a similar argument, precipitates should also be able to obtain the needed volume by ejecting an I, that is, by removing a silicon atom from the perimeter of the precipitate. This I would then diffuse off into the crystal. We can express the precipitate growth process as [33]

$$(1 + 2\gamma)\text{Si} + 2\text{O}_i + 2\beta\text{V} \leftrightarrow \text{SiO}_2 + 2\gamma\text{I} + \text{stress} , \tag{4.12}$$

where γ is the number of I that contribute to the precipitation process per O$_i$ atom joining the SiO$_2$, and β is a similar fraction for V. Here O$_i$ represents the interstitial oxygen atom and Si is simply a Si atom on a lattice site in the crystal. We have included a stress term on the right-hand side because it is unlikely that the point defects could provide all the necessary volume and so it is likely that the region surrounding the SiO$_2$ precipitates will be under stress, with the precipitate and the surrounding silicon under compression. From (4.12) we can observe that process environments that enhance the V population should promote precipitate growth, while those which enhance the I population should retard growth. We can also observe that, when precipitates are growing, they themselves should produce an excess of I and a deficiency of V in the surrounding crystal. We will see examples of these effects in later chapters when we see how point defect populations affect diffusion (Chapter 7) and other processes. A final point worth noting in connection with Figure 4.17 is that a stacking fault (see Figure 3.7) is shown associated with the SiO_2 precipitate. Such faults along with dislocation loops are often found in connection with such precipitates. They occur as a result of the high compressive stresses generated around the SiO_2 region because of volume expansion.

Such sites are ideal trapping sites for unwanted metal ions. The trick here is to cause this precipitation to occur in the bulk of the Si wafer, but not in the near-surface region where the devices are located. This turns out to be quite possible. In many cases, an epitaxial layer is grown on the silicon substrate as part of normal device fabrication. If this is the case, epitaxial silicon is normally very low in oxygen and so no precipitation would occur in the epilayer. Alternatively, the oxygen in the near-surface region can be out-diffused from the wafer as one of the first steps in wafer processing, creating the denuded zone shown in Figure 4.15. In either case, SiO_2 precipitation can then take place only in the wafer bulk, away from the active devices.

The SiO_2 precipitates and the resulting damage are created using a thermal cycle like that illustrated in Figure 4.18. This is normally done near the beginning of the wafer fabrication process flow and in some cases is actually done by the wafer supplier so that the as-supplied wafers already have an intrinsic gettering region in the wafers. If there is no epilayer on the wafer surface, an initial high-temperature step is used to out-diffuse the oxygen in the near-surface region. This creates the denuded zone shown in Figure 4.15.

The time required to create an "oxygen-free" surface region can be calculated using (4.11). The surface region does not actually have to be oxygen-free. What is required is that the oxygen concentration is reduced below the level at which precipitation will occur in the next part of the cycle. Normal oxygen concentrations in CZ silicon are 15–20 ppm ($\approx 10^{18}$ cm^{-3}). If the concentration is much higher than 20 ppm, too much precipitation will take place and there may be problems with wafer warpage or excessive defect formation. Precipitation does not readily occur below concentrations of about 10 ppm. A temperature cycle of several hours at 1100–1200 °C is sufficient to create a denuded zone many microns deep.

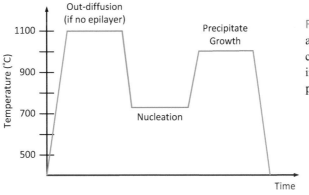

Figure 4.18 Processes and time–temperature cycle for a typical intrinsic gettering process.

The next step in the process is precipitation of the oxygen to nucleate small SiO_2 precipitates or embryos. Nucleation usually takes place over several hours. The embryos must be grown to a minimum critical size during this step so that, during the subsequent higher-temperature growth step, they will grow and not shrink. This critical size depends on process conditions (especially temperature), but is on the order of 1–3 nm. The density of embryos created also depends strongly on process conditions, but is on the order of 10^{11} cm^{-3}.

The final step in creating the intrinsic gettering region is growth of the embryo precipitates. This is accomplished by raising the temperature to approximately 1000 °C for several hours, which allows them to grow to 50–100 nm in size. It may be necessary to carefully control the temperature ramp rate at the start of the growth cycle, because the critical embryo size increases with temperature. If the temperature is ramped up too quickly, the embryos may be below the critical size at the growth temperature and therefore shrink rather than grow. Slow ramping gives them a chance to grow during the ramp up in temperature. An example of the result of these process steps is shown in Figure 4.19 [34]. A denuded zone about 25 μm deep is visible at the surface of the wafer. Longer precipitation and growth times result in a higher density and larger precipitates in the wafer bulk. Once the oxygen precipitates are formed, they are relatively stable throughout the rest of the IC manufacturing process, and as a result they serve as gettering sites throughout device fabrication.

Transition metal atoms preferentially reside at sites in the silicon lattice where "imperfections" exist. This is likely because these atoms do not fit well in the silicon lattice because of their very different atomic size, so disordered regions provide better sites for them. Thus creating such sites through oxygen precipitation in the bulk of the wafer or on the backside of the wafer provides places that can efficiently capture and trap metal contaminants. The third stage of the gettering process, trapping of the metal atoms, thus occurs when the fast-diffusing interstitial atoms reach either the intrinsic gettering sites due to SiO_2 formation in CZ wafers, or the extrinsic damage regions, usually on the wafer backside.

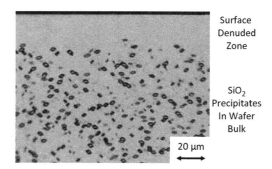

Figure 4.19 Cross-sectional optical image of a CZ wafer with a starting [O] of 1×10^{18} cm^{-3}. The wafer was heat-treated for 4 h at 1150 °C (out-diffusion), then for 8 h at 800 °C (nucleation), followed by 16 h at 1050 °C (precipitate growth), resulting in the surface denuded zone and the SiO_2 precipitates in the wafer bulk. Used with permission of IOP Publishing Ltd, from [34]; permission conveyed through Copyright Clearance Center Inc.

4.4.4 Extrinsic Gettering Trapping Sites in Silicon

Extrinsic gettering sites are obviously further away from the active devices in silicon ICs, so generally intrinsic gettering is preferred. However, there are situations in which intrinsic gettering is not possible. Examples include power devices (Si vertical MOSFETs, IGBTs, thyristors, etc.) in which current flow is vertically through the wafer. In many of these devices, float zone (FZ) silicon is used for the starting wafers, so the oxygen content is well below the levels needed to form the SiO_2 precipitates for intrinsic gettering. Other examples include photovoltaic devices in which the entire wafer thickness is again part of the active device. So, even in spite of the dominance today of intrinsic gettering, there remains interest in extrinsic gettering methods.

Extrinsic gettering sites have been formed in silicon by a wide variety of methods. Normally they are created on the wafer backside, which is an accessible surface far away from the devices on the top surface. Methods that have been used range from grinding and sandpaper abrasion, to ion implantation, laser melting, depositing amorphous or polycrystalline films (usually polysilicon), or high-concentration backside diffusions (usually phosphorus). Whatever the method, the objective is to create extended defects – dislocation loops, grain boundaries, precipitates – or other traps that are stable through subsequent high-temperature processing and hence can trap and hold metal atoms. Modern processes tend to use the "cleaner" methods from this list to produce backside damage (ion implantation, deposited polysilicon films or backside phosphorus diffusions), since it is clearly not desirable to introduce additional contamination into the wafer during the gettering step. Once the backside damage sites are created, any subsequent high-temperature processing step will allow the metal atoms to diffuse to the backside where they can be trapped.

The question of how they are trapped, however, is less clear. Physical damage to the wafer backside (ion implantation, laser or mechanical damage or polysilicon films containing many grain boundaries) is generally believed to trap the metal atoms at defect sites. This trapping process is most simply characterized by some binding energy E_B of the metal atom to the trap site. In that case, the metal atoms are kept trapped at modest temperatures. As T increases, more and more of them are able to escape the traps. Clearly we would want E_B as large as possible. We can express this as

$$\text{fraction bound} = 1 - K_1 \exp\left(-\frac{E_B}{kT}\right), \tag{4.13}$$

where K_1 represents some of the details of the trapping process and would be empirically determined in a specific case. Other models have been developed to explain the trapping process. These include segregation models in which the metals preferentially segregate to the damaged layer [35], enhanced solubility of metals like Au in heavily doped silicon, and ion pairing models in which large atoms like Au prefer to pair up with small atoms like P to minimize lattice strain [36].

Finally, we saw in Chapter 3 that the point defect concentrations are much higher in doped silicon than in intrinsic silicon. Au diffuses primarily by an interstitial mechanism, as we have seen, so when it arrives at the gettering site at the wafer backside, it needs to find a lattice site on which to sit, since, in equilibrium, the Au prefers to be predominantly on substitutional sites. One possible reaction is

$$Au_i + V^- \leftrightarrow Au_S^- . \tag{4.14}$$

The much higher V^- population in N^+ silicon would drive this reaction to the right, with the vacancies essentially providing lattice sites for the Au atoms. Whatever the exact trapping mechanism is (and it may vary with the type of damage produced on the wafer backside), extrinsic gettering has been found to work very effectively.

One final interesting point should be made with regard to extrinsic gettering. Suppose we have a wafer that is uniformly contaminated with, for example, Au. If we placed extrinsic gettering sites on the backside of the wafer and then heated the wafer to allow Au diffusion and trapping, the expected general shape of the Au profiles as diffusion to the backside occurs is illustrated in Figure 4.20 [37]. The mobile Au concentration goes to zero at the wafer backside where the Au atoms are trapped. We will see in Chapter 7 how to calculate the actual detailed shape of such profiles, but the general shape should be intuitive. As time increases, the depth over which the Au is gettered increases away from the wafer backside, until, at long enough times, the Au is removed entirely from the wafer.

Some actual experimental measurements of Au profiles are shown in Figure 4.21 [38]. Note that the shape is quite different from the expected simple out-diffusion profile in Figure 4.20. The experimental data can be understood in connection with the idea that Au_S needs to be converted to Au_i to make it mobile. The diffusivity of the I species indicated in Figure 4.16 is also important [28]. Note that this diffusion coefficient is much faster than dopant diffusivities, but significantly slower than interstitial metal atom diffusivities. Thus if the gettering process depends on the backside gettering region for injection of I to drive (4.7) as simulated in Figure 4.21(a), then the slowest (rate-limiting) part of the process would be the in-diffusion

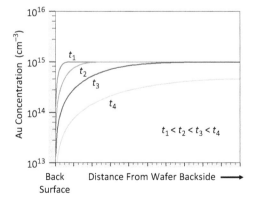

Figure 4.20 General shape of Au out-diffusion profiles from Athena simulations [37] if simple out-diffusion dominates the gettering process. Starting with a flat profile ($10^{15}\,cm^{-3}$), the concentration drops off with time through the wafer as the Au diffuses to the back surface.

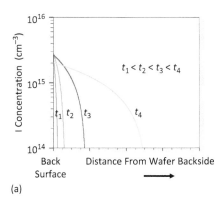
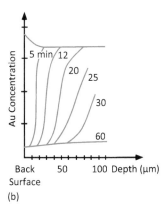

Figure 4.21 (a) Simulated interstitial injection from wafer back surface [37]. (b) Experimental 1000 °C Au out-diffusion profiles during gettering to the wafer backside. After [38].

of the I from the backside. As quickly as silicon interstitials reach a particular depth from the backside, the Au_S in that region are driven to Au_i that then quickly diffuse to the wafer backside where they are trapped. This qualitative explanation leads to Au profiles like those in Figure 4.21(b). The sharp drop in the Au profile, which moves away from the backside as time goes on, occurs at the depth corresponding to how far the silicon interstitials have diffused in from the backside at that particular time.

When gettering is performed at higher temperatures, the equilibrium concentration of I may be high enough that (4.7) can proceed without the excess I provided by the backside. In that case, the metal out-diffusion profiles should be more like Figure 4.20 and the rate-limiting process should be simply the metal atom diffusivity. The few metals which have diffusivities smaller than I (Ti, for example, in Figure 4.16) should also show out-diffusion profiles similar to Figure 4.20, since the rate-limiting step in this case is also metal diffusion at all temperatures.

4.4.5 Gettering in Compound Semiconductor Technologies

Gettering methods are generally not used in compound semiconductor chip manufacturing. Many of the applications involve direct-bandgap materials in which metal impurities play minor roles. Two examples will illustrate this point.

The only indirect-bandgap compound semiconductor of commercial interest today is SiC. Because of its wide bandgap (≈ 3.26 eV in the 4H polytype), SiC is currently being used for a variety of power devices, including vertical power MOSFETs, IGBTs and PIN diodes. In all of these applications, the current flow is vertically through the wafer, so that intrinsic gettering is not a viable approach. SiC crystal growth methods also do not introduce oxygen into the crystal in significant amounts, so, even if intrinsic or bulk gettering approaches were viable, the SiO_2 precipitation approach used in Si would not work.

Because SiC is an indirect-bandgap material, long carrier lifetimes are possible. Metal ions generally act as SRH recombination centers and reduce carrier lifetime and increase junction leakage currents just as they do in Si. It is known that Fe, Cr, Ti, Ni and other transition metals create deep levels in SiC [39]. Very little work seems to have been done to this point on gettering approaches in SiC. What is known is that the transition metals essentially do not diffuse at temperatures below 1500 °C. Above this temperature, they do diffuse, although the diffusion coefficients are very small (Cr is estimated to have a diffusivity of $\approx 5 \times 10^{-13}$ cm^2 s^{-1} at 1780 °C [39]). This is eight orders of magnitude smaller than the Cr value in Si shown in Figure 4.16 even though the temperature is much higher in the SiC data point. While 1780 °C may seem like an extraordinarily high temperature to those familiar with Si technology (it is higher than the melting temperature of Si), such temperatures are used in SiC processing to anneal implant damage, as we will see in Chapters 7 and 8.

There is some evidence that transition metals can be trapped in damage regions in SiC just as they are in Si, so in principle extrinsic gettering could work [39]. However, gettering seems to be basically unused in SiC manufacturing today, likely because the metal diffusivities are so small that getting them to diffuse to trapping sites is very difficult.

In GaAs devices, deep-level traps also play very important roles [40]. Unlike Si, in which deep-level traps are primarily caused by metal impurities, those in GaAs are primarily due to antisite defects such as As$_{Ga}$ (see Figure 3.12). Antisite defects can be grown into GaAs crystals by controlling the Ga/As ratio during the growth process. These defects are used to create semi-insulating material by pinning the Fermi level deep in the bandgap. They behave and are modeled just like SRH recombination centers in Si and can affect device characteristics both for DC and at high frequencies. A significant literature on these effects exists [40, 41]. Because GaAs is a direct-bandgap material, minority carrier lifetimes are very short and so the impact of metal impurities on lifetime is minimal. As a consequence, gettering methods are generally not used in GaAs IC manufacturing.

4.5   Lifetime Control in Indirect-Bandgap Semiconductors (Si and SiC)

In semiconductor devices that depend on minority carriers for their operation, controlling minority carrier lifetime is critical to device performance. This is generally only possible in indirect-bandgap materials because the intrinsic lifetimes are so short in direct-bandgap materials. The most important applications of these methods are in power semiconductor devices, IGBTs, thyristors and PIN diodes fabricated in Si and SiC. There are major markets for these devices, and so we include here a brief discussion of lifetime control methods. Interestingly, these methods involve introducing controlled amounts of deep levels that act as SRH recombination centers. This can be done using transition metals like Au or Pt, or it can be done by damaging the crystal in some other way to produce the deep levels. In many ways, these processes are the exact opposite of the gettering methods we discussed in the previous sections. There we were trying to remove the metal contaminants; here we purposely introduce them, but in a controlled manner.

Standard textbooks on semiconductor device physics derive the result from SRH theory, so that the rate of recombination (or generation) for a deep-level trap with an energy level E_T is

$$U = \frac{np - n_i^2}{\tau\{p + n + 2n_i \cosh[(E_T - E_i)/kT]\}},$$ (4.15)

where τ is the carrier lifetime. SRH theory is also discussed in Appendix A.1. In equilibrium, $np = n_i^2$ and U goes to zero. In this case, recombination and generation exactly balance each other. When excess carriers are present, $np > n_i^2$ and U is positive (net recombination). When $np < n_i^2$, then U is negative (net generation). Note that the rate of recombination (or generation) is maximized when the trap level is near the middle of the bandgap ($E_T \approx E_i$). The carrier lifetime is given by

$$\tau_R = \frac{1}{\sigma v_{th} N_t},$$ (4.16)

where τ_R is the recombination lifetime, σ is the capture cross-section of the trap (on the order of the atom cross-sectional area or about 10^{-16} cm^2), v_{th} is the minority carrier thermal velocity (about 10^7 cm s^{-1}) and N_t is the density of traps per cm^3. This is the same equation that we used in the example at the beginning of this chapter to illustrate the bad effects of metal atoms on DRAM devices.

The key parameter is N_t, the density of deep-level traps. A simple approach to creating these traps is to deposit a metal contaminant on the backside of the wafer and then let it diffuse into the wafer at high temperatures. An example is shown in Figure 4.22 in which Au was deposited on the backside of a 450 μm thick wafer and then diffused at 900 °C.

The Au diffuses rapidly as Au$_i$ and then becomes substitutional so that it can act as a deep-level trap. Because of the very high diffusivity of Au$_i$ (Figure 4.16), the Au profiles are relatively flat even for a 10 min diffusion. This produces a carrier lifetime according to (4.16) that is relatively constant throughout the wafer.

While controlled metal contamination was used in some early Si power devices, much more reproducible methods are used today [43]. These generally involve ion implantation of electrons, protons or He ions that create damage by knocking Si atoms off their lattice sites. This approach has two major advantages. First, the amount of damage is controlled by the dose of the implant, and this is very well controlled in modern ion implantation machines. Second, the energy of the implant controls where the damage is located, so it is possible to control the carrier lifetime spatially within the device. This provides enormous flexibility in optimizing the design of modern power devices. The damage is stable only up to a few hundred degrees Celsius, so it must be done at the end of the device fabrication process. We will study ion implantation in Chapter 8 and learn how energy and dose are controlled. These same methods can be used in SiC power devices although, because of the low diffusivities of metals in SiC, ion implantation methods would be preferred.

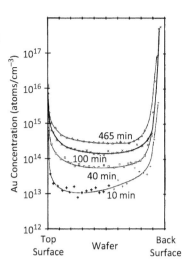

Figure 4.22 Au experimentally diffused through a 450 μm thick wafer at 900 °C from the wafer backside (right) for times between 10 and 465 minutes. Used with permission of IOP Publishing Ltd, from [42]; permission conveyed through Copyright Clearance Center Inc.

4.6 Modeling Yield in Chip Manufacturing

For companies involved in manufacturing semiconductor chips, yield is an extremely important parameter. Manufacturing yield is defined simply as the fraction of products that function correctly. In the case of ICs, since large numbers of chips are fabricated on a common substrate, yield is the fraction of the die that work at the end of the manufacturing process. Since wafer manufacturing costs are largely independent of how many chips function correctly, yield determines how many chips can be sold and hence what the revenue is for a company per wafer. Profitability of the company is thus directly tied to yield.

In this section, we will discuss functional yield, which is the fraction of die that do not have defects. In the next section on manufacturing process control, we will discuss parametric yield. Since all manufacturing processes have distributions of process parameters, the die that are produced will have parametric variations even when they are fully functional. Manufacturing process control is aimed at keeping these parametric variations within specified limits.

Functional yield is generally limited by particles that create defects. For this reason, yield models are based on a defect density D_0, such as defects/cm^2. The second basic parameter in yield modeling is the critical area A_C, which is the fraction of the die area in which a defect has a high probability of causing a fault. Generally, A_C is less than the total die area because defects can occur in a chip region where there is no device or where the particle will not cause a fault. We discussed this concept earlier in connection with Figure 4.3. Yield models can thus be formulated as follows [44]:

$$Y = f(D_0, A_C). \tag{4.17}$$

A variety of mathematical models have been developed to describe die yield. These models are differentiated by how the defect density D_0 is modeled. The simplest of these models is the

Poisson model, which assumes that the defects are spatially uncorrelated and uniformly distributed across the wafer. The model also assumes that each defect causes a faulty chip. The model can be derived in an elegant way, following [44]. Let C be the number of chips on a wafer and M be the number of possible defects. Then there are C^M unique ways to distribute the defects between the chips. If one chip is found to not have a defect, then the number of ways to distribute the defects among the remaining chips is

$$(C-1)^M. \tag{4.18}$$

Thus the probability that a circuit will contain zero defects is

$$\frac{(C-1)^M}{C^M} = \left(1 - \frac{1}{C}\right)^M. \tag{4.19}$$

Since $M = CA_C D_0$, the yield is the number of chips that have zero defects, which is given by

$$Y = \lim_{C \to \infty} \left(1 - \frac{1}{C}\right)^{CA_C D_0} = \exp(-A_C D_0). \tag{4.20}$$

The Poisson model generally gives reasonable results for small chips but underestimates the yield for large chips compared to actual experimental data. A variety of other yield models have been proposed [9, 44, 45]. Generally these models use different assumptions about the defect density distribution. The Poisson distribution assumes that the defect density is constant across the wafer, and this is generally not the case in practice. Defects are often clustered in specific regions and often the defect density is larger around the perimeter of the wafer.

Murphy first proposed that the defect density D should not be constant [46]. He suggested that D_0 should be summed over all circuits and substrates using a normalized probability density function $f(D)$ and that the yield should be calculated using the integral

$$Y = \int_0^\infty e^{-A_C D_0} f(D) \, dD. \tag{4.21}$$

Various forms for $f(D)$ thus are the basis for different analytical yield models. A detailed discussion of $f(D)$ functions that have been discussed in the literature is given in [44]. Two of the commonly used models are summarized below:

negative binomial model
$$Y = \left(1 + \frac{A_C D_0}{C}\right)^{-C}, \tag{4.22}$$

Seeds model
$$Y = \frac{1}{1 + D_0 A_C}. \tag{4.23}$$

In the negative binomial model, C is a measure of the particle spatial distribution called the clustering factor. When $C \to \infty$, which implies very little clustering of the defects, the particles become independent and (4.22) reduces to (4.20). When $C \to 1$, which implies a lot of clustering, the Seeds model results. In other words,

$$Y = \lim_{C \to \infty} \left(1 + \frac{A_C D_0}{C}\right)^{-C} = \exp(-A_C D_0), \tag{4.24}$$

$$Y = \lim_{C \to 1} \left(1 + \frac{A_C D_0}{C}\right)^{-C} = \frac{1}{1 + A_C D_0}. \tag{4.25}$$

Figure 4.23 plots chip yield versus chip area for several yield models. The negative binomial model is often used in industry with a C value of approximately 2. One can see in the figure that when $C \to \infty$ the negative binomial model approaches the Poisson model and when $C \to 0$ it approaches the Seeds model.

One might look at the numbers in Figure 4.23 and conclude that reasonable chip yields can be obtained even with defect densities on the order of 0.1 cm^{-2}. However, there is one very important additional factor that needs to be considered. Chip manufacturing consists of dozens

Figure 4.23 Plots of chip yield versus chip area for various yield models. The defect density is 5× smaller in plot (b) and the chip yields are correspondingly much higher.

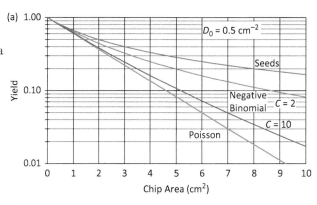

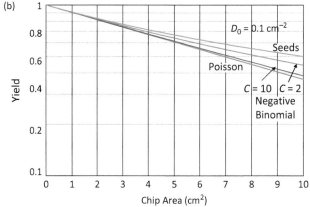

or hundreds of sequential steps, and defects introduced in any of these steps are cumulative. This means that the overall yield at the end of the process should actually be written as the product of the individual step yields. In other words, for the negative binomial model, for example,

$$Y = \prod_{i=1}^{\text{levels}} \left(1 + \frac{A_{C,i}D_{0,i}}{C_i}\right)^{-C_i} \cong \left\{1 + \left(\sum_{i=1}^{\text{levels}} A_{C,i}D_{0,i}\right)/C_i\right\}^{-C_i}. \tag{4.26}$$

Thus, if a particular manufacturing process consisted of 50 steps, each of which had a yield of 99%, the overall chip yield at the end would be on the order of $(0.99)^{50} \approx 60\%$. If the individual step yield were 95%, the chip yield at the end would be only a few percent. The very small numbers for defect densities in Table 4.1 (0.01 cm^{-2}) perhaps make more sense in light of these estimates. Of course, each process step in an IC manufacturing process is unique. As a result, one of the challenges in yield modeling in an actual manufacturing plant is determining the A_C and D_0 values for each process step. For readers interested in more depth on these topics, an excellent general reference is [44].

There are a number of interesting simulation tools that have been developed to help in functional yield modeling. Often the strategy is to take a particular circuit layout and then use Monte Carlo methods to randomly overlay defects on the layout. The simulator then calculates the impact of the defect distribution on circuit faults. One such example is the Very Large Scale Integration Layout Simulation for Integrated Circuits (VLASIC) simulation tool [44, 47].

4.7   Parametric Yield: Manufacturing Process Control

The previous section discussed functional yield, that is, what fraction of the chips on a wafer function correctly (have no defects) at the end of the manufacturing process. Because all manufacturing processes have tolerances and random variations, the "good" chips at the end of the process will exhibit a range of performance, where performance might be measured in terms of speed, power consumption or other parameters of interest. These variations in performance result from underlying variations in process parameters in the devices and other structures in the fabricated circuits. Figure 4.24 is a simple example. These types of measurements are usually assumed to be normally distributed so that they can be described by a mean and standard deviation, as also shown in Figure 4.24.

It is obviously important in manufacturing to control both the mean and the standard deviation of parameter distributions. Figure 4.25 shows examples. In Figure 4.25, the range of acceptable values for this parameter is defined by the lower specification limit (LSL) and the upper specification limit (USL). Three different examples of process distributions are shown. The uppermost one is centered but has a broad distribution. The middle one has a tighter distribution but is not centered. The lowest one is centered and has a tight distribution.

Two parameters that are widely used to describe how controlled a particular process step actually is are C_p (process capability) and C_{pk} (process capability index) defined below:

Figure 4.24 Typical distribution of a process parameter such as sheet resistance, oxide thickness, photoresist linewidth, etc., resulting from a manufacturing process.

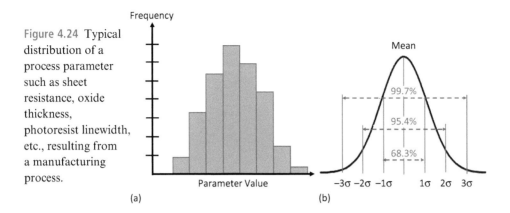

(a)   (b)

Figure 4.25 Normal parameter distribution examples shown with respect to the upper and lower specification or tolerance limits.

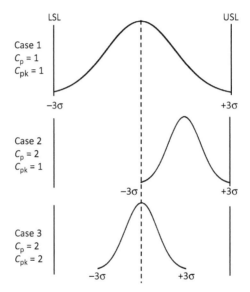

$$C_\mathrm{p} = \frac{\mathrm{USL} - \mathrm{LSL}}{6\sigma} \quad \text{and} \quad C_\mathrm{pk} = \min\left\{\frac{\mathrm{USL} - \overline{X}}{3\sigma}, \; \frac{\overline{X} - \mathrm{LSL}}{3\sigma}\right\}. \tag{4.27}$$

Thus C_p is primarily a measure of the width of the process parameter distribution and C_pk additionally adds in a measure of how well the distribution is centered. If the process is centered, $C_\mathrm{p} = C_\mathrm{pk}$. Generally, a process is considered to be under control if $C_\mathrm{p} > 1.33$. "Six sigma" high-quality processes have $C_\mathrm{p} \geq 2$.

Often manufacturing control is tracked using plots like that shown in Figure 4.26. In this example, the target parameter value is 100 with USL = 106 and LSL = 94. The actual data show a mean of 98.9 and a standard deviation $\sigma = 1.03$. This gives $C_\mathrm{p} = 1.94$ and $C_\mathrm{pk} = 1.60$.

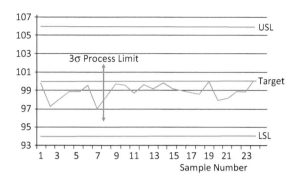

Figure 4.26 Example of process control in a manufacturing step. After https://en .wiki pedia.org/wiki/ Process_capability_ index.

The two parameters are not equal because the process is not centered, but the C_p value would indicate a well-controlled process.

There is also a time element to yield and process control issues. When a new technology is introduced, it often provides a competitive advantage for a company, since the products manufactured in that technology will often outperform older-generation products. It is not uncommon for companies to charge a premium price for these new products because they allow higher-performance systems to be built by their customers. As a result, getting a new technology to produce high yields in as short a time as possible can produce significant financial returns for a chip manufacturer. Balancing this, of course, is the fact that new technologies always bring new manufacturing challenges, and so there is a learning cycle associated with getting each new generation of technology to produce high yields.

An example of this is shown in Figure 4.27. Intel introduced its new 14 nm technology in 2013–2014. Their new microprocessors based on this technology offered performance advantages over older-generation products. However, the new technology carried with it significant manufacturing challenges and, as is always the case, chip yields were initially low. The time required to understand these yield challenges and to tune manufacturing processes to achieve high yields is a critical period that impacts net revenue to the company because, while yields are low, manufacturing costs are higher because more wafers must be processed to produce a given number of functioning chips.

When Intel was dealing with these yield issues in their 14 nm technology, Mark Bohr, one of the key technologists at Intel involved in technology development, was interviewed by *Semiconductor Technology*. His comments in September 2014 regarding Figure 4.27 are both interesting and insightful.

There was no one showstopper. It wasn't like we didn't know how to do the fins or we couldn't do the tight interconnects. It was just a variety of issues. Some were random. Some were systematic. Some were related to the transistors. Some were related to the interconnects. Because the features are so small and advanced, it was difficult to learn how to solve those problems. But we did that. It took us a little bit longer than what we would have liked.

Eventually, of course, Intel's 14 nm technology achieved high yields and became a dominant manufacturing platform. But Bohr's comments indicate how difficult this is with state-of-the-art silicon CMOS technologies.

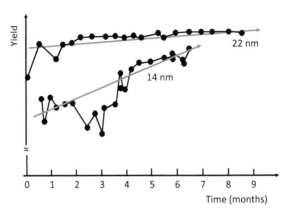

Figure 4.27 Yield versus time data presented by Intel in 2013 as their 14 nm process was ramped into full production. The vertical scale is not specified quantitatively.

Silicon IC manufacturing is among the most complex industrial processes practiced today. Because of this, there is a great need for computer tools for monitoring, controlling, simulating and analyzing manufacturing steps. Most modern IC manufacturing facilities employ sophisticated computer-integrated manufacturing (CIM) tools to monitor and control the status of machines, to download process recipes from central computer databases, to control wafer throughput in the fabrication facility (fab), to store information about factory performance and to improve operating efficiency. These topics are generally outside the scope of this book, but there are extensive references in the literature on these topics. See for example [48, 49].

4.8   Summary of Key Ideas

In this chapter, we have considered some of the "nuts and bolts" of IC manufacturing. Particle control, wafer cleaning and gettering sometimes do not receive the attention in the scientific community that more "high-tech" processes, like ion implantation, etching and lithography, do. However, the topics we have discussed in this chapter are crucial for the successful manufacturing of complex chips. In fact, it is likely the case that the economic success of any company in this business depends as much on its attention to particles, cleaning and gettering, as it does on the design features of a particular product. High yields, which come from attention to these issues, are the mark of successful manufacturers.

We have seen that a three-tiered approach is used to minimize contaminants in silicon wafer processing. The first step is to clean the environment around the wafers. Most commonly this is done through advanced air filtration systems to remove particles, careful protocols for workers in the plant, and the use of highly purified, filtered chemicals and gases. The second tier of contamination control is careful wafer cleaning. This is probably the most common step performed in wafer manufacturing. Cleaning needs to remove particulates that adhere to wafer surfaces, and it needs to strip trace levels of organic and metal contaminants. The RCA clean, which is most commonly used, is based on a solid scientific understanding of the chemical processes occurring in cleaning. Finally, gettering is the third level of contamination control, at least in advanced silicon technologies. This process collects metal atoms in regions of the wafer away from the active devices. Gettering occurs by releasing the metal atoms from whatever their current position is, allowing them to diffuse to the gettering site and then trapping them at those sites. While both extrinsic and intrinsic gettering are used in wafer manufacturing today, intrinsic gettering is the most common method used for silicon CMOS processing because the gettering sites are physically closer to the devices, minimizing the thermal cycling needed for gettering. In some types of devices, intrinsic gettering is not possible (vertical power devices, photovoltaic devices and SiC devices in which metal contaminant diffusivities are too small and oxygen is not typically incorporated in the wafers), so extrinsic gettering is the primary option, if gettering is used at all.

The bottom line is yield. A difference of a few percent in good die from each wafer can make a tremendous difference in the profitability of a company. Because of the batch fabrication processes used in IC manufacturing, with hundreds of die per wafer, the bad die are inherently manufactured along with the good die. There is generally no additional manufacturing cost when the yield goes up, so the incremental good die are essentially pure profit for the manufacturer. This fact should provide strong motivation for careful attention to the topics discussed in this chapter.

4.9   REFERENCES AND NOTES

[1] imec (Interuniversity Microelectronics Centre) is an international research and development and innovation hub in Leuven, Belgium, active in the fields of nanoelectronics and digital technologies. It is funded by semiconductor industry companies and by the local government. See https://www.imec-int.com/en.

[2] R. Guldi, T. Winter, N. Sridhar, *et al.*, "Systematic and random defect reduction during the evolution of integrated circuit technology," in *10th Annual IEEE/SEMI. Advanced Semiconductor Manufacturing Conference and Workshop. ASMC 99 Proceedings*, IEEE, pp. 2–7, 1999.

[3] P. Gupta and E. Papadopoulou, "Yield analysis and optimization," in *Handbook of Algorithms for Physical Design Automation*, eds. C. J. Alpert, D. P. Mehta and S. S. Sapatnekar, CRC Press, chap. 37, 2009.

[4] "International Technology Roadmap for Semiconductors," (ITRS). See http://www.itrs2.net/2012-itrs.html.

[5] C. Hu, *Modern Semiconductor Devices for Integrated Circuits*, Prentice Hall, 2010.

[6] W. Kern, "Overview and evolution of silicon wafer cleaning technology," in *Handbook of Silicon Wafer Cleaning Technology*, 3rd edn., eds. K. A. Reinhardt and W. Kern, Elsevier, pp. 3–85, 2018.

[7] S. M. Hu, private communication.

[8] D. K. Parasuraman, A. Kemps, H. Veeke and G. Lodewijks, "Prediction model for particle fallout in cleanrooms," *Journal of the IEST*, vol. 55, no. 1, pp. 1–9, 2012.

[9] C. H. Stapper, "Fact and fiction in yield modeling," *Microelectronics Journal*, vol. 20, no. 1–2, pp. 129–151, 1989.

[10] W. Kern, "Cleaning solution based on hydrogen peroxide for use in silicon semiconductor technology," *RCA Review*, vol. 31, pp. 187–205, 1970.

[11] S. Shwartzman, A. Mayer and W. Kern, "Megasonic particle removal from solid-state wafers," *RCA Review*, vol. 46, no. 81, pp. 81–105, 1985.

[12] R. H. Petrucci and J. W. Hill, *General Chemistry: An Integrated Approach*, Prentice Hall, 1999.

[13] M. Heyns, T. Bearda and I. Cornelissen, "Advanced cleaning strategies for ultraclean silicon surfaces," in *Proc. 6th International Symposium on Cleaning Technology in Semiconductor Device Manufacturing*, vol. 99, Electrochemical Society, pp. 2–15, 2000.

[14] T. Ohmi, "Total room temperature wet cleaning for Si substrate surface," *Journal of the Electrochemical Society*, vol. 143, no. 9, p. 2957, 1996.

[15] G. W. Gale, H. Cui and K. A. Reinhardt, "Aqueous cleaning and surface conditioning processes," in *Handbook of Silicon Wafer Cleaning Technology*, 3rd edn., eds. K. A. Reinhardt and W. Kern, Elsevier, pp. 185–252, 2018.

[16] D. W. Hess and K. A. Reinhardt, "Plasma stripping, cleaning, and surface conditioning," in *Handbook of Silicon Wafer Cleaning Technology*, 2nd edn., eds. K. A. Reinhardt and W. Kern, Elsevier, pp. 355–427, 2008.

[17] J. S. Song, Y. C. Choi, S. H. Seo, *et al.*, "Wet chemical cleaning process of GaAs substrate for ready-to-use," *Journal of Crystal Growth*, vol. 264, no. 1–3, pp. 98–103, 2004.

[18] J. Crites, W. Snodgrass and L. Luu, "Dynamics of surface treatments and pre-cleans for high volume wafer manufacturing," in *CS MANTECH Conference, May 18th–21st, 2015, Scottsdale, Arizona*, CS MANTECH, 2015.

[19] P. Raynal, M. Rebaud, V. Loup, *et al.*, "GaAs WET and Siconi cleaning sequences for an efficient oxide removal," *ECS Journal of Solid State Science and Technology*, vol. 8, no. 2, p. P106, 2019.

[20] J. Shenoy, G. Chindalore, M. Melloch, J. Cooper, J. Palmour and K. Irvine, "Characterization and optimization of the SiO_2/SiC metal-oxide semiconductor interface," *Journal of Electronic Materials*, vol. 24, no. 4, pp. 303–309, 1995.

[21] S. W. King, R. J. Nemanich and R. F. Davisa, "Wet chemical processing of (0001)Si 6H-SiC hydrophobic and hydrophilic surfaces," *Journal of the Electrochemical Society*, vol. 146, no. 5, p. 1910, 1999.

[22] L. Huang, Q. Zhu, M. Gao, F. Qin and D. Wang, "Cleaning of SiC surfaces by low temperature ECR microwave hydrogen plasma," *Applied Surface Science*, vol. 257, no. 23, pp. 10172–10176, 2011.

[23] P. Balk and J. Eldridge, "Phosphosilicate glass stabilization of FET devices," *Proceedings of the IEEE*, vol. 57, no. 9, pp. 1558–1563, 1969.

[24] G. Masetti and M. Severi, "Dependence of flat-band voltage of metal-oxide semiconductor structures on phosphosilicate-glass growing conditions," *Applied Physics Letters*, vol. 37, no. 2, pp. 226–228, 1980.

[25] C. Pearce, J. Moore and F. Stevie, "Removal of alkaline impurities in a polysilicon gate structure by phosphorus diffusion," *Journal of the Electrochemical Society*, vol. 140, no. 5, p. 1409, 1993.

[26] E. Snow and B. Deal, "Polarization effects in insulating films on silicon – a review," *Transactions of the Metallurgical Society of AIME*, vol. 242, no. 3, pp. 512–523, 1968.

[27] F. Frank and D. Turnbull, "Mechanism of diffusion of copper in germanium," *Physical Review*, vol. 104, no. 3, pp. 617–618, 1956.

[28] G. B. Bronner and J. D. Plummer, "Gettering of gold in silicon: a tool for understanding the properties of silicon interstitials," *Journal of Applied Physics*, vol. 61, no. 12, pp. 5286–5298, 1987.

[29] W. Beadle, J. Tsai and R. Plummer, *Quick Reference Manual for Silicon Integrated Circuit Technology*, Wiley-Interscience, 1985.

[30] E. R. Weber, "Transition metals in silicon," *Applied Physics A*, vol. 30, no. 1, pp. 1–22, 1983.

[31] W. Wijaranakula, "Solubility of interstitial oxygen in silicon," *Applied Physics Letters*, vol. 59, no. 10, pp. 1185–1187, 1991.

[32] J. Mikkelsen, "The diffusivity and solubility of oxygen in silicon," *MRS Online Proceedings Library*, vol. 59, no. 1, pp. 19–30, 1985.

[33] H. W. Kennel, "Physical modeling of oxygen precipitation defect formation and diffusion in silicon," Ph.D. Thesis, Stanford University, TR No. G710-2, 1992.

[34] C. Cui, D. Yang, X. Yu, X. Ma, L. Li and D. Que, "Effect of nitrogen on denuded zone in Czochralski silicon wafer," *Semiconductor Science and Technology*, vol. 19, no. 3, pp. 548–551, 2004.

[35] J. Kang and D. Schroder, "Gettering in silicon," *Journal of Applied Physics*, vol. 65, no. 8, pp. 2974–2985, 1989.

[36] R. Meek and T. Seidel, "Enhanced solubility and ion pairing of Cu and Au in heavily doped silicon at high temperatures," *Journal of Physics and Chemistry of Solids*, vol. 36, no. 7–8, pp. 731–740, 1975.

[37] Athena and Victory Process are Silvaco process simulation tools. See https://www.silvaco.com/products/tcad/process_simulation/process_simulation.html

[38] H. Higuchi and S. Nakamura, "Gettering of electrically active gold in silicon," *Journal of the Electrochemical Society*, vol. 122, no. 3, pp. C85–C85, 1975.

[39] K. Danno, H. Saitoh, A. Seki, *et al.*, "Diffusion of transition metals in 4H-SiC and trials of impurity gettering," *Applied Physics Express*, vol. 5, no. 3, p. 031301, 2012.

[40] N. Khuchua, L. Khvedelidze, M. Tigishvili, N. Gorev, E. Privalov and I. Kodzhespirova, "Deep-level effects in GaAs microelectronics: a review," *Russian Microelectronics*, vol. 32, no. 5, pp. 257–274, 2003.

[41] A. Zylberstejn, "The effects of deep levels in GaAs MESFETs," *Physica B + C*, vol. 117, pp. 44–49, 1983.

[42] F. Huntley and A. Willoughby, "The effect of dislocation density on the diffusion of gold in thin silicon slices," *Journal of the Electrochemical Society*, vol. 120, no. 3, pp. 414–422, 1973.

[43] J. Lutz, H. Schlangenotto, U. Scheuermann and R. De Doncker, *Semiconductor Power Devices: Physics, Characteristics, Reliability*, Springer, 2011.

[44] G. S. May and C. J. Spanos, *Fundamentals of Semiconductor Manufacturing and Process Control*, Wiley, 2006.

[45] D. Dance and K. Gildersleeve, "Estimating semiconductor yield from equipment particle measurements," in *Proceedings of the IEEE/SEMI International Semiconductor Manufacturing Science Symposium*, IEEE, pp. 18–23, 1992.

[46] B. T. Murphy, "Cost-size optima of monolithic integrated circuits," *Proceedings of the IEEE*, vol. 52, no. 12, pp. 1537–1545, 1964.

[47] W. Maly, "Realistic fault modeling for VLSI testing," in *24th ACM/IEEE Design Automation Conference*, IEEE, pp. 173–180, 1987.

[48] J. McGehee, J. Hebley and J. Mahaffey, "The MMST computer-integrated manufacturing system framework," *IEEE Transactions on Semiconductor Manufacturing*, vol. 7, no. 2, pp. 107–116, 1994.

[49] Y. Mizokami, "The total CIM system for semiconductor plants," in *IEEE/SEMI International Symposium on Semiconductor Manufacturing Science*, IEEE, pp. 24–25, 1990.

4.10   PROBLEMS

4.1 During the development of a new process technology, defect densities start out very high and are gradually reduced through running test structures through the manufacturing process line. For most of the development period, systematic defects are uncovered and eliminated. Once the technology is ready for manufacturing, it is usually dominated by random defects, and the defect density is much smaller. Give an example of a systematic and a random defect, and explain the difference between the two.

4.2 An IC manufacturing plant produces 1000 wafers per week. Assume that each wafer contains 100 die, each of which can be sold for $50 if it works. The yield on these chips is currently running at 50%. If the yield can be increased, the incremental income is almost pure profit, because all 100 chips on each wafer are manufactured whether they work or not. How much would the yield have to be increased to produce an annual profit increase of $10,000,000?

4.3 As MOS devices are scaled to smaller dimensions, gate oxides must be reduced in thickness.

   (a) As the gate oxide thickness decreases, do MOS devices become more or less sensitive to sodium contamination? Explain.

   (b) As the gate oxide thickness decreases, what must be done to the substrate doping (or alternatively the channel V_{TH} implant) to maintain the same V_{TH}? Explain.

4.4 The maximum minority carrier lifetime in Si observed experimentally is ≈ 10 ms. Many people believe that this implies that there must be some sort of intrinsic defect in Si that limits the lifetime even when sophisticated gettering methods are employed to remove trace impurities like Au which can also limit lifetime. One suggestion that has been made is that frozen-in point defects (vacancies and/or interstitials) are the intrinsic defects responsible. The reasoning is that, even though the equilibrium population of these defects at room temperature is roughly zero, there may not be time, as wafers are cooled from process temperatures, for all the point defects to recombine, and thus equilibrium is not reached at room temperature.

   (a) Assuming (4.2) holds, estimate the frozen-in concentration of point defects responsible for the 10 ms lifetime.

   (b) If the defect involved were neutral vacancies, what is the temperature at which the defects are frozen-in during wafer cooling?

4.5 In this chapter, we discussed a hierarchical approach to achieving high yields. Explain in your own words what this means and what the various levels and purpose of the hierarchy are.

4.6 In the discussion of cleaning procedures in this chapter, we saw the important role that H_2O_2 plays in modern cleaning chemistries. Explain in your own words, using simple chemical reactions, how the H_2O_2 contributes to effective wafer cleaning.

4.7 A new cleaning procedure has been proposed which is based on H_2O saturated with O_2 as an oxidant. This has been suggested as a replacement for the H_2O_2 oxidizing solution used in the RCA clean. Suppose a Si wafer, contaminated with trace amounts of Au, Fe and Cu, is cleaned in the new H_2O/O_2 solution. Will this clean the wafer effectively? Why or why not? Explain.

4.8 A new cleaning procedure has been proposed which is based on H_2SO_3 saturated with H_2SO_4 as an oxidant. This has been suggested as a replacement for the H_2O_2 oxidizing solution used in the RCA clean. Suppose a Si wafer, contaminated with trace amounts of Au, Fe and Cu, is cleaned in the new H_2SO_3/H_2SO_4 solution. Will this clean the wafer effectively? Why or why not? Explain.

4.9 In the discussion related to Figure 4.18, the importance of the temperature ramp rate between the nucleation and precipitation stages is described. Is the ramp rate coming back down to room temperature after the precipitation stage important in terms of gettering? Explain.

4.10 Many transition metals can cause problems in devices and therefore need to be gettered. These metals differ in how easily they can be gettered. For example, Ti is more difficult to getter than Au. Suggest a reason why Ti might be more difficult to getter than Au. Justify your reason with data if possible.

4.11 Figure 4.16 is a plot of the diffusion coefficients of various species in silicon. There are two curves for Si diffusion, one with very small D values (black line at the bottom), and another for silicon interstitials or interstitialcy defects (the shaded box region). How can a Si atom have two values of D that are different by 8–10 orders of magnitude? Explain.

4.12 A company is manufacturing a particular chip that is 1 cm^2. The manufacturing yield is only 40% and the company management claims the company is losing money on this product. A yield enhancement task force is told it must get the yield up to 95%. What reduction in defect density is required to accomplish this? You may assume Poisson statistics apply. Explain any other assumptions you make.

4.13 A semiconductor company is building chips that are 1 cm^2. There are 50 steps in the process flow that are critical in terms of defect densities. What does the defect density need to be in each of those 50 steps in order for the company to have a yield of 80% at the end of the process? Use Poisson statistics.

4.14 A company plans to manufacture a 1 cm^2 chip. With their current manufacturing process, they have a defect density of 1 defect cm^{-2}.

(a) Using both the Poisson model and the negative binomial model, estimate the chip yield based on their current defect density. (Assume $C = 2$ for the negative binomial model.)

(b) It turns out that the company is not satisfied with the yield calculated above because it is too low to be profitable. So they plan to make investments to further decrease the defect density. The goal is to improve the 1 cm^2 chip yield to 90%. If they could decrease the defect density by 25% per month, how long does it take to reach 90% chip yield? (Assume $C = 2$ for the negative binomial model.)

(c) The yield of another kind of 1 cm^2 chip at this company is 40%; thus they are losing money on this product. The company decides to improve the yield to 95%. What reduction in defect density is required to accomplish this? (Assume Poisson statistics apply.)

4.15 The yield crashes on a CMOS fabrication line to 1%. The factory manager sets a goal of reducing the defect density by a factor of 2 every month and claims that, if this can be done, the yield will be above 90% in six months. Is this correct? Calculate a numerical answer to show why you agree or disagree with the factory manager's claim. You may use Poisson yield statistics.

4.16 A semiconductor manufacturing company's factory typically achieves a defect density of 0.5 defect cm^{-2}. The company manufactures many products in this factory. Consider two products, one a 1 cm^2 chip that the company can sell for $50 per chip and a second product of 0.1 cm^2 that the company can sell for $1 per chip. In both cases, these products are manufactured on 300 mm (diameter) wafers and the wafer manufacturing costs are $2000 per wafer. Which chip should the company manufacture if it wants to maximize overall profits? Is either product profitable for the company? You can assume Poisson yield statistics apply in both cases.

4.17 A 1 cm^2 chip is introduced into manufacturing. Initially the defect density is 10 defects cm^{-2} and the yield is terrible. The company is not making any money on this chip and so a task force is set up to improve the defect density. The goal is to get the chip yield up to 90%. The task force is charged with decreasing the defect density by 25% per month until the yield is 90%, at which point the company will be making lots of money on this particular chip. How many months will this take? You may presume that Poisson statistics apply to the chip yield. State any other assumptions.

4.18 The two parameters C_p and C_{pk} are often used to describe whether a manufacturing process is "under control" or not. Use a sketch to describe in your own words what these parameters are and why they are so important in manufacturing.

4.19 You work for a large silicon chip manufacturing company. A potential customer comes to you and asks for a price quote on building chips for her company. She wants you to give her a price quote in dollars per square millimeter of silicon that the chip occupies. She says she already has bids around $0.25 mm^{-2} from other companies. You know that your costs for manufacturing a 12 inch (300 mm) wafer are about $5000. But, of course, you will have other costs to pursue her business as well, things like factory depreciation, design costs, etc. Your boss asks you to assess whether you should even continue to talk with this potential customer. What is your answer? Explain.

4.20 A restaurant's product is obviously food served to its customers. To meet quality standards, this food should be served between 38 °C and 49 °C. The restaurant has a system to keep food warm (warming lamps) that produces a mean T of 42 °C and a standard deviation of 2 °C. If we define a process being under control as a process having $C_p > 1$ and $C_{pk} > 1$, is this restaurant's process under control? Calculate a quantitative answer. If it is not under control, how would you fix this?

5   Lithography

5.1   Introduction

Lithography is arguably the most important process step in modern integrated circuit (IC) manufacturing. The ability to print patterns with features as small as 10–20 nm and to place those patterns on a substrate with a precision of a few nanometers is what makes today's chips possible. Virtually all ICs are manufactured today with deep-ultraviolet (DUV) optical lithography operating with 193 nm photons, the basic process introduced in Figure 1.7. Extreme ultraviolet (EUV) lithography ($\lambda = 13.5$ nm) is beginning to be used for some critical mask levels in the latest technology generations because the DUV systems have reached their practical limits in resolution. As we saw in Chapter 1, the concept is simple. A light-sensitive photoresist is spun onto the wafer, forming a thin layer on the surface. The resist is then selectively exposed by photons passing through a mask that contains the pattern information for the particular layer being printed. (In EUV lithography, the mask is actually a mirror, not transparent glass, because no materials are transparent at EUV wavelengths.) The resist is then developed, which completes the pattern transfer from the mask to the wafer. As we saw in the process flow in Chapter 2, the resist may then be used as a mask to etch underlying films or it may be used as a mask for an ion implantation doping step.

While the concept is simple, the actual implementation is very expensive and very complex, primarily because of the demands placed on this process for resolution, exposure field, placement accuracy, throughput and defect density. Resolution requirements result from the ever-increasing demand for smaller device structures. Exposure field requirements result from the need to fabricate large chips. Placement accuracy is an issue because generally each mask layer needs to be carefully aligned with respect to the existing patterns already on the wafer. Throughput and defect density are, of course, issues because of the competitive nature of the semiconductor industry. Throughput translates directly into manufacturing cost. Defects translate directly into yield loss and therefore less profitability in the finished chips. Defects introduced during the lithographic process are a significant contributor to final chip yields.

As we saw in Chapter 1, for many years, the "International Technology Roadmap for Semiconductors" (ITRS) guided the progression of new silicon technology generations on a 2–3 year cadence [1]. The feature size reduction in the ITRS corresponded to a factor of $0.7\times$ in linear dimension or a decrease in the area required per transistor by 50% approximately every 2–3 years. Not only did the minimum features decrease in average size with each technology generation, but the variation of these feature sizes also decreased with time in the ITRS. Generally, the feature size control is required to be about ±5% of the smallest feature size. This requirement is usually expressed in terms of the 3σ control. As an example, if the minimum feature size is 60 nm, the standard deviation of the distribution would be required to be $\sigma = 1$ nm, so that $3\sigma = 3$ nm. If the distribution is normal, this would imply that 99.7% of the minimum features printed on the wafer would be within 60 ± 3 nm (see Figure 4.24). The corresponding placement or alignment accuracy for these features (overlay) has generally been about one-third of the minimum feature size for each generation of technology.

The ITRS was discontinued in 2015 because individual companies by that time were following their own roadmaps, since it had become clear that there no longer was a universal roadmap. In one of the last versions of the ITRS (2013 edition), the resolution limit of the most advanced lithography tools in use at that time (193 nm immersion projection tools) was given as roughly 40 nm lines and spaces. It was also clear at that time that there would be no further significant advances in DUV lithography tools by, for example, using shorter-wavelength light systems, for reasons that we will discuss in this chapter. One might wonder, then, how smaller and smaller devices continue to be fabricated today. The answer is that a series of very interesting "tricks" were increasingly used, which enabled the industry to advance to today's 10 nm and smaller technology nodes. These "resolution enhancing" methods will be described later in this chapter. Today, however, those "tricks" have been pushed as far as they practically can, and so a radically new lithography approach – EUV lithography, with a much smaller wavelength – is now finally entering manufacturing.

Lithography approaches for compound semiconductor manufacturing are generally similar to those used for advanced silicon processes. However, the machines that are used in compound chip manufacturing are usually several generations behind those found in the latest silicon plants. SiC and GaN devices for power applications do not require nanometer-scale features, nor do most photonic circuits and light-emitting diodes (LEDs) manufactured in III–V semiconductors. So, much less expensive lithography tools can be used in manufacturing. Nevertheless, the same principles as discussed in this chapter for advanced silicon chips apply.

State-of-the-art 193 nm lithography tools today cost in excess of $50 million because of the precision required of these systems. Lithography accounts for at least one-third of the total wafer manufacturing costs. Thus, with a total wafer manufacturing cost of a few thousand dollars for a 12 inch (300 mm) silicon wafer, it is easy to understand why these machines must print patterns on a significant number of wafers per hour in manufacturing to justify their cost. Typically, 193 nm DUV systems can process as many as 250 wafers per hour. The radically different EUV lithography tools cost in excess of $100 million for each machine. As we will see, a major reason for their delayed introduction into manufacturing was that the throughput of these very advanced machines has not been adequate until recently to justify their cost.

In this chapter, we will explore two major areas associated with the lithographic process. The first is the exposure system used to print the patterns on the wafer. DUV optical exposure systems (193 nm projection systems) dominate the industry today. Typical systems today use "step-and-repeat" or "step-and-scan" reducing projection printing with complex lens systems. We will also discuss EUV systems that use a wavelength of 13.5 nm. This is a 14× reduction in the photon wavelength which allows a dramatic reduction in the size of printed features.

The second major area we will explore is the photoresist material itself. Here the primary issues are sensitivity, resolution and ruggedness. Typical resists today are hydrocarbon-based organic materials with one of the components sensitive to photons at the operating wavelength.

Powerful simulation tools are available today to predict quantitatively the results of the lithography process. We will use Optolith [2] extensively in this chapter to demonstrate concepts and to show what the printed images typically look like under various manufacturing conditions. Optolith implements many of the models we will discuss in this chapter.

5.2   Historical Development and General Concepts

The overall lithography process is conceptually illustrated in Figure 5.1. The patterns that comprise the various layers in an IC are designed using computer-aided design (CAD) systems. Today, these systems contain many advanced capabilities that greatly improve the efficiency of designing chips with many billions of components. Libraries of previous designs that are known to work are usually available, from which basic functions or circuits can be cut and pasted into new designs. Software tools are used to help route or wire the connections between functional blocks. Additional tools check the design to make sure that there are no violations of design rules. Design rules will be discussed more carefully in Section 5.7.4, but these rules are basically a set of constraints for things like feature size, feature spacing and feature shape that are intended to make the overall design more manufacturable. Finally, circuit- and system-level simulation tools are available to predict the performance of the new design.

Once the design is complete and ready to transfer to the fabrication facility, the information for each mask level is transferred to a mask-making machine that is either an electron beam or laser pattern generator. The pattern for each mask is literally written on a mask blank using a scanning electron or laser beam. The mask itself is usually a fused silica plate covered with

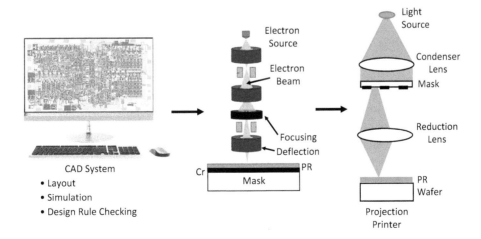

Figure 5.1 Lithography process from mask design to wafer printing.

a thin layer ($\approx$ 80 nm) of chrome and a layer of photoresist ("chrome" is the term used in the semiconductor industry as the mask material is likely not pure chromium). A thin antireflection coating (ARC) layer (10–15 nm) is also often used between the chrome and the resist to prevent reflections from the chrome layer that can degrade pattern resolution. The electron or laser beam exposes the resist, which is then developed and used as an etch mask to transfer the mask pattern into the chrome. The chrome layer is etched using dry etching methods (or wet etching if the dimensions are large). Tight dimensional control can be maintained because the chrome layer is so thin. Once the chrome is etched, the photoresist is removed. The fused silica substrate has highly polished surfaces, so that light is not scattered as it passes through the mask, and ideally has a small thermal expansion coefficient so that mask dimensions are stable over small temperature changes.

It is also obviously important that the clear areas of the mask where the chrome has been etched away have high transparency at the wavelength of light used in the wafer exposure system. For DUV systems operating at 193 nm wavelengths, SiO_2-based masks are reasonably transparent. There was considerable effort some years ago to develop 157 nm projection systems [3], but standard mask materials were found not to have sufficient transparency and stability at this shorter wavelength. CaF_2 was explored as a possible mask material, but it, too, proved not to be usable in a manufacturing environment. As a result, DUV projection systems have been "stuck" at 193 nm for many years. To construct systems below this wavelength, a major change involved the use of mirrors for the optical system (reflecting components) rather than the refracting (transmission) optical systems used until now. This is the approach used for EUV systems that we will discuss later in this chapter.

In DUV optical systems, usually the mask is fabricated with pattern dimensions 4× larger than the features actually desired on the wafer because the wafer exposure system reduces the image by the same factor. Demagnification in the wafer exposure system makes the mask easier to fabricate and easier to check for imperfections. This is a crucial step, because any defects on the mask will be directly imaged on every wafer exposed with the mask and contribute to yield loss. Since step-and-repeat exposure systems typically expose only a few chip areas at a time, a mask defect that prints on the wafer can easily make the yield zero.

Mask inspection is performed by comparing the mask chrome pattern with another identical pattern when there are two or more chip patterns on the mask. If the mask pattern is so large that only one chip pattern is present, the comparison can be with the database in the CAD system used to produce the mask. Defects can usually be repaired at this stage either by removing unwanted chrome areas with laser ablation or ion beams, or by depositing additional chrome to fill in pinholes. Once the mask is checked, and repaired if necessary, it is usually protected from later contamination by dust particles with a thin transparent membrane (pellicle) stretched over a metal frame above the chrome side of the mask. Pellicles are usually made from nitrocellulose a few microns thick. Since the pellicles are offset from the mask, dust particles that fall on the pellicle during use will be out of focus on the wafer and therefore will usually not print.

The actual mask writing is usually done with a laser or an electron beam (e-beam) pattern generator, as shown in Figure 5.1. These machines raster a laser beam or a beam of electrons in an X–Y pattern across the mask, with the beam blanked on and off as necessary to generate the appropriate mask pattern. As we will discuss later in this chapter, e-beam systems are capable of writing very small patterns. It is perhaps obvious that the e-beam system could be used directly to write images on a wafer simply by putting the electron-sensitive resist on the wafer rather than on the mask. The reason this is not done in high-volume manufacturing is simply because the wafer throughput would be too low. Typical e-beam pattern generators require tens of minutes to expose a full wafer. This is much slower than optical steppers, which typically have throughputs of more than 100 wafers per hour. However, in research laboratories and in universities, where throughput is not an issue, e-beam systems are widely used to directly write patterns on wafers.

The pattern information is transferred to the wafer by printing the mask pattern in a layer of photoresist on the wafer surface. Generally, today, this is done with a projection exposure system, as also illustrated in Figure 5.1. Light from a high-intensity source is collimated and passed through the mask. Each clear area on the mask transmits the light, which is then collected and focused by a second lens system. This second lens system also serves to reduce the image size typically by 4×. The field of view of such systems is typically only a few centimeters on a side, so that only a few chips are printed during each exposure. The wafer is physically moved ("stepped") to the next exposure field and the process repeated, so these systems are commonly called step-and-repeat tools or simply steppers. In some systems ("scanners") a narrow slit of light is produced by the light source system, and the mask and wafer are simultaneously scanned mechanically so that the mask image is scanned across the wafer. Note that, in systems of this type, the mechanical system must scan the mask 4× faster than the wafer, corresponding to the optical reduction factor. Scanners greatly increase the field size up to the limit of the mask dimension, at least in the scanning direction.

EUV exposure systems that operate at a wavelength of 13.5 nm are configured as shown in Figure 5.2. Reflecting optics are used (mirrors) and the system operates under high vacuum because the EUV photons would be absorbed by air. The mask itself is also made on a mirror and is typically 4× larger than the printed pattern on the wafer, which means that the mask is scanned 4× faster than the wafer. The purple color illustrates the path of the EUV photons, which are not visible.

It is convenient to separate the lithography process into three parts. The first is the light source used to generate the photons that ultimately expose the resist. While this might seem to

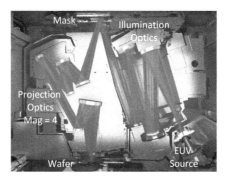

Figure 5.2 EUV exposure system [4]. Copyright © ASML Corporation, reprinted with permission.

be a simple task, higher-resolution lithography demands shorter-wavelength photons, and this makes the source issue more complex, especially in EUV systems in which very-short-wavelength, high-energy photons are required. The second part of the lithography process is the exposure system that images the mask on the wafer surface. This system produces what is known as the aerial image (Figure 5.3) at the top surface of the resist. This image is basically the pattern of optical radiation striking the top surface of the resist. In the example in Figure 5.3, a positive resist is illustrated, since this polarity of resist represents the vast majority of manufacturing applications today. The incoming photons (red) strike the resist in the light areas, changing its properties in those regions. The 3D optical intensity pattern through the photoresist layer produces a 3D latent image of the mask pattern that can then be developed. The exposed positive resist in this example would dissolve in the resist developer, allowing the unexposed resist to then be used as a mask to etch, for example, contact holes down to the source and drain regions in Figure 5.3. The third part of the lithography process contains all the issues associated with the resist itself, exposure, developing, baking, etc.

The aerial image thus serves as the dividing line between the major parts of the lithography system (the exposure tool and the resist). The job of the exposure tool is to produce the best aerial image possible. "Best" is defined in terms of resolution, exposure field, depth of focus, uniformity and lack of aberrations across the field, photon intensity, etc. The job of the photoresist is to translate this aerial image into the best thin-film 3D replica (latent image) of the aerial image possible. Here "best" is defined in terms of geometric accuracy, exposure speed and resist tolerance to subsequent processing (ruggedness). In the sections that follow, we will discuss in turn the various components of the lithography system, introducing a set of basic ideas about each that we can then apply to more detailed and more quantitative models and understanding later in the chapter.

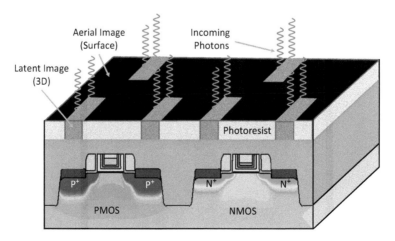

Figure 5.3 Separation of the lithography process into exposure and wafer features. The dividing line is the aerial image of the mask, which is the pattern of radiation that strikes the surface of the photoresist. This example corresponds to Figure 2.23 in Chapter 2.

5.2.1 Light Sources

We will discuss this in much more detail later in this chapter, but it is likely intuitively obvious that higher resolution can be achieved in lithography systems when shorter-wavelength light is used. To first order, the minimum feature size that can be printed with a lithography tool is directly proportional to the wavelength of the light used to print the feature. Lithography systems used for IC manufacturing are monochromatic (single-wavelength) because it is much easier to design optical systems when only one wavelength is used. Figure 5.4 summarizes the history of silicon IC manufacturing showing the light wavelength used in manufacturing for the various technology nodes. Recall from Figures 1.17 and 1.21 that the technology node names since about 2000 are not synonymous with the feature sizes actually being printed in a particular technology node. The red curve shows step-function changes in λ as new generations of lithography tools were introduced. Since the mid-1990s, feature sizes on silicon chips have been smaller than λ (subwavelength lithography) and today are more than 10× smaller than λ, something that might seem "impossible" given the general rule of thumb that printing features smaller than about $\lambda/2$ is "impossible." Clearly, it is *not* impossible, but the list in Figure 5.4 of tricks – OPC, phase shift, immersion, etc. – is a subject we will return to later in this chapter. EUV systems operating at 13.5 nm are shown with a dashed line since their introduction is just now starting. Today, 193 nm immersion optical systems are still the dominant tools used.

Prior to 1995, most lithography systems used arc lamps as the primary light source. These lamps usually contain Hg vapor inside a sealed glass envelope. Two conducting electrodes inside the envelope are separated by several millimeters. An arc is struck between the electrodes by applying a high enough voltage to ionize the gas (typically several kilovolts). Once the gas is ionized, it behaves like a plasma. In the arc lamp, the plasma is conducting and consists of ions, electrons and neutral species. Typical lamps used for photolithography consume about 1 kW of power. At room temperature, the Hg vapor pressure is on the order of 1 atm, but once the lamp is operating, power dissipation quickly raises the temperature and therefore the internal pressure to ≈20–40 atm.

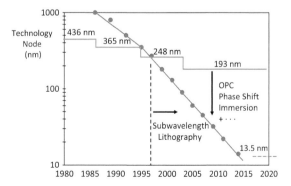

Figure 5.4 Light wavelengths used in silicon manufacturing over time (red). Technology nodes versus time are shown in blue. Note that node names are not equal to minimum feature sizes after 2000. After [5].

Light emission occurs through two processes. The free electrons in the plasma have an effective temperature on the order of 40,000 K and emit blackbody radiation according to Planck's law. The wavelength corresponding to this temperature is very deep in the ultraviolet and is above the bandgap of the fused silica used to form the glass envelope of the lamp. Thus this radiation is mostly absorbed before it exits the lamp.

The second source of emitted light comes from the Hg atoms themselves. Collisions in the gas between the free electrons and the Hg atoms provide energy to some of the electrons in the Hg atoms, raising them to higher energy levels. When these electrons drop back to their lower energy states, they radiate photons at specific energies (frequencies) characteristic of the allowed energy levels in the Hg atom. In the case of Hg, there is strong emission at a number of UV wavelengths. Two that were commonly used are 436 nm (g-line) and 365 nm (i-line) shown in Figure 5.4.

The brightest light sources in the deep-UV part of the spectrum are excimer lasers. Two that are of specific interest for lithography are KrF (248 nm) and ArF (193 nm). In excimer lasers, two elements are present that do not normally react in their unexcited state (often a noble gas and a halogen-containing compound). However, if these elements (Kr and NF_3, for example) are excited, a chemical reaction (forming, for example, KrF) is possible. When the excited molecule returns to its ground state, a photon is emitted in the deep UV and the molecule breaks up. Energy must be continually provided, usually from an internal pulsed discharge, to replenish the population of the excited species. Usually the laser source is pulsed at a frequency up to several kilohertz and modern excimer laser sources can produce pulses of several hundred watts. Multiple flashes are used at each exposure site on the wafer to minimize speckle noise. Speckle arises because of the use of a single wavelength in lithography systems and can be reduced by superimposing multiple images from multiple flashes of the light source. A total energy up to several hundred millijoules is provided to each exposure site. In steppers, this energy is focused on a wafer area of a few square centimeters, providing reasonable exposure times.

As is apparent in Figure 5.4, the first excimer laser systems were introduced in the mid-1990s and these used KrF laser sources (248 nm). In the early 2000s, those systems were gradually replaced with ArF systems (193 nm) and these systems have been used ever since. There are shorter-wavelength excimer lasers available, for example, F_2 lasers (157 nm), but, as discussed earlier, these systems have never been used in manufacturing, for a variety of reasons. The energy of a photon is given by

$$E(\text{eV}) = \frac{hc}{\lambda} = \frac{1.2398}{\lambda(\mu m)}.$$ (5.1)

Thus, in the excimer laser systems discussed above, the photon energy is about 5 eV in a 248 nm system and about 6.4 eV in a 193 nm system.

The next step in lithography systems is EUV, which uses 13.5 nm wavelength photons. These photons have an energy of about 92 eV, so generating such photons, particularly at the high intensities needed for an EUV lithography system, is a major challenge. Photons at these energies are absorbed by virtually all materials, which is the reason for the switch to reflecting optics and operating under a vacuum, as shown in Figure 5.2. We will discuss masks in more detail in the next subsection, but in EUV systems the masks and the system mirrors reflect less than half of the energy incident on them. Since there are multiple mirrors in these systems, as Figure 5.2 illustrates, approximately 98% of the EUV energy generated by the light source does not reach the wafer for exposure. Thus compared to a 193 nm excimer laser source with refracting optics, which do not absorb much of the light, an EUV source needs to be much brighter ($\approx 10\times$) to achieve reasonable throughput in manufacturing. Finally, the very high energy of EUV photons also changes how these photons interact with photoresist materials, as we will discuss later.

The dominant method used today to produce 13.5 nm photons for EUV systems is a laser-pulsed Sn plasma. Neutral atoms or condensed matter cannot emit EUV photons because the energy is simply too high. Ionization resulting from the thermal production of multi-charged positive ions is only possible in a hot, dense plasma. Generally a CO_2 pulsed laser is used to provide energy to small droplets of Sn. But these are very complex systems, for a number of reasons. First, the energy levels are very high to produce enough EUV photons. Sn "splattering" is a challenge because, if such deposits get on the parts of the lithography tool such as the mirrors, performance quickly degrades. Second, the plasma produced by the laser absorbs EUV radiation, as do the atoms in the plasma, so getting the photons out of the plasma is also a challenge. For the interested reader, the details are in [6]. There is also a quite good discussion of EUV systems in Wikipedia [7].

EUV lithography systems are just beginning to enter high-volume silicon manufacturing for the latest generation of chips (7 nm and 5 nm). These systems should make possible continued printing of smaller features for some years to come. However, they are enormously complex and expensive, and only a few leading-edge companies are likely to use them. Indeed, there is only one manufacturer of EUV exposure systems in the world today – ASML. Most silicon chips will continue to be manufactured with 193 nm lithography systems unless it is leading-edge technology. And virtually all compound semiconductor manufacturing will continue to use DUV or even i-line lithography systems.

5.2.2 Masks

Masks used in lithography systems used to be very straightforward. In simple optical lithography systems, masks are digital (opaque or transparent regions) on a glass (fused silica) substrate. If you were to look at such a mask, it would look exactly like the pattern on the CAD system used to design the chip, for one level of the chip. Beginning about 20 years ago when subwavelength lithography became important (Figure 5.2), masks began to be much more sophisticated. In this section we will deal only with simple digital masks for optical and EUV systems. Towards the end of this chapter we will return to masks and discuss the resolution enhancement techniques that have made subwavelength lithography possible. Much of that discussion will make more sense after we discuss the details of how lithographic systems actually work.

As discussed earlier, simple digital masks for optical lithography are made using direct-write e-beam or laser exposures of patterns on glass masks. The laser writing systems can write patterns down to about 0.5 µm, limited by diffraction effects due to the wavelength of the laser. Electron beam writing systems have a very small effective wavelength, as we will see later in this chapter, and can write patterns at the nanometer scale. The masks are coated with a thin chrome (Cr) layer with photoresist on top of the Cr. After exposure, the photoresist pattern is developed and then used as a mask to etch the Cr. Because the Cr is very thin, it can usually be wet etched, but, for very small geometries, dry etch methods can be used, which are much more anisotropic (no undercutting during the etching). We will discuss etching in Chapter 9. An example of a simple mask is shown in Figure 5.5. In this case 12 identical circuits have been written on the mask and would be printed simultaneously on the wafer by the lithography tool.

The concept of a "pellicle" is also shown in Figure 5.5. As described earlier in this chapter, a pellicle is a thin, transparent membrane that covers a mask in a manufacturing environment. They are typically ≈ 1 µm thick polymer films stretched across a frame that separates them from the mask by 5–10 µm. The pellicle is a dust cover, preventing particles and contaminants from falling on the mask. It also must be transparent enough to allow light to transmit from the lithography scanner to the mask. In DUV lithography tools, transparency is not generally a major issue since many materials are transparent at 193 nm. If dust particles do fall on the pellicle, they will be out of focus when printed on the wafer by the lithography tool, and hence there is more tolerance for particles on the pellicle than there would be if they fell directly on the mask. The importance of pellicles is perhaps illustrated by the fact that typical masks contain only a few circuits (one to 10, with four to six being typical), so if a single dust particle deposits on the mask itself that could reduce the yield by 25% if there are four chips on the mask.

EUV masks are reflecting masks because almost all materials absorb EUV radiation and hence a refracting or transmitting mask is not possible. The basic structure is shown in Figure 5.6. Between 40 and 50 alternating layers of silicon and molybdenum are deposited on the top of a very flat substrate to act as a Bragg reflector that maximizes the reflection of the 13.5 nm wavelength. The individual Mo and Si layers have nanometer thicknesses and must be atomically uniform over the large areas required for masks. This multilayer reflector is then capped to prevent oxidation, typically with a thin layer of ruthenium. A tantalum boron nitride

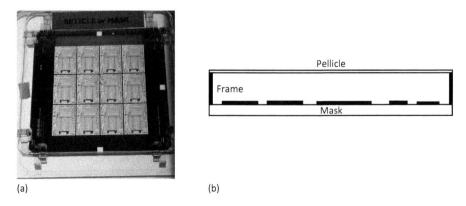

(a)   (b)

Figure 5.5 (a) Example of a relatively simple mask for an optical lithography system. (b) The use of a pellicle for mask protection.

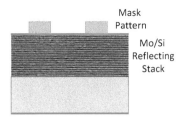

Figure 5.6 EUV mask structure. The Mo/Si stack shown is an actual TEM image of an EUV mirror or mask. The period of the stack is 6.8 nm. Courtesy of Purdue University.

film topped by a thin antireflective oxide acts as the EUV absorber. The absorber is typically dry etched after a pattern is written by an e-beam writer to produce vertical edges with very fine resolution. Typically the masks are 4× larger than the final printed pattern, as illustrated in Figure 5.2. Not surprisingly, EUV masks are very expensive. Just the substrates alone can cost over $100,000, which is 10× more than the glass substrates for optical photomasks. Because significant EUV energy is absorbed by the mask (or system mirrors which have a similar structure), materials must be chosen that have low thermal expansion coefficients. Otherwise thermal expansion would distort the pattern being printed.

A final issue with EUV masks is implementing pellicles. Pellicles protect the mask from unwanted particle deposition and are very important in increasing yield. In EUV systems, the issue is that essentially no materials are transparent to EUV photons, and so how can one implement a pellicle? Current technology involves using a 50 nm thick polysilicon or SiN_x pellicle mounted above the EUV mask. These extremely thin films must cover an area of hundreds of square centimeters, since they must protect the entire mask, which is obviously a challenging materials problem. Such pellicles transmit about 90% of the EUV photons because they are so thin. Nevertheless, with 250 W of EUV power passing through the pellicle, the absorption raises the temperature of the pellicle to as much as 1000 °C [8]. Successive heating and cooling of the pellicle as it is used creates difficult materials issues.

5.2.3 Wafer Exposure Systems

We now consider in more detail the systems used to create the aerial image at the photoresist surface. There are three general classes of wafer exposure tools, namely, contact, proximity and projection systems, although only the last is in widespread high-volume manufacturing use today. These are conceptually illustrated in Figure 5.7. While EUV systems (Figure 5.2) appear radically different than the optical systems illustrated in Figure 5.7, they are, in fact, simply projection printers that use reflecting rather than refracting optics.

Contact printing is the oldest and the simplest printing process. The mask is placed chrome side down in direct contact with the resist layer on the wafer. The exposure of the resist then takes place by shining light through the mask. The aligning of the mask to patterns already on the wafer takes place prior to exposure, by observing both the mask and wafer patterns through

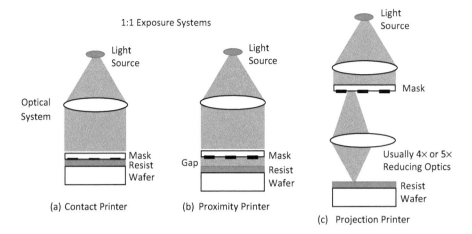

Figure 5.7 Three basic methods of wafer exposure in optical lithography systems.

a microscope with the mask slightly separated from the wafer. Contact printing systems are actually capable of fairly high-resolution printing because, with the mask and wafer in contact, diffraction effects are minimized. In addition, the machines to perform contact printing are relatively inexpensive since they are so simple. However, these types of systems cannot be used in high-volume manufacturing of complex chips for one very simple reason. The hard contact between the mask and the wafer results in damage to both the mask and the resist layer, and therefore results in high defect densities. The resulting chip yields are not compatible with economic manufacturing of today's chips. However, contact printing systems are used in some manufacturing applications where low volumes or small chip sizes make the economics of these systems more attractive, and they are also commonly used in universities and research facilities in which yield is not a primary issue.

Proximity printing largely solves the defect issues associated with contact printing, because the mask and wafer are kept separated. However, these systems are also not suitable for manufacturing most of today's chips because the separation of the mask and wafer degrades the resolution of the printed patterns due to diffraction effects. Interestingly, the resolution of these systems improves as the exposure wavelength decreases, and X-ray lithography systems can use proximity printing and achieve high resolution because of the very short exposure wavelength (1–2 nm). Such systems are, however, not commonly used today. There are a wide variety of contact/proximity lithography tools commercially available. Two examples are shown in Figure 5.8.

These systems are much simpler than projection aligners. They usually operate with broadband light sources, often from Hg lamps, so the light is not monochromatic. They are generally specified as having a resolution capability of a few microns in proximity mode and operate with a gap g between 0 and 100 μm. They usually provide an alignment capability on the order of 0.5 μm. The masks are 1:1 since there are no reducing optics and usually the exposure covers a full wafer, so the mask would contain hundreds of identical circuit patterns. For applications in which the lower resolution is adequate, these types of systems are a much lower-cost option.

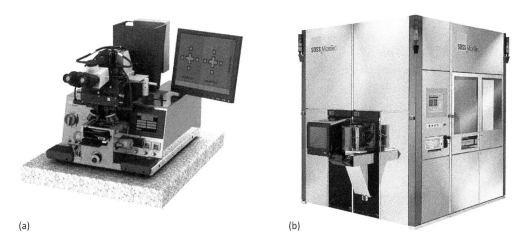

(a)   (b)

Figure 5.8 Examples of typical contact/proximity aligners. (a) SÜSS MJB4 system designed for research labs and small-scale production. (b) SÜSS MA300 system designed for production with up to 300 mm (12 inch) wafers. Photos from SÜSS MicroTec website.

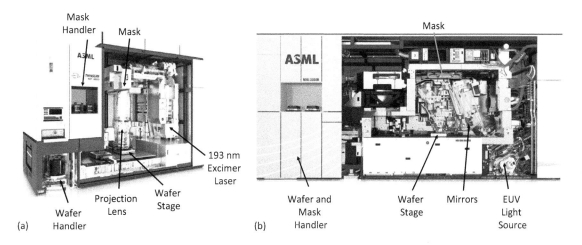

Figure 5.9 Examples of state-of-the-art lithography tools. (a) ASML 193 nm immersion tool. (b) ASML EUV exposure tool (yellow regions indicate the EUV path, see Figure 5.2). Photos copyright © ASML Corporation, reprinted with permission.

The dominant method of wafer exposure today is projection printing, examples of which are shown in Figure 5.9. These systems provide high resolution but without the defect problems of contact printing.

In projection exposure tools, the mask is physically separated from the wafer and an optical system is used to image the mask on the wafer. This obviously solves the defect issues associated with contact printing. The resolution of projection printers is limited by diffraction effects that we will describe in detail. Generally the optical system reduces the mask image by 4× to 5×, which means that only a small portion of the wafer is printed during each exposure. Systems designed for high-volume manufacturing cost approximately $50 million and are extremely high-precision machines.

Tools for 193 nm optical lithography like that shown in Figure 5.9(a) today can print features as small as 40 nm over a field of 26×33 mm^2 with an overlay accuracy of 1–2 nm on 275 (300 mm) wafers per hour. The latest EUV lithography tools like that shown in Figure 5.9(b) can print features as small as 13 nm with an overlay of 1–2 nm on 125 (300 mm) wafers per hour [9].

Some readers may have wondered when looking at Figure 5.1 why it is necessary to always make a physical mask. In other words, why not use the mask-making machine to directly write the pattern on the wafer itself rather than creating a mask, which is then used to transfer the pattern to the wafer? The answer is that "direct writing" lithographic tools do exist and are actually widely used in some applications. Direct writing essentially involves writing the pattern pixel by pixel either on a mask or on a wafer. This is a serial process. Printing through a mask is a parallel process in which many pixels are exposed in parallel. Thus modern projection printers like those shown in Figure 5.9 can print a 300 mm wafer in well under one minute. Creating the mask with an electron beam writing system or using that system to serially expose all the pixels on a wafer is a process that takes hours. Thus direct writing is not feasible for high-volume manufacturing. However, in research labs, in academic institutions and in some small-volume manufacturing plants, direct writing is entirely feasible, so we briefly discuss here some of these direct writing system options.

Direct-write systems using electron beams are illustrated in Figure 5.10. Under high vacuum, a beam of electrons is steered using magnetic or electromagnetic lenses to write a pattern in photoresist. The resist could be on a mask or on a wafer. The time required to write a pattern is simply

$$t = \frac{(\text{required dose})(\text{area})}{\text{beam current}}. \tag{5.2}$$

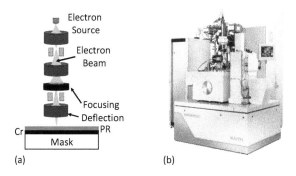

Figure 5.10 (a) Schematic of an electron beam lithography/mask-making tool. (b) A Raith machine EBPG5200 used for both direct writing on wafers and mask writing. (b) Photo courtesy Raith Corp. Reprinted with permission.

Example

A typical exposure dose for an electron beam resist is 500 μC cm^{-2} (C is coulomb). The Raith system shown in Figure 5.10 has a maximum beam current of 350 nA. How long would it take to write a 1 cm^2 area?

Answer

Substituting in the equation, we have

$$t = \frac{(500 \times 10^{-6} \text{ C cm}^{-2})(1 \text{ cm}^2)}{350 \times 10^{-9} \text{ C s}^{-1}} = 1428 \text{ s} = 24 \text{ min.}$$

Obviously to pattern an entire 300 mm (12 inch) silicon wafer would take a very long time!

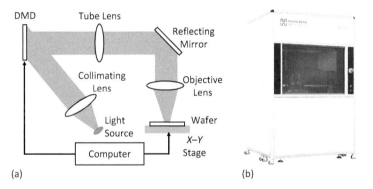

(a)   (b)

Figure 5.11 (a) Simple schematic of the DMD approach to laser direct writing. After [11]. (b) An example of a commercial laser direct-write system. This DWL 66+ laser lithography system is a high-resolution pattern generator for direct writing and low-volume mask making. (b) Photo courtesy Heidelberg Instruments. Reprinted with permission.

There are direct-write lithography tools available that use photons rather than electrons to write patterns. These systems are available commercially from a variety of companies. They often use laser light sources. The light is often controlled using mirrors to "steer" it to specific locations on a mask or on a wafer. Some systems use a "digital micromirror device" (DMD; Figure 5.11), which is a silicon chip containing thousands of controllable tiny mirrors which can individually direct light beams so that parallel processing with many light beams is possible [10].

Systems like that shown in Figure 5.11 are capable of writing submicron features and can pattern from a few to hundreds of square millimeters per minute, depending on the resolution of the pattern being written [12]. These types of systems are thus very useful in academic institutions, in R&D facilities and for small-scale production when they can meet the resolution required in a particular application.

5.2.4 Photoresists

Photoresist materials are designed to respond to incident photons by changing their properties when they are exposed to light. Many materials, of course, absorb light, but often the absorption results in electronic processes rather than chemical changes. Semiconductors, for example, absorb photons, and the energy is given to electrons and holes. Generally the free carriers will dissipate the absorbed energy through recombination or through phonon interactions (transferring the energy to heat). In some cases the energy can actually be collected, as in solar cells. None of these processes are useful in lithography because in photoresists we require a material that maintains a latent image of the impinging photons at least until the resist is developed. A long-lived response to light generally requires a chemical change in the material.

Almost all resists today are fabricated from hydrocarbon-based materials. When these materials absorb light, the energy from the photons generally breaks chemical bonds. After this happens, the resist material chemically restructures itself into a new stable form. Positive resists respond to light by becoming more soluble in the developer solution. Negative resists do the opposite. They become less soluble where they are exposed. Current practice in the semiconductor industry relies primarily on positive resists because they generally have better resolution than do negative resists.

Photoresists in use today are liquids at room temperature and are applied to the surface of wafers by placing the liquid on the wafer and then spinning the wafer at several thousand revolutions per minute (rpm). The spin speed and the viscosity of the resist determine the final resist thickness. The viscosity of the resist is controlled by a solvent, which is a constituent of the resist. Once the resist is spun onto the wafer, a baking step (pre-bake) is generally used to drive off the remaining solvent. The resist is then exposed using one of the lithography systems described in the previous subsection. Developing is done using liquid developers, either by immersion of the wafer in the liquid, or by spraying the developer onto the wafer. After developing is complete, the resist is generally baked again (post-exposure bake) to harden it and to improve its ability to act as an etch mask or ion implantation mask, depending on the particular step in the process flow. Finally, after the etching or implantation process, the resist is removed, usually in an oxygen plasma, although chemical stripping can also be used, as described in Chapter 4.

A number of important parameters determine the usefulness of a particular resist. Sensitivity is a measure of how much light is required to expose the resist. This is usually measured in millijoules per square centimeter (mJ cm^{-2}) and for g-line and i-line resists is typically 100 mJ cm^{-2}. DUV resists designed to work with 193 nm exposure systems achieve sensitivities of 20–40 mJ cm^{-2} because they use chemical amplification, a process we will describe shortly. Generally, high sensitivity is desired because this decreases the exposure time of the resist and therefore improves throughput in the lithography process. However, extremely high sensitivity is usually not desired because this tends to make the resist material unstable, makes it very sensitive to temperature and can also create problems with statistical variations due to shot noise during the exposure. Usually, higher contrast and more process latitude are achieved at lower sensitivities. Since the light used to expose the resist is monochromatic, it is also important that the sensitivity of the resist be optimized for the exposure wavelength.

Resolution is obviously important in resists. The quality of resist patterns today is generally limited by the exposure system (aerial image) and not by the resist itself. However, the resist materials and the process steps (exposure dose, baking and developing cycles) must be carefully controlled to maintain the aerial image resolution in the resist developed images.

The third parameter of importance has to do with the "resist" function of photoresists. The term "resist" describes the need for the photoresist to withstand etching or ion implantation after the mask pattern is transferred to the resist. Practical resists need to have reasonable robustness with respect to these processes. In practice, this means that resists must be able to resolve small features even when the resist is of reasonable thickness. Alternatively, we will see later that it is possible to use "hard masks," a material placed below the resist, in order to allow the very thin resist layers needed for high-resolution imaging.

We will briefly discuss three generations of photoresists in the remainder of this section. The g-line and i-line photoresists are designed to work with lithography tools using those wavelengths of light. While Figure 5.4 suggests that these systems became outdated in the mid-1990s, in fact they are still used today in situations in which they have adequate resolution, because these lithography systems are much less expensive than the newer DUV and EUV systems. The second generation of resists works with DUV systems, primarily at 193 nm today, and these resists introduced a new concept, chemical amplification to improve sensitivity. Finally, the newest resists are those designed to work with EUV exposure tools.

The g-line and i-line photoresists generally consist of three components: an inactive resin, which is usually a hydrocarbon that forms the base of the material; a photoactive compound (PAC), which is also a hydrocarbon; and a solvent, which is used to adjust the viscosity of the resist. Most of the solvent evaporates during spinning onto the wafer and during pre-bake processes before the resist is exposed, leaving a material that is about 1:1 base and active components.

The most commonly used g-line and i-line resists are diazonaphthoquinone (DNQ) materials. The base resin is generally novolac, which is a polymer material consisting of basic hydrocarbon rings with two methyl (CH_3) groups and one hydroxyl (OH) group attached. The basic ring structure may be repeated many times to form a long-chain polymer material. Novolac by itself will readily dissolve in the developer solution at a typical dissolution rate of about 15 nm s^{-1}.

The PACs in these resists are often diazoquinones. The photoactive part of the molecule is shown on the left-hand side of Figure 5.12. The role of the PAC is to inhibit the dissolution of the resist material in the developer. Diazoquinones are insoluble in typical developers and they reduce the overall dissolution rate of the resist to approximately 1–2 nm s^{-1}. Thus the DNQ material is essentially insoluble in the resist developer before it is exposed to light.

When the resist is exposed to light, the diazoquinone molecules chemically change, as illustrated in Figure 5.12. The N_2 molecule is weakly bonded in the PAC and the first part of the photochemical reaction involves the light breaking this bond. The 5–6 eV energy of the DUV photons is sufficient to break the N_2 bond. This leaves behind a highly reactive carbon site. The PAC structure can stabilize itself by moving a carbon atom outside the ring with the oxygen atom covalently bonded to it. This is known as the Wolff rearrangement, as shown in the center of Figure 5.12. The resulting ketene molecule finally transforms into carboxylic acid

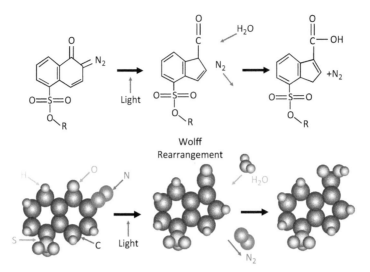

Figure 5.12 Process occurring in diazoquinones upon exposure to light. The top representation uses organic chemistry nomenclature; R refers to the rest of the molecule. The bottom shows the same reaction using atoms.

(as shown on the right-hand side of Figure 5.12) in the presence of water vapor from the room air. The carboxylic acid is now readily soluble in a basic developer – typically tetramethyl ammonium hydroxide (TMAH), potassium hydroxide (KOH) or sodium hydroxide (NaOH) dissolved in H_2O. The novolac matrix material is also readily soluble in this solution. The exposed resist material thus dissolves at a rate of 100–200 nm s^{-1}. The unexposed regions of the resist are essentially unaffected by the developer and so, if the exposed pattern accurately reproduces the mask pattern, the photoresist can produce a high-resolution image of the mask.

Conventional DNQ resist materials have two significant problems when shorter exposure wavelengths are used. The first is that, for wavelengths below i-line (365 nm), these resists strongly absorb the incident photons. Thus the incident radiation cannot penetrate through the full thickness of the resist. This is a significant issue in DUV exposure systems. The second problem has to do with resist sensitivity. When resists suitable for DUV applications were first being explored, the only viable light source was the Hg arc lamp described earlier in this chapter. These sources work well for g-line and i-line systems, but the intensity of their output in the DUV is much lower than at i-line. Thus it was believed at that time that, whatever resist was used for DUV applications, it would have to have improved sensitivity over the standard DNQ resists, in order to maintain manufacturing throughput. The development of bright excimer laser sources obviated this issue somewhat, but nevertheless higher sensitivity was still desirable in the DUV resists in order to improve throughput.

DUV resists are based on a completely new chemistry and make use of chemical amplification resists (CA resists or CARs) [13, 14, 15]. Standard DNQ resists achieve quantum efficiencies of about 0.3. This means that about 30% of the incoming photons interact with PAC molecules and are effective in exposing the resist. Thus the sensitivity improvement possible with these resists is at most a factor of about 3. CARs use a different exposure process in which the incoming photons react with a photo-acid generator (PAG) molecule, creating an acid molecule. These acid molecules then act as catalysts during a subsequent resist bake to change the resist properties in the exposed regions. Both positive and negative resist versions are possible. In the positive resist case, the PAG initiates a chemical reaction that makes the resist soluble in the developer; in a negative resist, the opposite happens. The key point in either case is that the reactions are catalytic; the acid molecule is regenerated after each chemical reaction and may thus participate in tens or hundreds of further reactions. Thus the overall quantum efficiency in a CAR is the product of the initial efficiency of the light–PAG reaction, multiplied by the number of subsequent reactions that are catalyzed. This product can be much larger than 1 and is responsible for the improvement in sensitivity of DUV resists compared to DNQ resists (20–$40\ \mathrm{mJ\ cm^{-2}}$ compared to $100\ \mathrm{mJ\ cm^{-2}}$).

Chemical amplification is a very powerful approach to create more efficient resists. Figure 5.13 illustrates the basic principle behind these resists. A typical example would be a polyhydroxystyrene (PHS) polymer with attached blocking groups that are insoluble in the developer [16, 17].

The incident DUV photons react with the PAG molecules to create an H^+ ion (proton or acid molecule). The spatial pattern of acid molecules in the resist after the exposure is thus a "stored" or latent acid 3D image of the mask pattern. After exposure, the wafer is baked at a temperature on the order of $120\ °C$ for a few minutes (post-exposure bake or PEB). The heat provides the energy needed for the reaction between the acid molecules and the insoluble fragments on the polymer chains to take place. It also provides mobility (through diffusion) for the acid molecules to find the insoluble fragments and to react with tens or hundreds of such fragments during the PEB. Thus, during the PEB, the insoluble blocked polymer is converted into an

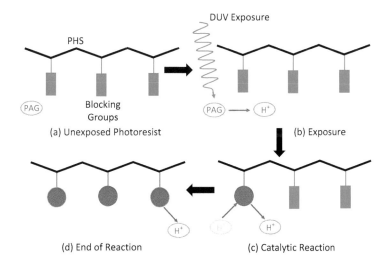

Figure 5.13 Basic operation of a chemically amplified (CA) resist. PAG is the photo-acid generator; the blocking groups are insoluble in the developer. Step (c) may repeat tens or hundreds of times during the post-exposure bake.

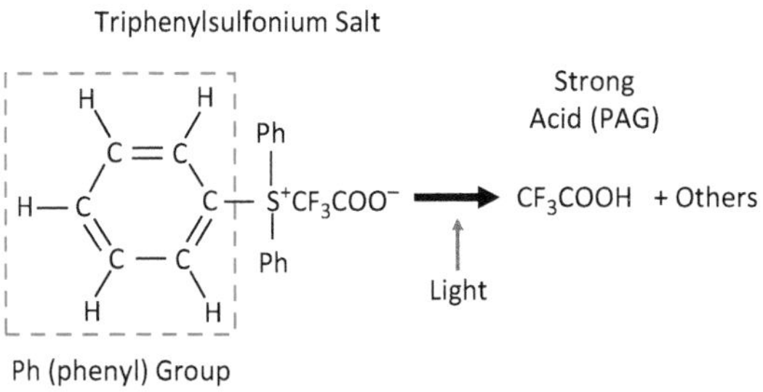

Figure 5.14 Creating the acid molecule in a CAR resist.

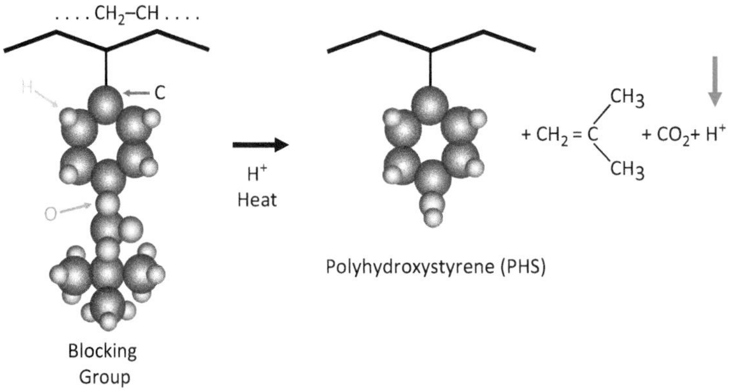

Figure 5.15 During the PEB, the H^+ reacts with the blocking group to produce OH-terminated PHS, which is very soluble in the developer.

unblocked polymer that is soluble in an aqueous alkaline developer. In negative working DUV resists, the PAG catalyzes a reaction that crosslinks the polymer chains, making the resist insoluble in the developer. The key mechanism in either process is the catalytic behavior of the acid molecules, which are regenerated after each reaction.

Figures 5.14 and 5.15 illustrate the reactions taking place in a chemically amplified resist. In Figure 5.14 a triphenylsulfonium salt molecule is struck by a DUV photon. This creates the strong acid molecule, trifluoroacetic acid (CF_3COOH) in this example. By strong acid, we mean that the H easily splits off from the PAG molecule as an H^+ ion and can then be used in the PEB to react with the blocking groups, as shown in Figure 5.15. Real CARs use more complicated acid molecules than this simple example, but the principle is the same.

In Figure 5.15, the chemistry occurring during the PEB is illustrated. The H^+ ion from the **PAG** reacts with the blocking group on the **PHS** backbone polymer. In this example, the blocking layer is a t-butoxycarbonyl (t-BOC) group, which makes the PHS much less soluble in the developer. The reaction cuts off the blocking group and terminates the PHS with an OH molecule, which makes the polymer soluble in the developer. Note that the H^+ ion is conserved (red arrow) in this reaction, so it is available to repeat the reaction multiple times on multiple blocking groups. In other words, the H^+ catalyzes the reaction. There are many other possible blocking groups that are used in CAR resists. The t-BOC example shown here was used in the very earliest CARs.

DUV resists require very careful control of the PEB conditions because that bake is used to drive the chemical reaction that completes the resist exposure. Chemical reactions and the diffusion associated with the acid molecules during the PEB generally depend exponentially on temperature, requiring careful temperature control during the PEB.

In EUV resists, there is a fundamental change in the absorption of the photons and in the exposure process. This occurs because of the energy difference between DUV and EUV photons. We saw in our discussion of (5.1) that the energy of DUV photons is in the range of 5–6 eV. Since the binding energy of electrons to the atoms and molecules of which they are part is generally greater than this, DUV photons interact with DUV resists either by raising the energy levels of electrons from a ground state to an excited state within the molecule of which they are part or by breaking chemical bonds. This results in chemical changes locally in the molecules, as we saw in the discussion in connection with Figures 5.12 and 5.13.

EUV photons have energies of approximately 92 eV, far greater than the binding energies of electrons to the atoms and molecules they are part of. As a result, when an EUV photon is absorbed, an electron is typically ejected from the molecule it is part of, with an energy of 70–80 eV, leaving behind a radical cation "hole" in the molecule. These ejected electrons have sufficient energy to ionize additional electrons in the photoresist, so that the quantum efficiency (number of electrons released per photon) can easily be higher than 1. The ejected electrons have been estimated to have a "travel distance" of 2–4 nm over which they can release their energy. It is thought that the electrons interact with PAG molecules, giving them the energy needed to produce the H^+ ion required for subsequent reaction with the insoluble "blocking groups" in the resist. However, while the physical mechanisms occurring in DUV resists are fairly well understood, as described above, the mechanisms operating in EUV resists are still subjects of active research [18].

EUV resists are CARs just like DUV resists. The EUV photon creates PAG molecules just as in DUV resists, even if the detailed mechanism is not fully understood. Whereas in DUV resists each photon generates at most one PAG molecule, several may be generated per photon in EUV resists because of the much higher photon energies. There is still significant research today in exploring chemistries for EUV resists, with Brainard *et al.* [18] providing a good summary of current efforts. Since essentially all materials absorb EUV photons, all of the molecules in the photoresist in addition to the PAG molecules absorb the EUV photons, so there are many choices to explore for EUV resists. Following the exposure process, a PEB (generally about 1 min at 100 °C) is used, as was the case with DUV resists to complete the exposure process and allow chemical amplification to occur.

A final point should be made with regard to the thickness of photoresist layers. We will discuss in the next section the idea that projection optical systems, whether DUV or EUV, have a finite depth of focus. As the wavelength decreases, this depth of focus also decreases. This means that the image the lithography system is producing is only in focus for a finite thickness of photoresist, a thickness that decreases as λ decreases. This generally means that the resist thickness must also decrease as λ decreases. In addition, since virtually all the components in EUV resists absorb EUV photons, the thickness of EUV resists must be fairly small in order to expose the resist through its entire thickness. This issue implies that, when EUV exposure tools are used, generally a hard mask below the EUV resist must be used in order to provide a more robust mask against subsequent etching or ion implantation or other process steps.

5.3   Optical System Basics: Ray Tracing and Diffraction

In order to understand and quantify the capabilities of modern wafer exposure systems, we will need to review some basic concepts about light and optical systems. Many standard references exist on these topics (see, for example, [19]) and the reader is referred to these texts for a more detailed treatment of the concepts described here. The application of basic optical principles to lithography systems is also described in detail in a number of books, for example, [20, 21]. In this section we will introduce the basic ideas in a somewhat qualitative way. In Section 5.5, we will be more quantitative and show how mathematical models can be implemented in simulation programs that provide significant insight into the performance of optical systems.

Light travels as an electromagnetic wave through space. When one is interested in the behavior of an optical system in which all of the dimensions are very large compared to the wavelength of light, the light can usually be treated as particles traveling in straight lines between the optical components. This simplifies the problem to one of "ray tracing." In projection lithography systems, for example, this approach can often be used in describing the optical source and condenser lens, but it fails when the light passes through the mask because the feature dimensions on the mask are comparable to the wavelength of the light. Ray tracing is accomplished simply by calculating the angle of refraction or reflection at each surface the ray encounters as it passes through the optical system. These calculations rely on three laws of geometrical optics:

1. Light travels in a straight line in a region of constant refractive index.
2. The angle of incidence equals the angle of reflection.
3. Snell's law, $n_1 \sin \theta_1 = n_2 \sin \theta_2$.

A simple example of ray tracing is shown in Figure 5.16 in which a "point" light source is passed through a convex lens to produce collimated light. At each air/glass interface, Snell's law is invoked to change the direction of the light ray. We will actually use this example shortly in the condenser portion of an optical lithography system.

Figure 5.16 Simple example of ray tracing applying Snell's law at each material interface. After [22].

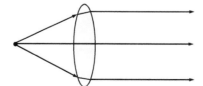

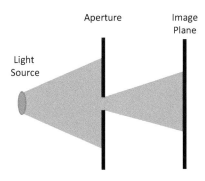

Figure 5.17 Simple example of diffraction effects. Light passes through a narrow aperture. The image formed covers a much larger area than can be explained based on simple straight-line ray tracing.

To understand the behavior of the light as it passes through the mask, the objective lens and on to the wafer, we must include the wave nature of light. The single most important effect that we must account for is diffraction. Diffraction effects occur because light does not in fact travel in straight lines. Many simple experiments can demonstrate this fact. Figure 5.17 illustrates the passage of light through a small aperture. The light pattern that strikes the screen (image plane) covers a much larger area than can be accounted for by simply drawing straight lines (ray tracing) to describe the light propagation. In fact, the smaller the aperture becomes, the more spread out the screen image becomes. This is very much like the situation we find in modern lithography, where light passes through a mask with apertures (clear regions) with dimensions smaller than the light wavelength. (We actually illustrated these effects in Figures 5.1 and 5.7 by showing the light spreading out after passing through the mask, but did not explain why this happens in connection with those figures.)

Consider Figure 5.18. A plane wave is shown propagating through space, unobstructed in Figure 5.18(a) and passing through an aperture in Figure 5.18(b). The Huygens–Fresnel principle can be used to construct the wavefront versus position as it propagates. This principle states that every unobstructed point of a wavefront at a given instant in time acts as a source of a spherical secondary wavelet of the same frequency as the primary wave. The amplitude of the optical field can be found by superimposing all these wavelets, considering their relative magnitudes and phases. In Figure 5.18(a) this superposition simply results in a propagating plane wave. In Figure 5.18(b), only the source points in the aperture serve as sources of Huygens wavelets, and the resulting propagating pattern beyond the aperture involves diffraction. As the arrows in Figure 5.18(b) illustrate, the light spreads out beyond the aperture. In fact, the smaller the aperture becomes, the more the light spreads out because there are fewer wavelets able to pass through the aperture and thus more spherical propagation results beyond the aperture.

Diffraction can be thought of simply as the "bending" of light when it passes through an aperture. The light that passes through the aperture carries with it the information on the size and shape of that aperture. If, for example, the aperture were part of a mask that we wished to print on a wafer, then the information about the aperture size and shape needs to be carried by the light to the photoresist on the wafer. The problem is that this information spreads out in

Figure 5.18 Propagation of a plane wave in (a) free space and (b) through a small aperture, illustrating the use of the Huygens–Fresnel principle to construct the wavefront as it propagates.

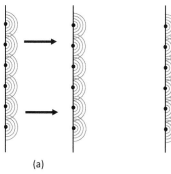
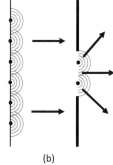

(a)   (b)

Figure 5.19 Qualitative example of a small aperture being imaged.

Point Source

Collimating Lens

Aperture

Focusing Lens

Image Plane

Diffracted Light

Collected Light

d

f

space because of diffraction, and it must all be collected to convey perfect information about the aperture to the resist on the wafer. Figure 5.19 illustrates this in a qualitative way.

Because of its finite size, the focusing lens collects only part of the total diffraction pattern associated with the light passing through the aperture. The light diffracted to wider angles carries the information about the finer details of the aperture, so it is those details that are lost first when a lens of finite size is used to collect and focus the light.

The aperture information carried by the extreme parts of the diffraction pattern is known as the "high-spatial-frequency information." Frequency does not imply that the photon frequency is changing, since all the light is monochromatic. What is meant is that information about the sharp corners of the aperture shape is carried in the extreme parts of the diffraction pattern. Thus, if that information is "lost" or not captured by the focusing lens, the printed pattern will show rounded corners rather than sharp square corners. We will see many examples of this later in this chapter.

The actual image produced in this simple example is shown in Figure 5.20 for a small circular aperture and is known as Airy's disk after Sir George Biddell Airy who first derived the expression describing the central intensity maximum. The central maximum approximates the image of the circular aperture. Because of diffraction effects, the image is composed of

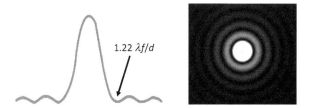

Figure 5.20 Image intensity of a circular aperture in the image plane (Fraunhofer diffraction pattern). The intensity is sketched along any diameter on the left. The pattern on the right illustrates the diffraction pattern, which is described more fully later in this chapter (Figure 5.33).

a bright center disk surrounded by a series of faint rings. The image intensity can be described mathematically by Bessel functions, and the approximate size of the image is given by

$$\text{distance to first minimum} = \frac{1.22\lambda f}{d},\tag{5.3}$$

where d is the focusing lens diameter, f is the focal length and λ is the wavelength of the light. Note that a point source only produces a point image if $d = \infty$ (or if λ or $f = 0$).

The propagation of light waves can be described mathematically using Maxwell's equations, and we will pursue a mathematical approach later in this chapter in Section 5.5. There are two approximations to a full electromagnetic wave treatment that are used in calculating diffraction patterns in semiconductor lithography applications. In Fresnel diffraction (often called near field), the image plane is close to the aperture (or mask). The light travels directly from the aperture to the plane where the image is formed with no intervening lens system. Fresnel diffraction is thus appropriate for modeling contact and proximity lithography tools (Figure 5.7(a,b)).

In Fraunhofer diffraction (often called far field), the image is far from the aperture and a lens is normally placed between the aperture and the image plane to capture and focus the image. Fraunhofer diffraction theory is thus appropriate for modeling projection lithography systems (Figure 5.7(c)). Mathematical descriptions and detailed models have been developed for both of these regimes. Powerful simulation tools based on these models have also been developed, which allow the calculation of the aerial image formed by the wafer exposure system. We will discuss the mathematics and these simulation tools in Section 5.5. Here, we will discuss qualitatively a set of ideas that allow estimates of the resolution, depth of focus and other parameters associated with modern lithography systems.

Figure 5.21 illustrates the situation for contact and proximity lithography systems. We assume that the mask and the wafer are separated by some small gap g. A plane wave is assumed to be incident on an aperture in a mask. The diffraction pattern on the opposite side of the mask can be constructed by imagining Huygens wavelets as emanating from each point in the aperture or mathematically using Fresnel diffraction theory. The resulting light intensity distribution striking the top surface of the resist is also shown in Figure 5.21.

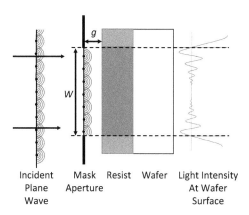

Figure 5.21 Basic contact or proximity (near-field) exposure system illustrating the use of Huygens wavelets emanating from an aperture in the mask. A mask feature size of W is assumed, along with a mask-to-resist separation of g. The resulting light intensity distribution (aerial image) at the resist surface due to Fresnel diffraction is shown on the right.

There are several features in the light distribution of interest. First, notice that the intensity drops off gradually outside the edges of the mask aperture. Because of diffraction effects, the light "bends" away from the aperture edges, producing some resist exposure outside the aperture edges. Second, note the "ringing" in the intensity distribution within the aperture dimension. This arises because of constructive and destructive interference between the Huygens wavelets emanating from the aperture. The ringing effects can be reduced in contact and proximity lithography systems through the use of multiple exposure wavelengths, and through using light sources that are not perfect plane waves. We will discuss some of these methods at the end of this chapter.

As the separation g increases between the mask and the resist, the quality of the aerial image produced at the resist surface will degrade because diffraction effects will become even more important. Contact printers minimize these effects by attempting to reduce g to zero. However, even in this case, the resist itself still has a finite thickness, and the resolution will still be limited by light scattering in the resist and by light reflection from surface features on the underlying wafer that scatter the light laterally into regions adjacent to the mask apertures. Contact printers can achieve resolutions of perhaps 0.1 μm under optimized conditions today, but generally they are used at much larger feature sizes.

Generally the aerial image can be calculated using Fresnel diffraction theory whenever the gap g falls within the limits

$$\lambda < g < \frac{W^2}{\lambda},$$ (5.4)

where W is the size of the mask aperture (feature size). The lower limit on g is certainly satisfied by proximity printing systems and is often satisfied by contact printing systems unless hard contact and thin resist layers are used. If g is below the wavelength of the light used for the exposure, the resulting light intensity distribution can still be calculated but only by a full numerical solution to Maxwell's equations. Such solutions are very complex, but fortunately are rarely needed [23]. The upper bound on g arises because, when g increases, Fresnel diffraction theory must be replaced by far-field (Fraunhofer) diffraction theory in order to calculate the aerial image. Within the Fresnel diffraction range, the minimum resolvable feature size is on the order of

$$W_{\min} \approx \sqrt{\lambda g}. \tag{5.5}$$

Therefore, a proximity exposure system operating with a 10 μm gap g and an i-line light source ($\lambda = 365$ nm) can resolve features slightly smaller than 2 μm. This is much larger than the dimensions used in modern silicon chips, so these systems are not useful for manufacturing such chips. However, proximity printers are much less expensive than DUV or EUV projection systems, so, for applications in which feature sizes are compatible with them, proximity printers are an economical solution.

Contact lithography has been projected to have ultimate resolution limits of $\approx \lambda/20$ [24], which for an i-line lithography tool would imply a resolution of ≈ 20 nm. However, such systems are very unlikely to be used in manufacturing complex chips because of the defect density issues associated with repeated hard contacts between the mask and the resist-covered wafers. (The one exception to this could be nanoimprint lithography (NIL), which is discussed in Section 5.8.) For the interested reader, there are online video demonstrations of calculated Fresnel diffraction patterns for a variety of simple shapes [25]. In this book we will primarily be interested in projection systems, since these are the most commonly used systems in most semiconductor manufacturing plants today.

The performance of projection printers is usually specified in terms of a number of basic parameters, such as resolution, depth of focus, field of view, modulation transfer function (MTF), alignment accuracy, throughput, etc. At least the first four of these are directly related to the basic properties of optical systems, which we will discuss in this section. The last two issues are more associated with the mechanical design of the system.

Consider Figure 5.22 where we now imagine that we have two point sources close together that we are trying to image. Qualitatively, when they are far apart, they are easily resolved. At some point, as they move closer, we would say that we cannot resolve two separate sources.

Rayleigh suggested that a reasonable criterion for resolution was that the central maximum of each point image lie at the first minimum of the adjacent point image, the distance given by (5.3). While this definition is somewhat arbitrary, it is useful and has been widely adopted. The Rayleigh criterion is the example in Figure 5.22(b). With this definition and using the dimensions shown in Figure 5.23, the resolution R of the lens is given by

$$R = \frac{1.22\lambda f}{d} = \frac{1.22\lambda f}{n(2f \sin \alpha)} = \frac{0.61\lambda}{n \sin \alpha}. \tag{5.6}$$

Here n has been included for generality and is the index of refraction of the material between the object and the lens (usually air with $n = 1$ in lithography systems); and α is the maximum half-angle of the diffracted light that can enter the lens or the acceptance angle of the lens. The angle α may be limited by the physical size of the lens itself, or by an entrance aperture or pupil in front of the lens.

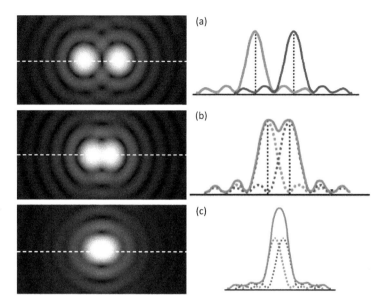

Figure 5.22 Two point sources moving closer together from (a) to (c). The Airy patterns are shown on the left (from Wikipedia). The light intensity on the white cutline through the center is shown on the right.

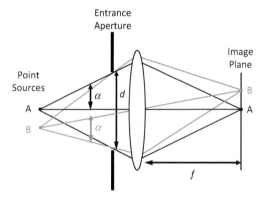

Figure 5.23 Illustration of the resolving power of a lens when two point sources are to be separated in the image. Note that d is the aperture size, or, if there is no aperture, it is the lens diameter.

The angle α is really a measure of the ability of the lens to collect diffracted light. This property was named the numerical aperture (NA) by Ernst Abbe:

$$NA = n \sin \alpha, \tag{5.7}$$

and therefore we have

$$R = \frac{0.61\lambda}{NA} = k_1 \frac{\lambda}{NA}. \tag{5.8}$$

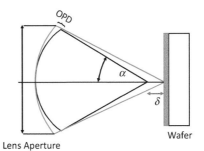

Figure 5.24 Geometry for estimating the depth of focus of an imaging system.

Equation (5.6) is based on the Fraunhofer diffraction pattern for an Airy disk and, as such, strictly applies only to point sources. Because of this, the 0.61 factor is often replaced by k_1 as in (5.8). In real lithography systems, the mask contains a variety of shapes. Also, k_1 depends in practice on the ability of the resist chemistry to distinguish closely spaced features, on the wafer structure below the resist (topography, reflectivity, etc.) and on defocusing at the image plane. Actual k_1 values achieved in simple optical lithography systems are 0.6 to 0.8, but k_1 can be much smaller when "tricks" are used in lithographic systems, as we will discuss later in this chapter.

It is obvious from (5.8) that shorter exposure wavelengths lead to better image resolution. It is also clear that lenses with higher numerical apertures also achieve better resolution, basically because they are able to capture more of the diffracted light and therefore construct a better image. Aside from the difficulty of building larger (higher-NA) lenses, there is also a significant drawback to using higher-NA lenses. This is the depth of focus of the lens, which can be estimated as shown in Figure 5.24.

The impact of focus errors on the resulting aerial image can be thought of as an aberration [21]. In Figure 5.24 the black lines represent a spherical wave converging or focusing on a point. In this example, the wafer is actually a distance δ away from the image plane and the image therefore is out of focus on the wafer. An ideal projection lens would create such a spherical wave coming out of the lens aperture. The image would be in focus on the wafer if the red spherical wave were produced by the projection lens. This is why we can think of the depth of focus problem as an aberration.

If δ is the on-axis path length difference at the limit of focus, then the path length difference or optical path difference (OPD) for a ray from the edge of the entrance aperture is simply $\delta \cos \alpha$ [21]. The Rayleigh criterion for depth of focus (DOF) is simply that these two lengths not differ by more than $\lambda/4$:

$$\frac{\lambda}{4} = \delta - \delta \cos \alpha \,. \tag{5.9}$$

Assuming that α is small, we have

$$\frac{\lambda}{4} = \delta \left[1 - \left(1 - \frac{\alpha^2}{2} \right) \right] \cong \delta \frac{\alpha^2}{2}, \tag{5.10}$$

$$\alpha \cong \sin \alpha = \frac{d}{2f} = \text{NA}, \tag{5.11}$$

and therefore

$$\text{DOF} = \delta = \pm \frac{\lambda}{2(\text{NA})^2} = \pm k_2 \frac{\lambda}{(\text{NA})^2}. \tag{5.12}$$

The factor of 1/2 is often replaced by k_2 in (5.12) because the 1/2 value is appropriate at the Rayleigh resolution limit but does not take into account the increase in depth of focus for larger features nor the dependence in practice on other parameters like the resist process.

Example

We can estimate the resolution and depth of focus of two lithography systems, a DUV stepper operating at $\lambda = 193$ nm and an EUV stepper operating at 13.5 nm. For simplicity, we shall assume both systems have a NA = 0.6, $k_1 = 0.75$ and $k_2 = 0.5$.

Answer

Using the equations above, we have

$$R_{193} = k_1 \frac{\lambda}{\text{NA}} = 0.75 \frac{193 \text{ nm}}{0.6} = 241 \text{ nm},$$

$$R_{13.5} = 0.75 \frac{13.5 \text{ nm}}{0.6} = 16.9 \text{ nm},$$

$$\text{DOF}_{193} = \pm k_2 \frac{\lambda}{(\text{NA})^2} = \pm 0.5 \frac{193 \text{ nm}}{(0.6)^2} = \pm 270 \text{ nm},$$

$$\text{DOF}_{13.5} = \pm 0.5 \frac{13.5 \text{ nm}}{(0.6)^2} = \pm 9.4 \text{ nm}.$$

The resolution of each system is on the order of the wavelength using these simple models. We will see later in this chapter that, with resolution enhancement methods, we can actually do much better than this. As expected, the EUV system has much better resolution (>10×), but it also has a very small depth of focus, which means that a very thin resist layer would need to be used and the mechanical position of the wafer would need to be very precise to produce an in-focus image.

There are two additional basic concepts regarding optical exposure systems that will be useful to us, the modulation transfer function (MTF) and spatial coherence (σ). The MTF is illustrated in Figure 5.25. This concept strictly applies only to coherent illumination, and as a result is not really applicable to modern lithography tools, which generally use partially coherent illumination, as described below. However, the basic idea is a useful one in understanding lithography issues.

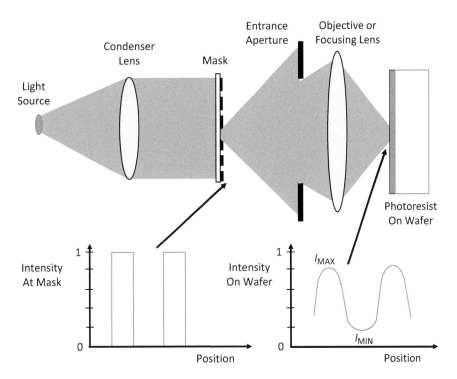

Figure 5.25 The modulation transfer function (MTF) concept. A generic lithography system is shown at the top, with a mask being imaged on photoresist on a wafer. The mask MTF is almost ideal ($M = 1$) since the feature sizes are 4–5× larger than those imaged in the resist and diffraction effects are minimal. The wafer aerial image MTF is much lower ($M \approx 0.6$) because of diffraction in the optical system.

Figure 5.25 shows a generic projection lithography system in which a reducing lens system is used to image a mask pattern in resist. Since diffraction effects are only important after the light passes through the mask, the optical intensity pattern as the light exits the mask will be almost an ideal representation of the mask. Because of diffraction effects and other non-idealities in the optical system, however, the aerial image produced at the resist plane will not be perfectly black and white. If the features are widely separated and large, the aerial image may approach the ideal shown on the left in Figure 5.25, but, as the features move closer together and the feature sizes become smaller compared to the light wavelength, the aerial image will look more like that sketched on the right (see Figure 5.22).

A useful measure of the quality of the aerial image is the modulation transfer function (MTF), which can be defined as

$$\mathrm{MTF} = \frac{I_{\mathrm{MAX}} - I_{\mathrm{MIN}}}{I_{\mathrm{MAX}} + I_{\mathrm{MIN}}}, \qquad (5.13)$$

where I is the intensity of the light. The MTF is really a measure of the contrast in the aerial image produced by the exposure system. Generally, an exposure system needs to achieve an MTF value of 0.5 or larger in order for the resist to properly resolve the features. DUV resists

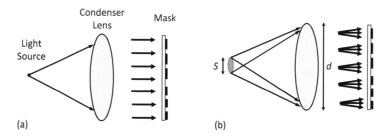

Figure 5.26 (a) An ideal point source produces coherent light at the mask. (b) A realistic light source produces partially incoherent light at the mask.

can work with somewhat smaller MTF values. It should be obvious that the MTF depends on the feature size being printed and that the MTF will be closer to 1 (ideal) for larger feature sizes.

Figure 5.26 illustrates the concept of spatial coherence. An ideal point source produces light in which the waves are in phase at all points along the emitted wavefronts. A condenser lens can then convert these waves to plane waves, all of which strike the mask at exactly the same angle, as illustrated in Figure 5.26(a). Such a source is an ideal coherent source. As the physical size of the source increases, as shown in Figure 5.26(b), light is emitted from a volume rather than a point and the waves will not be perfectly in phase everywhere. If the same condenser lens is used to convert the light to plane waves, the result will be light arriving at the mask from a variety of angles, as illustrated. Such a source is a partially coherent source. A useful definition of the spatial coherence of practical light sources, usually called σ in the lithography literature, is

$$\sigma = \frac{\text{light source diameter } S}{\text{condenser lens diameter } d}. \tag{5.14}$$

It might appear at first glance that we would choose to have an ideal coherent source ($\sigma = 0$) for optical lithography. However, this is not the case, for the following reasons. First, as σ approaches 0, the optical intensity also approaches zero, resulting in infinite exposure times to print the mask pattern in the resist. Second, the MTF is also affected by the value of σ and having $\sigma = 0$ is not the optimum choice.

To understand this latter point, we need to consider diffraction effects once again. If the light passing through the mask is partially coherent (i.e., coming in at a variety of angles), then the diffraction patterns resulting from mask features will be smeared out. If we refer again to Figure 5.19, this means that the diffraction pattern for a given feature will be spread over an angle larger than α. This initially seems like a bad thing, because it means that some information will be lost because the finite aperture of the objective lens will not collect it. However, this smearing also means that information from closely spaced features that would have been completely lost outside the aperture of the focusing lens is now partially collected because it is smeared inside the lens aperture. The result of these effects is that imaging is actually improved for small features when partially incoherent light is used. We will discuss this in more detail towards the end of this chapter (Section 5.7) when we consider resolution

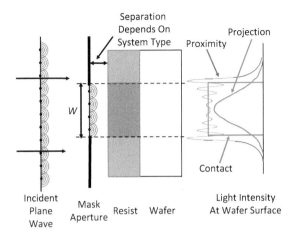

Figure 5.27 Aerial images produced by the three types of optical lithography tools. The mask and wafer would be in hard contact in a contact aligner, separated by a gap g in a proximity aligner, and far apart with an intervening focusing lens in a projection system.

enhancement techniques. In practice, a spatial coherence of 0.5–0.7 is often used in IC manufacturing, using the definition in (5.14).

Figure 5.27 summarizes the discussion on lithography systems. We imagine a plane wave passing through a mask aperture. The aperture is imaged on the resist on a wafer through one of the three types of exposure systems. In the contact printing case, a very-high-resolution image is produced because the mask and resist are assumed to be in hard contact. If the wafer and mask are separated slightly, as in a proximity printing system, the resolution degrades because of near-field Fresnel diffraction effects (the "ringing" shown in the proximity curve). Finally, if we place a lens between the mask and wafer and focus the aperture on the wafer, an image characterized by Fraunhofer (far-field) diffraction is produced. In this example, the resolution of the proximity system image is inferior to both of the other systems. This is usually the case in actual systems. Given the choice, therefore, between contact and projection systems to manufacture state-of-the-art chips, the choice is almost always projection systems because of the high defect densities associated with contact systems.

5.4   Photoresist Basics: Properties and Processing

We discussed earlier the types of resists used for g-line and i-line, DUV and EUV exposure systems. Regardless of which type of photoresist is being used in a particular lithography system, two basic parameters are often used to describe the properties of photoresists, namely, the contrast and the critical modulation transfer function (CMTF). We will define these terms and explain their importance in the following paragraphs.

Contrast is really a measure of the resist's ability to distinguish light from dark areas in the aerial image that the exposure system produces. As we have seen, diffraction effects and perhaps other imperfections in the exposure system result in an aerial image that does not

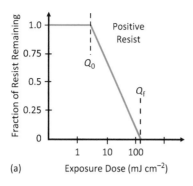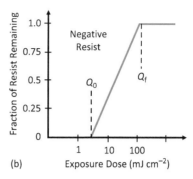

Figure 5.28 Idealized contrast curves for (a) positive and (b) negative resists.

have abrupt transitions from dark to light. An important question is how the resist responds to the "gray" region at the edges of features in the aerial image.

Contrast is an experimentally determined parameter for each resist, and its value is extracted from plots like those shown in Figure 5.28. The data from which these plots can be derived are obtained by exposing resist layers to a variety of exposure doses. Each of the samples is then developed for a fixed period of time, and the thickness of the photoresist remaining after developing is measured. For positive resists, samples receiving small exposure doses will not be attacked by the developer to any appreciable extent; those receiving large doses will completely dissolve in the developer. Intermediate doses will result in partial dissolution of the resist. For negative resists, the opposite behavior occurs.

Given data like those shown in Figure 5.28, the contrast is simply the slope of the steep part of the curve, defined as

$$\gamma = \frac{1}{\log_{10}(Q_f/Q_0)}, \tag{5.15}$$

where Q_0 is the dose at which the exposure first begins to have an effect and Q_f is the dose at which exposure is complete.

Typical g-line and i-line resists achieve contrasts of 2–3 and Q_f values of about 100 mJ cm^{-2}. DUV resists achieve significantly better contrast and better sensitivities than this. This is basically because the chemical amplification that occurs in DUV resists steepens the transition from the unexposed to the exposed condition. (Once the reaction is begun, the catalytic nature of the process carries it to completion, unlike DNQ resists where the PAC molecules in the resist must be exposed one by one by incoming photons during the exposure.) Thus DUV resists typically achieve γ values of 5–10 and Q_f values of about 20–40 mJ cm^{-2}.

It is important to note, however, that γ is not a constant for a particular resist composition. Rather, the experimentally extracted value of γ depends on process parameters like the development chemistry, bake times and temperatures before and after exposure, the wavelength of the exposing light and the underlying structure of the wafer on which the resist is spun. In general, it is desirable to have a resist with a high contrast because this produces better (steeper) edge profiles in the developed resist patterns (see Figure 5.29). Intuitively, this arises because

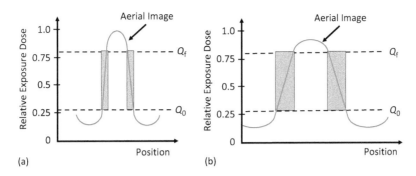

Figure 5.29 Two examples of how the quality of the aerial image and the resist contrast combine to produce the resist edge profile. (a) A sharp aerial image and narrow resist transition region (gray area). (b) A poorer aerial image and the resulting wider transition region in the resist profile.

a high contrast implies that the resist sharply distinguishes between dark and light areas in the aerial image. Thus resists with high contrast can actually "sharpen up" a poor aerial image.

The MTF of the aerial image was defined in (5.13) and Figure 5.25. The MTF is simply a measure of the "dark" versus "light" intensities in the aerial image produced by the exposure system. It is often useful to define a similar quantity for the resist, in this case called the critical modulation transfer function (critical MTF or CMTF). The CMTF is roughly the minimum optical transfer function necessary to resolve a pattern in the resist:

$$\mathrm{CMTF}_{\mathrm{resist}} = \frac{Q_{\mathrm{f}} - Q_0}{Q_{\mathrm{f}} + Q_0} = \frac{10^{1/\gamma} - 1}{10^{1/\gamma} + 1}. \tag{5.16}$$

Typical CMTF values for g-line and i-line resists are around 0.4. With their higher γ values, chemically amplified DUV resists achieve significantly smaller CMTF values ($\approx$0.1–0.2). The significance of this number is that the CMTF must be less than the aerial image MTF if the resist is going to resolve the aerial image.

A typical photoresist process sequence is shown in Figure 5.30. The first step in the lithography process is ensuring that the resist will adhere well to the wafer. Depending on the stage in the process flow, this may involve one or more operations. Normally the wafer is clean just before resist application, since resist is commonly deposited on wafers just after thin-film deposition or just after some other high-temperature process step. If this is not the case, the wafer may need to be chemically cleaned, using the procedures described in Chapter 4. It may also be necessary to heat the wafer to several hundred degrees Celsius to drive off any water vapor on the surface. These steps are represented by the first box in Figure 5.30.

Even with a clean, dry surface, adhesion of resists to ICs may not be as good as desired. The surface materials may include semiconductors, deposited dielectrics, metals and many other materials, and so it is common practice to use an adhesion promoter. Hexamethyldisilane (HMDS) is the most common substance used for this purpose. HMDS can be applied to the wafer in liquid form at room temperature by spinning under conditions similar to resist spinning (3000–6000 rpm for $\approx$30 s). However, in most cases, it is applied in vapor form, with

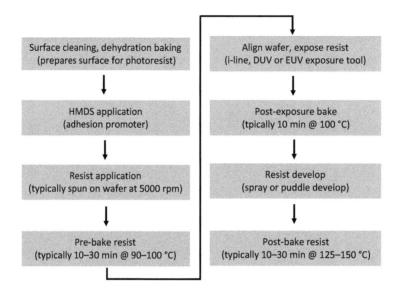

Figure 5.30 Typical photoresist process flow for i-line, DUV and EUV resists. Differences in the process flow for the various resist types are described in the text.

the HMDS being introduced as a vapor into a chamber containing the wafers, since this more easily produces the desired single monolayer on the wafers. In either case, one end of the HMDS molecule bonds readily with SiO_2 surfaces and the other bonds with the resist. Thus the HMDS acts an adhesion promoter. For surfaces other than SiO_2, other adhesion treatments may be used in addition to or instead of HMDS (see chapter 6 of [26]).

The next step is spinning the resist itself onto the wafer, and this should normally be done immediately after the HMDS application. The resist is dispensed onto the wafer and then the wafer is spun (3000–6000 rpm for ≈ 30 s) to produce a thin uniform layer (≈ 0.1–0.5 μm typically, but thinner in high-resolution imaging situations). If the resist is going to be used to mask an implant, it may be thicker than these values. While the procedure sounds simple, there are some subtleties associated with it. The dispensing of the resist can be done while the wafer is stationary or while it is spinning at slow speed to produce a uniform liquid layer on the wafer. The acceleration of the wafer to its final spin speed is also important, and is typically done rapidly (in a fraction of a second). The solvent in the resist begins to evaporate rapidly after the resist is dispensed and while it is being accelerated. Generally, more uniform films are obtained if the acceleration is as rapid as possible. During the first few seconds that the wafer spends at high spin rates, the film levels to a uniform thickness. During the remainder of the ≈ 30 s spin, the solvent continues to evaporate to produce the final resist thickness. The fluid dynamics of the spin coating process have been studied in some detail [26], but the exact process for a particular resist is usually determined experimentally. The viscosity of the resist in its liquid state (solvent content) and the spin speed are the primary factors affecting the final resist thickness.

The next step in the lithography process is the pre-bake. This is normally accomplished at 90–100 °C. Infrared or microwave heating can also be used, which are faster processes. The pre-bake step accomplishes several things. First, the remaining solvent in the resist is largely evaporated, reduced from $\approx 25\%$ to $\approx 5\%$ of the resist content. Second, adhesion of the resist is improved, since the heating strengthens the bonds between the resist and the HMDS and substrate. Finally, stresses present in the resist as a result of the spinning process are relieved through thermal relaxation. Chemical changes do take place in the resist during this high-temperature bake, with the result that required exposure times are increased as the bake temperature increases. The mechanism responsible for this is believed to be a decomposition of the light-sensitive molecules at high temperature, with the result that the resist sensitivity is degraded.

The exposure process creates a latent image in the resist that can be later developed. Most resists exhibit reciprocity, which means that light intensity and exposure time can be directly traded off with each other. Thus increases in the exposure system aerial intensity directly reduce exposure times. The required exposure time is also affected by the pre-bake thermal cycle, as was just discussed, and by the thickness of the resist. All of these parameters must therefore be carefully controlled. Exposure doses are normally designed to be well above Q_f (Figure 5.28), since this produces latent images in the resist that have the sharpest edges (limited, of course, by the quality of the aerial image). For typical DNQ resists, this corresponds to a dose >100 mJ cm^{-2}. For DUV and EUV resists, which use chemical amplification, the dose is typically >20–40 mJ cm^{-2}.

The next step illustrated in Figure 5.30 is a post-exposure bake (PEB). In g-line and i-line resists, this step is sometimes performed before development of the resist latent image, in order to minimize standing wave effects in the resist (discussed later in Figure 5.61). The mechanism is straightforward. At elevated temperatures, the PAC in these resists can diffuse. If the temperature and time are well controlled in this bake, the PAC molecules can diffuse far enough to "smear out" the standing wave effects along the edge of the resist features, but not diffuse far enough to significantly distort the image features themselves. We will see some examples of standing wave patterns in resists later in this chapter. A typical bake cycle might be ≈ 10 min at $100\,°C$.

In DUV and EUV resists, the PEB is a necessary and critical step in the process. This is the step in which the PAG reacts with the polymer chain to complete the exposure process. The time and especially the temperature must be very tightly controlled because chemical reaction rates and diffusion typically depend exponentially on temperature.

DNQ resists are developed in basic solutions (normally TMAH solutions diluted with H_2O, but NaOH or KOH solutions can also be used). The developer can be applied in a number of ways. The wafers may be immersed in the developer, the developer may be sprayed on a batch of wafers or on a single wafer at a time, or a puddle of the developer may be placed on the wafer. In each case, rinsing the wafers with H_2O stops the developing process. The rate at which developing proceeds is highly dependent on temperature, on developer concentration and on all the exposure and bake procedures used before developing. In DNQ resists, the developing proceeds by dissolving the carboxylic acid that results from the exposed PAC in the alkaline solution. The rate of developing is dependent on the local carboxylic acid concentration, which is proportional to the local exposure intensity in the resist. In positive working DUV resists, developing takes place in a similar fashion, since the unblocked polymer chains are soluble in the basic developer. We will consider detailed models in Section 5.6.

The final step in the photolithography process is the post-exposure bake. This step is done at higher temperature than the earlier bakes (typically 10–30 min at 125–150 °C), and is designed to harden the resist and improve its etch resistance. This bake process causes the resist to flow slightly and hence may also modify edge profiles. Any remaining solvents in the resist are driven off by this bake, and the adhesion of the resist to the underlying substrate is also improved. In some cases, when the resist must withstand a particularly harsh etching process or a high current ion implantation which may raise its temperature considerably, a further hardening and crosslinking of the resist with a blanket deep-UV exposure at elevated temperatures can improve the robustness of the resist layer.

In modern lithography processes, many of the steps we have described above are accomplished in a single integrated machine known as a wafer track system. These machines are usually tightly integrated with the exposure tool so that easy wafer transfer can occur between the two systems.

5.5 Modeling and Simulation of Wafer Exposure Systems (Calculating the Aerial Image)

Up to this point in this chapter, we have discussed lithography systems in somewhat qualitative terms in order to make clear what the basic concepts are that govern these systems. More quantitative lithography modeling and simulation relies on two principal areas of science. The first of these is optics, which provides a mathematical description of the behavior of light in the exposure systems used in modern lithographic tools. The second is chemistry, which provides the tools to treat exposure, baking and developing of the resists that are used to convert the aerial image created by the exposure system into a three-dimensional replica (latent image) of the mask patterns. Many of the basic ideas and models that optics provides about light propagation and diffraction effects have their origins in work done more than 100 years ago by Maxwell, Kirchhoff and others. And, of course, the basic ideas from chemistry regarding the rates of chemical reactions, catalysts and the like are equally old ideas.

Applying this science to lithography dates back to the mid-1970s. The pioneering work in this area began at IBM and resulted in a series of widely referenced papers written by Dill and his coworkers [27, 28, 29, 30]. Following that work, a number of other groups began work on lithography simulation [31, 32]. Today, several simulation tools for optical lithography are commercially available. Two examples are Athena and Victory Process (Silvaco) [33] and Sentaurus Lithography (Synopsys) [34]. See also [35]. In this section, we will describe the physics behind these simulators, and we will use the Athena simulation tool to illustrate the kinds of very useful insights and quantitative information that they can provide. In Section 5.6 we will discuss modeling and simulation of the latent image in the resist.

The science describing optical system performance comes from physics and is described by Maxwell's equations that describe propagating electric fields. Exact solutions of these equations are possible with today's computers, but they are quite complicated, and the usual approach is

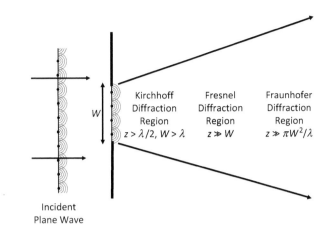

Figure 5.31 Diffraction regions where various approximations become accurate. Here W is the width of a slit that extends infinitely far into the page, λ is the light wavelength and z is the distance from the slit [21].

Example

Suppose we have an optical system operating at 193 nm and the light passes through an aperture (mask feature) 2 μm in size. Based on the definitions in Figure 5.31, the various diffraction approximate solutions would apply as follows for various distances beyond the mask.

Kirchhoff diffraction:	$z > 96$ nm (≈ 0.1 μm).
Fresnel diffraction:	$z \gg 2$ μm.
Fraunhofer diffraction:	$z \gg 65$ μm.

Thus, if we were imaging relatively large features in a projection lithography tool with a gap on the order of 10 μm, Fresnel or near-field diffraction would be an appropriate approximation. If we were using a projection lithography tool with much larger separations from the mask, Fraunhofer or far-field diffraction would be appropriate. However, if we were trying to image much smaller features, a similar calculation would show that essentially any practical separation would require Fraunhofer diffraction. Simulation tools like Athena implement both types of diffraction models, but the Fresnel model should only be used for large features (micron size).

to seek approximate solutions that apply under specific circumstances. This approach is shown in Figure 5.31 in which a plane wave is incident on a mask aperture. As we have seen, the light field propagating beyond the aperture is strongly affected by diffraction effects. Depending on the distance away from the mask, three approximations are illustrated in Figure 5.31.

There are excellent references that deal in detail with the mathematics of diffraction, for example [19, 21] and the interested reader is referred to these references for a complete mathematical treatment of these topics. Here we will outline the theory and discuss the implementation of this theory in modern lithography simulation tools. Since the vast majority of systems used in IC manufacturing today are projection systems (Fraunhofer diffraction), we will discuss these systems first.

5.5.1 Projection Systems

For the purposes of this discussion, we will consider a generic projection lithography system as illustrated in Figure 5.32. Modeling such a system requires a mathematical description of how light behaves. We begin with a basic description of light waves and their propagation.

Light travels as an electromagnetic wave. The electric and magnetic fields are perpendicular to each other and both are perpendicular to the direction of wave propagation. However, the materials that photons interact with in lithography are generally non-magnetic, so we can treat the light waves in such systems simply in terms of a traveling electric field $\mathcal{E}$. Thus a general description of a propagating monochromatic light wave at a point P in space is simply

$$\mathcal{E}(\mathrm{P}, t) = C(\mathrm{P})\cos(\omega t + \phi(t)), \tag{5.17}$$

where C is the amplitude, ω is the frequency and ϕ is the phase. We can also write this in the form of a complex exponential function, where j is the imaginary unit:

$$\mathcal{E}(\mathrm{P}, t) = \mathrm{Re}\{U(\mathrm{P})\mathrm{e}^{-\mathrm{j}\omega t}\} \quad \text{where} \quad U(\mathrm{P}) = C(\mathrm{P})\mathrm{e}^{-\mathrm{j}\phi(\mathrm{P})}. \tag{5.18}$$

The generic projection lithography system in Figure 5.32 consists of several components, each of which we have discussed in earlier sections of this chapter. The light source and condenser lens form the illumination system that needs to deliver light to the mask with the specified intensity, uniformity, spectral characteristics and spatial coherence. Once the light passes through the mask, diffraction effects come into play. The objective or projection lens is required to collect as much of the diffracted light as possible and focus it onto the resist layer on the wafer. Generally, demagnification also takes place through the objective lens. In Figure 5.32, the reduction ratio is given by the Abbe sine condition that holds for any good imaging system, i.e.,

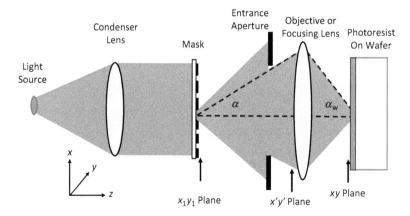

Figure 5.32 Generic projection lithography system. The shaded area inside the entrance aperture to the right of the mask illustrates the portion of the diffraction pattern that is collected by the projection lens and used to form the aerial image of the mask feature.

$$R = \frac{n_{\mathrm{w}} \sin \alpha_{\mathrm{w}}}{n \sin \alpha}, \tag{5.19}$$

where the w subscripts refer to the wafer side of the lens. Usually $R = 4$ in modern projection printers.

We will consider the mask to be transparent in some areas and completely opaque in others. For the chrome masks that are typically used today, this is a very good approximation, since the mask contrast is very high. Thus in the x_1y_1 plane, we can represent the transmission of the mask with a simple digital function:

$$t(x_1, y_1) = \begin{cases} 1 & \text{in clear areas,} \\ 0 & \text{in opaque areas,} \end{cases} \tag{5.20}$$

where $t(x_1, y_1)$ is the mask transmittance. If more advanced mask techniques such as phase shifting are used, then $t(x_1, y_1)$ will vary in both magnitude and phase, and will obviously be more complex than a simple digital function. We will consider such options in Section 5.7.

After the mask diffracts the light, it is described by the Fraunhofer diffraction integral in the far-field region. Thus, in the far-field diffraction region, the electric field intensity pattern of the light is given by [19, 21]

$$\mathcal{E}(x', y') = \int_{-\infty}^{+\infty} \int_{-\infty}^{+\infty} t(x_1, y_1) e^{-2j\pi(f_x x + f_y y)} \, dx dy. \tag{5.21}$$

In this expression, the terms f_x and f_y are known as the spatial frequencies of the diffraction pattern and are defined as

$$f_x = \frac{x'}{z\lambda} \quad \text{and} \quad f_y = \frac{y'}{z\lambda}, \tag{5.22}$$

where z is the distance from the mask to the lens. Equation (5.21) is simply the Fourier transform of the mask pattern. The light intensity pattern entering the objective lens can thus be calculated using well-known methods from the field of Fourier optics. In the shorthand of the Fourier transform, we can rewrite (5.21) as

$$\mathcal{E}(f_x, f_y) = F\{t(x_1, y_1)\}, \tag{5.23}$$

where F represents the Fourier transform. The terms f_x and f_y are simply scaled spatial coordinates in the $x'y'$ plane. The intensity distribution of the light is the square of the magnitude of the electric field, so that

$$I(f_x, f_y) = |\mathcal{E}(f_x, f_y)|^2 = |F\{t(x_1, y_1)\}|^2. \tag{5.24}$$

We now return to our generic lithography system in Figure 5.32. When the diffracted light passes the $x'y'$ plane, it enters the objective lens. Note that only a portion of the

Example

The Fourier transforms of simple shapes are generally described in standard texts on optical systems [19, 21]. Several examples that are relevant to lithography systems are illustrated here.

Long rectangular slit of width $w/2$:

$$\mathcal{E}(x') = F\{t(x)\} = \frac{\sin(\pi w f_x)}{\pi f_x}.$$

Square aperture of size $w/2$ by $w/2$:

$$\mathcal{E}(x'y') \propto \left(\frac{\sin(\pi w f_x)}{\pi f_x} \right) \left(\frac{\sin(\pi w f_y)}{\pi f_y} \right).$$

Circular aperture of radius b:

$$\mathcal{E}(x'y') \propto \frac{J_1(k\rho b/z)}{k\rho b/z} \text{ where } \rho = \sqrt{x_1^2 + y_1^2} \text{ (radial coordinate).}$$

The square diffraction pattern is simply the product of two slits perpendicular to each other. In the circle, J_1 is a Bessel function of the first kind. The intensities for each of these are shown in Figure 5.33 (using MATLAB). If a light detector were scanned over the $x'y'$ plane at some distance z from the mask, these computed images are what would be measured. Of course, these calculations assume that far-field diffraction holds. More generally, well-known computational methods exist for calculating the Fourier transform of arbitrarily shaped mask patterns, and this is the approach taken in simulators. We will consider a number of examples shortly.

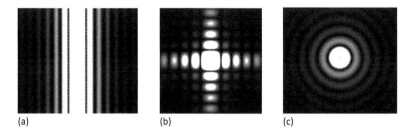

(a)   (b)   (c)

Figure 5.33 Far-field Fraunhofer diffraction patterns for (a) a slit, (b) a square and (c) a circle.

diffracted light is captured by the lens because of the entrance aperture or because of the limited physical size of the lens. We characterized this through the $\mathrm{NA} = \sin \alpha$ in our earlier discussion. The function of the objective lens is to reconstruct the diffraction pattern and to focus it on the wafer. Since the intensity pattern of the light at the $x'y'$ plane is simply the

Fourier transform of the mask pattern, the objective lens effectively needs to perform an inverse Fourier transform. This is, in fact, exactly what spherical lenses do. The lens can perform this operation only on the portion of the diffracted light that enters the lens, and in this sense we speak of the image that the lens forms as being "diffraction-limited."

We can characterize the portion of the diffracted light that the objective lens captures with a "pupil function" P defined as

$$P(f_x, f_y) = \begin{cases} 1 \text{ if } \sqrt{f_x^2 + f_y^2} < \dfrac{NA}{\lambda}, \\[2mm] 0 \text{ if } \sqrt{f_x^2 + f_y^2} > \dfrac{NA}{\lambda}. \end{cases} \tag{5.25}$$

Since, for small angles, the spatial frequencies f_x and f_y are given by

$$f_x = \frac{x'}{z\lambda} \cong \frac{\sin \alpha}{\lambda} = \frac{NA}{\lambda}, \tag{5.26}$$

The function P may be limited by the physical size of the objective lens, or by an aperture placed in front of the lens. Either way, the information passing through the objective lens is effectively low-pass-filtered by the pupil function P. Spatial frequencies higher than NA/λ are simply not passed on to the image on the wafer. The analogy in electric circuits, a sharp pulse passed through a low-pass filter which then ends up with rounded edges, may help to explain the optical degradation in the aerial image caused by P. A sharply defined object on the mask is imaged on the wafer with "smeared" edges, and sharp corners become rounded.

The objective lens now performs the inverse Fourier transform function on the portion of the diffracted light that it captures. Thus, at the image plane on the wafer (resist surface), the electric field associated with the light pattern is given by

$$\mathcal{E}(x,y) = F^{-1}\{\mathcal{E}(f_x,f_y)P(f_x,f_y)\} = F^{-1}\{F\{t(x_1,y_1)\}P(f_x,f_y)\}. \tag{5.27}$$

Finally, the intensity of the light distribution at the resist surface is simply given by

$$I_i(x,y) = |\mathcal{E}(x,y)|^2. \tag{5.28}$$

Equation (5.28) now represents the aerial image produced by the exposure system. Figure 5.34 summarizes the mathematical models that are used to describe the exposure system.

In real projection lithography systems, there are some additional issues that must be considered in calculating the aerial image. These include off-axis illumination, the use of a partially coherent light source, aberrations in the lens system, depth-of-focus issues and the finite thickness of the photoresist, which means that the aerial image must be determined over some volume, not just at the resist surface. All of these are typically modeled in modern lithography simulators like Athena and Victory Process (Silvaco) [33] and Sentaurus Lithography (Synopsys) [34]. We will describe in the following paragraphs the modifications needed in (5.28) in order to deal with these issues.

The first issues are off-axis illumination and partially coherent light sources. In both cases, the light that enters the mask does not have a direction normal to the mask. The result of this

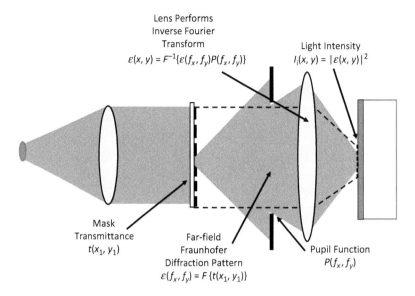

Figure 5.34 Mathematical models used to describe the projection exposure system.

(Figure 5.26) is that the position of the diffraction pattern is shifted in the $x'y'$ plane at the entrance to the objective lens. If the light enters the mask at an angle θ, then, in terms of the spatial frequencies f_x and f_y, the diffraction pattern is shifted by an amount $\sin\theta/\lambda$. The result then is that the electric field associated with the light pattern in the image plane is expressed as [21]

$$\mathcal{E}(x,y,f_x',f_y') = F^{-1}\left\{\mathcal{E}(f_x - f_x', f_y - f_y')P(f_x,f_y)\right\} \tag{5.29}$$

and

$$I_{\mathrm{i}}(x,y,f_x',f_y') = |\mathcal{E}(x,y,f_x',f_y')|^2 . \tag{5.30}$$

In the case of a partially coherent source, light enters the mask from a variety of angles. This is usually treated by superposition. The source is divided up into individual point sources and the intensities of the light at the image plane due to each point source are added. Expressed mathematically, if $S(f_x',f_y')$ is the function describing the source intensity versus position or angle, then [21]

$$I_{\mathrm{total}}(x,y) = \frac{\iint_{\text{source}} I(x,y,f_x',f_y') S(f_x',f_y')\mathrm{d}f_x'\mathrm{d}f_y'}{\iint_{\text{source}} S(f_x',f_y')\mathrm{d}f_x'\,\mathrm{d}f_y'}, \tag{5.31}$$

where I_{total} is the total light flux falling on the resist surface. Equation (5.31) thus allows the calculation of the aerial image for an arbitrary source illumination.

Aberrations or defects are always present in practical lens systems. They arise from manufacturing tolerances, from mechanical changes in a system during use (lenses out of alignment with each other), and from the fact that it is impossible to design a perfect lens that is completely free of aberrations. If these aberrations are known, they can also be incorporated in the modeling [21].

If the resist is not physically located in the plane of focus of the objective lens, further image degradation will occur. These effects can also be treated as a phase error in the wavefront emerging from the objective lens [19, 21]. Thus they can be treated mathematically in the same manner as lens aberrations. The interested reader is referred to the cited references, or to the user manuals for simulators like [33, 34].

We now turn to some examples from lithography simulators. We will use Silvaco's Athena, which includes Optolith as its lithography simulator. All of the models we have described above are implemented in this simulator. Figure 5.35 shows the optical system modeled in Optolith [2]. It is quite similar to the generic system we have discussed.

In the first example, in Figure 5.36, we use Optolith to calculate the aerial image of a technology that was state of the art around 1990 (see Figure 5.4). The feature size is 1 μm and the exposure system is an i-line projection printer with NA = 0.43. The color in Figure 5.36(a) corresponds to the calculated light intensity at each point in the aerial image. The features are fairly well resolved, although there is obvious rounding of sharp corners. In Figure 5.36(b), the calculated light intensity is plotted along a cutline at $y = 0$. Using the definition of MTF in (5.13), in this aerial image MTF = 1 since I_{MIN} is zero.

In the next example (Figure 5.37), we jump forward in time to a technology at the state of the art around 2005. The feature size now is 0.2 μm and the exposure system is a 193 nm KrF excimer laser projection printer with NA = 0.8. We used a more modern lithography tool and increased the numerical aperture so that smaller feature sizes can now be printed fairly well. The quality of the aerial image is roughly the same as the example in Figure 5.36. The MTF in this image is still approximately 1.

In Figure 5.38, the feature size is reduced to 0.1 μm with the same exposure system as in Figure 5.37. Clearly, the quality of this "subwavelength" aerial image is not great. The contact hole is barely resolved and the MTF has degraded to approximately 0.8. It is likely that the quality of this aerial image would not be acceptable in a manufacturing environment.

So, what can we do to improve the quality of the aerial image? If we look again at Figure 5.4, it seems that 193 nm lithography systems continue to be used today and that they are somehow printing features very much smaller than 0.1 μm. At least, in optical systems as we have

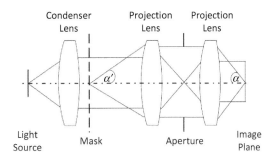

Figure 5.35 Optolith's lithography system model. See the Optolith user manual [2].

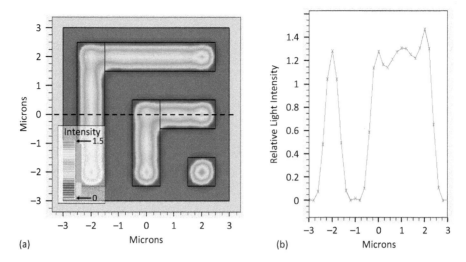

(a)

(b)

Figure 5.36 (a) Optolith simulation of the aerial image of two elbow shapes and a contact hole. The mask geometry is shown with the solid black lines. System: i-line (365 nm) stepper, NA = 0.43 and 1 µm feature sizes. (b) The intensity along a cutline at $y = 0$ (dashed line in (a)).

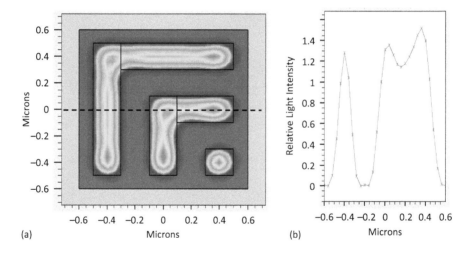

(a)

(b)

Figure 5.37 (a) Optolith simulation of the aerial image of two elbow shapes and a contact hole. The mask geometry is shown with the solid black lines. System: 193 nm stepper, NA = 0.8 and 0.2 µm feature sizes. (b) The intensity along a cutline at $y = 0$ (dashed line in (a)).

discussed them to date, the only "knobs" we have to turn are the NA (size of the lens) and perhaps spatial coherence. We will discuss these and other options in detail at the end of this chapter, since there obviously must be solutions if Figure 5.4 is to be believed.

The final simulation we will do here (Figure 5.39) increases the NA of the optical system to 0.93, which is the highest value currently available in simple "dry" optical systems [36]. We will

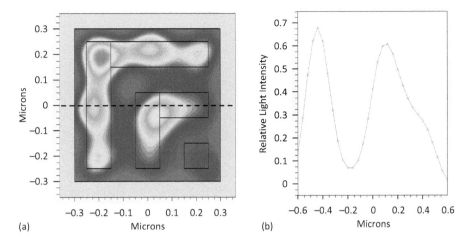

Figure 5.38 (a) Optolith simulation of the aerial image of two elbow shapes and a contact hole. The mask geometry is shown with the solid black lines. System: 193 nm stepper, NA = 0.8 and 0.1 μm feature sizes. (b) The intensity along a cutline at $y = 0$ (dashed line in (a)).

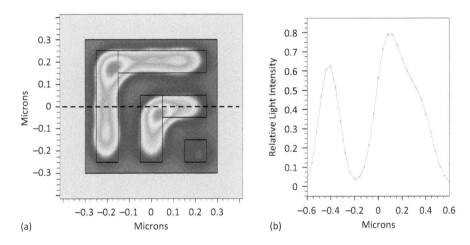

Figure 5.39 (a) Optolith simulation of the aerial image of two elbow shapes and a contact hole. The mask geometry is shown with the solid black lines. System: 193 nm stepper, NA = 0.93 and 0.1 μm feature sizes. (b) The intensity along a cutline at $y = 0$ (dashed line in (a)).

see later in the chapter that we can actually build systems with values NA > 1, but, for now, Figure 5.39 shows the impact of increasing NA to 0.93. Clearly, the aerial image is improved.

It should be clear from these examples that the last decade of progress illustrated in Figure 5.4 would not have happened without some breakthroughs in how lithography systems operate. We will return to this topic in Section 5.7 and discover how conceptual breakthroughs led to the ability to print aerial images with features 10× smaller than the wavelength of the light being used in the lithography tool.

Using (5.8) and (5.12), we estimated earlier in this chapter that the resolution and depth of focus of an EUV projection lithography tool should be on the order of 15–20 nm and ±10 nm, respectively. The resolution numbers are vastly superior to the 193 nm systems currently in widespread use because the wavelength of the EUV photons is 14× smaller. EUV systems are governed by the same physics as the 193 nm systems, and so the starting point for simulating the performance of EUV systems is Fraunhofer far-field diffraction, just as was the case above for the 193 nm systems.

Beyond the basic mathematics of Fraunhofer diffraction, there are a number of additional complications in predictive modeling for EUV systems. These areas are currently active research topics and we mention several of them here. First, there are several issues associated with the masks in EUV systems. These masks are reflective and we discussed them qualitatively in connection with Figure 5.6. In EUV systems, the photons incident on the mask arrive at an angle of 6 °, as shown in Figure 5.40. This results in shadowing effects, which can be visualized in the figure because the photons are reflected by the multilayer Mo/Si mirror and the absorber pattern has a finite thickness. In addition, not all of the photons are reflected by the mirror (typically 70%), so there is considerable energy absorbed by the mask. Finally, defects in the mirror structure can cause distortion or printing defects on the wafer. Modeling of all these effects is possible but generally requires full electromagnetic wave solutions [37, 38, 39]. Of course, all of the mirrors in EUV systems have the same physical structure as the mask, except that there is no absorber layer, so the same issues with energy absorption and defects exist in the mirrors as well.

Some of the commercially available lithography simulation tools do currently model EUV systems [34, 35], but these models are still being refined. It is to be expected that, as EUV tools

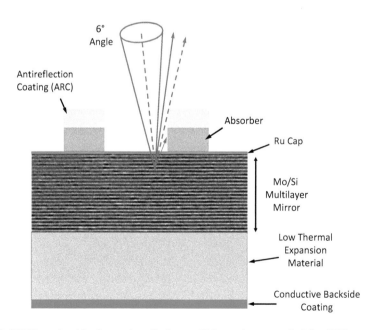

Figure 5.40 EUV mask with photon irradiation at 6° from the normal. After [37].

are introduced into manufacturing over the next few years, the simulation tools will become more robust and will include more of the effects that are unique to EUV systems.

5.5.2 Contact and Proximity Systems

Contact and proximity lithographic tools are still widely used today when their resolution limitations and higher defect densities, especially in the case of contact systems, are adequate for the intended application. They cannot, of course, be used for state-of-the-art silicon chips. Universities, R&D labs, applications such as sensors, LEDs and micro-electromechanical systems (MEMS) devices are typical examples of where these systems are used.

As we saw in Figure 5.31, when light passes through a mask, there are several regions beyond the mask in which the propagating diffracted light field can be approximated. In the previous section we considered far-field or Fraunhofer diffraction, which is an approximate solution that is useful for projection lithography systems. For contact and proximity systems, Fresnel diffraction is another approximation that may be useful depending on the mask feature sizes and on the distance between the mask and the resist-covered wafer. Commercially available simulators generally offer a Fresnel diffraction module in addition to a Fraunhofer diffraction module. We will outline here the mathematics behind the Fresnel approximation and then discuss some simulation results using this approximation.

Consider Figure 5.41. We have seen that g is the gap that exists in a projection printing system. In a contact printing system, g approaches zero. The diffracted light beyond the mask is described mathematically by the Fresnel diffraction integral if g and the mask features satisfy the constraints shown in Figure 5.31. The Fresnel equivalent to (5.21) is

$$\mathcal{E}(x,y,z) = \frac{1}{j\lambda z}\exp\left\{2\pi j\frac{z}{\lambda}\right\} \int\limits_{-\infty}^{\infty}\int\limits_{-\infty}^{\infty} t(x_1,y_1)\exp\left\{-\pi j\frac{(x-x_1)^2 + (y-y_1)^2}{\lambda z}\right\}\mathrm{d}x\mathrm{d}y. \qquad (5.32)$$

In this expression, $t(x_1, y_1)$ is defined exactly as it was in (5.20), and j is again the imaginary unit.

In contact and proximity lithography systems, there is no pupil usually and there is no focusing or objective lens, so simulating the aerial image basically comes down to solving (5.32). Analytic solutions to this equation are possible only in a few rare cases, so computational approaches are generally used. There are a number of such approaches to doing this, including

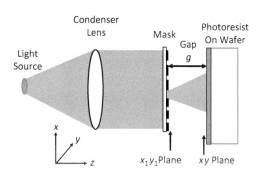

Figure 5.41 Generic proximity lithography system. The shaded area to the right of the mask illustrates the diffracted light that is incident on the resist.

using convolution and Fourier transforms [31, 33, 34]. In general, these approaches are more complex than the Fourier transform approach described earlier for far-field Fraunhofer diffraction. The interested reader is referred to the references for a discussion of these mathematical approaches. The Optolith simulator we will use below uses a beam propagation method to solve (5.32) [33].

Example

In Figure 5.31, we saw that Fresnel diffraction is a useful approximation to a full solution of the wave equations provided that $z \gg W$. Fraunhofer diffraction becomes the appropriate approximation when $z \gg \pi W^2 / \lambda$. So if we consider a typical proximity printing lithography tool that uses a Hg vapor lamp source, λ would be 436 nm if we consider g-line light. As discussed earlier, proximity printers often use multiple wavelengths from the Hg lamp source but we will consider g-line illumination in the examples below.

Assuming a feature size of 10 μm, Fresnel diffraction applies when $g \gg W = 10\,\mu m$ and $g \ll \pi W^2 / \lambda = 720\,\mu m$.

In order to illustrate the resolution capabilities of these kinds of systems, we include below several simulations using the Optolith simulator [33]. We will use the same simple mask pattern that we used earlier for projection system simulations. The first example is shown in Figure 5.42 where 10 μm features are imaged in a proximity printer using g-line light. The gap g in this case is 50 μm, which is well within the range where Fresnel diffraction should be appropriate. The aerial image shows that the features are quite well resolved. The intensity plotted along the $y = 0$ cutline illustrates some of the "ringing" that was conceptually illustrated in Figure 5.21.

Figure 5.43 shows a second simulation of the same mask pattern, this time with a 100 μm gap between the mask and the wafer. This is still well within the gap range appropriate for Fresnel diffraction, although a 100 μm gap is at the upper end of gap spaces typically used in modern proximity printing tools.

There is an important observation regarding Figures 5.42 and 5.43. The fact that the image quality is acceptable even when the gap increases by 50 μm implies that the depth of focus of proximity printing tools is very large compared to projection printers (see the example earlier in connection with Figure 5.24). There are manufacturing applications in which quite thick layers of photoresist are used, primarily in the MEMS industry. Provided the printed geometries are micron-scale, proximity printers do a very good job of exposing these thick resist layers, something high-resolution projection printers would find much more difficult.

In order to explore the limits of proximity printing systems, we now reduce the feature size to 1 μm in our final simulations. With this feature size, Fresnel diffraction is appropriate if $z \gg 1\,\mu m$ and $z \ll 7\,\mu m$ using the same bounds as we used above. Figure 5.44 shows the simulated results for 1 μm features and a gap of 5 μm. Clearly, the 1 μm features are not resolved with a 5 μm gap.

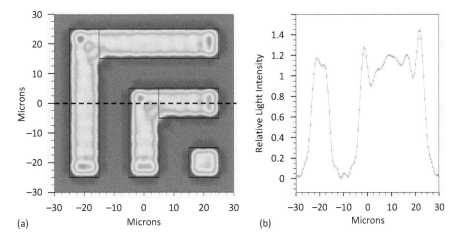

Figure 5.42 (a) Optolith simulation of the aerial image of two elbow shapes and a contact hole. The mask geometry is shown with the solid black lines. System: g-line (436 nm) stepper, 10 μm feature sizes and gap g = 50 μm. (b) The intensity along a cutline at y = 0 (dashed line in (a)).

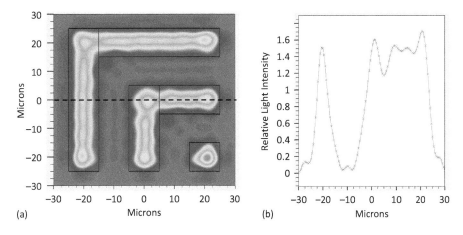

Figure 5.43 (a) Optolith simulation of the aerial image of two elbow shapes and a contact hole. The mask geometry is shown with the solid black lines. System: g-line (436 nm) stepper, 10 μm feature sizes and gap g = 100 μm. (b) The intensity along a cutline at y = 0 (dashed line in (a)).

The only options we have to improve the resolution are to reduce the exposing wavelength or to reduce the gap size. Figure 5.45 shows examples of both approaches. Reducing the gap from 5 μm to 2 μm (Figure 5.45(a)) clearly produces a much better aerial image, one that quite likely would be usable in manufacturing. Similarly, reducing the exposing wavelength to the DUV 193 nm wavelength used in modern projection systems also produces a much better image, even with the gap g left at 5 μm (Figure 5.45(b)). It should be pointed out that gaps as small as 2 μm are at the limits of what is generally possible in modern proximity printers.

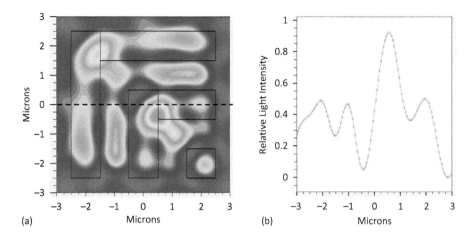

(a)

(b)

Figure 5.44 (a) Optolith simulation of the aerial image of two elbow shapes and a contact hole. The mask geometry is shown with the solid black lines. System: g-line (436 nm) stepper, 1 μm feature sizes and gap $g = 5$ μm. (b) The intensity along a cutline at $y = 0$ (dashed line in (a)).

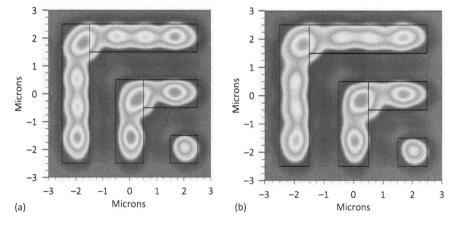

(a)

(b)

Figure 5.45 Optolith simulations of the aerial image of two elbow shapes and a contact hole with 1 μm feature sizes. The mask geometry is shown with the solid black lines. (a) The gap is reduced to 2 μm with a g-line (436 nm) stepper. (b) The exposing wavelength is reduced to 193 nm with the gap g at 5 μm.

The final observation regarding Figure 5.45 is that the improvements seen in reducing the gap g suggest that hard contact lithography systems could actually achieve quite good resolution, and this is indeed the case. However, the simulation tool used in these examples uses the Fresnel diffraction approximation and so it really cannot be used to simulate aerial images that contact aligners might produce.

To summarize the discussion in the last two sections, Figure 5.46 illustrates the diffraction pattern propagating away from a small slit. The sketches are the light intensity versus position that would be measured at each plane. In the region very close to the slit, the shape approximates a shadow of the slit but requires a full solution of the wave equations to calculate

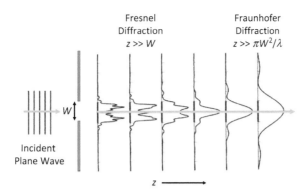

Figure 5.46 Summary of diffraction regimes.

quantitatively. For distances greater than the size of the slit ($z > W$), Fresnel diffraction provides a good approximation for calculating the image of the slit. For larger distances, $z > \pi W^2/\lambda$, Fraunhofer diffraction provides a good approximation.

5.6 Modeling and Simulation of Photoresists (Calculating and Developing the Latent Image)

5.6.1 Optical Intensity Pattern in the Photoresist

In Section 5.5 we discussed models and simulation methods for calculating the aerial image produced by modern lithography tools. The aerial image is the optical intensity in the air immediately above the surface of the photoresist. This is $\{\mathcal{E}_i(x,y)\}^2$, as we saw in (5.28). The second step in simulating the performance of lithography tools is to calculate the optical intensity $\{\mathcal{E}_{PR}(x,y,z)\}^2$ pattern through the photoresist layer, as illustrated in Figure 5.47. Once this is done, we can then calculate the chemical response of the resist and therefore the three-dimensional latent image produced in the resist by the incoming photons. Note that, while the aerial image is 2D (x and y), the resist light intensity and the latent image are 3D (x, y and z).

Simulation tools use different mathematical approaches to calculating $\{\mathcal{E}_{PR}(x,y,z)\}^2$. The simulation tool we will use (Optolith [33]) computes the field in the photoresist by numerically solving the Helmholtz equation, which is a time-independent form of the wave equation. The Helmholtz equation is often used in electromagnetic problems and is

$$\nabla^2 \mathcal{E} + k^2 n^2 \mathcal{E} = 0 \,, \tag{5.33}$$

where k is the wavenumber and $n(x, y)$ is the complex index of refraction of the various materials.

Because the incoming photons interact with the photoresist components by creating PAC molecules or PAG molecules, as we discussed earlier in Section 5.2.4, the resulting field and therefore light intensity is generally a function not only of x, y and z, but also of time during the

Figure 5.47 Representation of the electric field distribution in the photoresist layer.

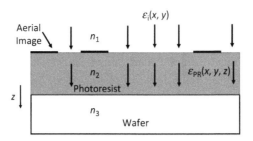

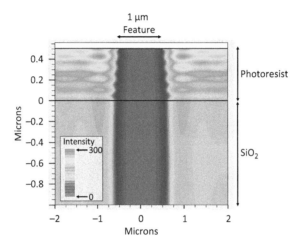

Figure 5.48 Cross-sectional view of calculated optical intensity pattern for a 1 μm feature exposed in an i-line lithography system with NA = 0.43 using the Optolith simulator [33].

exposure process. This is certainly true in DNQ resists that have fairly low quantum efficiencies. In chemically amplified resists used for DUV and EUV systems, it is often the case that the light intensity in the resist is fairly constant during the exposure, and this simplifies the situation. In DNQ resists, $n(x, y)$ is also often a function of time because this parameter is used to model some of the chemistry occurring in the resist. So the calculation of $\{\mathcal{E}_{PR}(x, y, z)\}^2$ is normally done in multiple time steps for DNQ resists during the exposure simulation, with material parameters updated after each iteration. The details of how this is done are described in the user manuals for the various simulation tools [33, 34, 35].

An example of a calculated optical intensity pattern is shown in Figure 5.48. The structure is a 0.5 μm layer of AZ 1350J resist on an SiO_2 substrate that is exposed in an i-line lithography tool with NA = 0.43. The use of an SiO_2 substrate is simply because this is the most common situation in device processing. Of course, underneath the SiO_2 would be other materials, including the silicon or semiconductor substrate, but these are not needed in the simulation. The photons are incident on the left and right sides. The central 1 μm feature is supposed to be unexposed.

There are several features of note in this simulation. First, note that the optical intensity decreases with depth into the photoresist. Second, note that there seem to be periodic increases and decreases of intensity in the vertical direction. These are actually due to the fact that some of the light that penetrates to the photoresist/SiO$_2$ interface reflects back up into the photoresist, setting up a standing wave pattern as the incoming and outgoing waves interfere. We will discuss this issue more carefully in Section 5.6.3 but will ignore it in this section. Finally, note that the light penetrates into the SiO$_2$ substrate beyond the photoresist. These photons obviously do not help to expose the resist.

5.6.2 Exposure and Latent Image in the Photoresist

We will consider in this section the chemistries and the models that are used to simulate exposure in simple resists (non-chemically amplified, novolac-based resists used for g-line and i-line), and also in chemically amplified resists used for DUV and EUV lithography systems. These models are used to provide the material parameters needed along with the photon intensity $\{\mathcal{E}_{PR}(x,y,z)\}^2$ to calculate the latent image in the resist at the end of the exposure process.

The light incident on a DNQ photoresist is primarily absorbed by the PAC component of the resist. The PAC is assumed to be uniformly distributed throughout the resist. The basic equation describing the light absorption is known as Lambert's law and is given by

$$\frac{\mathrm{d}I}{\mathrm{d}z} = -\alpha I, \tag{5.34}$$

where z is the direction into the resist as shown in Figure 5.47 or downwards in Figure 5.48. This equation basically says that the probability that a photon will be absorbed is proportional to the light intensity, with α, the absorption coefficient, as the proportionality constant. If α is a constant, integration of this equation results in (5.35), which simply says that the light intensity falls off exponentially with distance z into the resist with a decay length of α, that is,

$$I = I_0 \exp(-\alpha z). \tag{5.35}$$

This simple relationship will turn out to hold in chemically amplified resists (discussed shortly); but in DNQ resists, α will turn out to be related to the concentration of the PAC that is unexposed in the resist at a particular time and location. Thus in DNQ resists, α is not a constant, but rather is a function of position and time. At any time during the exposure, the light intensity profile in the resist will be given by

$$I(z) = I_0 \exp\left(\int_0^z \alpha(z')\mathrm{d}z'\right), \tag{5.36}$$

where the integral in the exponent is the absorbance of the resist down to a depth z.

Before proceeding further, we need to consider physically why the absorption coefficient α in DNQ resists depends on time and position within the resist. Imagine in connection with Figure 5.47 that the exposure process has just begun. The PAC concentration is assumed uniform throughout the resist. As PAC molecules near the resist surface absorb photons, those photons are not available for exposing the deeper layers of the resist. Thus the light concentration falls off with depth and the resist is first exposed near the top surface.

Fortunately, a process known as "bleaching" occurs in most DNQ resists. As the resist is exposed, the PAC is altered, absorbing less and less light, and therefore it becomes more and more transparent. Thus, as the top layers of the resist are exposed, they transmit more of the light to the deeper layers, which are subsequently exposed. The bleaching process is not surprising since, as the PAC component of these resists reacts and converts to carboxylic acid, more of the light will pass through to the deeper layers.

We now need to relate α to the material properties of the resist. If the PAC component of the resist is dilute, then the absorption coefficient of the resist is simply

$$\alpha_{\text{resist}} = \alpha_{\text{PAC}}[\text{PAC}]\,, \tag{5.37}$$

where [PAC] is the concentration of the PAC constituent in the resist and α_{PAC} is the absorption coefficient of the PAC material. Equation (5.37) simply says that the resist absorption is determined by the amount of PAC in the resist. The dilute approximation required for (5.37) has been shown to hold for practical resist formulations [40].

DNQ resists consist of several constituents, as we saw in Section 5.2.4. In general, such resists will have a resin R (usually novolac), a photoactive component PAC, a solvent S and, as the exposure proceeds, exposure products P (principally the carboxylic acid produced by the PAC). More generally, then, (5.37) can be written as

$$\alpha_{\text{resist}} = \alpha_{\text{PAC}}[\text{PAC}] + \alpha_{\text{R}}[\text{R}] + \alpha_{\text{S}}[\text{S}] + \alpha_{\text{P}}[\text{P}]\,. \tag{5.38}$$

Since the exposure products result from the PAC, we may write that

$$[\text{P}] = [\text{PAC}]_0 - [\text{PAC}]\,. \tag{5.39}$$

Following the approach of [28], we rewrite (5.38) in the following form:

$$\alpha_{\text{resist}} = Am + B\,, \tag{5.40}$$

where

$$\left.\begin{aligned} A &= (\alpha_{\text{PAC}} - \alpha_{\text{P}})[\text{PAC}]_0, \\ B &= \alpha_{\text{P}}[\text{PAC}]_0 + \alpha_{\text{R}}[\text{R}] + \alpha_{\text{S}}[\text{S}], \\ m &= [\text{PAC}]/[\text{PAC}]_0. \end{aligned}\right\} \tag{5.41}$$

In (5.40) A and B are experimentally measurable parameters for a given photoresist, and are known as the first two Dill resist parameters [28]: A is the absorption coefficient of the bleachable components of the resist and B is the absorption coefficient of the non-bleachable components of the resist. We will describe how these parameters can be measured for a given resist shortly. Note that when the resist is unexposed, $m = 1$ and $\alpha_{\text{resist}} = A + B$. When the resist is fully exposed, $m = 0$ and $\alpha_{\text{resist}} = B$.

We can now substitute (5.40) into (5.34) and express the light intensity through the resist layer as

$$\frac{\mathrm{d}I}{\mathrm{d}z} = -(Am + B)I \,. \tag{5.42}$$

Unfortunately, solving this equation is complicated by the fact that m is a function of time and position, so that (5.42) really should be written as

$$\frac{\mathrm{d}I}{\mathrm{d}z} = -\{Am(z,t) + B\}\,I, \tag{5.43}$$

where m is the fraction of the PAC component of the resist that is unexposed at time t. If we again assume first-order reaction kinetics, then

$$\frac{\mathrm{d}m}{\mathrm{d}t} = -CIm \,. \tag{5.44}$$

This equation states that the exposure rate of the PAC is proportional to the remaining unexposed PAC concentration and to the light intensity. The constant of proportionality is C, which is the third Dill resist parameter.

Equations (5.43) and (5.44) are coupled equations that must be solved simultaneously. The photon intensity I is also in these equations, and we know that it depends on time and position, so the third equation that must be solved is (5.33), which gives $\{\mathcal{E}_{\text{PR}}(x,y,z)\}^2$, which is I. Simulators that are available today [33, 34, 35] solve this system of equations to calculate the time evolution of the exposure of the resist. An iterative procedure is used, which basically operates as follows.

At $t = 0$, $m = 1$ and the light intensity pattern can be calculated in the resist using (5.33). This intensity distribution is then used in (5.44) to calculate the exposure versus position. The result is $m(x,y,z,t_0)$ in the general three-dimensional case. The result is then used in (5.43) and (5.33) to calculate the light intensity for the next time step. This result is again used in (5.44) to calculate the new light intensity after this next time step. Through this iterative procedure, the time evolution of the exposure process is simulated. The overall result is that the aerial image produced by the optical system is converted into a latent 3D image in the photoresist. We will show and discuss some examples shortly.

Calculation of the latent image in the resist requires knowledge of the Dill resist parameters A, B and C. These parameters can be measured experimentally for a particular resist material, as illustrated in Figure 5.49. In their series of classic papers in 1975, Dill and his coworkers showed that a simple measurement of the time-dependent light transmission through an exposing resist film could be used to extract the A, B and C parameters.

As illustrated in Figure 5.49, light at the exposing wavelength is used to illuminate a resist-coated substrate. In the simplest case, a transparent substrate is chosen to have the same index

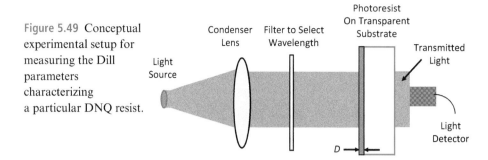

Figure 5.49 Conceptual experimental setup for measuring the Dill parameters characterizing a particular DNQ resist.

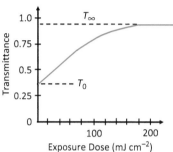

Figure 5.50 Typical experimental result from a measurement like that shown in Figure 5.49.

of refraction as the resist, so that there are no reflections at the resist/substrate interface. An antireflection coating is used on the backside of the substrate to avoid reflections at that interface. A typical measurement result is shown schematically in Figure 5.50. For this simple case, Dill *et al.* [28] showed that

$$A = \frac{1}{D} \ln\left(\frac{T_\infty}{T_0}\right), \tag{5.45}$$

$$B = -\frac{1}{D} \ln(T_\infty), \tag{5.46}$$

$$C = \frac{A+B}{AT_0(1-T_0)T_{12}} \left.\frac{\mathrm{d}T}{\mathrm{d}E}\right|_{E=0}, \tag{5.47}$$

where

$$T_{12} = 1 - \left(\frac{n_{\text{resist}} - 1}{n_{\text{resist}} + 1}\right)^2. \tag{5.48}$$

Here T_0 is the transmitted light intensity at the start of the exposure, T_∞ is the transmitted light intensity at the end of the exposure, $\mathrm{d}T/\mathrm{d}E$ is the slope of the red curve in Figure 5.50 at $E = 0$, D is the resist thickness and T_{12} is the transmittance through the air/resist interface, with n_{resist}

the index of refraction of the resist. More sophisticated parameter extraction methods can be used if the experimental situation is not quite as ideal as Figure 5.49 [28, 31].

Unlike DNQ resists, chemically amplified resists (CARs) require two chemical reactions in order to change their solubility in the developer. The first reaction caused by incoming photons creates the PAG molecules and stores a latent image of the photon intensity pattern in the resist. The second reaction occurs during the post-exposure bake (PEB) in which the PAG molecules diffuse and react with blocking molecules that make the resist insoluble in the developer. These reactions convert the blocking molecules into a soluble form (in a positive resist) and are catalytic, since the PAG molecules are not used up during this reaction. Thus, as we saw earlier in this chapter, CARs can have quantum efficiencies larger than 1, and as a result require lower exposure doses than do DNQ resists. For readers interested in the details of the chemical reactions occurring in CARs both during exposure and during the PEB, a good reference is [21].

Chemically amplified DUV resists consist of a polymer resin (blocked to make it insoluble in the developer in the case of a positive resist), a PAG and other additives. For most CARs, the bleachable component of the resist is negligible, and so the light intensity in the resist is independent of time during the exposure. This means that the light intensity pattern resulting from (5.33), which is a function $f(x, y, z)$, only needs to be calculated once rather than at each time step as in the case of DNQ resists. The kinetics of the exposure reaction are assumed to be first order, i.e.,

$$\frac{d[PAG]}{dt} = -CI[PAG],$$
(5.49)

where [PAG] is the time- and position-dependent PAG concentration, I is the exposure intensity and C is the exposure rate. Note the similarity to (5.44). If I is constant during the exposure, then

$$[PAG] = [PAG]_0 e^{-CIt} .$$
(5.50)

Since the acid concentration [H] results directly from the PAG exposure, [H] is given by

$$[H] = [PAG]_0 (1 - e^{-CIt}),$$
(5.51)

where, in general, [H] and I will be functions $f(x, y, z)$. If the light intensity is also a function of time, then an iterative technique like that described earlier for DNQ resists can be used to calculate $[H](x, y, z, t)$, but this is generally not necessary in CARs. The aerial image is thus converted into a latent 3D image in the resist, which is "stored" as $[H](x, y, z)$ at the end of the exposure.

The fact that CARs do not bleach means that the experimental setup shown in Figures 5.49 and 5.50 does not work to extract experimental information about the parameter C in (5.51). There are other approaches to determining this value. The interested reader is referred to [21]. The final point to be made about CAR exposure is that, since bleaching does not occur, if the light intensity is to be reasonably uniform through the resist thickness, then C must be small enough that many of the photons actually propagate completely through the resist without interacting with a PAG molecule. If the material below the resist is reflective, this can set up standing waves in the resist as the incoming and outgoing waves interfere. We will consider this more carefully in Section 5.7.3, but these effects can be mitigated by placing an antireflection coating (ARC) below the resist.

An example of the exposure process is shown in Figure 5.51 using the Optolith simulator [33]. In this example, a 1 μm feature is exposed using an i-line lithography tool. The three simulations show the PAC concentration in the i-line resist (AZ 1350J) at various times during the exposure process. The substrate in these simulations is SiO$_2$, as it was in Figure 5.48. The simulations in Figure 5.48(a)–(c) are after 50 mJ cm^{-2} (25% of the full exposure dose), after 100 mJ cm^{-2} (50% of the full dose) and after 200 mJ cm^{-2} (full exposure), respectively. Note that, while the colors are similar in all three simulations, the scale of remaining PAC concentration changes, so that in Figure 5.48(c) the PAC concentration everywhere in the exposed region is approaching zero. The progression of the exposure from top to bottom in the resist layer is evident in each of the simulations, as is the presence of standing wave effects due to reflections from the photoresist/SiO$_2$ interface.

5.6.3 Post-Exposure Bake (PEB)

We saw in the photoresist process flow in Figure 5.30 that, following resist exposure, the next step is usually a post-exposure bake (PEB). In DNQ resists, this is intended to smooth out effects due to standing waves before development. It has been found that a simple thermal bake between exposure and development can have a dramatic effect on standing wave effects [41]. Typically, this bake is a few minutes at 100–125 °C (similar to the resist pre-bake step).

The effect of the PEB on DNQ resist chemistry is usually modeled as a simple diffusion process. The exposed PAC (P in our terminology below) is assumed to be able to diffuse in the resist with some diffusivity given by

$$D_P = D_0 \, \exp(E_A/kT) \,. \tag{5.52}$$

Figure 5.51 Simulation cross-section of the exposure of a 1 μm feature in 0.5 μm AZ 1350J resist on an SiO$_2$ substrate using an i-line stepper and the Optolith simulator [33].

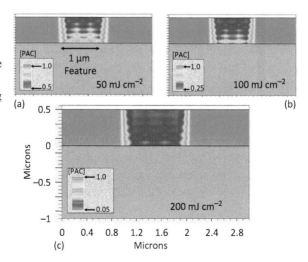

This is of the same form as dopant diffusion coefficients that we will discuss in Chapter 7. The effect of the PAC diffusion is simply to "smear out" the standing wave effects. Such a diffusion process is generally modeled in lithography simulators. However, this is more complicated than might appear, first, because D_P has not been measured for most resists, and second, because it is likely that the diffusivity is not as simple as (5.52) might suggest. For example, it is believed that D_P is concentration-dependent.

In some lithography simulators like Optolith, default or user-specified values for D_P are used along with (5.52). The PEB is then modeled as a diffusion process. It is also possible in these simulators to take an even simpler approach because of the lack of D_P data. In this approach, the distance over which the PAC "smearing" occurs during the bake is simply given by a diffusion length,

$$\sigma = \sqrt{2D_P t_{bake}}, \tag{5.53}$$

where t_{bake} is the bake time. The diffusion length σ has units of distance and is generally a user-supplied parameter in simulators. A typical value is $\approx 0.05\ \mu m$ (50 nm) [33].

An example of such a PEB simulation in a DNQ resist is shown in Figure 5.52. This is the same structure as simulated in Figure 5.51. In Figure 5.52(a) is the same output for the fully exposed resist as shown in Figure 5.51. In Figure 5.52(b) that structure has been subjected to a 45 s, 115 °C PEB. The simulation uses default parameters for AZ 1350J photoresist and models the PEB as a diffusion process. The smearing out of the standing wave pattern in the exposed photoresist is apparent in the simulation in Figure 5.52(b). Note particularly the edges of the 1 μm wide exposed region. The "scalloped" edges that appear before the PEB can cause non-vertical or non-straight edges in developed photoresist images, which can cause process issues when the photoresist is subsequently used to mask implants or mask etching processes. The PEB largely eliminates these issues. We will discuss standing waves more fully in Section 5.7.3.

In DUV resists, the PEB is a critical step because it completes the exposure process by driving the reaction between the PAG molecules and the blocking molecules in the polymer chains. In positive DUV resists, this reaction unblocks the polymer chain, making it soluble in the

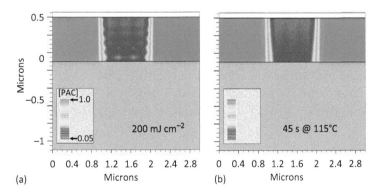

Figure 5.52 Simulation cross-sections of the post-exposure bake (PEB) of a 1 μm feature in 0.5 μm AZ 1350J resist on an SiO$_2$ substrate using an i-line stepper and the Optolith simulator [33]: (a) before PEB and (b) after a 45 s, 115 °C PEB.

developer. In negative resists, the PAG drives a crosslinking reaction that makes the resist insoluble in the developer. In either case, the acid molecules are not consumed by the reactions and hence, to first order, [H], given by (5.51), remains constant during the PEB.

If [M] represents the concentration of the reactive sites on the resist polymer chains, then

$$\frac{d[M]}{dt} = -C[M][H] , \qquad (5.54)$$

where C is the reaction rate constant. This expression assumes that the reaction between the acid H and the reactive sites M is first order. During the PEB, M begins with a starting concentration $[M]_0$ and decreases with time:

$$[M]_0 - [M] = [M]_0(1 - \exp(-C[H]t)). \qquad (5.55)$$

In general, [H] and therefore [M] are functions $f(x, y, z, t)$. The PEB provides the thermal energy for this reaction to take place because C in (5.54) is generally exponentially dependent on temperature.

The second process driven by temperature is the diffusion of the acid molecules, since they must physically move from reactive site to reactive site to complete the resist exposure (recall Figure 5.13). The diffusivity D_H of the acid molecules generally depends exponentially on temperature, as in (5.52). But it may also depend on the acid concentration, making the modeling problem more difficult. Simulation of the PEB effects on DUV resists would look similar to Figure 5.52, although the degree of exposure would increase during the PEB rather than just the smearing that Figure 5.52 illustrates.

5.6.4 Photoresist Developing

A number of models have been used to describe photoresist developing. The earliest models came from Dill and his coworkers and were empirically based [28]. Additional models have also been developed [31]. In all cases, the process is basically described as a surface controlled etching process. The developer solution is assumed to etch the surface of the resist isotropically at a point (x, y, z) and at a rate that is determined by the resist inhibitor concentration at that point. The photoactive compound (PAC) in DNQ resists or the blocked polymers in DUV resists are insoluble in the developer, whereas, when the PAC is converted to carboxylic acid (P) in DNQ resists, or when the PAG unblocks the polymer chains in DUV resists in the exposure process, the resist becomes quite soluble in the developer. We will describe the developing process below in terms of DNQ resist parameters. However, the same models apply to DUV resists.

The development or etching rate at point (x, y, z) thus depends on the [P] or [PAC] at that point. This is exactly what is calculated by the exposure models described in the previous section, and what is shown in the examples in Figures 5.51 and 5.52. What is needed, then, is a mathematical relationship between [P] or [PAC] and the development rate.

The first such relationship was described by Dill *et al.* [28] and is implemented in simulators today typically in a form like the following expression, which is used in Optolith [33]:

$$R(x,y,z) = \begin{cases} \exp(E_1 + mE_2 + m^2E_3) & \text{if } m > 0.4, \\ \exp(E_1 + 0.4E_2 + 0.16E_3) & \text{if } m \leq 0.4, \end{cases} \quad (5.56)$$

where $R(x, y, z)$ is the local development or etching rate (µm min^{-1}), $m(x, y, z)$ was defined in (5.41) and is the local [PAC] after exposure, and E_1, E_2 and E_3 are empirical parameters obtained from fitting (5.56) to experimental data for a particular resist.

Another development model is due to Mack [31]. This model is based on a physical model that involves diffusion of the developer from the solution to the resist surface, chemical reaction of the developer with the resist, and finally diffusion of the reaction products from the surface back into the liquid developer. Mathematically, it is expressed as

$$R(x,y,z) = R_{\max} \frac{(a+1)(1-m)^n}{a+(1-m)^n} + R_{\min}, \quad (5.57)$$

where $R_{\max}$ is the etch rate of fully exposed resist ($m = 0$), $R_{\min}$ is the etch rate of unexposed resist, and a and n are constants that are determined experimentally for a given resist material.

Commercially available simulators generally implement a number of models for resist development, including the two described above. The time evolution of the developing resist profile is calculated by setting up a two- or three-dimensional grid in which each grid point in the resist has a specific value of $m(x, y, z)$. The developing pattern is then allowed to evolve over time, with the developer moving into the resist at a local rate determined by the development model.

Figure 5.53 shows an example using Optolith of the development of a resist pattern. The structure is the same as in Figure 5.52(b), which includes the post-exposure bake. The Dill development model was used in the simulation with default model parameters for AZ 1350J resist. The simulation in Figure 5.53(a) is part way through the development cycle. The simulation in Figure 5.53(b) shows the structure after the develop cycle is complete. Note in Figure 5.53(b) that the developed image is actually about 1.15 µm wide (dashed lines) rather than the designed 1.0 µm. This simulation used a large exposure dose (200 mJ cm^{-2}) and a long development time (60 s), which resulted in some degree of overexposure. In a real manufacturing environment, additional

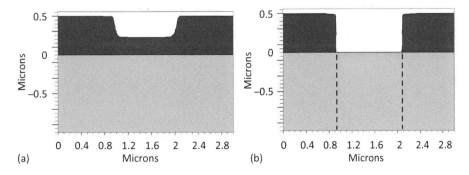

Figure 5.53 Simulation cross-sections of the develop cycle of a 1 µm feature in 0.5 µm AZ 1350J resist on an SiO$_2$ substrate using an i-line stepper and the Optolith simulator [33]: (a) part way through and (b) on completion.

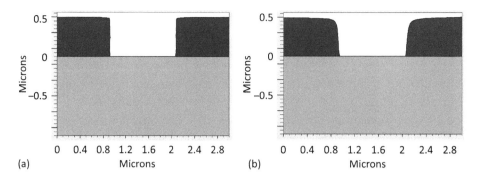

Figure 5.54 Simulation cross-sections of the post-bake (5 min at 125 °C) of a 1 μm feature in 0.5 μm AZ 1350J resist on an SiO$_2$ substrate using an i-line stepper and the Optolith simulator [33]: (a) before and (b) after the post-exposure bake.

simulations could be run at smaller doses and development times to better optimize the final result. Alternatively, a better resolution lithography tool could be used that would produce a sharper aerial image with less grayscale at the edges of the 1 μm feature.

5.6.5   Photoresist Post-Exposure Bake

As shown in Figure 5.30, the final step in a typical photoresist process is a post-exposure bake, typically done at 100–140 °C for 10–30 min. This step hardens the resist by evaporating any remaining solvents and also improves adhesion to underlying materials. Since the resist is often used to mask a plasma etching step, or to block an ion implantation step, this hardening process is important in improving the robustness of the resist. The resist will generally flow somewhat during this post-exposure bake, rounding the edges of the profile. Obviously, the degree of flowing depends on the time and temperature of the post-exposure bake, as well as on the resist material properties.

Lithography simulators generally model the post-exposure bake process using models that are similar to glass reflow models. The interested reader is referred to the user manuals of these simulation tools. Figure 5.54 shows an example of the simulated effects of the post-exposure bake on the resist profile. In this case, the developed resist structure from Figure 5.53 was post-exposure baked at 125 °C for 5 min. The rounding of the corners and general smoothing of the profile is evident.

5.7   Resolution Enhancement Methods

Given what we have discussed to this point regarding modern lithography systems, perhaps the most interesting question we have not yet answered is how did the feature size reduction shown in Figure 5.4 (or more accurately in Figure 1.21) after the year 2000 actually occur? For close to two decades, lithography has been based on 193 nm exposure systems, and simulations such as Figure 5.39 would suggest it might be extremely difficult to go much below 0.1 μm (or 100 nm) features with these systems. Yet Figure 5.4 shows that, today, features more like 20 nm are being routinely produced in high-volume manufacturing. How can this be?

We will first consider in the sections below a variety of improvements to 193 nm projection systems and photoresist materials (in Sections 5.7.1–5.7.3) that have made some of this progress possible. Then, in Sections 5.7.4–5.7.7 we will discuss several conceptual breakthroughs that account for most of the feature size reduction in the past 20 years. The key conceptual breakthrough was the following: the mask does not have to look like the pattern we are trying to print! We will explain this in detail below.

5.7.1   Off-Axis and Köhler Illumination

In our discussion of partially coherent light in connection with Figure 5.26, we saw that, in modern lithography systems, the light waves illuminating a mask are generally not all perpendicular to the mask. This is usually referred to as "off-axis illumination" and it can actually be a good thing rather than a problem.

Consider Figure 5.55. In the upper part, coherent or normally incident light is used, and some of the diffracted light is lost because of the finite size of the projection lens, or because of the presence of an entrance aperture. As we have seen, this degrades the quality of the aerial image. If off-axis illumination is used, as shown in the lower part of Figure 5.55, some of the diffracted light lost in the normally incident case will be captured by the lens. The diffraction patterns shown on the right of Figure 5.55 correspond to a slit pattern on the mask (Figure 5.33). The off-axis illumination shifts the diffraction pattern within the objective lens entrance aperture so that some of the higher-order diffraction information is now captured by the lens. If an illumination source with light incident at ± off-axis angles were used, we could do even better at capturing the diffracted light. The net result is that the resolution can be somewhat improved.

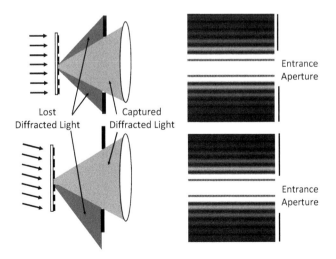

Figure 5.55 Illustration of the idea of on-axis illumination (top) and off-axis illumination (bottom). The shaded areas emanating from the mask regions represent the diffracted light pattern.

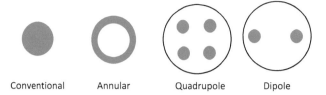

Figure 5.56 Illustration of light source geometries used in modern projection lithography tools.

In (5.8), k_1 can be improved by as much as a factor of 2 by using off-axis illumination [21]. In actual modern lithography tools, a variety of light source geometries are used, as illustrated in Figure 5.56. In some cases, very complex source designs are used [42].

An arbitrarily shaped illumination source produces partially coherent light and can be described as an extended source, where the source shape is divided into a collection of individual point sources. Each point is a source of spatially coherent illumination, producing a plane wave striking the mask at one angle, and resulting in an aerial image given by (5.28). Two point sources from the extended source, however, do not interact coherently with each other. Thus, the contributions of these two sources must be added to each other incoherently, e.g., the intensities are added together. Illumination systems for lithographic tools are carefully designed to have this property. Simulators like Optolith compute the resulting aerial image by adding together the contributions of each point source in the illumination source, and they generally allow the user to specify the shape of the illumination source [21, 33].

Although we did not discuss this in connection with Figure 5.34, it should be obvious that the system pictured there will produce higher-resolution images for features at the center of the mask than it will for features at the edges of the mask. The diffraction patterns for features in the center will be more completely captured by the objective lens than will the diffraction patterns produced by features at the mask edges. This is clearly not a good thing, and it can be mitigated by an additional change in the projection lithography system architecture known as a Köhler illumination source.

Figure 5.57 illustrates the idea behind a Köhler illumination source. Named for August Köhler, developer of illumination systems for Zeiss microscopes in the late 1800s, Köhler illumination uses the condenser lens to form an image of the source at the entrance pupil of the objective lens, rather than having collimated light pass through the mask, as illustrated conceptually in the top part of Figure 5.55. The reason for using a Köhler source arrangement is so that the projection lens can capture the diffracted light from any of the features on the mask equally well. This is illustrated in Figure 5.57 by the triangles emanating from the mask apertures that represent the angular spread of the diffracted light. If collimated light were used through the mask, it is apparent that much of the diffracted light would be lost from mask features near the outside edge of the mask.

Off-axis and Köhler illumination system concepts are used in virtually all modern lithography systems and they do help to improve the resolution of these systems. By themselves, however, they are not sufficient to explain the remarkable progress that Figure 5.4 illustrates.

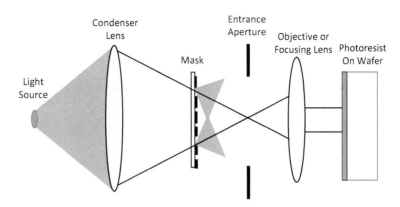

Figure 5.57 Köhler illumination system. The triangles emanating from the mask regions represent the diffracted light pattern.

5.7.2 Immersion Lithography

Immersion lithography refers to the idea of changing the medium between the final lens element in the objective or focusing lens from air to water. This permits improved resolution in the projection lithography tool, as we will see. The idea is actually not new. It was first introduced in the 1880s by Carl Zeiss to increase the resolving power of optical microscopes. The idea was proposed for lithography tools in the 1980s, and it was introduced into semiconductor manufacturing in about 2005. Such systems are often referred to as 193i lithography tools.

To understand how the introduction of water changes the optical system, consider Figure 5.58. We considered Snell's law earlier in Section 5.3. This law states that light rays transitioning from one medium to another with different indices of refraction obey the law $n_1 \sin \theta_1 = n_2 \sin \theta_2$. Thus in Figure 5.58(b), if the medium is air after the lens, light rays are bent such that $\theta_2 > \theta_1$ as shown. In the case of a water medium, $\theta_3 \approx \theta_1$ since the indices of refraction of water and the lens material are similar. Thus the focal planes are different in the two cases, as shown.

The resolution in the two cases shown in Figure 5.58 would be identical. The simplest way to think about this is that the lens gathers the diffracted light information after it passes through the mask and then performs an inverse Fourier transform to print the aerial image. There is no difference in the dry and wet systems illustrated in Figure 5.58 in terms of their ability to gather the diffracted light and so both systems should print similar images although in different focal (image) planes. So what have we accomplished by introducing the water?

Even if nothing is changed in the optical system, we actually still have accomplished something. Refer back to Figure 5.24 where we calculated the depth of focus (DOF) of a projection lithography system. If we were to redraw Figure 5.24 with a water medium after the objective lens, the angle α would change and as a result δ would become larger. In other words, the DOF would increase. The derivation of the increase in the DOF follows our derivation of (5.12) (see [21] if interested in the details), with the result that

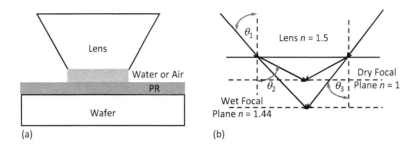

Figure 5.58 Dry versus wet (immersion) lithography systems. After [43].

$$\frac{\text{DOF(immersion)}}{\text{DOF(dry)}} = \frac{1 - \sqrt{1 - (\lambda/p)^2}}{n - \sqrt{n^2 - (\lambda/p)^2}}, \tag{5.58}$$

where n is the index of refraction of the material between the lens and the wafer, λ is the exposure wavelength and p is the pitch of the features being printed. The DOF improves at least by the index of refraction of the water and improves further for smaller feature sizes. Increasing the DOF can be a major improvement in lithography tools because of the finite thickness of photoresist (PR) layers, as we saw earlier when we derived (5.12).

The goal in using immersion lithography is, of course, to improve resolution and, as we have seen, simply replacing air with water between the final lens element in the objective or focusing lens does not accomplish this. However, changes in the optical system along with immersion *can* do this. According to (5.8) we can improve the system resolution by increasing the NA, which basically means using larger lenses to capture more of the diffracted light information. Suppose we try to do this as shown in Figure 5.59. A larger lens will bring light rays to the bottom lens/air or lens/water interface at larger angles. However, in order for light to be refracted (pass through to the second material), it must arrive at less than the critical angle or it will be reflected, as shown in Figure 5.59(b). The critical angle from optics is given by

$$\theta_{\text{crit}} = \sin^{-1}(n_2/n_1), \tag{5.59}$$

where n_1 and n_2 are the indices of refraction of the air/water material and the lens material, respectively. For a lens/air boundary, $\theta_{\text{crit}} \approx 41°$; for a lens/water boundary, θ_{crit} is much larger because n_1 and n_2 are similar.

Immersion systems can therefore capture higher-order diffraction information and use this information to print higher-resolution images. Water is a great choice for the liquid in these systems because it transmits more than 90% of 193 nm light, it has a relatively high index of refraction, it is environmentally safe and semiconductor factories are used to dealing with highly purified water because it is widely used in cleaning procedures and elsewhere in the fab. Today, 193i immersion lithography systems are designed with NA values of about 1.35. This produces a significant improvement in resolution, as can be seen from (5.8). The upper bound on the NA in these systems is set by the index of refraction of the medium between the lens and

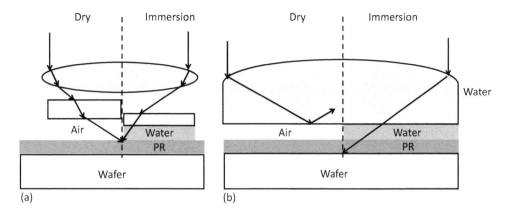

Figure 5.59 Projection lithography systems with (a) NA < 1 and (b) NA > 1.

the wafer, which for water is 1.44 [21, 43]. However, an NA of 1.35 seems to be a practical upper bound for these systems.

The engineering challenges associated with bringing immersion lithography into a manufacturing environment should not be underestimated. Highly purified water must be introduced in a narrow region (≈1 mm) above the wafer surface and must be maintained there while the wafer stage moves to expose successive chips. Water interacts with some photoresist components (PAG molecules, for example) and so the chemistries of the photoresist and the stack of photoresist materials need to be optimized for a water environment. In some cases a "topcoat" layer above the resist must be added to prevent the water from interacting with the photoresist. There are possible liquids with even higher indices of refraction than water. However, the complications in using these materials in a practical lithography machine have so far precluded their use in manufacturing. For readers interested in additional details about immersion lithography systems, numerous references exist in the literature [21, 43].

5.7.3 Multilayer Resists and Antireflection Coatings

There are several issues associated with the photoresist layers used in lithography that we have not yet discussed in detail. Figure 5.60 illustrates some of these issues.

Notice in Figure 5.60(a) that the resist thickness tends to be non-uniform across the wafer because it is spun on as a liquid and therefore tends to fill in the "hills and valleys" of the underlying topography. The exposure process that patterns the resist therefore must contend with exposing different regions with different thicknesses of resist. Effectively, the resist is thinner on top of high structures and thicker over low-lying structures. The result of these effects is that the resist can be underexposed where it is thicker and overexposed where it is thinner. This can result in linewidth variations, particularly where the photoresist features cross steps in the underlying structures. In the case in Figure 5.60(a), only a low-resolution pattern is being exposed and the resist is only being used to block the As^+ implant, so the resist thickness variation would likely not be a problem. In Figure 5.60(b), a high-resolution pattern is being

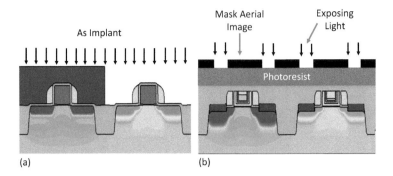

Figure 5.60 Illustration of the impact of surface topography on thickness variations in deposited photoresist (PR) layers using (a) Figure 2.12 and (b) Figure 2.25.

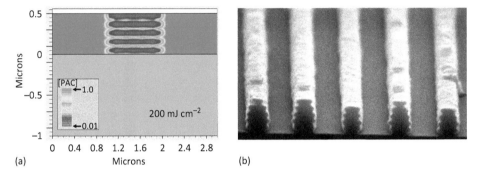

Figure 5.61 Illustration of the impact of standing waves. (a) The simulation is identical to Figure 5.51 except with a Si substrate. (b) SEM image. Photo courtesy P. Rissman.

printed and here planarization using chemical–mechanical polishing (CMP) is used prior to the photoresist deposition, so that a thin uniform resist layer results. This would allow for higher-resolution printing.

If there are highly reflective layers below the photoresist, light that passes all the way through the resist without being absorbed will be reflected by these underlying layers and pass back up through the resist again. While this may speed up the exposure process, it also has the potential for setting up standing wave light patterns in the resist because of constructive and destructive interference between the incoming and outgoing waves. We saw an indication of this problem in Figures 5.48 and 5.51 in which the photoresist was placed on top of an SiO_2 layer that is not very reflective. In Figure 5.61(a), the simulation from Figure 5.51 is repeated, but this time with the photoresist on a silicon substrate, which is much more reflective. The standing wave pattern is now much more obvious in the exposed resist. If the underlying reflective surface were not perfectly flat, light could be scattered sideways, which could degrade the latent image quality. The cross-sectional scanning electron microscope (SEM) image in Figure 5.61(b) shows a developed resist pattern with "scalloped" edges that result from the standing wave pattern during the exposure.

The final issue associated with resists is that the depth of focus achievable in modern high-resolution lithography tools is very limited, especially in EUV machines, which means that only thin layers of resist can be used. Such thin layers may not be adequate for subsequently masking implants, masking against etching processes, etc.

Two changes to the resist layer have been widely adopted to deal with all of the above issues. The first is the adoption of a multilayer resist in which a thin top layer is used for imaging and the lower thicker layer provides the full thickness needed for subsequent steps. An alternative to this approach is to use a "hard mask," which is basically a material like SiO_2 or Si_3N_4 that is deposited below a thin resist layer. The thin resist produces the desired pattern and it is then used as a mask to etch the "hard mask," which then serves as a mask for the layers below it for whatever the subsequent process step may be. We saw a number of examples of this in the process flow in Chapter 2, and Figure 5.62 illustrates several additional examples. In each case, the red layer has been deposited on a substrate that is nonplanar and the lithography step is supposed to result in selective etching of the red layer.

The process flow in Figure 5.62(a) is simply a single layer of resist. In Figure 5.62(b) a thick organic layer (such as another photoresist not sensitive to the exposing wavelength) planarizes the structure and provides a flat surface for the thin imaging photoresist layer. In Figure 5.62(c), an organic layer provides planarization and an oxide hard mask is deposited above this layer and below the thin resist imaging layer. Generally, the organic layer in these processes is a liquid at room temperature and is applied just like a photoresist. The hard mask layers could be deposited by a chemical vapor deposition (CVD) process or using a spin-on liquid source that is

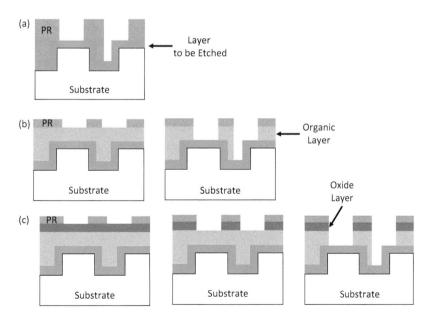

Figure 5.62 Examples of multilayer resist process flows. In each case, the final step (not shown) would be to etch the red layer.

basically a solvent containing the SiO_2. After the solvent evaporates, a solid hard mask remains. There are many variations on these processes, but the core ideas are to planarize the structure using a liquid, provide for a very thin imaging layer to deal with depth-of-focus issues, and to provide a more rugged (hard) mask when this is required. Obviously, the process complexity increases with these additional steps. The final imaging photoresist (PR) layer is often less than 100 nm thick. Of course, CMP, as we saw in several instances in the CMOS process in Chapter 2, can also be used to planarize the surface before a thin layer of resist is deposited. This is illustrated in Figure 5.60(b).

The second significant change in the resist layer structure has been the adoption of antireflection coatings. The most common approach is to deposit a layer underneath the imaging photoresist layer called a bottom antireflection coating (BARC), as illustrated in Figure 5.63. The purpose of this layer is to minimize standing waves or, more generally, reflections from material layers beneath the BARC back into the photoresist. The BARC material is characterized by its refractive index, which is a complex number, by its absorption and by its thickness. In the simplest case, when the materials are not absorbing, then the thickness of the BARC layer is chosen such that the reflections from the n_2/n_3 interface that propagate back up to the n_1/n_2 interface are $180°$ out of phase with the reflections occurring at the n_1/n_2 interface. This requires that the BARC thickness be a quarter wavelength or

$$D = \frac{\lambda}{4n_2}. \tag{5.60}$$

If the light is normally incident on the stack, then the reflection that occurs at the interface when light travels through material i and strikes material j is given by [21]

$$\rho_Y = \frac{n_i - n_j}{n_i + n_j}, \tag{5.61}$$

where n_i and n_j are the complex indices of refraction for the two layers. In Figure 5.63, the total reflectivity at the top of layer 2 (the BARC layer) is given by

$$R_{total} = |\rho_{total}|^2 = \left| \frac{\rho_{12} + \rho_{23}\tau_D^2}{1 + \rho_{12}\rho_{23}\tau_D^2} \right|, \tag{5.62}$$

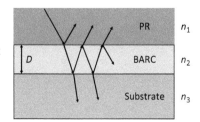

Figure 5.63 Illustration of the use of a BARC to minimize reflections back into the PR layer.

where $\tau_D = e^{-i2\pi D/\lambda}$ is the internal transmittance or the change in the electric field as it travels through layer 2 from top to bottom. If the BARC layer is designed to completely eliminate reflections, then we would require that

$$\rho_{12} + \rho_{23}\tau_D^2 = 0. \tag{5.63}$$

In the simplest case, when the BARC layer is not absorbing, $\tau_D^2 = -1$ if D satisfies (5.60). We can then make (5.63) true if $\rho_{12} = \rho_{23}$. This will be true if

$$n_2 = \sqrt{n_1 n_3}. \tag{5.64}$$

The above simplified analysis is only true if all three materials are transparent, i.e., the imaginary parts of their refractive indices are zero. In this ideal case, there is no reflection back into the photoresist and all of the light striking the BARC will be transmitted into layer 3. While this simple solution is actually used in ideal situations like camera lenses or lithography tool projection lenses, it generally fails in the situation pictured in Figure 5.63 for a number of reasons. Photoresist is not transparent. The thicknesses of the various layers vary in a real manufacturing environment. The underlying layers are often also not transparent. And, most importantly, the light coming into the photoresist layer is not normal to the surface, especially in high-NA projection tools. So the design of real BARC layers in semiconductor manufacturing generally uses numerical methods that can account for these factors. For the interested reader, there is much more detail available in [21] and other similar references.

5.7.4 Design Rule Restrictions

If we revisit the aerial image simulations discussed earlier (Figures 5.36–5.39), it is apparent that the quality issues in the aerial image occur primarily where we have sharp corners. This should not be surprising because these are the areas in the pattern that contain the highest spatial frequencies, and it is exactly these frequencies that lithographic systems have the hardest time printing. So, one strategy for helping to mitigate these issues is simply to restrict chip designers from using the kinds of patterns that are difficult to print. Designers, of course, do not like these restrictions because it constrains their designs (and likely makes their job more difficult). But, in fact, design rule restrictions are very common today in state-of-the-art technologies. They have become progressively more restrictive as device dimensions have decreased. Figure 5.64 illustrates how these rules have evolved over generations of technologies. Obviously, the single-size, single-orientation approach in the latest generations of technologies is very restricting to designers.

5.7.5 Optical Proximity Correction, Phase Shifting Masks and Computational Lithography

In all of the examples that we have considered so far, the pattern on the mask has been an exact replica of the pattern designed by the chip designer. To this point, we have focused our efforts on optimizing the lithography system so that it could do the best possible job of printing this

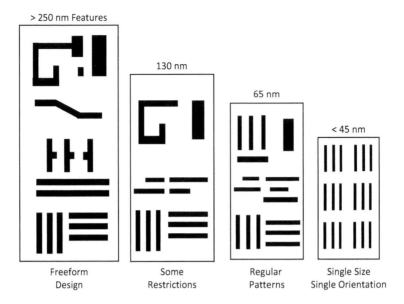

> 250 nm Features

130 nm

65 nm

< 45 nm

| Freeform Design | Some Restrictions | Regular Patterns | Single Size Single Orientation |

Figure 5.64 Evolution of design rule restrictions as feature size has decreased. After [44].

pattern on the wafer. When subwavelength lithography began to be used about 20 years ago (Figure 5.4), it was soon realized that something radically different would need to be done. The conceptual breakthrough that occurred was the realization that we actually do not care what the mask pattern looks like. What we care about is what the printed pattern on the wafer looks like. So, what might be possible if we asked the question: "What is the best pattern to use on the mask in order to get a printed pattern that best resembles the designed feature?"

We will briefly discuss several approaches to doing this: optical proximity correction (OPC), phase shift masks (PSMs) and computational lithography. These approaches might be considered "mask engineering." These methods are also sometimes called "wavefront engineering" [45].

The Optolith simulation in Figure 5.65(a) is similar to Figure 5.39(a). The parameters used in the simulation are at the limits of what is possible in dry DUV projection systems (λ = 193 nm, NA = 0.93). If we look at the quality of the aerial image in Figure 5.65(a), we can observe that diffraction effects are contributing too much light in regions like A and too little light in regions like B. In the simulation in Figure 5.65(b), the mask pattern has been modified to add serifs on outer corners like C and cutouts or notches on inner corners like D. Effectively, these changes allow more light where we need it and reduce the light where we do not want it. The net result is a higher-quality aerial image. This process of adding serifs, cutouts and making other changes in the mask geometry is called optical proximity correction (OPC).

While the concept of using OPC to improve aerial image quality may make sense, an obvious question is how one goes about designing the OPC pattern to optimize the aerial image. There are two basic approaches that are used. The first involves the use of look-up tables that contain pre-computed OPC geometries based on the designed patterns. This is often called "rule-based" OPC. A second approach is to use models to dynamically simulate the aerial image and use this process to move edges, add serifs, etc., until the

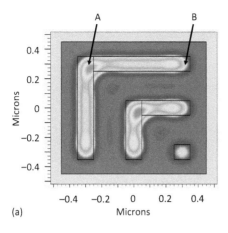
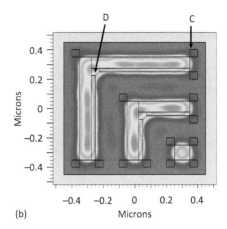

Figure 5.65 (a) Optolith simulation of the aerial image of two elbow shapes and a contact hole. The mask geometry is shown with the solid black lines. System: 193 nm stepper, NA = 0.93 and 0.1 μm feature sizes, similar to Figure 5.39(a). (b) Here OPC using serifs has been used to improve the quality of the aerial image.

best solution is found, usually by iteration. This is often called "model-based" OPC. The models that are used basically follow the approach discussed in connection with Figure 5.34, although they may be simplified to run faster when very complex designs are involved. As one might expect, the rules-based approaches generally are much faster, especially when complex designs are involved. Commercially available simulation tools usually offer OPC design as an option [46]. In addition, for the interested reader, there are many technical references available, e.g., [47, 48].

Phase shift masks (PSMs) are an additional strategy that can be used to modify the mask pattern in order to produce a better aerial image. Conventional masks only make use of light intensity information to form an image. Conventional masks are "binary," i.e., black or white. PSMs make use of phase information as well as intensity information to improve the aerial image. The basic concept was suggested by Levenson *et al.* [49] and is illustrated in Figure 5.66.

In this example, a periodic mask with equal lines and spaces (diffraction grating) is used as the mask. Figure 5.66(a) shows the electric field, $\mathcal{E}$, associated with the light just after it passes through the mask, and also at the wafer after the diffracted light has passed through the optical system, without any phase shifting in the mask. The period of the grating is chosen in this case so that the lines on the mask are barely resolved on the wafer. The photoresist responds to the intensity of the light or the electric field squared, $\mathcal{E}^2$, and hence the intensity pattern at the bottom is barely sufficient to resolve the two lines.

In the example in Figure 5.66(b), a material whose thickness and index of refraction are chosen to phase shift the light by exactly 180 ° is added to the mask. The thickness of this layer is given by

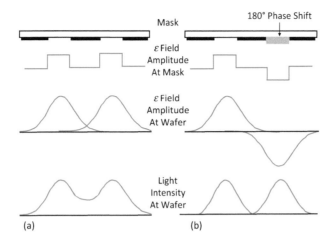

Figure 5.66 Example of the use of phase shifting techniques on masks to improve the resolution of the aerial image. After Levenson *et al.* [49].

$$d = \frac{\lambda}{2(n-1)}, \tag{5.65}$$

where n is the index of refraction of the phase shift material. The period of this added pattern is half the period of the original grating. The corresponding $\mathcal{E}$ fields at the mask and wafer are also shown. Since the light intensity at the aerial image is the square of the field intensity, $\mathcal{E}^2$, the quality of the resulting aerial image is significantly improved, as illustrated, because of the destructive interference of the two patterns.

There are a variety of ways to implement PSM, several of which are illustrated in Figure 5.67. In the example in Figure 5.67(a), the quartz mask substrate is etched to produce the 180° phase shift. In Figure 5.67(b), a material (such as $MoSi_2$) that provides the 180° phase shift but also allows a small amount of transmission of light is used. This approach is called an attenuated phase shift mask. Typically only about 5–15% transmission is used. This approach does not generally provide the full benefit of alternating phase shift masks (in Figure 5.67(a)), but it is simpler and cheaper to apply and is used in some manufacturing situations. The two examples in Figure 5.67(c) and (d) are top views of the mask and illustrate two additional approaches to phase shift masks.

Application of this principle to arbitrary mask shapes is quite complex and generally requires the addition of features on the mask smaller than the minimum feature size to be printed. OPC can be combined with PSM, although this further complicates the mask design. Simulation tools provide a very powerful approach to exploring the benefits of phase shifting masks and they can also help in the optimum design of mask patterns. An Optolith simulation in Figure 5.68 illustrates the use of phase shift elements to sharpen an aerial image. Three 0.1 μm contact holes are simulated with phase shifting elements added to the middle and right contact holes. Both the aerial image in Figure 5.68(a) and the intensity cutline in Figure 5.68(b) show the sharpening of the contact hole as a result of the phase shifting regions.

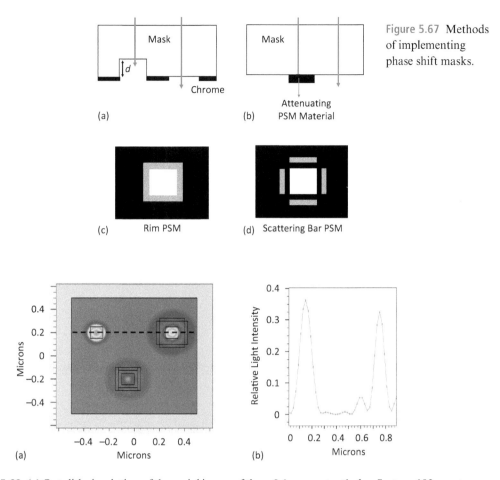

Figure 5.67 Methods of implementing phase shift masks.

Figure 5.68 (a) Optolith simulation of the aerial image of three 0.1 μm contact holes. System: 193 nm stepper, NA = 0.93, similar to Figure 5.65. In the middle and right contact holes, scattering bar PSM elements have been added with 180° phase shifts. (b) The intensity cutline in panel (a) is at $y = 0.2$ μm.

In Section 5.5.1 we discussed the mathematics used in calculating the aerial image in projection lithography tools. The inverse problem – "Given a desired aerial image, what should the mask look like?" – combines many of the approaches to resolution enhancement that we have discussed in this section. This field is often called computational lithography or inverse lithography, or sometimes pixelated lithography since the mask is broken up into individual pixels whose properties are designed through inverse modeling to achieve the best aerial image. Companies fabricating chips in state-of-the-art technologies have made major investments in the field of computational lithography. Designers create designs that appear fairly traditional on their CAD systems (although constrained by many design rules). Those designs are then processed through very complex software to create the actual masks that are used in building those designs. There is an extensive literature on this topic [50, 51, 52, 53]. The computationally computed mask pattern may

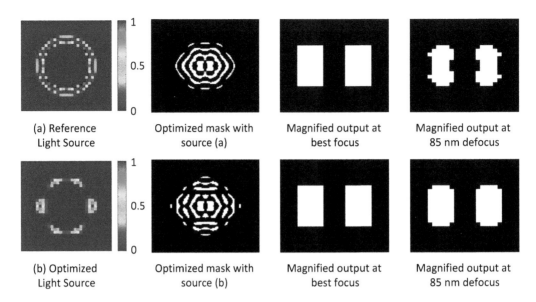

(a) Reference
Light Source

Optimized mask with
source (a)

Magnified output at
best focus

Magnified output at
85 nm defocus

(b) Optimized
Light Source

Optimized mask with
source (b)

Magnified output at
best focus

Magnified output at
85 nm defocus

Figure 5.69 Example of optimizing the light source and the mask to print a simple two-rectangle pattern. In this example, the primary effect of the optimization is to improve the depth of focus of the image. In these examples, the pixel size is 10 nm × 10 nm. The imaging system uses $\lambda = 193$ nm and NA = 1.35. The rectangular features are 60 nm × 110 nm. Reprinted/adapted with permission from [52]. © The Optical Society.

involve a chrome mask with OPC, PSM and other elements, or it may even involve only a glass mask in which individual pixels have been etched to control the phase of the light passing through that pixel.

Perhaps the best way to illustrate the radical changes that have been made in masks in recent years is through two examples. In Figure 5.69, from [52], two simple rectangles 60 nm × 110 nm are to be printed. In Figure 5.69(a), an annular light source is used, with the intensity varying with position as shown. Given that light source, the mask is optimized to produce the best possible image (magnified). With this source and mask, the image is quite poor in a plane 85 nm away from the plane of best focus. These are calculated areal images that used models similar to those described in this chapter. In Figure 5.69(b), computational methods using inverse lithography are used to optimize the light source and mask to improve the depth of focus of the calculated areal image. The improvement is significant.

The second example is shown in Figure 5.70 from a paper presented by Intel [53]. The goal in this example is to print the pattern in Figure 5.70(c), which is the first metal layer in a microprocessor chip. The pixelated mask, designed using inverse lithography, is shown in Figure 5.70(a) using an atomic force microscope (AFM) to illustrate the etched glass. The mask pattern is in Figure 5.70(b). Obviously, the mask pattern is considerably more complex than a simple digital representation of the desired pattern. The pattern in Figure 5.70(c) was printed with a 193 nm lithography tool with NA = 1.35.

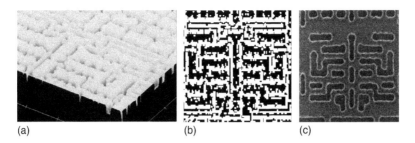

Figure 5.70 (a) Example of a pixelated mask. The glass has been etched pixel by pixel to control the phase of the light passing through each pixel. (b) The mask pattern. (c) The printed pattern in the photoresist. © 2013 IEEE. Reprinted with permission from [53].

5.7.6 Multi-Patterning

Consider Figure 5.71. In this simulation, a series of 0.1 μm lines and spaces are simulated in a lithography tool with the specifications given in the figure caption. It is clear from the aerial image (Figure 5.71(a)) and from the intensity profile (Figure 5.71(b)) that the optical system cannot resolve these closely spaced lines and spaces, or perhaps it can barely resolve them. The quality of the aerial image would not be acceptable in a manufacturing environment.

Now consider Figure 5.72. This simulation is identical to Figure 5.71 except that only alternate lines are simulated. It is clear that the quality of the aerial image is greatly improved and the intensity cutline shows that the contrast in the aerial image is much better. Physically, the reason for this is simply that the interference caused by the diffracted light from adjacent lines in Figure 5.71 disappears when those lines are not part of the mask pattern. This example illustrates the concept of multi-patterning. The original mask pattern is split into two masks in which the features for each mask are chosen so that there is greater spacing between the features. The cost is obviously twice as many patterning steps, and the two masks need to be precisely aligned with each other on the wafer. Nevertheless, for at least the last decade, multi-patterning has been common in state-of-the-art manufacturing plants. Separating the mask features into two different masks also permits more space to use other resolution-enhancing techniques like OPC and PSM.

There are a variety of options for how to actually implement multi-patterning. One example is shown in Figure 5.73. In this case, the desired mask pattern is split into two patterns. The first is exposed, developed and the underlying film is etched. Then a second photoresist layer and second exposure process is used for the second half of the pattern, followed by a second etch. It is possible to use two lithography steps using the same resist layer, followed by a single development and etch cycle, which would be less expensive. The particular approach chosen depends on the manufacturing facility and the pattern being printed. It is quite possible to extend double patterning to triple patterning or even quadruple patterning, that is, use as many as four masks to print a single layer in manufacturing. The latest generations of 193 nm based lithography (10 nm, 7 nm and 5 nm technologies) all use multi-patterning on critical levels and often more than two masks per layer [54, 55, 56].

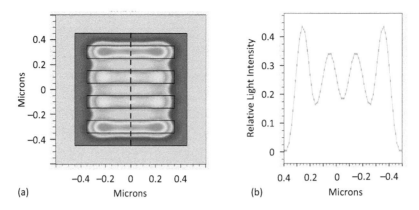

(a)

(b)

Figure 5.71 (a) Optolith simulation of 0.1 μm lines and spaces using a 193 nm lithography tool with NA = 0.8. (b) The cutline shows the light intensity along the dashed line in panel (a) at $x = 0$.

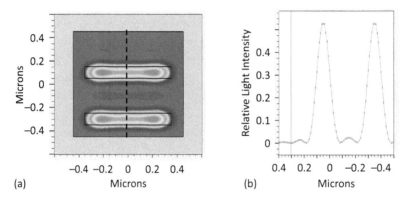

(a)

(b)

Figure 5.72 (a) Same Optolith simulation of 0.1 μm lines and spaces as in Figure 5.71 but with only alternate lines simulated. (b) The cutline shows the light intensity along the dashed line in panel (a) at $x = 0$.

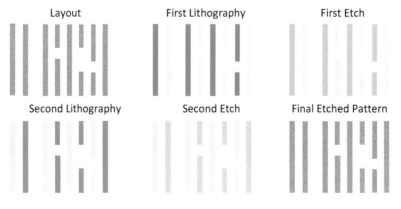

Figure 5.73 Typical process sequence for double patterning.

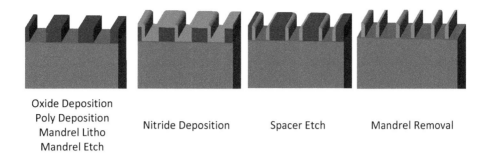

Oxide Deposition
Poly Deposition   Nitride Deposition   Spacer Etch   Mandrel Removal
Mandrel Litho
Mandrel Etch

Figure 5.74 Process steps in the self-aligned double patterning (SADP) process flow. The simulation was done using Silvaco's Victory Process 3D simulation tool [33].

5.7.7 Sidewall Spacer Methods (Self-Aligned Double Patterning)

There is one final resolution enhancement technique that is widely used today in state-of-the-art manufacturing. This is the self-aligned double patterning (SADP) technique illustrated in Figure 5.74. The method makes use of sidewall spacers that were introduced in the CMOS process flow in Chapter 2.

The process begins with the deposition of an expendable material (the mandrel), in this case polysilicon (poly) deposited on a thin SiO_2 layer. The poly is then patterned using lithography. Next, a Si_3N_4 layer is conformally deposited using a deposition method such as low-pressure chemical vapor deposition (LPCVD), which we will discuss in detail in Chapter 10. The Si_3N_4 layer is then etched anisotropically without using a mask. Because the nitride is vertically much thicker on the sidewalls of the poly stripes, an anisotropic etch will leave the sidewall spacers shown in Figure 5.74 after the nitride is removed from all flat surfaces. If the poly is then removed, the sidewall spacers remain. They have a spacing that is half of the poly spacing and a width that is determined by the thickness of the deposited nitride film.

Figure 5.75 illustrates how the SADP process can implement a mask pattern. We used sidewall spacers in the CMOS process in Chapter 2 for a different purpose, in that case to form the lightly doped drain (LDD) regions in the transistors.

All state-of-the-art process flows use various combinations of SADP, multi-patterning, OPC and PSM to achieve pattern resolutions that are much smaller than the 193 nm wavelength light used in the lithography tools. Table 5.1 summarizes the approaches taken in various technology generations to achieve higher resolution.

The k_1 factors listed in Table 5.1 refer to the Rayleigh resolution limit defined in (5.8). The values in the table are calculated using the NA, pitch and λ values given in the table. Note that the actual minimum pitch (line plus space) in each generation is much larger than the technology node "name." The names today actually have little or no meaning in terms of the physical sizes of the features, as described in Chapter 1 (Figure 1.23).

Values of k_1 smaller than ≈ 0.6 reduce manufacturing yield, and so the various "tricks" described above and listed in Table 5.1 must be used to compensate for the small k_1 values. It is generally agreed that these "tricks" have reached their limits with current 10 nm or 7 nm technologies. The only option to continue to reduce feature sizes is to reduce the exposure

Table 5.1 **Summary of approaches to subwavelength lithography used with 193 nm systems. After [55].**

Technology node name (nm)	130	90	65	45	32	22	14	10	7	5
Minimum pitch (nm)	350	250	200	140	100	80	64	48	≈40	≈30
Lithography λ (nm)					193 nm					
NA	0.5	0.75	0.85	1.2			1.35			
Litho "tricks"	Rayleigh factor, $k_1 = 0.5 \times$ pitch $\times (NA/\lambda)$ (equation (5.8))									
Conventional litho	0.6									
OPC		0.5								
Off-axis illumination			0.44	0.44	0.35					
Mask optimization, double exposure						0.28				
Double/triple patterning							0.22	0.16		
Regular arrays, trim exposure, SADP									0.12	0.10

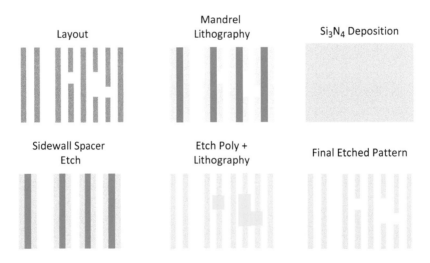

Figure 5.75 Process steps in the self-aligned double patterning (SADP) process flow, including patterning the sidewall spacers.

wavelength, and EUV systems are now beginning to be used in next-generation technologies (7 nm and 5 nm). This is not explicitly shown in Table 5.1 but, in fact, the last two columns in this table (7 nm and 5 nm) do use EUV for some of the lithography steps. However, it should be noted that EUV has a wavelength of 13.5 nm. This is much smaller than the 193 nm currently used, but even EUV systems have resolution limits of 10–15 nm with single exposures and the currently available NA in these systems. Thus it is very likely that higher-NA systems and/or the resolution enhancement methods described above will soon need to be incorporated into EUV lithography processes as well. It is likely that double patterning will need to be used in EUV systems within the next technology generation or two [55, 56].

5.8   Nanoimprint Lithography

The combination of incredibly expensive lithography tools (which cost more than $100 million for EUV systems) and very clever resolution enhancement methods has enabled state-of-the-art chips with printed features as small as 10–20 nm. But the very high cost of these methods means that only a few very large companies can afford them, and only a relatively small number of high-volume chip designs can use them. Is there any feasible alternative?

Imprint lithography has existed literally for centuries, with examples dating back to 500 BC in China and many examples from Europe dating to the Middle Ages. As an example, carved woodblocks were used to transfer images to paper and other materials by inking the woodblock and rubbing the paper on the woodblock. Modern applications of these methods to produce features at the nanoscale were introduced by Chou in 1996 [57], as illustrated in Figure 5.76.

In the thermal nanoimprint lithography (NIL) process introduced by Chou (see Figure 5.76(a)), a mold is first prepared, often using a silicon substrate, since the thermal expansion coefficient will then be matched to that of the silicon wafer that is to be patterned. The mold contains the pattern to be printed, and, since no optical system is involved, the pattern is 1:1 in size with the image desired on the wafer. For nanometer-scale features, the mold would generally be created using e-beam lithography, similar to Figure 5.1. If the mold etch depth is significant, a hard mask and perhaps metal lift-off, as will be discussed in connection with Figure 5.78, would be used to etch the mold. Alternatively, an SiO_2-on-Si structure can be used in which the SiO_2 is etched after e-beam lithography to produce the mold.

A resist material such as poly(methyl methacrylate) (PMMA) is then spin-coated on the substrate to be patterned. The substrate and the resist are then heated above the glass transition temperature of the resist (140–180 °C for PMMA). The mold is then pressed into the resist as shown

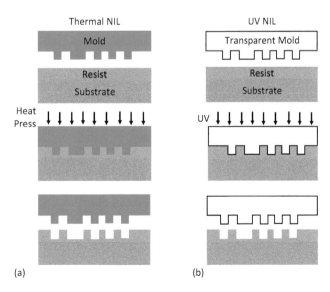

Figure 5.76 Nanoimprint lithography. (a) Thermal NIL: temperature and pressure are used to create the image in the resist. (b) UV NIL: UV light is used to cure the resist.

using a typical pressure of a few hundred psi (a few MPa). The resist material softens above its glass transition temperature and flows into the mold pattern. The temperature is then reduced, which "freezes" the mold pattern in the resist, and the mold is separated from the substrate as illustrated.

An alternative to the thermal NIL process is shown in Figure 5.76(b). In this case, a transparent mold is used, perhaps made using a quartz substrate. A liquid photopolymer resist is spin-coated onto the substrate and the mold is then pressed into the resist. This is done at room temperature, since a low-viscosity resist is used and much lower pressures are required. This process is generally much faster than the thermal NIL process because no temperature cycling is required. Once the resist has flowed into the mold, UV light is used to expose the resist as shown. This hardens the resist, after which the mold can be removed. Because the UV NIL process has much higher throughput, it is generally preferred today.

Figure 5.77 adds the next step in the process for the UV NIL technology. The photoresist must be removed down to the underlying substrate. This can be achieved using an O_2 plasma etch, often called "descumming," which thins the photoresist everywhere, as shown in Figure 5.77(a). Alternatively, if a UV blocking layer is incorporated on the bottom of the mold pattern, as shown in Figure 5.77(b), the resist will be unexposed in the thinned regions and can be simply developed away. The NiCr absorber is often simple to implement, since such a layer is part of normal e-beam mask making, as shown in Figure 5.1.

Figure 5.78 shows two examples of how the pattern would be transferred into the substrate following the nanoimprint process. In Figure 5.78(a), the photoresist itself is used as a mask for dry etching the substrate. This would work if the resist is thick enough and robust enough to withstand the dry etching process. If this is not the case, then a hard mask could be formed that would be more resistant to the substrate etching process. This is illustrated in Figure 5.78(b), in which a hard mask

Figure 5.77 Two approaches in UV NIL for clearing the photoresist down to the substrate. (a) An O_2 plasma descumming is used, which thins the resist everywhere. (b) A NiCr absorber blocks the UV light so the photoresist can simply be developed without thinning where it was exposed.

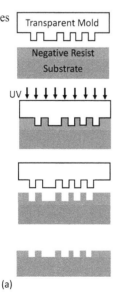

(a)

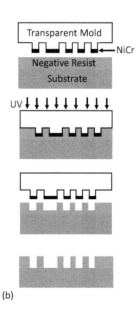

(b)

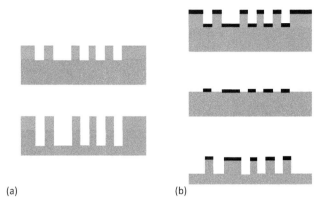

(a)   (b)

Figure 5.78 Two approaches for etching the mold pattern into the substrate. (a) The photoresist is used as the mask and dry etching transfers the pattern. (b) A hard mask is deposited, then lift-off is used, followed by dry etching.

(the black material) is deposited and then lift-off is used to leave the hard mask on the substrate surface where it could serve as a mask for the substrate etching. The lift-off process is described in detail in Chapter 11 (Section 11.7). Note that this process forms a negative image compared to the original mold, which would have to be accounted for in the mold design.

Modern machines that are designed for chip production using NIL are considerably more sophisticated than the simple conceptual drawings in the above figures. Printing patterns over a 300 mm silicon wafer usually requires a step-and-flash process since molds cannot likely be made large enough to print an entire wafer in one operation. Conceptually, this is done as illustrated in Figure 5.79. A planarization process is first used to provide a flat surface on which to do the printing. We will discuss planarization methods in detail in Chapter 11 since they are critical in back-end technologies. A low-viscosity imprint fluid (resist) is dispensed with an inkjet printer in a pattern that is determined by the mold geometry, since the resist thickness after it flows into the mold will depend on this pattern. Alignment with existing patterns on the chip is accomplished through the transparent mold in a similar manner to conventional optical lithography tools. At each position of the mold, the resist is first dispensed, then the mold is moved into position, pressure is applied to allow the resist to flow into the mold and then UV light flashes to expose the resist. The mold is then lifted off the substrate, leaving the molded pattern. The process then repeats. This is often called "jet-and-flash imprint technology" or J-FIL because of the use of inkjet technology to dispense the resist.

An example of a production system operating in this way is shown in Figure 5.80, taken from the Canon website. Production tools based on the ideas in this section are much less expensive than the equipment used to print nanometer-scale features with EUV lithography tools and the resolution enhancement methods described in this chapter. Both methods are capable of printing features with 10–20 nm sizes.

The challenges associated with NIL include defect densities, mold robustness over many printing operations, achieving reliable separation of the mold from the resist after pattern

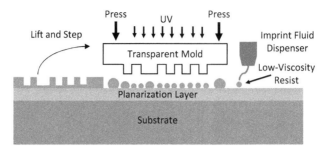

Figure 5.79 Step-and-flash approach to using NIL on large-diameter silicon wafers.

Figure 5.80 Example of production NIL tool using the concepts shown in Figure 5.78. Copyright © Canon Corporation, reprinted with permission.

transfer, and overall manufacturing economics. As an example, if the mold size is approximately 2 cm × 3 cm (on the order of one or a few chips), the inkjet process would need to dispense 10^4–10^5 pl (picoliter) drops at each site, followed by pressing the mold into the resist and exposing with UV light. All of this needs to be accomplished in a time on the order of 1 s, in order to provide an overall throughput of ≈ 20 wafers per hour in manufacturing. In fact, the process is even more complicated than this, because usually the mold needs to be coated with a release layer to make certain that, after exposure, the mold can be cleanly separated from the resist, and during the contact process the mold may need to be bowed using mechanical pressure so that the center of the mold contacts the resist first. For the interested reader, a number of review papers in the literature have summarized the current state of the art, such as [58]. To this point in time, all of the companies involved in manufacturing the latest-generation silicon chips have chosen the EUV option for patterning. However, the NIL approach conceptually offers economic advantages, and these systems could play a role in the future. It is likely that the first applications of NIL in silicon chip manufacturing will be in memory products, because these products are less sensitive to defect densities (because of redundancy) than are logic products like microprocessors.

5.9 Summary of Key Ideas

Lithography is one of the fundamental technologies on which modern semiconductor technologies are based. The two key elements of modern optical lithography systems are the exposure tool and the photoresist. Exposure tools today generally use projection optics with diffraction-limited refracting lenses. These systems can print an area on the order of a few square centimeters on the wafer, so that a step-and-repeat or step-and-scan approach must be used to print an entire wafer. Today, g-line and i-line resists are largely based on DNQ materials, which are used for manufacturing down to the 0.35 μm generation. DUV resists that use chemical amplification are used for all shorter wavelengths, including 193 nm and EUV systems. Lithography simulation tools are based on two scientific fields: Fourier optics to describe the performance of exposure tools; and resist chemistry to describe the formation of the mask pattern in the resist. These tools are very useful today in understanding and optimizing the performance of lithographic systems, and we have used them extensively in this chapter to illustrate concepts and to simulate achievable resolutions in current lithographic tools.

Computational lithography, sometimes called inverse lithography, has become a very powerful tool today in designing both the mask pattern and the other properties of the optical system in order to achieve the highest resolution in aerial images. Many "tricks" have been invented to push these systems far beyond the conventional limit of "half of the exposing wavelength." We have seen how OPC, PSM, source illumination shape optimization, immersion lithography, SADP and other techniques have achieved resolutions of a small fraction of the exposing wavelength. Collectively these "tricks" are often referred to as design technology co-optimization (DTCO) since they effectively couple the design process with the manufacturing technology.

Nevertheless today's 193 nm exposure tools have reached their practical limits, and we are now finally seeing the introduction of EUV tools into manufacturing with an exposure wavelength of 13.5 nm. These tools are enormously expensive and will really test the economic model of continuing to push feature sizes in semiconductor chips. Many semiconductor devices today are manufactured with much larger feature sizes and can therefore use much less sophisticated lithography tools. If we have learned anything over the past several decades of "Moore's law," it is that, when economic drivers exist to push technologies, the ingenuity of engineers and scientists in designing and building machines to create ever smaller device structures should never be underestimated.

5.10 REFERENCES AND NOTES

[1] "International Technology Roadmap for Semiconductors," (ITRS). See http://www.itrs2.net/2012-itrs.html.
[2] Optolith is the lithography simulation tool portion of Silvaco's Athena process simulator. It implements many of the models we discuss in this chapter. See https://www.silvaco.com/products/vwf/athena/optolith/optolith_br.html.
[3] A. K. Bates, M. Rothschild, T. M. Bloomstein, *et al.*, "Review of technology for 157-nm lithography," *IBM Journal of Research and Development*, vol. 45, no. 5, pp. 605–614, 2001.

[4] ASML is the only manufacturer of EUV lithography systems. See https://www.asml.com.

[5] M. Bohr, "The new era of scaling in an SoC world," in *2009 IEEE International Solid-State Circuits Conference. Digest of Technical Papers*, IEEE, pp. 23–28, 2009.

[6] H. Mizoguchi, H. Nakarai, T. Abe, *et al.*, "Development of 250 W EUV light source for HVM lithography," in *High-Power Laser Materials Processing: Applications, Diagnostics, and Systems VI*, SPIE Proc., vol. 10097, International Society for Optics and Photonics, p. 1009702, 2017.

[7] Extreme Ultraviolet Lithography – Wikipedia. See https://en.wikipedia.org/wiki/Extreme_ultraviolet_lithography.

[8] J. Hong, C. Park, C. Lee, *et al.*, "Development of full-size EUV pellicle with thermal emission layer coating," in *International Conference on Extreme Ultraviolet Lithography 2018*, SPIE Proc., vol. 10809, International Society for Optics and Photonics, p. 108090, 2018.

[9] ASML Technical Specifications for Twinscan NXT:2000i and Twinscan NXE 3400B lithography systems.

[10] Digital micromirrors are manufactured by Texas Instruments. See www.ti.com/lit/an/dlpa008b/dlpa008b.pdf.

[11] J. Zhu, M. Li, J. Qiu and H. Ye, "Fabrication of high fill-factor aspheric microlens array by dose-modulated lithography and low temperature thermal reflow," *Microsystem Technologies*, vol. 25, no. 4, pp. 1235–1241, 2019.

[12] Heidelberg Instruments DWL 66+Data Sheet.

[13] H. Ito, "Chemical amplification resists: history and development within IBM," *IBM Journal of Research and Development*, vol. 41, no. 1–2, pp. 119–130, 1997.

[14] D. Seeger, "Chemically amplified resists for advanced lithography: road to success or detour?," *Solid State Technology*, vol. 40, no. 6, pp. 115–119, 1997.

[15] G. M. Wallraff, R. D. Allen, W. D. Hinsberg, *et al.*, "Single-layer chemically amplified photoresists for 193-nm lithography," *Journal of Vacuum Science & Technology B: Microelectronics and Nanometer Structures Processing, Measurement, and Phenomena*, vol. 11, no. 6, pp. 2783–2788, 1993.

[16] Integrated Micro Materials, "Chemistry and processing of DUV chemically amplified photoresists," *Integrated Micro Materials Technical Bulletin*, Integrated Micro Materials, 2013. See http://www.imicromaterials.com/technical/duv-photoresist-processing.

[17] H. Ito and C. G. Willson, "Applications of photoinitiators to the design of resists for semiconductor manufacturing," in *Polymers in Electronics*, ACS Symposium Series, vol. 242, American Chemical Society, pp. 11–23, 1984.

[18] R. Brainard, M. Neisser, G. Gallatin, A. Narasimhan and V. Bakshi, "Photoresists for EUV lithography," in *EUV Lithography*, 2nd edn., SPIE, chap. 8, 2018.

[19] J. W. Goodman, *Introduction to Fourier Optics*, Roberts and Co., 2005.

[20] H. J. Levinson, *Principles of Lithography*, SPIE, 2005.

[21] C. Mack, *Fundamental Principles of Optical Lithography: The Science of Microfabrication*, Wiley, 2008.

[22] lumenlearning, Image Formation by Lenses, https://courses.lumenlearning.com/austincc-physics2/chapter/25-6-image-formation-by-lenses/.

[23] B. J. Lin, "Electromagnetic near-field diffraction of a medium slit," *Journal of the Optical Society of America*, vol. 62, no. 8, pp. 976–981, 1972.

[24] S. J. McNab and R. J. Blaikie, "Contrast in the evanescent near field of $\lambda/20$ period gratings for photolithography," *Applied Optics*, vol. 39, no. 1, pp. 20–25, 2000.

[25] Video of Fresnel diffraction for a square shape, https://www.youtube.com/watch?v=3hoyxTbDA-Y.

[26] W. M. Moreau, *Semiconductor Lithography: Principles, Practices, and Materials*, Springer, 2012.

[27] F. H. Dill, "Optical lithography," *IEEE Transactions on Electron Devices*, vol. 22, no. 7, pp. 440–444, 1975.

[28] F. H. Dill, W. P. Hornberger, P. S. Hauge and J. M. Shaw, "Characterization of positive photoresist," *IEEE Transactions on Electron Devices*, vol. 22, no. 7, pp. 445–452, 1975.

[29] F. H. Dill, A. R. Neureuther, J. A. Tuttle and E. J. Walker, "Modeling projection printing of positive photoresists," *IEEE Transactions on Electron Devices*, vol. 22, no. 7, pp. 456–464, 1975.

[30] K. L. Konnerth and F. H. Dill, "In-situ measurement of dielectric thickness during etching or developing processes," *IEEE Transactions on Electron Devices*, vol. 22, no. 7, pp. 452–456, 1975.

[31] C. A. Mack, "PROLITH: a comprehensive optical lithography model," in *Optical Microlithography IV*, SPIE Proc., vol. 538, International Society for Optics and Photonics, pp. 207–220, 1985.

[32] W. G. Oldham, S. N. Nandgaonkar, A. R. Neureuther and M. O'Toole, "A general simulator for VLSI lithography and etching processes: Part I – Application to projection lithography," *IEEE Transactions on Electron Devices*, vol. 26, no. 4, pp. 717–722, 1979.

[33] Athena and Victory Process are commercially available simulation tools from Silvaco. They include lithography simulation tools. See https://www.silvaco.com/products/tcad/process_simulation/process_simulation.html.

[34] Sentaurus Lithography is a commercially available simulation tool from Synopsys. See https://www.synopsys.com/silicon/mask-synthesis/sentaurus-lithography.html.

[35] PROLITH is a lithography simulation tool available from KLA Corp. See https://www.kla-tencor.com/products/chip-manufacturing/patterning-simulation.

[36] ASML Twinscan 1460 K 193 nm step and scan system. See https://www.asml.com/en/products#lithography-systems.

[37] V. Luong, V. Philipsen, E. Hendrickx, *et al.*, "Ni-Al alloys as alternative EUV mask absorber," *Applied Sciences*, vol. 8, no. 4, p. 521, 2018.

[38] A. Erdmann, P. Evanschitzky, F. Shao, *et al.*, "Predictive modeling of EUV-lithography: the role of mask, optics, and photoresist effects," in *Physical Optics*, SPIE Proc., vol. 8171, International Society for Optics and Photonics, p. 81710, 2011.

[39] M. D. Smith, T. Graves, J. Biafore, *et al.*, "Comprehensive EUV lithography model," in *Extreme Ultraviolet (EUV) Lithography II*, SPIE Proc., vol. 7969, International Society for Optics and Photonics, p. 796906, 2011.

[40] C. A. Mack, "Absorption and exposure in positive photoresist," *Applied Optics*, vol. 27, no. 23, pp. 4913–4919, 1988.

[41] E. J. Walker, "Reduction of photoresist standing-wave effects by post-exposure bake," *IEEE Transactions on Electron Devices*, vol. 22, no. 7, pp. 464–466, 1975.

[42] Y. Zou, Y. Deng, J. Kye, *et al.*, "Evaluation of lithographic benefits of using ILT techniques for 22nm-node," in *Optical Microlithography XXIII*, SPIE Proc., vol. 7640, International Society for Optics and Photonics, p. 76400L, 2010.

[43] Y. Wei and D. Back, "193nm immersion lithography: status and challenges," *SPIE Newsroom*, SPIE, 2007.

[44] S. Hotta and S. Okazak, "Layout design and lithography technology for advanced devices," *Hitachi Review*, vol. 57, no. 3, p. 117, 2008.

[45] M. D. Levenson, "Extending the lifetime of optical lithography technologies with wavefront engineering," *Japanese Journal of Applied Physics*, vol. 33, no. 12S, p. 6765, 1994.

[46] "Performing Optical Proximity Correction (OPC) in Athena," Silvaco. See https://www.silvaco.com/tech_lib_TCAD/simulationstandard/1997/feb/a1/a1.htm.

[47] A. Awad, A. Takahashi and C. Kodama, "A fast manufacturability aware optical proximity correction (OPC) algorithm with adaptive wafer image estimation," in *2016 Design, Automation & Test in Europe Conference & Exhibition (DATE)*, IEEE, pp. 49–54, 2016.

[48] A. Awad, A. Takahashi and C. Kodaman, "Optical proximity correction (OPC) under immersion lithography," in *Micro/Nanolithography – A Heuristic Aspect on the Enduring Technology*, IntechOpen, 2017.

[49] M. D. Levenson, N. Viswanathan, and R. A. Simpson, "Improving resolution in photolithography with a phase-shifting mask," *IEEE Transactions on Electron Devices*, vol. 29, no. 12, pp. 1828–1836, 1982.

[50] Y. Borodovsky, W.-H. Cheng, R. Schenker and V. Singh, "Pixelated phase mask as novel lithography RET," in *Optical Microlithography XXI*, SPIE Proc., vol. 6924, International Society for Optics and Photonics, p. 69240E, 2008.

[51] X. Ma, C. Han, Y. Li, L. Dong and G. R. Arce, "Pixelated source and mask optimization for immersion lithography," *Journal of the Optical Society of America A*, vol. 30, no. 1, pp. 112–123, 2013.

[52] N. Jia and E. Y. Lam, "Pixelated source mask optimization for process robustness in optical lithography," *Optics Express*, vol. 19, no. 20, pp. 19384–19398, 2011.

[53] R. Schenker and V. Singh, "Foundations for scaling beyond 14nm," in *Proceedings of the IEEE 2013 Custom Integrated Circuits Conference*, IEEE, pp. 1–4, 2013.

[54] Multiple Patterning in Wikipedia provides a good introduction to many of these methods. See https://en.wikipedia.org/wiki/Multiple_patterning.

[55] P. McLellan, "The Rosetta Stone of lithography," SemiWiki.com, November 20, 2013.

[56] M. LaPedus, "Single vs. multi-patterning EUV," Semiconductor Engineering, March 25th, 2019. See https://semiengineering.com/single-vs-multi-patterning-euv/.

[57] S. Y. Chou, P. R. Krauss and P. J. Renstrom, "Nanoimprint lithography," *Journal of Vacuum Science and Technology B*, vol. 14, no. 6, pp. 4129–4133, 1996.

[58] S. V. Sreenivasan, "Nanoimprint lithography steppers for volume fabrication of leading-edge semiconductor integrated circuits," Microsystems & Nanoengineering, vol. 3, p. 17075, 2017.

5.11   PROBLEMS

5.1 Calculate and plot versus exposure wavelength the theoretical resolution and depth of focus for a projection exposure system with NA = 0.6. Assume k_1 = 0.6 and k_2 = 0.5. Consider wavelengths between 100 nm and 1000 nm (DUV and visible light). Indicate the common exposure wavelengths being used on your plot (g-line, i-line, KrF and ArF). Is an ArF source adequate for the 0.13 μm and 0.1 μm technology generations according to these simple calculations?

5.2 In a particular i-line positive resist process, it is sometimes noticed that there is difficulty developing away the last few hundred ångströms of resist in exposed areas. This sometimes causes etching problems because the resist remains in the areas to be etched. Suggest a possible cause of this problem and therefore propose a solution.

5.3 Ray tracing, illustrated below, is a simple concept that works very well for describing and understanding many optical systems. Explain why this method cannot be used in modern optical lithography systems, at least for the focusing lens system.

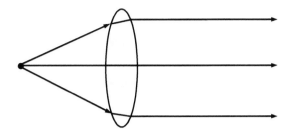

5.4 The MTF of an aerial image was defined as MTF $= (I_{MAX} - I_{MIN})/(I_{MAX} + I_{MIN})$. This parameter depends on the features being printed. Explain using a sketch why the MTF depends on the features being printed.

5.5 An X-ray exposure system operating in the proximity printing mode uses photons with an energy of 1 keV. If the separation between the mask and wafer is 20 μm, estimate the diffraction-limited resolution that is achievable by this system.

5.6 In this chapter, we considered several examples of mask patterns and looked at their Fourier transforms (far-field Fraunhofer diffraction patterns). This function describes the spatial variation of the light intensity that the objective lens must capture. Another example of the effects of diffraction is given by a mask pattern consisting of a periodic pattern of lines and spaces as shown here.

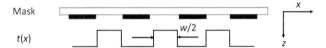

Find the Fourier transform for this mask function and plot it, as was done in the examples in this chapter. The objective lens must capture at least the first diffraction order if it is going to resolve features with a pitch p. Use this criterion to derive a result like (5.8) for the resolution of such a system.

5.7 Several examples of far-field Fraunhofer diffraction patterns for simple geometries were discussed in this chapter (Figure 5.31). Choose an additional possible mask shape of interest to you and do an internet search to find the mathematical expression that describes the diffraction pattern. If you have access to MATLAB and want to calculate the diffraction pattern (analogous to Figure 5.31), please do so, but, alternatively, you can likely find the diffraction pattern for most simple shapes on the internet. Either by calculating or searching, show the diffraction pattern for your shape. Discuss qualitatively the general features of this diffraction pattern. Your discussion does not need to be more than a paragraph or two. Please make sure to list any references or internet sites you use in your solution.

5.8 In discussing the mathematics of projection lithography systems, we saw that the aerial image that is produced by such systems is described by the equation

$$I(x,y) = \left[F^{-1}\left\{F\{t(x_1,y_1)\}P(f_x,f_y)\right\}\right]^2,$$

where $I(x, y)$ is the light intensity at each point x, y in the aerial image. Explain what the physical meaning of each of the terms on the right-hand side of the equation is, and how this equation describes what the various parts of the optical system do.

5.9 The numerical aperture for a given lithography system is NA = 0.5. To ensure the quality of exposure, we want the depth of focus to cover the whole thickness of the photoresist. Also, the minimum resolution required is 500 nm. Assume $k_1 = 0.75$ and $k_2 = 0.5$. The photoresist is uniformly coated on the wafer with a thickness of 0.7 μm.

(a) If the focal plane of the projection system is positioned at the bottom of the resist (the boundary of the photoresist and the wafer), calculate the wavelength range that is suitable for this exposure process. Comment whether you could successfully design this wavelength based on your result.

(b) If the focal plane of the projection system is positioned in the middle height of the photoresist, calculate the wavelength range that is suitable for this exposure process.

5.10 A projection lithography system uses an ArF laser source to produce a 193 nm wavelength and an exposure lens with NA = 0.9. Assume values of $k_1 = 0.75$ and $k_2 = 0.3$ in answering the following.

(a) Estimate the height variation (surface roughness) tolerance on the wafer surface of this projection system. (Use the depth of focus, DOF, as a measure of tolerance.)

(b) To improve the system DOF, an immersion lithography system is used with water, which has a refractive index of 1.44. Calculate the improvement of the DOF if this system is being used to create lines with a pitch of 200 nm.

5.11 A student plans to make a grating pattern on his wafer. After trying an exposure on an ordinary lithography system, he realizes this system cannot achieve the resolution needed for the grating, which is less than 100 nm. So he plans to try e-beam lithography instead. An experienced user tells him that a 600 nm thick e-beam resist, with a dose of 500 μC cm^{-2}, should work well for the desired resolution. The pattern he plans to write is a 1 cm^2 area. The e-beam system has a current of 9000 pA. Estimate how long it will take for this e-beam exposure.

5.12 This problem should give you a feeling for why e-beam tools are not used today for exposing wafers. An e-beam lithography system (like the mask-making system illustrated in Figure 5.1) has a beam current of 1 nA, an accelerating voltage for the e-beam of 100 kV and a spot size of 50 nm × 50 nm square. (This size corresponds to the pixel size the machine can write in photoresist.)

(a) If the e-beam resist the machine is exposing requires a dose of 10 mJ cm^{-2}, how long would it take to expose each pixel? (Remember: 1 J = 1 V C.)

(b) How long would it take to write the pattern on an entire 12 inch diameter wafer?

5.13 In projection lithographic systems, an ideal point illumination source can produce a spatial coherence S = 0. Real sources have finite physical dimensions and therefore produce an S larger than 0, explain in your own words why this is actually a good thing.

5.14 DUV resists typically have higher contrast than the older g-line and i-line resists. Explain in your own words physically why this is the case.

5.15 A particular DUV photoresist has a contrast of 4. The post-exposure bake (PEB) time is doubled in an experiment. (The PEB is the heat cycle after exposure which allows the acid molecules created by the exposure to react with the resist molecules.) Estimate quantitatively the contrast that the resist would exhibit with this doubled PEB time. State any assumptions.

5.16 A student goes to work in the semiconductor industry and realizes that there are major issues with getting complex EUV lithography systems to work in manufacturing. Since if they cannot be made to work, the industry could be in serious trouble, the student has the interesting idea of using an EUV source with a wavelength of 13.5 nm in a proximity printing system, rather than a projection printing system. This would imply a much simpler lithography tool and should provide high resolution. Is this a good idea or not? Why?

5.17 For conventional g-line and i-line resists, the exposure process can be experimentally measured using a setup like that shown below. The curve on the right allows extraction of the "Dill parameters" that are used in modeling photoresist exposure. Explain in your own words how this measurement works and why the experimental curve on the right has the shape that it does. Your explanation should include the starting and final values, and what is happening chemically in the resist between T_0 and T_∞.

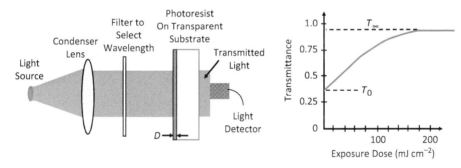

5.18 Assume that Figure 5.50 in the text was experimentally determined for a 0.6 μm thick resist with an index of refraction of 1.68. Estimate the Dill resist parameters from the data in that figure.

5.19 Why is the quantum efficiency of DNQ resists fundamentally limited to ≤1?

5.20 In the Dill resist exposure model, the parameters can be measured experimentally for a resist using the setup shown in Figures 5.47 and 5.48. The C parameter in Dill's model is extracted from such data using the following equation:

$$C = \frac{A+B}{AT_0(1-T_0)T_{12}} \frac{dT}{dE}\bigg|_{E=0}.$$

In this expression $dT/dE|_{E=0}$ is the slope of the experimental curve at the beginning of the exposure. Explain physically why C should be proportional to this slope.

5.21 As described in the text, the exposure process for DNQ resists consists of incoming photons reacting with PAC molecules to change them through the Wolff rearrangement process so the PAC molecules are soluble in the developer. This process is time- and position-dependent throughout the resist volume. Consider the top layers of the resist film, where the incoming light intensity I is relatively constant during the exposure. The PAC concentration during the exposure starts at some value $[PAC]_0$ and decreases with time during the exposure. We described this process with the parameter m, which starts at 1 and goes to 0 as the exposure goes to completion. Given a resist with a Dill parameter $C = 0.01 \text{ cm}^2 \text{ mJ}^{-1}$, and a light intensity $I = 1000 \text{ mJ cm}^{-2} \text{ s}^{-1}$, how long does it take to expose 90% of the resist PAC molecules in the top layers of the resist?

5.22 In Figure 5.49 we discussed the use of the following experimental method to extract the Dill resist parameters for the exposure of g-line and i-line resists. This method does not work to characterize more modern DUV and EUV resists. Explain in your own words why this simple method does not work for these newer resists.

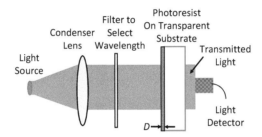

5.23 Lithography often has to be done over underlying topography on a silicon chip. This can result in variations in the resist thickness as the underlying topography goes up and down. This can sometimes cause some parts of the photoresist image to be underexposed and/or other regions to be overexposed. Consider an i-line resist. Explain in terms of the chemistry of the resist exposure process why these underexposure and overexposure problems occur.

5.24 One of the techniques we discussed to improve the resolution of optical lithography systems is to use Köhler illumination, which basically focuses the mask image in the center of the projection optics aperture. Explain physically in your own words why this improves resolution.

5.25 EUV systems use a shorter exposure wavelength to achieve higher resolution. They also generally require planarization techniques to provide "flat" substrates on which to expose the resist layers. Explain why "flat" substrates are critical for these exposure systems.

5.26 Modern DUV resists can achieve arbitrarily high quantum efficiencies simply by extending the post-exposure bake time. This has the advantage of improving the throughput of lithography tools because the exposure required is reduced when the quantum efficiency is higher. What limits how much you might want to take advantage of this trade-off? Explain.

5.27 Current optical projection lithography tools produce diffraction-limited aerial images. A typical aerial image produced by such a system (Figure 5.37) is shown below, where a square and rectangular mask regions produce the image shown. (The mask features are the black outlines; the calculated aerial image is the colored area inside the black rectangles.)

The major feature of the aerial image is its rounded corners compared to the sharp square corners of the desired pattern. Explain physically in your own words why these features look the way they do, using diffraction theory and the physical properties of modern projection optical lithography tools.

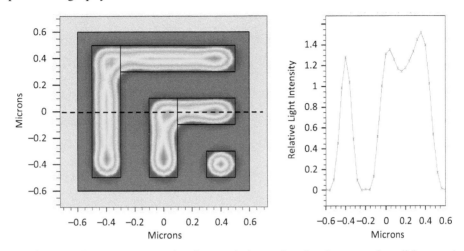

5.28 The diagram shows a rectangular feature being printed using a modern lithography tool. The aerial images are on the right. The top images show the aerial image with no OPC; the bottom images show the effect of adding OPC to the mask. Explain in your own words why the aerial image using OPC looks so much better. Include in your explanation the concepts of high spatial frequencies and diffraction.

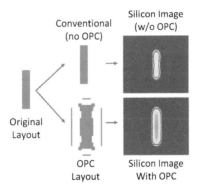

5.29 The highest-resolution 193 nm optical lithography systems are immersion systems, that is, they use a layer of water between the final optical lens and the wafer. Explain why this approach enables higher-resolution lithography.

5.30 A 193 nm lithography tool is being used to print features that are much smaller than the wavelength of the light being used. As a result, the mask pattern (the rectangle) below is distorted when it is printed. The printed aerial image is shown superimposed on the rectangular mask pattern. Using OPC, sketch how you would redesign the mask so that a better quality aerial image is produced. Explain your sketch.

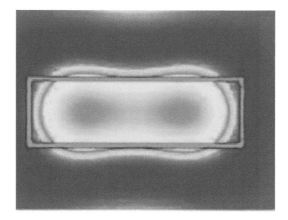

5.31 The minimum feature that a 193 nm lithography tool can directly print in photoresist is on the order of 40 nm. Yet features much smaller than this are routinely used in the latest generations of technology. For example, the fin width in FinFET technologies is much smaller than 40 nm. Suggest one method by which these smaller features can be achieved in manufacturing.

5.32 A circuit designer is constructing a physical layout for a circuit using a CAD system. A small portion of the design is shown on the left below. Once the full design is finished, the designer runs her design through a computer program that automatically corrects the layout using OPC software, so that a better replica of the desired design will be printed on the wafer during lithography. Two of the corrections made by the software are shown on the right below. (Other corrections were also made.) Explain physically why these corrections were made.

5.33 In the latest generations of silicon chips, multi-patterning has become widely used as a method for printing smaller features. Explain in terms of diffraction why this works to improve resolution.

6 Semiconductor/Insulator Interfaces

6.1 Introduction

The interface between a semiconductor and an insulator often determines the viability of the material combination in device structures. Silicon is unique in nature, at least among the semiconductors, for having a robust, reliable oxide that can be grown on its surface. The interface between Si and SiO_2 is perhaps the most carefully studied of all material interfaces, and is probably the principal reason why silicon has been the dominant semiconductor material. The fact is that silicon naturally oxidizes in the sense that it can be simply placed in a furnace at high temperature with oxygen or water vapor and one obtains a nice, stable dielectric material that is essentially electrically perfect. This distinguishes silicon from all the other simple column IV semiconductor materials. Germanium can be oxidized, but its oxide is soluble in water, which makes it very hard to do any sort of chemical processing. Silicon carbide can be oxidized to produce SiO_2, but it also forms a gaseous oxide, CO or CO_2, and the SiC/SiO_2 interface is not nearly as good as the Si/SiO_2 interface. In the III–V or II–VI materials, any of these compound semiconductors generally have the problem that the oxide or dielectric that forms on them is some combination of the underlying elements and is not homogeneous. The electrical properties of these interfaces are not nearly as good as those of the Si/SiO_2 interface.

The perfection of the Si/SiO_2 interface caused the transition from germanium to silicon in the early days of semiconductor processing. This interface is rather unique among semiconductor/insulator interfaces. SiO_2 can be selectively etched, it is a good masking material against the diffusion of dopants, it has a very wide bandgap so is an excellent insulator, and it can tolerate very high electric fields. The interface between silicon and its oxide is virtually perfect on the atomic scale, has very few mechanical or electrical defects and is stable over time. These properties make metal-oxide–semiconductor (MOS) structures easy to build in silicon, and it means that silicon devices of all types are generally reliable and stable.

We saw a number of applications of SiO_2 layers in the complementary metal-oxide–semiconductor (CMOS) technology example in Chapter 2. These included using the thin oxide as the gate dielectric layer, as a spacer mask against implantation, as an isolation oxide in the shallow trench and as the insulator between metal layers in back-end processing. These and some other uses are illustrated in Figure 6.1. In other semiconductor systems, insulating layers are not as perfect as the canonical Si/SiO_2 interface and often require deposited oxides because their oxides cannot be thermally grown. The perfection of the silicon/oxide interface

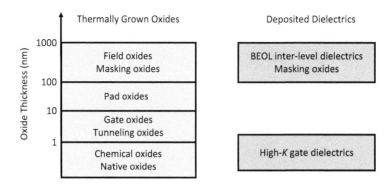

Figure 6.1 Approximate range of oxide and dielectric thicknesses in silicon technology.

is due to careful growth conditions and processing, and understanding how this was achieved is useful for optimizing other semiconductor/insulator combinations.

Even at room temperature, a bare silicon surface will begin to oxidize in air and will form a thin native oxide, which grows slowly and effectively stops after a few hours with a final thickness on the order of 1–2 nm. The Standard Clean SC-1 and SC-2 cleaning procedures discussed in Chapter 4 create chemical oxides on the silicon surface due to H_2O_2 oxidation, which can be less than 1 nm thick. Because of the unique perfection of the silicon/oxide interface, even when high-K dielectrics are used as the gate insulator, a thin interfacial oxide is often used. If layers are to be deliberately grown on silicon by epitaxy, for example, silicon, gallium arsenide or gallium nitride, this native oxide or chemical oxide must first be desorbed from the interface in a H_2 ambient at 700–900 °C.

Looking at Figure 6.1, thin oxide layers in the range of 1–10 nm are used as gate oxides or passivation oxides in silicon devices. Aside from the highest-performance logic devices using high-K dielectrics, most silicon devices use these thermally grown gate oxides because of their high quality. Such oxides would be used in biosensors, micro-electromechanical systems (MEMS) driver devices and power devices, for example. For the thinner oxides, the term "insulator" begins to lose its meaning, because these thin films will conduct finite amounts of current due to quantum mechanical tunneling of electrons or holes. The process occurs because of the wave nature of particles, which basically allows them to pass through an otherwise impenetrable barrier if it is thin enough. These tunneling currents increase exponentially as the barrier becomes thinner, and were once thought to be a fundamental limit on Moore's law scaling before the introduction of high-K dielectrics. A positive aspect of these tunneling currents is that they can form the basis for a non-volatile memory element by trapping charge in the polysilicon gate under electrical bias. These electrically erasable and programmable memories take advantage of the tunneling currents to store charge (see Figure 1.24). In some applications, even minute tunneling currents can be problematic, such as in the pass transistor that isolates the storage capacitor in a dynamic random access memory (DRAM; see Figure 4.4). This leakage current can be a major problem because it directly affects data retention.

Slightly thicker thermal oxides in the range of 10–50 nm are often used under silicon nitride layers as a stress relief or "pad" oxide during LOCOS (local oxidation of silicon) isolation processes. LOCOS was the dominant technology used to laterally isolate transistors before the STI (shallow trench isolation) process we described in Chapter 2 became dominant around 1990. These oxides can also be used to passivate the sidewalls of trenches during the STI process. Thicker thermal oxides often serve as masks for ion implantation or gas-phase doping steps. We will consider these applications more carefully in Chapters 7 and 8. It is also possible to deposit oxide layers using chemical vapor deposition (CVD) or low-pressure chemical vapor deposition (LPCVD) techniques. Deposition usually involves a much smaller "thermal budget" compared to thermal oxidation and is preferred even in front-end processes whenever it is important to minimize the temperature cycle to which the wafers are subjected. The interface between a deposited oxide and the underlying silicon is not as perfect electrically as that formed by a thermal oxide. It is possible to anneal a deposited oxide to improve its properties and approach those of a thermally grown oxide. The simplest way to "anneal" such an interface is simply to expose the deposited oxide to oxygen or water vapor at high temperatures. Because the oxidant diffuses to the interface to grow new thermal oxide, the interface improves and the high-temperature anneal densifies the oxide, improving its bulk properties.

Deposited oxides are essential in the back-end-of-line (BEOL) interconnect processes and in this case a low-temperature deposition processes is required. In addition, modifications to the oxide deposition process to achieve a "low-K" material to minimize resistive–capacitive (RC) delay are needed. We will discuss these issues in detail in Chapter 11.

This chapter will focus primarily on the thermal oxidation process and the properties of the silicon/oxide interface. The measurement methods used to characterize the oxide are broadly applicable to all semiconductor/insulator interfaces. The only other thermal oxidation process that is widely used in semiconductor technologies is the thermal oxidation of SiC, which also produces SiO_2. The carbon forms CO or CO_2, which is carried away as a gas. But, as noted above, the SiC/SiO_2 interface is not nearly as perfect as the Si/SiO_2 interface.

6.2   Basic Model for Oxide Growth

The basic equipment used to oxidize silicon is very simple compared to the sophisticated tools used for processes like lithography, etching and deposition. Figure 6.2 shows a horizontal furnace with Si wafers stacked side-by-side in the furnace. The oxidant gases are introduced at the rear of the furnace, O_2 for dry oxidation and $H_2 + O_2$ for wet oxidations. In the wet case, the H_2 is ignited in the back end of the furnace to produce H_2O.

We will see shortly that the fact that the wafers can be stacked closely together in such a furnace implies that the transport of the oxidizing species from the gas phase to the wafer surface is not a rate-limiting step in the oxidation process. If it were, grown oxide thicknesses would vary across a wafer with the arrangement shown in Figure 6.2. Vertically oriented furnaces and single-wafer oxidation systems are also common today.

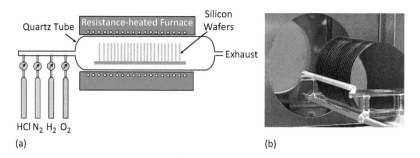

Figure 6.2 (a) Conceptual picture of a thermal oxidation furnace. (b) Example of Si wafers being loaded into such a horizontal furnace system. Photo courtesy of R. Carranza.

Figure 6.3 Basic process for oxidation of silicon where oxygen diffuses through the growing layer to complete a chemical reaction at the Si/SiO$_2$ interface.

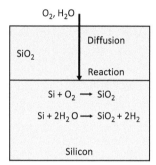

In silicon oxidation, the oxygen or water vapor diffuses through the oxide to the silicon/oxide interface where the new growth occurs. Silicon oxidation therefore occurs by the inward diffusion of the oxidant rather than the outward diffusion of silicon, as depicted in Figure 6.3. This was demonstrated by growing an SiO$_2$ layer using one isotope of oxygen (say ^{16}O) and then continuing the growth with a second isotope (say ^{18}O). By profiling through the resulting composite oxide with a mass-sensitive technique, one can determine whether the second oxide grows at the top or the bottom surface. This kind of experiment has shown that the new oxide grows at the Si/SiO$_2$ interface.

Conceptually, one can think of this as oxygen atoms in-diffusing to the Si/SiO$_2$ interface, breaking Si–Si bonds and inserting oxygen atoms to form Si–O bonds. The process involves a volume expansion because of the room needed for the oxygen atoms. As shown in the center of Figure 6.4, the oxide would like to expand by 30% in all three directions to accommodate the oxygen atoms. However, because the silicon substrate constrains the oxide on two sides, as shown on the right-hand side of Figure 6.4, the only option is for the oxide to expand both down into the silicon as it grows and above the silicon to accommodate the volume expansion. The volume expansion is thus accommodated by a 2.2× expansion of the oxide compared to the volume of silicon oxidized.

Figure 6.5 illustrates this process for a classical LOCOS structure where a thick isolation oxide is grown next to a nitride mask. In the LOCOS process, the nitride acts as diffusion

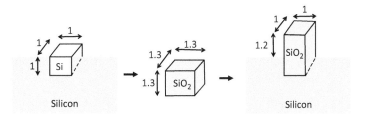

Figure 6.4 Schematic of the volume expansion that occurs during silicon oxidation where a 30% volume expansion on converting silicon to silicon dioxide is constrained by the silicon substrate to give a 2.2 unit thickness expansion.

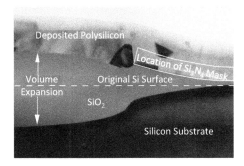

Figure 6.5 A cross-section scanning electron micrograph of a simple LOCOS structure. The 2.2× volume expansion of the grown oxide above and below the original silicon interface is clearly shown. Photo courtesy J. Bravman.

barrier to the O_2 or H_2O oxidant and prevents oxidation underneath the Si_3N_4 layer. Thus the silicon is oxidized locally, and hence the name LOCOS. The cross-section image clearly shows that the volume expansion on the planar surface is accommodated by the oxide both growing down towards the silicon interface and expanding upwards beyond the original interface for the 2.2× volume expansion. It is also clear that the nitride does not act as a perfect mask for the oxidation, as the oxidant diffuses and reacts laterally under the mask edge, thereby lifting the nitride to produce the characteristic "bird's beak" shape. This is a very useful and simple isolation structure but, because of the lateral encroachment, it has been replaced in advanced devices by the more compact STI structure that we discussed in Chapter 2.

Because the oxide is grown on a crystalline silicon substrate, one might assume that the oxide is in a crystalline form like a quartz crystal. However, the lattice constant of quartz is so far from that of silicon that this forces the growth of the oxide to be amorphous. A central silicon atom is tetrahedrally coordinated to four oxygen atoms, one of which is shared by an adjacent silicon atom, as depicted in Figure 6.6. These SiO_4 tetrahedra are the basic units from which SiO_2 forms. They bond together by sharing the bridging oxygen atoms, and these bridging atoms allow one tetrahedron to rotate with respect to its neighbors and lose long-range order and thus become amorphous.

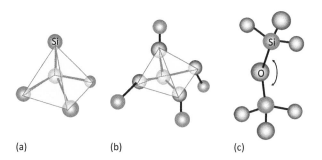

(a)   (b)   (c)

Figure 6.6 A silicon tetrahedron (a) is oxidized by inserting O atoms between each Si atom. The volume expands to accommodate the oxygen atoms. This forms SiO_4 tetrahedra (b), which are the basic units that link up to make SiO_2 as shown (c).

The requirement for volume expansion during oxide growth means that the oxide is in compressive stress. As shown in Figure 6.5, the growing oxide can only expand upwards and not laterally, resulting in compressive stresses that can be as large as 5×10^9 dyne cm^{-2}. So even on a flat, planar, unpatterned surface, there are large compressive stresses in thermally grown oxides. At temperatures above about 1000 °C, the oxide can relieve some of the stress by viscous flow, so that oxides grown at higher temperatures introduce less stress in the silicon. The oxide viscosity is nonlinear, so even a 100 °C change in temperature can change the profile in a trench corner and consequently affect the dielectric breakdown and reliability of the oxide. These stresses can give rise to defects especially at corners or if they are interacting with damage produced by ion implants. We will look at these 2D and 3D effects later in this chapter. Another source of stress is the large difference in the thermal expansion coefficients between Si and SiO_2. As the wafers are cooled down after a thermal oxide growth, this results in an additional compressive stress in the oxide that can be as high as a few $\times 10^9$ dyne cm^{-2}. In the early days of semiconductor manufacturing, random defects or dislocations would occur in devices and were eventually traced to these stress fields. Careful control of the growth temperature and the cooling rates after thermal oxidation have eliminated these problems. As we will see later in this chapter, simulation tools are capable of predicting these stresses, which are important in designing new structures with different corners and edges.

The high-resolution transmission electron micrograph (TEM) image in Figure 6.7 shows the crystal structure in the silicon and an amorphous oxide layer, with an abrupt transition between the two. The interface appears to be atomically abrupt, though there is some evidence of steps of a single atomic layer occurring occasionally. The quality of the interface can be inferred by measuring the mobility of electrons or holes in the inversion layer of a MOS device, and it is known that the interface is a little rougher for oxides that are rapidly grown (water vapor versus O_2 oxidation, for example) or grown at lower temperatures.

The growth of oxides has been carefully studied since the early 1960s given their importance in semiconductor manufacturing. One of the earliest explanations of the growth rate, known as the Deal–Grove model, captured most of the physics of oxidant transport and reaction during oxidation [1]. The key ideas behind the original Deal–Grove model are illustrated in Figure 6.8. The thickness of the oxide on the silicon surface, x_o, is a function of time, and we assume that

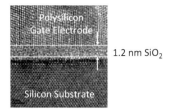

Figure 6.7 High-resolution TEM image of the abrupt interface between silicon and silicon dioxide, showing crystalline silicon and amorphous oxide. Photo © Intel Corporation. Reprinted with permission.

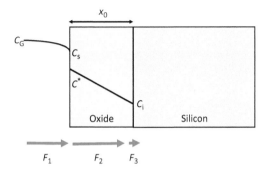

Figure 6.8 Oxidant flux from the gas phase to the silicon/oxide interface during thermal oxidation. The bold line represents the O_2 or H_2O concentration.

the growth is one-dimensional on a planar flat substrate, so we do not have to worry about two-dimensional effects and viscous flow.

The oxide grows by in-diffusion of the oxidizing species to the oxide/silicon interface, where a simple chemical reaction occurs with O_2 or H_2O:

$$\text{Si} + O_2 \rightarrow SiO_2, \tag{6.1}$$

$$\text{Si} + 2H_2O \rightarrow SiO_2 + 2H_2. \tag{6.2}$$

Note that neutral molecular species are assumed to be involved in the reaction, although there was some controversy in the early work as to whether charged species played a role. In the original Deal–Grove model, three separate fluxes were postulated, as shown in Figure 6.8: a transport flux of oxidant through the gas phase to the outer interface, F_1; a diffusive flux of oxidant through the oxide layer, F_2; and a chemical reaction flux at the interface, F_3. Deal and Grove suggested that, at steady state, the fluxes of each of these components must be equal, otherwise you would build up concentrations of oxidant; in other words, the flux in each phase must balance.

However, a simple experiment shows that gas-phase transport and adsorption do not limit the rate at which oxidant enters the oxide and F_1 is therefore not a limiting process in the oxidation of silicon wafers. If we take three wafers and stack two closely together vertically and place another flat in a high-temperature furnace with oxidation occurring, the thickness of the oxide layers on all the exposed surfaces will be equal if the temperature in the furnace is uniform. Essentially, the experiment says that the gas-phase flux is very fast, with the gas being an infinite supplier of oxidizing reactant. If F_1 were important, indicating that getting the oxygen to the surface of the wafer was limiting, then the experiment would not show the same oxide thickness at the bottom of the stacked wafers or the same thickness on a stacked wafer versus a flat wafer. This shows that the gas-phase transport is negligible in terms of being a limiting process for oxidation.

When we discuss CVD systems in Chapter 10, the gas-phase processes will be very important. Many CVD systems have important gas-phase limitations, and we will return to this model in Chapter 10 and include F_1. Because F_1 is not rate-limiting in oxidation, the wafer stacking arrangement shown in Figure 6.2 produces uniform oxide thickness across wafers and allows many more wafers to be loaded into an oxidation furnace than would be possible if the wafers had to be loaded horizontally, side-by-side.

The basis of the Deal–Grove model can thus be simplified to the diffusion and reaction fluxes F_2 and F_3. Oxidant is available at the SiO$_2$ surface at an equilibrium concentration C^* that is the solubility of the oxidant in SiO$_2$. For wet oxides, the H$_2$O solubility C^* is approximately 3×10^{19} cm^{-3} and for dry oxides the O$_2$ solubility C^* is 5×10^{16} cm^{-3}. The oxidant diffuses through the existing oxide until it reaches the silicon interface, where it reacts with the silicon to form SiO$_2$. Because the volume of the reactants is greater than the volume of silicon consumed, there is a volume expansion where the new oxide expands both downwards and upwards from the original silicon interface.

The flux F_2 represents the diffusion of the oxidant through the oxide to the Si/SiO$_2$ interface. Using Fick's law (which will be described in more detail in Chapter 7), we can express this flux as

$$F_2 = -D\frac{dC}{dx} = D\left(\frac{C^* - C_i}{x_o}\right),\tag{6.3}$$

where D is the oxidant diffusivity in the oxide, C^* and C_i are the concentrations at the two interfaces and x_o is the oxide thickness. In writing this expression, we have assumed that the process is in steady state (not changing with time) and that there is no loss of oxidant as it diffuses through the oxide. Under these conditions, the flux F_2 is constant through the oxide and hence the derivative can be replaced by a simple linear gradient.

The rate at which the oxidant is consumed at the interface is proportional to the concentration there, so the reaction flux is given by

$$F_3 = k_s C_i,\tag{6.4}$$

where F_3 is the number of oxidant molecules reacting per unit area of the interface per unit time, k_s is the reaction rate and C_i is the concentration of oxidant at the interface. The oxide growth rate is equal to the rate at which oxidant molecules are reacting at the interface, divided by the number, N_1, of oxide molecules that must be incorporated to form a unit volume of oxide. For O$_2$ oxidation, $N_1 = 2.2 \times 10^{22}$ cm^{-3} and it is twice this value for H$_2$O oxidation assuming regular oxygen and water are the reactant species.

Under steady-state conditions, these fluxes representing the oxidation process must be equal since they occur in series with each other, and the overall process must proceed at the rate of the slowest process. Thus $F_2 = F_3$, giving

$$D\left(\frac{C^* - C_i}{x_o}\right) = k_s C_i ,$$

(6.5)

so that

$$C_i = \frac{C^*}{k_s x_o / D + 1} .$$

(6.6)

The growth rate is then F_3/N_1, so from (6.4)

$$\frac{dx_o}{dt} = \frac{F_3}{N_1} = \frac{k_s C^* / N_1}{k_s x_o / D + 1}$$

(6.7)

or equivalently

$$\frac{dx_o}{dt} = \frac{B}{2x_o + A} ,$$

(6.8)

where we have defined $B/A = k_s C^* / N_1$ as the linear growth coefficient and $B = 2DC^* / N_1$ as the parabolic growth coefficient, representing limiting forms of the growth law as we will see below. The B and B/A terminology follows the definitions Deal and Grove used in their original paper.

It is now simple to integrate this equation to obtain the oxide thickness x_o from

$$\frac{x_o^2}{B} + \frac{x_o}{B/A} = t + \tau ,$$

(6.9)

where

$$\tau = \frac{x_i^2}{B} + \frac{x_i}{B/A} .$$

(6.10)

In these expressions, x_i represents any initial oxide present at the start of the oxidation and τ represents the time taken to originally grow that initial oxide. Solving the parabolic equation (6.9) leads to the following explicit expression for the oxide thickness in terms of the growth time:

$$x_o = \frac{A}{2}\left\{\sqrt{1 + \frac{t + \tau}{A^2/(4B)}} - 1\right\} .$$

(6.11)

There are two limiting forms of the linear–parabolic growth law that can be seen from (6.9) and that are illustrated in Figure 6.9. These occur when one of the two terms dominates, leading to

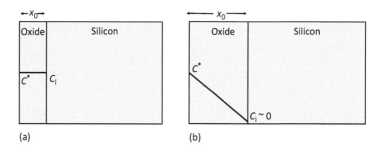

Figure 6.9 Limiting cases of the Deal–Grove oxidation model. (a) The interface reaction is the limiting step in the thin oxide. (b) The diffusive transport of oxidant through the existing thick SiO_2 is the limiting step.

$$x_o = (B/A)(t + \tau) \quad \text{or} \quad x_o^2 = B(t + \tau), \tag{6.12}$$

making it clear that the linear term dominates for small x_o values, while the parabolic term dominates for larger x_o values. SiO_2 growth on a bare silicon wafer thus starts out with a linear thickness versus time characteristic, which becomes parabolic as the oxide gets thicker.

For very thin oxides ($x_o \approx 0$) the linear rate constant is the maximum oxide growth rate attainable at that temperature and has the dimensions of velocity. Given that this term models oxidation from the bare silicon surface, it is perhaps no surprise that it is sensitive to the crystal orientation, and it has the ratios 1.68:1.4:1 for the (111):(110):(100) orientations. The parabolic growth coefficient gives rise to the parabolic growth behavior for longer times, and is related to the diffusion through the oxide. It is not a physical diffusivity, being the product of the actual diffusivity through the oxide and the dimensionless parameter $2C^*/N_1$.

When an oxidation is performed on a flat, lightly doped substrate, the growth kinetics are well described by the linear parabolic law. Highly doped substrates show faster oxidation rates, and we will discuss the reasons later. The main limitation is for thin dry oxides, where the initial 20 nm is not well modeled. Adding a fictitious 20 nm initial oxide brings the data for dry oxidation into agreement with the model described above. The reasons for this initial, anomalous rapid oxidation are also discussed later. Over a remarkable range of times and temperatures in both O_2 and H_2O ambients, the Deal–Grove model does extremely well in modeling the experimental growth rate data. Experimentally, when we fit the growth data, we find that both B and B/A are well described by Arrhenius expressions of the form

$$B = C_1 \exp(-E_1/kT), \tag{6.13}$$

$$B/A = C_2 \exp(-E_2/kT), \tag{6.14}$$

where E_1 and E_2 are the activation energies associated with the physical processes that B and B/A represent, and C_1 and C_2 are the pre-exponential constants. Although these are empirically extracted by fitting (6.13) and (6.14) to experimental data, the fact that the model follows well-behaved Arrhenius relationships over a wide range of temperatures indicates that there are underlying physical reasons for the relationships. Figure 6.10 shows a plot of the constants and Table 6.1 lists experimentally extracted values for the parameters.

Table 6.1 **Experimentally extracted rate constants describing the Deal–Grove (111)**
oxidation kinetics at 1 atm. For (100) silicon, the *B/A* values should be divided by 1.68.

Ambient	B/A (μm min^{-1})	B (μm^2 min^{-1})
Dry O$_2$	$1.04 \times 10^5 \exp\left(-\dfrac{2.0\ \text{eV}}{kT}\right)$	$12.87 \exp\left(-\dfrac{1.23\ \text{eV}}{kT}\right)$
H$_2$O	$2.95 \times 10^6 \exp\left(-\dfrac{2.05\ \text{eV}}{kT}\right)$	$7.0 \exp\left(-\dfrac{0.78\ \text{eV}}{kT}\right)$

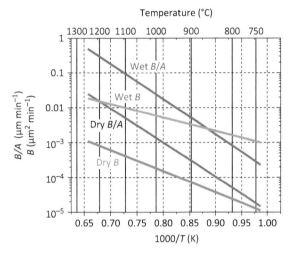

Figure 6.10 Graph showing B and B/A for O$_2$ and H$_2$O oxidation of (111) silicon. Values are taken from the parameters in Table 6.1.

The activation energy for the parabolic constant B in Table 6.1 is quite different for O$_2$ and H$_2$O ambients. Since N_1 is a constant and C^* does not vary much with temperature, this seems to indicate that the origin of the difference is related to the different diffusivities of the O$_2$ and H$_2$O oxidant species in SiO$_2$. Indeed, independent measurements of the diffusion coefficients of O$_2$ and H$_2$O in SiO$_2$ show that these parameters vary with temperature with activation energies similar to those in Table 6.1.

The activation energy for the linear constants B/A are all close to 2 eV, regardless of the oxidation ambient. This suggests that the activation energy for the linear growth coefficient is related to interface effects. Traditionally, the 2 eV activation energy has been associated with the Si–Si bond breaking process. It is also independent of the substrate orientation, suggesting that it is somehow a fundamental part of the oxidation physics. The dependence of the rate on the substrate orientation is subsumed into the C_2 pre-exponential constant, indicating an effective increase in the number of reaction sites for the different orientations.

The magnitudes of the values for B and B/A are much larger for H$_2$O than for O$_2$ ambients, and this is due to the much larger solubility C^* for the H$_2$O oxidant. For this reason, O$_2$ ambients are generally useful for producing thin oxides, while H$_2$O ambients can produce much thicker films in an equivalent time.

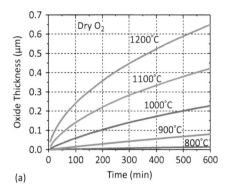
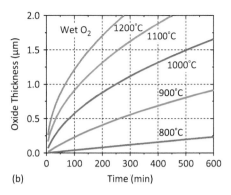

Figure 6.11 Plots of thickness versus time for oxides grown using (a) dry O_2 and (b) wet O_2 (H_2O) for (100) silicon using the parameters from Table 6.1, scaled appropriately for the orientation dependence (B/A values for dry O_2 in Table 6.1 divided by 1.68).

There is actually a slight "break" in the slope of the B and B/A values for wet oxidation below around 900 °C, with the linear constant being about a factor of 2 larger and the parabolic constant a factor of 2 smaller than calculated from Table 6.1 at 750 °C. We ignore this variation for simplicity, but modern simulation programs take these into account. Thickness versus time data for O_2- and H_2O-grown oxides are shown in Figure 6.11 for (100) silicon.

These physical insights into the oxidation process are part of the reason that the Deal–Grove model was so widely embraced and accepted. Other models have been proposed, each claiming to be an improvement over the linear–parabolic model. Examples are a power-law model with two fitting parameters, models involving two parallel oxidation streams like O and O_2, and models involving space-charge effects. An argument can be made that none of these more complicated models fit the data better than the simple Deal–Grove model when a large range of oxide thicknesses is considered. The one exception is that something must be done to bring the Deal–Grove model into agreement with experimental data for dry oxides less than 20 nm thick (the Deal–Grove model works well for thin oxides grown in H_2O ambients).

6.3   Thin Oxide Growth Kinetics in Dry O_2 Ambients

Deal and Grove recognized that their model did not fit experimental data for very thin oxides grown in dry O_2, and they proposed a mathematical "correction" to their model to "fix" this anomaly. Nevertheless, this empirical "fix" drove many investigators to try to understand the physical reasons for the modeling problem. Even after a vast number of academic papers on thin oxidation kinetics, there is still today no definitive answer to the anomaly observed by Deal and Grove, where the initial 20 nm of oxide growth in dry O_2 was not accurately modeled. Their "fix" invoked a simple "fudge factor" by adding a fictitious 23 nm initial oxide x_i and then the linear–parabolic theory agreed with experiment. The same x_i seemed to be present at any temperature, indicating that it was some property of the initial oxide. The anomalous time $\tau = (x_i^2 + Ax_i)/B$ can be calculated, and τ changes with temperature because A and B change

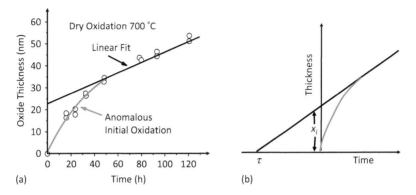

Figure 6.12 Experimental dry oxide data at 700 °C showing (a) the anomalous initial oxidation regime and (b) the definition of the anomalous initial thickness x_i and the equivalent initial time τ.

with temperature. This can be added as a "correction factor" to the actual oxidation time to obtain a more accurate estimate of the oxide thickness, as shown in Figure 6.12.

At the time Deal and Grove developed their model (1965), this was a minor adjustment, but as scaling advanced, these thin oxides became important for device applications. From an industrial perspective, growing the thin oxides reproducibly and reliably was the key to progress, but basing an essential component of a technology on poorly understood fundamentals is always risky. Despite the enormous academic interest, there is still not a good physical understanding of the process even though many of the models can reproduce the experimental data for thin oxides.

Perhaps the model that has been most accepted (at least in simulation programs) is based on both extensive experimental data and also a simple modification to the Deal–Grove physics. In what is termed the Massoud model, Massoud *et al.* [2] reported that growth in the thin oxide regime could be fitted by a simple decaying exponential term added to the Deal–Grove model, so that

$$\frac{dx_o}{dt} = \frac{B}{2x_o + A} + C \exp\left(-\frac{x_o}{L}\right), \tag{6.15}$$

where

$$C = C_0 \exp\left(-\frac{E_A}{kT}\right), \tag{6.16}$$

with $C_0 \approx 6.5 \times 10^6$ µm min^{-1}, $E_A = 2.37$ eV and $L = 7$ nm. These numbers do not vary much with different orientations of the silicon substrate, suggesting that the initial anomalous oxidation process is inherently related to the properties of the thin oxide itself. The first term in (6.15) is the Deal–Grove model, while the second term represents an additional oxidation mechanism that is important initially but decreases exponentially with a decay length of 7 nm. All the attractive features of the Deal–Grove model are retained with the addition of this temporary decaying fitting factor, which allows the data to be modeled over the whole range of thicknesses.

6.4   Dependence of Growth Kinetics on Pressure

In chemistry, Henry's law states that the solubility of a gas in a liquid is proportional to its partial pressure above the liquid. A similar effect occurs in silicon oxidation, where the solubility of the oxidant C^* at the gas/SiO$_2$ interface is proportional to the pressure, so that both B and B/A are also proportional to the pressure, because the solubility C^* is included in both B/A and B. The oxide growth rate should then be proportional to the pressure.

Experimental measurements have shown that in H$_2$O oxidation this prediction is correct. From pressures below atmospheric to well above atmospheric, the pressure dependence in the linear–parabolic model accurately captures the H$_2$O growth. However, the situation is less clear for dry O$_2$ oxidation. The data have consistently shown that, in order to model dry O$_2$ results with the linear–parabolic equation, the parabolic rate constant is proportional to pressure, as expected, but the linear rate constant has a sublinear dependence on pressure, so that

$$B/A \propto P^n \text{ where } 0.5 < n < 1. \tag{6.17}$$

One can only conclude that the linear–parabolic model is incomplete. There have been a number of attempts to modify the Deal–Grove model to account for these results. One model included both O and O$_2$ in separate streams reacting at the Si/SiO$_2$ interface. This model could explain the pressure dependence because an O-based reaction would vary as $P^{0.5}$ while an O$_2$-based reaction would vary as $P^{1.0}$. Most of the models proposed have been based on a modified Si/SiO$_2$ interface reaction, but none have found widespread acceptance.

Instead, in most simulation programs, an empirical approach to modifying the linear and parabolic coefficients is used to model a wide range of data with the following pressure dependences:

$$\frac{B}{A} = \left(\frac{B}{A}\right)^i P \text{ and } B = (B)^i P \text{ for H}_2\text{O}, \tag{6.18}$$

$$\frac{B}{A} = \left(\frac{B}{A}\right)^i P^n \text{ and } B = (B)^i P \text{ for O}_2, \tag{6.19}$$

with $n \approx 0.7$–0.8. The i superscripts refer to the values at 1 atm. A similar nonlinear pressure dependence has been noted in the anomalous thin oxide regime in dry O$_2$ (<20 nm).

6.5   Dependence of Growth Kinetics on Crystal Orientation

As we saw in Chapter 3, the (100) crystal orientation is almost exclusively used for building silicon integrated circuits (ICs) because the electrical properties of the Si/SiO$_2$ interface are superior to those of other crystal orientations. However, other crystal orientations can easily be exposed on trench sidewalls or in building structures other than conventional logic devices. For example, microfluidic channels might wind around, exposing different channel sidewalls. Some wet etchants expose the (111) silicon planes, producing triangular or pyramidal etch pits (see Figure 9.3 in Chapter 9). MEMS devices such as accelerometers, pressure sensors, oscillators and micromirrors expose different crystalline planes. In many applications, the precise oxide thickness grown may not be critical, but if devices are integrated into these structures, knowing the gate oxide thickness would determine the device performance.

Even before the development of the linear–parabolic model, it had been observed that the crystal orientation affected the oxidation rate. It was postulated that the effect might be caused by differences in the surface density of silicon atoms in the various crystal faces. Since silicon atoms are required for the oxidation process, crystal planes that have higher densities of atoms should oxidize faster. Ligenza [3] argued that it was not just the number of silicon surface atoms that was important, but also the number of bonds, since it was necessary for Si–Si bonds to be broken for the oxidation reaction to occur. He calculated the number of "available" bonds on the various silicon surfaces and concluded that the oxidation rate on (111) surfaces should be faster than on (100) surfaces. For most oxidation conditions of interest in silicon technology, these two orientations represent the upper and lower bounds for oxidation rates.

These effects are simple to incorporate in the linear–parabolic model in the following way. The oxide that grows on silicon is amorphous; that is, it does not incorporate any information about the underlying single-crystal structure. Because of this, the parabolic rate constant B should not be orientation-dependent since B represents the diffusion of oxidant through the oxide layer. If the oxide structure is amorphous and unrelated to the underlying substrate, there should be no crystal orientation effect on B. This is what is found experimentally when the growth data are analyzed in the context of the linear–parabolic model and the rate constants B and B/A are extracted. Conversely, B/A should be orientation-dependent because it involves the reaction at the Si/SiO$_2$ interface. The breaking of Si–Si bonds and the insertion of O atoms should be dependent on the number of available reaction sites. The orientation affects are simply accounted for by scaling the B/A linear rate constant as follows:

$$\left(\frac{B}{A}\right)_{111} = 1.4 \left(\frac{B}{A}\right)_{110} = 1.68 \left(\frac{B}{A}\right)_{100}. \tag{6.20}$$

An example of a simulation showing orientation affects in a shallow trench is shown in Figure 6.13. A trench etched parallel or perpendicular to the flat or notch in a (100)-oriented silicon wafer will have a sidewall with a (110) surface. The simulation shows the thicker oxide on the trench sidewall compared to the trench bottom. It also shows the stress affects and 2D effects in the silicon corner, which we will discuss in Section 6.8. Note also the rounding of the bottom corner caused by the oxidation process.

6.6   Substrate Doping Effects

It has been known for many years that highly doped substrates oxidize more rapidly than do lightly doped wafers. The effect is particularly noticeable at lower temperatures and for thinner oxides. Under these conditions, the difference in oxidation rates between heavily doped and lightly doped regions can be three to four times [4]. The difference is more pronounced for N$^+$ regions than for P$^+$ regions. Analysis has indicated that the mechanism associated with the faster oxidation is occurring at the Si/SiO$_2$ interface, in other words, with the linear rate

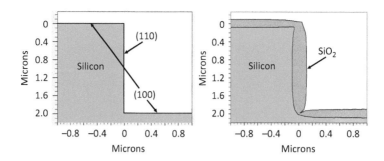

Figure 6.13 Athena simulation of shallow trench oxidation (20 min at 1000 °C in H_2O), where both the orientation effects of the sidewall (110) and base (100) are visible, as well as retardation in the corners due to stress effects.

coefficient B/A in the Deal–Grove model. The parabolic coefficient B is not affected, which is perhaps not too surprising since the diffusion of oxidant through the oxide would not be expected to depend on the doping concentration in the silicon.

The most widely accepted model relies on the fact that oxidation requires volume expansion (as shown in Figure 6.4). One way to accommodate the volume expansion is to consume vacancies at the growing Si/SiO_2 interface. We saw in Chapter 3 that the total number of vacancies V is much higher in extrinsic, highly doped regions than in intrinsic, lightly doped material. The vacancies exist in a number of charge states, and the total equilibrium number depends on the doping concentration (through the Fermi level). Fortunately, the charge states are such that there are more vacancies in N^+ regions than in P^+ regions for the same doping concentration, thereby explaining one of the key experimental observations.

If these vacancies can help provide sites for the volume expansion, they can enable more rapid oxide growth in N^+ regions than in P^+ regions and, because the linear constant is affected, it explains why thinner oxides and lower temperatures are most affected by the substrate doping. The effect is modeled by a multiplicative correction to the intrinsic B/A value as follows:

$$\left(\frac{B}{A}\right)_{doping} = \left(\frac{B}{A}\right)_i \left(1 + 2.6 \times 10^3 \exp\left(-\frac{1.1 \text{ eV}}{kT}\right)\left(\frac{V^*}{V_i^*} - 1\right)\right), \tag{6.21}$$

where V^*/V_i^* is the ratio of the total vacancy concentration in the doped region V^* to the vacancy concentration in lightly doped intrinsic regions V_i^*. The other factor of $2.6 \times 10^3 \exp(-1.1 \text{ eV}/kT)$ is an empirical fit to the data that presumably represent the interaction of the vacancy mechanism with the other components of B/A.

An example of the effect that doping can have on oxidation is shown in the simulation in Figure 6.14. A high-dose phosphorus implant is done on the right side of the structure, which raises the surface doping concentration to above 10^{20} cm^{-3}. A 5 min wet oxidation at 1000 °C grows an oxide over twice as thick on the right as on the undoped left side. In the heavily doped N^+ region, the concentrations of V^- and $V^=$ would be much higher than on the intrinsic left side (see (3.11) and (3.13) in Chapter 3). Thus the model in (6.21) would predict a significantly larger B/A value on the right side of Figure 6.14. The color contours in the figure represent the phosphorus doping levels.

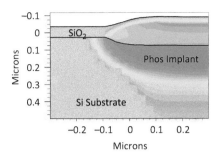

Figure 6.14 Athena simulation of heavily doped oxidation. A 3×10^{15} cm^{-2}, 50 keV phosphorus implant was done on the right side. The surface oxidation was 5 min, wet at 1000 °C.

Several points should be made with regard to this model. The first is that, since it is based on V populations and these depend only on T and E_F, it makes no difference if the N$^+$ regions are formed with arsenic or phosphorus. The same electrically active surface concentration should produce the same E_F and hence the same B/A. Second, the relevant concentration that should be used in calculating E_F and hence V^* is the surface concentration, since that is where the oxidation reaction is taking place.

Segregation and pile-up effects may increase this concentration above bulk values. We introduced the idea of segregation in Chapter 3 in connection with crystal growth. In that case, we were concerned with the distribution of an impurity across the solid/liquid interface of the growing crystal. In the oxidation process, we have an Si/SiO$_2$ interface where dopant segregation also occurs. N-type dopants tend to segregate into the silicon whereas boron prefers to be in the SiO$_2$. As a result, the surface concentration for N-type dopants will tend to build up as the oxidation proceeds, and it will tend to decrease during oxidation for P-type dopants (see Figure 7.22 in Chapter 7). Because of this, the B/A values calculated from (6.21) will change with time during the oxidation. The oxidation rate will therefore change with time. Experiments that carefully measured the dopant profiles near the surface clearly showed the time dependence of the oxidation rate and its correlation with the dopant pile-up process [5]. The model also predicts that P$^+$ substrates should not show nearly as large an increase in oxidation rate as do N$^+$ regions, because, according to the models we discussed in Chapter 3, there are fewer total vacancies present in P$^+$ regions at a given doping level. This also agrees with experiments.

6.7 Point Defects Associated with the Oxidation Process

The model described above is a point defect-based model to explain the local oxidation kinetics on doped substrates. However, there are even more interesting point defect effects associated with a growing oxide interface, and these have to do with interactions between oxidation and other process steps like diffusion.

One of the earliest indications that more was actually going on during oxidation than the effects we have considered to this point were observations that surface oxidations changed the

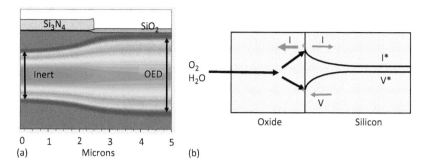

Figure 6.15 (a) Simulation of oxidation-enhanced diffusion (OED). The colors correspond to boron concentration (red being highest concentration to blue being lowest). (b) Point defect reactions taking place at the Si/SiO$_2$ interface during oxidation. Oxidation reactions (*) can both consume V and generate I in order to help provide the volume needed for the reaction to take place. Most generated I are injected into the SiO$_2$.

rate of diffusion of dopants in the underlying silicon substrate. This phenomenon is now known as oxidation-enhanced diffusion (OED) in cases in which dopant diffusion rates are increased by surface oxidation, or oxidation-retarded diffusion (ORD) in cases in which diffusion is slowed down by surface oxidation. Figure 6.15(a) shows a simulated example of OED and Figure 6.15(b) shows the underlying physical mechanisms.

First, a boron implant is performed uniformly in the wafer beneath the surface. The peak of the implant is at the position of the red buried layer on the left side of Figure 6.15(a). The processing then is similar to the experimental LOCOS structure shown in Figure 6.5. Si$_3$N$_4$ is deposited on the left, which prevents oxidation on that side and a LOCOS oxide is grown on the right. During the high-temperature oxidation, the buried boron diffuses up and down, but it diffuses significantly faster on the right side where a surface oxidation is occurring. We will consider these cases in more detail in Chapter 7. However, we will introduce here some additional ideas about oxidation that will be needed to explain OED, ORD and other related effects.

All of the oxidation mechanisms we have considered to this point are purely local phenomena. That is, they help to explain why oxides grow at the rates they do, but they offer no suggestions as to why surface oxidation should, for example, perturb dopant diffusion rates tens of microns away from the surface. Clearly, some nonlocal phenomenon must be involved. We have seen that there is a very large volume expansion that occurs during oxidation. This volume expansion results in very large compressive stresses in the oxide layer and tensile stresses in the silicon in the near-surface region. Dobson was among the first to suggest that there might be mechanisms at the atomic level available to the oxidizing interface to help relieve these stresses [6]. These mechanisms in essence provide some of the "volume" that is needed for the oxide to grow. Figure 6.15(b) illustrates these ideas. At a microscopic level, Si–Si bonds are broken at the Si/SiO$_2$ interface, O atoms are inserted between the Si atoms, and Si–O bonds are formed. The volume expansion requirement comes from the room needed for the added O atoms.

The two major types of point defects present in silicon are vacancies (V) or missing Si atoms and interstitials (I) or excess silicon atoms. These are the two reactions illustrated in Figure 6.15(b). The reaction forming SiO_2 consumes vacancies and generates interstitials because the vacancies provide some of the volume needed for the oxidation to proceed, as can removing Si atoms as I. The red arrows indicate that the generated interstitials could flow either into the SiO_2 or into the bulk silicon. The effects we will focus on here are due to the latter.

The generation of I is thought to explain the interactions between oxidation and other processes like diffusion (i.e., OED and ORD) because the I that are generated and injected into the silicon can diffuse far away from the Si/SiO_2 interface and change the diffusivity of dopants that interact with I. We will discuss this further in Chapter 7. The consumption of V as a path to provide the volume required for the oxidation is favored whenever there are large numbers of V present. This is the case in heavily doped silicon regions (as we saw in the model in the previous section), and this reaction has been used to explain why oxidation proceeds at a much higher rate on heavily doped substrates than it does on lightly doped substrates.

It is quite clear that the mechanisms proposed do not provide all the volume required at the oxidizing interface. The fact that grown oxides are under very high compressive stresses is clear evidence for this. Estimates of the number of I that are injected into the substrate during oxidation suggest that fewer than 1 in 10^3 Si atoms does this. The rest are consumed by the oxidation process and end up in the SiO_2 layer. In other words, the oxidizing interface is a very active surface, with many Si–Si bonds being broken and lots of I being generated, but most of those I diffuse back into the SiO_2 layer where they are consumed by the oxidation reaction itself and only a relatively small number are injected into the silicon. Nonetheless, the effects that the few point defects that do diffuse into the substrate have on process phenomena in the underlying silicon are quite remarkable. We will discuss oxidation-enhanced diffusion and oxidation-retarded diffusion more carefully in Chapter 7, as these effects were largely responsible for the detailed physical understanding of how dopants diffuse during processing.

6.8 Oxide Flow, Stress and Two-Dimensional Growth

It is often the case in real device structures that oxidation takes place on nonplanar (shaped) surfaces. Figure 6.13 was one example of this, in which a trench structure was oxidized. The Deal–Grove oxidation model was developed and explains planar (1D) oxidation, but oxidation of shaped structures is more complex, and we consider some of those effects in this section.

We start by considering the oxidation of a silicon cylinder, as shown in Figure 6.16. Such a cylinder could easily be made by applying a circular mask and anisotropically etching the silicon to produce a vertical standing cylinder. A circular trench or hole could also be made in the same fashion. This would provide an "outside" corner and an "inside" corner to observe stress effects. Such cylinders and holes have actually been used [7] to investigate the two-dimensional effects during oxide growth.

During the oxidation of the curved surface of a cylinder, the expansion of the interfacial layer lifts the old oxide away from the interface, forcing it to stretch along the outside edge, as shown

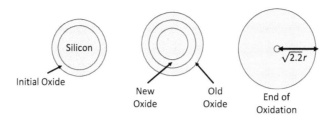

Figure 6.16 Complete oxidation of a silicon cylinder leads to a 2.2× increase in volume. Note the schematic ignores the crystal orientation dependence around the cylinder.

in Figure 6.16. If we consider the complete oxidation of a cylinder as in Figure 6.16, where the silicon oxidizes to a volume of oxide 2.2× larger, the radius must increase by $\sqrt{2.2}$ and the circumference must also be $\sqrt{2.2}$ longer than before. This 48% tangential extension of the surface oxide is quite a large strain. Similar large strains also occur under the bird's beak formation during LOCOS, as can be inferred by comparing the length of the interface under the nitride before and after growth in Figure 6.5. These strains are clearly related to the two-dimensional character of the oxide growth.

We have mentioned that the oxide is also in compressive stress during planar growth (see Figure 6.4). If the oxide is stripped from one side of an oxidized wafer, the wafer curves or bows away from the oxide-coated surface, indicating a compressive stress in the oxide. This wafer curvature can be measured by a laser scanning and reflection technique. If measured during growth, the curvature measures the intrinsic stress at growth temperature. The curvature increases as the oxide film grows thicker and is also larger for oxides being grown at low temperature rather than high temperature. If the curvature is measured at room temperature, it includes the extra stress due to the thermal expansion mismatch between the silicon and the oxide.

A simple mechanical calculation of the stress from this level of strain is given by

$$G\varepsilon \approx 10^{11} \text{ dyne cm}^{-2}, \tag{6.22}$$

where G is the shear modulus (3.5×10^{11} dyne cm^{-2} and ε is the strain (30–48%)). This estimate is almost two orders of magnitude larger than the measured stress. A likely explanation is that Mott's model of oxidation at kink sites lowers the stress [8]. We mentioned that atomic-level steps are sometimes observed in the otherwise abrupt interface between oxide and silicon (Figure 6.7). If the oxidation proceeds at kink sites moving along the interface like a zipper, then the oxidation front advances parallel to the wafer and the perpendicular strain can be reduced. So these one-dimensional stresses have an atomistic origin and, since the planar growth coefficients (B and B/A) already have the effects of these 1D stresses built into them, we account for the 2D stresses by adding some fitting parameters to the standard Deal–Grove model to account for the 2D oxide growth results on curved surfaces.

First, we examine the experimental results from Figure 6.17, where a large range of curvatures for both outside oxide growth on silicon cylinders and inside oxide growth on holes in silicon were experimentally measured. By cross-sectioning the structures, i.e., by chemical–mechanical

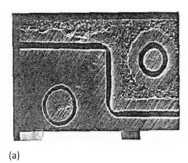 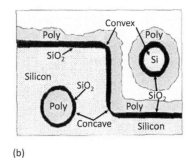

(a)   (b)

Figure 6.17 Experimental results on oxidation on inside (concave) and outside (convex) corners, cylinders (convex) and holes (concave) in silicon. The structures are anisotropically etched into the screen, oxidized and then cross-sectioned by CMP for viewing by microscopy. The polysilicon is deposited after oxidation to provide a contrast layer to more easily measure the oxide thickness by microscopy. © 1987 IEEE. Reprinted with permission from [7].

polishing (CMP), Kao was able to observe the oxide thickness as a function of curvature in a microscope [7]. This elegant set of experiments allowed a quantitative understanding of the physical mechanisms responsible for stress-retarded growth on curved surfaces to be developed.

Typical results from Kao's work are shown in Figures 6.17 and 6.18 for a nominal 500 nm planar oxide, which acted as a built-in reference for the measurements. In Figure 6.18 the horizontal axis is $1/r$, so that progressively sharper corners are shown to the right. The value $1/r = 0$ corresponds to planar oxide growth, which should be well modeled by the Deal–Grove model. Several interesting observations can be made from the data in Figure 6.18.

1. No cracking of the oxide layer occurred for any radius of curvature.
2. Retardation occurs on both inside (holes) and outside (cylinders) corners.
3. Retardation is more significant at lower temperatures.
4. Retardation is more significant for holes than for cylinders.

The fact that the oxide does not crack means that it must flow at the growth temperature. It is somewhat surprising that the outside corners of cylinders show reduced oxidation, as they are more exposed to incoming oxidant. Lower temperatures increase the resistance to flow (increased viscosity), so it is not surprising that lower temperatures exhibit more retardation. Smaller radii have more strain, more stress and more retardation. Kao suggested stress-dependent modifications to the Deal–Grove coefficients and oxidant solubility to account for the results. Later, Rafferty and Dutton [9] showed that the experimental results could be quantitatively modeled by accounting for a stress dependence in the reaction and diffusion coefficients and by including a nonlinear oxide viscosity:

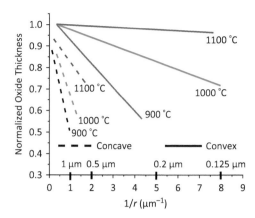

Figure 6.18 Experimental wet oxide thickness measurements on convex (outside corner) and concave (inner corner) structures of various radius of curvature, normalized to the thickness on a planar surface. After [7]. The normalized thickness of 1.0 corresponds to a 500 nm planar oxide thickness.

- stress-dependent reaction rate

$$\frac{B}{A} = \left(\frac{B}{A}\right)_0 \exp\left(-\frac{\sigma_n V_r}{kT}\right), \tag{6.23}$$

- stress-dependent diffusion rate

$$D = D_0 \exp\left(-\frac{p V_D}{kT}\right), \tag{6.24}$$

- nonlinear viscosity

$$\mu = \mu_0 \frac{\sigma_t V_\mu/(2kT)}{\sinh(\sigma_t V_\mu/(2kT))}. \tag{6.25}$$

The stress-dependent reaction rate is needed to account for the reduced oxidation thickness on the outside corners or cylinders, and reduces the reaction rate proportional to the normal stress σ_n along with a fitting volume V_r. The stress-dependent diffusion rate is needed to provide the extra reduction in inside corners or holes, and reduces the oxidant diffusivity proportional to the hydrostatic pressure p that builds up due to deformation along with a fitting volume V_D. Finally, the oxide viscosity has a strong nonlinear dependence that depends on the tangential stress σ_t with a volume fitting parameter V_μ. The $x/\sinh(x)$ function peaks at $x = 0$ and is symmetric about $x = 0$ and falls off very sharply for larger values of x. It is a common nonlinear function seen in the mechanics of materials. Modern simulation tools use these parameters to solve the nonlinear viscoelastic equations for the oxide properties to predict both the stress induced by the oxidation and the oxidation thickness in shaped structures.

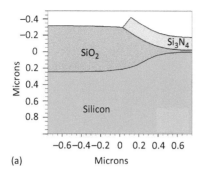
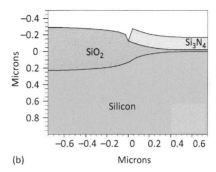

Figure 6.19 A simple LOCOS structure simulated, using Athena, (a) without and (b) with stress-dependent effects. The oxidation was 90 min at 1000 °C in H_2O in both cases.

An example of the difference between accounting for only simple Deal–Grove kinetics and the stress-dependent oxidation model is shown in Figure 6.19. Here, a simple LOCOS structure is simulated with both models.

Using the stress-dependent oxidation model shows that the nitride is a stiff material that does not bend easily at the oxidation temperature and imposes stress under the nitride edge that limits encroachment of the "bird's beak." This is a good thing if one is trying to make compact isolation structures, but comes at the expense of creating stresses in the silicon. It is important to keep these stresses to tolerable levels so as not to introduce defects like stacking faults or dislocations in the substrate. The simulation tools do not predict when a dislocation will form, but the simulations allow different structures to be compared and choices made to minimize the stress levels. Another example in Figure 6.20 shows the stress-dependent models being used to simulate a range of structures composed of oxide, nitride and polysilicon spacer or buffer layers.

This study grew from an attempt to find a compact, inexpensive isolation structure for 256 Mb DRAM devices at Samsung. The shallow trench isolation option was also considered, but is a more expensive option in devices meant for commodity markets. The poly spacer LOCOS (PSL) structure provided the most benefit and was subsequently used in manufacturing the DRAMs [10].

6.9 Electrical Defects in the Silicon/Oxide System

Although we have said that silicon is the dominant semiconductor because of the perfection of the silicon/oxide interface, it is also because intensive research has shown how to minimize any defects and create the near-ideal structure. In 1980, in an effort to unify research in this field, Deal [11] (of the Deal–Grove model) suggested the nomenclature shown in Figure 6.21 and Table 6.2 to represent the various types of electrical defects that are found experimentally at the Si/SiO_2 interface and in SiO_2 layers. There are four basic types of defects or charges that exist, listed below.

The first, Q_f, is known as the fixed oxide charge. Experimentally, we find that a sheet of positive charge (usually 10^9–10^{11} cm^{-2}) exists in the oxide, very close to the interface. It seems to be located within 2 nm of the interface (perhaps closer) and is likely associated with the transition from Si to SiO_2, perhaps involving partially oxidized silicon SiO_x. Most physical

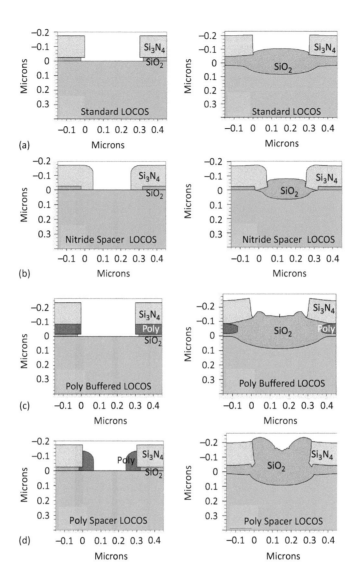

Figure 6.20 A range of isolation structures, (a) standard LOCOS, (b) nitride spacer LOCOS, (c) poly buffered LOCOS and (d) poly spacer LOCOS, tested by simulation using the nonlinear viscous models discussed in the text. The poly spacer LOCOS produced the best oxide isolation with acceptable encroachment and the least stress in the silicon. The oxidation was 24 min at 1050 °C in H_2O.

explanations for Q_f suggest that it is due to incompletely oxidized Si atoms that have a net positive charge. These are sometimes represented by $\equiv$Si·, where the three bars indicate three silicon bonds and the dot indicates a missing fourth bond or unpaired valence electron (the so called "dangling bond"). This charge Q_f has a similar dependence on orientation as B/A, so is thought to be related to the interface reactions. Note that Q_f is called the fixed oxide charge because it cannot be charged and discharged by varying the silicon surface potential. It is positive and invariant under normal conditions.

Table 6.2 **Charges associated with the silicon/oxide interface, based on the standardized nomenclature suggested in [11]. Note that we define Q as the number of electronic defects per unit area, so we would need to multiply by the electronic charge q to get coulombs per square centimeter (C cm^{-2}).**

Fixed oxide charge	Q_f (cm^{-2})
Mobile ionic charge	Q_m (cm^{-2})
Oxide trapped charge	Q_{ot} (cm^{-2})
Interface trapped charge	Q_{it} (cm^{-2})
Interface trap density	D_{it} (cm^{-2} eV^{-1})

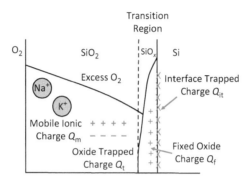

Figure 6.21 Schematic of the position and type of charges in silicon dioxide. After Deal [11].

But the amount of Q_f does vary with processing conditions. Oxidation at high temperatures minimizes Q_f, as presumably the faster interface reactions consume the partially oxidized silicon in the transition region shown in Figure 6.21. Oxidation at lower temperatures leads to higher Q_f values. The green arrows in Figure 6.22 indicate dry O$_2$ oxidations or anneals done at various temperatures and the resulting impact on Q_f. The experiments behind the data in Figure 6.22 were done by oxidizing a wafer at the indicated temperature and then pulling the wafer from the furnace fairly rapidly (a few seconds) to "freeze in" the Q_f corresponding to that temperature. Annealing in an inert ambient like N$_2$ or Ar drops the Q_f to a low value. To minimize Q_f, a final high-temperature Ar anneal is used towards the end of a device fabrication sequence. As shown in Figure 6.22, the equilibrium values for Q_f after an N$_2$ or Ar anneal are relatively independent of temperature. However, a longer anneal time is required at lower temperatures to achieve the low final Q_f value. Less than a minute may be required at 1100 °C or higher, whereas 30 min may be necessary at much lower temperatures.

The second type of charge, Q_m, is the mobile oxide charge, which may be located anywhere in the oxide, and was a serious problem in the early days of MOS transistor development. At that time, it was often the case that MOS structures were unstable after fabrication. We discussed this briefly in Chapter 1 and again in Chapter 4, and observed there that this instability in transistor threshold voltage was traced to the presence of mobile ions like Na$^+$ and K$^+$ in the

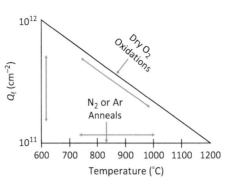

Figure 6.22 The Q_f triangle, showing the variation of the fixed oxide charge with oxidation temperature and ambient. Values shown are for (111) silicon; (100) values are ≈ 3× lower.

gate oxides. With proper attention to cleanliness in wafer fabrication facilities, this problem largely disappeared by the 1970s. It is well understood and easy to measure by electrical means and to eliminate by careful processing. In the example we considered in Chapter 4, we saw that even parts per million of Q_m due to Na^+ or K^+ cause a shift in the V_{TH} of MOS devices that is inversely proportional to C_{ox} (see (4.1)). Thus, as the gate oxide thickness is scaled down, larger Q_m values can be tolerated.

The oxide trapped charge, Q_{ot}, which may be located anywhere in the oxide, is due to defects that are likely broken Si–O bonds in the bulk of the oxide, well away from the Si/SiO$_2$ interface. Such bonds can be broken by ionizing radiation or by some of the process steps used in manufacturing ICs today. Plasma etching, for example, exposes oxides to energetic ions, electrons and other neutral species. Ion implantation is often done through an oxide layer. These and other processes can damage oxides resulting in traps in the bulk oxide, Q_{ot}. Such traps are normally repaired by a high-temperature anneal before device fabrication is complete, which allows the broken bonds to repair themselves. If they exist in the oxide because they are not fully annealed, or because the device is exposed to ionizing radiation, these traps can capture holes or electrons that may be injected into the oxide during device operation, resulting in trapped charge. Hence the name Q_{ot}.

The oxide trapped charge Q_{ot} has taken on increased importance in recent years because of the high electric fields present in scaled devices. These higher fields result in more energetic or "hot" carriers that can achieve energies high enough to be injected into the gate oxides of modern MOS devices. If oxide traps are present, or if they are created by the energetic carriers themselves, charge trapping can occur. This results in device threshold shifts with time (see (4.1)) and reliability concerns. Charge trapping in SiO$_2$ insulators is also a major issue in programmable devices like erasable programmable read-only memories (EPROMs) in which current is purposely passed through the gate oxide as part of the write operation.

Like Q_f, the second type of charged defect present near or at the Si/SiO$_2$ interface is the interface trapped charge, Q_{it}. The physical origin of these Q_{it} charges is often suggested to be similar to the origin of Q_f. That is, Q_{it} is likely due to some type of incompletely oxidized silicon atom with unsatisfied or dangling bonds located in the oxide, but very close to the interface. However, there is a very important difference between Q_{it} and Q_f. The charge associated with Q_f is fixed and positive. The charge associated with Q_{it} may be positive, neutral or negative and, in fact, may change during normal device operation because of the capture of holes or electrons. These are often called fast

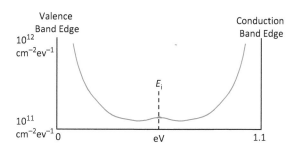

Figure 6.23 Typical distribution of fast interface traps D_{it} in the silicon bandgap after thermal oxidation in dry O_2. Note that D_{it} measurement accuracy decreases away from mid-gap.

states because of their ability to quickly trap holes or electrons. These traps behave very much like the bulk deep-level traps we discuss in Appendix A.1 and Chapter 4. Energy levels associated with Q_{it} exist throughout the forbidden band, although usually there are more traps at energy levels near the conduction and valence band edges than there are in the middle of the bandgap. For this reason, it is common to express the density of these interface traps in terms of both the number per unit area and position in the bandgap, giving the special symbol D_{it} with units of number per square centimeter per electronvolt ($cm^{-2}\,eV^{-1}$). A typical plot of the fast interface traps D_{it} in the bandgap is shown in Figure 6.23. Often, the D_{it} value at mid-gap is given as an approximate measure of the interface trap density Q_{it}.

Oxidizing a silicon surface usually results in a density of Q_{it} on the order of 10^9–$10^{11}\,cm^{-2}$, about the same density as is found for Q_f. In fact, it is usually the case that a process that results in a high value of Q_f will also result in a high density of Q_{it}. This correlation is one of the experimental results that suggests a common origin for the two charges.

However, another very important distinction between the two charges is that Q_{it} can be passivated through hydrogen annealing at fairly low temperatures (300–500 °C) whereas Q_f is unaffected by such an anneal. By "passivated" we mean that, once the Q_{it} traps are bonded with H atoms, they are no longer electrically active and no longer trap carriers. The mechanism of passivation is believed to be simply the diffusion of hydrogen through the SiO_2 layer to the Si/SiO$_2$ interface and reaction there to form Si–H bonds. Since the process takes place readily at temperatures that metal layers can tolerate, this "forming gas" anneal step is normally done at the end of a chip process.

Because of the importance of this passivation process, it has been carefully studied over many years in order to understand not only the origins of the interface states themselves, but also how the passivation works. Most of the atomic-level models for interface states are based on excess silicon or incompletely bonded silicon atoms, with the trivalent silicon defect ≡Si· that we discussed above being a leading candidate. The annealing or passivation process is usually modeled with molecular hydrogen diffusion to the Si/SiO$_2$ interface followed by hydrogen dissociation to form atomic hydrogen and then chemical bonding between the ≡Si· defect and H [12]. This is often expressed simply as

$$H_2 \leftrightarrow 2H \text{ followed by } \equiv Si\cdot + H \leftrightarrow \ \equiv SiH. \tag{6.26}$$

The final annealed value is below $10^{10}\,cm^{-2}$ for anneal times on the order of 30 min at 400 °C.

All four of the charges in Figure 6.21 can have deleterious effects on device operation. As a result, great care is normally taken during fabrication to choose process sequences that will minimize these charges. Generally, this is accomplished by high-temperature inert anneals in Ar or N_2 late in the process flow, and by a final moderate-temperature (400 °C) anneal in H_2 or forming gas (N_2/H_2) at the end of the process.

6.10 Characterization of the Silicon/Insulator Interface

The simplest way to characterize a thermally grown oxide is to look at it – your eye is remarkably sensitive to any irregularities in the thickness of the oxide layer. On a flat substrate, it will look perfectly uniform and takes on a color that depends on the thickness, which can be read from Figure 6.24. A more complete listing of SiO_2 colors is given in Appendix A.7. The colors occur because of standing wave interference between the top of the oxide and the reflective silicon surface. We discussed these issues in Chapter 5 in connection with standing waves in photoresist films (see Figure 5.61, for example). These ideas form the basis for a host of optical measurement tools which scan wavelength and measure reflectance to characterize deposited thin films. More sophisticated techniques like ellipsometry can measure both the dielectric constant and the film thickness by measuring the change in polarization as an incident wave reflects off the substrate at an angle.

Removing the deposited layer from a region and scanning over the step with a stylus-type instrument (or a more sophisticated atomic force microscope) provides an alternative physical method to measure the film thickness. However, the method that is of most utility relies on the electrical characterization of the deposited film, because it is extremely sensitive.

Virtually all of the experimental techniques that are available to investigate interface charges are electrical in nature. The problem is easily visualized in connection with the TEM photo in Figure 6.7. The normal densities of Q_f and Q_{it} are around 10^9–10^{12} cm^{-2} or about one charge for every 10^3–10^6 silicon atoms at the interface. Physical techniques such as TEM are simply not sensitive enough to "find" such charges. It is quite literally like "looking for a needle in

Oxide thickness (nm)		Color
50		tan
70		brown
100		dark violet to red violet
120		royal blue
150		light blue to metallic blue
170		metallic to very light yellow-green
200		light gold or yellow – slightly metallic
220		gold with slight yellow orange
250		orange to melon

Figure 6.24 Oxide thickness in nm and layer color under normal incident daylight fluorescent lighting. After 250 nm (2500 Å), the colors begin to repeat, starting with the dark violet, though they become somewhat muted for the thicker oxides (see Appendix A.7).

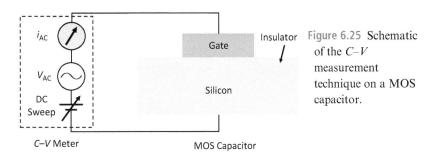

Figure 6.25 Schematic of the C–V measurement technique on a MOS capacitor.

a haystack." Electrical methods do have the required sensitivity and are commonly used to measure Q_f and Q_{it}.

Electrical methods measure parameters that are of direct interest to semiconductor devices, such as capacitances, threshold voltages, insulator quality and electrical charges. By far the most dominant electrical measurement technique is the capacitance–voltage or C–V method, which provides a large amount of information about the insulator/semiconductor properties. We describe how this technique works for the oxide/silicon interface, but it is widely applicable to any insulator/semiconductor interface.

The basic structure used to make the measurement is a MOS capacitor, consisting of a metal or doped polysilicon gate electrode, a dielectric film and a semiconductor substrate connected to a sweeping DC voltage with a small superimposed AC capacitance measurement voltage, as shown in Figure 6.25. In principle, this can be made with one mask, by taking the silicon wafer, growing some oxide on it, depositing some metal (say, aluminum) and patterning the metal into some dots of known area that can then be probed electrically between the dot on top and the back of the wafer. In practice, it might be more complicated because a large backside contact may be needed to minimize series resistance, though modern C–V instruments attempt to account for series resistance.

By definition, capacitance is the change in charge when there is a change in voltage:

$$C \equiv \frac{\Delta Q}{\Delta V}. \tag{6.27}$$

The small AC measurement voltage of a millivolt or so at some fixed frequency is used to allow us to measure with a current meter the AC current that is flowing into the structure, which allows us to measure the capacitance according to (6.28). The frequency of the AC voltage may be important as it "wiggles" the charge carriers in the semiconductor and depending on their response rate gives rise to high-frequency and low-frequency C–V curves, as we will see later:

$$i_{AC} = C \frac{dV_{AC}}{dt}. \tag{6.28}$$

One practical way to implement this is to apply a small AC voltage and measure the resulting AC current. Then integrate the current over time and use the ΔQ and ΔV values to calculate the capacitance. The C–V curves are measured by using two series voltage sources, the small AC voltage source and a DC voltage that is swept in time. The purpose of the DC voltage is to allow the capacitance to be sampled at different depths in the device. The AC signal allows the capacitance to be measured at each depth.

When we say we sweep the DC voltage, what we really mean is that we take a capacitance measurement at one DC voltage, then change the DC voltage and take another measurement when everything is stabilized. In other words, the sweep rate is slow enough to approximate a series of steady-state or equilibrium measurements at a range of DC voltages, say +3 V to −3 V.

The reason this structure is so important is that it mimics the gate of a MOS transistor, minus the source/drain regions, so any electrical measurements are directly relevant to the semiconductor device operation. Consider first an N-type silicon substrate with a positive voltage $+V_G$ on the gate, as shown in Figure 6.26(a). In N-type material, the dominant carriers are electrons, which are negatively charged and therefore attracted to the surface if they see the positive voltage applied to the metal. The electrons form an "accumulation" layer at the silicon surface.

There is an assumption here, and it is that the electric field lines pass through the insulator and into the substrate. The assumption is that the insulator and the insulator/semiconductor interface (SiO_2/Si in this case) have only small numbers of charges or defects. In other words, the four types of charges described in Figure 6.21 are assumed to be relatively small in density. If this is the case, electric field lines that start on charges in the gate will extend through the insulator and into the substrate where there are negative charges available to terminate the field lines. If there were large numbers of charges or defects in the insulator or at the insulator/semiconductor interface, the electric field lines could terminate on these charges before they reached the underlying semiconductor. The result would be that applying voltages to the gate would have little effect on the semiconductor and the interface could not be electrically controlled. This is in fact what happens in many insulator/semiconductor systems, but the SiO_2/Si interface is close to electrically perfect. Presuming that the insulator is close to perfect, the capacitance we measure is simply the oxide capacitance, C_{ox}, which will be independent of the gate voltage, as shown in Figure 6.26(c).

Now, imagine applying a negative DC voltage, as in Figure 6.27. Negative voltages will repel the majority carrier electrons from the surface, creating a depleted region. Any negative charge

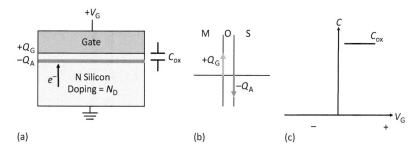

Figure 6.26 The C–V curve under accumulation conditions. A positive charge on the gate attracts majority carrier electrons in the substrate and the oxide capacitance is measured. By overall charge neutrality, $|+Q_G| = |-Q_A|$, where Q_G is the charge on the gate and Q_A is the electron charge in the accumulation layer.

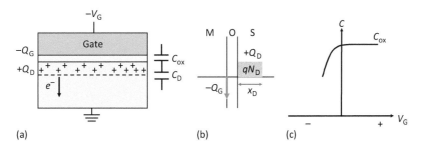

Figure 6.27 The *C–V* curve under depletion conditions. A negative gate voltage depletes majority carrier electrons under the gate and uncovers positive donor charges to a depth X_D required to balance the gate voltage.

we place on the gate must be balanced by a corresponding positive charge in the substrate to maintain charge neutrality. The positive charge is provided by the substrate donor atoms, which have a net positive charge when the mobile electrons that the donors provide are forced away from the surface. A depletion region is much like an insulator – to first order there are no free carriers in a depleted region, so it looks like a series combination of the oxide capacitance and the depletion capacitance, which results in a smaller overall capacitance.

The gate charge is balanced by the depletion charge as

$$|Q_G| = |Q_D| = N_D x_D \, , \tag{6.29}$$

where Q_G and Q_D are in units of number of charges per square centimeter, and N_D is the doping concentration in the substrate, which is assumed here to be uniform. The depletion region forms to a depth x_D until enough donor charge is uncovered to balance the gate charge. The depletion region has a capacitance per unit area associated with it, which is given by

$$C_D = \frac{\varepsilon_S}{x_D} \, , \tag{6.30}$$

where ε_S is the permittivity of silicon. However, x_D is a function of the gate voltage, increasing as V_G increases. The overall capacitance that is measured for the structure is now C_{ox} in series with C_D, and capacitances in series sum to a smaller overall capacitance,

$$\frac{1}{C} = \frac{1}{C_{ox}} + \frac{1}{C_D} \, , \tag{6.31}$$

which reduces with an increasing depletion region.

As we apply more negative voltage, the carriers are pushed further away, the depletion region grows and the capacitance decreases. The rate at which it goes down the curve depends on how fast the depletion-layer thickness changes, which depends on the doping in the substrate, so we can extract the doping in the substrate from this plot. The plot in Figure 6.27(c) shows the

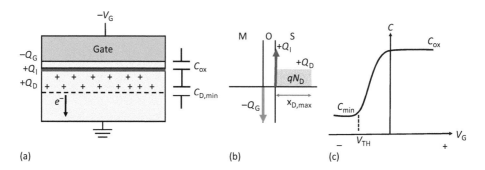

Figure 6.28 The C–V curve under inversion conditions. A large negative gate voltage "inverts" the substrate and attracts mobile holes to the interface. The gate charge is balanced by the charge in the inversion layer and the fixed donor charge in the substrate out to the maximum depletion layer thickness $x_{D,max}$. Further increases in gate charge are balanced by increased charge in the inversion layer.

capacitance decreasing as V_G becomes more negative, because x_D grows as V_G becomes more negative.

At some point, this expansion of the depletion layer stops and the capacitance bottoms out, and that happens when we form an inversion layer, as illustrated in Figure 6.28. That occurs when there is enough field in the substrate that the minority carriers in the substrate, the holes, are attracted to the surface in sufficient numbers to create the inversion layer. By definition, surface inversion occurs when the number of minority carriers equals the number of donor atoms, essentially the beginning of the inversion of the surface electrical layer to mobile holes. This is also the point that corresponds to the turn-on of a MOS transistor, the threshold voltage or the voltage at which the device transitions from OFF to ON. Once the inversion layer forms, x_D stops expanding and reaches the maximum value of $x_{D,max}$.

For any condition, we need overall charge balance. If we have negative gate charge, it must be balanced by charge in the inversion layer or charge in the depletion layer. If we apply a sufficiently high negative voltage on the gate, then the inversion layer has positive charge (holes) and there is positive charge (donors) in the depletion layer, and they must add up to equal the gate charge. That is,

$$Q_G = N_D x_D + Q_I, \tag{6.32}$$

where Q_I is the charge density (number of charges per cm^2) in the inversion layer; Q_I is negligible during the depletion condition, so that x_D expands during depletion to balance Q_G. Once the inversion layer forms, additional gate charge is balanced by Q_I rather than Q_D, so that x_D reaches its maximum value of $x_{D,max}$. The reason this occurs is because small changes in the surface potential result in an exponential increase in mobile holes, which means that it is easy for Q_I to balance additional Q_G. Increasing Q_G after inversion is reached will result in a very small increase in x_D and hence in Q_D, so we speak of x_D being pinned at $x_{D,max}$ once inversion occurs. This means that the C–V curve will reach a minimum, as shown in Figure 6.28(c).

Whether the capacitance remains at a minimum with increasing negative gate voltage or reverts to the oxide capacitance depends on whether the AC measurement is a high-frequency

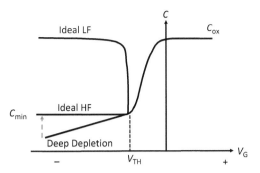

Figure 6.29 Ideal MOS C–V curves at high (HF) and low (LF) frequencies. Also shown is the deep depletion C–V curve, where the DC bias sweeps faster than equilibrium.

(HF) or low-frequency (LF) signal, as illustrated in Figure 6.29. For any condition, we need overall charge balance. The AC signal that is superimposed on $-V_G$ modulates Q_G very slightly. This change in Q_G must be balanced by a change in either Q_I or Q_D. Does the ΔQ show up in the inversion layer or in the depletion layer? This depends on whether the AC measurement is an HF or LF signal. What you measure depends entirely on whether or not there is a source of holes (Q_I) that can follow the AC signal. If there is a source of holes that can follow the signal, then you measure the oxide capacitance. If there is not, and you cannot generate or produce them at the frequency of the applied signal, then the only option the system has is to move the edge of the depletion layer and uncover charge. If this happens, then the charge balancing that occurs is between the gate and the bottom of the depletion region. The bottom edge of the depletion region moves up and down along with the gate voltage, providing the ΔQ_D needed to balance the ΔQ_G. Hence the capacitance that is measured is between these two plates (C_{ox} in series with C_D). At high frequencies, you measure the capacitance all the way to the bottom of the depletion layer.

Now why is it that Q_I cannot change as fast as the HF signal in the gate? The answer is simply that there is no source of additional holes available in the substrate to provide the ΔQ_I. There are very few holes present in an N-type semiconductor. The only mechanism usually available to create or remove holes is thermal generation and recombination, and these processes are quite slow in materials like silicon that have long carrier lifetimes. As a result, Q_I cannot change rapidly and the HF C–V curve measures a constant minimum after inversion.

Now suppose that we slow down the AC signal on the gate. If we reduce the frequency to a low enough value (typically less than 1 Hz for silicon), then generation and recombination processes can keep up with the AC signal and Q_I can follow the changes in Q_G. If Q_I follows Q_G, then the capacitance measured is just the oxide capacitance, because the top and bottom plates of the capacitor where Q is changing are on the two edges of the oxide. So the C–V curve jumps back to the oxide capacitance after inversion. Both the ideal HF and LF C–V curves are shown in Figure 6.29. Note that where the curves reach their minimum value corresponds to the onset of inversion and so is the threshold voltage for the MOS structure.

Figure 6.29 also shows the deep depletion condition that can be observed for MOS capacitors. Deep depletion occurs if the DC bias voltage on the gate is swept too rapidly into the inversion region. If this is done, the holes needed for the inversion region cannot be generated rapidly enough to follow the applied DC sweep voltage. Thus a deeper depletion than would be present in equilibrium (beyond $x_{D,max}$) forms to provide overall charge balance. A smaller capacitance than the equilibrium value is measured, which gradually relaxes back to C_{min} as hole generation takes place, illustrated by the broken red arrow in Figure 6.29.

The sweep rate of the gate voltage to maintain equilibrium conditions depends on the generation rate of minority carriers. Since there are no mobile charges in the depletion region (n and p are negligible), the generation rate is simply

$$U = -\frac{n_i}{\tau_G},\tag{6.33}$$

where n_i is the intrinsic concentration and τ_G is the generation lifetime. Assuming that this generation rate holds throughout the volume of the depletion region, the current density that results is

$$J_{gen} = \frac{qn_i\, x_{D,max}}{\tau_G}.\tag{6.34}$$

This current will flow to the surface, providing the holes for the inversion layer. In order to avoid deep depletion, the rate at which the DC bias voltage is swept on the gate of the MOS capacitor must be less than a value given by

$$\frac{dV}{dt} \leq \frac{J_{gen}}{C_{ox}} = \frac{qn_i x_{D,max}}{\tau_G C_{ox}}.\tag{6.35}$$

In practice, this means that the DC sweep voltage must be less than about 0.1 V s^{-1}.

The "ideal" C–V curves can be used to measure the oxide capacitance, the threshold voltage and the doping in the substrate, and to estimate the lifetime in the silicon. A host of other more subtle shifts and distortions in the C–V curves can shed light on the charges and defects in the insulator and at the interface. We saw in Figure 6.21 the four charge types generally associated with insulator/semiconductor interfaces. These charges cause shifts and distortions in the C–V curves because they terminate field lines from the gate charges and prevent the gate charge from influencing the substrate charge. Figure 6.30 illustrates these effects on the "ideal" HF C–V curve.

In Figure 6.30, Q_f is a fixed positive charge (usually) that is present in the oxide and positioned near the SiO_2/Si interface. Such a positive charge close to the interface will induce a corresponding negative image charge in the silicon, effectively making the silicon surface more N-type and thus requiring more voltage to invert to P-type. This shows up as a lateral shift in the C–V curve, with the magnitude of the shift given by qQ_f/C_{ox}. Measurement of this shift from the ideal C–V curve (which can be calculated) allows the calculation of Q_f. An additional lateral shift occurs because of the difference in the work functions between the metal and the semiconductor, $\varphi_{MS} = \varphi_M - \varphi_S$. The work function is the energy required to move an electron from the Fermi level E_F to just outside the material. This material property can be measured, so the lateral offset due to φ_{MS} is known. These two terms, φ_{MS} and qQ_f/C_{ox} together account for an increase in the gate voltage required to cause inversion and cause the lateral shift in the C–V curve shown in Figure 6.30.

The voltage at the cusp of the transition between accumulation and depletion, where there is no charge on the gate or in the semiconductor, is referred to as the flat-band voltage V_{FB}. In

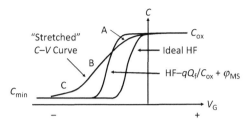

Figure 6.30 High-frequency (HF) C–V curves illustrating some of the non-idealities that may be present in actual experimental structures. The "ideal" C–V curve is laterally shifted by the fixed charge and the metal–semiconductor work function. A stretch in the C–V curve occurs because of interface states with different energy levels in the silicon bandgap.

theory, it is just the voltage to overcome the difference in the semiconductor and metal work functions. The metal work function φ_M is known and the semiconductor work function depends on the doping as

$$\varphi_S = \frac{kT}{q} \ln \frac{N_D}{n_i}. \tag{6.36}$$

In practice, you must add the shift due to Q_f at the interface as described above to obtain the flat-band voltage. The threshold voltage is the additional amount of voltage beyond V_{FB} required to form the depletion layer and the added voltage necessary to invert the surface to mobile holes. These terms show up in the turn-on voltage of the MOS transistor.

The "distorted" or "stretched" C–V curve shown in Figure 6.30 illustrates what happens when interface traps Q_{it} are present. We saw in Figure 6.23 that these traps can have energy levels throughout the forbidden band. As the applied gate voltage causes the semiconductor surface to move from accumulation through depletion to inversion, the surface potential (or Fermi level E_F) sweeps from one band edge to the other (from E_C to E_V in this N-type substrate we have been considering). As a result, the Q_{it} traps will fill and empty as E_F moves through their various energy levels. As this happens, the C–V curve will distort from its ideal shape, as the gate field lines terminate on the trapped charges instead of the substrate charges. The form of the distortion depends on the density and energy level of the Q_{it} traps. In Figure 6.30, the region labeled A corresponds to interface trap states near E_C, region B corresponds to states near the middle of the bandgap and region C corresponds to states near E_V. The assumption is that the interface states cannot follow the HF signal used to measure the capacitance. That is, they do not charge and discharge as the HF signal changes. But these traps will respond to the DC voltage applied to the gate, since we assume that this signal is applied for a long enough time to reach equilibrium. Thus, some of the gate charge terminates on the Q_{it} traps and the C–V curve is stretched out, requiring more gate voltage to achieve the same effect on the substrate charge.

A number of methods have been developed to experimentally extract the interface state density from C–V measurements on MOS capacitors. Indeed, these measurements and methods to minimize or passivate the interface traps by hydrogen annealing have been instrumental in

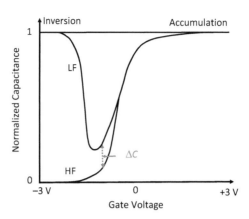

Figure 6.31 Schematic of the "quasi-static" measurement of the interface traps Q_{it}.

the progress of MOS transistors [12]. The most common method is the "quasi-static" measurement that is illustrated in Figure 6.31.

In this method, HF and LF C–V curves are superimposed from measurements on the same MOS capacitor. It is assumed that the Q_{it} traps cannot respond to the HF signal but can charge and discharge during the LF measurement. In this case, the LF measurement will see an additional capacitance, which will be proportional to the trap density at the DC voltage corresponding to the measurement point. The difference between the HF and LF plots shown in Figure 6.31 can be used to generate a plot like that previously shown in Figure 6.23, where the interface state density can be shown as a function of the applied DC voltage, or, equivalently, as a function of the position in the energy gap. The U-shape shown in Figure 6.23 is typical of what is observed for the Si/SiO$_2$ interface, where the density of interface trap states increases towards each band edge.

These interface traps are normally passivated by a "forming gas" or an anneal in a hydrogen-containing ambient at around 400 °C after finishing silicon processing [12]. At mid-gap, the density of traps typically reduces from $10^{12}\text{cm}^{-2}\text{eV}^{-1}$ to $10^{10}\text{cm}^{-2}\text{eV}^{-1}$ after hydrogen passivation. Given that the silicon atomic surface atom density is $\approx 10^{15}\text{cm}^{-2}$, the traps are rare and far apart even when the trap density is high. Because these traps are right at the interface and affect mobility and lifetimes, a significant research effort has been undertaken to understand the nature of the traps. It is particularly important to minimize the traps in solar cell applications, where high lifetimes are necessary to increase efficiency.

If mobile charges (Q_m) or oxide trapped charges (Q_{ot}) are present, they produce similar lateral shifts like Q_f (rather than stretching) of the C–V curves. The only real difference is that these charges are not located directly at the SiO$_2$/Si interface, so the magnitude of the shift they cause is reduced in proportion to their distance away from the interface. For example, a charge mid-way between the gate and the substrate would induce equal and opposite charges at each interface and balance out to a null effect.

Mobile charges (Q_m) from ions such as sodium (Na$^+$) or potassium (K$^+$) have largely been eliminated from MOS processing due to extra care with contamination levels introduced by humans or chemicals, as described in Chapter 4. But these mobile ions were part of the reason that MOS devices took so long to reach a reliable state. For seemingly random reasons (a change of soap in a restroom, a cleanroom near a seashore, a different bottle of chemicals), MOS devices produced unknown drifts of threshold voltage during operation, long after being fabricated, a severe problem for a chip maker. However, C–V measurements provide a simple way to measure whether mobile charges are present. Bias temperature stressing (BTS) involves making an initial C–V measurement at room temperature. The MOS capacitor is then heated to a low temperature around 200 °C with a DC bias applied on the gate. At this temperature, Na^+ or K^+ ions are highly mobile and easily drift upwards or downwards in the oxide layer. The wafer is then cooled to room temperature and the capacitance measured again. The extent of the lateral shift is proportional to Q_m. A second measurement with the opposite DC voltage at 200 °C moves the charges to the other interface, and the lateral difference between the two C–V curves corresponds to the total Q_m present, regardless of where it was originally located in the insulator. Because powerful C–V measurements could track Q_m, the origins of mobile charges in silicon processing were largely eliminated.

Oxide trapped charge (Q_{ot}) does not show up often in silicon oxides unless they are externally stressed. This may not be the case in deposited oxide or dielectrics. But silicon oxide is essentially perfect after Q_m, Q_f and Q_{it} are eliminated by either care in processing or annealing conditions. The oxide trapped charge can be measured by forcing current through an insulator for a period of time and making C–V measurements before and after current stressing. The trapped charge will show up as a lateral shift in the C–V curve, the magnitude of which is proportional to Q_{ot}. These kinds of measurements are essential for some devices that rely on charge trapping for a memory element (electrically erasable programmable read-only memories, EEPROMs).

Simply measuring the current–voltage (I–V) characteristic of the insulator in the MOS capacitor structure also provides useful information. For oxides thicker than about 10 nm, the current is too small to measure; but if a high enough voltage is used, the oxide will eventually break down destructively. A large number of such measurements on different capacitor structures can provide a histogram of the measured breakdown fields, which provides statistical information on the microscopic defect densities in the oxides.

Another interesting I–V measurement to make is of the tunneling current in thin (<5 nm) oxides illustrated in Figure 6.32. In the simple inverter circuit, $I_{G,N}$ is the gate tunneling current. As the oxide thickness decreases, this current starts to dominate normal leakage currents like $I_{D,off,P}$.

Tunneling is a purely quantum mechanical effect that can provide a way to non-destructively pass current through an oxide layer when the barrier is thin enough. Such currents can then place charge in the floating gate of a memory element (flash memory devices), for example. Mead [13] was first to recognize that tunneling currents would eventually limit the scaling of MOS transistors in 1972, just a few years after Moore made his scaling observations that became Moore's law. Mead basically realized that the same transistors could be made to work until the oxide thickness got to 5 nm and tunneling currents became an issue. At that point, leakage through the gate insulator would rise alarmingly due to tunneling currents, and power dissipation due to leakage would begin to dominate device performance. At the time, scaling that far into the future seemed impossible, and it was considered a crazy prediction by most people. Mead revisited the calculations more than 20 years later in 1994 [14] and predicted

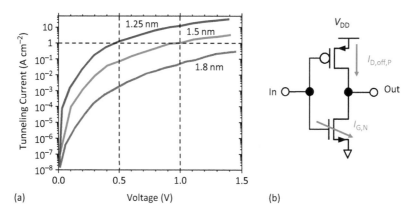

Figure 6.32 Simulated tunneling currents for very thin silicon dioxide layers on silicon. The simulations take quantum mechanical corrections into account.

a further order of magnitude of scaling was likely, leading to 30 nm devices with 1 nm gate oxides. Both predictions were largely ignored, but were among the most prescient made of the scaling limits of MOS transistors.

In summary then, C–V and I–V measurements are powerful characterization tools for determining the insulator quality and the insulator/semiconductor interface properties. These electrical measurements are used to monitor the quality of the insulators and optimize the processing steps to minimize defects in the layers, enabling the continued scaling of transistors. Although we have described the measurements for N-type substrates, P-type substrates produce similar C–V plots except that the horizontal voltage axis is flipped (mirror imaged). As mentioned earlier, C–V measurement can be performed on any arbitrary semiconductor/ insulator structure. In general, the distortions and shifts in the ideal C–V curve illustrated in Figure 6.30 would be greatly magnified in systems other than the Si/SiO$_2$ system.

6.11 High-K Dielectrics

The tunneling current limits described above came into sharp focus in 1999 when it appeared that there were no known solutions to the tunneling problem [15]. Combined with the limited solubility of dopants and statistical fluctuations of dopants in small devices [15] (which we discuss in Chapters 7 and 8), it seemed to indicate that Moore's law was in serious danger. The severity of the tunneling problem is shown in Figure 6.32, which shows the exponential dependence of the current on the insulator thickness, leading to leakage currents through the gate oxide of 1 A cm^{-2} at a 1 V supply voltage. Since the gate dielectric in the 90 nm node was already 1.2 nm thick, this did indeed seem to be a formidable problem.

At first sight, silicon nitride (Si$_3$N$_4$), with its dielectric constant of 7 versus 3.9 for silicon oxide, seemed like a good choice for a gate insulator, as it can be thermally grown by exposing silicon to an ammonia ambient in a nitridation reaction, similar to the oxidation reaction. Si$_3$N$_4$ is amorphous, has a dielectric strength similar to that of SiO$_2$ and has a large bandgap. In spite of extensive research, Si_3N_4 never was able to perform as a gate insulator because of high defect levels, making the search for a more exotic high-*K* material seem even more daunting. However, it did provide a stop-gap solution at the 90 nm and 65 nm nodes, as nitriding SiO_2 by exposing it to a nitrogen-containing ambient to form an oxynitride (SiON) film did increase the dielectric constant somewhat. It could be followed by a short oxidation to improve the interface properties, as we have seen that these are determined by the in-diffusion of oxygen to form the interface layer. Efforts to improve device performance at the 65 nm node did not rely on gate oxide scaling, but focused instead on introducing strain in the channels. It was therefore a big surprise when Intel introduced a hafnium-based high-*K*/metal gate stack at the 45 nm node, demonstrating that high-*K* insulators in MOS devices were possible.

The conceptual advantage of a high-*K* dielectric can be best understood by considering the simple parallel-plate MOS capacitor in accumulation. The measured capacitance per unit area is

$$C = \frac{\varepsilon}{t} = \frac{\kappa \varepsilon_0}{t}, \tag{6.37}$$

where ε is the permittivity of the insulator, t is the thickness of the insulator, and κ is the relative permittivity or dielectric constant (3.9 for SiO_2) compared to the permittivity of free space ε_0. For proper device operation in smaller devices, all vertical and lateral dimensions must be scaled equally to maintain equivalent electric fields in the device [16]. When silicon dioxide was used as the gate insulator, this meant scaling down the oxide thickness t to provide the necessary increased capacitive control over the channel. At some point, the thickness is so thin that the tunneling currents dominate the leakage power and prevent scaling. But if we relax the assumption that the insulator is silicon dioxide, then the other way to provide the necessary capacitive control over the channel is to increase the dielectric constant of the insulator. Thus the physical thickness of a high-*K* dielectric employed to achieve the equivalent capacitance density of an oxide of thickness t_{eq} is

$$\frac{t_{eq}}{\kappa_{ox}} = \frac{t_{hi-K}}{\kappa_{hi-K}}. \tag{6.38}$$

Since the device engineer does not care particularly about the material used, it is convenient to define an electrical oxide thickness of a new insulator in terms of the equivalent silicon dioxide thickness (or equivalent oxide thickness, EOT) as

$$t_{eq} = EOT = \frac{3.9}{\kappa_{hi-K}} t_{hi-K}. \tag{6.39}$$

A dielectric with a dielectric constant of 16 can use a physical thickness of ~4 nm to obtain an equivalent oxide thickness of $t_{eq} = 1$ nm. The increased physical thickness of the gate insulator means that the tunneling leakage current is virtually eliminated and explains why the introduction of a high-*K* dielectric was such an important event. There are minor corrections to the EOT calculation above due to quantum mechanical effects in the inversion layer position in the channel, depletion effects in the gate near the interface (avoided by going to metal gates) and the generally lower bandgaps of high-*K* materials compared to SiO_2. These corrections to the EOT amount to fractions of a nanometer but lead to the true electrical thickness of the insulator, which is what is seen by the device and is the electrical thickness that is measured by the capacitance measurements.

The introduction of high-K gate dielectrics, while surprising, did not come totally out of the blue. The first application to make use of high-K dielectrics was in DRAM capacitor memory elements in the 2001 timeframe. DRAM capacitors store charge and, as dimensions shrink, the requirement to maintain the capacitance of the cell required thinner insulators or higher-K dielectrics. DRAM manufacturers used a variety of high-K dielectric materials, including Al_2O_3, Ta_2O_5, HfO_2, ZrO_2 and laminates of the above. Higher dielectric constant options such as TiO_2, $SrTiO_3$ and Al-doped TiO_2 were also considered. But, because the high-K dielectric was sandwiched between two metal plates to make a metal–insulator–metal capacitor stack, the requirements were much relaxed compared to what is required for a gate insulator in a MOS device. Even small numbers of defects in the insulator of a MOS device can cause mobility degradation, reliability problems and device variations over time that affect circuit performance.

Some key early papers [17, 18] identified several requirements for the successful introduction of high-K oxides, namely these:

1. A K value high enough to enable continued scaling to lower EOTs.
2. Having an adequate bandgap.
3. Thermodynamically stable with the silicon channel.
4. Form a good electrical interface and limit the loss of carrier mobility in the silicon channel.
5. Prevent threshold instabilities caused by high levels of electrically active defects.

A desire to have a high-K oxide that could tolerate high-temperature processing and enable a gate-first process (like the polysilicon gate process) was too stringent to meet and led to the adoption of the gate-last process discussed in Chapter 2.

The first requirement is that the K value be large enough that the high-K development could be used economically for a reasonable number of scaling nodes. But there is a trade-off, because there is an empirical relationship where the oxide's K value tends to vary inversely with its bandgap, as shown in Figure 6.33. There are oxides with very high K, such as TiO_2 (80) or $SrTiO_3$ (2000), but their bandgaps are too small. The band offsets with the Si valence and conduction bands need to be at least 1 eV to prevent carrier injection over the insulating barrier rather than through the barrier. These very-high-K materials can be considered for DRAM capacitors, where the metal–insulator–metal system has much less stringent demands.

From Figure 6.33, it would seem that ZrO_2, HfO_2 or LaO_2 might be reasonable choices with a fair trade-off between high K value and bandgap. The requirement for thermodynamic stability arises from the condition that the high-K oxide must not react with silicon to form either SiO_2 or a silicide according to either of these reactions:

$$MO_2 + Si \Leftrightarrow M + SiO_2 \,, \tag{6.40}$$

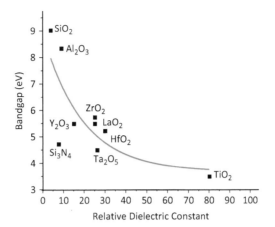

Figure 6.33 Bandgap and relative dielectric constant for a range of insulators.

$$MO_2 + 2Si \Leftrightarrow MSi + SiO_2 ,\tag{6.41}$$

where M is Zr, Hf or La. The first is because the resulting SiO_2 layer would increase the EOT and negate the effect of using the high-*K* dielectric. The second is because a metallic silicide layer would short out the gate field and prevent gate control of the channel. This requires that the high-*K* oxide has a higher heat of formation than SiO_2. As we have seen in Chapter 4, a chemical reaction proceeds towards the side with the higher heat of formation, which would prevent the formation of SiO_2 or silicide if this were true for the high-*K* oxide. The high heat of formation correlates with a wide bandgap in ionic compounds, so both problems are solved with high-*K* materials with good thermodynamic stability. This restricts the range of possible oxides to a very few, and our candidates are among those. Figure 6.34 summarizes some of the options.

It turns out that ZrO_2 is slightly unstable and reactive with silicon, potentially forming the silicide $ZrSi_2$. And La_2O_3 is more hygroscopic (absorbs water) than HfO_2. For these reasons, industry has concentrated on hafnium-based high-*K* materials. Obviously, this enables silicon technology to be scaled further. But the main reason silicon became the dominant semiconductor material was because of the perfection of the silicon dioxide insulating layer. It is simple to make by thermal oxidation, has very few defects, is amorphous with a high bandgap, and forms a smooth, abrupt interface with silicon. Most other semiconductors (Ge, GaAs, SiC, GaN, etc.) have unstable oxides, defective oxides or interfaces with poor qualities.

The actual incorporation of a Hf-based dielectric in place of SiO_2 is quite complex, as the example in Chapter 2 shows (see Figures 2.18 and 2.19). The actual composition of the dielectric layers and the process conditions used to deposit and anneal these films are closely guarded proprietary secrets in the companies that use these films. So the example shown in Chapter 2 is meant only to be illustrative and not a specific recipe.

Once SiO_2 was replaced by a high-*K* dielectric, silicon lost its key performance advantage and other materials with similar high-*K* layers can be considered for devices [18]. However, they are difficult to grow in large wafers and have brittle mechanical properties, so they must be integrated onto silicon substrate wafers if they are to achieve their full potential. We address some of these integration issues in Chapter 10 on deposition and epitaxial growth.

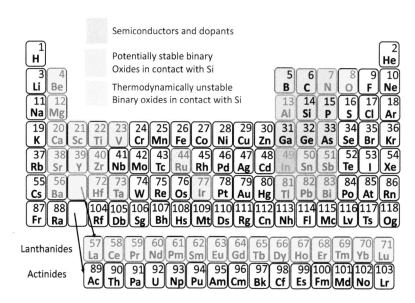

Figure 6.34 Periodic table of elements with potentially stable and unstable binary oxides with silicon indicated.

Silicon will remain the dominant semiconductor material (at least as a substrate) for the foreseeable future.

6.12 Oxidation of Silicon Carbide

Silicon carbide (SiC) is unique in that it is the only compound semiconductor that can be thermally oxidized to provide a high-quality oxide SiO_2 layer. These oxides can be used to passivate the SiC surface and to provide the gate oxide for SiC power MOSFETs. The kinetics of SiC oxidation are similar to those of Si oxidation, with the added complication that C is a component in the reaction through the formation of CO. The reaction for the thermal oxidation of SiC is given by

$$SiC + 1.5O_2 \rightarrow SiO_2 + CO .$$ (6.42)

This means that the flux of O_2 is 1.5× the flux of CO, i.e., $F_{O_2} : F_{CO} = 1.5{:}1$. The thermal oxide of SiC is SiO_2 and the interface seems to be atomically abrupt [19], similar to the oxide/silicon interface. Based on the Si density in SiC, the amount of SiC consumed by the thermal oxidation can be calculated to be 46%, similar to the value for the thermal oxidation of Si. For example, to grow 100 nm of SiO_2 on SiC, 46 nm of SiC are consumed [20]. During thermal oxidation, most of the C atoms diffuse out to the surface as CO molecules, with a small number of C interstitials being injected into the substrate, reducing the number of carbon-vacancy defects [21]. This is very reminiscent of the case of silicon interstitial injection causing oxidation-enhanced diffusion during the oxidation of silicon.

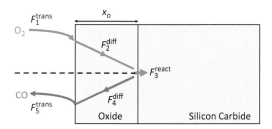

Figure 6.35 Schematic of the O_2 and CO fluxes involved in the oxidation of SiC. The transport, diffusion and reaction fluxes are the same as for silicon oxidation, with the addition of a CO out-diffusion flux and a transport flux of CO back to the gas phase.

Just as in the case of silicon oxidation, we can examine the major fluxes involved, shown schematically in Figure 6.35. The transport fluxes are given by

$$F_1^{O_2} = h_{O_2}(C_{O_2}^g - C_{O_2}^s),$$ (6.43)

$$F_5^{CO} = h_{CO}(C_{CO}^s - C_{CO}^g),$$ (6.44)

where the superscript s represents the concentration of O_2 or CO at the solid/gas interface and the superscript g represents the concentrations in the gas phase. The gas-phase transport coefficient $h = D/\delta$ represents the diffusive transport across the boundary layer δ (cm s^{-1}). Just as in the case of silicon oxidation, these transport fluxes are fast and are not limiting for the growth kinetics.

The reaction flux at the SiO$_2$/SiC interface is more complex, corresponding to both a forward reaction of oxidant creating SiO$_2$ and a reverse reaction where carbon from SiC is oxidized and out-diffuses as CO [20]:

$$F_3^{react} = K_f C_{O_2} - K_r C_{CO},$$ (6.45)

where K_f and K_r are the forward and reverse reaction coefficients for the O_2 and CO reactions at the SiC interface. Thus, the only fluxes that matter are the in-diffusion flux of oxidant, the reaction flux at the interface (these are the same two fluxes as silicon) and the out-diffusion flux of CO. During oxidation, the oxidant concentration at the oxide surface $C_{O_2}^*$ is much greater than the out-diffusing concentration of CO at the oxide interface C_{CO}^*, i.e., $C_{O_2}^* \gg C_{CO}^*$. Thus, the oxidant flux determines the Deal–Grove linear reaction coefficient, just as in silicon oxidation:

$$B/A \approx K_f \frac{C_{O_2}^*}{N_1}.$$ (6.46)

The Deal–Grove parabolic reaction coefficient B, which is determined by the rate-controlling diffusion step, is more complex, because both the O_2 in-diffusion flux and the CO out-diffusion flux are involved, giving [20]

$$B = \frac{(K_f C_{O_2}^* - K_r C_{CO}^*)/N_1}{1.5 K_f / D_{O_2} + K_r / D_{CO}}. \tag{6.47}$$

The factor of 1.5 comes from the difference in the O_2 and CO fluxes due to the interface reaction, as mentioned above. This equation has two limiting regimes. If oxygen diffusion is the rate-controlling step, i.e., $K_f / D_{O_2} \gg K_r / D_{CO}$, then

$$B \approx \frac{C_{O_2}^*}{1.5 N_1} D_{O_2}, \tag{6.48}$$

and B will have the same activation energy as that for the oxidation of silicon. If the CO out-diffusion is the rate-controlling step, i.e., $K_f / D_{O_2} \ll K_r / D_{CO}$, then

$$B \approx \frac{C_{O_2}^* K_f}{N_1 K_r} D_{CO}, \tag{6.49}$$

where the diffusivity of CO to escape the oxide is the limiting step. Note that, strictly speaking, the wet oxidation of silicon should look very similar to the oxidation of SiC, where the out-diffusion of H_2 is considered as part of the kinetics:

$$Si + 2H_2O \rightarrow SiO_2 + 2H_2. \tag{6.50}$$

However, since hydrogen is such a small molecule with high diffusivity, the H_2 diffusion is not a rate-limiting step in silicon oxidation and the simple Deal–Grove reaction suffices.

These results support a Deal–Grove model for SiC oxidation, with a possible modification to the parabolic coefficient B in cases when CO out-diffusion controls the kinetics, as described by the equation:

$$\frac{x_o^2}{B} + \frac{x_o}{B/A} = t + \tau. \tag{6.51}$$

This model fits the experimental results, as shown in Figure 6.36 for different faces of SiC.

Perhaps the most surprising observation from Figure 6.36 is that there is almost an order-of-magnitude difference in the oxidation rates, with the $(000\bar{1})$ C-terminated face oxidizing much faster than the (0001) Si-terminated face. The activation energies for oxidation on the different faces are given in Table 6.3 based on the data in Figure 6.36.

The activation energies in Table 6.3 do not provide much insight into the physical mechanisms of oxidation, with the exception of the large activation energy for the diffusion-limited rate constant B for the Si face compared to the C face. This might indicate that the in-diffusion of oxidant is limiting in one case and the out-diffusion of CO is limiting in the other. Otherwise we might expect that diffusion in oxide would have the same activation energy. The available SiC oxidation data are not extensive enough to go beyond the phenomenological model and determine how the activation energy for B corresponds to the products of the activation energies for K_f, K_r and D_{CO} in (10.49).

Table 6.3 **The B and B/A values for SiC oxidation on the C face and Si face. Note that some activation energies are different from [20] perhaps due to different fitting procedures.**

Ambient dry O_2 on SiC	B/A (μm min^{-1})	B (μm^2 min^{-1})
$(000\overline{1})$ C face	$5.91 \times 10^2 \exp\left(-\dfrac{1.01 \text{ eV}}{kT}\right)$	$7.68 \times 10^4 \exp\left(-\dfrac{1.87 \text{ eV}}{kT}\right)$
(0001) Si face	$2.39 \times 10^3 \exp\left(-\dfrac{1.34 \text{ eV}}{kT}\right)$	$1.08 \times 10^6 \exp\left(-\dfrac{2.67 \text{ eV}}{kT}\right)$

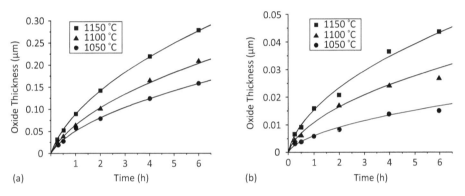

Figure 6.36 (a) Oxide thickness for dry O_2 oxidation of the $(000\overline{1})$ C-terminated face of 4H-SiC. (b) Oxide thickness for dry O_2 oxidation of the (0001) Si-terminated face of 4H-SiC. The solid lines are Deal–Grove fits from (10.51) with $\tau = 0$, using values from Table 6.3. Data are from [19, 20].

What is clear from the data is that the oxidation rates are very different on the Si face and the C face, and a possible mechanism is indicated in Figure 6.37. The surface component is clearly an important factor: for the $(000\overline{1})$ C-terminated face, the first layer is 100% C atoms, while for the (0001) Si-terminated face, the first layer is 100% Si atoms. For the C face, there is a large driving force to break the Si–C bonds and form CO, a very stable molecule. After the first layer removal, the Si atoms in the second layer can react rapidly with the oxygen since there is only one Si–C bond connected to the third C-face layer. However, on the (0001) Si face, it is more difficult to break the Si–C bonds since there are three Si–C bonds connected to the second C layer. This provides at least a qualitative explanation for the vastly different oxidation rates [20].

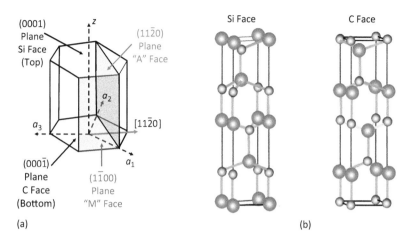

Figure 6.37 (a) Hexagonal SiC crystal structure showing the Si and C faces. The M and A faces are of interest in SiC devices. (b) The 4H unit cell showing the Si and C faces along with the bonding structure within the cell. (See Figures 3.5 and 3.6, which discuss how directions and planes are defined in hexagonal unit cells.)

6.13 Summary of Key Ideas

Thermal oxidation has been a cornerstone of silicon technology since its inception. In fact, it is really the ability to easily grow SiO_2 layers that distinguishes silicon technology from all other semiconductors. SiO_2 layers adhere well to silicon, are relatively chemically inert, mask dopant diffusion, block ion implantations and are easily patterned. However, the real strength of SiO_2 lies in the almost perfect electrical interface it makes with silicon. Electrical charges are reproducible and small, and the interface is almost atomically abrupt. These features allow junctions to be almost ideally passivated at silicon surfaces, and they allow a variety of device structures that critically depend on surface properties to be easily built in silicon (MOS structures in particular). Even devices which are not primarily surface-controlled, such as bipolar transistors, do not have their electrical characteristics degraded by parasitic surface effects. No other semiconductor/insulator interface has all these desirable properties.

Since SiO_2 is such an important part of silicon technology, the growth kinetics of these layers on silicon have been carefully studied over many years. The basic growth mechanism is oxidant transport through the SiO_2 layer to the Si/SiO_2 interface, where a simple chemical reaction produces the new layers of oxide. The growth generally follows a simple linear–parabolic law. This basic model of oxidation has been extended in recent years to handle many situations for which the original formulation falls short. These include heavily doped substrates, 2D effects and thin oxides. The same basic ideas have been applied to other materials like polysilicon and silicides, which are common in silicon technology and which are often oxidized. We introduced in this chapter the concept of the interaction of oxidation with other process steps such as diffusion. Oxidation-enhanced diffusion (OED) or oxidation-retarded diffusion (ORD) and impurity redistribution and segregation during oxidation are important effects. We will discuss them in more detail in Chapter 7 after we describe the basic ideas of dopant diffusion.

We also carefully studied $C-V$ measurement methods in this chapter because they are widely used and extremely useful in characterizing the charges and defects that exist in all semiconductor/insulator interfaces. While the discussion was primarily focused on the Si/SiO_2 interface, which is close to perfect, these same measurement methods are widely applied to other materials systems and provide the same kind of very useful information in those systems.

With the end of SiO_2 thickness scaling in gate dielectrics in MOS devices around 2005, a massive effort was launched to find a high-K dielectric that could be used in place of SiO_2. Starting in about 2008, HfO_2-based dielectric structures were found to be a solution and were introduced into manufacturing. We used such a process in the CMOS process flow we considered in Chapter 2.

Finally, many of the models and physical mechanisms described in this chapter have been implemented in process simulators. These simulators today are capable of accurately predicting the oxidation processes in modern very-large-scale integrated (VLSI) structures. As new understanding develops, particularly with respect to very small structures, new models are incorporated in these simulators so that their capability is constantly improving. These simulation tools are especially powerful when coupled process phenomena such as OED occur because these are able to track the complexities of interacting phenomena and interacting models for these effects.

6.14   REFERENCES AND NOTES

[1]   B. E. Deal and A. Grove, "General relationship for the thermal oxidation of silicon," *Journal of Applied Physics*, vol. 36, no. 12, pp. 3770–3778, 1965.

[2]   H. Z. Massoud, J. D. Plummer and E. A. Irene, "Thermal oxidation of silicon in dry oxygen: growth-rate enhancement in the thin regime I. Experimental results," *Journal of the Electrochemical Society*, vol. 132, no. 11, pp. 2685–2693, 1985.

[3]   J. R. Ligenza, "Effect of crystal orientation on oxidation rates of silicon in high pressure steam," *Journal of Physical Chemistry*, vol. 65, no. 11, pp. 2011–2014, 1961.

[4]   C. Ho and J. Plummer, "Si/SiO₂ interface oxidation kinetics: a physical model for the influence of high substrate doping levels I. Theory," *Journal of the Electrochemical Society*, vol. 126, no. 9, pp. 1516–1522, 1979.

[5]   E. Biermann, H. H. Berger, P. Linke and B. Müller, "Oxide growth enhancement on highly n-type doped silicon under steam oxidation," *Journal of the Electrochemical Society*, vol. 143, no. 4, pp. 1434–1442, 1996.

[6]   P. Dobson, "The effect of oxidation on anomalous diffusion in silicon," *Philosophical Magazine*, vol. 24, no. 189, pp. 567–576, 1971.

[7]   D.-B. Kao, J. P. McVittie, W. D. Nix and K. C. Saraswat, "Two-dimensional thermal oxidation of silicon – I. Experiments," *IEEE Transactions on Electron Devices*, vol. 34, no. 5, pp. 1008–1017, 1987.

[8]   N. Mott, "On the oxidation of silicon," *Philosophical Magazine B*, vol. 55, no. 2, pp. 117–129, 1987.

[9]   C. Rafferty and R. Dutton, "Plastic analysis of cylinder oxidation," *Applied Physics Letters*, vol. 54, no. 18, pp. 1815–1817, 1989.

[10]   D. Ahn, S. J. Ahn, P. B. Griffin, *et al.*, "A highly practical modified LOCOS isolation technology for the 256 Mbit DRAM," in *Proceedings of 1994 IEEE International Electron Devices Meeting*, IEEE, pp. 679–682,1994.

[11]   B. E. Deal, "Standardized terminology for oxide charges associated with thermally oxidized silicon," *IEEE Transactions on Electron Devices*, vol. 27, no. 3, pp. 606–608, 1980.

[12] M. L. Reed and J. D. Plummer, "Chemistry of Si–SiO$_2$ interface trap annealing," *Journal of Applied Physics*, vol. 63, no. 12, pp. 5776–5793, 1988.

[13] B. Hoeneisen and C. A. Mead, "Fundamental limitations in microelectronics – I. MOS technology," *Solid-State Electronics*, vol. 15, no. 7, pp. 819–829, 1972.

[14] C. A. Mead, "Scaling of MOS technology to submicrometer feature sizes," *Analog Integrated Circuits and Signal Processing*, vol. 6, no. 1, pp. 9–25, 1994.

[15] P. A. Packan, "Pushing the limits," *Science*, vol. 285, no. 5436, pp. 2079–2081, 1999.

[16] R. H. Dennard, F. H. Gaensslen, V. L. Rideout, E. Bassous and A. R. LeBlanc, "Design of ion-implanted MOSFET's with very small physical dimensions," *IEEE Journal of Solid-State Circuits*, vol. 9, no. 5, pp. 256–268, 1974.

[17] E. Gusev, E. Cartier, D. A. Buchanan, *et al.*, "Ultrathin high-*K* metal oxides on silicon: processing, characterization and integration issues," *Microelectronic Engineering*, vol. 59, no. 1–4, pp. 341–349, 2001.

[18] J. Robertson, "High dielectric constant oxides," *European Physical Journal – Applied Physics*, vol. 28, no. 3, pp. 265–291, 2004.

[19] T. Kimoto and J. A. Cooper, "Device processing of silicon carbide," in *Fundamentals of Silicon Carbide Technology: Growth, Characterization, Devices and Applications*, Wiley, chap. 6, 2014.

[20] Y. Song, S. Dhar, L. C. Feldman, G. Chung and J. Williams, "Modified Deal–Grove model for the thermal oxidation of silicon carbide," *Journal of Applied Physics*, vol. 95, no. 9, pp. 4953–4957, 2004.

[21] T. Hiyoshi and T. Kimoto, "Reduction of deep levels and improvement of carrier lifetime in n-type 4H-SiC by thermal oxidation," *Applied Physics Express*, vol. 2, no. 4, p. 041101, 2009.

6.15   PROBLEMS

6.1 A spherically shaped piece of silicon is cut and polished from a Czochralski single-crystal ingot and oxidized in a thermal oxidation furnace. Upon pulling the silicon sphere from the furnace, the color is observed to vary significantly over the surface. Why?

6.2 Why is steam oxidation more rapid than dry O$_2$ oxidation?

6.3 According to the Deal–Grove model, oxidation kinetics start out linear and become parabolic as the oxidation proceeds. The transition from linear to parabolic growth is often taken to be when $k_s x_o / D \approx 1$. This can be seen from (6.6) since, for $k_s x_o / D \ll 1$, Figure 6.9(a) applies, and, for $k_s x_o / D \gg 1$, Figure 6.9(b) applies. Calculate the oxide thickness at which this transition takes place and plot this versus oxidation temperature for both O$_2$ and H$_2$O oxidation.

6.4 Does the oxide thickness at which there is a transition from linear to parabolic rates change if we perform an oxidation at a pressure of 20 atm rather than at 1 atm?

6.5 A gate oxide for a MOS device needs to be grown thermally and it must be controlled to 2 nm ± 0.1 nm. The oxide is grown in dry O$_2$ at 800 °C. You can assume that the Deal–Grove model holds and you may neglect the anomalous thin oxide regime (Massoud model). You may also assume that, since the oxide is so thin, the growth is entirely in the linear regime of the Deal–Grove linear–parabolic mode. What temperature control is required in the furnace in order to achieve the ±0.1 nm control? Calculate a quantitative answer.

6.6 In the Deal–Grove oxidation model, the flux F_1 is normally neglected. Explain what this flux is and why it is OK to ignore it in developing the linear–parabolic model.

6.7 In the Deal–Grove oxidation model, explain why the oxidant flux through the SiO_2 (F_2) is drawn as a straight line.

6.8 In an experiment, an oxidation furnace is set up as shown below. The two wafers on the left are lying flat, while those on the right are standing vertically and are closely spaced. Using the Deal–Grove oxidation model, explain what you think would result in this experiment in terms of the thicknesses of the oxides grown on the wafers.

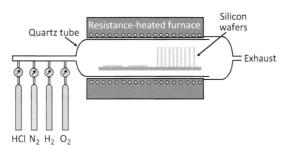

6.9 In the Deal–Grove oxidation model, B/A changes with the crystal orientation of the silicon substrate. Explain physically why this is not also true for B in the Deal–Grove model.

6.10 A (111) silicon wafer is covered by an SiO_2 film 0.3 μm thick.
(a) What is the time required to increase the thickness by 0.5 μm by oxidation in H_2O at 1200 °C?
(b) Repeat for oxidation in dry O_2 at 1200 °C.

6.11 Suppose an oxidation process is used in which (100) wafers are oxidized in O_2 for 3 h at 1100 °C, followed by 2 h in H_2O at 900 °C, followed by 2 h in O_2 at 1200 °C. Use Figure 6.11 to estimate the resulting final oxide thickness. Explain how you use these figures to calculate the results of a multi-step oxidation like this.

6.12 For a (111) silicon wafer, what is the oxide thickness after a 100 min dry O_2 oxidation followed by a 35 min H_2O oxidation at 900 °C? Calculate your answer. Do not use the charts to read off values. You can ignore the anomalous initial oxidation regime in dry O_2.

6.13 A new semiconductor material (X) is being investigated for possible use in making semiconductor devices. Many experiments are done to characterize the material, including oxidation experiments. In the oxidation experiments, it is found that an oxide with the composition XO_2 is grown and that the growth kinetics are purely parabolic with time. What can you conclude about the physical mechanisms involved in the oxidation of X?

6.14 The structure shown below is formed by oxidizing a (111) silicon wafer (x_o = 500 nm), and then using standard masking and etching techniques to remove the SiO_2 in the center region. An N^+ doping step is then used to produce the structure shown. The structure is next placed in an oxidation furnace and oxidized at 900 °C in H_2O. The oxide will grow faster over the N^+ region than it will over the lightly doped substrate, as discussed in the text. However, the enhancement will decrease over time as the N^+ region diffuses and the surface concentration drops. Assume B/A is enhanced by 4× for the first hour of oxidation, then by 2× for the second hour and then drops to a normal value for longer times.

Will the growing oxide over the N^+ region ever catch up in thickness to the other oxide? If so, when and at what thickness. Use the Deal–Grove model for the oxidation kinetics.

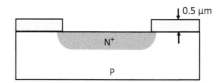

6.15 Highly doped N^+ regions (extrinsic regions) oxidize significantly faster than lightly doped regions (intrinsic regions). In an experiment, the oxidation rates of N^+ versus N regions on a wafer are compared for oxidations done at a relatively high temperature (say 1000 °C) versus the rates at a lower temperature (say 800 °C). The N^+ region is doped heavily enough that it is extrinsic in both of these experiments. What result would you expect from such an experiment? Would the N^+/N rate difference be larger at 1000 °C or at 800 °C? You do not have to calculate a numerical answer, but explain your answer in terms of the physical mechanisms occurring.

6.16 A 1 μm thick dielectric layer is deposited on a (100) silicon wafer. This wafer is then placed in an oxidation furnace at 1000 °C in an O_2 ambient. It is known from other experiments that the diffusivity of O_2 in the deposited dielectric material is one-third of its value in SiO_2.
  (a) How long does it take to thermally grow a 50 nm SiO_2 layer underneath the deposited dielectric?
  (b) Suppose this experiment is actually done and the experimental time is half of the value you calculated in (a). How would you explain this result?

6.17 A 1 μm wide trench is etched in a (100) silicon wafer, so that the sides of the trench are (110) planes. An angled implant is performed, doping only one of the sidewalls N^+ and thereby enhancing the linear rate constant by a factor of 4. The structure is then oxidized in H_2O at 1100 °C. At what time during the oxidation will the groove be filled with SiO_2? You may assume that the oxidation only occurs on the trench sidewalls; ignore the effect of oxidation on the trench bottom. Assume the appropriate oxidation coefficients scale as (111:110:100) = (1.68:1.2:1.0).

6.18 A uniform oxide layer of 0.4 μm thickness is selectively etched to expose the silicon surface in some locations on a wafer surface. A second oxidation at 1000 °C in H_2O grows 0.2 μm on the bare silicon.
  (a) Calculate the final oxide thicknesses and sketch a cross-section of the SiO_2 in all locations on the wafer and the position of the Si/SiO_2 interface. Assume (111) silicon and use the Deal–Grove oxidation model.
  (b) Would your picture be the same if the second oxidation grew the 0.2 μm at a different temperature? Explain.

6.19 For a local oxidation experiment, a thin SiO_2 layer is first thermally grown on (100) silicon substrate, followed by Si_3N_4 mask layer deposition. To form the local oxide structure, a local silicon trench is etched with depth of 30 nm, as shown below. Now the wafer is put into the oxidation furnace, at 1000 °C in H_2O for 200 min. If we want the top surface of the local oxide structure to be exactly the same height as the top surface of the thin oxide, what is the initial oxide thickness needed for this requirement?

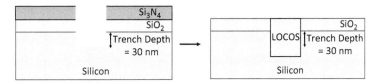

6.20 A semiconductor company is evaluating a new shallow trench isolation (STI) technology illustrated below. The transistors are fabricated first and then shallow trenches are etched to laterally isolate the transistors. The trenches are thermally oxidized to fill them with SiO_2. The etched width of the trenches is 100 nm. The oxidation is done in H_2O at 900 °C. The trench sidewalls are (100) planes. Assume the N^+ heavily doped regions enhance B/A in the Deal–Grove model by 4× over the lightly doped value. Please take this into account in your answer.

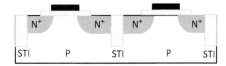

(a) What oxidation time is required to close the trenches?
(b) What would the shape of the oxide in the trenches actually look like? You only need to sketch this and explain your sketch.

6.21 A traditional LOCOS structure is shown below. Here we consider a modification to this LOCOS process in an attempt to create a structure with a more planar surface after oxidation. In the modified process, a 40 nm thin oxide is thermally grown on a (100) Si wafer as part of the masking structure. Following nitride deposition, patterning and etching of the nitride and the thin oxide, the LOCOS oxide is ready to grow. In the new process, the Si is etched before the LOCOS oxidation just enough so that, after the LOCOS oxide is grown, the top surface of the LOCOS oxide on the unmasked side (left) will be at exactly the same height as the top surface of the thin masking oxide. Assuming that the LOCOS oxidation is 90 min at 1000 °C in an H_2O ambient, how deeply should the silicon be etched prior to the LOCOS growth to achieve this planar surface?

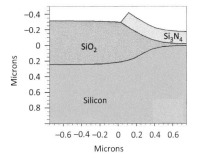

6.22 A process technician plans to grow a 500 nm thick thermal oxide. Her plan was to grow the oxide in H_2O at 1000 °C. She calculated the time needed and started the process. However, when the process is done, she noticed that the oxide thickness was much thinner than expected. Then she noticed that she had forgotten to turn on the "H_2" torch for the H_2O oxidation. Calculate the extra time needed to get the final oxide thickness equal to 500 nm (in H_2O at 1000 °C). Assume (100) silicon.

6.23 In a particular CMOS process, it is desired grow a 2 nm thick gate oxide. The oxide is grown in dry O_2 at 900 °C. Because of other constraints, the oxidation time must be 30 min. Assume that the Deal–Grove model holds for this thin oxide, and both rate constants B and B/A are proportional to the O_2 partial pressure. What partial pressure of O_2 should be used in this oxidation? Assume (111) silicon.

6.24 An experimental oxidation process is performed in which both O_2 and H_2O are used. You can assume that the two oxidation processes act in parallel and independently. Based on the Deal–Grove model, write a mathematical expression that could be used to model the overall growth rate dx/dt which includes both the O_2 and H_2O oxidation processes. You do not have to integrate this equation to derive an overall growth law, just show what dx/dt would look like. Based upon what you know about O_2 and H_2O oxidation rates, explain how you would simplify your dx/dt expression so that it could be easily integrated into a simple growth law.

6.25 Silicon on insulator (SOI) is a substrate material that is used for some integrated circuits. The structure, shown below, consists of a thin single-crystal silicon layer on an insulating (SiO_2) substrate. The silicon below the SiO_2 provides mechanical support for the structure. One of the reasons this type of material is used is because junctions can be diffused completely through the thin silicon layer to the underlying SiO_2. This reduces junction capacitances and produces faster circuits. Isolation is also easy to achieve in this material, because the thin Si layer can be completely oxidized, resulting in devices completely surrounded by SiO_2. A LOCOS process is used to locally oxidize through the silicon, as shown on the right. Assuming the LOCOS oxidation is done in H_2O at 1000 °C, how long will it take to oxidize through the 0.3 μm (111) silicon layer? Calculate a numerical answer using the Deal–Grove model.

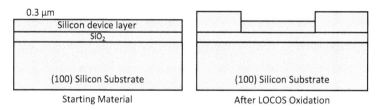

6.26 An SiO_2 layer is thermally grown at 1000 °C using a 10–20–30 min dry–wet–dry oxidation cycle. Upon pulling the (100) wafer from the furnace, the oxide color is observed to be tan. Is this right? If not, suggest what might have gone wrong in the experiment.

6.27 For this problem, we are interested in why the Deal–Grove model does not work well for

thin oxide. In one experiment, the oxidation is done in dry O_2 at 850 °C on (111) substrate. Calculate the growth rate when the oxide thickness is 3 nm, based on
(a) the Deal–Grove model and
(b) the Massoud model.

Hint: The rate constant C in the Massoud model is

$$C = C_0 \exp\left(-\frac{E_A}{kT}\right),$$

where $E_A = 2.35$ eV and $C_0 = 5.8 \times 10^6 \mu\text{m min}^{-1}$. Assume $L = 7$ nm.

6.28 In the simulation example in Figure 6.15, the difference in oxidation rate between the heavily doped and lightly doped regions would be more pronounced at low temperatures than at high temperatures. Explain physically why this is the case using the mechanisms in the Deal–Grove model and the behavior of point defect (V) concentrations versus doping.

6.29 A standard OED-type test structure is used in an experiment as shown in Figure 6.14. The LOCOS oxide is grown at 1100 °C in H_2O. The substrate is (111).
(a) What is the instantaneous oxidation rate when the oxide has grown to 0.1 μm?
(b) Assume that Si lattice planes are separated by 2.5 Å (0.25 nm) in the substrate and the surface area density of Si atoms is 10^{15} cm^{-2}. If 1% of the oxidized silicon atoms are injected as interstitials into the substrate, what is the flux of injected interstitials?

6.30 The structure shown below is implanted with oxygen using a 1×10^{18} cm^{-2} implant at 200 keV. At this energy, the average depth of the implanted ions is $R_P = 0.35$ μm. The left-hand side is masked from the implant. Following the implant, a high-temperature anneal is performed, which forms stoichiometric SiO_2 in a buried layer on the right-hand side. Calculate the structural dimensions on the right-hand side following this anneal (oxide thickness, distance from the surface and all other important dimensions). You can assume the silicon atomic density is 5×10^{22} atoms cm^{-3}, and the Si lattice planes are 0.25 nm apart. State any other assumptions you make.

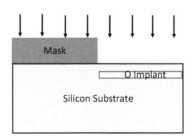

6.31 As part of an IC process flow, a CVD SiO_2 layer 1.0 μm thick is deposited on a $\langle 100 \rangle$ silicon substrate. This structure is then oxidized at 900 °C for 60 min in an H_2O ambient. What is the final SiO_2 thickness after this oxidation? Calculate an answer; do not use the oxidation charts in the text.

6.32 The basic concept of the MOS transistor was invented in 1926 by Lilienfeld, so why did it take almost 50 years before MOSFETs were introduced in manufacturing and why did bipolar devices that were invented 25 years after FETs dominate the semiconductor industry in the 1950s and 1960s?

6.33 An experimental metal–insulator–semiconductor (MIS) structure is fabricated by depositing Si_3N_4 (silicon nitride) on a silicon substrate. The nitride is deposited using silane and ammonia in a CVD system:

$$3SiH_4 + 4NH_3 \rightarrow Si_3N_4 + 12H_2.$$

A metal electrode is deposited and a C–V plot is made as shown below. A representative C–V plot is also shown for an identical structure, except with thermally grown SiO_2 as the insulator. Explain the lateral shift in the C–V curve of the Si_3N_4.

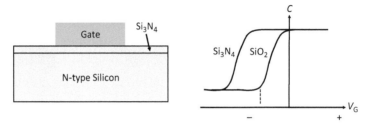

6.34 Construct a HF C–V plot for a P-type silicon sample, analogous to Figures 6.26 to 6.28. Explain your plot based on the behavior of holes and electrons in the semiconductor in a similar manner to the discussion in the text for Figures 6.26 to 6.28.

6.35 A MOS structure is fabricated to make C–V measurements as shown below. The C–V plot shows the result if the P^+ diffusion is NOT present. Sketch the expected shape of the C–V plot with the P^+ diffusion. Explain.

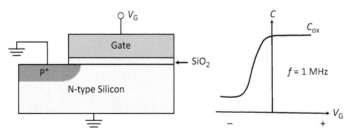

6.36 High-frequency C–V measurements are made on N-type silicon substrates. The doping concentrations in the substrates are 10^{14}, 10^{15} and 10^{16} cm^{-3} as shown on the plot. The minimum capacitance under inversion increases with doping as seen in the plot. Explain physically why this should happen by discussing the processes of depletion and inversion.

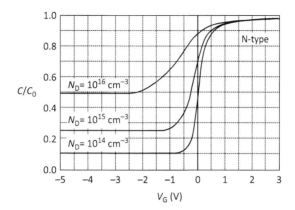

6.37 A student builds a MOS capacitor structure and then measures the C–V curve associated with the capacitor. She learned in class that a typical high-frequency (HF) measurement frequency is 1 MHz, so she chooses that frequency for her measurement. Surprisingly, she measures a low-frequency (LF) curve. Suggest two possible reasons why this might have happened.

6.38 In a small MOS device, there may be a statistical variation in V_T due to differences in Q_F from one device to another. Suppose, in a 0.13 µm technology, that the minimum device gate oxide area is 0.1 µm × 0.1 µm with a 2.5 nm gate oxide. What would the difference in threshold voltage be for devices with zero or one fixed charge in the gate oxide?

6.39 SiC is a semiconductor material of interest for power devices because it has a wide bandgap and therefore can support high voltages more easily than silicon can. SiC can be oxidized just like silicon to create an SiO_2 dielectric according to the following chemical reaction:

$$SiC + 3H_2O \rightarrow SiO_2 + 3H_2 + CO.$$

The H_2 and the CO byproducts are gases and are removed by out-diffusion just as the H_2 byproduct is in silicon oxidation. A Deal–Grove linear–parabolic model is found to model SiC oxidation as it does in the silicon oxidation case. Experimental data give a value for B of $C \exp(-0.78 \text{ eV}/kT)$, where C is a constant. What can you conclude about the physical processes involved in SiC oxidation from these data? Explain your answer.

6.40 We know that thermal oxidation of silicon can cause OED. If the SiO_2 layer were deposited rather than created by thermal oxidation, and the wafer were subsequently heated in an N_2 atmosphere to allow dopant diffusion, would you expect OED to occur? Explain.

7 Doping in Semiconductors

7.1 Introduction and Basic Concepts

One of the main challenges in designing a front-end process for building a device is accurate control of the placement of the active doping regions. Understanding and controlling diffusion and annealing behavior are essential to obtaining the desired electrical characteristics. Consider a cross-section of a state-of-the-art MOS transistor and imagine what happens when it gets scaled down to smaller dimensions (Figure 7.1). In "ideal" or Dennard scaling, as described in Chapter 1, everything shrinks down linearly from one generation to the next. This means that not only do the lateral dimensions scale, but the vertical dimensions, such as the deep source/drain contacting junctions and the shallower tip or extension junctions, also scale. This maintains the same electric field patterns (assuming the operating voltage also scales proportionally). With the same $\mathcal{E}$-field patterns, the device operates in the same manner as before, except that the shorter channel length allows for faster switching speeds [1]. Thus there is a continuous drive to decrease junction depth with each new technology generation.

There are two things that become very important and place constraints on junction doping profiles: the junctions are getting shallower; and the doping concentrations are going up. The doping concentration increase is related to the parasitic resistances inherent to the device structure. Those resistances have to scale as the device gets smaller because the intrinsic channel resistance is designed to get smaller to produce higher-performance devices. So if the parasitic resistances are not going to dominate and limit the current through the device, they have to get smaller as well, or at least not get any bigger. As the depth of the junctions decreases, there is less area available to carry the current and to reduce the resistance; hence the need to increase the doping concentration in the doped regions. Dopant concentrations in MOS transistors have increased more than 100-fold over the past 20 years and are on the order of 1% of the silicon lattice density for current device technologies, which is at the solubility limit of dopants in the lattice [2]. The diffusion cycles required to electrically activate the implanted dopant atoms are often the limiting factor on the junction depth that can be obtained.

We introduced the concept of sheet resistance in Chapter 2 when we silicided the source/drain regions to reduce resistance. Since this concept is central to doped regions, we revisit it here. Consider a region that is uniformly doped where we want to obtain the resistance between

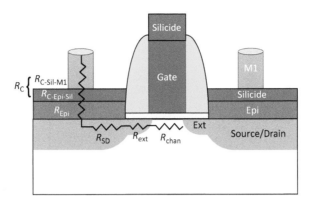

Figure 7.1 Schematic of a MOS device cross-section showing the shallow extension (or tip) region to control short-channel effects and the deeper source/drain region (contact junction) to allow good silicide contacts.

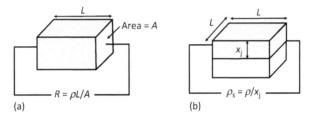

Figure 7.2 The relationships between resistance, resistivity and sheet resistance. (a) Resistivity ρ (Ω cm) is the resistance that would be measured between the sides of a cube of material having edges of unit length (1 cm). The resistance R (Ω) that would be measured between the sides of a more general material shape is $\rho L/A$. (b) Sheet resistance (Ω/square) is the resistance that would be measured between the sides of a square of any dimension L with junction depth x_j.

contacts at either end of the region, as shown in Figure 7.2(a). Three different types of resistance measurements are commonly used in semiconductor technology, conventional resistance R (Ω), resistivity (Ω cm) and sheet resistance ρ_s (Ω/square). Each is illustrated and defined in Figure 7.2. In each case, the current is assumed to be distributed uniformly over the area.

In general, the ability of the region to carry current is given by the number of carriers, to first order equal to the doping concentration N_D or N_A, times the charge q, times the velocity v at which they are moving. By definition, the velocity of the carriers is defined to be the mobility μ times the electric field $\mathcal{E}$. Carriers are accelerated by the field and move with a velocity that is proportional to the field. The constant of proportionality is the mobility, which really measures how well the carriers can respond to the electric field, and it is determined by the scattering mechanisms and defects and doping in the material.

The conductivity σ of the doped region is then defined as the relationship between the current density J and electric field $\mathcal{E}$,

$$J = nqv = nq\mu\mathcal{E} = \sigma\mathcal{E} = \frac{1}{\rho}\mathcal{E}, \tag{7.1}$$

where $\mathcal{E}$ has the units of V cm^{-1} and J has units of A cm^{-2}. The resistivity ρ defines the relationship between the electric field and the current density,

$$\rho = \frac{1}{nq\mu} = \frac{\mathcal{E}}{J} \ (\Omega \text{ cm}), \tag{7.2}$$

and can be thought of as the resistance that would be measured between opposite sides of a cube of length 1 cm, with units of Ω cm, as illustrated in Figure 7.2. If, instead of a cube, the resistance was measured between the shallow edges of a square with depth x_j (Figure 7.2(b)), the resistance would be higher and would measure

$$\rho_s \equiv \frac{\rho}{x_j} \ (\Omega/\text{square}). \tag{7.3}$$

The definition of the sheet resistance ρ_s is simply the resistivity divided by the junction depth. The units of sheet resistance are interesting. Dimensionally, the sheet resistance is expressed in ohms but has the geometrical significance that the sheet resistance is the same for any square. A smaller square has less area for the current to flow through and thus higher current density, but proportionally a higher field, giving the same measured resistance. What that means is that, given a square, it does not matter whether the edge dimension is a millimeter or a centimeter, the resistance from one end to the other is the same, a certain number of ohms/square. If the resistor were part of an integrated circuit, one could calculate the resistance by figuring out the number of squares and that would give the resistance between the two ends. It is extremely useful to be able to specify the resistance of a doped region, without having to specify the dimensions of the region or the depth of the junction, once it is known that the resistance applies to any surface square. The sheet resistance of a junction can be measured by using the four-point probe technique or the van der Pauw method.

The above derivation only applies when the doping is constant throughout the junction, otherwise the resistivity is not as simple as $1/nq\mu$, because n is now a function of depth. We can consider taking the material and slicing it up into small regions and adding the resistance of each of the slices, since these parallel slices all contribute to the conductivity. In the general case, the mobility also is a function of the doping concentration. Qualitatively, in a doped semiconductor with carriers moving through it, there will be a number of mechanisms that cause the electrons or holes to scatter, the most obvious one being carriers bouncing off silicon atoms. The carriers are accelerated by the field, bounce off a silicon atom, then get accelerated again. The mobility is really a measure of how far the carriers go before one of these scattering events takes place.

One reason the mobility is a function of position is that, in addition to scattering off silicon atoms, the electrons or holes can also scatter off doping atoms. If one of the atoms were an arsenic atom, scattering could take place off the arsenic atom, and that scattering off an arsenic atom could be quite different than scattering off a silicon atom, because the arsenic atom is charged and may be a more effective scatterer than a neutral silicon atom. So the higher the percentage of doping atoms, the more frequent the scattering and the lower the mobility. In a doping profile, both n and μ vary as a function of depth. Given an analytic expression for the doping profile, and knowledge of how the mobility varies with doping, one can integrate to find the sheet resistance. Irvin's curves incorporate the doping dependence of the mobility for simple analytical shapes. We discuss how these are used later in this chapter.

If the doping varies in the conducting region, an integral of the doping and mobility (which varies with doping) must be performed from the surface to the junction depth x_j. An accurate value for the sheet resistance of a diffused layer can be calculated from the integral

$$\rho_s = \frac{1}{\overline{\sigma} x_j} = \frac{1}{q \int_0^{x_j} [n(x) - N_B] \, \mu[n(x)] \, dx}. \tag{7.4}$$

The junction depth occurs where the impurity profile meets the background doping concentration N_B. We will see examples of such calculations later in the chapter.

The general guideline when designing modern transistors is that the resistance of the source or drain region ($R_C + R_{Epi} + R_{SD} + R_{ext}$ in Figure 7.1) should not amount to more than a small percentage of the channel resistance (R_{chan}). It seems obvious from the definition of sheet resistance that a simple way to reduce the sheet resistance of a layer is to simply increase the junction depth to allow more room for the current to flow. However, this causes an immediate problem in MOS devices – the deeper junctions make it easier for the voltages on the drain to affect the current flow from the source. Ideally, only the gate voltage causes current to flow from the source to drain by inverting the surface of the channel if $V_G > V_{TH}$. But in a submicron MOSFET, the two-dimensional spreading of the electric field from the drain can attract carriers from the source even when the device is supposed to be off with $V_G < V_{TH}$. This off-current in the device is a key design parameter and can be minimized by keeping the junctions shallow. The change in off-current with drain voltage is called drain-induced barrier lowering (DIBL) and is a key parameter that describes how well a transistor is functioning. The challenge in designing MOS devices is thus to keep the junctions shallow, so DIBL is reduced, and at the same time keep the resistance of the source and drain regions small so that the drive current is maximized. These are conflicting requirements.

Very shallow junctions with very high doping concentrations are required to simultaneously meet DIBL and ρ_s requirements. Of course, devices also use deeper, less-critical doped regions for wells and shallow, more lightly doped regions in the channel region as described in the complementary metal-oxide–semiconductor (CMOS) process flow in Chapter 2. These regions are generally easier to produce than the shallow, highly doped source and drain regions. We will see examples of all these types of doping applications later. In this chapter, we will explore the basic process of dopant diffusion, which is one of the fundamental processes used to form junctions in modern semiconductor devices. We will be concerned with the fundamental mechanisms that cause diffusion and methods to accurately predict dopant profiles. The requirements in the future will require knowledge of dopant positions with almost atomic-scale accuracy.

In Chapter 1 we described some of the techniques that have been used to produce doped regions in silicon devices. Formation methods, such as alloy junctions, mesa junctions and the like, were used only until the invention of the planar process because of the far superior

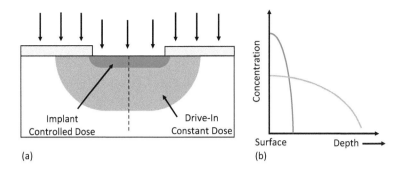

Figure 7.3 (a) Two-step process for producing a junction at the desired depth. The implant or predeposition step introduces a controlled number of impurity atoms, while the drive-in step thermally diffuses the dopant to the desired junction depth. (b) The doping concentration versus depth shown qualitatively for both profiles along the dashed cutline.

manufacturing capability and the improved electrical characteristics of planar junctions. Thus, by 1960, the basic junction formation method shown in Figure 7.3 was in widespread use, and these techniques continue to dominate today.

The only fundamental change that has occurred in this process in the past 40 years is the preferred method for "predeposition." This step is designed to controllably introduce a desired dose of dopant atoms into the silicon crystal. Throughout the 1960s, the dominant predeposition methods were solid-phase diffusion from glass layers deposited on the wafer surface, or high-temperature gas-phase depositions in which the wafers were placed in a furnace with a gas containing the desired doping species. Typical examples included B_2H_6, PH_3 and AsH_3, all of which are gases. Aside from the safety issues associated with using these gases (many of them are toxic), both of these deposition methods have significant limitations in practice. In the gas-phase doping technique, the only way to make it reproducible was to pin the surface concentration at the solid solubility level. The reason this was reproducible was that one could always introduce an overpressure of dopant gas, and the silicon would accept dopant up to its solubility and no more, so the boundary condition was fixed at the solubility. Running the gas concentration below that was never really reproducible. There are many situations in IC fabrication in which relatively small doses of dopant are required (the threshold-adjust steps in the CMOS process we described in Figure 2.8 in Chapter 2 are one example). These kinds of processes are almost impossible to perform with solid-phase or gas-phase predeposition.

Ion implantation was studied extensively in the 1960s as an alternative predeposition method. Because it provides much better control of the predeposition dose, it became the dominant doping method by the mid-1970s and continues to be so today. We will discuss this process in detail in Chapter 8. The concept is simple: if you accelerate ions to high energy and shoot them into silicon, they come to a stop by "billiard ball" collisions with the silicon atoms. So, damage is the big issue with ion implantation, and it must be annealed at high temperatures to repair it, and during this step diffusion and redistribution of dopant atoms occur. For

Table 7.1 **Comparison of ion implantation versus solid- or gas-phase doping methods.**

	Ion implantation and annealing	Solid- or gas-phase diffusion
Pro:	Room-temperature mask (usually)	No damage created by doping
	Precise dose control	Batch fabrication
	10^{11}–10^{16} atoms cm^{-2} doses	
	Accurate depth control	
Con:	Implant damage enhances diffusion, causes amorphization	Usually limited to solid solubility
	Dislocations caused by damage may cause junction leakage	Low surface concentration hard to achieve without a long drive-in
	Implant channeling may affect profile	Low-dose predeposition very difficult

relatively deep junctions, this is of little consequence because the drive-in step illustrated in Figure 7.3 diffuses the junctions deeper than any transient effects due to implantation damage. However, for today's very shallow junctions, the anomalously high diffusion rate observed during damage annealing is a real issue. This transient enhanced diffusion (TED) is beginning to limit how shallow junctions can be made and is a subject of intense study today. We will discuss this topic in detail later in Chapter 8.

In fabricating shallow junctions, the challenge is thus to minimize the redistribution of dopants during any subsequent drive-in or anneal following implantation. Because of TED effects, there is a resurgence of interest in solid- or gas-phase diffusion techniques today, driven by the need to form shallow junctions without any damage. However, it is likely that ion implantation will remain the dominant predeposition method in the future because of its large installed base in current manufacturing. Table 7.1 summarizes the advantages and disadvantages of these doping techniques.

7.2   Dopant Solid Solubility

The concept of solid solubility comes from asking a simple question: "If you start introducing dopant into a silicon lattice, how much can you introduce before the dopant won't sit on silicon lattice sites and instead forms separate phases?" As you get to doping concentrations that high, not all the dopant atoms are isolated and bound by four neighboring silicon atoms and thus able to donate an electron. Above the solid solubility limit, the dopant atoms interact with each other as a result of electrochemical interactions and the strain fields caused by the atomic size mismatch of the dopant atoms and the silicon lattice. This leads to the formation of clusters of dopant atoms, such as As–As clusters, or other structures which are not located on silicon lattice sites and do not contribute to electrically active sites. The doping concentrations needed for current process technologies are already at the solid solubility limit.

Some of the numbers for solubility (arsenic, for example) in Figure 7.4 are quite high, several atomic percent (recall that the silicon lattice concentration is 5×10^{22} cm^{-3}). At

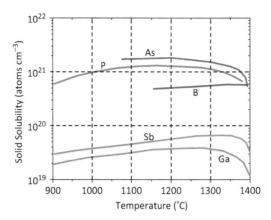

Figure 7.4 Solid solubility curves for various dopants in silicon. These values are the equilibrium chemical solubilities at each temperature and may be different from the electrically active concentrations. After [3].

these levels, one in 20 sites are arsenic atoms rather than silicon atoms, so the dopants are not beside each other but are not very far apart either on average. Unfortunately, this is not the parameter we really care about, because, when we put dopants into silicon, we want the dopant atom to become ionized and produce a free carrier to alter the conductivity of the region. This number is the electrical solubility and tells us how many free electrons or free holes we get as a result of the doping process. From a practical point of view, the maximum electrically active doping concentration is around 2×10^{20} cm^{-3}, roughly similar for **B**, **P** and As in equilibrium. For boron, it is close to the solid solubility, but the electrical solubility is an order of magnitude below the solid solubility for phosphorus and arsenic.

The origin of this discrepancy is of enormous practical interest. Typically, dopants above the electrical solubility limit form an inactive complex that is electrically neutral and does not contribute free carriers. Techniques such as laser melting of the silicon can introduce arsenic into silicon in metastable electrically active concentrations near the solid solubility limit. However, there is an enormous driving force that tends to inactivate the arsenic during any subsequent thermal cycling. Upon annealing, some of the arsenic, while not strictly forming a separate precipitate phase, forms an electrically inactive structure.

One such proposed structure, shown in Figure 7.5, which is consistent with the experimental evidence, is that of several arsenic atoms surrounding a vacancy. The arsenic atoms remain on substitutional sites but adjoin a vacancy, which leaves the arsenic three-fold coordinated with the silicon lattice while retaining two electrons in a dangling bond, for a full shell of eight electrons. Thus the As atoms are on lattice sites so they are chemically soluble, but are not electrically active in this form and do not contribute free electrons to the crystal. There are attempts today to use metastable doping concentrations, perhaps by doing millisecond anneals, laser anneals or epitaxial growth. Unfortunately, these ultra-high mobile charge concentrations are fragile and are extremely difficult to maintain during device processing [2].

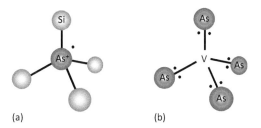

Figure 7.5 Representation of (a) arsenic in a substitutional position versus (b) four arsenic atoms around a vacancy in an inactive configuration.

7.3   Diffusion from a Macroscopic Viewpoint

Diffusion can be discussed from either a macroscopic or a microscopic viewpoint. The macro-scopic viewpoint considers the overall motion of a dopant profile and predicts the amount of motion by solving a diffusion equation subject to some boundary conditions. This is a very useful approach for the practical problem of designing dopant profiles in devices. Diffusion can also be examined by considering the motion of the dopant on an atomic scale, where an attempt is made to relate the overall motion of the whole profile to the motion of unseen individual atoms based on interactions of atoms and point defects in the lattice. This second approach is needed to explain the complex behavior exhibited by dopants diffusing in modern devices and forms the physical basis for the models used in today's simulation tools. There is a lot to be gained by looking at the overall macroscopic diffusion process first, because analytical solu-tions are available for some simple cases.

There are many examples of diffusion processes – if you take a glass of water and add a drop of ink to it, the ink spreads over time, quicker at the start and more gradually later as it spreads to produce a homogeneous distribution of ink in the water. The driving force is the gradient of the concentration, with the higher-concentration regions wanting to diffuse faster and equili-brate to a lower background concentration. The assumptions made in this section were first used by Fick to describe the diffusion process. It is a tribute to the power of the approach that, even today, the most sophisticated diffusion models can trace their origins to an intelligent application of Fick's laws. Fick's first law relates the diffusion of dopant atoms to the concen-tration gradient.

If we consider a block of material in which the concentration varies in different places, as shown in Figure 7.6, this law says there will be a flow of material because of these concentration variations, and postulates that the flow is proportional to the concentration gradient. The proportionality constant is the diffusivity. In essence, this law says that the amount of flow is proportional to the effect causing the flow. Intuitively, this seems to make sense. When the concentration gradient goes to zero and everything is homogeneously distributed, the flow goes to zero. Fick's first law is mathematically described by the equation

$$F = -D\frac{\partial C}{\partial x},$$

(7.5)

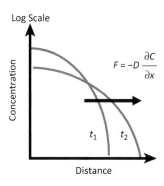

Figure 7.6 Fick's first law states that diffusion is driven by the concentration gradient.

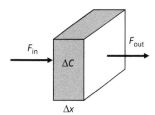

Figure 7.7 Flux into and out of a volume element, used to derive Fick's second law.

where F is the flux or flow (atoms cm^{-2} s^{-1}), D is the diffusivity (cm^2 s^{-1}) and $\partial C/\partial x$ is the concentration gradient.

In a diamond lattice (Si, Ge, GaAs), which has cubic symmetry, D has the same value in all directions. Fick's first law is similar to Fourier's law of heat flow that we used in Chapter 3 in discussing crystal growth, which states that the flow of heat is proportional to the temperature gradient, or to Ohm's law, which relates the current to the potential gradient. The negative sign indicates that the flow is down the concentration gradient.

A more useful practical description of the concentration profile is given by Fick's second law, which relates the concentration to both time and space variables. It is obtained by examining the flux into and out of a volume element, as shown in Figure 7.7. Fick's second law is a fundamental conservation law for matter, which says that the increase in concentration in a cross-section of unit area with time is simply the difference between the flux into the volume and the flux out of the volume:

$$\frac{\Delta C}{\Delta t} = \frac{\Delta F}{\Delta x} = \frac{F_{\text{in}} - F_{\text{out}}}{\Delta x}. \tag{7.6}$$

In other words, "what goes in and does not go out, stays there." Fick's first law is valid at any instant, even if the concentration and concentration gradient are changing with time. Therefore, substituting from (7.5) and taking the limit gives

$$\frac{\partial C}{\partial t} = \frac{\partial}{\partial x} \left(D \frac{\partial C}{\partial x} \right). \tag{7.7}$$

If D is a constant (at a particular temperature, perhaps), then the diffusivity can be taken outside the differential to give

$$\frac{\partial C}{\partial t} = D \left(\frac{\partial^2 C}{\partial x^2} \right). \tag{7.8}$$

This equation has some very useful analytical solutions. However, there is a key assumption made in going from (7.7) to (7.8). What it says is that, if we have a profile that is evolving with time, then to pull D outside the derivative is to say that D is not a function of position, i.e., D is a constant. In many device situations, this is not correct. For example, suppose we have boron or phosphorus diffusing into silicon and we hypothesize that the mechanism of diffusion is through interactions with vacancies or interstitials. To say the dopant diffusivity is constant with position means that the vacancy or interstitial profiles are also constant with position. But we already know that point defect populations depend on the position of the Fermi level, because of the charged defects in the material. And if they depend on Fermi level, then they depend on the doping level. If we have a high doping concentration profile that falls off in the bulk, the population of defects will be higher in the high-concentration region than in the lower-concentration region, so the diffusion coefficient will change with position.

The analytical solutions with constant diffusivity are usually valid for concentrations below n_i (e.g., at 1000 °C, $n_i = 7.14 \times 10^{18}$ cm^{-3}), because the intrinsic population of electrons and holes dominates the mobile charges donated by the dopants. At concentrations above n_i, the diffusion coefficient is concentration-dependent (or position-dependent) and numerical solutions for the diffusion must be used.

The three-dimensional generalization of the diffusion equation is

$$\frac{\partial C}{\partial t} = \nabla \cdot F = \nabla \cdot (D \nabla C). \tag{7.9}$$

This is often referred to as Fick's second law of diffusion. It states that the divergence of the flux F gives the rate at which the concentration in a unit volume is being depleted. The flux of material in turn is proportional to the gradient of the concentration, where the proportionality constant is the diffusivity D.

7.4   Analytic Solutions of the Diffusion Equation

7.4.1   Steady-State Solution ($C \neq f(t)$)

The simplest solution of the diffusion equation occurs when a steady-state condition applies and there is no variation in the concentration with time. Then,

$$D \left(\frac{\partial^2 C}{\partial x^2} \right) = 0. \tag{7.10}$$

Integrating twice gives

$$C = a + bx. \tag{7.11}$$

This steady-state solution gives a linear concentration profile over distance. This solution was of particular interest when we solved the diffusion equation for the oxidant in the oxide during the oxidation of silicon. (Note the linear oxidant profiles in Figure 6.9.)

There are two other particular analytic solutions of Fick's second law that are of interest to diffusion problems in silicon and other semiconductor technologies. These arise from differences in boundary conditions.

7.4.2 Gaussian Solution in an Infinite Medium

Consider first the case where we introduce a spike or delta function of dopant in the middle of a lightly doped region, as illustrated in Figure 7.8. To build such a structure, we could perhaps use a low-temperature epitaxial growth of single-crystal silicon on a silicon wafer and introduce dopant gas into the growth ambient for a very short time. Or, we might implant a very narrow peak of dopant at a particular depth, which approximates a delta function.

Taking the origin to be at the delta function, the boundary conditions are

$$\begin{aligned} C &\to 0 \ \text{ as } \ t \to 0 \ \text{ for } |x| > 0, \\ C &\to \infty \ \text{ as } t \to 0 \ \text{ for } |x| = 0, \end{aligned} \tag{7.12}$$

and

$$\int_{-\infty}^{+\infty} C(x,t) \ \ \mathrm{d}x = Q, \tag{7.13}$$

where Q is the total quantity or dose of dopant that is contained in the spike. The key here is that the initial profile can be approximated as a delta function and that a fixed, constant dose is introduced and remains at all times.

The solution of Fick's second law that satisfies these boundary conditions is

Figure 7.8 A delta function of dopant containing a dose Q is introduced into an infinite medium and subsequently diffused.

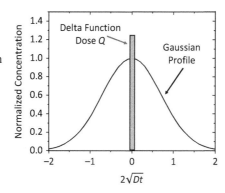

$$C(x,t) = \frac{Q}{2\sqrt{\pi Dt}} \exp\left(-\frac{x^2}{4Dt}\right) = C(0,t) \exp\left(-\frac{x^2}{4Dt}\right). \tag{7.14}$$

This equation is provable by differentiation to be a solution which also satisfies the boundary conditions. This equation describes a Gaussian profile, which evolves with time and retains the same Gaussian form. It is clearly symmetrical about the origin, so that the profile is mirrored about a plane at $x = 0$.

There are several immediate consequences of the solution. The peak concentration decreases as $1/\sqrt{t}$ and is given by $C(0, t)$. When the distance from the origin is $x = 2\sqrt{Dt}$, the peak concentration has fallen by $1/e$, which is easily seen by substituting $2\sqrt{Dt}$ in (7.14).

An approximate measure of how far the dopant has diffused is thus given by $2\sqrt{Dt}$. This factor is such a convenient measure of the extent of the profile motion that it is often termed the "diffusion length." The time evolution of a Gaussian profile is plotted in Figure 7.9 on both linear and logarithmic scales. Note that a Gaussian solution remains Gaussian when more diffusion time is added. Thus, the effect of successive Dt cycles on a Gaussian profile can be easily calculated. Since ion-implanted profiles are, to first order, Gaussian profiles because of the statistical nature of the ion stopping process, (7.14) can often be used to make predictions of the evolution of these profiles during subsequent heat cycles. We will discuss this in more detail in Chapter 8.

7.4.3 Gaussian Solution near a Surface

The symmetry represented by Figure 7.9 and (7.14) allows another very useful solution of Fick's diffusion equation to be easily derived when one considers a dopant dose Q that is introduced near a surface, as shown in Figure 7.10. In practice, this might be done by low-energy ion implantation, with a profile so shallow that it approximates a delta function when compared with the final diffused profile. One assumption is that no dopant is lost through evaporation or segregation during the anneal, so the dopant dose is again fixed and constant.

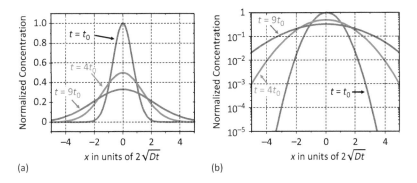

Figure 7.9 Time evolution of a Gaussian diffusion profile on (a) linear and (b) log scales.

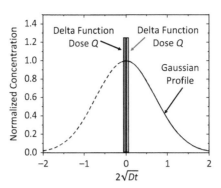

Figure 7.10 A surface Gaussian **Figure 7.10** A surface Gaussian diffusion can be treated as a Gaussian diffusion with dose $2Q$ in an infinite bulk medium.

As Figure 7.10 makes clear, a surface at $x = 0$ can be treated as a reflecting or mirror boundary condition, so that effectively a dose of $2Q$ is introduced into a (virtual) infinite medium, giving the following solution for diffusion of a dopant near a surface:

$$C(x, t) = \frac{Q}{\sqrt{\pi D t}} \exp\left(-\frac{x^2}{4Dt}\right) = C(0, t) \exp\left(-\frac{x^2}{4Dt}\right). \tag{7.15}$$

The surface concentration is given by

$$C(0, t) = \frac{Q}{\sqrt{\pi D t}} \tag{7.16}$$

and falls with time. These surface boundary conditions may be unrealistic in practice because generally there is segregation into a deposited or growing oxide layer or evaporation into the ambient, requiring the use of numerical simulation to accurately characterize the profile. However, (7.15) represents a useful analytical solution when a dose Q is introduced "near" the surface (generally by ion implantation) and annealed for long enough that the initial distribution is reasonably approximated by a delta function, i.e., the spatial extent of the initial profile is much less than the final $2\sqrt{Dt}$.

7.4.4 Error-Function Solution in an Infinite Medium

In addition to the Gaussian profile, the other solution that is useful in silicon processing is the case where we consider the diffusion from an infinite source of dopant. This might correspond to putting a heavily doped epitaxial layer on a lightly doped wafer, as in Figure 7.11. The question is: "How far does the dopant from the heavily doped region diffuse into the lightly doped region?" In this case, the boundary conditions are

$$\begin{aligned} C(x, t) = 0 \text{ at } t = 0 \text{ for } x > 0 \ , \\ C(x, t) = C \text{ at } t = 0 \text{ for } x < 0 \ . \end{aligned} \tag{7.17}$$

It is instructive to consider that the problem is made up by a sum of the previous Gaussian solutions. Consider a series of slices, each of width Δx and of unit cross-section, as shown in Figure 7.11. Each slice initially contains a dose $C\Delta x$ of dopant atoms, which in the absence of

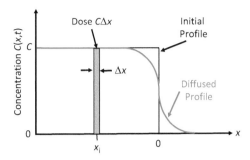

Figure 7.11 An infinite source of material for $x < 0$ can be considered to be made up of a sum of delta functions. The diffused solution is then given by a sum of Gaussians, known as the error-function solution.

the rest of the structure is exactly analogous to the previous condition and would diffuse according to (7.14). To obtain the solution for the present case, we can make use of a simple linear superposition of solutions for each of the thin slices to give

$$C(x,t) = \frac{C}{2\sqrt{\pi Dt}} \sum_{i=1}^{n} \Delta x_i \exp\left(-\frac{(x-x_i)^2}{4Dt}\right). \qquad (7.18)$$

In the limit of thin slices, the sum becomes an integral and the solution is

$$C(x,t) = \frac{C}{2\sqrt{\pi Dt}} \int_{0}^{\infty} \exp\left(-\frac{(x-\alpha)^2}{4Dt}\right) d\alpha. \qquad (7.19)$$

Letting

$$\frac{(x-\alpha)}{2\sqrt{Dt}} = \eta \qquad (7.20)$$

and substituting gives

$$C(x,t) = \frac{C}{\sqrt{\pi}} \int_{-\infty}^{x/(2\sqrt{Dt})} e^{-\eta^2} d\eta. \qquad (7.21)$$

A related integral is tabulated because it is common in the solution of diffusion equations and is not easy to calculate otherwise. This function is called the error function and is defined by

$$\operatorname{erf}(z) = \frac{2}{\sqrt{\pi}} \int_{0}^{z} e^{-\eta^2} d\eta. \qquad (7.22)$$

This function is plotted and tabulated in Appendix A.9, where some additional properties of the error function (erf) and complementary error function (erfc) are also given. Thus the solution of the diffusion equation from an infinite source becomes

$$C(x, t) = \frac{C}{2} \left[1 - \text{erf}\left(\frac{x}{2\sqrt{Dt}}\right) \right]. \tag{7.23}$$

To simplify the notation, the complementary error function (erfc), defined as

$$\text{erfc}(x) = 1 - \text{erf}(x), \tag{7.24}$$

can be used, so that

$$C(x, t) = \frac{C}{2} \left[\text{erfc}\left(\frac{x}{2\sqrt{Dt}}\right) \right]. \tag{7.25}$$

The time evolution of the error function profile is shown in Figure 7.12(a).

7.4.5 Error-Function Solution near a Surface

The error function also describes the diffusion kinetics when the profile is characterized by a constant surface concentration at all times. Such a diffusion profile might occur if the diffusion occurred from a gas ambient with a concentration above the solid solubility of the dopant in the solid.

That the error-function-type solution is valid for diffusion from a constant surface concentration can easily be seen by a symmetry argument. Because of symmetry, it is clear that the mid-point in Figure 7.12(a), i.e., $C(x, t) = C/2$ at $x = 0$, must remain stationary in the solution of the diffusion equation. The solution on either side of $x = 0$ is also equivalent by symmetry. Using this property, if a source concentration is held constant at a value of C_S, we can write

$$C(x, t) = C_S \left[\text{erfc}\left(\frac{x}{2\sqrt{Dt}}\right) \right]. \tag{7.26}$$

The time evolution of the profile thus looks exactly the same as the $x > 0$ portion of Figure 7.12. The time evolution of the constant C_S profile is shown in Figure 7.12(b) on a log scale.

The error-function solution is (very approximately) triangular on a linear scale, so the dose introduced can be approximated by the area of a triangle of height C_S and a base equal to the diffusion distance $2\sqrt{Dt}$, giving $Q = C_S\sqrt{Dt}$. More accurately,

$$Q = \int_0^\infty C_S \left[\text{erfc}\left(\frac{x}{2\sqrt{Dt}}\right) \right] dx = \frac{2C_S}{\sqrt{\pi}} \sqrt{Dt}. \tag{7.27}$$

The spatial forms of the error-function and Gaussian solutions are similar when normalized by the same surface concentration, as shown in Figure 7.13. The major

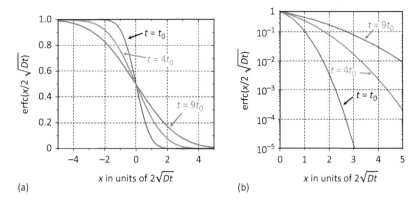

(a)   (b)

Figure 7.12 Time evolution of erfc diffused profiles on (a) linear and (b) log scales. (a) The time evolution of (7.25). (b) The time evolution of (7.26). The horizontal axis units are in terms of the diffused distance $2\sqrt{Dt}$, a measure of how far the profile has diffused.

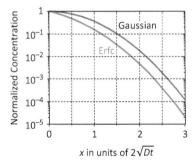

Figure 7.13 The spatial forms of the Gaussian and error-function solutions are similar for a particular normalized value of Dt.

difference between these solutions is that the error-function solution applies when there is an infinite supply of dopant, which implies that an increasing dose of dopant is introduced into the substrate during the diffusion process to maintain a constant surface concentration. The Gaussian solution applies when the initial dose of dopant is fixed and, consequently, the surface concentration must drop as the dopant diffuses deeper into the bulk. Thus, while a snapshot at a particular time may show similar profiles, the time evolutions of the Gaussian and error-function profiles are vastly different (compare Figures 7.9 and 7.12).

With the exception of the Gaussian and erfc analytical solutions to the diffusion equations, most practical problems in silicon technology today require numerical solutions because the simple boundary conditions required for an analytical solution are not usually satisfied. Modern structures employ doped regions in which concentration-dependent diffusion, electric field effects, dopant segregation and complicated point-defect-driven diffusion processes take place. All of these effects generally require numerical methods to calculate the resulting dopant profiles. We will return to these issues later in this chapter.

7.5   Intrinsic Diffusion Coefficients of Dopants in Silicon

Diffusion coefficients are a measure of how fast dopants move in a solid. The diffusion coefficients of common impurities in silicon are found to go exponentially faster at higher temperatures, perhaps not surprisingly since, if dopants move by interacting with vacancies or interstitials, then we know that the population of the native point defects increases exponentially with temperature. If you plot measured diffusion coefficients on a semi-log scale, you get a straight line with the slope given by E_A, so they have the form

$$D = D_0 \exp\left(-\frac{E_A}{kT}\right), \tag{7.28}$$

where k is the Boltzmann constant and T is the temperature in kelvins.

The activation energy E_A has units of electronvolts, and is typically 3.5–4.0 eV for impurity diffusion in silicon. Plots of the intrinsic diffusion coefficient of common dopants in silicon are shown in Figure 7.14, corresponding to the prefactors and activation energies in Table 7.2. This table represents an Arrhenius fit to the diffusivity under intrinsic conditions. We will update this table later to account for high-concentration diffusion effects in extrinsic regions. The reason the impurity diffusion clusters into groups of slow and fast diffusers is not obvious. For example, the physical size of the dopant atoms does not seem to be responsible, because the ionic radius of As is almost a perfect fit in the silicon lattice, though it diffuses slowly, while Sb is large and diffuses slowly, and In is large and diffuses quickly. Charge state effects are also not the solution, because there are both fast (P) and slow (As, Sb) N-type diffusers and fast P-type (B, In) diffusers. The microscopic mechanisms discussed later provide some hints into this behavior based on interactions with interstitials or vacancies.

Another interesting observation is that, while the activation energy E_A of all the dopants are essentially the same, the E_A for silicon self-diffusion is 1 eV higher than for the dopant atoms even though it is obviously perfectly matched in its native lattice. There is an argument for why dopants as a class should have a lower energy for diffusion than silicon, because there might be

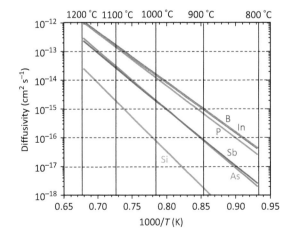

Figure 7.14 Arrhenius plot of the intrinsic diffusivity of the common dopants in silicon along with the self-diffusion coefficient of silicon itself (boron and indium are indistinguishable on this scale).

Table 7.2 Intrinsic diffusivity (cm^2 s^{-1}) for silicon self-diffusion and of common dopants in single-crystal silicon, fitted to an Arrhenius expression.

	Units	Si	B	In	As	Sb	P
D_0	cm^2 s^{-1}	560	1.0	1.2	9.17	4.58	4.70
E_A	eV	4.76	3.5	3.5	3.99	3.88	3.68

interesting pairings between charged vacancies and interstitials and charged dopants on a microscopic scale that might be different from neutral silicon. Lattice strain may also enhance interactions with point defects. This coupling between the dopants and the point defects becomes especially important at high concentrations, and we discuss it in more detail later.

It is important to recall that the intrinsic carrier concentration in silicon is quite high at normal diffusion temperatures (see Figure A1.4). Thus there are many practical conditions in which the silicon is intrinsic, and dopant diffusion is described by the data in Figure 7.14. For example, at 1000 °C, $n_i = 7.14 \times 10^{18}$ cm^{-3}, so that for all N_D, $N_A < n_i$, the material behaves as an intrinsic material. This also means that the analytical solutions given above can be valid if the doping is not extrinsic at the diffusion temperatures. (The analytic solutions, of course, also require specific boundary conditions in order to be valid.)

As junctions have become shallower in recent years, the need for "slow" diffusers has become more important. For N-type regions, this has led to the dominance of As as a dopant because it has both a small D and a high solid solubility. For P-type regions, B is unfortunately the only dopant with a high solid solubility, and its higher diffusivity means that fabricating shallow P-type regions is usually more difficult than forming shallow N-type regions. In addition, as we saw earlier in this chapter, modern devices often require both shallow and heavily doped junctions. When dopant concentrations are very large, diffusivities are not well described by the intrinsic values given in this section. Under extrinsic conditions, all dopants show much higher D values than those given here. We will discuss these issues later in the chapter.

7.6   Effect of Successive Diffusion Steps

Since there are often multiple diffusion steps in a full IC process, they must be added in some way before the final profile can be predicted. It is clear that, if all the diffusion steps occurred at a constant temperature where the diffusivity is the same, then the effective Dt product is given by

$$(Dt)_{\text{eff}} = D_1(t_1 + t_2 + \cdots) = D_1 t_1 + D_1 t_2 + \cdots. \tag{7.29}$$

In other words, doing a single step in a furnace for a total time of $t_1 + t_2$ is the same as doing two separate steps, one for time t_1 and one for time t_2. We often consider the total Dt product to be a measure of the thermal steps or thermal budget that is used in a process. It is this time–temperature product that occurs in the diffusion equations above, so the final profile is unique not to a particular time or temperature but rather to a time–temperature product.

Consider a dopant that is diffused at a temperature T_1 with diffusivity D_1 for time t_1 and then diffused at temperature T_2 with diffusivity D_2 for time t_2. We can write

$$(Dt)_{\text{eff}} = D_1 t_1 + D_2 t_2 + \cdots \tag{7.30}$$

to derive a formula for the total effective Dt. The total effective Dt is given by the sum of all the individual Dt products. Because the diffusion coefficient is exponentially activated, the highest-temperature steps in the process generally dominate the thermal budget (or the effective Dt product). Some of the steps in the process may thus be negligible in determining the overall amount of diffusion.

These concepts also allow us to calculate the equivalent time needed at temperature T_2 to diffuse a profile as far as a time t_1 at a temperature T_1. The time needed is related to the ratio of the diffusivities at each temperature:

$$t_2^{\text{equiv}} = t_1 \times \frac{D_1}{D_2}. \tag{7.31}$$

In many cases, in real device structures, dopant diffusivities are not constant during a particular process step even if the temperature is constant. In fact, in many cases, D is a function of both time and position. This can result from effects like TED following an ion implantation, from concentration-dependent diffusivities and from many other effects that we discuss later. In these cases, numerical simulation is the only viable way to accurately calculate the final impurity profile. Equations like (7.31) are only really useful if D is constant over time and position during each process step.

7.7   Design and Evaluation of Diffused Layers

The key parameters that are important in designing a diffused layer are the sheet resistance, the surface concentration and the junction depth. These three parameters are interdependent and any two of them fully defines a simple erfc or Gaussian profile. There are useful design curves known as Irvin's curves that have numerically integrated (7.4) for the sheet resistance of simple Gaussian and erfc profiles and plotted the average conductivity of these layers versus the surface concentration. An example of these curves is shown in Figure 7.15 for P-type Gaussian diffusions. Additional curves given are in Appendix A.8.

When designing a process, a key step is to use any available information to decide which (if any) analytic solution to Fick's laws applies. For example, a surface concentration at the solid solubility limit might imply that the profile is likely to be described by an error-function-type solution for a constant surface concentration. A low surface concentration might mean that the profile was driven-in with a long heat cycle, indicating a Gaussian-type solution is applicable. It is always important to check these initial assumptions later to verify that the boundary conditions implicit in the solutions of the equations are valid.

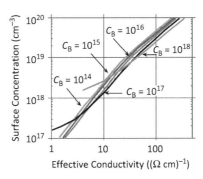

Figure 7.15 Example of Irvin's curves, in this case for P-type Gaussian profiles in an N-type background of concentration C_B. After [4].

Example

Design a boron diffusion process (say, for the well or tub of a CMOS process), such that $\rho_s = 900\ \Omega/\text{square}$, $x_j = 3\ \mu\text{m}$ and $C_B = 1 \times 10^{15}\ \text{cm}^{-3}$ (substrate concentration). These specifications might correspond to the P well in Figure 2.7 in the CMOS process we considered in Chapter 2, except that the junction depth specification given here would likely correspond to an older-generation process flow. The well is deep, so a device can fit at the surface and there is room for the junction depletion regions to spread without punching through the well. And the well sheet resistance helps determine the threshold voltage of the device and the junction capacitance of the surface layers.

Answer

The average conductivity of the layer is given by

$$\bar{\sigma} = \frac{1}{\rho_s x_j} = \frac{1}{(900\ \Omega/\text{sq})(3 \times 10^{-4}\text{cm})} = 3.7\ (\Omega\ \text{cm})^{-1}.$$

We know that $\sigma = nq\mu$, but we cannot calculate n or μ directly because both are functions of depth. Since CMOS N and P wells are moderate-concentration profiles, we make the assumption that the P profile is Gaussian. Thus we can use Irvin's curve in Figure 7.15, from which we obtain

$$C_S \approx 4 \times 10^{17}\text{cm}^{-3}.$$

This surface concentration is well below the solid solubility, so perhaps our original assumption that the profile is Gaussian is reasonable. We surmise that the profile was driven-in to a deep junction depth from an initial dose that was introduced into the silicon probably by ion implantation. Given that the profile is a Gaussian, we can calculate the extent of the thermal anneal (the Dt product) used to diffuse it to the required junction depth, from (7.15):

$$C_B = \frac{Q}{\sqrt{\pi D t}}\ \exp\left(-\frac{x_j^2}{4Dt}\right) = C_S\ \exp\left(-\frac{x_j^2}{4Dt}\right),$$

so that

$$Dt = \frac{x_j^2}{4\ln(C_S/C_B)} = \frac{(3 \times 10^{-4})^2}{4\ln(4 \times 10^{17}/10^{15})} = 3.7 \times 10^{-9}\text{cm}^2.$$

Given the thermal budget for the anneal, we can choose a diffusion temperature (which fixes the diffusivity) and see if the time required at that temperature makes sense, given the constraints of typical manufacturing equipment. If the drive-in is done at 1100 °C, then D_B from Figure 7.14 or using Table 7.2 is $D = 1.5 \times 10^{-13}$ cm^2 s^{-1}. The drive-in time is therefore

$$t_{\text{drive-in}} = \frac{3.7 \times 10^{-9}\,\text{cm}^2}{1.5 \times 10^{-13}\,\text{cm}^2\,\text{s}^{-1}} = 6.8\ \text{h}.$$

Thus, to form the deep, low-concentration boron well requires 6.8 h at 1100 °C. Such long, high-temperature steps need to occur early in the process to avoid the effects of this high thermal budget step on other more sensitive and shallower junction profiles. The well process was in fact performed near the beginning of the CMOS process described in Chapter 2 (see Figure 2.7 and the associated text).

Given both the surface concentration and the Dt product, the initial dose can be calculated for this Gaussian profile as

$$Q = C(0,t)\sqrt{\pi Dt} = (4 \times 10^{17})\sqrt{\pi \times 3.7 \times 10^{-9}} = 4.3 \times 10^{13}\,\text{cm}^{-2}.$$

This dose could easily be implanted in a narrow layer close to the surface, justifying the implicit assumption in the Gaussian profile that the initial distribution approximates a delta function.

Alternatively, a gas/solid-phase predeposition step might be used to deposit the required initial dose at the surface. We assume that the dopant is introduced at the solid solubility limit because of manufacturing control issues, and the surface concentration is maintained constant at the solid solubility. We assume a reasonable temperature like 950 °C for the predeposition. We choose a relatively low temperature because of the small dose involved. For boron, the solid solubility at 950 °C is approximately 2.5×10^{20} cm^{-3} and the diffusivity is 4.2×10^{-15} cm^2 s^{-1} from Figure 7.14 or using Table 7.2. The time required to obtain the required dose of 4.3×10^{13} cm^{-2} for an erfc profile can be found from (7.27) as

$$Q = \frac{2C_S}{\sqrt{\pi}}\sqrt{Dt},$$

so that the time required for the predeposition is

$$t_{\text{pre-dep}} = \left(\frac{4.3 \times 10^{13}}{2.5 \times 10^{20}}\right)^2 \left(\frac{\sqrt{\pi}}{2}\right)^2 \frac{1}{4.2 \times 10^{-15}} = 5.5\ \text{s}.$$

(Note that the use of this boron diffusivity is really not valid in this example because Figure 7.14 gives intrinsic diffusivities and the boron is certainly not intrinsic at this temperature at the solid solubility limit.)

We must now check that the delta-function approximation is valid for this combination of predeposition step and drive-in, so that our initial assumption of a Gaussian profile is reasonable. We have

$$Dt_{\text{pre-dep}} = 2.3 \times 10^{-14} \ll Dt_{\text{drive-in}} = 3.7 \times 10^{-9},$$

which is completely adequate in this particular case. In other words, the erfc predeposition profile sufficiently approximates a delta function when compared with the Gaussian drive-in profile.

However, in this particular case, the predeposition time is too short to be "reasonable" in a manufacturing environment, where the dopant is introduced in a furnace, and this example shows why ion implantation is the technique of choice for introducing the low dose of dopant required in this process. This is, in fact, what was done in the process flow in Chapter 2 (Figure 2.6). This simple example shows how the analytic solutions of the diffusion equation can be used to design a simple process, provided the underlying assumptions are valid.

7.8   Models and Numerical Simulation

Fick's first and second laws have formed the basis for understanding and predicting diffusion profiles for many years. However, there are only a few simple analytic solutions to these equations because such solutions require a time- and position-independent dopant diffusivity. These constraints are rarely met in modern structures. Thus, while first-order designs of profiles can be accomplished analytically, accurate design today requires numerical solutions to Fick's equations.

Modern process simulators use numerical methods, along with sophisticated physical models of the diffusion process. In this section, we will focus on such models and use simulation tools to illustrate the effects of various physical models. The simulators we will use are based on derivatives of the Stanford University program SUPREM IV [5], which has been implemented in commercially available versions like Sentaurus [6] and Athena and Victory Process [7]. These programs implement many of the models we will discuss. These 2D and 3D simulators are generally available and fairly widely used in the semiconductor industry. The simulation examples in this chapter were run on commercially available versions of the Athena and Victory Process programs.

After briefly discussing some issues associated with the numerical implementation of diffusion models, we will consider a number of modifications to Fick's first and second laws which account for additional physical effects beyond those which can be included in analytic solutions. Recall from our discussion earlier in the chapter that analytic solutions generally require that the dopant diffusivity is constant over space and time during a particular process step. Neither of these is generally true in real very-large-scale integrated (VLSI) fabrication sequences. We will thus consider modifications to Fick's laws which broaden their applicability, and are capable of incorporating better physics.

It is interesting to think about how to solve the diffusion equation numerically, because the numerical approach will not be restricted to the simple initial and boundary conditions we have examined. For example, the diffusion of an arbitrary initial profile that is non-Gaussian could easily be examined numerically, but would have no tractable analytic solution.

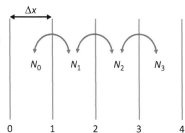

Figure 7.16 Atomic planes Δx apart, with N_i atoms at each plane. Atoms at a plane have an equal probability of jumping left or right.

Following Crank [8], let us consider the physical situation where we have atoms hopping from one plane to another in the crystal, as shown in Figure 7.16. Assume we divide up the crystal into planes Δx apart, labeled 0, 1, 2, ..., and we have a number of atoms at each plane N_0, N_1, N_2, ..., i.e., the planar density is N_i atoms cm^{-2}. The average volume concentration at any point is then $C_i = N_i \Delta x$ cm^{-3}. The atoms remain relatively fixed in the lattice, but vibrate about their equilibrium position at the Debye frequency, which is about 10^{13} s^{-1} in silicon. Sometimes, an atom will be able to surmount the energy barrier keeping it in place and will hop to an adjacent plane if a free site exists (a vacancy) or exchange directly with an adjacent atom. The frequency of hopping is given by

$$v_b = v_d \exp\left(-\frac{E_b}{kT}\right),$$ (7.32)

where v_d is the Debye frequency and E_b is the energy barrier. The jump frequency is v_b. Any given atom will have an equal probability of jumping to the left or to the right, so there will be $(v_b/2)N$ atoms jumping right and $(v_b/2)N$ atoms jumping left per unit time.

The number of atoms crossing the plane with N_2 atoms on one side and N_1 atoms on the other is simply

$$F = -\frac{v_b}{2}(N_2 - N_1) = -\frac{v_b}{2}\Delta x(C_2 - C_1) = -\frac{v_b}{2}(\Delta x)^2 \frac{\Delta C}{\Delta x} = -D\frac{\Delta C}{\Delta x},$$ (7.33)

where we have defined

$$D = \frac{v_b}{2}(\Delta x)^2.$$ (7.34)

This provides some atomic-scale insight into the origins of Fick's first law. It is our first indication that diffusion processes have an origin in atomic-scale jumps of atoms, which we will return to in a later section. It is, at first sight, surprising that independent hops with equal probability of going left or right for each individual atom in a dilute mixture leads to a flow of atoms down the concentration gradient.

Consider, for a moment, all the atoms hopping to or from plane i only: atoms at plane $i - 1$ can jump to plane i, atoms at plane i can jump to plane $i - 1$ or $i + 1$, and atoms at plane $i + 1$

can jump back to plane i. Summing all the atom hops at plane i, we have, at the end of an interval of time Δt,

$$N_i^+ = N_i + \frac{v_b}{2}\Delta t(N_{i-1} - 2N_i + N_{i+1}).\tag{7.35}$$

In terms of concentrations, we have

$$C_i^+ = C_i + \frac{v_b}{2}\Delta t(C_{i-1} - 2C_i + C_{i+1}).\tag{7.36}$$

Making the substitution for diffusivity defined above, we get

$$C_i^+ = C_i + \frac{D\,\Delta t}{(\Delta x)^2}(C_{i-1} - 2C_i + C_{i+1}).\tag{7.37}$$

This forms the core of our numerical solution. Given initial concentrations at different places, we can calculate the subsequent concentrations after a small interval of time Δt. We have considerable flexibility in choosing the magnitudes of Δt and Δx. These parameters should be chosen such that they are relevant to the problem we are attempting to solve. For example, Δx should be such that the profile is uniformly divided up into a sufficient number of distance intervals that the profile can be fitted in a reasonable manner by these piecewise-linear approximations. Similarly, Δt should be chosen so that the time interval is divided into a sufficient number of time steps to resolve the diffusion process. Each iteration of the numerical solution for C_i^+ advances the solution of the diffusion equation at each of the distance intervals Δx by an amount of time Δt.

What are the limits on the values of Δt and Δx that we choose? We can investigate this in a little more detail by considering a simplification of this equation that is mathematically clever if we take

$$\frac{D\,\Delta t}{(\Delta x)^2} = \frac{1}{2}.\tag{7.38}$$

Then, our numerical solution becomes

$$C_i^+ = \frac{1}{2}(C_{i-1} + C_{i+1}),\tag{7.39}$$

which simply relates the new value at a node to the average of the values at the adjacent nodes. This may be considered a mathematical trick to simplify the equation, but it contains some interesting physics and numerics. For example, if we were to take the increment in the numerical solution $D\Delta t/(\Delta x)^2 > 0.5$ in (7.37), we would find that the solution becomes unstable – oscillations in the values of the concentration increase for each succeeding numeric interval. Thus, $D\Delta t/(\Delta x)^2 = 0.5$ represents the maximum value of the numeric interval that can be used before the numerical solution becomes unstable. Physically, this means that we are asking more than the available number of atoms at a plane to jump within a time Δt.

This very simple derivation allows us to solve the diffusion equations for any arbitrary initial dopant profile. It provides an introduction to the numerical solution of diffusion equations, which takes on a heightened importance when the diffusivity is no longer constant and when the diffusion of one species affects the motion of another. It also provides an introduction to the numerical issues that must be tackled in a process simulator.

7.9   Modifications to Fick's Laws to Account for Electric Field Effects

When the doping concentrations exceed the intrinsic carrier concentration at the diffusion temperature, electric fields set up by the doping atoms can affect the diffusion process. This is an example of an extrinsic effect that is not observed below the intrinsic carrier concentration and means that the diffusivity cannot be taken outside the derivative in (7.7).

The electric field effect is driven by the fact that, if you take an arsenic atom (for example) and introduce it into silicon, it can ionize and become a donor and we end up with two profiles inside the silicon. We end up with an As$^+$ profile and an electron profile, which are different in general, because the mobility of electrons is much higher than that of the atoms. Some separation of the profiles occurs as the fast-moving electrons diffuse into the silicon. If there were not something to stop that process, we would end up with the electrons homogeneously distributed throughout the crystal and we would not be able to use the local doping technique to create local N- and P-type regions. What stops the process is that we end up with a balance between diffusion and drift, diffusion being the tendency for the two species to separate in space, and drift being the electric field due to the charge separation that opposes the tendency, as shown in Figure 7.17. The built-in field results in macroscopic effects like a change in the arsenic diffusivity, because the field is in a direction to pull the charged arsenic atoms deeper into the bulk.

If all of the dopant is ionized $C_A^+ \cong C_A$, the electric field adds an enhancement factor h to the flux,

$$F = -hD\frac{\partial C}{\partial x}, \qquad (7.40)$$

Figure 7.17 Schematic of the electric field. This develops because the electron or hole mobility exceeds the dopant mobility, so that electrons or holes diffuse ahead until they reach a steady-state condition where the drift flux from the internal electric field balances the diffusion flux.

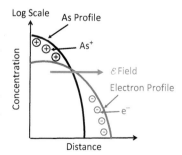

which increases the apparent diffusivity, with h given by (not derived here)

$$h \equiv 1 + \frac{C}{\sqrt{C^2 + 4n_i^2}}.$$

(7.41)

The factor h has an upper bound of 2, which means that this electric field enhancement term can enhance the diffusivity of the dopant causing the field by a maximum factor of 2, when the doping concentration term is much greater than the intrinsic electron concentration.

The other point that needs to be made is that, in silicon at process temperatures, there is the parameter n_i, which is the background concentration of electrons and holes that comes from thermal bond breaking. It shows up in the device equations at room temperature, and at process temperatures it is much higher, around 8×10^{18} cm^{-3} at 1000 °C. If we have an arsenic profile with the peak less than 8×10^{18} cm^{-3}, then there would be no electric field effects, because the intrinsic concentration would set the electron concentration at 8×10^{18} cm^{-3} uniformly flat with position, so these effects show up only at high concentrations.

However, in a case where there are species at different concentrations, the field term can cause even bigger changes in the diffusivity of a low-concentration dopant in the vicinity of the field, as illustrated in Figure 7.18.

In this simulation, the initial phosphorus and boron profiles are shown as dashed lines. Anneals were then done at 1000 °C for 15 min with and without $\mathcal{E}$-field effects included in the simulations. Whatever species is dominant at a given location determines the magnitude of the $\mathcal{E}$-field. In the surface region, this is phosphorus. Below 0.1 μm initially and later, after some diffusion, below 0.2 μm, boron dominates and sets the $\mathcal{E}$-field. Below n_i ($\approx 8 \times 10^{18}$ cm^{-3} at 1000 °C), there are no electric field effects.

To illustrate that (7.41) provides a reasonable estimate of electric field effects, the phosphorus diffusivity with no $\mathcal{E}$-field in the simulation (green curve) was doubled. This roughly matches the red curve for phosphorus which includes $\mathcal{E}$-field effects. The field thus roughly doubles the flux of the high-concentration species.

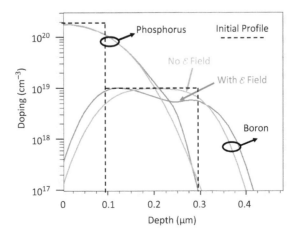

Figure 7.18 Athena simulation of phosphorus and boron profiles with and without the $\mathcal{E}$-field effect. The phosphorus diffusivity was increased by 2× when there was no $\mathcal{E}$-field, almost matching the effect of the field (see text). Diffusion was 15 min at 1000 °C.

However, the magnitude of this term can completely dominate the diffusion flux of a lower-concentration dopant. This is because the high-concentration dopant profile determines the $\mathcal{E}$-field. Note in Figure 7.18 that the direction of the field pulls the B^- atoms into the N^+ region and depletes significant amounts of boron from the region below the phosphorus. In the region below 0.3 μm, boron dominates and determines the field. This increases the boron diffusivity at least until the doping drops below n_i and the electric field effects disappear.

7.10 Modifications to Fick's Laws for Concentration-Dependent Diffusion

When the doping concentration exceeds the intrinsic electron concentration at the diffusion temperature, another extrinsic diffusion effect in addition to electric field effects is seen, and it can be a much bigger effect than the electric field effect. Fick's first law is really based on the premise that the flux or flow is directly proportional to the concentration gradient. However, if we examine the actual concentration profile of many diffused dopants, we find that the diffusion profile is box-like – that is, the diffusion appears to be faster in the higher-concentration regions. An example is shown in Figure 7.19.

If we consider a low-concentration constant surface concentration diffusion profile, below n_i, which is 8×10^{18} cm^{-3} at 1000 °C, we get an erfc profile assuming constant D as shown in Figure 7.12. If we now consider a higher-concentration constant surface concentration profile ($C_S = 10^{20}$ cm^{-3}), again with constant D, such erfc profiles often do not match experimental results. Instead, experimental profiles are highly non-Gaussian and much more box-like, as shown in Figure 7.19. To match these experimental profiles, the diffusivity would have to vary as n or n^2, where n is the electron concentration, or, in other words, the doping concentration. Qualitatively, in the high-concentration regions, the atoms are moving around very quickly, which tends to produce a flat profile, and, once the concentration begins to drop, we observe a very sharp fall-off because the diffusivity is a lot lower at the diffusion front and atoms cannot move as rapidly in the tail part.

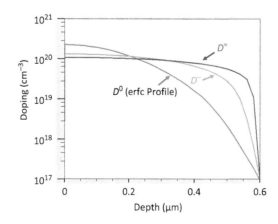

Figure 7.19 Effect of concentration-dependent diffusion coefficients on the profile shapes. The labeling of the D^- and $D^=$ curves is explained below.

This observation represents a major revision of the original intent of Fick's laws, where the flux depended only on the concentration gradient, not the absolute value of the concentration. All of the common dopants in silicon exhibit this behavior and tend to have more box-shaped profiles than would be expected from simple error-function or Gaussian solutions.

If we define the diffusivity as being a function of concentration, then Fick's formulation with non-constant D still represents a very useful way of describing dopant diffusion in silicon. The modified form of Fick's law where the diffusivity is a function of concentration cannot usually be directly integrated, so that the solution of (7.42) must be computed numerically:

$$\frac{\partial C}{\partial t} = \frac{\partial}{\partial x}\left(D_A^{\text{eff}} \frac{\partial C}{\partial x}\right).$$

(7.42)

Experimentally, it often appears that the profile can be well modeled by assuming that the diffusivity is directly proportional to the free carrier profile, as illustrated in Figure 7.19, where $D \propto (n/n_i)$ or $D \propto (n/n_i)^2$. The box-shaped profiles characteristic of concentration-dependent diffusion are evident. Why does the diffusivity behave in this manner? If we believe that dopants move by interacting with point defects, then one obvious reason is that the concentrations of the vacancies or interstitials change with the background doping or Fermi level.

Information on the concentration dependence of the diffusivity can be obtained from iso-concentration experiments, which indicate the dependence of the diffusion coefficient on the background concentration. Boron is a particularly interesting element on which to perform isoconcentration experiments, because there are two common isotopes ^{10}B and ^{11}B. If a high-concentration background doping is set by one isotope, say ^{10}B, then this sets p/n_i. If a profile of the other isotope, say ^{11}B, is introduced into the ^{10}B background by ion implantation, the diffusivity of the ^{11}B can be measured as a function of the background ^{10}B concentration. This is a particularly nice experiment, because it is all boron diffusion, albeit different isotopes. The ^{11}B profiles remain Gaussian because there is no spatial variation in the background concentration, but diffuse with a constant but faster diffusion coefficient than normal, making the analysis simple. Any complications that might occur due to dopant–dopant interactions are removed. Secondary ion mass spectrometry (SIMS) is a particularly convenient analysis tool because its mass sensitivity allows a ^{11}B profile to be measured inside a uniform ^{10}B background, for example. This kind of experiment gives the diffusivity of boron as a function of concentration.

Based on many results like these, the diffusivities of the common dopants in silicon have been characterized and found to depend linearly or sometimes quadratically on the carrier concentration, as shown in Figure 7.19. The effective diffusivity to be used in (7.42) can be written as

$$D_A^{\text{eff}} = D^0 + D^-\left(\frac{n}{n_i}\right) + D^=\left(\frac{n}{n_i}\right)^2 \text{ for N-type dopants,}$$

(7.43)

$$D_A^{\text{eff}} = D^0 + D^+\left(\frac{p}{n_i}\right) \qquad\qquad \text{ for P-type dopants.}$$

(7.44)

The form of these equations is similar to (3.11)–(3.13), which showed the dependence of the charged point defect concentrations on n/n_i, p/n_i or $(n/n_i)^2$. Higher charged point defect concentrations in extrinsic material thus provide the physical basis for the profile shapes shown in Figure 7.19.

Table 7.3 Concentration-dependent diffusivities of common dopants in single-crystal silicon.

	Si	B	In	As	Sb	P
$D^0(0)$ (cm^2 s^{-1})	560	0.05	0.6	0.011	0.214	3.85
$D^0(E)$ (eV)	4.76	3.5	3.5	3.44	3.65	3.66
$D^+(0)$ (cm^2 s^{-1})		0.95	0.6			
$D^+(E)$ (eV)		3.5	3.5			
$D^-(0)$ (cm^2 s^{-1})				31.0	15.0	4.44
$D^-(E)$ (eV)				4.15	4.08	4.0
$D^=(0)$ (cm^2 s^{-1})						44.2
$D^=(E)$ (eV)						4.37

The superscripts on D^0, D^+, etc. are chosen because, on an atomic level, these different terms are thought to occur due to interactions with neutral and charged point defects. The diffusivity under intrinsic conditions for an N-type dopant (when $p = n = n_i$) is

$$D_A^* = D^0 + D^- + D^= . \tag{7.45}$$

Each of these individual diffusivities is exponentially activated, moving faster at higher temperatures, and each individual diffusivity can be written in Arrhenius form (below) with a pre-exponential factor $D(0)$ and an activation energy $D(E)$:

$$D = D(0) \ \exp\left(-\frac{D(E)}{kT}\right). \tag{7.46}$$

The diffusion coefficients that are used in process simulators (where they may be written as D.0 and D.E, respectively) are given in Table 7.3. From this table, we can see that boron diffuses with both neutral and positively charged defects. Phosphorus diffuses with a combination of neutral, negative and doubly negatively charged point defects.

7.11 Segregation and Interfacial Dopant Pile-Up

Interfaces between different materials occur frequently in silicon processing. Dopants have different solubilities in different materials, and so redistribute at an interface until the chemical potential is the same on both sides of the interface. The ratio of the equilibrium doping concentration on each side of the interface is defined as the segregation coefficient. We used this same concept in Chapter 3 in connection with dopant behavior during Czochralski (CZ) crystal growth (Section 3.8). This difference in a dopant's solubility in each phase drives a diffusion flux until the chemical potential equalizes, and for this reason is an important boundary condition.

For example, consider two separate phases A and B (which might be oxide and silicon). We want to derive the transfer of a component X between A and B having concentrations C_A in

Example

Calculate the effective diffusion coefficient at 1000 °C for two different box-shaped arsenic profiles grown by silicon epitaxy, one doped at 1×10^{18} cm^{-3} and the other doped at 1×10^{20} cm^{-3}.

Answer

At 1000 °C, the intrinsic electron and hole concentrations are 7.14×10^{18} cm^{-3}, so, for dopant concentrations less than this, the profile appears intrinsic. For the 1×10^{18} cm^{-3} profile:

$$D_{As} = 0.011 \exp\left(-\frac{3.44}{k(1000+273)}\right) + 31.0 \exp\left(-\frac{4.15}{k(1000+273)}\right)$$

$$= 2.67 \times 10^{-16} + 1.17 \times 10^{-15} = 1.43 \times 10^{-15} \, \text{cm}^2\text{s}^{-1}.$$

As a sanity check, we can calculate the value from the Arrhenius fit in Table 7.3, which was obtained by fitting a single activation energy to diffusion profiles under intrinsic conditions, giving

$$D_{As} = 9.17 \exp\left(-\frac{3.99}{k(1000+273)}\right) = 1.48 \times 10^{-15} \text{cm}^2 \, \text{s}^{-1}.$$

For the 1×10^{20} cm^{-3} profile:

$$D_{As} = 0.011 \exp\left(-\frac{3.44}{k(1000+273)}\right) + 31.0 \exp\left(-\frac{4.15}{k(1000+273)}\right)\left(\frac{1 \times 10^{20}}{7.14 \times 10^{18}}\right)$$

$$= 2.67 \times 10^{-16} + 1.63 \times 10^{-14} = 1.66 \times 10^{-14} \text{cm}^2\text{s}^{-1}.$$

Note that the highly doped layer has a 10-fold higher diffusion coefficient in the extrinsic material. In a simulation of a diffusion profile, the computer keeps track of the diffusion coefficient at every point in the profile and then uses these local values in calculating the doping profile time evolution.

A and C_B in B. The basic first-order assumption is that the species interchange at the interface according to the chemical reaction

$$X_A \underset{k_2}{\overset{k_1}{\rightleftharpoons}} X_B, \tag{7.47}$$

so that the rate of flow of material from A to B is represented by a first-order chemical reaction

$$R_{AB} = k_1 C_A - k_2 C_B. \tag{7.48}$$

The flux across the interface is modeled by

$$F_{AB} = h_{A \to B} \left(\frac{C_A}{m_{A \to B}} - C_B \right),$$ (7.49)

where $h_{A \to B}$ is the mass transport velocity across the interface and $m_{A \to B}$ is the segregation coefficient at the interface. The speed with which equilibrium is achieved across an interface is determined by the magnitude of h. Both h and m may be exponentially activated. In steady state, F_{AB} goes to zero and there can be a discontinuous concentration on each side of the interface.

For thick oxides, the segregation coefficient can be determined from careful SIMS profiles. For thin oxides, SIMS does not have adequate resolution, and the electrical effects of segregation provide a better monitor of the dopant segregation. Measurements such as threshold voltage or capacitance measurements are sensitive indications of the segregation of dopants at interfaces. Process simulators contain experimentally determined values for m for the common dopants in silicon across the Si/SiO$_2$ interface. These parameters are often temperature-activated but are in the range of the values given below:

$$m = \frac{C_{Si}}{C_{SiO_2}} \approx \begin{cases} 0.3 \text{ for boron,} \\ 10 \text{ for arsenic,} \\ 10 \text{ for antimony,} \\ 10 \text{ for phosphorus.} \end{cases}$$ (7.50)

Boron segregates into oxide layers, tending to deplete the boron concentration in the silicon near the oxide/silicon interface. Figure 7.20(a) shows a two-dimensional simulation of the depletion of boron by a local oxidation at the silicon surface, with a detailed cross-section shown in Figure 7.20(b).

Phosphorus, on the other hand, tends to "pile-up" on the silicon side of the interface. This leads to a "snow-plow" effect when a phosphorus-doped substrate is oxidized and there is

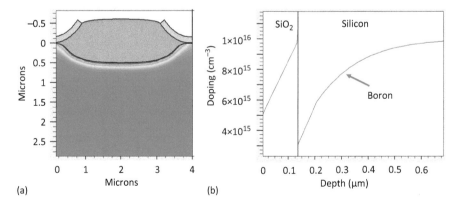

(a)   (b)

Figure 7.20 (a) Athena plot of boron contours after a LOCOS (local oxidation of silicon) oxidation of an initially uniform (10^{16} cm^{-3}) boron-doped substrate. (b) The boron depletes from the bulk into the oxide layer due to segregation.

a moving interface which attempts to reject the phosphorus from the growing oxide into the silicon. Figure 7.21 illustrates these effects for As, P and B in silicon. The As profile is steeper than the P profile in the silicon because of arsenic's smaller diffusivity compared to phosphorus.

As junctions become shallower, it has been observed that dopants may pile-up in a very narrow interfacial layer between SiO_2 and Si, as illustrated in Figure 7.22. This pile-up is separate from the equilibrium segregation that we discussed above. The interfacial layer may be as thin as a monolayer and can act as a sink for dopant atoms and is able to trap on the order of an atomic layer of dopant at the oxide/silicon interface. The dopants are inactive in the interfacial layer and can be removed if the oxide layer is stripped in dilute hydrofluoric acid. A simulation of the dose loss model is shown in Figure 7.23(a) and an actual SIMS profile of the dose loss from an implanted arsenic layer is shown in Figure 7.23(b), before and after an anneal. The "lost" dose in this example has been incorporated into a thin layer just at the interface, which is not resolved in the SIMS data.

Because the amount of interfacial dose loss may be a significant fraction of the dose of dopants that are used in the tip or extension regions of small MOS devices, this effect can take on a dominant role in determining the electrical characteristics of these devices. TED of the implanted dopant atoms (see Chapter 8) can allow many of the dopant atoms to reach the interface, where they can be trapped. Some of the dopant can return to the silicon upon subsequent annealing.

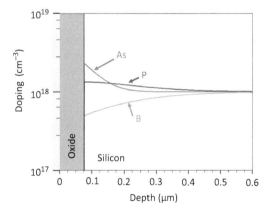

Figure 7.21 Athena simulation of initially uniform (10^{18} cm^{-3}) arsenic, phosphorus and boron profiles after an oxidation of the silicon surface. Arsenic and phosphorus pile-up in front of the moving interface, while boron is depleted.

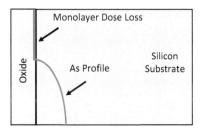

Figure 7.22 Schematic diagram of interfacial dopant dose loss.

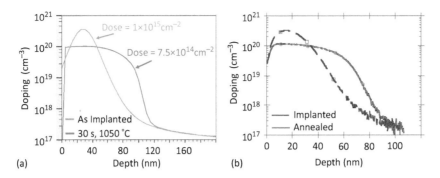

Figure 7.23 (a) Simulation showing 25% arsenic dose loss after a 30 s 1050 °C rapid thermal anneal (RTA). The dose loss results in the areas under the two curves being different by 25%. (b) Experimental SIMS data showing similar dose loss for the same RTA. After [9, 10].

This dose loss is important because it occurs on the same time scale as the rapid thermal cycles that are used to anneal implants. Understanding this process is important for designing the source/drain profiles in MOS devices [9, 10]. At present, trapping models in programs like Athena account for a flux of dopant to unfilled traps at the interface and consider a trapping flux

$$F = h \left(C \left(f + r\frac{\sigma}{\sigma_{\max}} \right) - k\sigma \right), \tag{7.51}$$

where h is a transport coefficient, C is the dopant concentration near the interface, f is the fraction of unfilled interface traps, σ and $\sigma_{\max}$ are the available and maximum trap density, respectively, and r and k are rate constants for trapping and detrapping. In general, this dose loss effect is poorly understood and only preliminary models are available in process simulators. While it seems like a subtle effect, it can dramatically increase the sheet resistance in ultra-shallow source/drain tips in modern MOS devices.

7.12 The Physical Basis for Diffusion at an Atomic Scale

We have discussed many of the issues related to macroscopic dopant diffusion in silicon. Darken and Gurry [11] make the point that Fick's first law is a fundamental physical law, in the sense that it appears to properly describe diffusion behavior in the limit of sufficiently low concentrations or for small concentration differences. We have modified the original intent of Fick's first law because experimental data suggested that it was not adequate to describe the evolution of doping profiles, by adding electric field effects and concentration-dependent diffusivities. It is at this stage that we run out of steam in dealing with a macroscopic description of the diffusion process. The complexities from these *ad-hoc* descriptions begin to outweigh their usefulness. In order to progress further, we need to understand a little more about how dopants diffuse on an atomic scale. This atomic-scale understanding forms the physical basis for the models employed in most process simulation programs used today. We will see later that Fick's laws, intelligently applied to the mobile diffusing species on an atomic level, still provide a very useful description of the full range of anomalous dopant diffusion in silicon.

Point defects (vacancies and interstitials) and dopant diffusion are intimately linked at the atomic scale. Understanding the behavior of point defects can help to provide a unified explanation of the atomic mechanisms underlying dopant diffusion. This provides a clear picture of many effects that previously seemed anomalous. The basic concepts relating to point defects were introduced in Chapter 3. The basic dopant diffusion mechanisms by interacting with point defects are illustrated in Figure 7.24.

Intuitively, it seems clear that a vacancy adjacent to a dopant atom provides a mechanism for the dopant atom to hop to the adjacent site, thus effecting a single migration jump for the dopant atom, as shown in Figure 7.24(a). This is precisely the mechanism of diffusion that occurs in metals. In fact, the vacancy concentrations in metals are so high that changes in the lattice constant can be observed directly by X-ray diffraction (XRD). If the metal temperature is changed and the lattice constant is measured as a function of temperature, the change in lattice constant can be used to extract the vacancy concentration versus temperature. A similar experiment in silicon does not reveal a measurable change in the lattice constant, which implies that the vacancy concentrations are below the detection limit of this technique. The change in the length of a specimen is related to the change in volume by $\Delta L/L = \Delta V/3V$, so that the length of a silicon specimen changes by only 0.7×10^{-7} for a vacancy concentration of 1×10^{16} cm^{-3}. This is below the detection limit of the XRD technique.

The success of the "vacancy-only" theory of impurity diffusion in metals was initially applied with great enthusiasm to diffusion in silicon. It had some early dramatic successes, in that it allowed many of the observations of high-concentration dopant diffusion to be explained in a consistent manner using a single set of vacancy energy levels that allowed the vacancy concentrations to depend on the Fermi level [12]. Much of the current success in modeling dopant diffusion in process simulators stems directly from this idea that an atomic-level understanding of the process is both possible and powerful.

From Figure 7.24, it is clear that, if a vacancy simply exchanges sites with a dopant, there is a very high probability that it switches back again, leading to no long-range motion of the dopant atom. But if the vacancy is loosely bound to the dopant atom, then perhaps the

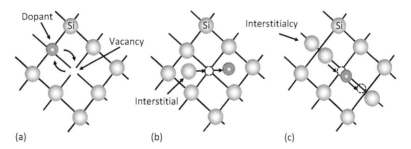

Figure 7.24 Schematics of (a) vacancy-assisted diffusion mechanism, (b) interstitial-assisted kick-out diffusion mechanism and (c) interstitialcy-assisted diffusion mechanism.

vacancy can wander some distance away, so that it can return along a different path and affect a diffusion jump for the dopant atom. In a square lattice like that shown in Figure 7.24, corresponding to a body-centered cubic (BCC) or face-centered cubic (FCC) crystal structure, the vacancy needs to migrate only to a second-nearest-neighbor site to complete one diffusion jump. In a diamond lattice, the vacancy must diffuse to a third-nearest-neighbor site. The binding energy that keeps the vacancy in the vicinity of the dopant atom makes it easier for a dopant atom to diffuse. The Coulombic interaction between an ionized dopant atom and a charged point defect contributes to the binding energy. So do non-Coulombic interactions caused by the size effects or elastic distortions of dopants in proximity with point defects. Because the local state of the lattice may be quite different when point defects are in intimate contact with dopants than when they are both isolated, simple treatments based on the tetrahedral covalent radius or Coulombic potentials are generally inadequate.

It is also easy to imagine how a silicon interstitial might "kick out" a substitutional dopant atom from its lattice site and enable it to quickly diffuse down the relatively open channels in the silicon lattice, as illustrated in Figure 7.24(b). Eventually, the mobile dopant interstitial will regain its place on a substitutional site by displacing a silicon atom or perhaps by finding a vacant site. The interstitial-assisted diffusion of the dopant then takes place by a series of these hops or episodes of mobility. Each hop of the diffusing dopant is actually a random walk through the channels of the silicon lattice, interspersed with long periods in a stable, substitutional site.

It is perhaps less easy to imagine that a dopant and a silicon interstitial could diffuse as a bound pair, as illustrated in Figure 7.24(c), but this could correspond to diffusion along the bond directions rather than through the open channels in the lattice. A dopant and a silicon atom could share a lattice site, and, by moving in the bond directions, could effectively migrate as a bound pair through the lattice. Once again, a free silicon interstitial is temporarily bound with a dopant atom while the pair migrates as a mobile species, before the pair breaks up, leaving the dopant atom on a substitutional site and releasing the silicon interstitial. To distinguish this process from the kick-out process, the mechanism in Figure 7.24(c) is known as the interstitialcy mechanism. In either case, a silicon interstitial initiates the diffusion event and a silicon interstitial is released when the dopant reoccupies a substitutional site. A strong interaction between a silicon interstitial and a substitutional dopant atom enhances the probability of formation of a mobile dopant interstitialcy, just like a strong binding energy between a dopant and an interstitial enhances the kick-out diffusion jump. Mathematically, they are very similar and, in the literature, both the kick-out and interstitialcy processes are often simply referred to as "interstitial"-assisted diffusion. We will adopt this terminology in this text.

7.13 Dopant Diffusion Occurs by Both Interstitials and Vacancies

It is known that some dopants (e.g., P and B) have their diffusion coefficients enhanced when the surface of the silicon is oxidized, while others (Sb) appear to have their diffusion coefficient reduced. These experimental observations on oxidation-enhanced diffusion (OED) and oxidation-retarded diffusion (ORD) were important in elucidating the microscopic mechanisms of diffusion, because the variations were postulated to be due to perturbed point defect concentrations caused by the surface oxidation. The fact that the diffusion of one dopant was enhanced while that of another was retarded is evidence that two different diffusion mechanisms are operating on an atomic scale.

The kind of test structure that was used to understand the microscopic diffusion mechanisms is illustrated in Figure 7.25. It looks very much like the LOCOS structure discussed in Chapter 6, where oxidation occurs in one area and a thin oxide covered by a deposited nitride masks the other area. Buried dopant layers are often used, to avoid complications from dopant segregation at interfaces. The reason for having two regions is to have a built-in reference on the same sample, where the reference is the area where inert or normal diffusion occurs under the masked region. The active surface of the oxidizing area injects interstitials and enhances the diffusion of dopants like phosphorus and boron. Of course, it is not obvious that interstitials are injected under an oxidizing region, though it makes intuitive sense, because, when a cube of silicon is oxidized, there is an expansion of 30% to form the SiO_2 structure (recall Figure 6.4 in Chapter 6). Some of the resulting compressive stress could be relieved by injection of a silicon interstitial to make space at the silicon surface where oxidation is occurring.

The key original insight that led to understanding the role of point defects in mediating dopant diffusion arose from recognizing that the growth of oxidation stacking faults and the OED of dopants always happened together. This meant that they were different manifestations of the same underlying phenomena, namely the injection of interstitials during oxidation [13]. Oxidation-induced stacking faults are non-equilibrium defect structures which have been shown by transmission electron microscopy (TEM) image contrast to be composed of an extra partial plane of interstitials (see Figures 3.7 and 3.8 in Chapter 3). These stacking faults grow during oxidation of the silicon surface, indicating that oxidation is injecting interstitials.

OED of boron or phosphorus is seen under the same conditions that cause stacking faults to grow (Figure 7.26(a)). Antimony, on the other hand, undergoes ORD, i.e., the diffusion of antimony is slower than normal during an oxidation process (Figure 7.26(b)). How could this happen? The oxidation-injected interstitials must recombine with vacancies in the bulk, depressing the vacancy concentration and therefore retarding Sb diffusion. This implies that B and P prefer to diffuse with interstitials, and Sb prefers to diffuse with vacancies. Physically, the preference by antimony for a vacancy mechanism may be related to its large size in the

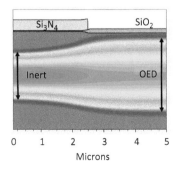

Figure 7.25 Two-dimensional simulation showing boron OED under a growing oxide.

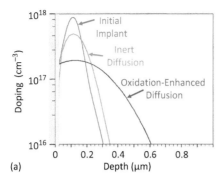
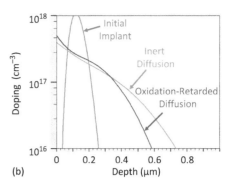

Figure 7.26 Athena simulations of (a) oxidation-enhanced diffusion (OED) of boron and (b) oxidation-retarded diffusion (ORD) of antimony during the growth of a thermal oxide on the surface of silicon. The initial implants were $1 \times 10^{13}\,\mathrm{cm}^{-2}$, the B at 30 keV and the Sb at 300 keV. The inert drive-in for boron was 300 min at 900 °C in each case. The oxidation was 300 min in dry O_2 at 900 °C. For Sb, the drive-in was 3000 min at 1000 °C since it diffuses slowly.

silicon lattice. The elastic interaction between a point defect and a dopant atom depends on the mismatch in size, so that a large dopant atom may prefer to migrate with a vacancy, while a small dopant atom may prefer to migrate with an interstitial.

Further experiments found that nitridation of the silicon surface in an ammonia ambient has exactly the opposite effects to oxidation. If silicon is exposed to an ammonia ambient at high temperatures, the surface grows a thin silicon nitride layer that is in tension. Pre-existing stacking faults also shrink in a nitriding ambient. And B and P exhibit retarded diffusion, while the diffusion of Sb is enhanced. Thus the conclusion is that nitridation injects vacancies and creates an interstitial undersaturation through recombination of the injected vacancies with interstitials. This complementary set of observations "weaves a seamless logic" for the dual roles of interstitials and vacancies in assisting dopant diffusion in silicon [14].

The idea that both I and V contribute to dopant diffusion in silicon is now generally accepted based on both experimental observations and theoretical calculations. The theoretical calculations have only recently become possible based on a direct quantum mechanical description of solid-state systems on powerful computers. The experimental results that dramatically capture the existence of a dual diffusion mechanism are based on the observation that, under certain identical conditions, such as oxidation of the silicon surface, one dopant has its diffusion enhanced while another dopant has its diffusion retarded.

It is important to note that this information could not be obtained by observing the diffusion of the dopants in an inert (argon or nitrogen) atmosphere. In that case, the only observation that could be made is that one dopant diffused faster or slower than another one. Instead, it is because the point defect levels are perturbed from their equilibrium values by oxidation that some inference can be made about dopant diffusion mechanisms.

These observations can be formalized by saying that dopants diffuse with a fraction f_I interstitial-type diffusion mechanism and with a fraction $f_V = 1 - f_I$ vacancy-type diffusion mechanism. Under perturbed conditions, when the interstitial (C_I) or vacancy (C_V)

concentrations are different from their equilibrium concentrations (C_I^* and C_V^*), the diffusivity of a dopant can be written as

$$D_A^{eff} = D_A^* \left(f_I \frac{C_I}{C_I^*} + f_V \frac{C_V}{C_V^*} \right), \qquad (7.52)$$

where D_A^{eff} is the effective diffusivity of the dopant measured under conditions where the point defect populations are perturbed and D_A^* is the normal, equilibrium diffusivity measured under inert conditions. By definition $f_I + f_V = 1$. This kind of diffusion model, where interstitials, vacancies and dopants interact, is often referred to as a three-stream diffusion model.

It is interesting to note that, if two different diffusion mechanisms are operative (i.e., interstitial- and vacancy-assisted mechanisms) and if they have very different activation energies (E_A), then a plot of D_A versus $1/T$ would show distinct regions with different slopes. For dopants in silicon, there has never been such an observation in the intrinsic D_A versus $1/T$ plot (Figure 7.14). We might expect that this slope change or curvature would be most obvious for arsenic, which has approximately equal I and V components of diffusion. This indicates that the activation energies of dopant diffusion via both I and V mechanisms are comparable at processing temperatures.

If an independent measure of the vacancy or interstitial supersaturation can be made, the fraction of diffusion by a vacancy or interstitial mechanism can be directly computed from (7.52). Such an estimate of the supersaturation might be made by observing the growth rate of stacking faults, for example. However, these methods tend to be less reliable than the self-consistent approach that relies on observing enhanced and retarded dopant diffusion in the same experiment as described above. Remember that this problem of inferring the mechanism of dopant diffusion occurs because there is no reliable way to directly observe the interstitial or vacancy populations themselves at processing temperatures because of their low concentrations. In spite of this limitation, a self-consistent picture of the atomic-scale mechanisms of dopant diffusion in silicon can be deduced by following the trail of physical evidence. These kinds of experiments have been widely done, resulting in the following f_I and f_V values for the common dopants in silicon, shown in Table 7.4.

This is a much more rigorous description of the diffusion process, as we have now "split" the overall dopant diffusivity into components dominated by interstitial- and vacancy-assisted mechanisms. Though this contains a lot more insight than a purely macroscopic description of the diffusion process, it is still a crude description of the atomic-level mechanisms. Consider

Table 7.4 **Approximate values for f_I and f_V for silicon self-diffusion and for the common dopants in silicon.**

	f_I	f_V
Silicon	0.6	0.4
Boron	1.0	0
Phosphorus	1.0	0
Arsenic	0.4	0.6
Antimony	0.02	0.98

what really happens: A substitutional dopant does not diffuse in any average sense. Instead, an individual dopant atom might by chance interact with a vacancy or interstitial, producing an episode of mobility for that particular dopant atom, after which it sits back on a substitutional lattice site. The overall or macroscopic effect of many of these individual hops is that the substitutional dopant appears to move with a macroscopic diffusion coefficient described above. These physical effects of the mobile species are not yet captured in our description of the diffusion process. In the following sections, we will consider how these mobile species form on an atomic level and how Fick's laws can be applied to these mobile diffusing species to give a detailed description of the diffusion process on an atomic scale.

The mobile species come about via interactions of individual dopant atoms and individual point defects. We can consider a dopant atom A interacting with an interstitial I as follows:

$$A + I \leftrightarrow AI, \qquad (7.53)$$

where AI is the actual interstitial-assisted mobile species on an atomic scale. The substitutional dopant is immobile by itself, unless it interacts with a point defect. (For a vacancy mechanism, I is replaced by V in this equation and the subsequent discussion.) This equation might represent either a migrating bound pair or a substitutional dopant "kicked out" of a lattice site by an interstitial to become mobile in an interstitial site.

Equation (7.53), although simple, contains a surprising amount of physics. It is clear that an interstitial supersaturation will drive more dopant atoms into a mobile state by shifting the equation to the right, thus enhancing the dopant diffusivity. This explains the phenomenon of OED, where the interstitial supersaturation during oxidation enhances the dopant diffusivity. A suppression of I, by nitridation of the silicon surface, which introduces vacancies, can occur through the following reaction:

$$I + V \leftrightarrow Si_S, \qquad (7.54)$$

where Si_S represents a silicon atom on a lattice site. In other words, the I and V annihilate each other, leaving a substitutional silicon behind. If there is an excess of vacancies, this will lead to retarded diffusion for dopant species that diffuse with interstitials. Interstitial–vacancy (I–V) recombination is discussed more fully in Section 7.14.

Even under inert conditions where we consider the diffusion of a dopant in from the surface (say, from a shallow phosphorus implant), the atomic-level description says that the interior of the sample soon contains a lot of AI species, which drives (7.53) to the left, releasing interstitials in the interior when the mobile dopant regains a substitutional lattice position. In this way, silicon interstitials are "pumped" from the surface region into the interior region by the dopant diffusion. If there is a strong source or sink of interstitials at surface kinks and ledges which maintain near-equilibrium levels of interstitials in the surface region, this leads to a supersaturation of interstitials in the interior of the sample. But a supersaturation of interstitials causes enhanced dopant diffusion. Thus, the tail of a phosphorus profile can diffuse much faster than expected, which is exactly what is observed experimentally.

This full coupling between the point defects and the dopant diffusion is illustrated in Figure 7.27. A simulation is shown for a high-concentration phosphorus diffusion profile that is diffusing into silicon at 800 °C. Two diffusion models are shown in Figure 7.27(a), the typical concentration-dependent model and the fully coupled diffusion model. The box-shaped concentration-dependent model is typical of what would be observed for a slow-diffusing species, such as arsenic. But because the phosphorus has very high solubility and very high diffusivity, it carries a lot of point defects (interstitials or interstitialcies) with the profile as it diffuses into the bulk. These are released deeper in the bulk when the phosphorus atoms regain substitutional sites and cause an enhanced interstitial concentration in the tail region of the profile.

The simulation in Figure 7.27(b) shows that there is a 40× enhancement in the interstitial supersaturation, which causes the enhanced phosphorus tail diffusion. This feedback effect from the coupling between the dopants and the point defects produces the characteristic "kink and tail" phosphorus diffusion profile, where the tail diffusion is much faster than expected. The effect is much more pronounced at lower temperatures, where the point defect flux and the dopant fluxes diverge because of the higher activation energy for point defect diffusion. The point defects are not able to diffuse fast enough to smooth out the interstitial supersaturation generated by the in-diffusing phosphorus. These fully coupled effects occur to a lesser extent for the slower-diffusing dopants, producing small modifications in their tail diffusion.

The fully coupled model in the simulation in Figure 7.27(a) incorporates the concentration dependence of the dopant diffusion, but there is a complex interplay between the dopant fluxes and the point defect fluxes that depends in subtle ways on the individual dopant diffusivities and solubilities and the diffusion temperature. Only a sophisticated simulator can combine these effects to predict what will actually happen in devices.

This description of "chemical pumping" of point defects also explains an anomalous inter- action between emitter and base diffusions in some bipolar processes. In bipolar devices, the

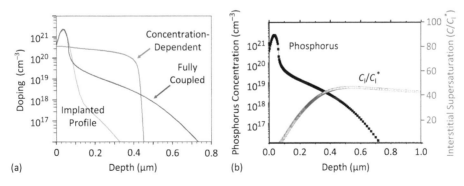

Figure 7.27 (a) Athena simulation using the Fermi-level (concentration-dependent) model (red curve) and the fully coupled model (blue curve). The initial implant (green curve) is a 30 keV, 1×10^{16} cm^{-3} phosphorus implant. (b) The fully coupled model illustrates the chemical or diffusion pumping of native point defects from the surface region into the bulk region by dopant–defect coupling. A supersaturation of 40× in the interstitial concentration causes enhanced tail diffusion in the phosphorus profile, producing the characteristic "kink and tail" profile.

current flows vertically through N–P–N or emitter–base–collector regions (or the P–N–P equivalent). These were the original logic devices, because they did not rely on the poorly understood MOS surface properties at the time (1960s), as discussed in Chapter 1. Now, variations on the bipolar device are very common in power devices because large amounts of current can flow through the bulk of the semiconductor versus along the surface in a MOS device. There was the same push to scale down the base thickness and increase the speed of the devices. A simple way to do this seemed to be to diffuse the emitter profile some more, thus narrowing the base. But this did not seem to work in practice. The "emitter-push" or "emitter-dip" effect refers to the enhanced diffusion of the base profile below a phosphorus emitter, as shown in Figure 7.28. This manifests itself in a much wider base width than expected. It is now understood to be caused by the pumping of interstitials by the diffusion of phosphorus, which builds up a high supersaturation of these point defects in the interior of the sample, well beyond the phosphorus diffusion front. The interstitial supersaturation then enhances the base (typically boron) diffusion.

In beginning this section, we indicated qualitatively how a diffusing dopant atom transported interstitials into the bulk of the silicon, where it could enhance the tail diffusion of the diffusing dopant. We can now understand the origin of this effect if we properly account for the full coupling between the defects and the dopants when we consider the diffusion equation for the defects.

To obtain an equation describing the full coupling between defects and dopants, we should apply Fick's law to the mobile species, giving

$$F_{AI} = d_{AI} \left[\frac{\partial C_{AI}}{\partial x} \right], \tag{7.55}$$

where d_{AI} is the diffusivity of the actual mobile species causing the migration and C_{AI} is the concentration of the mobile species. At this point, our description of the diffusion flux is in terms of the unseen mobile species parameters d_{AI} and C_{AI}, the diffusivity and concentration of the mobile species in the lattice. These mobile species are not readily observable. Thus, the description as it stands is not very helpful. If we had numbers for the diffusivity of the mobile

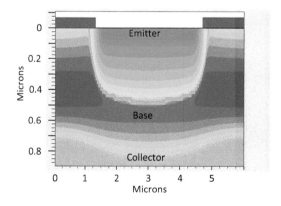

Figure 7.28 Schematic of the emitter-push effect in a bipolar transistor. The diffusion of the emitter region (phosphorus) affects the base region (boron) underneath the emitter and causes the base to "push-out" further than expected [15]. This interaction is now known to be due to the chemical pumping of interstitials from the emitter region to the deeper bulk regions.

species itself and the concentration of the mobile species, we would have a useful description of the evolution of the profile shape. To make this description useful, we must relate these parameters to what can be observed, namely the overall dopant concentration C_A and the apparent or effective diffusion coefficient D_A of the total dopant concentration (measured, say, by SIMS). The one thing we know is that the flux of the mobile species must equal the flux of the total dopants, since that is what we observe. Thus,

$$D_A C_A = d_{AI} C_{AI} . \tag{7.56}$$

Note that this does not mean that $D_A = d_{AI}$, but rather that the "DC" product for each must be equal (i.e., to make the fluxes equal). The overall profile motion that is measured for a substitutional dopant C_A with an apparent measured diffusion coefficient D_A is in fact generated by a very small number of species on the atomic scale that become mobile intermittently (C_{AI}) that migrate with a high diffusivity d_{AI} in the silicon lattice.

All of the previous macroscopic diffusion effects, such as concentration-dependent diffusion, now have a clearer physical foundation. At high dopant concentrations, the higher Fermi level gives rise to more interstitials (because of their charge states in the bandgap), and this creates more mobile species as (7.53) is pushed to the right. Other effects such as OED, ORD and chemical pumping of defects causing enhanced tail diffusion are naturally accounted for by increased or decreased interstitial numbers.

But there is another more subtle effect from (7.53) ($A + I \leftrightarrow AI$) that provided a surprising explanation for a previous anomalous device observation known as the "reverse short-channel effect." As channels get shorter, the influence of the drain voltage is to partially turn-on the device (the DIBL effect), leading to lower threshold voltages. Experimentally, it was observed that the threshold voltages began to increase for shorter channels, a phenomenon that is very hard to explain from a device physics perspective. The short-channel effects are shown in Figure 7.29.

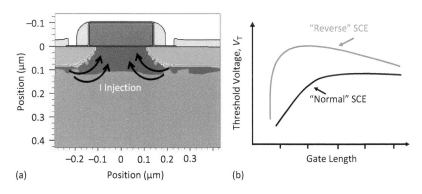

(a)   (b)

Figure 7.29 (a) Athena simulation of short-channel MOS devices. The doping profiles are shown, with arrows that indicate the interstitial fluxes emanating from the diffusing heavily doped source/drain regions. (b) Qualitative plot of the reverse short-channel effect (RSCE), where the threshold voltage actually increases as the gate length reduces.

The explanation is simple (in hindsight). We can take (7.53) and apply the law of mass action to the solid-state system and relate the concentrations on each side of the equation by

$$C_{AI} = kC_A C_I. \tag{7.57}$$

This particular description of the diffusion process is in "chemical equilibrium" in the sense that the mobile pairs are directly related to the reacting species by the equilibrium reaction constant. Experimentally, this is justified because large enhancements in the diffusivity when excess interstitials are present can be seen for short diffusion times and remain for longer anneals. This indicates that the mobile species forms quickly and that the equilibrium assumption is valid.

Applying Fick's law to the mobile species C_{AI} gives

$$F_{AI} = -d_{AI} \frac{\partial C_{AI}}{\partial x}. \tag{7.58}$$

Substituting from (7.57), differentiating with respect to x and using the chain rule for differentiation gives

$$F_{AI} = -d_{AI} \left(kC_I \frac{\partial C_A}{\partial x} + kC_A \frac{\partial C_I}{\partial x} \right). \tag{7.59}$$

The first term can be thought of as the "normal" diffusion, where the gradient in the doping concentration determines the diffusion flux and more interstitials increase the flux. But the second term, which has a gradient in the interstitials driving the diffusion flux, is what explains the anomalous reverse short-channel effect (RSCE), first realized by Rafferty [16]. Implant damage from the implants in the source/drain regions (discussed in detail in Chapter 8) combined with the local sink for point defects at the surface under the gate creates large gradients in the point defect profiles in the channel region, as illustrated in Figure 7.29(a).

But from (7.59), we know that the gradient of either the dopant or the point defects can cause diffusion. It is not just the dopant gradient (expected), but also the interstitial gradient (unexpected, but based on these atomic-level insights) that determines the flux. Thus, it is not just that more interstitials create more diffusion, it is also the case that a gradient of interstitials creates more diffusion. This means that dopant diffusion can be driven by strong sources and sinks of point defects that create defect gradients. These large interstitial gradients drive the overall dopant diffusion "uphill" towards the surface. Higher doping under the channel region results in a higher threshold voltage, and the effect is more pronounced as the channel length reduces, as shown in Figure 7.29.

This previously anomalous RSCE showed not only that the microscopic diffusion mechanisms provide a clearer and more physical picture for dopant diffusion, but also that they could explain new and unconsidered effects. Process simulators are now capable of predictively simulating these kinds of effects in small devices. The details of how the various physical models we have discussed are implemented in process simulation programs like Athena are discussed in the user manuals for those simulators.

7.14 Point Defect (Interstitial–Vacancy) Recombination

We have seen several examples in which non-equilibrium concentrations of point defects can be present in silicon processing. OED is caused by oxidation injecting excess interstitials. The emitter-push effect is caused by high-concentration diffusion injecting excess interstitials as the phosphorus–interstitial (PI) pairs break up at lower concentrations. And, finally, the RSCE was explained by the flux of excess interstitials out of ion-implanted regions to recombination sites at the silicon surface. In Chapter 8 we will see that the ion implantation process introduces very large excess concentrations of both I and V because the implanted ions slow down partly by knocking silicon atoms off lattice sites, creating one I and one V each time this happens.

In all these cases, dopant diffusion is affected by the non-equilibrium point defect concentrations. Recall (7.52):

$$D_A^{\text{eff}} = D_A^* \left(f_I \frac{C_I}{C_I^*} + f_V \frac{C_V}{C_V^*} \right).$$

If $C_I/C_I^* \neq 1$, for example, because a surface oxidation is increasing C_I, then, in order to calculate what happens to a particular dopant's diffusivity, we need to know what happens to C_V/C_V^* under these conditions. In (7.54) it was suggested that I–V recombination could take place through simple recombination, $I + V \leftrightarrow Si_S$. If this happens, then when C_I/C_I^* increases, we should expect C_V/C_V^* to decrease because the excess I will reduce the V concentration through recombination.

It is generally assumed in silicon that a steady-state condition given by

$$C_I C_V = C_I^* C_V^* \tag{7.60}$$

will be established whenever C_I and C_V deviate from their equilibrium values. Thus an interstitial supersaturation will produce a vacancy undersaturation inversely proportional to the interstitial supersaturation. However, the time required to reach this steady-state condition will depend on the rate at which I and V recombine.

There is both theoretical and some experimental evidence in the literature that suggests that there is a barrier to I–V recombination, which would imply that recombination is not instantaneous. However, we will see in Chapter 8 that, for the case of ion implantation damage, which produces the largest non-equilibrium concentrations of I and V, recombination appears to be very fast (Section 8.8). In fact, very fast recombination will be one of the key assumptions we make in discussing TED in Section 8.9. In order to calculate values in (7.52), there are two limiting cases that are generally useful. If we assume that I–V recombination is very fast, then (7.60) will hold and $C_I/C_I^* = C_V^*/C_V$.

In the other limiting case, we assume that I–V recombination is very slow. In this case, a supersaturation of one type of point defect will have no effect on the concentration of the other type of point defect. Numerical simulators can, of course, handle intermediate cases in which the time scale of I–V recombination is comparable to the time scale of the simulated experiment.

7.15   Why Silicon Self-Diffusion Matters

It is found experimentally that the activation energy (E_A in (7.28)) is about 1 eV lower for dopant diffusion in silicon than it is for self-diffusion (3–4 eV versus 4–5 eV). This raises a fundamental question regarding this difference. Because dopants have a lower activation energy for either I or V or mixed mechanisms of diffusion, there must be a lowering of the activation barrier specifically related to group III and V elements that is not present for group IV elements (Si or Ge). Coulombic effects can contribute to some but not all of this barrier lowering. (The dopant atoms are charged on substitutional lattice sites and the Si atoms are not.) If Coulombic effects alone dominated dopant-defect interactions, then all N-type or P-type dopants would have the same activation energy. Since this not the case, there must be additional dopant–defect interactions to be considered.

One such possibility is the relaxation of the silicon lattice around a dopant, related to the size of the dopant atom. For example, theoretical calculations indicate that a substitutional boron atom causes the lattice to relax inwards by about 12%. It is easy to imagine that lattice distortions when both dopants and defects are in close proximity could significantly alter the activation energy for diffusion. Another possibility relates to the electronic interactions due to charge transfer between the dopant and the defect. These effects lead to a higher probability of a point defect binding to a dopant atom or interacting with a dopant atom to kick it out of its normal, substitutional lattice site. It is this interaction between dopant atoms and point defects that is the key to understanding dopant diffusion on an atomic scale. The difference in activation energies between self-diffusion and dopant diffusion plays a key role in how native point defects and mobile dopants interact as the temperature is changed, generating complex diffused profiles, as described in the following sections.

The difference in activation energies between self-diffusion (4.8 eV) and dopant diffusion (3–4 eV) explains many otherwise strange or anomalous results. Self-diffusion measures the flux or flow of silicon atoms in the lattice and is surprisingly easy to measure because of the availability of silicon isotopes. The tendency of silicon atoms to move by an interstitial or vacancy mechanism can also be determined by OED and ORD experiments, just as for dopants. Such measurements show that silicon moves with almost equal components of I and V mechanisms with similar activation energies [17]. We know that the absolute numbers of interstitials or vacancies are very low, so in spite of the presumed high diffusivity for interstitials and vacancies, the silicon self-diffusion flux is lower than dopant diffusion fluxes. The "DC" product for a dopant diffusing with interstitials ($D_A^{\mathrm{eff}} C_A$ or equivalently $d_{AI}C_{AI}$) is larger than the "DC" product for the native silicon self-diffusion ($d_I C_I^*$ or $d_V C_V^*$) and this difference increases at lower temperatures because of the higher activation energy for silicon diffusion.

This explains why the emitter-push effect is worse at lower temperatures. On an atomic scale, a mobile dopant "carries" an interstitial with it and releases the interstitial when it regains a substitutional site deeper in the bulk. Obviously, the faster a dopant moves, the more it perturbs the interstitial concentration deeper in the bulk, which is why dopants with a fast diffusivity and high solubility like phosphorus have larger fully coupled effects. It also explains why the effect is larger at lower temperatures, because the flux of interstitials cannot match the flux of dopants and there is a larger pile-up of interstitials deeper in the bulk. Once again, this detailed understanding of the full coupling between mobile dopant atoms and native point defects on an atomic scale provides answers to questions that would otherwise remain a puzzle.

The mathematical origin came from the application of the chain rule to the gradient in the mobile species, while the physical interpretation is that areas with high point defect populations have faster diffusion, so a gradient in point defects can drive a diffusion flux. This cross-term is of paramount importance in the neighborhood of point defect sources and sinks, where point defect gradients are large. In these cases, it can dominate the overall diffusion of the dopant.

7.16   Diffusion in Polysilicon

Diffusion in polycrystalline materials is important for several device applications. In a CMOS process, the requirement is to dope the gate with a uniform heavy concentration of dopant, but still not penetrate the thin gate oxide, as that would affect the threshold voltage. The polysilicon gate needs to be doped right up to the interface to avoid the effects of poly depletion when gate voltage is applied, making the gate capacitance seem less than desired. Heavily doped polysilicon in direct contact with silicon can also be used as a dopant source to form shallow junctions or buried contacts or the emitters of bipolar transistors. The substrate doping profile is influenced by the diffusivity in the polysilicon. Lightly doped polysilicon can be used to make high-value resistors, and the performance is limited by the diffusion of dopant along the polysilicon grain boundaries from the highly doped contacts to the resistor.

The structure of polysilicon depends on the deposition and annealing conditions. The most common method for depositing polysilicon is to decompose silane SiH_4 gas in a low-pressure ambient between 575 and 650 °C, a process known as low-pressure chemical vapor deposition (LPCVD). At the lowest temperature, the deposited film is amorphous or microcrystalline, with grains less than 5 nm. At the higher temperatures, the films are composed of monocrystalline grains of different orientations, often with a strong preferred crystalline texture and a columnar orientation, with grains as large as 50 nm that extend through the thickness of the film. Annealing causes further morphological changes in the films, with grain growth occurring. Doping of the films tends to enhance both the initial grain size and the grain growth rates.

Modeling diffusion in polysilicon is obviously a complicated problem because it is not possible to account for the exact microstructure of the grains and grain boundaries. But the average behavior of dopants in polysilicon can be accounted for by considering the following four major mechanisms:

1. fast dopant diffusion along grain boundaries;
2. dopant diffusion in the grain interiors;
3. segregation of dopants between the grain interior and grain boundaries; and
4. transport of dopants during grain growth.

The experimental results show extraordinarily fast dopant diffusion in polysilicon. The concentration profile in the polysilicon can be explained by the low diffusivity of dopant in the interior of the grains and the much higher diffusivity along the grain boundaries. Consider an arsenic implant into polysilicon followed by an anneal at 800 °C in nitrogen. After the implant, the arsenic is mostly in the interior of the grains. During annealing, it moves from the bulk of the grain to the grain boundary, where it preferentially segregates. The weak or irregular bonding between different grain orientations provides a natural sink for dopants where they have a fast diffusion pathway along the boundary. The amount of dopant that reaches the grain boundary is limited by the slow diffusion in the grain interior. The diffusion in the grain interior, which has a regular silicon lattice, is the normal concentration-dependent diffusion seen in bulk silicon. Larger-grained polysilicon tends to have larger peaks near the surface and less diffusion deep into the polysilicon, because the dopant has to diffuse a greater distance in the grain before segregating to the grain boundary. The diffusivity in the grain boundaries is approximately 10,000 times faster than in the single-crystal silicon grains. Because of the interaction between diffusion in the interior of grains and grain boundaries, very simplistic models for dopant diffusion in polysilicon simply enhance the bulk diffusion rate by a factor of 100.

More sophisticated models track two streams of dopant using the following equations:

$$\frac{\partial C_A^{grain}}{\partial t} = \frac{\partial}{\partial x}\left(D_A^{grain}\frac{\partial C_A^{grain}}{\partial x}\right) - k_{eff}\left(C_A^{grain} - \frac{C_A^{gb}}{s}\right), \tag{7.61}$$

$$\frac{\partial C_A^{gb}}{\partial t} = \frac{\partial}{\partial x}\left(D_A^{gb}\frac{\partial C_A^{gb}}{\partial x}\right) + k_{eff}\left(C_A^{grain} - \frac{C_A^{gb}}{s}\right), \tag{7.62}$$

where k_{eff} is the effective transfer rate between grains and grain boundaries, and s is the effective segregation coefficient between grains and grain boundaries. The grains are assumed to increase in size due to diffusion of silicon atoms to grains with favorable energetics, giving a time-dependent grain size described by

$$L_g(t) = \sqrt{L_{g,0}^2 + 2\sigma t}, \tag{7.63}$$

where $L_{g,0}$ is the initial grain size and σ is an effective diffusion constant of atoms to grain boundaries. Numerical simulations based on these two-stream diffusion models do a remarkably good job of modeling the complex diffusion profiles in polysilicon.

7.17 Dopant Diffusion in Compound Semiconductors

As in silicon, dopant diffusion in compound semiconductors is mediated by native point defects. One should expect that the physical mechanisms for diffusion in silicon would transfer to other semiconductors with some modifications, and indeed that is the case. These defects, as well as substitutional and interstitial dopants, can be charged, leading to Fermi-level-dependent effects. Apparent anomalous diffusion behavior can be understood if the fundamental charge and point defect mechanisms are understood and modeled.

We briefly discuss diffusion in two important compound semiconductors, GaAs and SiC, to show both the similarities and differences between diffusion in these compounds versus Si. Similar considerations would hold for the other binary compound semiconductors. Diffusion in GaAs is more complex than in Si because the Ga and As sublattices behave differently. Diffusion in SiC is simpler than in Si because the strong bonds in SiC give rise to minimal diffusion of dopants, leaving dopant activation as the main problem.

7.17.1 Diffusion in GaAs

The different behaviors of the Ga and As sublattices are captured by the differences in the self-diffusion of Ga and As measured by isotope experiments using the stable isotopes of Ga and by radioactive tracer diffusion for As. Just as in Si, the motion of the host-lattice ("self") atoms in compounds is mediated by the native point defects such as vacancies and interstitials. The difference is that GaAs has four native point defects, the Ga vacancy, the Ga interstitial, the As vacancy and the As interstitial. In addition, it has two antisite defects, a Ga atom sitting on an As site, and an As atom sitting on a Ga site. The native antisite defects are important for the electrical properties of the bulk GaAs, but not so much for diffusion. A schematic diagram of these point defects was shown in Figure 3.12 for a binary compound semiconductor. The concentrations of these defects depend on the Fermi level, just as in silicon. Diffusion of the self-atoms (and also impurity atoms) is mediated by the native point defects, and measurements of the self-diffusion of the Ga and As give [18]

$$
\begin{aligned}
D_{Ga} &= 0.1 \ \exp^{-3.2/kT} \ cm^2 s^{-1}, \\
D_{As} &= 0.7 \ \exp^{-5.6/kT} \ cm^2 s^{-1}.
\end{aligned}
\tag{7.64}
$$

Other well-respected researchers [19] have compiled data and fitted the gallium diffusivity to the expression

$$
D_{Ga} = 2.9 \times 10^8 \ \exp^{-6.0/kT} \ cm^2 s^{-1} .
\tag{7.65}
$$

The origin of the difference may be due to poor control over the As vapor pressure in different sets of experiments, often a problem when trying to interpret diffusion results in GaAs. In spite of the very different values, the trends are clear and are shown in Figure 7.30. The large difference between Ga and As diffusion in GaAs suggests that the self-diffusion of the gallium and arsenic atoms occurs independently along their own sublattices. Similar large differences between gallium and antimony self-diffusion are seen in GaSb [20]. Impurities in GaAs tend to follow similar trends, with P-type acceptors (the column II dopants) moving along the gallium sublattice and N-type donors (the column VI dopants) moving along the arsenic sublattice. There is an interesting group of dopants from column IV, such as Si, Ge and Sn, that can occupy sites on either sublattice.

To compare and contrast with diffusion in silicon, we will examine the diffusion behavior of two important dopants in GaAs, the P-type dopants and the N-type column IV dopants. These dopants undergo many of the same types of diffusion behavior as dopants in the silicon lattice.

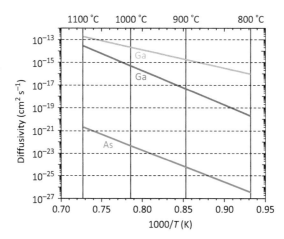

Figure 7.30 Self-diffusion data for Ga and As in GaAs or GaAs superlattices. After [18, 19].

At low concentrations, the diffusion coefficient is small and the behavior satisfies the simple analytic erfc or Gaussian formulas. At higher concentrations, Fermi-level-dependent diffusion is obvious, giving rise to the characteristic box-shaped profiles. At even higher concentrations, some anomalous diffusion profiles occur, such as the "kink and tail" profiles characteristic of interactions between the dopant flux and the native point defect flux. In general, diffusion in GaAs is more complex than diffusion in Si because of the abundance of native point defects, but also because the low vapor pressure of As can cause arsenic evaporation, generating As vacancies, which can dramatically affect dopant profiles. This can be understood if all of the point defect, extended defect and Fermi-level effects are modeled. Even implanting dopants can dislodge As atoms, giving rise to a striking implant energy dependence and implant temperature dependence for dopant diffusion [21, 22]. These effects can be understood using the same point defect concepts and equations as for diffusion in silicon, as we will see in the next subsections.

7.17.2 P-Type Dopant Diffusion in GaAs

Zinc and beryllium are both common column II P-type acceptor dopants in GaAs. Zn is very well studied, though Be is a useful dopant because it is a lighter element that causes less implant damage. Both have very similar diffusion mechanisms [19, 23, 24]. Zn came to be extensively studied because its diffusion was seen to cause intermixing in GaAs/AlAs heterostructures [25, 26]. To understand why this is important, recall that GaAs and AlAs have identical lattice constants but different bandgaps. This makes it very easy to grow alternating layers of each material and confine carriers in the lower-bandgap material, thereby enhancing carrier recombination, which can produce efficient light-emitting diodes or laser action because of the direct bandgap of these materials. The emission wavelength can be further tuned by creating lattice-matched GaAs/Al$_x$Ga$_{1-x}$As layer heterostructures grown by epitaxy. To confine light emission to a region or to create optical waveguides, it is useful to be able to "destroy" the heterostructures in defined regions, which is why intermixing by Zn diffusion became technologically important. Figure 7.31 shows a GaAs/AlAs heterostructure, where Zn is allowed to diffuse in

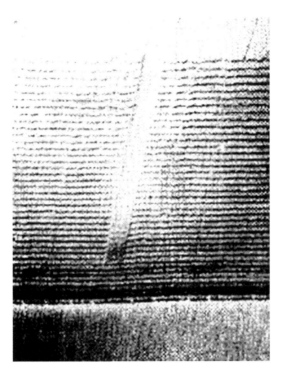

Figure 7.31 A beveled image of GaAs/AlAs heterostructures disrupted by Zn diffusion causing impurity-induced lattice disordering (IILD) in the center of the image. Reproduced from [25] with the permission of AIP Publishing.

the center of the structure through an opening in a deposited Si_3N_4 layer. The structure under the Zn diffusion profile remains crystalline, but it takes on the average alloy composition of the Ga, Al and As layers and acts to confine the heterostructure to a localized region.

The conventional wisdom is that the P-type dopants in GaAs diffuse by an interstitial-assisted mechanism. Some evidence for this comes from the TED of Be after ion implantation, a process that is discussed in more detail in Chapter 8. Here, we simply state that implantation introduces extra atoms that generate interstitials when they occupy a substitutional site. These extra interstitials cause a signature transient enhancement in diffusion for dopants that diffuse by an interstitial-type mechanism [27]. Just as in silicon, the interstitial-assisted dopant diffusion mechanism can be either a dopant–interstitial pair (the interstitialcy) or a pure dopant interstitial. For Be, this would correspond to the reactions for the pair and pure dopant interstitial

$$Be^{-1} + I_{Ga}^{+2} \Leftrightarrow (Be\text{--}I_{Ga})^{+1},$$
$$Be^{-1} + I_{Ga}^{+2} \Leftrightarrow Be_I^{+1}, \tag{7.66}$$

where I_{Ga}^{+2} is a gallium interstitial in the +2 charge state, $Be - I_{Ga}$ corresponds to the dopant–interstitialcy pair and Be_I^{+1} represents the Be interstitial. The issue of which particular charge state dominates P-type dopant diffusion in GaAs remains open and confused. Some authors propose that Be diffusivity depends linearly on the hole concentration p [28], while others found that the data for both implanted and grown-in Be can be modeled by a square dependence on p [24, 27].

What is clear is that P-type dopant diffusion in GaAs proceeds by interacting with point defects, producing the familiar concentration-dependent diffusion resulting in box-shaped profiles and more subtle "kink and tail" diffusion where the dopant flux affects the native point defect flux. In the following sections, we show some examples of these effects.

For example, Zn diffusion profiles show box-shaped profiles characteristic of concentration-dependent diffusion that is very well modeled by a single term of the form

$$D_{Zn}^{eff} = D_{Zn}^{++} \left(\frac{p}{n_i} \right)^2 , \tag{7.67}$$

where D_{Zn}^{++} is the intrinsic diffusivity corresponding to a defect with a $(p/n_i)^2$ dependence. Note that we do not have D_{Zn}^0 and D_{Zn}^+ terms like we often have in silicon – the data seem to be well modeled by a single term that depends only on $(p/n_i)^2$. This equation is a simplification of the analogous equation (7.44) in silicon, where all of the three terms occur. Since Zn and Be are both shallow acceptors, the hole concentration mostly corresponds to the dopant profile, unless precipitation takes place at very high concentrations. A SIMS plot showing exquisite fitting to a square-law diffusivity is shown in Figure 7.32.

At lower anneal temperatures, the Zn profile in the tail region is not quite so box-shaped and gets "kicked out" by the supersaturation of interstitials due to the dopant injecting interstitials as it regains substitutional sites deeper in the bulk. This is the fully coupled effect that is also seen in silicon as the dopant flux becomes larger than the self-diffusion flux and the self-diffusion profile cannot quickly equilibrate. An example is shown in Figure 7.33 for a 750 °C anneal versus the 850 °C anneal shown in Figure 7.32.

The intrinsic diffusivities for Be and Zn are plotted in Figure 7.35 (discussed later), for the single D^{++} intrinsic charge state that appears to be the dominant one in GaAs. The values from [23, 27] are given by

Figure 7.32 SIMS data (blue squares) and fitted results (red line) for an 850 °C, 120 min anneal of implanted zinc in GaAs. The box-shaped profile is a result of the square-law concentration-dependent diffusivity. Silvaco Athena simulation to data from [23].

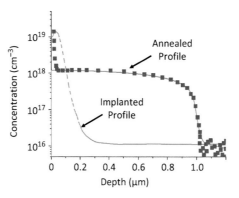

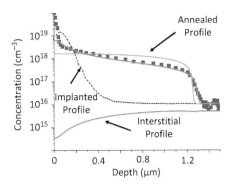

Figure 7.33 SIMS data (blue squares) and simulated results for a 750 °C, 120 min anneal of implanted zinc in GaAs. The green line shows the box-shaped profile, a result of the square-law concentration-dependent diffusivity (similar to Figure 7.32). The red line shows the extended tail of the dopant profile, as the fully coupled model captures the interaction between the dopant and self-interstitial fluxes. Silvaco Athena simulation to data from [23]. The pink line shows the interstitial profile, after interstitials are "pumped" into the bulk by the dopant flux, enhancing the dopant diffusion in the tail. This is analogous to the "kink and tail" phosphorus diffusion profile in silicon (Figure 7.27).

$$D_{\text{Be}}^{++} = 35 \, \exp\left(-\frac{3.86}{kT}\right) \quad \text{and} \quad D_{\text{Zn}}^{++} = 0.6 \, \exp\left(-\frac{3.21}{kT}\right). \tag{7.68}$$

In reference [23], the product of the gallium interstitial diffusion coefficient and gallium interstitial concentration was found to have an activation energy of 5.61 eV, considerably higher than the activation energy of 3.21 eV for the Zn diffusion. This would imply that, at lower temperatures, the "DC" product for the native point defects would be much lower than the "DC" product for the diffusing Zn, so that the flux of the diffusing Zn (which depends on the DC product) would exceed the ability of the native point defects to maintain equilibrium and would result in a supersaturation of native point defects beyond the tail of the Zn profile. This is exactly what causes the kick-out of the tail of the Zn profile at lower temperatures that is captured by the fully coupled model which captures the interaction between the dopant and self-interstitial fluxes, shown in Figure 7.33. Although many details remain unclear for P-type diffusion in GaAs, the basic trends are well captured by the same point defect mechanisms that produce the box-shaped concentration-dependent profiles and the fully coupled "kink and tail" diffusion profiles in the silicon lattice.

7.17.3 N-Type Dopant Diffusion in GaAs

While it is very simple to dope GaAs with acceptors (as shown above), N-type doping is much more difficult to achieve. For example, chromium (Cr) is a column VI dopant and can sit on a column V arsenic site and donate an electron. However, it is a deep level and is more often used to alter the bulk resistivity of GaAs and produce semi-insulating GaAs. Of more practical interest are the column IV dopants (Sn, Ge and Si, in particular), which are termed "amphoteric dopants," meaning they can act as either a donor or acceptor. For example, if Si sits on a column III Ga site, it acts as a donor, and if it sits on a column V As site, it acts as an acceptor.

As expected, this can give rise to complex doping behavior. At low temperatures ($T \leq 700\,°C$), Si prefers to sit on As sites and acts as an acceptor with an energy level about 35 meV above the valence band edge. At high temperatures ($T \geq 900\,°C$), Si tends to sit on Ga sites and acts as a donor with an energy level about 6 meV below the conduction band edge. Amphoteric doping tends to become less efficient at donor concentrations greater than about 3×10^{18} cm^{-3} and tends to saturate at levels slightly above 10^{19}cm^{-3}. In other words, larger donor concentrations do not produce larger electron concentrations above about 10^{19}cm^{-3}. It has been argued that, at high doping levels, the dopants begin to occupy both donor and acceptor sites and begin to compensate each other. While this is a plausible explanation, it is now thought that interactions with vacancies cause the compensation because the vacancy concentrations increase with the Fermi level [29]. The gallium vacancies compensate the donors and n becomes a sublinear function of N_D.

It is thought that these same vacancies mediate column IV dopant diffusion in GaAs. Two leading models [30, 31] proposed that three species were involved in Si diffusion – a Si atom on an As site (Si_{As}), a Si atom on a Ga site (Si_{Ga}) and a pair composed of Si atoms on adjacent Ga and As sites ($Si_{Ga}\cdots Si_{As}$). However, the models made strikingly opposite assumptions about which were the mobile and immobile species. In spite of this, the assumption that the mobile species interacted with vacancies allowed effects like the Fermi-level-dependent concentration-dependent box-shaped profiles to be modeled. A more recent paper [32] calculated the energy barriers for the three species interacting with gallium vacancies and found that all three species were likely mobile with relatively small energy barriers. Since the Si prefers to undergo jumps on the Ga sublattice, it will often be found migrating with a vacancy in the form of a ($Si_{Ga}\cdots V_{Ga}$) complex, which also helps provide another mechanism for the compensation mentioned above.

For practical purposes of estimating diffusion profiles, it is important to know what the concentration dependence of the dopant diffusion is. A comparison of the column IV dopants in GaAs found that good fits to the experimental profiles for Sn, Ge and Si could be obtained using only the quadratic term for the effectivity diffusivity [33]:

$$D_{\text{eff}} = D^=\left(\frac{n}{n_i}\right)^2. \tag{7.69}$$

The quadratic dependence on n suggests that the dominant vacancy is the doubly charged one. There are some conflicting theoretical calculations and Al–Ga superlattice interdiffusion data that suggest a triply charged Ga vacancy. However, a $(n/n_i)^3$ concentration dependence for the diffusion coefficient gives far more abrupt profiles than are actually observed. There are many open questions in the GaAs literature, but it is obvious that the basic physical ideas regarding point defect and dopant interactions are central to an understanding of diffusion in GaAs. Many of the apparent anomalous effects can be explained by understanding the evolution of point defect profiles, e.g., anomalous uphill diffusion due to gradients for point defects [21] or the effect of implant temperature on diffusion of Si [22]. An example of the fit to Si implants is shown in Figure 7.34, showing the characteristic box-shaped profiles due to concentration-dependent diffusion. A compilation of the diffusion data is shown in Table 7.5 and in Figure 7.35.

Table 7.5 **Diffusivities of the common N-type and P-type dopants in GaAs [23, 27, 33].**

N-type dopants in GaAs	$D^=$ (cm^2 s^{-1})	E_A (eV)
Sn	8×10^2	4.1
Ge	2×10^{-3}	2.9
Si	4×10^{-2}	3.3
P-type dopants in GaAs	D^{++} (cm^2 s^{-1})	E_A (eV)
Zn	0.6	3.21
Be	35	3.86

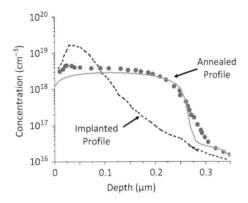

Figure 7.34 SIMS profiles (blue dots) and square-law-dependent fit to the profile based on (7.69) for a 1×10^{14} cm^{-2}, 40 keV Si implant into a semi-insulating GaAs substrate, annealed for 45 min at 900 °C. Based on Silvaco Athena simulations and data from [33].

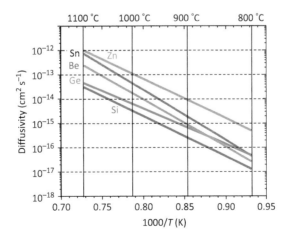

Figure 7.35 Compilation of diffusivity for N-type (Sn, Ge, Si) and P-type (Zn, Be) dopants in GaAs (from Table 7.5).

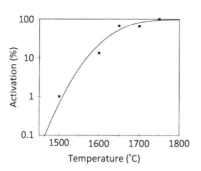

Figure 7.36 Plot of percentage activation for implanted boron versus anneal temperature, showing the very high temperatures required for substantial activation. After [34].

7.17.4 Diffusion and Activation in SiC

Because of very strong bonding in SiC, making it one of the hardest compound materials, it is practically impossible to diffuse anything into SiC. Instead, dopants need to be implanted or grown into the material. Aluminum and boron are commonly used for P-type dopants, with acceptor levels approximately 230–300 meV above the valence band. Nitrogen is commonly used as an N-type dopant, with an energy level 50–100 meV below the conduction band.

Very high annealing temperatures are needed to activate a substantial fraction of the dopants in SiC, as shown in Figure 7.36. Typically, temperatures around 1700 °C are commonly used. At higher temperatures, the SiC begins to sublime and the surface becomes rough. Even at these extremely high temperatures, there is negligible dopant diffusion in SiC, and the dopant profile is determined by the implanted profile. However, this does allow for selective area doping, which is essential for producing devices in SiC.

7.18 Doping Methods

Most workers believe that the dominant doping method used today – ion implantation – will continue to dominate in the future. Ion implantation affords tremendous flexibility in terms of the doping species, in the physical location of the dopants and in manufacturing throughput. Deeper doped regions in device structures are straightforward to form by moderate- or high-energy implants. The very shallow junctions required in future scaled devices can be formed by very-low-energy implants. The only real drawback to this technology is the damage produced by the implantation process, and the transient diffusion and extended defects that result when implants are annealed. In order to minimize these effects, rapid thermal processing is likely to dominate annealing and diffusion processing in the future. We will discuss ion implantation and some of the issues associated with extending it to future device fabrication in Chapter 8.

A number of alternative doping methods are currently being investigated. These methods range from very old methods, like out-diffusion from deposited thin films, to very new processes, like laser-assisted annealing or doping. Because the fins in the FinFET structure described in Chapter 2 are so thin and so closely spaced, some of the older techniques such as doping from deposited silicon layers are coming back into fashion.

7.19 Summary of Key Ideas

Doping of semiconductors is the fundamental process by which devices are fabricated. To make dopants electrically active and therefore useful in devices, they must sit on substitutional lattice sites. Both the concentration and the spatial variation (profile) of these dopants are critical to device performance. Diffusion is thus the basic process by which we produce these doping profiles.

Ion implantation is the dominant method used today to introduce dopants into silicon. This process allows precise control of the numbers of doping atoms and their physical position. Annealing and sometimes additional diffusion are required following implantation in order to allow the dopants to diffuse to their final, substitutional lattice sites. For short times, this process is dominated by transients associated with the annealing of implantation damage. During this period, the dopant diffusivity is much larger than its equilibrium value because of the high concentration of point defects present. Following this transient, the dopant diffusivity drops to a more normal value.

Dopants diffuse in silicon by interacting with point defects through a number of possible atomic-scale mechanisms. The dopant diffusivity is basically proportional to the number of such defects present in a crystal. Point defect concentrations depend exponentially on temperature and, as a result, dopant diffusivities do also. However, point defect concentrations are also changed by Fermi-level effects, by ion implantation damage and by surface processes like oxidation. The result is that dopant diffusivities may vary spatially and with time during the diffusion of a particular doped region. This greatly complicates the modeling of dopant profiles and generally requires that numerical simulation tools be used to calculate such profiles. Only in the simplest cases in which D is constant can simple analytical solutions be used to calculate impurity profiles.

Much progress has been made in recent years in developing physically based models for dopant diffusion and in implementing these models in simulators. These simulators today allow full coupling of the point defects and the dopants and can accurately predict very complex experimental dopant profiles in silicon. Many complex insights such as point defect gradient-driven uphill diffusion and chemical pumping of point defects appear as straightforward consequences of these physically based, atomistic models of dopant diffusion. The same basic diffusion physics applies to other compound semiconductor materials, and we have shown some results for GaAs that have analogies to diffusion in silicon.

While we have focused our discussion on the issues associated with CMOS technology, these tools are also used in many other types of silicon device structures. In many cases, they are crucial in reducing the number of experiments required to develop new processes. They are also very useful in understanding process sensitivities and manufacturing tolerances.

7.20 REFERENCES AND NOTES

[1] R. H. Dennard, F. H. Gaensslen, V. L. Rideout, E. Bassous and A. R. LeBlanc, "Design of ion-implanted MOSFET's with very small physical dimensions," *IEEE Journal of Solid-State Circuits*, vol. 9, no. 5, pp. 256–268, 1974.

[2] P. A. Packan, "Pushing the limits," *Science*, vol. 285, no. 5436, pp. 2079–2081, 1999.

[3] F. A. Trumbore, "Solid solubilities of impurity elements in germanium and silicon," *Bell System Technical Journal*, vol. 39, no. 1, pp. 205–233, 1960.

[4] J. C. Irvin, "Resistivity of bulk silicon and of diffused layers in silicon," *Bell System Technical Journal*, vol. 41, no. 2, pp. 387–410, 1962.

[5] SUPREM-IV (Stanford University Process Engineering Model) is a computer program developed by M. Law and C. Rafferty.

[6] Sentaurus Process is a simulator from Synopsys Inc.

[7] Athena and Victory Process are simulators from Silvaco Inc.

[8] J. Crank, *The Mathematics of Diffusion*, Oxford University Press, 1979.

[9] P. Griffin, S. Crowder and J. Knight, "Dose loss in phosphorus implants due to transient diffusion and interface segregation," *Applied Physics Letters*, vol. 67, no. 4, pp. 482–484, 1995.

[10] R. Kasnavi, P. B. Griffin and J. D. Plummer, "Dynamics of arsenic dose loss at the SiO_2 interface during TED," in *Simulation of Semiconductor Processes and Devices 1998*, Springer, pp. 48–50, 1998.

[11] L. Darken and R. Gurry, *Physical Chemistry of Metals*, McGraw-Hill, p. 443, 1953.

[12] R. Fair, "Concentration profiles of diffused dopants in silicon," in *Impurity Doping Processes in Silicon*, ed. F. F. Y. Wang, Materials and Processing Theory and Practices, vol. 2, North-Holland, 1981.

[13] S. Hu, "Formation of stacking faults and enhanced diffusion in the oxidation of silicon," *Journal of Applied Physics*, vol. 45, no. 4, pp. 1567–1573, 1974.

[14] P. M. Fahey, P. Griffin and J. Plummer, "Point defects and dopant diffusion in silicon," *Reviews of Modern Physics*, vol. 61, no. 2, p. 289, 1989.

[15] R. B. Fair, "Explanation of anomalous base regions in transistors," *Applied Physics Letters*, vol. 22, no. 4, pp. 186–187, 1973.

[16] C. Rafferty, H.-H. Vuong, S. Eshraghi, M. Giles, M. Pinto and S. Hillenius, "Explanation of reverse short channel effect by defect gradients," in *Proceedings of IEEE International Electron Devices Meeting*, IEEE, pp. 311–314, 1993.

[17] A. Ural, P. B. Griffin and J. D. Plummer, "Self-diffusion in silicon: similarity between the properties of native point defects," *Physical Review Letters*, vol. 83, no. 17, p. 3454, 1999.

[18] S. K. Ghandhi, *VLSI Fabrication Principles: Silicon and Gallium Arsenide*, Wiley, 2008.

[19] T. Tan, U. Gösele and S. Yu, "Point defects, diffusion mechanisms, and superlattice disordering in gallium arsenide-based materials," *Critical Reviews in Solid State and Material Sciences*, vol. 17, no. 1, pp. 47–106, 1991.

[20] H. Bracht, S. Nicols, W. Walukiewicz, J. P. Silveira, F. Briones and E. Haller, "Large disparity between gallium and antimony self-diffusion in gallium antimonide," *Nature*, vol. 408, no. 6808, pp. 69–72, 2000.

[21] Y. M. Haddara, C. C. Lee, J. C. Hu, M. D. Deal and J. C. Bravman, "Modeling diffusion in gallium arsenide: recent work," *MRS Bulletin*, vol. 20, no. 4, pp. 41–50, 1995.

[22] H. Robinson, T. Haynes, E. Allen, C. Lee, M. Deal and K. Jones, "Effect of implant temperature on dopant diffusion and defect morphology for Si implanted GaAs," *Journal of Applied Physics*, vol. 76, no. 8, pp. 4571–4575, 1994.

[23] M. P. Chase, M. D. Deal and J. D. Plummer, "Diffusion modeling of zinc implanted into GaAs," *Journal of Applied Physics*, vol. 81, no. 4, pp. 1670–1676, 1997.

[24] S. Yu, T. Tan and U. Gösele, "Diffusion mechanism of zinc and beryllium in gallium arsenide," *Journal of Applied Physics*, vol. 69, no. 6, pp. 3547–3565, 1991.

[25] W. D. Laidig, N. Holonyak Jr., M. D. Camras, *et al.*, "Disorder of an AlAs–GaAs superlattice by impurity diffusion," *Applied Physics Letters*, vol. 38, no. 10, pp. 776–778, 1981.

[26] D. G. Deppe and N. Holonyak Jr., "Atom diffusion and impurity-induced layer disordering in quantum well III–V semiconductor heterostructures," *Journal of Applied Physics*, vol. 64, no. 12, pp. R93–R113, 1988.

[27] J. Hu, M. Deal and J. Plummer, "Modeling the diffusion of implanted Be in GaAs," *Journal of Applied Physics*, vol. 78, no. 3, pp. 1606–1613, 1995.

[28] M. D. Deal and H. G. Robinson, "Diffusion of implanted beryllium in gallium arsenide as a function of anneal temperature and dose," *Applied Physics Letters*, vol. 55, no. 10, pp. 996–998, 1989.

[29] W. Walukiewicz, "Intrinsic limitations to the doping of wide-gap semiconductors," *Physica B: Condensed Matter*, vol. 302, pp. 123–134, 2001.

[30] M. E. Greiner and J. F. Gibbons, "Diffusion of silicon in gallium arsenide using rapid thermal processing: experiment and model," *Applied Physics Letters*, vol. 44, no. 8, pp. 750–752, 1984.

[31] S. Yu, U. M. Gösele and T. Y. Tan, "A model of Si diffusion in GaAs based on the effect of the Fermi level," *Journal of Applied Physics*, vol. 66, no. 7, pp. 2952–2961, 1989.

[32] M. Reveil and P. Clancy, "Resolving the mystery of the concentration-dependence of amphoteric dopant diffusion in III–V semiconductors," *Acta Materialia*, vol. 186, pp. 555–563, 2020.

[33] E. Allen, J. Murray, M. Deal, J. Plummer, K. Jones and W. Rubart, "A comparison of the diffusion behavior of ion-implanted Sn, Ge, and Si in gallium arsenide," *Journal of the Electrochemical Society*, vol. 138, no. 11, p. 3440, 1991.

[34] M. Capano, S. Ryu, M. Melloch, J. Cooper and M. Buss, "Dopant activation and surface morphology of ion implanted 4H- and 6H-silicon carbide," *Journal of Electronic Materials*, vol. 27, no. 4, pp. 370–376, 1998.

7.21 PROBLEMS

7.1 A resistor for an analog integrated circuit is made using a layer of deposited polysilicon 0.5 μm thick, as shown below.

(a) The doping of the polysilicon is $1 \times 10^{16}\,\mathrm{cm}^{-3}$. The carrier mobility $\mu = 100\ \mathrm{cm^2\ V^{-1}\ s^{-1}}$ is low because of scattering at grain boundaries. If the resistor has $L = 100$ μm and $W = 10$ μm, what is its resistance in ohms?

(b) A thermal oxidation is performed on the polysilicon for 2 h at 900 °C in H_2O. Assuming B/A for polysilicon is two-thirds that of $\langle 111 \rangle$ silicon, what is the polysilicon thickness that remains.

(c) Assuming that all of the dopant remains in the polysilicon (i.e., does not segregate to oxide), what is the new value of the resistance in (a). Assume the mobility does not change.

7.2 A resistor is made as part of a high-frequency analog integrated circuit, as shown below. The N^- epilayer forms the body of the resistor. If the width of the resistor in the direction into the paper is 2.5 μm, what should the length X be to give a resistor of approximately

50 kΩ. The epilayer is doped with phosphorus at a concentration of $1 \times 10^{15}\,\text{cm}^{-3}$ and is 30 µm thick.

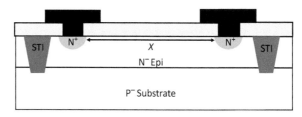

7.3 A P-type (boron) diffusion is performed as follows: predeposition, 30 min, 900 °C, solid solubility; drive-in, 60 min, 1000 °C.
  (a) What is the deposited Q?
  (b) If the substrate is doped $1 \times 10^{15}\,\text{cm}^{-3}$ phosphorus, what is x_j?
  (c) What is the sheet resistance of the diffused layer?
  Use the analytic solutions to Fick's law to solve this problem but state any assumptions you make and explain why they are reasonable (or not).

7.4 Suppose we perform a solid-solubility-limited predeposition from a doped glass source which introduces a total of Q impurities per square centimeter.
  (a) If this predeposition was performed for a total of t minutes, how long would it take (total time) to predeposit a total of $3Q$ impurities/cm^2 into a wafer if the predeposition temperature remained constant.
  (b) Derive a simple expression for the $(Dt)_{\text{drive-in}}$ which would be required to drive the initial predeposition of Q impurities/cm^2 sufficiently deep so that the final surface concentration is equal to 1% of the solid solubility concentration. This can be expressed in terms of $(Dt)_{\text{predep}}$ and the solid solubility concentration C_S.

  Use the analytic solutions to Fick's law to solve this problem but state any assumptions you make and explain why they are reasonable (or not).

7.5 A diffused region is formed by an ultra-shallow implant followed by a drive-in. The final profile is Gaussian. Derive a simple expression for the sensitivity of x_j to the implant dose Q. Is x_j more sensitive to Q at high or low doses?

7.6 From a process control point of view, predeposition times >10 min are required. From an economic point of view, times <10 h are required. Equipment limitations restrict 700 °C < T < 1200 °C. You need only consider simple Gaussian- and erfc-type profiles in this problem.
  (a) Is it possible to dope a MOS channel region by predeposition to shift threshold voltages? The required dose is $5 \times 10^{11}\,\text{cm}^{-2}$ boron. Assume that boron's electrically active solubility is $\approx 4 \times 10^{19}\,\text{cm}^{-3}$ at the lowest allowed temperature.
  (b) What would be a reasonable schedule (T, t) for a MOS source/drain predeposition? The required dose is $1 \times 10^{15}\,\text{cm}^{-2}$ arsenic. The junction depth cannot be deeper than 0.2 µm for device reasons. Assume the substrate doping is $1 \times 10^{18}\,\text{cm}^{-3}$.

Use the analytic solutions to Fick's law to solve this problem but state any assumptions you make and explain why they are reasonable (or not).

7.7 A boron diffusion is performed in silicon such that the maximum boron concentration is $1 \times 10^{18}\,\mathrm{cm}^{-3}$. For what range of diffusion temperatures will electric field effects and concentration-dependent diffusion coefficients be important?

7.8 An N^+ region is formed in a P^- substrate ($10^{16}\,\mathrm{cm}^{-3}$ doping) with a junction depth x_j as shown below. For the device being fabricated, it is important to minimize the sheet resistance of the N^+ region.

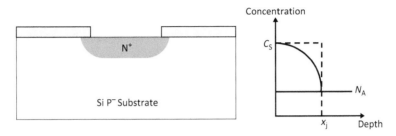

(a) If the ideal "box" profile shown (dashed line) could be obtained, it would provide an absolute minimum limit on ρ_s. Derive an approximate expression for this lower bound on ρ_s.

(b) If As had been used as the dopant and a metastable doping at the solid solubility of $2 \times 10^{21}\,\mathrm{cm}^{-3}$ was obtained in a box-shaped profile by, for example, laser melting, estimate the absolute minimum ρ_s for an x_j of 0.1 μm if the limiting carrier mobility in N^+ silicon is 85 cm^2 V^{-1} s^{-1}.

(c) If a normal error-function profile was used with the surface at the metastable solid solubility, what value of ρ_s would be realized with an x_j of 0.1 μm?

(d) Repeat the calculation in (c) if the arsenic had deactivated to its normal "electrical solubility" in silicon.

7.9 A (111) silicon wafer which has an existing 0.5 μm SiO$_2$ layer on its surface is placed in an oxidation furnace at 1000 °C to grow a thicker oxide in an H$_2$O oxidizing ambient. We are interested in how long it takes for the oxidation to reach its steady-state value at 1000 °C.

(a) You may assume that the oxidant concentration (H$_2$O) is fixed at $C^* = 3 \times 10^{19}\,\mathrm{cm}^{-3}$ at the outer surface of the oxide, which means that F_1 in the Deal–Grove model can be neglected. When the oxidation first starts, the H$_2$O needs to diffuse through the existing SiO$_2$ layer to raise the H$_2$O concentration at the oxidizing interface to C_1. While this transient is occurring, the oxidation rate will be slower than the Deal–Grove model predicts. Using the Deal–Grove model, calculate the value of C_1 that must be reached to achieve the full oxidation rate at 1000 °C.

(b) Presuming that the H$_2$O concentration at the top surface of the oxide is fixed at $C^* = 3 \times 10^{19}\,\mathrm{cm}^{-3}$, and assuming that the diffusivity of the H$_2$O in SiO$_2$ is 2×10^3 μm^2 h^{-1} at 1000 °C, how long does it take after the wafer is placed in the furnace and the H$_2$O is switched on, before steady-state oxidation is achieved? You will have to choose one of the analytic diffusion profiles discussed in this chapter even though this problem does not really satisfy the boundary conditions of those solutions.

7.10 A bipolar transistor is fabricated by implanting the boron base with a dose of $Q_B = 2 \times 10^{13}$ cm^{-2}. Then an *in-situ* arsenic-doped polysilicon emitter region doped at 10^{20} cm^{-3} is deposited on the surface. A drive-in is performed for 60 min at 1100 °C. Assume that the implant can be treated as a delta function. The substrate is N-type doped at 10^{15} cm^{-3}. Calculate the base width of the final NPN bipolar transistor. You can neglect all second-order effects like concentration-dependent diffusion, $\mathcal{E}$-field effects, segregation, rapid diffusion in polysilicon, etc.

7.11 A special twin-well CMOS technology requires that the wells have precisely the same depth at the substrate concentration of 1×10^{15} cm^{-3}, with arsenic used for the N well and boron used for the P well. A shallow implant dose of 1×10^{14} cm^{2} is used for both, and the slow-diffusing arsenic is introduced first and partially driven-in. Then the boron is introduced, and the rest of the anneal is performed until both junctions reach 2.5 μm. Calculate both drive-in times, assuming 1100 °C is used for both drive-ins. Use analytic solutions to Fick's law and assume intrinsic diffusion coefficients apply.

7.12 A process engineer on the day shift started a boron isolation diffusion for a structure in which the boron diffusion needs to penetrate completely through a 6 μm thick N-type epitaxial layer that is lightly doped with phosphorus ($N_D = 1 \times 10^{15}$ cm^{-3}) on a P-type substrate ($N_A = 1 \times 10^{14}$ cm^{-3}). The purpose of the diffusion is to provide isolation between different N-type regions. The day shift engineer left no information on what he did. On a monitor wafer, you put probes on adjacent N-type regions that are supposed to be isolated by a P$^+$ boron diffusion between them. But they are still connected, so isolation has not been achieved. The P$^+$ region is 100 μm long by 5 μm wide, and you measure a resistance underneath this P region of 400 kΩ. The sheet resistance of the P$^+$ diffusion measured using a four-point probe is 250 Ω/square. If an additional drive-in is done at 1100 °C, what time is needed to complete the isolation diffusion?

7.13 Some advanced device structures have an epitaxial Ge layer as a buried channel in a silicon substrate (as shown below) because the mobility of holes in Ge is higher than that in Si. Unfortunately, if Ge atoms diffuse to the gate oxide interface, they would create traps at the interface, therefore degrading the device characteristics. In this device, the epitaxial Ge layer is 2 nm thick. If the diffusivity of Ge in silicon is $1535 \exp(-4.7 \text{ eV}/kT)$ cm^2 s^{-1}, how far must the buried layer be from the gate oxide interface to ensure that, after a 900 °C anneal for 30 min, not more than one atom in 100,000 at the silicon–gate oxide interface is a Ge atom? (The atomic density of Ge is 4.42×10^{22} cm^{-3}).

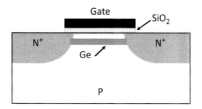

7.14 In a silicon semiconductor manufacturing plant, a cleaning solution is accidentally contaminated with Au. Not knowing this has happened, technicians in the factory continue to process wafers using this contaminated solution. The result of wafers being exposed to this cleaning solution is that each wafer has a monolayer of Au ($1 \times 10^{15}\,\text{cm}^2$) deposited on the top wafer surface. The wafers are 500 μm thick. The wafer processing includes using backside gettering, which means that the wafer backside incorporates traps that will capture any Au atoms that manage to diffuse to the backside. The subsequent wafer processing steps are equivalent to a thermal cycle of 60 min at 1000 °C. You may assume that, at this temperature, Au has a diffusivity of $10^{-5}\,\text{cm}^2\,\text{s}^{-1}$. What fraction of the Au deposited on the top surface will be trapped in the backside gettering region during the subsequent wafer processing? State any assumptions you make.

   Hint: Use the equation

$$\frac{1}{\sqrt{\pi D t}} \int_{x_1}^{\infty} \exp\left(-\frac{x^2}{4Dt}\right) dx = \text{erfc}\left(\frac{x_1}{2\sqrt{Dt}}\right).$$

7.15 A P$^+$ boron-doped Si substrate has a 2 μm thick epitaxial layer grown on it, as shown. A shallow surface boron implant ($1 \times 10^{14}\,\text{cm}^{-2}$) is performed. The energy is low, so this implant can be regarded as a delta function at the surface. The structure is then diffused for 30 min at 1000 °C. You may neglect any anomalous effects such as electric fields, concentration-dependent diffusion, etc. What is the remaining width of the N layer after the diffusion? Calculate a quantitative answer.

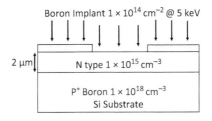

7.16 A bipolar transistor is fabricated by implanting the base and emitter regions and then driving them in together. Boron is used for the base and arsenic for the emitter. The implants are $Q_B = 2 \times 10^{13}\,\text{cm}^{-2}$ and $Q_E = 2 \times 10^{15}\,\text{cm}^{-2}$. The drive-in is 60 min at 1100 °C. Assume that the implants can be treated as delta functions at the surface. The substrate is $10^{15}\,\text{cm}^{-3}$ N type.
   (a) Calculate and plot the resulting impurity profiles. Assume that the B and As diffusivities are the intrinsic values from Table 7.2. You can neglect all second-order effects like concentration-dependent diffusion, $\mathcal{E}$-field effects, etc.
   (b) A parameter of interest in bipolar transistors is the base Gummel number, which is the total dose (atoms cm^{-2}) in the base region under the emitter. Estimate this parameter for the device in part (a).

7.17 All semiconductor devices are really "unstable" at device operating temperatures because dopants will continue to diffuse and hence, given long enough, profiles will change and devices will stop functioning correctly. How long would it take at a 150 °C operating temperature for the base width in the device in problem 7.10 to double? Assume the As does not move, just the boron moves.

7.18 A silicon wafer is uniformly doped with boron $(2 \times 10^{15}\,\text{cm}^{-3})$ and phosphorus $(1 \times 10^{15}\,\text{cm}^{-3})$ so that it is net P type. This wafer is then thermally oxidized to grow about 1 µm of SiO_2. The oxide is then stripped and a measurement is made to determine the doping type of the wafer surface. Surprisingly it is found to be N type. Explain why the surface was converted from P to N type.

   Hint: Consider the segregation behavior of dopants during oxidation.

7.19 An engineer wants to use analytical solutions to diffusion equations in a programmable calculator to make rapid estimates for process changes on junction depths. Consider the following possible diffusion regimes:
   (a) high temperatures,
   (b) low temperatures,
   (c) long times,
   (d) short times.

   Which of them are most appropriate for analytical solutions (i.e., which would minimize E-field or concentration-dependent effects)? Explain.

7.20 A shallow phosphorus implant with a dose of 10^{14} cm^2 is covered in some regions with a deposited layer of inert nitride. An anneal is performed at 1000 °C in dry O_2, resulting in local oxidation of the silicon surface. After the anneal, the junction depth of the phosphorus region below the original surface is measured in the inert region to be 0.5 µm and under the oxidizing region to be 1.2 µm. What is the diffusivity enhancement that the phosphorus in the oxidizing region experiences? If f_I for phosphorus is 0.9 and I–V recombination is fast at 1000 °C, what is the interstitial supersaturation that is generated by the oxidation? The silicon substrate doping is 10^{15} cm^3.

7.21 Rapid thermal annealing (RTA) systems are becoming common for activating dopants. Silicon has quite a high thermal diffusivity of 0.88 cm^2 s^{-1}, which describes how fast heat flows through silicon. Calculate the time required for a silicon wafer of thickness 500 µm to reach a constant temperature if a thin surface layer absorbs all the incident light and quickly reaches a steady-state temperature of T_S, i.e., when does the center reach $T/T_S = 0.5$, since the wafer is heated from both sides.

7.22 We know that thermal oxidation of silicon can cause OED. The mechanism was proposed to be injection of silicon interstitials into the silicon. If the SiO_2 layer were deposited by CVD rather than created by thermal oxidation, and the wafer were subsequently heated in an N_2 atmosphere to allow dopant diffusion, would you expect OED to occur? Explain.

7.23 An OED experiment is done with two different phosphorus buried layers. In the first wafer, the phosphorus doping is 10^{15} cm^{-3}. In the second wafer, the doping is 5×10^{19} cm^{-3}. On both wafers, oxide is grown at 1000 °C. The first wafer shows the expected OED, as illustrated below. What would you expect the second wafer to show (the one with the much higher phosphorus doping concentration)? Explain.

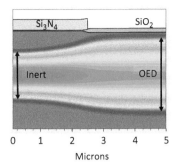

7.24 A simulation of OED is shown below using a standard LOCOS test structure. Assume that, in this simulation, a uniformly doped epitaxial layer was the starting point, and that the dopant in this layer diffused downwards during the OED experiment to produce the result shown. Calculate quantitatively the average diffusivity enhancement due to OED that this simulation shows. Explain any assumptions.

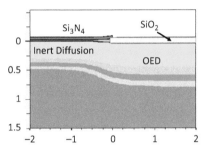

7.25 An enterprising graduate student designs an experiment to answer the basic question of how many interstitials are injected into the silicon substrate during oxidation. An SOI structure is used in which the silicon above the buried oxide is grown as ^{29}Si for most of the 10 μm SOI layer and a thin layer of ^{30}Si is grown at the surface, as shown. The idea is that, when the surface is oxidized, ^{30}Si atoms will be injected as interstitials. These will diffuse through the SOI layer to the buried oxide, where they will recombine by growing additional atomic layers of ^{30}Si on top of the buried oxide. By knowing how much SiO_2 was grown on the top surface and how many atomic layers of ^{30}Si formed on the buried oxide, the graduate student hopes to determine what fraction of the oxidized silicon atoms on the top were injected as interstitials.

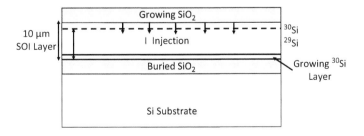

(a) Estimate the transit time for the injected interstitials across the SOI layer at 1000 °C.

(b) You should have found in part (a) that the interstitial transit time is fast compared to normal oxidation times. Suppose a surface oxidation is performed in dry O_2 at 1000 °C for 3 h. A measurement of the ^{30}Si layer grown on top of the buried oxide shows that 10 atomic planes of ^{30}Si were grown. Estimate the fraction of silicon atoms that are oxidized that end up injected as silicon interstitials. You can assume that silicon atomic planes are spaced 2.5 Å apart and that the silicon is (100).

7.26 In Figure 4.16 in Chapter 4, we saw that the Si self-diffusion coefficient is quite small (much smaller than dopant diffusion coefficients, for example). We also saw that the silicon interstitial diffusion coefficient is very fast, much faster than dopant diffusion coefficients. Explain physically why these two Si diffusion coefficients should be so different.

7.27 We used the equation $D_A^{\text{eff}} = D_A^* (f_I C_I / C_I^* + f_V C_V / C_V^*)$ to describe how dopant diffusivities are impacted when C_I or C_V changes. Consider OED, which we know injects interstitials and changes C_I. Thus the first term in the equation would be determined by the oxidizing surface. Describe two options for how we might deal with the second term in an OED situation. Explain physically and mathematically each of these options.

7.28 An experiment involving high-concentration phosphorus diffusion is done in order to better understand phosphorus diffusion mechanisms. In the experiment, the phosphorus profile is carefully analyzed to extract the effective phosphorus diffusion coefficient at each point on the profile. What is discovered is that D_P is proportional to n/n_i at each point along the profile. Given what you have learned about phosphorus diffusion, what can you say about the atomistic diffusion process that was likely dominant in this experiment? Explain your answer.

7.29 Figure 7.28 considered the example of the kinds of profile shapes resulting from concentration-dependent diffusion and the fully coupled diffusion model for a phosphorus implant. Explain physically in terms of point defects, in your own words, why the shapes shown in the diagram result from these two models.

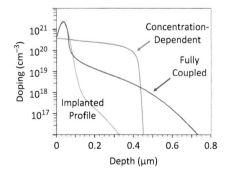

7.30 A special test structure is created as shown below. As and B implants are performed into a highly Sb-doped silicon wafer. The wafer is then annealed at 1000 °C. What would the Sb concentration need to be in order to have the As and B regions exhibit the same diffusion coefficient?

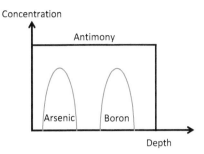

7.31 The diffusivity of a certain buried dopant is only 30% of its normal diffusivity in an inert ambient after a phosphorus implant is performed at the surface of a silicon wafer. The phosphorus diffusion coefficient is itself enhanced by a factor 4 (400%). Presumably, these effects are caused by the phosphorus injecting interstitials as it diffuses, as discussed in the text. Assuming that the phosphorus diffuses completely by an interstitial mechanism and that interstitial–vacancy recombination is very fast, calculate how much of the buried dopant diffusion is mediated by interstitials and how much by vacancies.

7.32 An OED experiment is done using the test structure shown below. The buried layer is a new dopant for which $f_I = f_V = 0.5$. The oxidation increases the interstitial concentration by a factor 2. Consider two cases: one in which interstitial–vacancy recombination is very fast, in which case $C_I C_V = C_I^* C_V^*$; the other in which there is no interstitial–vacancy recombination. Which case results in more OED enhancement of the buried layer? Calculate a quantitative answer.

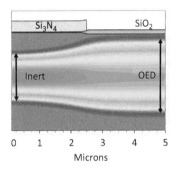

7.33 This problem concerns an OED experiment using the structure described in this chapter and shown below. In OED, interstitials are generated by the oxidation process. These interstitials diffuse into the silicon and enhance the diffusivity of any nearby dopants. Models for these processes have been developed, and in their simplest form can be represented as $C_I/C_I^* = K\sqrt{dx/dt}$, where K is a constant and dx/dt is the instantaneous oxidation rate. Suppose an oxidation is performed in dry O_2 on a (111) substrate at 1000 °C. Calculate and plot the time-dependent boron diffusivity in the buried marker layer. The total diffusivity would be the intrinsic diffusivity plus the OED term. You may assume $K = 10 \; (\text{min}/\mu\text{m})^{1/2}$.

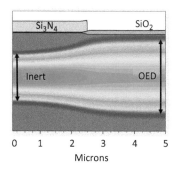

7.34 An NMOS transistor with a 0.1 μm channel length (see below) is operated in a very unusual environment. The transistor is operated at 0 °C and in a radiation environment that creates a steady-state interstitial concentration of 10^3 cm^{-3} throughout the silicon substrate. A student who has studied diffusion mechanisms is very worried that the excess interstitials will cause measurable dopant diffusion even at the 0 °C temperature and hence cause the transistor to fail in a short time. If the source/drain profiles can be regarded as As box profiles with a concentration of 10^{20} cm^{-3}, calculate the time required for the source/drain dopants to diffuse half-way through the channel from each side. If this happened, the entire channel would become N type and clearly the transistor would fail. The channel is P type, doped at 10^{17} cm^{-3}.

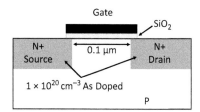

7.35 An experimental structure is fabricated as shown below. The boron and arsenic regions are marker layers to enable measurement of B and As diffusivities. The structure is heated to 1000 °C for 30 min and then the diffused profiles are measured to extract the dopant diffusivities. You may assume that there is no implant damage. Calculate the expected diffusivities for

(a) the boron,
(b) the As inside the phosphorus region, and
(c) the As beyond the phosphorus profile.

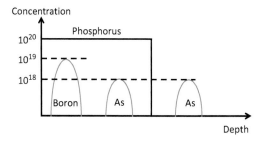

7.36 An unusual substrate doping profile is created by epitaxially growing P, B and then As layers with the doping concentrations shown below (the dashed straight-line structure). Three phosphorus implants are then done at different energies to position the phosphorus implants as shown. Each of the implants has a peak doping concentration of 10^{18} cm^{-3}. If the structure is heated to 1000 °C, the phosphorus implants may have different diffusivities. What is the ratio of the largest to smallest phosphorus implant diffusivity?

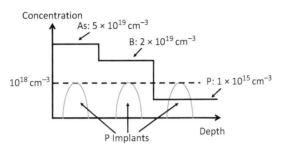

7.37 The simulations below show the channel doping profiles in a MOS transistor with and without electric field effects turned on during the source/drain diffusion. Clearly there is a big change in the boron channel region dopant profile. Electric field effects are due to the separation of the carrier and dopant distributions. The separation distance is actually very small and, in this example, the electric field effects extend for only a very small distance outside the black source/drain regions. So why do the changes in the boron dopant profile extend throughout the entire channel region? You do not need to calculate anything in this problem, but explain physically why the electric field effects affect the doping profile throughout the channel region.

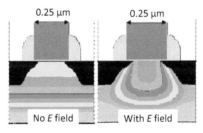

7.38 A diffusion experiment is carried out with the dopant profiles shown below. The As profile has a peak concentration of 10^{20} cm^{-3} and decreases as shown on either side of the peak. (Such an As profile could be obtained by epitaxially growing a Si layer and varying the doping during growth.) The phosphorus implant peaks at 10^{18} cm^{-3} and can be assumed to initially be a Gaussian profile. Consider only concentration-dependent diffusion effects in this problem (i.e., neglect electric field effects, full point defect coupling effects, etc.).

(a) A 1000 °C, 60 min annealing is done. At the beginning of the annealing, what is the largest value of the phosphorus diffusivity, and where does this value occur on the phosphorus profile?

(b) Sketch what you expect the shape of the As and P profiles will look like after the 60 min annealing. You do not have to calculate anything but you must explain your sketch.

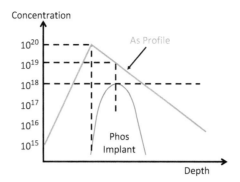

7.39 In a particular process flow, As and P regions are implanted into a Si wafer, as shown below. The implant dose is $1 \times 10^{13} \, \text{cm}^{-2}$ in both regions, and can be regarded as shallow compared to the final junction depth. The implants are then driven-in at $1100 \, °\text{C}$. Normally the P diffusion depth would be significantly deeper than the As diffusion depth because of the higher P diffusivity. In this particular application, it is important that the two junction depths are equal. An engineer at the company realizes that she can make the junction depths equal if she can adjust the interstitial concentration in the wafer during the diffusion. Assume interstitial and vacancy recombination is fast. Calculate what the required ratio of actual to equilibrium interstitial concentration (C_I/C_I^*) would need to be to make the junction depths the same.

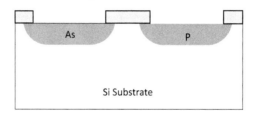

7.40 In a particular fabrication process, it is important that the diffusivities of As and B in the structure below are equal to each other. We have seen in this chapter that we can modify diffusivities by controlling surface injection of point defects. In this case, what type of surface injection would you use? If the As and B regions are diffused at $1000 \, °\text{C}$, calculate quantitatively what would be needed in terms of non-equilibrium I and V populations to make the D values match. Assume for this problem that recombination between I and V is very slow, i.e., does not happen.

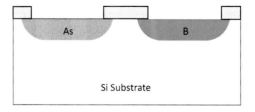

7.41 A bipolar junction transistor (BJT) is fabricated using a phosphorus emitter and a boron base. The emitter diffusion is at 1100 °C. As discussed in connection with Figure 7.28, the phosphorus emitter injects interstitials while it is diffusing, causing the "emitter-push" effect shown. The tail of the phosphorus diffusion will also be "pushed" by this same effect. Make a quantitative estimate as to whether the base width of the transistor will increase or decrease with time during the diffusion.

7.42 The emitter-push effect was illustrated in Figure 7.28 and discussed above in problem 7.41. Suppose you were making a PNP transistor with the base made with Sb as a dopant and the emitter made with B as a dopant. Would you expect a similar emitter-push effect? Explain.

8   Ion Implantation

8.1   Introduction

Ion implantation has been the dominant doping technique for silicon integrated circuits (ICs) and most other semiconductors for the past 45 years. It is expected to retain this position of dominance for the foreseeable future. In this process, dopant ions are accelerated to 0.1–1000 keV of energy and smashed into a crystalline semiconductor substrate, creating a cascade of damage that may displace hundreds or thousands of lattice atoms for each implanted ion. In this chapter, we will seek to understand how such an energetic and violent technique has become the dominant and preferred method of doping semiconductor wafers in manufacturing. At first glance, it seems that the technique would not be of much use in the precise art of fabricating integrated circuits. Indeed, although the original patent for ion implantation was issued to William Shockley in 1954, it was not until the late 1970s that ion implantation was used in manufacturing. To appreciate the power of the technique, we will investigate its ability to precisely control the distribution and dose of various dopants in semiconductors and examine how the implant damage is annealed and how the dopant is introduced to active sites in the silicon lattice.

8.2   Historical Development and Basic Concepts

Ion implantation provides a very precise means to introduce a specific dose or number of dopant atoms into a semiconductor substrate. Figure 8.1 shows the basic elements of a modern ion implantation machine. The system operates under high vacuum.

The basic requirement is a source of ions of sufficiently high density to be useful. Either a solid source that is vaporized or a gas source are conventionally used to deliver the ions to the ion implanter. The gas from the feed source is ionized by energetic electrons boiled off a hot filament or by a plasma discharge. The ions are extracted by a voltage bias on a grid and mass-analyzed to select only one ion species and perhaps even a single isotope of an ion species. This mass analysis is necessary because a gas like BF_2, for example, will dissociate into many ions such as B^{++}, B^+, F^+, BF^+ and BF_2^+ for both the boron-10 and boron-11 isotopes. The mass analysis relies on balancing the force exerted on the charged ions in a magnetic field with their centrifugal force, so that the path of the ion is given by the solution of the equation

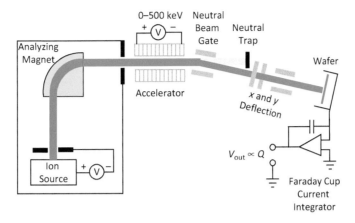

Figure 8.1 Schematic of an ion implanter.

$$\frac{mv^2}{r} = qvB,\qquad(8.1)$$

where m is the mass of the ion, q is the charge on the ion, v is the ion velocity, r is the radius of curvature and B is the magnetic field intensity which is perpendicular to the ion velocity. The ion velocity is related to the extraction voltage by

$$v = \sqrt{\frac{2E}{m}} = \sqrt{\frac{2qV_{\text{ext}}}{m}}.\qquad(8.2)$$

The field can be tuned to allow an ion of particular mass m to exactly follow an arc of radius r and exit the analyzer through a narrow slit.

For most situations, the selected ions are further accelerated in a small linear accelerator to the final energy of implantation. Some beam neutralization may occur during this acceleration phase and this is a problem, because neutral atoms cannot be electrostatically scanned or counted as ion current. They implant in the center of the wafer as an uncontrolled dose. For this reason, the ion path typically undergoes an electrostatic deflection from the linear path just before final implantation, which acts to trap the neutrals which continue undeflected. The ion beam may then undergo an x–y electrostatic deflection, which scans the beam onto the wafer, perhaps combined with a mechanical translation and scan of the wafer holder. Because charged ions are being implanted into the substrate, charging effects can become an issue, especially if part of the wafer is covered with insulating layers like SiO_2. To mitigate these issues, ion implanters typically include a supplemental flux of low-energy electrons from an electron flood system like a filament or plasma. Many practical issues in implanter design are discussed in [1].

The implant dose is measured by locating the sample at the end of a deep Faraday cup, which collects the current and integrates it over time. The normalized dose is given by

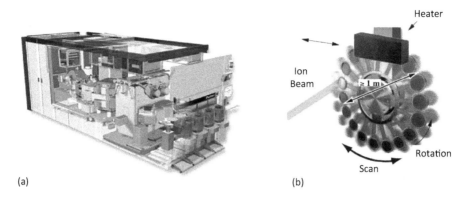

Figure 8.2 (a) Modern ion implanter (Axcelis Purion H Series). (b) Typical configuration for the end station wafer mounting system. Reprinted with permission of Hitachi Ltd., from [2].

$$Q = \frac{1}{A} \int \frac{I}{q} dt \quad \text{ions cm}^{-2}, \tag{8.3}$$

where I is the collected beam current, A is the implant area, t is the integration time and q is the charge on the ion. Modern ion implanters are designed to produce beam currents as high as 25 mA and to operate with accelerating voltages as high as 500 keV. Figure 8.2 shows an example of such a machine and a typical end station that holds the wafers. Especially in high-current implanters, the end station normally includes mechanical rotation of the wafers in order to decrease the average power density incident on each wafer, to reduce wafer heating. Figure 8.3 illustrates many of the applications of ion implantation in complementary metal-oxide–semiconductor (CMOS) processing.

In spite of the precision with which the dose can be controlled, at its heart ion implantation is a random process, because each ion follows a random trajectory, scattering off the lattice silicon atoms before losing its energy and coming to rest at some location, as illustrated in Figure 8.4. The ion tracks in Figure 8.4 were calculated using Monte Carlo methods that we will discuss in more detail in Section 8.7. Since large numbers of ions are implanted, an average depth in the z direction for the implanted dopants, which is called the projected range R_P, can be calculated.

The first step is to investigate the distribution of ions implanted at a given energy. Heavy ions like antimony do not travel as far in the crystal as light ions like boron. If different ions are implanted with the same energy, the heavy ions stop at a shallower depth, as shown in Figure 8.5. (We will understand the physics of this more deeply when we discuss nuclear and electronic stopping in Section 8.6.) The projected range R_P depends on the energy that is used for the implant, with higher energies giving a deeper range. While each implanted ion itself follows a random trajectory with a range R, on average the distribution of a large group of ions will peak at a projected depth R_P below the surface of the wafer. Because of the random nature of the process, some ions will stop sooner because of more collisions, while some will travel further. The spread of the ions depends on the range traveled, with deeper ranges allowing for more random stopping events. This gives rise to

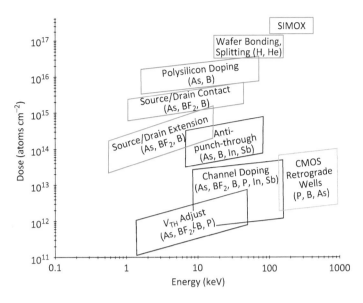

Figure 8.3 Typical ranges of implant dose and energy used in CMOS processing. Red indicates a high-current implanter. Blue indicates a high-voltage implanter. After [3].

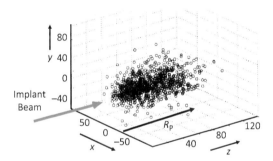

Figure 8.4 Monte Carlo simulations (using the UT-MARLOWE code [4]) of the 3D distribution of 1000 phosphorus ions implanted at 35 keV. The implant beam is centered at $(x, y) = (0, 0)$ and the axes are in units of nanometers.

a distribution of ions, where most of the ions are within a standard deviation $\pm\Delta R_P$ of the projected range. The heavy ions with a smaller range have a more narrow distribution than the light ions.

Since the number of ions implanted is usually greater than 10^{11} cm^{-2}, the distribution can be statistically described and is often modeled to first order by a symmetric Gaussian distribution given by

$$C(x) = \frac{Q}{\sqrt{2\pi}\Delta R_P} \exp\left(-\frac{(x - R_P)^2}{2\Delta R_P^2}\right) = C_P \exp\left(-\frac{(x - R_P)^2}{2\Delta R_P^2}\right), \tag{8.4}$$

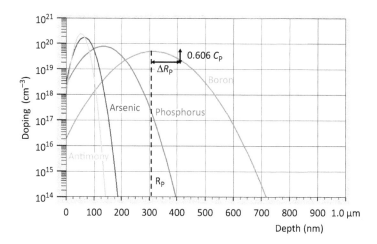

Figure 8.5 Athena simulations of 1×10^{15} cm^2, 100 keV implants into silicon [6]. The Si substrate is assumed amorphous; the profiles are Gaussian.

where R_P is the projected range normal to the surface, ΔR_P is the standard deviation or straggle about that range, Q is the dose and C_P is the peak concentration where the Gaussian is centered. Simulation tools such as Silvaco's Athena use look-up tables for R_P and ΔR_P when they are asked to construct a Gaussian profile for an implant such as shown in Figure 8.5. Table 8.1 lists values for these parameters for common dopants in silicon [5].

The total number of ions implanted Q is defined as the dose, and is simply

$$Q = \int_{-\infty}^{\infty} C(x) \, dx. \tag{8.5}$$

Making use of the fact that the sum (or integral) of Gaussian functions is an error function (Chapter 7), and using the formula that defines the error function gives

$$\int_{-\infty}^{\infty} \exp(-u^2) \, du = \frac{\sqrt{\pi}}{2}[\text{erf}(\infty) - \text{erf}(-\infty)], \tag{8.6}$$

and therefore

$$Q = \sqrt{2\pi}\Delta R_P C_P. \tag{8.7}$$

This provides a useful relationship between the dose and the peak concentration of the implant.

Modern simulation tools have built-in look-up tables for a variety of implanted ions, including Al, Sb, As, Be, B, BF$_2$, C, Cr, …. These ions can be implanted into a variety of substrate materials, including Si, Ge, GaAs, InP, InGaAs, polysilicon, SiC, SiO$_2$, photoresist, etc. [6]. Implanted profiles can also be calculated in multi-material substrates, as we will see later in this chapter.

Table 8.1 **Projected range R_P and standard deviation ΔR_P for common dopants in silicon [5]. The substrate is assumed to be amorphous.**

	Phosphorus		Arsenic		Antimony		Boron	
Energy (keV)	R_P (nm)	ΔR_P (nm)	R_P (nm)	ΔR_P (nm)	R_P (nm)	ΔR_P (nm)	R_P (nm)	ΔR_P (nm)
10	19.9	6.4	8.4	4.3	12.1	5.8	47.3	24.9
20	34.2	12.5	15.6	7.5	21.9	10.0	82.6	38.4
30	47.3	17.9	22.6	10.2	30.6	13.3	114	48.3
40	59.8	22.9	29.4	12.8	38.5	16.2	143	56.2
50	71.7	27.5	36.2	15.2	45.9	18.7	171	62.8
60	83.3	31.7	42.9	17.6	52.8	20.9	198	68.5
70	94.7	35.6	49.5	19.8	59.4	22.9	223	73.6
80	105	39.3	56.1	22.0	65.6	24.8	248	78.0
90	116	42.8	62.6	24.1	71.6	26.5	272	82.1
100	127	46.1	69.2	26.1	77.3	28.0	296	85.7
120	148	52.2	82.1	30.1	88.3	30.9	341	92.2
140	169	57.9	95.0	33.9	98.5	33.4	385	97.8
160	189	63.0	107	37.5	108	35.7	428	102
180	210	67.8	120	41.1	117	37.8	469	107
200	229	72.3	133	44.6	126	39.7	509	110

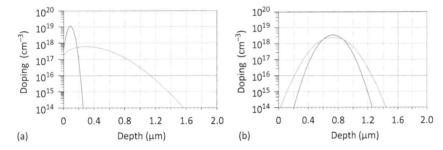

(a)

(b)

Figure 8.6 Broadening of an implanted profile by diffusion. (a) A shallow implant (1×10^{14} cm^{-2} at 25 keV) broadens into a one-sided Gaussian profile after diffusing (60 min at 1100 °C). (b) A deep Gaussian profile (1×10^{14} cm^{-2} at 300 keV) remains Gaussian after diffusion (60 min at 1000 °C). Athena simulations [6].

After ion implantation, it is common to diffuse the implanted ions to create the final implant profile desired. Based on the Gaussian approximation to the implanted profile, there are two simple analytic methods for calculating the shape of the final diffused profile. These are illustrated in the simulations in Figure 8.6.

In Figure 8.6(a), the implant is shallow enough and the diffusion is long enough that the implant can be treated as a delta function at the surface. As we saw in Chapter 7, a delta function broadens into a one-sided Gaussian profile, so that the profile is described by

$$C(x,t) = \frac{Q}{\sqrt{\pi Dt}} \exp\left(-\frac{x^2}{4Dt}\right) = C(0,t) \exp\left(-\frac{x^2}{4Dt}\right). \tag{8.8}$$

It should be noted in this simulation that the near-surface region of the profile after the diffusion (green curve in Figure 8.6(a)) is not exactly a one-sided Gaussian. This is because of surface segregation effects that affect the near-surface profile, as discussed in Chapter 7 (Section 7.11). Thus (8.8) should be used with some caution. Numerical simulators such as were used in Figure 8.6 can, of course, easily include these effects.

In Figure 8.6(b), the implant is deep enough that it can be described by a two-sided Gaussian after the implant. Such a profile when it broadens due to diffusion will remain a two-sided Gaussian and is described mathematically by

$$C(x,t) = \frac{Q}{\sqrt{2\pi(\Delta R_P^2 + 2Dt)}} \exp\left(-\frac{(x - R_P)^2}{2(\Delta R_P^2 + 2Dt)}\right). \tag{8.9}$$

For the implanted profile, $Dt = 0$, and the width of the profile is described by ΔR_P. With diffusion, the width increases $(\Delta R_P^2 + 2Dt)$ and the peak concentration decreases in the same way. We can simply add successive Dt cycles to predict how a diffused profile continues to evolve with time (Section 7.6 in Chapter 7).

We conclude this introductory section by briefly returning to the issue of ion implantation damage. The basic issue is illustrated in Figure 8.7 in a simple 2D picture. The implantation process in Figure 8.7(a) damages the crystalline substrate because the implanted ions slow down partially through collisions with substrate atoms. We will study this process in detail later in this chapter, but this collision process typically displaces the lattice atoms from their lattice sites and, if the damage is severe enough, can completely amorphize the substrate. This creates a material that is unsuitable for semiconductor devices. As a result, we will pay careful attention later in the chapter to how we can repair this damage and at the same time activate the

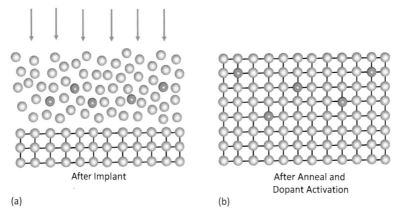

Figure 8.7 Illustration showing the impact of implantation damage and the need for an anneal and repair process.

implanted dopants by having them sit on lattice sites as shown in Figure 8.7(b). The fact that this repair and activation process works well has been critical to the widespread adoption of ion implantation.

A final point is perhaps best made by referring back to Figure 3.12 in Chapter 3. In compound materials, including semiconductors, the implant damage process also affects local stoichiometry because the compound elements can be scattered differently by the incoming ions. This can complicate damage repair in these materials, as we will see later in this chapter.

8.3   More Accurate Analytic One-Dimensional Implant Profiles

A large amount of experimental data has clearly shown that a simple symmetric Gaussian model of implanted profiles is an inadequate representation of real implanted profiles. We will discuss in this section the reasons why this is true and explore more sophisticated 1D representations of implanted profiles.

The first issue we will consider is that experimental implant profiles are often asymmetric. Light ions tend to have profiles skewed towards the surface, and heavy ions tend to have profiles skewed towards the deeper part of the profile. We can understand this physically by considering how light ions like boron scatter off silicon atoms compared with how heavy atoms like antimony scatter off silicon atoms. There is a greater tendency for light ions to backscatter and fill in the front side of the distribution and vice versa for heavy ions. We will discuss stopping mechanisms in more detail in Section 8.6.

An arbitrary probability distribution can be described by a series of moments. The normalized first moment is the projected range and is given by

$$R_P = \frac{1}{Q} \int_{-\infty}^{\infty} x C(x) \, dx.$$

(8.10)

As we saw in the Gaussian distribution discussed earlier, this is simply the mean value of the distribution, R_P in our case.

The second moment is the straggle or standard deviation, given by

$$\Delta R_P = \sqrt{\frac{1}{Q} \int_{-\infty}^{\infty} (x - R_P)^2 C(x) \, dx}.$$

(8.11)

Here ΔR_P is a measure of the width of the probability distribution, as we also saw in the example of the Gaussian distribution discussed earlier.

The third moment describes the skewness and is given by

$$\gamma = \frac{\int_{-\infty}^{\infty} (x - R_P)^3 C(x) \, dx}{Q \Delta R_P^3}. \tag{8.12}$$

Here γ is a measure of the symmetry of the distribution around the mean value. In the example we considered earlier, in which light ions are backscattered, the distribution would have negative skewness. Heavy ions scattering forward produce a positive skewness in the distribution.

The fourth moment is the kurtosis, given by

$$\beta = \frac{\int_{-\infty}^{\infty} (x - R_P)^4 C(x) \, dx}{Q \Delta R_P^4}. \tag{8.13}$$

This moment is a measure of how much of the distribution is contained in the tail regions of the distribution. The kurtosis is positive if the tail regions are less populated than the Gaussian distribution; it is negative if the tail regions are more populated than the Gaussian distribution.

These four moments can be used to get a more accurate description of a wide range of implanted species. The integral representation of the moments is not particularly convenient, so in practice the equivalent moments are described by coefficients in a functional form known as Pearson's equation [7]. Look-up tables of the moments can then be used to generate a profile. This is the approach taken in simulation programs (see, for example, the Athena User Manual [6]). An example is shown in Figure 8.8 in which the B and Sb 100 keV implants discussed in Figure 8.5 are simulated again using both the Gaussian model and the Pearson model.

In the B case (light ion) in Figure 8.8, the Pearson distribution is clearly skewed towards the surface, as suggested above when we argued that light ions could "bounce off" the heavier lattice atoms in the substrate and therefore more heavily populate the shallow portion of the profile. The Sb profile shows the opposite behavior, with the Pearson distribution skewed towards the deeper part of the profile. This is again consistent with the simple idea that heavy ions will tend to scatter forwards as they slow down in the target material.

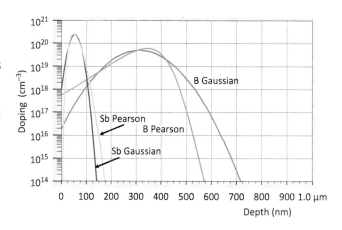

Figure 8.8 Athena simulation of 100 keV, 1×10^{15} cm^{-2} Sb and B implants modeled with Gaussian and Pearson distributions [6]. The Si substrate is assumed amorphous.

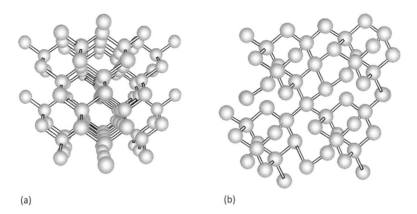

Figure 8.9 Structure of a silicon crystal looking down the 110 axial channels (a) and with a tilt and rotation to simulate a "random" direction (b).

In addition to the effects described above, there is an additional critical property of semiconductor substrates that we have ignored to this point in modeling implant profiles. Semiconductor substrates are crystalline, with a regular array of lattice atoms. This leads to the existence of planar and axial channels that can have quite dramatic effects on the implant profile. Figure 8.9 shows two images of a silicon crystal. In Figure 8.9(a), the crystal is aligned so that the 110 axial channels are apparent. In Figure 8.9(b), the crystal has been tilted and rotated so it appears to be a random arrangement of atoms. It is easy to imagine that an incoming ion beam normal to these images might produce very different profiles in the two cases.

We will see in Section 8.6 that most of the scattering events that do occur are through relatively small angles, so that a tilt and rotation can minimize the number of ions that immediately find the relatively open channels in the silicon lattice. However, once an ion does enter a channel, the small-angle scattering events from the atoms that line the walls of the channel mean that the ion can be steered quite a long distance along the channel before coming to rest from the electronic drag forces or from a sharp collision that causes the ion to exit the channel. The effect of channeling on the implant profile is to cause a tail that continues much further than expected.

Figure 8.10 shows simulations of a series of P implants (various doses at 100 keV) into a silicon substrate that is aligned with the incoming beam (0° tilt and rotation, like the image in Figure 8.9(a)). The yellow curve in Figure 8.10 is the Gaussian model for the lowest dose implant $(1 \times 10^{13}$ cm$^{-2})$. The modeled profile including channeling effects is the red curve, a vastly different prediction! While 0° tilt and rotation is the worst possible case, the prediction is that a significant fraction of the P ions will channel to depths well beyond the Gaussian profile prediction. Clearly, such a result would have major consequences on device performance.

Simulation tools like Athena model channeling using a dual Pearson model. The profile is constructed by adding two Pearson distributions, each with its own moments determined from a look-up table. The two distributions describe the random part of the distribution similar to Figure 8.8 and the channeled part of the distribution. The moments in the look-up tables are

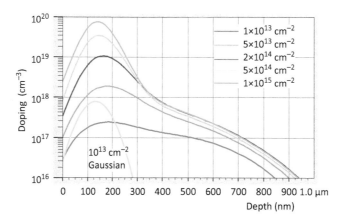

Figure 8.10 Athena simulations [6] of phosphorus profiles with different doses implanted at 100 keV into ⟨100⟩ crystalline silicon wafers with 0 ° tilt and rotation.

generally derived from experimental results, so they depend strongly on the tilt and rotation, which must be specified in the simulation.

Another important observation from the simulations in Figure 8.10 is that the impact of channeling seems to decrease as the implanted P dose increases. This agrees with experiments and is a result of the fact that, as the implant dose rises, the substrate Si atoms that are dislodged from their lattice sites by collisions gradually clog up the open channels and prevent further channeling from occurring. Simulators like Athena model this by using look-up table parameters that depend on dose.

Figure 8.11 shows a series of B implants simulated in Athena with the ion beam tilted at various angles with respect to the crystalline substrate. The black curve is a simulation using the Pearson distribution with the substrate assumed amorphous (no channeling). The black curve shows the skewness towards the surface expected for a light ion implant. Even with a tilt of 10°, there is still a significant effect of the channeling component on the final boron profile.

To attempt to minimize channeling, a thin screen oxide which is amorphous is often used, causing some randomization of the incident beam before it enters the lattice. But the results in Figure 8.11 suggest an alternative approach. If the substrate could be preamorphized before the implant, channeling should be essentially eliminated. In fact, this can be done by using a high-dose Si implant before the dopant implant. This option is actually used in manufacturing in situations in which it is critical to control the dopant profile and to eliminate channeling. We will consider preamorphization more carefully in Section 8.7 when we discuss Monte Carlo modeling of implant profiles. Of course, this is only possible if we really have a means to repair the damage in the substrate, a topic we will consider in detail later in this chapter.

Though tabulating all possible parameters that affect implant profiles is a tedious task, it does provide simulation programs with the ability to use a look-up table of moments and to simulate the shape of a profile for various implant conditions. Because of extensive characterization using secondary ion mass spectrometry (SIMS), most implanted profiles can be well modeled in silicon and other materials today, although, not surprisingly, the experimental data are more extensive in silicon compared to most other materials.

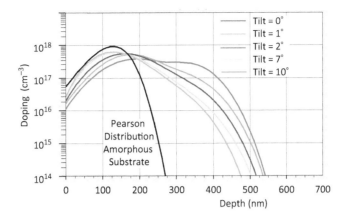

Figure 8.11 Athena simulations [6] of boron profiles (1×10^{13} cm^2 at 35 keV) implanted with the beam at various tilt angles with respect to the substrate.

8.4   Multilayer Implants

Most applications of ion implantation involve implanting into a multilayer structure. Implants into silicon or other semiconductors, such as the examples we have considered up to this point, are usually done with a thin screening SiO$_2$ or other layer on the surface in order to minimize channeling. Most implants are also masked in parts of the target wafer using photoresist or deposited thin films. We saw many examples of this in the CMOS process flow in Chapter 2.

The general situation is illustrated in Figure 8.12. The right-hand side corresponds to the situation we have discussed to this point, direct implantation into the substrate. The analytic methods we have discussed (Gaussian, Pearson, dual Pearson) can also be applied on the left-hand side. However, in general, the moments for these distributions will be different in the different materials, and the overall profile will have to be constructed with this in mind.

Simulation tools like Athena use a variety of methods to construct multilayer profiles, but the simplest and most common is the dose matching method [8]. This method works as follows. First, the profile in material 1 is calculated using any of the analytic methods we have discussed and tabulated moments for material 1. The number of ions that stop in material 1 is then given by (see (8.5))

$$Q_1 = \int_0^{t_1} C_1(x)\ \mathrm{d}x. \tag{8.14}$$

Next, a profile is calculated in material 2 assuming that the implant went directly into material 2. This is equivalent to replacing material 1 by material 2 in the first step. Then the thickness of material 2 that would contain the dose Q_1 is calculated. Assume this thickness is x'. The actual profile in material 2 is then calculated by assuming that material 2 extends from x' to $(x' + t_2)$. Of course, the appropriate moments for material 2 would be used in calculating this profile.

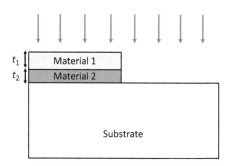

Figure 8.12 Implantation into a multilayer target.

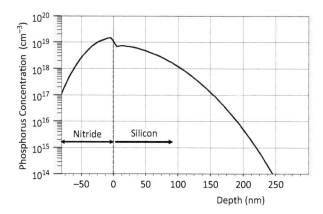

Figure 8.13 Athena simulation [6] of a 1×10^{14} cm^{-2}, 100 keV P implant into a multilayer structure (80 nm Si$_3$N$_4$ on a Si substrate). Substrate assumed amorphous.

This procedure can then be repeated for any other materials below material 2, including the substrate.

An example of a simulation using the dose matching method is shown in Figure 8.13. A P implant into a multilayer structure is shown. A dual Pearson analytic profile was calculated using moments appropriate for the two materials. Note that the dose matching approach can result in discontinuities at material interfaces, as seen in this example.

A multilayer implant situation of considerable importance involves masking of implants. The basic question is how large do t_1 and/or t_2 need to be in Figure 8.12 to block an implant intended to dope the substrate on the right-hand side of the figure? Any layer thick enough to capture the implanted ions can be used as a masking layer. For example, photoresist is often used as a convenient mask, because implants are usually performed at room temperature. To estimate the required mask thickness, we would obviously need to know the moments of the distribution function in the mask material. But even if we know these, how do we define masking?

Suppose that we need to do a 1×10^{15} cm^{-2}, 100 keV B implant into Si and we wish to mask this implant with a deposited SiO$_2$ layer. Figure 8.14 simulates this situation using Athena. An SiO$_2$ thickness of 600 nm (0.6 μm) was chosen for this example. In the Athena simulation,

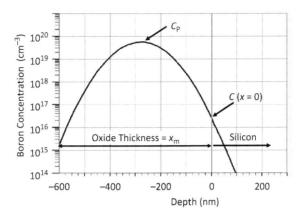

Figure 8.14 Athena simulation [6] of a $1 \times 10^{15}\ \text{cm}^{-2}$, 100 keV B implant into an SiO_2 mask on a Si substrate. Gaussian profile, Si assumed amorphous.

Gaussian profiles were assumed and the substrate Si was assumed amorphous. The tail region of the Gaussian decreases exponentially with distance and does not go to zero until $x = \infty$. So we need a definition of what "masking" means.

One reasonable definition could be that the thickness of the mask should be large enough that the concentration in the tail of the implant profile in the silicon does not exceed some specified background concentration, as illustrated in Figure 8.14. Then, the criterion for masking, assuming that the profile is Gaussian, is that C at $x = 0$ or $C(x_m)$ is

$$C(x_m) = C_P \exp\left(-\frac{(x_m - R_P)^2}{2\Delta R_P^2}\right) \leq C_B, \tag{8.15}$$

where x_m is the mask thickness, R_P and ΔR_P refer to the SiO_2 mask moments and C_B is some specified concentration in the substrate. Setting $C(x = 0) = C_B$ and solving for the mask thickness gives

$$x_m = R_P + \Delta R_P \sqrt{2\ln\left(\frac{C_P}{C_B}\right)} = R_P + m\,\Delta R_P, \tag{8.16}$$

where the parameter m indicates that the mask thickness should be equal to the range plus some multiple m times the standard deviation in the masking material.

In some situations, a different definition of "masking" is used. If the total number of ions that reach the substrate rather than the peak concentration is critical, then the dose Q_P that penetrates the mask is given by

$$Q_P = \frac{Q}{\sqrt{2\pi}\Delta R_P} \int_{x_m}^{\infty} \exp\left[-\left(\frac{x - R_P}{\sqrt{2}\Delta R_P}\right)^2\right]\ dx, \tag{8.17}$$

where Q is the total implanted dose, and the moments R_P and ΔR_P refer to the mask material. Since the integral or sum of Gaussian functions can be described as an error function, where

$$\int_{d}^{\infty} \exp(-u^2) \, du = \frac{\sqrt{\pi}}{2} \mathrm{erfc}(u), \tag{8.18}$$

the dose that penetrates is

$$Q_P = \frac{Q}{2} \mathrm{erfc}\left(\frac{x_m - R_P}{\sqrt{2}\Delta R_P}\right). \tag{8.19}$$

This masking criterion could be important if the region beneath the mask is the channel region of a metal–oxide semiconductor field effect transistor (MOSFET), since the total dose would affect the threshold voltage of the device.

Example

A process engineer building an N-channel metal-oxide–semiconductor (NMOS) device wants to dope the polysilicon gate at the same time as doing the arsenic source/drain diffusion. The source/drain implant dose is 2×10^{15} cm^{-2} at an energy of 50 keV.

(a) For these implant conditions, what is the minimum polysilicon thickness that can be used if the implant is not to exceed the channel doping, which is 1×10^{16} cm^{-3} near the surface (assume the gate oxide is negligibly thin compared to the polysilicon)?

(b) Assuming that this polysilicon thickness is actually used, how much of the implant dose will penetrate the polysilicon mask if the process engineer decides to change the implant energy to 80 keV?

Answer

(a) From Table 8.1, 50 keV arsenic has a range of 36.2 nm and a standard deviation of 15.2 nm, so the implant profile in the polysilicon is equal to the channel profile when

$$C(x_m) = 1 \times 10^{16} = \frac{Q}{\sqrt{2\pi}\Delta R_P} \exp\left[-\frac{(x_m - R_P)^2}{2\Delta R_P^2}\right]$$

$$= 1 \times 10^{16} = \frac{2 \times 10^{15}}{\sqrt{2\pi}(152 \times 10^{-8})} \exp\left[-\frac{(x_m - 362 \times 10^{-8})^2}{2(152 \times 10^{-8})^2}\right],$$

therefore giving

$$x_m = 10.7 \times 10^{-6} \mathrm{cm} = 0.107 \ \mu\mathrm{m}.$$

(b) If the implant energy is raised to 80 keV, the range from Table 8.1 is 56.1 nm and the standard deviation on the implant is 22 nm. The dose that penetrates the mask of thickness $x_m = 0.107$ μm is

$$Q_P = \frac{Q}{2} \, \text{erfc} \left[\frac{x_m - R_P^*}{\sqrt{2} \, \Delta R_P^*} \right]$$

$$= \frac{2 \times 10^{15}}{2} \, \text{erfc} \left[\frac{0.107 \times 10^{-4} - 561 \times 10^{-8}}{\sqrt{2}(220 \times 10^{-8})} \right]$$

$$= 1 \times 10^{15} [1 - \text{erf}(1.636)];$$

that is,

$$Q_P = 2.1 \times 10^{13} \, \text{cm}^{-2}.$$

8.5   Two- and Three-Dimensional Implants

Most applications of ion implantation involve masked implants which introduce the dopants only into localized regions of the wafer (see Figure 8.12). In these situations, the 2D or 3D profile that is produced is of great interest. One example is shown in Figure 8.15 in which As implants at 35 keV (Figure 8.15(a)) and 120 keV (Figure 8.15(b)) are shown, masked by the edge of a polysilicon gate structure. This is the kind of structure often encountered in MOS device fabrication, and we saw similar examples in Chapter 2.

The analytic descriptions we have discussed (Gaussian, Pearson, dual Pearson) can be extended to 2D or 3D profiles if we know the moments of these distributions in the lateral directions. For example, if we consider a Gaussian profile, the lateral distribution around the peak can also be statistically described by a Gaussian distribution with a lateral straggle ΔR_{lat} replacing the vertical straggle. The 2D distribution for each incoming vertical beam (red arrows in Figure 8.15) is often assumed to be composed of just the product of the vertical and lateral distributions, so that

$$C(x,y) = C_{\text{ver}}(x) \frac{1}{\sqrt{2\pi} \Delta R_{\text{lat}}} \exp\left(-\frac{y^2}{2 \Delta R_{\text{lat}}^2} \right). \tag{8.20}$$

This is often called the point spread function. Each of the incoming beams (red arrows) contribute to the total lateral profile under the mask edge, so the total profile is really a summation of equations like (8.20). We saw in Chapter 7 and in (8.19) earlier that the sum of Gaussian functions is an erfc profile, so, for a mask opening extending from $y = -a$ to $y = +a$, the lateral distribution under the mask edge is given by

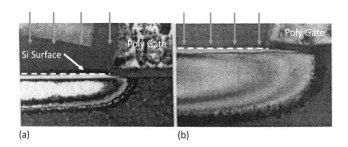

(a)   (b)

Figure 8.15 Cross-sectional images of As implants at (a) 35 keV and (b) 120 keV masked by the edge of a polysilicon gate region. The doped regions were delineated by etching the silicon. Reproduced from [9] with the permission of the American Vacuum Society.

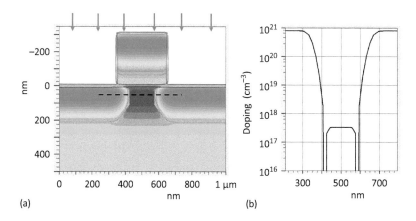

(a)   (b)

Figure 8.16 (a) Athena simulation of an As source/drain implant in a 0.3 μm gate-length MOS device. The implant was 5×10^{15} cm^{-2} at 100 keV and 0 ° tilt (red arrows). (b) The cutline shows the lateral doping profile at approximately the peak of the As vertical profile (dashed line in panel (a)).

$$
\begin{aligned}
C(x,y) &= C_{\mathrm{ver}}(x) \int_{-a}^{+a} \exp\left(\frac{y - y'}{2\,\Delta R_{\mathrm{lat}}}\right)\, \mathrm{d}y \\
&= C_{\mathrm{ver}}(x)\left[\mathrm{erfc}\left(\frac{y - a}{\sqrt{2\,\Delta R_{\mathrm{lat}}}}\right) - \mathrm{erfc}\left(\frac{y + a}{\sqrt{2\,\Delta R_{\mathrm{lat}}}}\right)\right].
\end{aligned}
\tag{8.21}
$$

Simulators like Athena use a variety of methods to construct 2D profiles from the analytic distributions we have discussed, including the simple erfc profile given by (8.21) (see [6], for example). An example is shown in Figure 8.16. In this simulation, a Pearson distribution was used, including channeling (crystalline substrate). The blue color contours in the device channel region in Figure 8.16(a) are the result of a channel B implant that was done in the simulation prior to the As implant.

Figure 8.17 shows a second example in which the same device structure as in Figure 8.16 is simulated, except that the As source/drain implant is done at an angle of 30 ° to the substrate.

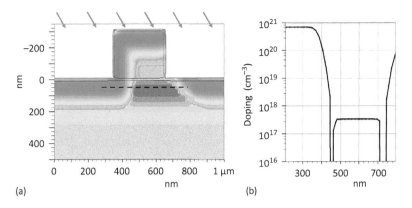

Figure 8.17 (a) Athena simulation of an As source/drain implant in a 0.3 μm gate-length MOS device. The implant was 5×10^{15} cm^{-2} at 100 keV and 30 ° tilt (red arrows). (b) The cutline shows the lateral doping profile at approximately the peak of the As vertical profile (dashed line in panel (a)).

The asymmetry of the implant is obvious in the simulation. This transistor structure would not work as shown because there is a separation between the drain N$^+$ region and the edge of the gate. Tilted implants like this are actually used in device fabrication. One example was discussed in Chapter 2 in which a "halo," i.e., a tilted implant, was used to locally boost the deeper channel concentration in order to reduce source/drain punch-through (Figure 2.11). Such a halo implant would be a lower dose than the example in Figure 8.17. It is obvious from Figure 8.17 that, if tilted implants are used, tilts of ±30 ° at a minimum would have to be used and perhaps other angles as well, to obtain symmetrical device characteristics, if a variety of device orientations were used in the layout.

8.6 Nuclear and Electronic Stopping

While the analytic descriptions of ion distributions we have discussed to this point are very useful, a complete description of the detailed atomic-scale events that take place along an ion trajectory is a more complete and often more useful way to describe implant profiles. In addition, atomistic mechanisms are required to understand and model implant damage, damage annealing and transient enhanced diffusion (TED). So, at this point, we turn our attention to a more atomistic view of the implantation process. Much of the theoretical description of the stopping of charged ions by matter came from original work on the structure of the nucleus. Following Ziegler [10], we provide a brief review of some of the interesting historical highlights.

Rutherford, in 1911, observed that about 0.01% of alpha particles were backscattered from thin aluminum foils, and he proceeded to show that these came from a single collision with a positively charged nucleus, thereby altering the prevailing view of the nature of matter. In the terminology of ion implantation, the alpha particle is just a doubly ionized helium atom, though the source of Rutherford's particles came from radioactive decay. Soon after, Bohr attempted one of the first unified theories of ion stopping by matter, and made the hypothesis that the energy loss of ions could be divided into two components: the nuclear energy loss to the positive atomic cores of the target, and the electronic energy loss to the free electrons in the target. The major unknown in calculating the energy loss was the effective charge state of a moving ion in the target. Bohr assumed that the ion's electrons with an orbital velocity less than the ion velocity would be stripped off, leaving a highly charged particle interacting with the orbiting target electrons, which could be treated as harmonic oscillators. Bethe and Bloch later introduced a quantum mechanical treatment of the electronic energy loss of a charged particle to a free electron gas, while Fermi considered how the charged particle would polarize the electron gas and alter the interaction. Fermi and Teller extended this work and found that the electronic energy loss would be directly proportional to the particle's velocity. The nuclear energy loss was treated as an elastic collision of the charged particle with the Coulomb field of a positively charged nucleus, partially screened by the outer valence electrons.

This historical work culminated in 1963 when Lindhard, Scharff and Schiøtt developed what has become known as LSS range theory [11]. This brought all the pieces together and allowed the stopping powers to be calculated for any arbitrary atomic species and elemental targets to within a factor of 2. The improvements in range calculations since then have come about from considering the shell structure of the atoms using quantum mechanical atomic structure calculations, by numerically calculating the path of the ion trajectory during interactions with a target atom rather than making asymptotic collision approximations, by improved effective charge approximations [12] and by considering the lattice structure of crystalline materials.

Computers today can follow the atomic-scale trajectory of a representative sample of tens or hundreds of thousands of ions through the lattice, accurately accounting for an arbitrary screen oxide thickness together with any tilt and rotation used during implantation. By combining the information on the final resting positions of the ions, a picture of the final distribution can be obtained. Because of the random nature of the starting condition, this approach is often called a Monte Carlo simulation. We will consider it in more detail in Section 8.7. In amorphous silicon, the position of the silicon atoms is also randomly chosen so the overall density is correct. Given the atomic density for the target, the average distance between atoms in an amorphous target is $(1/N)^{1/3}$, which gives the average path length between nuclear interactions in amorphous material.

In crystalline materials, the trajectory is completely determined by the structure of the lattice after the ion is set in motion. The ion scatters deterministically from target atoms through angles determined from classical two-body collision theory, slowing down by an additional drag force from electronic interactions. The rate at which an ion loses energy depends on both the nuclear and electronic stopping power of the target, giving

$$\frac{\mathrm{d}E}{\mathrm{d}x} = -N[S_\mathrm{n}(E) + S_\mathrm{e}(E)]\,, \tag{8.22}$$

where N is the target atom density ($5 \times 10^{22}\,\mathrm{cm}^{-3}$ for silicon) and $S_\mathrm{n}(E)$ and $S_\mathrm{e}(E)$ are the nuclear and electronic stopping powers (eV cm^2). Both are, in general, functions of energy. Note that, even though $S_\mathrm{n}(E)$ is called nuclear stopping, it has nothing to do with nuclear forces in atoms; rather, it is due to elastic "billiard ball" collision processes.

If $S_n(E)$ and $S_e(E)$ are known, then (with dx from above) the range of the ion can be calculated using

$$R = \int_0^R dx = \frac{1}{N}\int_0^{E_0}\frac{dE}{S_n(E) + S_e(E)}. \tag{8.23}$$

Note that R is the total range, and what we are generally interested in is R_P, the distance normal to the surface. These are related by [11]

$$R_P = \frac{R}{1 + [m_2/(3m_1)]}, \tag{8.24}$$

where m_1 is the mass of the implanted ion and m_2 is the substrate atomic mass. The challenge then is to develop physically based models for the nuclear and electronic stopping powers. We will develop simple models, useful for analytic calculations, below and then discuss Monte Carlo simulation methods in Section 8.7 that use more sophisticated models.

8.6.1 Nuclear Stopping

When an ion with mass m_1, atomic number Z_1 and kinetic energy E_0 collides with a stationary target atom with mass m_2 and atomic number Z_2, it loses energy by interacting with the electric field of the nucleus of the target. Not all collisions are head-on collisions, so the nuclear energy loss depends on the distance of closest approach, often called the impact parameter (b in Figure 8.18).

As the ion interacts with the electric field of the nucleus, it exchanges its kinetic energy for potential energy, reaching a maximum transfer at the distance of closest approach. This potential energy is partitioned between the ion and target atom in accordance with their masses, and the ion continues on a deflected path while the lattice atom recoils. The velocities and trajectories can be found from the conservation of momentum and energy in a classical treatment of two colliding particles. The ion has a momentum of m_1v_1 and a kinetic energy

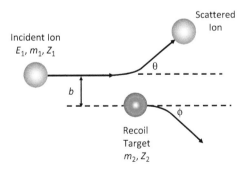

Figure 8.18 Schematic of a nuclear collision or scattering encounter between an ion and a stationary target atom.

of $\frac{1}{2} m_1 v_1^2$. After scattering, the two atoms still have the same total energy and momentum. Thus the nuclear energy loss is elastic, in that the energy lost by the ion is transferred to the lattice atom, providing the atomic-scale basis for the damage created by the incoming ions. These nuclear interactions give rise to scattering and deflected trajectories, with the deflection angle depending on the impact parameter b.

From classical mechanics, the energy lost by the ion in such a scattering event is

$$\Delta E = E_1 \frac{4m_1 m_2}{(m_1 + m_2)^2} \cos^2\phi , \qquad (8.25)$$

where the angles depend on b, the masses of the atoms and the interaction potential between the ion and the lattice atom. The energy loss ΔE is at a maximum in the event of a head-on collision, in which case $\phi = 0\,°$.

If the ion and target atoms were bare nuclei, the interaction or scattering potential would be given by the simple Coulomb potential:

$$V(r) = \frac{q^2 Z_1 Z_2}{4\pi\varepsilon r} . \qquad (8.26)$$

Because of the electrons that surround the target nucleus, the full effect of the positive core potential is screened from the incoming ion. Modifying the potential by an exponential screening function gives the Thomas–Fermi model of the atom,

$$V(r) = \frac{q^2 Z_1 Z_2}{4\pi\varepsilon r} \exp\left(-\frac{r}{a}\right) , \qquad (8.27)$$

where a is the screening distance. To predict the scattering angle, the interaction of the ion with the screened Coulomb potential is integrated over the path length where the nuclear force is important, as shown in Figure 8.18. Since the scattering is deterministic in the binary collision approximation, tables of precalculated scattering angles for each impact parameter and energy speed up the calculations in practice, and this is the strategy typically used in Monte Carlo simulators.

The nuclear stopping power $S_n(E)$ depends on the ion energy. The nuclear energy loss is small at very high energies because fast particles have less interaction time with the scattering nucleus (reduced cross-section). Thus, the nuclear energy loss tends to dominate towards the end of the range when the ion has lost much of its energy and where the nuclear collisions produce most of the damage. While more sophisticated models are used in modern computer simulation tools, as we will see in Section 8.7, $S_n(E)$ can sometimes be usefully approximated as a constant value at energies below where electronic stopping is important by [13]

$$S_n(E) = 2.8 \times 10^{-15} \frac{Z_1 Z_2}{(Z_1^{2/3} + Z_2^{2/3})^{1/2}} \frac{m_1}{m_1 + m_2} \text{ eV cm}^2 , \qquad (8.28)$$

where m_1 and Z_1 refer to the ion, and m_2 and Z_2 are the substrate atom mass and atomic number. Note that this simple expression does not include any energy dependence in $S_n(E)$, which means that it models the nuclear stopping as independent of position along the ion's path. We will see shortly that this is not really correct, but (8.28) is a useful approximation that allows us to calculate a numerical value given an ion and a substrate, which can then be used in (8.23) to estimate the range.

In compound semiconductors, or other materials like SiO_2, there are two or more types of atoms to which the ion can transfer energy as it slows down. The simplest way to model this is known as the Bragg rule [14], which states that the interactions with individual target atoms are independent of the surrounding target structure. Thus, if the stopping powers of elements A and B are $S_{n,A}(E)$ and $S_{n,B}(E)$, respectively, then, for a compound $A_m B_n$, the total stopping power is given by

$$S_{n,AB} = mS_{n,A} + nS_{n,B} .$$ (8.29)

Experimental stopping powers for compound materials usually differ somewhat from this simple relationship, but simulation programs can tabulate experimental results to provide better prediction of actual implant profiles.

8.6.2 Electronic Stopping

Energy loss to electrons in the target material is actually more complicated to model than is nuclear stopping. This is mainly because there are multiple energy loss mechanisms. There can be direct energy transfer from electrons surrounding the ion to electrons in target atoms because of collisions. Target electrons can be excited to higher energy levels in the atoms they surround. Target atoms can be ionized by exciting conduction band electrons to high enough energies to free them from the atoms to which they are bound. And, finally, treating the target material as a dielectric leads to a drag force on the ion because of polarization fields set up in the substrate. This last mechanism is usually regarded as a nonlocal mechanism modeled as a drag force. The others are more local mechanisms and depend on the particular atom with which the ion is interacting.

Many of the electronic loss mechanisms depend directly on the ion velocity, and so the simplest model for electronic energy loss is

$$S_e(E) = c v_{ion} = kE^{1/2} ,$$ (8.30)

where c and k are parameters that depend on the ion, the substrate and the particular electronic stopping process being modeled. In the simplest case, in which a Si substrate is assumed to be amorphous, k is roughly independent of the ion being implanted and is given by [13]

$$k \cong 0.2 \times 10^{-15} \, eV^{1/2} cm^2 .$$ (8.31)

Electronic stopping is inelastic, with the transferred energy generally being converted to heat in the substrate, as opposed to nuclear stopping, which is elastic.

8.6.3 Total Stopping Powers

Both the nuclear and electronic stopping powers depend on energy and contribute to the total stopping, as seen in Figure 8.19 [15]. There are a number of observations that we can make in looking at this figure.

1. For all ions, the importance of nuclear stopping versus electronic stopping increases at low energies.
2. The approximation that nuclear stopping is independent of energy (in (8.28)) is approximately true for heavy ions like As, but is increasingly wrong as the mass of the ion decreases.
3. The approximation that electronic stopping is independent of the ion being implanted (in (8.30) and (8.31)) is approximately true.
4. If the implant energy is, for example, 200 keV, As will be dominated by nuclear stopping over its whole range. Near the surface, P will have similar contributions from nuclear and electronic stopping, but will be dominated by nuclear stopping once it slows down to 100 keV and below. B will be dominated by electronic stopping until the very end of its range ($\approx$10 keV and below).
5. Since damage is caused by nuclear collisions, As will create damage along its entire path. Light ions like B create most of their damage near the end of their paths.

These observations will be important when we consider damage and damage annealing later in this chapter.

We can also perhaps now begin to understand more deeply the implant profiles in Figure 8.5. In that plot we saw that light ions have much larger R_P values than do heavy ions. We can now see from Figure 8.19 that the light ions have a much smaller energy loss rate than do heavy ions because nuclear stopping plays almost no role in their stopping until they are almost stopped. Stated another way, for the same energy, a light ion will have a much higher velocity ($E = \frac{1}{2} mv^2$), and nuclear stopping is much less effective at high velocities. Therefore, the light ions travel much further in the substrate.

Figure 8.19 Nuclear (superscript N) and electronic (superscript e) stopping powers for common dopants in silicon. The vertical scale is $NS_n(E)$, where N is the silicon density. After [15].

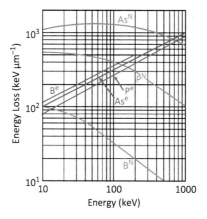

The analytic expressions for $S_n(E)$ and $S_e(E)$ in (8.28) and (8.30) can be substituted into (8.23), which can then be integrated to provide simple estimates of the range for an arbitrary ion implanted into silicon. If the substrate is not silicon, (8.29) can be used for nuclear stopping, provided the component stopping powers are known. A much more powerful approach, however, is to use simulation tools that implement more complete physical models for $S_n(E)$ and $S_e(E)$ and individually track ions through their stopping process. This is the Monte Carlo approach we discuss next.

8.7   Monte Carlo Implant Profiles

Computers today are powerful enough to follow the atomic-scale trajectory of a representative sample of tens or hundreds of thousands of ions through the lattice, accurately accounting for an arbitrary screen oxide thickness together with any tilt and rotation used during implantation. By combining the information on the final resting positions of the ions, a picture of the final distribution can be obtained. Because of the random nature of the starting condition, this approach is often called a Monte Carlo simulation.

The first Monte Carlo simulation tools to be developed treated the substrate as amorphous [16]. The "target" in this case is a random arrangement of substrate atoms with an average spacing set to give the correct material density $(1/N)^{1/3}$. In the simulation, ions are individually implanted and lose their energy through nuclear and electronic stopping. The basic ideas in the previous section regarding the physics of these processes are used in the simulations, although, since computational models are being used, the models can be more sophisticated. The simulators use the binary collision approximation (BCA) (see Figure 8.18).

As Monte Carlo simulators became more sophisticated, they were used to predict damage, basically by keeping track of the number of displaced target atoms [17]. These calculations are the starting point for damage annealing that we will discuss in Section 8.8. Today, Monte Carlo simulators can also model crystalline substrates, including Si and other semiconductors [6, 18]. The starting point in this case is a model of the atomic structure of the material and knowledge of the tilt and rotation of the incoming ion beam with respect to that structure.

While Monte Carlo methods require much longer computation times than the simple analytic approaches described earlier, they are the most flexible approach to modeling implant profiles in multilayer structures and in situations in which there are few experimental results that can provide the moments for the analytic approaches. There is an extensive literature on the physical models used in these simulators, and the interested reader is referred to that literature (e.g., [19]) and to the User Manuals for commercial simulation tools like Silvaco's Athena and Victory Process [6].

Figure 8.20 shows an example of a Monte Carlo implant profile simulation. In this case, 500,000 As^+ ions were implanted to simulate a 1×10^{13} cm^{-2}, 100 keV implant. With no preamorphization and $0°$ tilt, significant channeling is present, which is included in the Monte Carlo simulation. For the blue and green curves, Si^+ implants at 70 keV and $7°$ tilt were used to preamorphize the substrate, and then the same As^+ implants were done. The channeling is obviously mostly eliminated for the 1×10^{14} cm^{-2} Si^+ implant, and is essentially gone with a fully amorphizing 1×10^{15} cm^{-2} Si^+ implant.

Figure 8.20 Monte
Carlo simulations using
Silvaco's Victory
Process [6]. Here,
500,000 As$^+$ ions were
implanted in each case
with varying degrees
of Si$^+$ preamorphization.

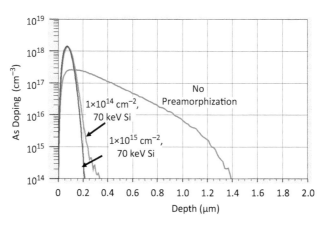

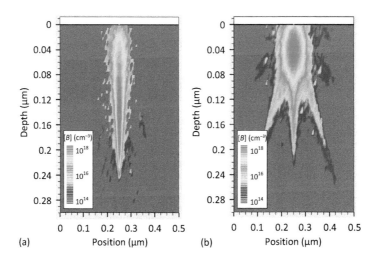

Figure 8.21 Sequential 1×10^{13} cm^{-2}, 10 keV B and As implants: (a) B first and (b) As first. Victory Process Monte Carlo simulations, tilt $= 0\,°$ with respect to $\langle 100 \rangle$ surface.

Figure 8.21 shows another interesting simulation, again using Victory Process [6]. In these simulations, sequential B and As implants were done into a narrow, 10 nm wide window. In the simulation in Figure 8.21(a), the B implant is done first and then the As implant. The colors show the calculated B concentration profile. When the B implant is done first, the substrate is undamaged and there is clear evidence of significant channeling down the $\langle 100 \rangle$ channels that are aligned with the implant beam (tilt $= 0\,°$). A significant number of B ions penetrate down to 0.2 μm.

In the example in Figure 8.21(b), the As implant is done first. Even though this is a relatively light As$^+$ implant, it does damage the crystal, blocking the direct $\langle 100 \rangle$ channels. As a result, the B$^+$ ions see more scattering near the surface. This results in a shallower B profile overall, but it also results in some B$^+$ ions being scattered into different channels in the silicon. This simple change in the sequential ordering of the two implants obviously has a dramatic effect on the B profile. Although not shown, the As profiles in the two cases are quite similar because, even when the B implant is done first, it does little damage near the surface because B is a light ion.

To illustrate the discrete nature of Monte Carlo simulations, Figure 8.22 shows the same simulation as Figure 8.20 except that only 5000 As$^+$ ions are simulated (100× fewer). The simulated profiles are qualitatively similar but there is clear evidence of the discreteness of the profiles in Figure 8.22. Monte Carlo simulations usually ask the user to specify the number of ions to be simulated. The trade-off is obviously simulation time versus accuracy of the calculated profile.

The same Monte Carlo simulation methods can also be used in compound semiconductor materials. An example is shown in Figure 8.23. SiC is a material of growing interest for power semiconductor devices. It has a hexagonal crystal structure (see Figure 3.6 in Chapter 3) and is available in a number of polytypes (the hexagonal Si and C layers are stacked and rotated to form a number of different crystal structures). The 4H polytype is commonly used for power devices with a wafer surface orientation of [0001]. Often wafers supplied by manufacturers are deliberately cut at a small angle to the [0001] direction in order to facilitate device processing. In Figure 8.23, the Al$^+$ implants are done normal to the SiC wafer surface, which effectively means that the tilt is the miscut angle shown. As the miscut angle increases, the amount of channeling simulated in the profiles decreases. This is similar to the effect seen in Figure 8.11 for B implants in Si.

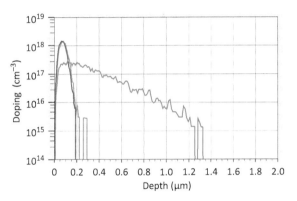

Figure 8.22 Same simulation as Figure 8.20 except that only 5000 ions were simulated.

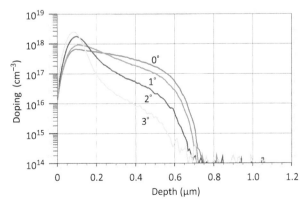

Figure 8.23 Monte Carlo simulations of Al$^+$ (2×10^{13} cm^2, 65 keV) implants into 4H-SiC using Victory Process [6]. The substrates were miscut from the [0001] axis by 0 ° to 3 °.

8.8   Damage Production and Damage Annealing

The major disadvantage of ion implantation is that nuclear stopping creates damage that must be repaired. If we consider a crystalline Si wafer as the substrate, the energy required to displace a silicon atom just far enough from its lattice site to create a stable separated interstitial and vacancy (a Frenkel pair) is the displacement energy E_d and is approximately 15 eV in silicon. Because the implant energy is often tens or hundreds of kiloelectronvolts, a large number of displacements will be caused.

Example
How many displaced Si substrate atoms are created by a one 30 keV arsenic atom?

Answer
For a heavy ion like arsenic, this energy is mainly dissipated due to nuclear collisions (see Figure 8.19). From Table 8.1, a 30 keV arsenic implant has a range of 23 nm, which means it traverses ≈ 90 planes of atoms (0.25 nm average inter-plane spacing), so that on average it loses 333 eV in a nuclear collision at each plane, as illustrated in Figure 8.24. This is much more than the Si displacement energy, so the lattice atom that is displaced by the incident ion (the primary knock-on) can continue and itself lose energy and displace other secondary lattice atoms. This is essentially equivalent to a silicon implant with 333 eV of energy at each point where a nuclear collision occurred. When the energy of the incident ion or the secondary knocked-on atoms reach E_d, they can be considered stopped, because if they do manage to transfer all their energy to a lattice atom they can cause a single displacement but remain at rest in the lattice position themselves. Thus, an energetic particle can only increase the number of moving particles if it has energy greater than $2E_d$. This allows an estimate of the number of displaced atoms created by an energetic particle to be made (the Kinchin–Pease formula) [20]:

$$n = \frac{E_n}{2E_d} = \frac{30,000}{2 \times 15} = 1000 \text{ displaced atoms},$$ (8.32)

where E_n is the energy lost in nuclear collisions. Thus, each incident arsenic ion creates a trail or cascade of 1000 displaced lattice atoms.

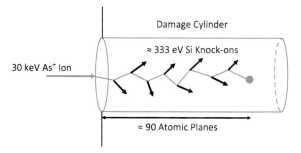

Figure 8.24 Schematic of damage caused by a 30 keV arsenic ion in silicon. With a range of 23 nm, the ion crosses approximately 90 atomic planes. After [19].

What is the time scale on which this damage production occurs? We can estimate that the time it takes the ion to come to rest is given by the range divided by its velocity:

$$t = \frac{R_P}{\sqrt{2E/m}} \approx 10^{-13} \text{ s} .$$

(8.33)

Some recombination occurs within the cascade due to elementary diffusion hops on the time scale of 10^{-11} s, after which the primary damage generated by the incident ion can be considered stable. This damage is primarily small defect clusters and dopant–defect complexes, and some isolated interstitials and vacancies. Such a detailed picture of the damage evolution has been obtained from molecular dynamics calculations of ions impacting a crystalline silicon lattice.

Molecular dynamics simulations are based on atomistic models of materials, as are the Monte Carlo methods we discussed earlier. However, Monte Carlo methods use randomness to model a system that might be deterministic in principle. As we saw, a Monte Carlo simulation involves simulating the random paths that tens of thousands of implanted ions take in order to build up a probability distribution and hence the implanted profile. Molecular dynamics simulations track individual particles deterministically and simulate a detailed picture of the scattering events, including secondary damage by recoiled Si atoms, of each particle. Generally, molecular dynamics methods are used for individual or small numbers of particles because of the computational complexity.

An example is shown in Figure 8.25 in which a single Si$^+$ ion is implanted at 5 keV into a crystalline silicon substrate. The black dots are the displaced Si atoms that result from nuclear scattering events. Two observations are important. First, the total number of displaced Si atoms is very large, roughly 800 atoms in this example. This is several times larger than what

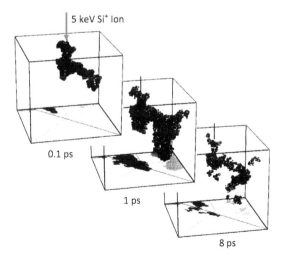

Figure 8.25 Molecular dynamics simulation of a single 5 keV Si$^+$ ion implant into silicon, showing the damage evolution and stabilization. Only displaced atoms are shown. The simulation space is a 15.2 nm square [100] surface. The depth is 13.5 nm. Reprinted with permission from [21]. © 1995 by the American Physical Society.

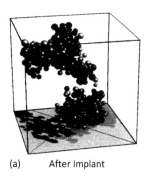 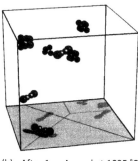

(a)   After Implant   (b)   After 1 ns Anneal at 1025 °C

Figure 8.26 Molecular dynamics simulation of a 1 ns 1025 °C anneal of a portion of the damage structure shown in Figure 8.24. Reprinted with permission from [21]. © 1995 by the American Physical Society.

would be estimated from (8.32). However, after only 8 ps, when the ion has stopped, the number of displaced atoms is much smaller. Clearly, some of the displaced Si atoms have locally diffused (jumped) back onto lattice sites.

Figure 8.26 shows a further molecular dynamics simulation of a very short anneal (1 ns at 1025 °C) of a small region of the damage simulated in Figure 8.25. Clearly, much of the damage is removed even in this very short-time anneal. Molecular dynamics simulations have proven very helpful in understanding damage and damage annealing, but the computational costs are too large to simulate real device structures and real anneal times. Generally, these kinds of simulations are limited to structures with a few million atoms and a few nanoseconds of time. However, we should take away from Figures 8.25 and 8.26 that a lot of the damage seems to "heal" and disappear pretty quickly. In order to develop models useful for real device structures and real processing times, we now discuss different approaches at a more macroscopic level.

Since the number of displacements n produced by an ion depends on the amount of energy deposited into nuclear collisions, the damage production is a maximum where the nuclear energy losses are highest. In Figure 8.19, we saw that, for heavy ions like arsenic, whose stopping is dominated by nuclear collisions, this means that the damage profile is relatively flat over the whole range up to R_P. For lighter ions like boron, which have an appreciable component of electronic stopping at higher energies, Figure 8.19 also suggested that the damage accumulation should be concentrated near R_P where the ion energy is lower. For light ions, then, the damage will accumulate first at the peak of the profile near R_P and expand on both sides of this depth as the implant dose is increased. If the damage is high enough to create an amorphous layer, it will form first near R_P and then gradually expand in the case of light ions.

Experimental measurements of the amorphous layer thickness as a function of implant dose support this picture and indicate a U-shaped distribution for the amorphous layer thickness depending on implant dose, as shown in Figure 8.27. In these experiments, Si^+ ions were implanted at 300 keV into Si substrates at increasing doses. The red double-headed arrows on the figure illustrate the approximate bounds of the amorphous region. A buried amorphous layer first forms, and it may take a considerably higher implant dose before a continuous amorphous layer extending down from the surface forms [22]. Other experimental results

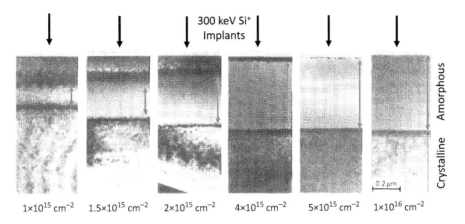

Figure 8.27 Cross-sectional transmission electron microscope (TEM) images of the formation of an amorphous layer for increasing 300 keV Si⁺ implant doses. The layer begins to form where the nuclear damage is highest, spreading with higher doses. Reproduced from [22] with the permission of AIP Publishing.

indicate that it is easier to form an amorphous layer at low temperatures (liquid nitrogen) rather than at room temperature or higher. This observation is easily explained by the larger fraction of ions that recombine within a cascade at higher implant temperatures.

The above examples of molecular dynamics simulations and the experimental results in Figure 8.27 demonstrate that the damage introduced by ion implantation is quite complex. Yet, in order to use this doping process, we must find a way to remove this damage. The goal of damage annealing is to remove the primary damage created by the implant and restore the semiconductor lattice to its perfect crystalline state, leaving the dopants on active, substitutional sites. This damage removal can be divided into two distinct regimes, one below and the other above the amorphous threshold.

Figure 8.25 suggests that the damage actually begins to anneal at quite low temperatures, since those simulations were done at room temperature. Many experiments, mainly in the 1980s and 1990s, showed that the damage remaining after implantation begins to annealing at relatively low temperatures of about 400 °C. Figure 8.28 explores this annealing in more detail [23]. These simulations used molecular dynamics to create the initial damage distribution after implantation, as in Figure 8.26. That profile was then used in a Monte Carlo diffusion simulator to allow the defects to diffuse and react over longer time scales, in this case at 815 °C.

Note the very short time scale on which the Frenkel pairs recombine in this simulation through bulk I–V recombination. This is consistent with the simulations in Figure 8.26. Some vacancies recombine independently at the surface. After 10^{-2} s, only interstitials remain in small clusters, and these slowly dissolve by recombining at the surface. Though this picture is simple, it captures much of the essential physics that occurs during implant damage annealing.

Figure 8.28 would seem to suggest that the implant damage anneals very quickly (<1 s at 815 °C), restoring the substrate to crystalline form. While this is true, at least for the relatively low dose considered in this simulation, there is something we have forgotten to this point. The interstitials

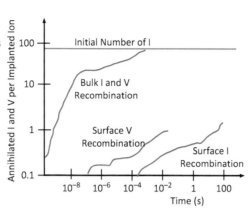

Figure 8.28 Molecular dynamics and Monte Carlo simulation of the recombination of I and V damage generated by a 40 keV, 5×10^{13} cm^{-2} Si$^+$ implant into a Si substrate. The anneal shown was at 815 °C in the simulation. After [23].

and vacancies generated by the implanted ions are formed in exactly equal numbers, and so, if they all recombined, we would be left with a perfect crystal. But we must also consider the implanted ion, which is typically a dopant ion. It, too, needs to sit on a lattice site, and when it does so, it will create one additional Si interstitial. This gives rise to the "+1" model for residual damage due to implants that we will discuss in detail in Section 8.9 [24]. This model states that, to a good approximation, all of the original damage recombines, leaving only one interstitial created when the implanted ion finally occupies a lattice site. The profile of the excess "+1" interstitials will mirror the implanted dopant profile.

The +1 model was a conceptual breakthrough that tremendously simplified the modeling of ion implant damage and its effects, as we will see when we discuss TED in the next section. But, before doing so, we need to understand what these +1 interstitials do in the crystal. They are very mobile and have no local vacancies with which to recombine, so it would seem that they would diffuse away to the surface or into the bulk to restore the crystal to an equilibrium population of I. However, many experiments in the 1980s and 1990s showed that this is actually not what they do.

In fact, the silicon interstitials generated by this "+1" amount of damage very quickly condense into characteristic rod-shaped defect clusters upon annealing at temperatures above 400 °C. These rod-shaped defects are composed of ribbons of silicon atoms that lie on {311} planes, as illustrated in Figure 8.29(a). An actual TEM image of such a defect is also shown in Figure 8.29(b). After a 5 s anneal at 900 °C, there may be a very high concentration ($\approx 10^{11}$ cm^{-2}) of these {311} defects, which are only 10 nm long.

We will see when we discuss TED in the next section that, over the course of a high-temperature implant anneal, the interstitials "stored" in these {311} defects are gradually released. While this is happening, they raise the interstitial concentration well above the equilibrium level. Anything that is affected by interstitials will be affected, including, for example, dopant diffusion rates. These interstitials are the origin of TED and the high diffusion rates will exist for the time required for the {311} defects to dissolve.

The basic picture described above for relatively low-dose implants has been confirmed by molecular dynamics and Monte Carlo simulations. Figure 8.30 shows one example [25]. The implant is 1×10^{14} cm^{-2} Si$^+$ at 5 keV. This dose causes significant damage but does not

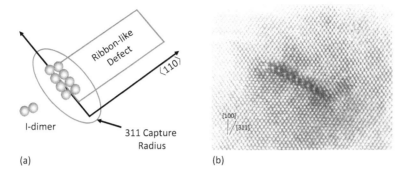

Figure 8.29 (a) Schematic and (b) cross-section TEM of {311} clusters that form by capturing a row of interstitial dimers that lie on the {311} plane and grow by extending in the ⟨110⟩ direction. Photo reprinted from [25] with the permission of AIP Publishing.

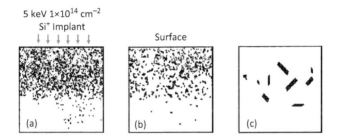

Figure 8.30 Molecular dynamics and kinetic Monte Carlo simulations of the defect structure resulting from a 5 keV, 1×10^{14} cm^{-2} Si$^+$ implant. Black and gray points represent silicon I and V, respectively. (a) The damage at room temperature after the implant. (b) The damage during the beginning of the anneal (ramp up in temperature to 800 °C). (c) After the anneal at 800 °C. Reproduced from [26] with the permission of AIP Publishing.

amorphize the substrate. A molecular dynamics simulation was used to create the initial damage profile shown in Figure 8.30(a). In Figure 8.30(b) the beginning of the anneal is shown as the temperature is ramped up to 800 °C. In this case, a kinetic Monte Carlo simulation approach was used, which follows the defects but not all of the lattice atoms, so anneals of minutes or longer can be simulated. After the 800 °C anneal in Figure 8.30(c), all of the damage is gone, except for the {311} defects that have formed. These are stable at this low anneal temperature, but we will see in Section 8.9 that they can be easily removed by a higher-temperature anneal. The basic process we are seeing in Figure 8.30 is similar to the plots in Figure 8.28 which show the I and V defects recombining.

If the damage is larger (higher-dose implant), some of the {311} defects grow at the expense of others that shrink. The larger {311} defects can turn into stable dislocation loops, which are much more difficult to remove. These stable dislocation loops are called secondary defects and remain when the primary damage is completely annealed. The dislocation loops can be

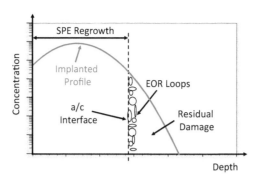

approximated by an extra circular atomic layer of silicon atoms precipitated on a {111} plane of silicon. They can be shown from TEM images to be extrinsic loops, that is, they are composed of an extra plane of silicon interstitials.

Stable dislocation loops are a much more dominant feature of higher-dose implants, which drive the silicon amorphous. The largest concentration of these loops is seen at the interface between amorphous and crystalline silicon after an anneal which regrows the amorphous region by solid-phase epitaxy (SPE), which we discuss below. These defects at the original amorphous/crystalline (a/c) interface are referred to as end-of-range (EOR) defects, as illustrated in Figure 8.31. They occur because there is a large amount of damage that is just below the threshold of amorphization beyond the a/c interface. By definition, just beyond the a/c interface is the maximum possible amount of damage that can exist in the crystal at the implant temperature without it being amorphous. This damage is sufficiently high to be able to nucleate a mixture of {311} defects and a band of dislocation loops in a very narrow region just beneath the a/c interface on the crystalline side of the interface.

Because the annealing of implant damage is very different when the material is amorphous versus more lightly damaged, we need to consider how to define the a/c interface shown in Figure 8.31. There are a number of models that have been proposed in the literature to estimate when the material becomes amorphous and where the a/c boundary is. An excellent review of this field is in [25]. The simplest model assumes that the transition to amorphous material occurs when a critical energy density is deposited in the material. This energy density is on the order of $E_{cr} \approx 5 \times 10^{23}$ eV cm^{-3} [26]. We can write this in terms of the implant dose required to amorphize the substrate:

$$E_{cr} = \Phi N S_n(E) . \tag{8.34}$$

Here, Φ is the dose required for amorphization and $S_n(E)$ is the nuclear stopping discussed earlier, and is the rate at which energy is deposited per unit length in creating the damage.

While not reflected directly in (8.34), the required dose to create an amorphous region depends on many parameters in reality. This is because, during the implant, damage is being created, but the damage is also repairing itself, as we saw in Figure 8.25. So, for example, if the implant is done at reduced temperature, less repair will take place during the implant and a smaller dose would be required to amorphize the substrate. If the implant is done in

Example
What As^+ dose is required to amorphize the substrate in a 100 keV implant?

Answer
For As, the nuclear stopping from Figure 8.19 is $NS_n(E) \approx 1200$ keV μm^{-1}. Thus

$$\Phi = \frac{E_{cr}}{NS_n(E)} = \frac{5 \times 10^{23} \text{ eV cm}^{-3}}{1.2 \times 10^{10} \text{ eV cm}^{-1}} \cong 4.2 \times 10^{13} \text{ ions cm}^{-2}.$$

Note from Figure 8.19 that $S_n(E)$ for a light ion like B is $\approx 50\times$ smaller, so that the required dose would be $>10^{15}$ cm^2.

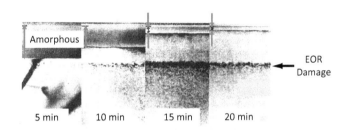

Figure 8.32 Cross-sectional TEM images of the regrowth of an amorphous layer at 525 °C as a function of time. The red arrows indicate the amorphous region. The dark band at the original a/c interface consists of EOR dislocation loops. Reproduced from [27] with the permission of AVS: Science and Technology of Materials, Interfaces and Processing.

a higher-current implanter, there is less time for repair to take place before more damage is created, and a lower dose would generally be required [26]. So values calculated using (8.34) should be regarded only as estimates.

If the substrate is amorphized, then a rather remarkable process called solid-phase epitaxy (SPE) can take place to repair the damage. This process is shown in experimental data in Figure 8.32. Even though there is significant damage at the a/c interface, the material regrows from that depth upwards, plane by plane to produce crystalline material.

The SPE process shown experimentally in Figure 8.32 is shown in simulation in Figure 8.33(a). This simulation is similar to Figure 8.30 except that the Si^+ dose is increased to 1×10^{15} cm^{-2}, sufficient to amorphize the substrate. In Figure 8.33(b), the simulated temperature is only 550 °C and the amorphized region can be seen regrowing plane by plane from the bottom up towards the surface. After the SPE is complete at 550 °C, the simulation in Figure 8.33(c) shows an 800 °C anneal which causes the {311} defects and the EOR damage dislocation loops to form.

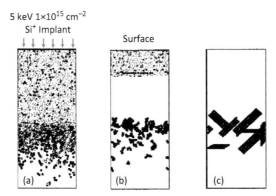

Figure 8.33 Molecular dynamics and kinetic Monte Carlo simulations of the defect structure resulting from a 5 keV, 1×10^{15} cm^{-2} Si$^+$ implant. Black and gray points represent silicon I and V, respectively. (a) The damage at room temperature after the implant. (b) The SPE regrowth process occurring at 550 °C. (c) After the anneal at 800 °C. Reproduced from [26] with the permission of AIP Publishing.

The SPE process is similar to the crystallization process that occurs from the melt onto a single-crystal seed or from a gas phase onto a crystalline substrate, except that it is now occurring in a solid phase rather than a liquid or gas phase. It occurs quite quickly, with regrowth rates in undoped silicon at 600 °C of 50 nm min^{-1} for $\langle 100 \rangle$ orientations, 20 nm min^{-1} for $\langle 110 \rangle$ orientations and 2 nm min^{-1} for $\langle 111 \rangle$ orientations. The activation energy for the regrowth process on any orientation is 2.3 eV, which is indicative of a silicon–silicon bond breaking and formation process. The regrowth rate is given by

$$v = A \, \exp\left(-\frac{2.3 \text{ eV}}{kT} \right), \tag{8.35}$$

where v is the regrowth rate and A is an experimentally determined parameter.

The regrowth rate can be enhanced by a factor of 10 for high doping concentrations characteristic of the source and drain in MOS devices [27]. The SPE process quickly eliminates all of the primary damage in the amorphous region, so that no anomalous dopant diffusion occurs there. Most of the dopant atoms in the amorphous regions are incorporated onto substitutional lattice sites during regrowth, so that high levels of activation are possible even at these low temperatures. Indeed, were it not for the residual band of EOR defects just below the a/c interface and the {311} defects below the a/c interface, SPE would provide an ideal method for simultaneously removing all the implant damage and obtaining high levels of dopant activation without any noticeable dopant diffusion.

Figure 8.34 summarizes what we have learned to this point about damage issues in crystalline substrates during an implant. Figure 8.34(a) applies to implant doses low enough that they do not amorphize the substrate. In that case, the damage anneals readily at modest temperatures ($\approx$ 800 °C), leaving behind only the {311} clusters due to the +1 interstitials created by the implanted atoms taking lattice sites.

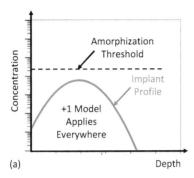 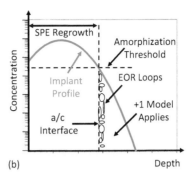

Figure 8.34 Summary of damage modeling for implant doses (a) below the amorphization threshold and (b) above the threshold.

In Figure 8.34(b), the implanted dose is large enough to create a surface amorphous layer. This layer is readily repaired by SPE at quite low temperatures ($\approx 500\,^\circ$C). Beyond the a/c interface, we are left with the EOR dislocation loops, which are difficult to anneal out completely, and a tail region in which the +1 model again applies. Note that in Figure 8.34(a) the number of +1 interstitials stored in {311} defects corresponds to the full implant dose. In Figure 8.34(b) only the tail of the implant profile contributes to the stored +1 magnitude. The +1 interstitials created in the SPE region are incorporated into the regrown epitaxial layer simply by extending the crystalline substrate upwards by an atomic plane or two (there are roughly 10^{15} Si atoms cm^{-2} in each atomic plane).

In both panels of Figure 8.34, the implanted dopant atoms are incorporated on lattice sites as part of the SPE or moderate-temperate annealing. So the main issue we are left with is the {311} stored interstitials in both cases. The EOR loops in Figure 8.34(b) may be an issue if they are physically located in sensitive regions like depletion regions, where they can reduce lifetime and increase leakage currents. But if they are located inside heavily doped regions, they may be relatively benign.

8.9 Transient Enhanced Diffusion

The {311} defects generally need to be removed from the silicon, not only because they are crystallographic defects that could affect device performance, but also because the interstitials (I) they store are gradually released during any subsequent thermal cycles that the wafers see. As we saw in Chapter 7, dopants generally diffuse with I and with V, and so an increase in the I population due to the dissolving {311} defects should be expected to increase dopant diffusivities. This mechanism is known as transient enhanced diffusion (TED), and it can be a very important issue especially when device performance relies on tightly controlled impurity profiles.

TED is anomalous because there are conditions where the dopant can diffuse more at lower temperatures than at higher temperatures. It might seem that only the Dt product should

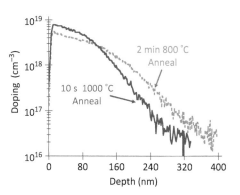

determine the extent of profile motion, i.e., a longer time at low temperature could be chosen to be equivalent to a short time at high temperature. A longer time at lower temperature would allow a controllable anneal to be performed in a furnace environment. However, when implant damage is present, the result is very surprising. An example is shown in Figure 8.35 [28]. Here, the profile annealed at the lower temperature diffused further than the one annealed at high temperature. Using normal D values, the Dt in this experiment at 1000 °C is significantly larger than the Dt at 800 °C.

The reason is that, at lower temperatures, the {311} defects can stay around longer and enhance the dopant diffusion, while, at higher temperatures, the {311} defects annihilate faster. Historically, long anneals at high temperatures masked this effect, because eventually normal thermal diffusion will occur as the interstitial concentration returns to equilibrium values. However, for the small-thermal-budget anneals used in present-day devices, anomalous TED has become the dominant effect that determines how far an implanted dopant moves during an anneal.

The magnitude of TED can be quantified by examining how large the enhancement in dopant diffusivity is at each temperature. The results are astonishing, with enhancements ranging from 10,000 at 700 °C to 800 at 1000 °C. Our goal below is to quantify this behavior and develop predictive models that can be used in process simulators for process optimization.

Consider the buried marker layer experiment shown in Figure 8.36. A lightly doped boron layer is shown that might be grown using epitaxial silicon growth, followed by a lightly doped intrinsic silicon layer. A high-dose arsenic implant amorphizes the surface region of the substrate, and SPE regrowth removes much of the implant damage. But the region below the a/c interface in the As profile will create {311} defects and EOR defects, as illustrated in Figure 8.34(b).

If this structure is heated to a modest temperature, say 900 °C, for a short time (say 30 s), the {311} defects will dissolve and the tail region of the As profile will become defect-free (except for remaining EOR defects). In 30 s at 900 °C, the boron buried layer would not be expected to diffuse very much because $\sqrt{Dt}$ is very small for normal boron diffusion. But what is observed experimentally is a large amount of diffusion in the boron profile, as illustrated. The mechanism is simply that the dissolving {311} defects release I that diffuse over large distances. Since the

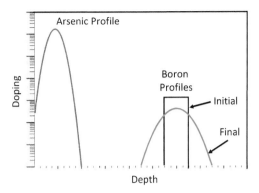

Figure 8.36 Buried marker layer experiment structure.

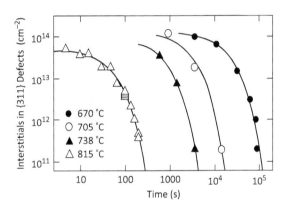

Figure 8.37 Plot of the number of interstitials contained in {311} defect clusters as a function of anneal time at various temperatures. The {311} defects dissolve and provide the interstitial supersaturation that controls TED. Reproduced from [31] with the permission of AIP Publishing.

diffusivity of B is proportional to C_I, as we saw in Chapter 7, then D_B is much larger than normal as long as the {311} defects are present. The arsenic profile itself also shows some slight but significant diffusion, but it is less for arsenic than for boron because of the lower component of arsenic diffusion that takes place by an interstitial mechanism and also because of the lower intrinsic diffusivity of arsenic.

Figure 8.36 is based on hundreds of experimental results (e.g., [29]) using similar structures and illustrates the effect that TED can have in determining how the damage introduced by one dopant affects its own diffusion and also that of a nearby dopant. This structure decouples the damage itself from the effects of the damage on a nearby region.

Atomic-level understanding of TED has come from key experimental [30] and theoretical studies. An important experimental finding was that the point defects responsible for TED are interstitial-type extended defects in the form of {311} clusters, whose evaporation rate matches the kinetics of TED [31]. Figure 8.37 shows the experimentally measured dissolution rate of the {311} defects.

In these experiments, 5×10^{13} cm^{-2} Si$^+$ implants were done at 40 keV. Cross-sectional and plan-view TEMs were used to count and measure the {311} density and to calculate the number of I contained in the {311} defects. Note that the number of I in {311} defects in Figure 8.37 starts at approximately the +1 value (equal to the implant dose), is relatively constant for some time period and then declines precipitously after a time that decreases with temperature. Data like these were critical in developing the quantitative TED model we discuss below. The same time dependence is seen experimentally in TED [31]. Key theoretical results come from molecular dynamics simulations of the collision cascade process in silicon and from Monte Carlo simulations of the point defect evolution and clustering, as discussed earlier in connection with Figures 8.30 and 8.33.

The microscopic explanation of the energy dependence of TED is even more interesting for the insights it provides into the nature of the residual implant damage. In a clever experiment, Giles *et al.* [32] implanted a dopant at the same depth below the surface using both a normal angle of incidence for the ion beam and using a higher energy at a tilted angle of incidence. One expects that the higher energy creates more damage, but the amount of TED in both profiles is identical. This indicates that, even though more primary damage is created, the bulk recombination process is efficient enough that essentially all the interstitials and vacancies produced by the implant annihilate in the early stages of the anneal, leaving only a dose of interstitials due to the extra ions introduced into the lattice. This provides a remarkable confirmation of the "+1" approach to implant damage modeling.

Given these atomic-level insights, we can now quantitatively model the impact of TED on dopant diffusion [33]. The fact that the {311} defects form remarkably quickly implies that the TED transient is controlled by the evaporation of interstitials from these aggregated defects. The process of cluster growth and evaporation can be atomistically described by

$$I + CI_n \Longleftrightarrow CI_{n+1}, \tag{8.36}$$

where the subscript n indicates a cluster CI with n interstitials. If we assume that the properties of the cluster remain the same with size, we can write an equation [33] for the time rate of change of the cluster growth as

$$-\frac{\partial C_1}{\partial t} = \frac{\partial CI}{\partial t} = k_f C_1 CI - k_r CI$$
$$= \text{growth} - \text{shrinkage}, \tag{8.37}$$

where the forward reaction rate for {311} cluster growth is given by k_f and the reverse reaction rate for cluster shrinkage is given by k_r. We can assume that the growth of the clusters is diffusion-limited, so that the forward reaction rate is given by

$$k_f = 4\pi a d_I, \tag{8.38}$$

where a is an encounter distance equal to the nearest-neighbor distance in silicon and d_I is the interstitial diffusivity. The reverse reaction constant, which describes the rate of evaporation, will be given by the attempted hop frequency of the bound interstitial, d_I/a^2, times a Boltzmann factor to account for the binding energy to the cluster, so that

$$k_r = \frac{d_I}{a^2} \exp\left(-\frac{E_b}{kT}\right). \tag{8.39}$$

For longer anneal times, the growth of the clusters is not important, leaving only the second term

$$\frac{\partial C_I}{\partial t} = -k_r C I, \tag{8.40}$$

which is the origin of the exponential decay kinetics of the clusters and the consequent change in the dopant diffusivity for longer-time anneals. As the experimental data in Figure 8.37 shows, for much of the TED time period, C_I is approximately constant because there is a balance between the growth and shrinkage of the {311} defects. This balance between growth and evaporation when $\partial C I / \partial t \cong 0$ gives rise to a constant interstitial concentration during the steady-state period of

$$C_I^{\max} = \frac{k_r}{k_f} = \frac{1}{4\pi a^3} \exp\left(-\frac{E_b}{kT}\right). \tag{8.41}$$

It is more useful to write this as a constant interstitial supersaturation during the steady-state period, since this will determine the enhancement in the dopant diffusivity, giving

$$\frac{C_I^{\max}}{C_I^*} = \frac{1}{4\pi a^3 C_I^0} \exp\left(-\frac{E_b - E_F}{kT}\right), \tag{8.42}$$

where we used the formula for the equilibrium interstitial concentration from Chapter 3,

$$C_{I^0}^* = N_{Si} \exp\left(-\frac{S^f}{kT}\right) \exp\left(-\frac{H^f}{kT}\right) = C_I^0 \exp\left(-\frac{E_F}{kT}\right). \tag{8.43}$$

The enhancement in diffusivity observed during TED is determined by the balance between cluster growth and evaporation. As long as there are a sufficient number of clusters around, some will shrink and others will grow, maintaining a steady-state interstitial supersaturation. It is during this steady-state condition that most of the profile motion occurs because the interstitial supersaturation is large and constant. This supersaturation has been measured experimentally and is as large as 10,000 at 750 °C [28]. It reduces to a factor of 800 at a typical rapid thermal anneal (RTA) temperature of 1000 °C.

A second critical parameter is how long this steady-state condition lasts. If we assume the "+1" model for damage, then an implanted dose Q introduces an equivalent dose Q of excess interstitials which rapidly form {311} defects, and we wish to determine how long these excess defects survive. The interstitials in the clusters will eventually diffuse into the bulk or recombine at the surface. Since the damage is in general near the surface, the surface is the dominant sink for the excess interstitials, as shown in Figure 8.39.

Example

Calculate the interstitial supersaturation versus temperature using the above equations. Assume the binding energy of an interstitial to a {311} cluster is 1.77 eV.

Answer

From Table 7.5, we have the self-diffusion coefficient for silicon

$$D_{\text{self}} = 560 \ \exp\left(-\frac{4.76}{kT}\right) \ \text{cm}^2 \text{s}^{-1}.$$

This is the overall measured diffusion coefficient for silicon atoms, so the self-diffusion flux is given by

$$D_{\text{self}} C_{\text{Si}} = 560 \ \exp\left(-\frac{4.76}{kT}\right) \ \text{cm}^2 \text{s}^{-1} (5 \times 10^{22}) \ \text{cm}^{-3}$$

$$= 2800 \times 10^{22} \ \exp\left(-\frac{4.76}{kT}\right) \ \text{cm}^{-1} \ \text{s}^{-1}.$$

The interstitial fraction of the self-diffusion (using Table 7.4) is

$$d_{\text{I}} C_{\text{I}}^* = (0.6)(2800 \times 10^{22}) \ \exp\left(-\frac{4.76}{kT}\right)$$

$$= 1680 \times 10^{22} \ \exp\left(-\frac{4.76}{kT}\right) \ \text{cm}^{-1} \ \text{s}^{-1}.$$

This equation describes the magnitude and temperature dependence of the interstitial flux. Note that the activation energy is large compared to the activation energy for dopant diffusion. This will become important in the later discussion. A reasonable value for the diffusion coefficient of interstitials is given by

$$d_{\text{I}} = 51 \ \exp\left(-\frac{1.8}{kT}\right) \ \text{cm}^2 \ \text{s}^{-1},$$

allowing C_{I}^* to be calculated as

$$C_{\text{I}}^* = 33 \times 10^{22} \ \exp\left(-\frac{2.96}{kT}\right) \ \text{cm}^{-3}.$$

The jump distance can be taken as half the lattice constant, so that

$$a = \frac{5.41 \times 10^{-8}}{2} = 2.7 \times 10^{-8} \, \text{cm}.$$

The binding energy of interstitials to {311} clusters is 1.77 eV, which was obtained by fitting the decay kinetics of the {311} defects in Figure 8.37 [33]. It is a pure coincidence that this is so similar to the activation energy for interstitial diffusion. Using these values in (8.42) gives

$$\frac{C_I^{max}}{C_I^*} = \frac{1}{4\pi(2.7 \times 10^{-8})^3(33 \times 10^{22})} \exp\left(-\frac{1.77 - 2.96}{kT}\right)$$

$$= 1.22 \times 10^{-2} \exp\left(\frac{1.19}{kT}\right),$$

which is plotted in Figure 8.38.

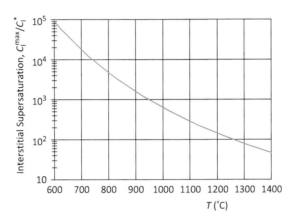

Figure 8.38   The maximum interstitial supersaturation as a function of anneal temperature.

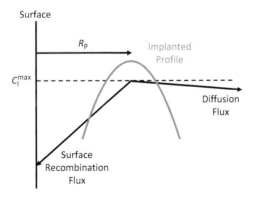

Figure 8.39   A schematic of the recombination flux at the surface which removes the +1 damage introduced by an implant at a range R_P [33]. Using R_P for the diffusion distance presumes that the implant dose is below the amorphization level.

The flux of interstitials towards the surface can be estimated from Fick's first law as

$$\text{flux} = D\frac{\partial C}{\partial x} = \frac{d_\mathrm{I} C_\mathrm{I}^{\max}}{R_\mathrm{P}}, \tag{8.44}$$

where R_P is the projected range of the implant. The time to dissolve the clusters is given by the dose divided by the flux, i.e.,

$$\tau_\text{enh} = \frac{4\pi a^3 R_\mathrm{P} Q}{d_\mathrm{I}} \exp\left(\frac{E_\mathrm{b}}{kT}\right). \tag{8.45}$$

To clearly see the activation energy for the steady-state period, we can substitute for the self-interstitial diffusivity,

$$d_\mathrm{I} = d_\mathrm{I}^0 \exp\left(-\frac{E_\mathrm{m}}{kT}\right),$$

and therefore

$$\tau_\text{enh} = \frac{4\pi a^3 R_\mathrm{P} Q}{d_\mathrm{I}^0} \exp\left(\frac{E_\mathrm{b} + E_\mathrm{m}}{kT}\right). \tag{8.46}$$

Note that the use of R_P in the equation for τ_enh assumes that the implant dose is below the amorphization threshold and hence SPE does not repair any of the damage. In a higher-dose implant, only the region below the SPE region needs to be repaired and so a more appropriate diffusion distance to use in (8.44) and (8.46) would be the depth of the SPE layer.

We can estimate the overall amount of profile motion after TED by combining the above information and taking the product of the dopant diffusivity, the steady-state interstitial supersaturation and the steady-state time to give

$$D_\mathrm{A}^\text{eff} t_\text{eff} = \left(D_\mathrm{A} \frac{C_\mathrm{I}^{\max}}{C_\mathrm{I}^*}\right)\tau_\text{enh} = D_\mathrm{A}\left(\frac{Q R_\mathrm{P}}{d_\mathrm{I} C_\mathrm{I}^*}\right)$$
$$= \frac{D_\mathrm{A}}{f_\mathrm{I}^\text{self} D^\text{self} C_\text{Si}} Q R_\mathrm{P}, \tag{8.47}$$

which immediately makes clear the anomalous temperature dependence of TED. Though the dopant diffusivity has an activation energy of approximately 3.5 eV, it is overwhelmed by the activation energy of self-diffusion of $\approx$ 4.8 eV, which causes more profile motion at lower temperatures. Thus, this equation neatly explains the single most anomalous observation about TED – the fact that profiles diffuse more at lower temperatures than at higher temperatures, even for similar Dt values, as shown in Figure 8.35. Though puzzling, this observation can be rationalized by considering that a fixed amount of damage is introduced by the implant, while the background point defect concentration and interstitial self-diffusion coefficient fall sharply with temperature. Thus, the supersaturation in the point defect levels rises sharply with falling temperature and the excess defects remain around longer due to the lower interstitial diffusion coefficient, leading to greater overall motion in the dopant profile.

Example

Calculate and plot how the duration of TED depends on temperature for a 100 keV, 5×10^{13} cm^{-2} phosphorus implant into silicon.

Answer

From Table 8.1, a 100 keV phosphorus implant has a range of 127 nm. Therefore,

$$\tau_{enh} = \frac{4\pi(2.7 \times 10^{-8} \text{cm})^3(127 \times 10^{-7}\text{cm})(5 \times 10^{13}\text{cm}^{-2})}{51 \text{ cm}^2\text{s}^{-1}} \exp\left(\frac{1.77 + 1.8}{kT}\right)$$

$$= 3.08 \times 10^{-15} \exp\left(\frac{3.57}{kT}\right) \text{ s},$$

which is plotted in Figure 8.40.

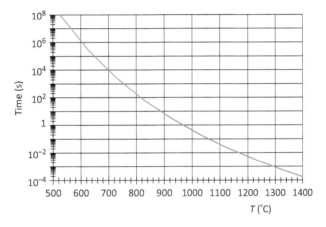

Figure 8.40 Duration of TED versus temperature for a 100 keV, 5×10^{13} cm^{-2} phosphorus implant into silicon.

We can make another observation from looking at the form of (8.42) and (8.46). Assuming that we are below the amorphization threshold, if we did two experiments, implanting a dopant at the same energy in both cases, but doubling the dose in the second case, the interstitial supersaturations would be identical in the two cases according to (8.42). Thus the diffusivity enhancement would be the same in both cases. However, (8.46) predicts that the TED time duration of the higher-dose implant would be twice as long and thus there would be more overall motion in the higher-dose case. Similarly, if the two experiments used the same dose, but different energies, the higher-energy implant would create more TED through the R_P dependence in (8.46) even though, again, the interstitial superstation would be the same in the two cases. Physically, the higher-energy implant will place the dopant and the {311} defects further from the surface, decreasing their rate of surface recombination and extending the TED time.

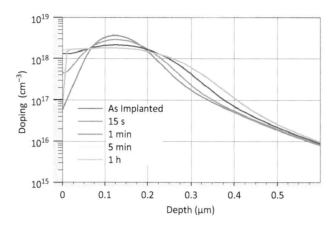

Figure 8.41 Silvaco Athena simulation of TED. A 5×10^{13} cm^{-2}, 100 keV P$^+$ implant with a tilt of 7° into a silicon substrate was used. A Pearson distribution (red curve) was used to model the implant. The other curves show the TED during an 800 °C anneal.

Commercial process simulation programs like Silvaco's Athena and Victory Process implement models very much like those we have discussed. These simulators include an amorphization threshold, use the {311} model and base the magnitude of TED on a +1 model. Figure 8.41 shows an example of a TED simulation. A 5×10^{13} cm^{-2}, 100 keV phosphorus implant with a tilt of 7° into a silicon substrate was used. A Pearson distribution (red curve) was used to model the implant. The remaining curves show the diffusion during an anneal at 800 °C. The 5 min curve is essentially identical to the 1 h curve, indicating that the TED is over within the first 5 min. Beyond that time, D_P returns to its normal value, which is small at 800 °C and so little further diffusion takes place.

Figure 8.42 provides a deeper look into the simulation. The calculated {311} defect concentrations (dashed lines) and the free interstitial concentrations (solid lines) are shown. The {311} defects mirror the implanted P profile, as would be expected for a low-dose implant which does not amorphize the substrate. The {311} defects completely disappear after 5 min at 800 °C. Note the change in depth scale in Figure 8.42, which is necessary in order to track the fast-diffusing I species. The I profiles are relatively flat over several microns into the substrate, reflecting the high diffusivity of I. In the analytic treatment above, (8.42) assumes that the interstitial supersaturation is constant everywhere. We can see in Figure 8.42 that this is an approximation but holds reasonably well in the region of the P$^+$ implant. Numerical simulators can account for the varying I concentrations and adjust the P diffusivity locally given the local I value. The length of the TED we derived analytically, in (8.46) and also shown in Figure 8.40 for this same dose P$^+$ implant, is a few minutes at 800 °C, in the same range as the numerical simulation results.

From (8.46) or from Figure 8.40, we could calculate that, if the anneal temperature were 1200 °C rather than 800 °C, the {311} defects would disappear in a much shorter time, roughly 5 ms. Figures 8.43 and 8.44 repeat the simulation of the same implant at this higher anneal temperature.

We can see in Figure 8.44 that the {311} defects disappear in <25 ms, slightly longer than the simple analytic estimate, but in reasonable agreement. Notice also that the free interstitial

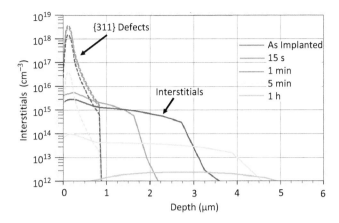

Figure 8.42 The {311} defect concentrations and free I concentrations calculated in the simulation shown in Figure 8.39.

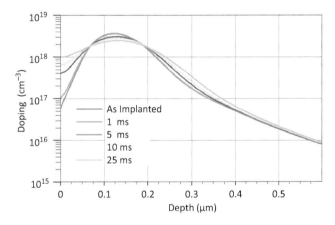

Figure 8.43 Silvaco Athena simulation of TED. A 5×10^{13} cm^{-2}, 100 keV P^{+} implant with a tilt of 7° into a silicon substrate was used. A Pearson distribution (red curve) was used to model the implant. The other curves show the TED during a 1200 °C anneal.

concentration is much higher in Figure 8.44 than in Figure 8.42. This is because C_I^* is much higher at 1200 °C ($\approx 1.5 \times 10^{16}$ cm^{-3}) than it is at 800 °C. In Figure 8.38, C_I^{max}/C_I^* is about 5000 at 800 °C and only about 150 at 1200 °C. But the much higher value of C_I^* at 1200 °C means that C_I^{max} is higher at 1200 °C, which is what is seen in Figure 8.44. The orange curve in Figure 8.44 would come down to the C_I^* value if the simulation were run a little longer. Even though the {311} defects produce a smaller supersaturation at the higher T (Figure 8.38), the magnitude of C_I^{max} is larger at higher T.

Most importantly, however, the amount of TED is significantly smaller in the 1200 °C anneal than it is in the 800 °C anneal, consistent with what we saw in Figure 8.34 when we began this discussion of TED. The very short time required to dissolve the {311} defects at 1200 °C means

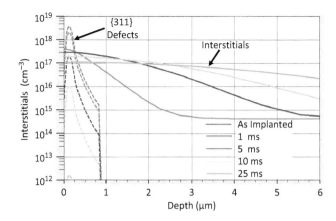

Figure 8.44 The {311} defect concentrations and free I concentrations calculated in the simulation shown in Figure 8.41.

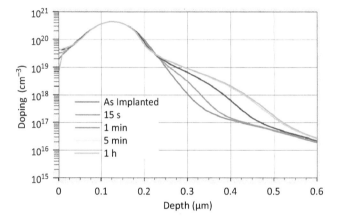

Figure 8.45 Silvaco Athena simulation of TED. A 5×10^{15} cm^{-2}, 100 keV P$^+$ implant with a tilt of 7° into a silicon substrate was used. A Pearson distribution (red curve) was used to model the implant. The other curves show the TED during an 800 °C anneal.

that a very short anneal can be used at the higher temperature, minimizing the amount of TED and therefore overall dopant diffusion. The physics we have discussed thus provides the motivation to do very-short-time, very-high-temperature implant anneals if the goal is to minimize TED.

The simulation results in Figures 8.45 and 8.46 show what happens when a much higher-dose implant is used. Here the dose is 5×10^{15} cm^{-2}, 100× higher than the previous simulations and large enough to amorphize the top part of the profile.

In Figure 8.46 we see that the {311} defects are only created in the tail of the implanted profile. The simulator assumes that the amorphized part of the profile recrystallizes through SPE and does not create {311} defects. The tail of the profile does create

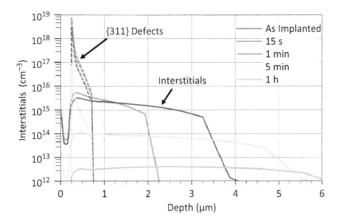

Figure 8.46 The {311} defect concentrations and free I concentrations calculated in the simulation shown in Figure 8.45.

{311} defects and these anneal out and cause TED just as in the earlier non-amorphizing example. The simulation predicts very little TED in the SPE part of the profile, as expected. In the tail region, TED is observed. The amount of TED is a little larger than in Figure 8.41. This is because the effective +1 dose in the tail region is larger than the +1 dose in Figure 8.41 and hence the duration of the TED is a little longer. The EOR defects that would likely be present (Figure 8.34) in an experiment like this are not modeled in the simulations in these figures.

While we have relied on simulation results to illustrate the essential features of TED modeling, there are many experimental results in the literature that support these models (e.g., [27, 28, 29, 30, 31, 32, 33] and many others). The models that are incorporated in simulators like Athena are based on these experiments and the theoretical understanding they made possible. The key ideas are encapsulated in Figure 8.47. The supersaturation of interstitials depends only on T (equation (8.42)). The duration of the TED depends on the effective +1 dose and on the implant energy through R_P (equation (8.46)).

Figure 8.47 provides a summary of what we have learned about TED. The key points follow.

1. Implantation produces significant damage in a crystalline substrate. Typically hundreds of lattice atoms are displaced for every ion implanted.
2. For regions that are damaged beyond the amorphization threshold, SPE easily repairs this damage at temperatures around 500 °C. Dopants are activated and no excess interstitials are produced during SPE.
3. In regions damaged below the amorphization threshold, excess point defects recombine quickly at moderate temperatures, typically 800 °C, leaving behind +1 interstitials when the implanted ions sit on lattice sites.
4. The +1 interstitials quickly form {311} defects at modest temperatures, typically 800 °C, that store the +1 excess interstitials.

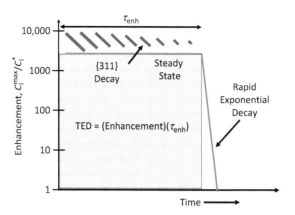

Figure 8.47 Schematic of the magnitude and duration of TED, determined from (8.42) and (8.47). The blue-shaded area is primarily responsible for the profile motion observed during TED.

5. Annealing the implant to remove the {311} defects can be done at temperatures from 800 °C to >1200 °C. While the {311} defects are dissolving, they maintain a constant supersaturation of interstitials given by (8.42). This is the flat red curve (steady state) in Figure 8.47.

6. Once the {311} defects are gone, the interstitial supersaturation quickly returns to equilibrium values (C_I^*). This is the decreasing red curve in Figure 8.47.

7. The τ_{enh} value is much shorter at higher temperatures (8.47), leading to smaller TED effects in high-T anneals.

8. Higher implant doses lead to longer τ_{enh} and more TED.

Thus, two parameters control the magnitude and duration of TED for practical purposes – the critical supersaturation level, and the time that the steady-state condition lasts, as shown schematically in Figure 8.47. This has been validated experimentally for As, P and Si implant damage [28]. Boron TED appears to be a bit more complex, perhaps because of the enormous strain that boron atoms introduce in the lattice, leading to the preferential formation of boron–interstitial complexes (BICs) [34]. There are fewer {311} defects seen in high-concentration boron layers, but there is still substantial TED. Small dots that appear on TEM images may be related to BICs. There appears to be competition between BICs and {311} defects for the excess interstitials. The BICs appear to have a higher binding energy for interstitials, so retain them longer during an anneal, giving rise to a longer time transient for boron TED than for the other dopants. However, the general picture holds, in that the magnitude and duration of TED is controlled by the evaporation of interstitials from microscopic defect aggregates.

8.10 Flash Annealing

The examples in the previous section indicated that TED can be minimized by using very short, very-high-temperature implant anneals. Once this was understood, there was a major effort in industry to develop manufacturing tools that could actually accomplish such anneals. The problem is not simple, because 300 mm silicon wafers need to be heated

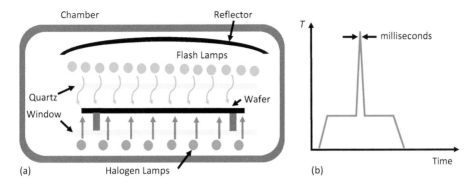

Figure 8.48 (a) Schematic of flash annealing system and (b) the temperature profile in the wafer.

to >1200 °C for a few milliseconds and then rapidly cooled down. To accomplish this, the wafers are normally supported on a very-small-thermal-mass holder (a few "pins" supporting the wafer) and heated with a bank of xenon lamps that are powered by a bank of capacitors which can discharge large amounts of power in a very short time. Such systems are called "flash annealing" systems and they are widely used today for implant annealing [35, 36]. A typical flash annealing system is shown schematically in Figure 8.48 along with a typical temperature versus time heat cycle.

Halogen lamps or other heating systems are used to heat the wafer to an intermediate temperature, which could be used, for example, to accomplish SPE regrowth of the amorphous part of the implanted profile. Then flash lamps are used to provide millisecond heating up to 1200 °C or higher to remove the {311} defects and to make certain the implanted dopant ions occupy lattice sites. In addition to minimizing TED, this high-temperature anneal also allows higher active dopant concentrations because the solubility of dopants increases as T increases. This helps to reduce the sheet resistance of heavily doped implanted regions. The dual heating system provides great flexibility in optimizing the T versus time cycle. One of the challenges in designing such a machine is that the temperature uniformity across the Si wafer must be kept within tight bounds because otherwise thermal gradients can cause significant issues with defects that can be introduced in the substrate if thermal stresses become too large.

8.11 Implantation and Damage in Semiconductors Other than Silicon

As we saw in Figure 8.23, process simulation tools like Athena can simulate implants in compound semiconductors as well as in silicon. For analytic profile calculations, the appropriate moments are needed in order to construct the profiles. For Monte Carlo methods, the nuclear and electronic stopping powers are needed. While data for these parameters are not as plentiful as for silicon, there are data for most of the commonly used compound materials, and so simulation is usually available.

In many compound semiconductor applications, especially in GaAs, the active layers are grown epitaxially and are not produced by ion implantation. Often this is because the atomic-scale precision and sharpness of these layers required for proper device operation are beyond the capability of ion implantation. However, ion implantation may still be used in these device structures to create isolation regions [37]. In these applications, the goal is often to introduce damage that makes the material insulating, and O^+ or He^+ or other ions may be used. Obviously, damage annealing is not an issue in these instances since the damage is the goal.

In cases in which implantation is used to create active regions in GaAs device structures, the same kinds of effects and physical mechanisms as described for Si have been found to occur [38, 39]. Amorphization and SPE take place at high doses. In less damaged regions, annealing is generally done at moderate temperatures (800–900 °C). Fairly high dopant activation can be achieved in P-type implants under these conditions; N-type implants are often more difficult to anneal and activate. The amount of experimental data is much less than in the case of Si and, as a result, accurate quantitative models for annealing and dopant activation generally are not available for GaAs. Nevertheless, optimization of particular processes follows the guiding principles described above for implants in silicon. There is also a significant literature on implantation and damage in materials like InP, GaSb, InGaAs, etc. [40]. These papers generally describe mechanisms similar to those we discussed for Si wafers.

SiC is another compound semiconductor of current interest, primarily for power devices because of its wide bandgap and large critical electric field. Ion implantation is used in fabricating many SiC device structures. We saw an example of a simulation of a SiC implant in Figure 8.23 that demonstrated channeling and other features similar to implants into Si. There are some important differences in SiC, however. First, if the implant dose is high enough to create an amorphous region, it is very difficult to anneal out this damage, likely because the substrate Si and C atoms are recoiled differently, creating stoichiometry issues. As a result, if high doses are required, often high-temperature implants are used in order to reduce the damage. Second, even if the SiC is not amorphized, extraordinarily high temperatures (1600–1700 °C) are required to repair the damage. Interestingly, essentially no dopant diffusion takes place during these anneals, so that there is no concern with TED or other anomalous effects. We discussed dopant diffusion in SiC briefly in Chapter 7 (see Figure 7.36). For the interested reader, an excellent reference on this topic is [41].

8.12 Other Applications of Ion Implantation

Because ion implantation has the flexibility of implanting almost any species into any substrate at almost any dose and energy, many other applications have been invented for the technology aside from doping semiconductors. In the previous section, we mentioned using implantation to produce damaged regions that provide lateral isolation between devices in some compound semiconductor technologies. Here we briefly mention two other applications that have

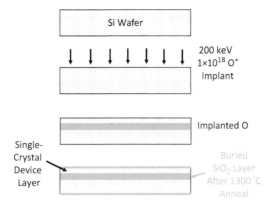

Figure 8.49 Process flow for the SIMOX process.

significant commercial importance in the semiconductor industry – SIMOX wafers and the "Smart Cut" process. These two applications are shown in the high-dose part of Figure 8.3.

The SIMOX (separation by implanted oxygen) process is used commercially to produce silicon on insulator (SOI) wafers. Many device structures and ICs can show improved performance using SIMOX starting wafers rather than bulk silicon wafers. This is primarily because SIMOX wafers put a true insulator (SiO_2) underneath the active devices, reducing leakage current and capacitances [42, 43, 44, 45].

The basic SIMOX process is shown in Figure 8.49. A very-high-dose ($\approx 10^{18}$ cm^{-2}) O$^+$ implant is done at relatively high energy (≈ 200 keV). Usually, this is an unmasked implant, since the goal is to create a wafer with a buried oxide layer everywhere. Normally, an implant dose this high would completely amorphize the silicon surface region, making it very difficult to repair the Si without a crystalline substrate below the surface region. In this case, there is no crystalline substrate below the Si surface layer because the implanted O converts the buried region into SiO_2. So the O$^+$ implant is normally done at high temperature (≈ 600 °C) in order to prevent the Si from being amorphized.

Following the implant, a very-high-temperature anneal is performed (≈ 1300 °C for several hours). This anneal is often done in an oxidizing ambient in order to prevent surface evaporation of Si. This forms a buried, stoichiometric SiO_2 layer and recrystallizes the surface Si layer. With optimization of the implant and annealing conditions, the buried Si/SiO_2 interface can be made abrupt and of high electrical quality (low interface state density).

Typically, the buried oxide (BOX) layer is 0.1–0.5 μm thick and the single-crystal device layer above it has a thickness on the same order. It is actually quite remarkable that a perfect single-crystal layer forms under these conditions. The silicon is heavily damaged by the O$^+$ implant, but, because a single crystal is the lowest-energy form for the silicon layer, there is a significant driving force to convert this damaged material into a single crystal, and a very-high-temperature anneal is used to allow this to happen. There are many commercially available ICs that use SIMOX substrates, and SOI processes based on these wafers are one of the options offered by commercially available chip foundry services. CMOS process flows like the example in Chapter 2 can be used with SOI substrates in place of the bulk silicon wafers used in that example.

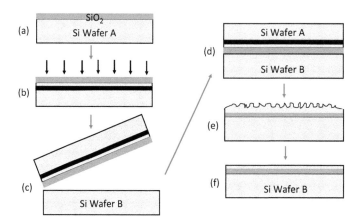

Figure 8.50 Process flow for the "Smart Cut" process. In (b) an H$^+$ implant is performed. In (c) and (d) the two wafers are bonded. In (e) the H-implanted layer splits wafer A. The surface roughness is exaggerated. In (f) CMP produces an atomically flat surface.

The second example of an interesting ion implantation application is the "Smart Cut" process, which is an alternative method for forming SOI wafers, but is also one of the methods used more generally in wafer bonding. The basic process flow is illustrated in Figure 8.50. The process was originally suggested by Bruel [45], but many others have since worked on turning it into a commercially viable technology [46, 47].

In this process, an oxidized silicon wafer is implanted with a high dose of H$^+$ or He$^+$ (other inert gases have also been used). Typically, a dose of 10^{16}–10^{17} cm^{-2} is used. The energy is chosen to place R_P at the desired depth below the surface of the wafer but is typically in the range of 10–100 keV. Because H and He are light ions, most of the damage they create is near the end of the range (near R_P), but the primary effect that they have is not so much the damage but rather simply the very high concentrations of these elements around R_P. As shown in Figure 8.50, the next step is careful cleaning of the wafer and a second wafer that will become the substrate of the completed SOI structure. Typically an RCA cleaning process is used, as discussed in Chapter 3. After cleaning, the two wafers are bonded, which involves carefully placing them in contact in a clean environment that ensures that no particles are present on either surface so that an intimate contact between the two is possible. The hydrophilic surfaces of the two wafers cause them to initially be held together by van der Waals' forces.

The next step is a heat treatment that is done in two steps. In the first step, an intermediate temperature ($\approx$ 500 °C) is used. This strengthens the bond between the two wafers. During this anneal, the H atoms diffuse to form microscopic cavities ("bubbles") near R_P and these cause the silicon wafer to split at that point, as shown in Figure 8.50. The original wafer can be reused at this point. The new composite wafer is further annealed at a higher temperature ($\approx$1100 °C), which strengthens the bond between the two wafers and makes the bonding energy essentially as high as the rupture strength of a silicon wafer. The SOI wafer can then be polished using chemical–mechanical polishing (CMP) to produce an atomically flat surface suitable for device fabrication. The thickness of the thin Si layer on top of the BOX SiO_2 layer is determined by the original energy of the H^+ implant.

It should be obvious that this process is more general than the SOI example shown in Figure 8.49, and the Smart Cut method has indeed been applied to a number of other structures. Thin Si layers have been bonded to other substrates like quartz, and other semiconductor materials have been bonded to a variety of substrates using the same process. There are clearly opportunities for interesting structures and applications, and this remains an interesting research area.

8.13 Plasma Immersion Ion Implantation

We have seen in this chapter that ion implantation using standard beam line machines (Figure 8.1) has become the dominant doping method used today. Especially when very high doses are required, the cost associated with conventional ion implantation can be quite high. An alternative is the plasma immersion ion implantation (PIII) system illustrated in Figure 8.51. These systems are much simpler than the beam line systems in Figure 8.1 and have found some use in high-volume manufacturing in applications in which their low cost and reduced capabilities make economic sense [48, 49].

In PIII systems, gases such as PH_3, B_2H_6 or BF_3 are used to create the doping ions to be implanted into the wafer. A plasma is a mixture of ions, electrons, neutral molecules or atoms, excited molecules and free radicals (neutral species that are incompletely bonded and very reactive), all existing in an ionized gas mixture. Plasmas are sometimes called the fourth state of matter and are the result of adding energy to a gas mixture to effectively raise its temperature to very high levels. Plasma systems are widely used in semiconductor processing, and we discuss these systems extensively in Chapters 9 and 10 because the most common uses of plasmas are in etching and deposition systems. The energy is commonly added to the gas using a microwave or RF power source, and this energy rips apart molecules such as BF_3, creating species such as B^+,

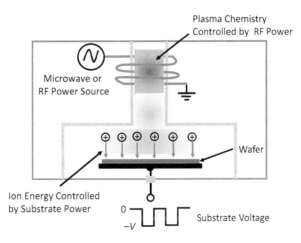

Figure 8.51 Plasma immersion ion implantation (PIII) system.

F^+, BF_2, etc. The chemistry is quite complex, but basically the more power added to the plasma, the more ions and other reactive species are created.

In PIII systems, a second power source biases the wafer negatively, which attracts the ions towards the wafer, resulting in implantation. The wafer power source could be a DC negative voltage if the surface is conducting, but more commonly an AC or pulsed source is used, as illustrated in Figure 8.51, to capacitively produce a negative voltage on the wafer even if insulating films are present on the wafer surface. We discuss the details of these systems in Chapter 9. For now it is sufficient just to understand that the wafer is biased negatively at a voltage controlled by the substrate voltage, which results in ion implantation.

The advantages of PIII systems include the fact that very high ion fluxes can be achieved simply by using higher plasma power levels. Thus high-dose implants can be achieved in much shorter times than in beam line systems. Under appropriate conditions, the wafer is essentially immersed in the plasma. Since electric field lines are always perpendicular to equipotential lines, and since the wafer is at a constant voltage, the ions are implanted perpendicular to the surface even if the surface is three dimensional, as in FinFETs or trench structures, for example. This allows uniform doping over complex surface topography, which is another major advantage of PIII systems. Dynamic random access memory (DRAM) capacitor structures and solar cells with textured surfaces are examples of applications in which PIII has been explored. Also, as illustrated in Figure 8.51, the entire wafer is typically exposed to the plasma, so the whole wafer is implanted at the same time without the need for complex mechanical or electrostatic scanning as is used in beam line implantation systems.

The disadvantages of PIII systems are a direct result of the fact that mass selection is not part of the system, so many ion types can be implanted (B^+ and F^+ in the example given above). In addition, depending on the pressure at which the PIII system operates, there may be considerable ion/neutral scattering that occurs in the plasma, with the result that the implanted ions will have a range of energies. This means that most applications of PIII involve low-energy implants at high doses, so that the exact implant energy is not critical.

8.14 Summary of Key Ideas

Ion implantation is a technique that introduces a precise number of dopant atoms into the silicon substrate. The ions slow down by a combination of nuclear and electronic interactions with the lattice atoms. Ion implantation energies range from hundreds to millions of electron-volts, with higher energy producing a deeper implant profile. This energy is much larger than the ≈ 15 eV needed to displace a lattice atom from the lattice and create a stable interstitial and vacancy pair. As a result, large numbers of displaced atoms are created by a single implanted ion. The damage can accumulate and eventually produce a completely disordered or amorphous layer where the implant has occurred. Damage produced by an implant causes a large enhancement in the diffusion coefficient of dopants, until such time as the damage is completely repaired. This effect is known as transient enhanced diffusion (TED), and often limits how shallow a junction can be made.

Simple models for ion-implanted profiles rely on analytic distribution functions such as the Gaussian and Pearson distributions. Generally, higher-order moments beyond R_P and ΔR_P are needed to characterize profiles even in amorphous material. The relatively open structure of crystalline semiconductor lattices leads to channeled profiles at low doses, where ions that enter a planar or axial channel travel further. The most general modeling approach to constructing implanted profiles involves Monte Carlo simulation. This capability is generally available in process modeling tools.

Because reliable methods have been discovered for annealing the damage caused by ion implantation, this doping method has become dominant in the silicon industry. It is also widely used in compound semiconductor fabrication and has found a number of uses beyond simple doping. It is one of the foundational technologies of modern chip fabrication.

8.15 REFERENCES AND NOTES

[1] P. H. Rose and G. Ryding, "Concepts and designs of ion implantation equipment for semiconductor processing," *Review of Scientific Instruments*, vol. 77, no. 11, p. 111101, 2006.

[2] K. Mera and H. Tomita, "High-current ion implanter for 300-mm SIMOX wafer production," *Hitachi Review*, vol. 51, no. 4, p. 113, 2002.

[3] L. Rubin and J. Poate, "Ion implantation in silicon technology," *Industrial Physicist*, vol. 9, no. 3, pp. 12–15, 2003.

[4] Simulation performed with computer code UT-MARLOWE, distributed by the University of Texas, Austin, Texas, USA.

[5] J. F. Gibbons, W. S. Johnson and S. W. Mylroie, *Projected Range Statistics. Semiconductors and Related Materials*, 2nd edn., Wiley, 1975.

[6] Athena and Victory Process are Silvaco process simulation tools. See https://silvaco.com/tcad/.

[7] D. Ashworth, R. Oven and B. Mundin, "Representation of ion implantation profiles by Pearson frequency distribution curves," *Journal of Physics D: Applied Physics*, vol. 23, no. 7, p. 870, 1990.

[8] G. Amaratunga, K. Sabine and A. Evans, "The modeling of ion implantation in a three-layer structure using the method of dose matching," *IEEE Transactions on Electron Devices*, vol. 32, no. 9, pp. 1889–1890, 1985.

[9] R. Alvis, S. Luning, L. Thompson, R. Sinclair and P. Griffin, "Physical characterization of two-dimensional doping profiles for process modeling," *Journal of Vacuum Science & Technology B: Microelectronics and Nanometer Structures Processing, Measurement, and Phenomena*, vol. 14, no. 1, pp. 231–235, 1996.

[10] J. F. Ziegler, "The stopping and range of ions in solids," in *Ion Implantation: Science and Technology*, ed. J. F. Ziegler, Elsevier, pp. 51–108, 1984.

[11] J. Lindhard, M. Scharff and H. E. Schiøtt, "Range concepts and heavy ion ranges (notes on atomic collisions, II)," *Matematisk-fysiske Meddelelser udgivet af det Kongelige Danske Videnskabernes Selskab*, vol. 33, no. 14, pp. 1–42, 1963.

[12] W. Brandt and M. Kitagawa, "Effective stopping-power charges of swift ions in condensed matter," *Physical Review B*, vol. 25, no. 9, p. 5631, 1982.

[13] J. F. Gibbons, "Ion implantation in semiconductors – Part I: Range distribution theory and experiments," *Proceedings of the IEEE*, vol. 56, no. 3, pp. 295–319, 1968.

[14] W. H. Bragg and R. Kleeman, "XXXIX. On the α particles of radium, and their loss of range in passing through various atoms and molecules," *London, Edinburgh, and Dublin Philosophical Magazine and Journal of Science*, vol. 10, no. 57, pp. 318–340, 1905.

[15] T. E. Seidel, "Ion implantation," in *VLSI Technology*, ed. S. M. Sze, McGraw-Hill, p. 219, 1983.

[16] J. P. Biersack and L. Haggmark, "A Monte Carlo computer program for the transport of energetic ions in amorphous targets," *Nuclear Instruments and Methods*, vol. 174, no. 1–2, pp. 257–269, 1980.

[17] S. Tian, S. Morris, M. Morris, B. Obradovic and A. Tasch, "Monte Carlo simulation of ion implantation damage process in silicon," in *International Electron Devices Meeting. Technical Digest*, IEEE, pp. 713–716, 1996.

[18] S. Tian, "Monte Carlo simulation of ion implantation in crystalline SiC with arbitrary polytypes," *IEEE Transactions on Electron Devices*, vol. 55, no. 8, pp. 1991–1996, 2008.

[19] K. Suzuki, T. Yoko, Y. Kataoka and T. Nagayama, "Monte Carlo simulation of ion implantation profiles calibrated for various ions over wide energy range," *Journal of Semiconductor Technology and Science*, vol. 9, no. 1, pp. 67–74, 2009.

[20] J. F. Gibbons, "Ion implantation in semiconductors – Part II: Damage production and annealing," *Proceedings of the IEEE*, vol. 60, no. 9, pp. 1062–1096, 1972.

[21] T. D. De La Rubia and G. Gilmer, "Structural transformations and defect production in ion implanted silicon: a molecular dynamics simulation study," *Physical Review Letters*, vol. 74, no. 13, p. 2507, 1995.

[22] W. Maszara and G. A. Rozgonyi, "Kinetics of damage production in silicon during self-implantation," *Journal of Applied Physics*, vol. 60, no. 7, pp. 2310–2315, 1986.

[23] M. Jaraiz, G. Gilmer, J. Poate and T. D. De La Rubia, "Atomistic calculations of ion implantation in Si: point defect and transient enhanced diffusion phenomena," *Applied Physics Letters*, vol. 68, no. 3, pp. 409–411, 1996.

[24] M. D. Giles, "Transient phosphorus diffusion below the amorphization threshold," *Journal of the Electrochemical Society*, vol. 138, no. 4, p. 1160, 1991.

[25] D. Eaglesham, P. Stolk, H. J. Gossmann and J. Poate, "Implantation and transient B diffusion in Si: the source of the interstitials," *Applied Physics Letters*, vol. 65, no. 18, pp. 2305–2307, 1994.

[26] L. Pelaz, L. A. Marqués and J. Barbolla, "Ion-beam-induced amorphization and recrystallization in silicon," *Journal of Applied Physics*, vol. 96, no. 11, pp. 5947–5976, 2004.

[27] J. Narayan, O. W. Holland and B. R. Appleton, "Solid-phase-epitaxial growth and formation of metastable alloys in ion implanted silicon," *Journal of Vacuum Science & Technology B: Microelectronics Processing and Phenomena*, vol. 1, pp. 871–887, 1983.

[28] S. W. Crowder, "Processing physics in silicon-on-insulator material," Ph.D. Thesis, Stanford University, 1995.

[29] H. Chao, S. Crowder, P. Griffin and J. Plummer, "Species and dose dependence of ion implantation damage induced transient enhanced diffusion," *Journal of Applied Physics*, vol. 79, no. 5, pp. 2352–2363, 1996.

[30] P. Packan and J. Plummer, "Transient diffusion of low-concentration B in Si due to [29]Si implantation damage," *Applied Physics Letters*, vol. 56, no. 18, pp. 1787–1789, 1990.

[31] P. A. Stolk, H.-J. Gossmann, D. J. Eaglesham, *et al.*, "Physical mechanisms of transient enhanced dopant diffusion in ion-implanted silicon," *Journal of Applied Physics*, vol. 81, no. 9, pp. 6031–6050, 1997.

[32] M. D. Giles, S. Yu, H. W. Kennel and P. A. Packan, "Modeling silicon implantation damage and transient enhanced diffusion effects for silicon technology development," in *Defects and Diffusion in Silicon Processing Symposium, San Francisco, 1997*, eds. T. Diaz de la Rubia, S. Coffa, P. A. Stolk and C. S. Rafferty, MRS Online Proceedings Library, vol. 469, Materials Research Society, 1997.

[33] C. Rafferty, G. Gilmer, M. Jaraiz, D. Eaglesham and H. J. Gossmann, "Simulation of cluster evaporation and transient enhanced diffusion in silicon," *Applied Physics Letters*, vol. 68, no. 17, pp. 2395–2397, 1996.

[34] K. S. Jones, J. Liu and L. Zhang, "Evidence of two sources of interstitials for TED in boron implanted silicon," in *Proceedings of the Fourth International Symposium of Process Physics and Modeling in Semiconductor Technology*, Electrochemical Society, 1996.

[35] C. H. Poon, A. See, Y. Tan, M. Zhou and D. Gui, "Improved boron activation with reduced preheating temperature during flash annealing of preamorphized silicon," *Journal of the Electrochemical Society*, vol. 155, no. 2, p. H59, 2007.

[36] S. Prucnal, L. Rebohle and W. Skorupa, "Doping by flash lamp annealing," *Materials Science in Semiconductor Processing*, vol. 62, pp. 115–127, 2017.

[37] T. Kazior, "Isolation implant studies in GaAs," *Journal of the Electrochemical Society*, vol. 137, no. 7, p. 2257, 1990.

[38] T. Haynes and O. Holland, "Comparative study of implantation-induced damage in GaAs and Ge: temperature and flux dependence," *Applied Physics Letters*, vol. 59, no. 4, pp. 452–454, 1991.

[39] I. Naik, "Annealing behavior of GaAs ion implanted with p-type dopants," *Journal of the Electrochemical Society*, vol. 134, no. 5, p. 1270, 1987.

[40] N. Rahimi, M. Behzadirad, E. J. Renteria, *et al.*, "Beryllium implant activation and damage recovery study in n-type GaSb," in *Physics, Simulation, and Photonic Engineering of Photovoltaic Devices III*, SPIE Proc., vol. 8981, International Society for Optics and Photonics, p. 89811Q, 2014.

[41] T. Kimoto and J. A. Cooper, *Fundamentals of Silicon Carbide Technology: Growth, Characterization, Devices and Applications*, Wiley, 2014.

[42] A. Auberton-Herve, A. Wittkower and B. Aspar, "SIMOX – a new challenge for ion implantation," *Nuclear Instruments and Methods in Physics Research B: Beam Interactions with Materials and Atoms*, vol. 96, no. 1–2, pp. 420–424, 1995.

[43] S. Krause, M. Anc and P. Roitman, "Evolution and future trends of SIMOX material," *MRS Bulletin*, vol. 23, no. 12, pp. 25–29, 1998.

[44] A. Ogura and H. Ono, "Evaluation of buried oxide formation in low-dose SIMOX process," *Applied Surface Science*, vol. 159, pp. 104–110, 2000.

[45] M. Bruel, "Silicon on insulator material technology," *Electronics Letters*, vol. 31, no. 14, pp. 1201–1202, 1995.

[46] B. Aspar, M. Bruel, H. Moriceau, *et al.*, "Basic mechanisms involved in the Smart-Cut® process," *Microelectronic Engineering*, vol. 36, no. 1–4, pp. 233–240, 1997.

[47] H. Moriceau, F. Mazen, C. Braley, F. Rieutord, A. Tauzin and C. Deguet, "Smart Cut™: review on an attractive process for innovative substrate elaboration," *Nuclear Instruments and Methods in Physics Research B: Beam Interactions with Materials and Atoms*, vol. 277, pp. 84–92, 2012.

[48] N. Cheung, "Plasma immersion ion implantation for semiconductor processing," *Materials Chemistry and Physics*, vol. 46, no. 2–3, pp. 132–139, 1996.

[49] S. Felch, F. Torregrosa, H. Etienne, Y. Spiegel, L. Roux and D. Turnbaugh, "PULSION® HP: tunable, high productivity plasma doping," *AIP Conference Proceedings*, vol. 1321, no. 1, pp. 333–336, 2011.

8.16   PROBLEMS

8.1 Arsenic is implanted into a lightly doped P-type Si substrate at an energy of 100 keV. The dose is 1×10^{14} cm^{-2}. The Si substrate is tilted 7° with respect to the ion beam to make it appear amorphous. The implanted region is assumed to be rapidly annealed so that complete electrical activation is achieved. What is the peak doping concentration produced?

8.2 A graduate student is interested in using Ga as a P-type dopant in Si in some experiments. She wants to implant the Ga into Si and then study Ga diffusion in Si. In looking at the properties of Ga, the student discovers that it has two naturally occurring isotopes [69]Ga and [71]Ga. The student wonders whether the range statistics for the two isotopes will be significantly different and therefore she should select only one isotope in the implanter. Would the profiles of the two isotopes be within 5% of each other, so selecting only one would not be necessary?

8.3 The graph below shows plots of the projected range for boron and arsenic. The data are taken from Table 8.1. The dashed lines are straight lines and they show that the As data are well fitted by a straight line. The B data, on the other hand, are sublinear. Show mathematically why this behavior should be expected for As and for B.

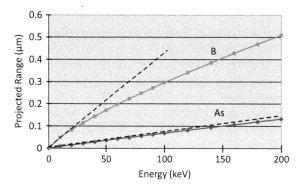

8.4 Tin ($Z = 50$, $M = 118$) is implanted into a silicon ($Z = 14$, $M = 28$) substrate at an energy of 10 keV.
(a) Calculate the expected projected range. State any assumptions you make.
(b) Assume that all of the damage caused by the Sn implant is located between the surface and R_P. If 25% of the Si atoms must be displaced to "amorphize" the silicon, estimate the dose required to amorphize the Si crystal to a depth of R_P.

8.5 A 120 keV phosphorus ion implantation is performed into the structure shown below. Following the implant, a heat treatment sufficient to achieve a flat phosphorus profile in the 0.1 μm Si layer is performed. You may assume that, during the heat treatment, there is no transfer of dopant between the SiO_2 layers and the 0.1 μm Si layer. If the goal is to achieve a flat phosphorus profile with a concentration of 10^{20} cm^{-3} in the 0.1 μm Si layer, what dose Q should be implanted?

SiO_2	0.1 μm
Si	0.1 μm
SiO_2	0.1 μm
Si Substrate	

8.6 The simulations below show calculated 2D implanted profiles for (a) boron, (b) phosphorus and (c) arsenic, with the energies adjusted to provide roughly the same range in each case. The differences in the calculated lateral straggle are striking, under the mask edge. Explain why these should be so different.

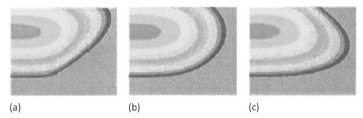

(a)   (b)   (c)

8.7 A 50 keV boron implant is done into a silicon substrate. Parts of the silicon substrate are masked with a 0.4 µm SiO_2 layer. Suppose it is important that not more than 1 cm^{-2} boron ion on average penetrates the SiO_2 layer. What is the maximum boron dose that can be implanted into the silicon and meet this criterion? You may assume that the range statistics are the same in Si and SiO_2.

8.8 In the CMOS process flow we discussed in Chapter 2, a boron source/drain implant is done as shown below. Suppose B^+ is implanted at 70 keV with a dose of 1×10^{15} cm^{-2}. What is the minimum thickness of the polysilicon gate in order that the threshold voltage of the PMOS transistor not be modified by this implant by more than 0.1 V? The gate oxide in the transistor is 2 nm thick. State any assumptions you make.

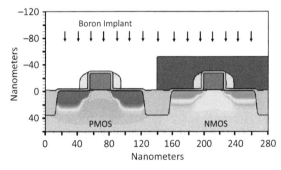

8.9 A 1×10^{13} cm^{-2} phosphorus implant is done at 100 keV (R_P = 0.127 µm, ΔR_P = 0.0428 µm). An SiO_2 layer covers regions on the wafer where the implant is supposed to be masked. How thick does the SiO_2 layer need to be to ensure that on average not more than 1000 cm^{-2} phosphorus ions penetrate the SiO_2 and get into the Si beneath the oxide? Calculate a numerical answer.

8.10 An engineer worried about avoiding punch-through in an NMOS device decides to perform a deep punch-through implant using boron while keeping the surface concentration at a maximum value of 1×10^{17} cm^{-3}. What is the maximum implant dose that would be suitable if the peak of the implanted profile is at 0.2 µm?

8.11 How thick does a mask have to be to reduce the peak doping of an implant by a factor of 10,000 at the mask/substrate boundary. Provide an equation in terms of R_P and ΔR_P.

8.12 In a particular application, it is important to produce a fairly flat profile over an extended distance by ion implantation, as indicated below. One way to do this is to superimpose several implants at different energies. If phosphorus implants with energies chosen to produce projected ranges of $0.5R_P$, R_P and $2R_P$ are used with a dose of Q in the middle peak, approximately what doses should be used in the adjacent peaks if the initial peak concentrations are to be the same.

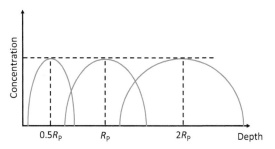

8.13 Estimate the range of 500 keV boron and arsenic implants using Table 8.1. The table only goes up to 200 keV, so you will have to extrapolate. Explain your reasoning.

8.14 The equations A–F provide a reasonable analytical description for some of the diffusion processes indicated schematically in the diagrams that follow:

A $C(x, t) = \dfrac{C}{2}\left[\mathrm{erfc}\left(\dfrac{x}{2\sqrt{Dt}}\right)\right]$,

B $C(x, t) = \dfrac{Q}{\sqrt{\pi(D_1 t_1 + D_2 t_2)}}\exp\left(-\dfrac{x^2}{4(D_1 t_1 + D_2 t_2)}\right)$,

C $C(x, t) = \dfrac{Q}{2\sqrt{\pi Dt}}\exp\left(-\dfrac{x^2}{4Dt}\right)$,

D $C(x, t) = C\left[\mathrm{erfc}\left(\dfrac{x}{2\sqrt{Dt}}\right)\right]$,

E $C(x, t) = \dfrac{Q}{\sqrt{\pi Dt}}\exp\left(-\dfrac{x^2}{4Dt}\right)$,

F $C(x, t) = \dfrac{Q}{\sqrt{2\pi(\Delta R_P^2 + 2Dt)}}\exp\left(-\dfrac{(x - R_P)^2}{2(\Delta R_P^2 + 2Dt)}\right)$.

For each of the diagrams (a)–(f) below, which equation A–F is the best match? Equations may be reused, or multiple equations may describe the same figure. A brief explanation is required for each figure.

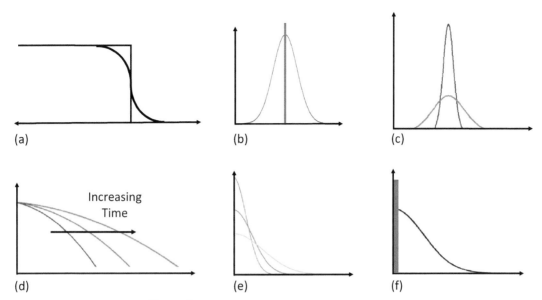

8.15 An 80 keV, 5×10^{13} cm^{-2} boron implant is performed into bare silicon. A subsequent anneal at 950 °C ($D_B = 3.82 \times 10^{-15}$ cm^2 s^{-1}) is performed for 60 min.

(a) Can the annealed profile evolution be described by the following formula?

$$C(x,t) = \frac{Q}{\sqrt{\pi D_B t}} \exp\left(-\frac{x^2}{4D_B t}\right)$$

(b) Assume that all the dopant remains in the silicon and that none evaporates to the ambient. By considering a virtual or imaginary image profile on the ambient side of the interface (a reflecting boundary), calculate the surface concentration after a 60 min, 950 °C anneal.

8.16 A 1×10^{14} cm^{-2} phosphorus implant through a 200 nm SiO$_2$ mask layer is performed so the peak concentration is at the Si/SiO$_2$ interface. An anneal is then performed for 30 min at 1000 °C. Calculate the location of the junction with the substrate doped at 1×10^{15} cm^{-3}. Assume no diffusion in the masking layer and ignore any segregation effects. Assume the same range statistics for SiO$_2$ and Si.

8.17 Phosphorus is implanted at 50 keV with a dose of 1×10^{14} cm^{-2}. Calculate the junction depth where the phosphorus meets the substrate (P type, 1×10^{15} cm^{-3}) after:

(a) a 900 °C, 3 h anneal and
(b) a 1200 °C, 3 h anneal.

8.18 An implant machine for 300 mm wafers is required to have a throughput of 60 wafers per hour. What beam current is required in order to implant a source/drain region in a CMOS device with a dose of 1×10^{15} cm^{-2}?

8.19 In the ion implantation process, positively charged ions impact on the semiconductor surface. Normally, these ions are neutralized by capturing an electron from the conducting substrate. However, when the mask is an insulator like SiO$_2$, the charge on the ions may not be neutralized as easily. Consider the case where a dose Q is implanted into the surface of an SiO_2 layer (assume all the charge resides at the oxide surface). Further assume that the oxide can withstand an electric field of 10^7 V cm^{-1} before it breaks down. What implant dose Q is required to cause electrical failure of the mask? That is, what dose will cause a field of 10^7 V cm^{-1} across the oxide?

8.20 An engineer investigating SPE regrowth after amorphizing ion implants of various species (P, B, Si, Ge, As and Sb) makes the following observations.

(a) N-type dopants of very different size or atomic radius (e.g., antimony versus phosphorus) show identical regrowth rates, approximately an order of magnitude faster than the regrowth of silicon implanted and amorphized with silicon ions.

(b) P-type and N-type ion-implanted regions have much faster regrowth rates than Si or Ge implanted regions, although they are not identical.

(c) B-doped regions compensated with an equal dose of an arsenic implant show identical regrowth rates to Ge implanted and amorphized regions.

Construct a unified physical explanation of all three phenomena, considering possibilities such as size or stress effects, dopant charge or electric field effects, or point-defect-based effects (no calculation required).

8.21 We discussed the ion implantation damage production model shown below, in the text. We estimated the total number of displaced Si atoms caused by the primary As ion as 1000 atoms.

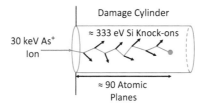

(a) Using the Kinchin–Pease formula, estimate the number of displaced Si atoms caused by each of the Si knock-ons in the figure above.

(b) Estimate the same number as in part (a), but find this number by calculating the range of the knock-on Si atoms. Presume that they create one displaced Si atom for each lattice plane that they cross. Do not use online look-up tables for the range of the Si knock-ons. You must calculate this number.

(c) Why is your answer in (b) smaller than your answer in (a)? Even if you did not get answers to parts (a) and (b), you can still reason about this question without numerical answers.

8.22 Suppose boron is implanted at 100 keV into a silicon substrate. While most of the stopping initially is through electronic energy loss mechanisms, the nuclear stopping component for boron is not zero (see the main text on ion implantation). Estimate a lower bound on how many atomic planes each boron ion passes through before it dislodges a Si atom through a nuclear collision.

8.23 In two separate experiments, As and then B are implanted at the same energy through a thin SiO_2 layer into the underlying substrate. As a result of the implantation, some of the oxygen atoms in the SiO_2 layer are knocked into the silicon substrate. Would you expect the As or the B to produce more oxygen knock-ons? Why?

8.24 In a particular ion-assisted etching process, a CF_4/H_2 plasma is used to etch a silicon substrate. (We will discuss etching in detail in Chapter 9.) The etching system produces many molecular and ionic species in the plasma, including CF_3, CF_3^+, H^+ and F^+. The etching system generates a potential of 1000 eV between the plasma and the electrode on which the wafer sits. Since energetic species are involved in the etching process, some damage can be created in the silicon substrate during the etching. You can consider this problem to be a low-energy ion implantation problem. Considering just the four species listed above, over what depth into the silicon would you expect the etching to have an impact on the silicon material? Explain your reasoning and calculate a quantitative answer.

8.25 A 30 keV As implant is done into a silicon wafer ($R_P = 25$ nm). If we define "amorphized" to mean that 10% of the Si atoms in the substrate are displaced from lattice sites, quantitatively estimate the As dose needed to amorphize the substrate in the region of the implant.

8.26 A boron implant is performed into silicon at 100 keV. The boron beam is aligned with the silicon crystal so that channeling is present. Estimate the range of the channeled boron profile, by considering that electronic stopping is the only mechanism for slowing the boron ions.

8.27 In 1993, Giles published an interesting experimental result. He did implants into two wafers. In the first case, a dose Q of the dopant was implanted normal to the surface at an energy E. In the second case, the same dose and dopant were implanted at significantly higher energy and at an angle such that the two implanted profiles were quite similar. The difference between the two was that the higher-energy implant should have produced more damage for the same implanted dose. When the wafers were annealed to eliminate the implant damage, very similar amounts of TED were observed in the two wafers. Explain how this result is consistent with current atomic-level understanding of TED.

8.28 In an experiment, two 10^{13} cm^{-2} boron implants are done on two separate wafers, one at 50 keV and one at 200 keV. A buried marker layer well below these implants is used to measure the TED that occurs during the implant damage anneal. Each of the wafers is annealed at the same temperature for a time just sufficient to repair the damage. Estimate quantitatively the difference in TED seen in the two wafers (i.e., the TED Dt product seen by the buried marker layer). Explain physically why there is a difference, if there is a difference.

8.29 An amorphizing implant is performed using a high dose of arsenic (5×10^{15} cm^{-2}, 200 keV) and TEM cross-section images indicate complete amorphization to a depth of 50 nm. If SPE regrowth instantly removes all of the damage in the amorphous region, calculate the fraction of the "+1" implanted dose that is available for TED.

8.30 Calculate the change in junction depth for a 40 keV boron threshold adjust implant of 5×10^{13} cm^{-2} annealed at 750 °C in a furnace or at 1000 °C in an RTA for a time just long enough to remove all the damage that causes TED. Assume a uniform well doping (background doping) of 5×10^{16} cm^{-3}.

8.31 The diffusion of a buried antimony layer is only 20% of its normal diffusion in an inert ambient after a phosphorus implant is performed at the surface of a silicon wafer. The phosphorus diffusion coefficient is itself enhanced by a factor of 4 (400%).

(a) Suggest a qualitative reason for this observation.

(b) Assuming that the phosphorus diffuses completely by an interstitial mechanism and that interstitial vacancy recombination is fast, calculate how much of the antimony diffusion is mediated by interstitials and how much by vacancies.

8.32 A phosphorus implant is performed into a bare P-type silicon wafer with a background doping of 1×10^{15} cm^{-3}, at an energy of 10 keV and a dose of 1×10^{14} cm^{-2}. After an anneal at 900 °C for 10 min, the junction depth is measured to be 0.25 μm.

(a) During the 10 min anneal, the phosphorus diffusivity would be enhanced by a variety of effects, including TED, concentration-dependent D, electric field effects, etc. Given the final junction depth, calculate the average enhancement in the phosphorus diffusion coefficient during the 10 min anneal that was caused by the combined effects of all these things, compared to the intrinsic diffusivity of phosphorus. You may assume the phosphorus profile is Gaussian.

(b) The answer you calculated in part (a) is the average diffusivity enhancement during the 10 min anneal. Without doing any calculations, what would you expect the actual time dependence of the enhancement would really be during the anneal? Explain.

8.33 In the CMOS process that we described in Chapter 2, we did V_{TH} adjust implants without annealing the damage from those implants, and then we grew a temporary gate oxide of 1–2 nm, and then deposited and oxidized polysilicon to serve as the "dummy" gate while we finished the fabrication steps for the transistors. The V_{TH} implants actually will have at least some of their damage annealed by the high-temperature steps even though no specific implant anneal was done at this stage. Suppose the V_{TH} implant was 1×10^{13} cm^{-2} B at 10 keV ($R_P = 47.3$ nm). The subsequent steps were gate oxidation (800 °C for 10 min in dry O_2), poly deposition (LPCVD, 600 °C for 30 min) and poly oxidation (800 °C for 30 min in dry O_2). What fraction of the V_{TH} implant damage is actually removed by these subsequent steps?

8.34 As discussed in this chapter, there is great interest in using millisecond annealing of implanted junctions because these high-temperature, very-short-time anneals repair implant damage while minimizing TED.

(a) Using the TED models developed in the text, calculate the actual time required to anneal an As source/drain implant at 1200 °C. The As implant is 5×10^{15} cm^{-2} at 100 keV. You may neglect any impact of SPE. State any assumptions that you make.

(b) If SPE had been included in this calculation, would your answer have changed significantly? Why or why not?

8.35 A TED experiment is done using a standard marker layer (shown below). The implant is 10^{14} cm^{-2}, 50 keV Si and the buried marker layer is Sb. The Si implant damage is annealed at 1350 °C for a few milliseconds (just long enough to anneal the damage).

(a) Calculate the change (enhancement or retardation) in the buried layer Sb diffusivity during TED. You can assume interstitial/vacancy recombination is instantaneous.

(b) If the TED anneal had been done at a much lower temperature, say 1000 °C, you would have calculated that the Sb diffusivity was greatly enhanced over its normal value. Explain why this result would have been found.

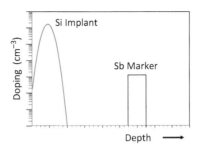

8.36 A special test structure is fabricated as illustrated below. A Si wafer with a P epi-layer on an N substrate is patterned to have a Si_3N_4 layer on the left side. The right side is oxidized to produce the SiO_2 layer shown.

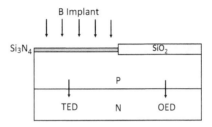

Prior to the oxidation, a 1×10^{13} cm^{-2}, 100 keV B implant on the left side is performed. You may assume the nitride is thin enough so that the entire B implant passes through it and goes into the Si epi-layer and also that the B implant is shallow compared to the thickness of the P epi-layer. Thus, during the LOCOS oxidation, OED will occur on the right side and TED will occur on the left side. The oxidation is 30 min at 1000 °C. After the oxidation is complete, it is observed that the location of the P–N junction is the same on the left and right sides. What was the average OED diffusivity enhancement during the LOCOS process? State any assumptions.

8.37 A TED experiment is done using a standard marker layer. The implant is a 1×10^{13} cm^{-2}, 50 keV As implant. The buried marker layer is also As. The implant damage is annealed at 800 °C for a time long enough to anneal out the damage.
   (a) Calculate the enhancement in the buried layer diffusivity during TED. You can assume interstitial–vacancy recombination is fast.
   (b) Repeat your calculation if the buried marker layer is Sb. Is the Sb retarded or enhanced by the TED?

8.38 A TED experiment is done using a standard marker layer (shown below). The implant damage is annealed at an unknown temperature. The buried marker layer is Sb and the result of the experiment is the rather surprising observation that the diffusivity of the Sb during the implant anneal is exactly the normal (intrinsic) Sb diffusivity. Explain how this could happen. What must the temperature of the implant anneal have been? You may assume $f_I = 0.02$ and $f_V = 0.98$ for Sb and you may assume that I–V recombination is very fast. State any assumptions you make.

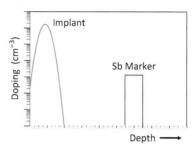

8.39 An arsenic implant with dose of 10^{13} cm^{-2}, 200 keV is performed into a silicon wafer.
   (a) Assuming the profile is Gaussian, sketch the resulting profile. Label your axes quantitatively to show the key parameters associated with the profile.
   (b) The implant is then annealed at 800 °C for 60 min. Assuming that the profile remains Gaussian, calculate the resulting profile, including TED effects. State any assumptions.
   (c) Before the implantation, if it is required to coat photoresist to totally "mask" the implantation, what is the minimum thickness of the resist needed? Assume the silicon wafer has P-type doping of 10^{15} cm^{-3} and the criterion for masking is that the implant cannot exceed this concentration anywhere.

8.40 An engineer wants to form a shallow boron-doped source/drain junction for an advanced technology ($Q = 1 \times 10^{15}$ cm^{-2} and $E = 40$ keV). The manager wants to know whether the company should buy an inexpensive, batch furnace and achieve the required junction depth using a low-thermal-budget anneal (1 h at 800 °C) or an expensive, single-wafer, RTA using a high-temperature anneal.
   (a) Calculate the time required to achieve the target junction depth if the annealing temperature is 1050 °C in the RTA and make a simple estimate ($2\sqrt{Dt}$) of how far the dopants move during these anneals.
   (b) Now, consider that the boron is introduced using an implant and that TED due to the implant will be important. Using the TED charts in the text of this chapter for the expected enhancement in diffusivity and the time TED lasts, calculate how far the dopants move at each temperature. Which anneal would you recommend?

8.41 In a particular CMOS process, it is important to minimize the anneal time needed to repair implant damage, so a preamorphization implant is done to amorphize the Si. This implant is a 1×10^{16} cm^{-2} Si implant, for which $R_P = 0.12$ μm and $\Delta R_P = 0.06$ μm. You may assume this implant amorphizes the Si from the surface down to a depth where the Si implant concentration drops to 10^{20} cm^{-3}. A boron implant is then done (70 keV, 5×10^{13} cm^{-2}) for which $R_P = 0.223$ μm and $\Delta R_P = 0.0736$ μm. An anneal is then done at 1200 °C to repair the implant damage. How long does this anneal need to be to completely repair the damage?

8.42 Germanium is starting to be of interest as a semiconductor substrate for MOS devices because of its higher carrier mobility and hence higher device performance. Thus basic processing issues in Ge are being investigated. In a series of experiments, As and B diffusion in Ge are studied. Contrary to silicon, As is found to be a fast diffuser and boron a slow diffuser. Obviously, diffusion mechanisms may be different in Ge. In a TED-type experiment, implant damage is created through a Ge implant. The diffusion of As is found to be retarded when the implant damage is present. What would you propose as the basic mechanism for As diffusion in Ge? Why?

8.43 An experimental test structure shown below is used to test the usefulness of a preamorphization Si implant in minimizing TED. On one wafer, a 20 keV, 1×10^{14} cm^{-2} boron implant is done and then subsequently annealed at 1100 °C for long enough to repair the implant damage. On a second wafer, the same boron implant is done after a 5×10^{15} cm^{-2} Si implant is done to amorphize the substrate. For the Si implant you may assume that $R_P = 0.1$ μm and $\Delta R_P = 0.04$ μm. You may also assume that the Si implant amorphizes the substrate above the point where the Si implant concentration equals 1×10^{20} cm^{-3}. On the second wafer, following the Si and B implants, the wafer is annealed at 1100 °C, for long enough to repair the implant damage. Determine quantitatively whether the Si implant actually reduces the TED seen by the boron marker layer or not. State any assumptions.

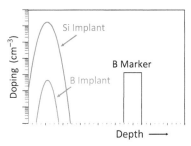

9 Etching Technologies and Chemical–Mechanical Polishing

9.1 Introduction

As we have seen throughout this book, material deposition and material removal are critical steps in integrated circuit (IC) fabrication. A wide variety of materials, insulators, semiconductors and conductors must be deposited at various stages in chip manufacturing. Usually, these materials are deposited in blanket form covering the entire wafer surface, although there are some deposition methods which are selective and deposit materials only in specific locations on the wafer surface. We will discuss deposition methods in detail in Chapter 10. Selective removal of material is usually accomplished using a lithography-defined mask followed by etching. We will discuss a variety of etching methods in this chapter.

Material removal can also be accomplished using chemical–mechanical polishing (CMP). This process is usually not selective but uses a combination of chemical etching and mechanical polishing to remove materials. The original motivation for developing CMP was to planarize wafer surfaces in back-end structures, since the polishing produces a flat surface. We discuss this history in Chapter 11 and illustrate the evolution of back-end structures as CMP became manufacturable. However, as we saw in the CMOS process in Chapter 2, CMP is widely used today, even in front-end processing for planarization. Shrinking device geometries and higher-resolution lithography imply smaller depth of focus in printing patterns, as we saw in Chapter 5. CMP has made this work in manufacturing by producing perfectly flat surfaces on which to do lithography. So in this chapter we will also discuss CMP as a general material removal method, although most of the chapter will focus on etching.

Etching is the process that turns designs into reality. It takes design information from masks and transfers it into regions on the semiconductor where doped regions, isolation regions, gates and interconnects are formed. It could be argued that, without precise etching, the patterns on the mask could not be faithfully transferred into the substrate. Modern etch machines are among the most sophisticated second-tier machines in semiconductor processing, where lithography machines occupy the first tier.

But we can begin with the simpler concepts of isotropic and directional etching, sometimes referred to as wet and dry etching (though there are some important qualifiers). Isotropic etching means the etching proceeds at the same rate in all directions (vertically and laterally) and is typical of most wet etchants. Directional etching means that the process only etches in the vertical direction and does not etch laterally. A simple schematic is shown in Figure 9.1.

For the most part, wet etchants create isotropic profiles while dry (or plasma) etching creates vertical profiles. First, we discuss wet etching and some of the caveats that apply where wet etch profiles can be quite directional. The rest of the chapter focuses on the much more prevalent and flexible dry or plasma etching processes.

9.2   Wet Etching

Wet etching is simple and selective. First, simplicity – it just requires dipping a wafer into a wet chemical bath, and the exposed material is etched if the chemical is chosen properly. Second, selectivity – because it is a chemical process, it can be very selective between the material etched and the underlying material, i.e., it can etch only the target material and stop etching when it reaches the underlying material. This makes it a very uniform and reproducible process [1].

A common etchant for SiO_2 is hydrofluoric acid (HF), which etches the oxide selectively compared to silicon or silicon nitride. The overall reaction is

$$SiO_2 + 6HF \rightarrow SiH_2F_6 + 2H_2O \,. \tag{9.1}$$

A buffering agent such as NH_4F is often added to HF to help prevent depletion of the fluoride ions in the oxide etch and maintain uniform etch rates. This is called "buffered oxide etch" (BOE) and is very common because it decreases the etch rate and lifting of photoresist masks during oxide etching. As seen in Table 9.1, BOE has effectively infinite selectivity against silicon or silicon nitride while etching thermal oxide at 100 nm min^{-1}. For very thin oxides or for

Table 9.1 **Etch rates (nm min^{-1}) for some common materials and etchants used in silicon and micro-electromechanical systems (MEMS) processing.**

	Isotropic Si etch	5:1 BOE	Phosphoric acid
Silicon	150	≈0	0.17
Thermal SiO_2	8	100	0.18
Fused quartz	12	130	0.23
Pyrex 7740	140	43	3.7
Si_3N_4	≈0	≈0	4.5

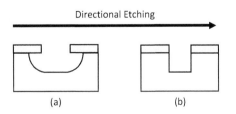

Directional Etching

(a)   (b)

Figure 9.1 Schematic of (a) isotropic etching and (b) directional or anisotropic etching.

etching the native oxide that grows on silicon, a very dilute 100:1 HF solution is often used with an etch rate of ~1 nm min^{-1}.

In order to get silicon to etch in HF, an oxidizer such as nitric acid (HNO$_3$) is added, which generates a thin chemical oxide that is then etched by the HF. Buffering agents such as acetic acid (CH$_3$COOH) are added to limit the dissociation of the nitric acid and create uniform chemical oxidation rates in these isotropic silicon etchants. This combination is the "isotropic silicon etch" in Table 9.1.

Table 9.1 shows the exquisite selectivity that can be obtained with wet chemical etches. The selectivity S of an etchant for two materials is the ratio of their etch rates, giving

$$S = \frac{r_1}{r_2}, \tag{9.2}$$

where r_1 is the etch rate of the film being etched and r_2 is the etch rate of the masking material or the underlying material. For example, the simple LOCOS (local oxidation of silicon) isolation structure consists of a silicon nitride film on a thin pad oxide on top of silicon. After growing the LOCOS oxide, the silicon nitride is removed in hot phosphoric acid. Using the etch rates in Table 9.1, we see that the selectivity for etching nitride over oxide is

$$S = \frac{4.5}{0.18} = 25. \tag{9.3}$$

Thus, the nitride can be safely over-etched to achieve uniform removal without worrying about the pad oxide being etched. Even if the pad oxide is etched, the selectivity against the underlying silicon is equally good.

While the selectivity of wet etching is excellent, the major problem is that it etches laterally (sideways) just as fast as it etches vertically. As shown in Figure 9.2, this means that the profile under the mask is semicircular if the film is just etched to its full thickness. Many processes cannot tolerate this lateral etching under the mask, and this provides the incentive for dry etching, which is more directional. Note, however, that, if the film is over-etched, then the sideways profile appears more vertical, as the radius of curvature is larger. So, just looking at the etched profile after the mask is removed does not tell you much about the etch process, without knowing the position of the mask edge.

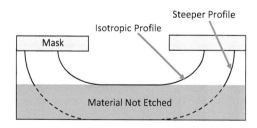

Figure 9.2 Isotropic (semicircular) etch profile at a mask edge, where the lateral and vertical etch rates are equal, and apparent steeper lateral etch profile after over-etch.

Table 9.2 **Crystallographic etch rates of silicon and common hard mask materials in potassium hydroxide (KOH; strongly anisotropic) and tetramethyl ammonium hydroxide (TMAH; weakly anisotropic). After [1].**

Etchant	Si (110) (nm min^{-1})	Si (100) (nm min^{-1})	Si (111) (nm min^{-1})	Si$_3$N$_4$ (nm min^{-1})	SiO$_2$ (nm min^{-1})	Si ratio 110:100:111
KOH (44%, 85 °C)	2800	1400	47	~0	1.4	600:300:1
TMAH (25%, 80 °C)	919	500	13.5	~0	0.2	68:37:1

Between the two extremes of fully isotropic and fully anisotropic etching illustrated in Figure 9.1, there are many intermediate profiles that are partially anisotropic. Such profiles can be considered a mixture of isotropic and anisotropic etching, and many etching systems operate in this intermediate mode, as we will see. A useful definition of the degree of anisotropy is

$$A = 1 - \frac{r_{lat}}{r_{vert}}, \tag{9.4}$$

where r_{lat} is the etch rate in the lateral direction and r_{vert} is the etch rate in the vertical direction. In Figure 9.2, the isotropic profile has an anisotropy $A = 0$ since chemical etching produces the same etch rate in all directions. For times longer than required to etch the isotropic profile in Figure 9.2, r_{vert} may go to zero if there is perfect selectivity in the etching process, or r_{vert} may have a lower value than during the isotropic etching if the selectivity is finite. Note that A could become negative if r_{vert} is small.

9.2.1 Directional Wet Etching

Some wet etchants have the interesting property that they etch some crystal planes much faster than others. For example, potassium hydroxide (KOH) at 85 °C etches the silicon crystal planes (110) > (100) > (111) with relative values of 600:300:1 (see Table 9.2). Essentially, this means that the etching stops on (111) planes. This can produce interesting pyramidal or truncated pyramidal structures, which are often useful for generating membranes or suspended cantilevers in micro-electromechanical systems (MEMS) structures, as shown in Figure 9.3. The cantilever structure shown in Figure 9.3 would be deposited and defined on the top surface before the pyramidal cavity is etched out underneath it.

Note from Table 9.2 that the selectivity against SiO$_2$ is so high that a brief 100:1 HF dip may be needed to remove native oxide before starting to etch with these anisotropic etches. Both KOH and tetramethyl ammonium hydroxide (TMAH) attack photoresist, so hard masks such as Si$_3$N$_4$ or SiO$_2$ must be used. Aluminum can also be used as a mask with TMAH.

9.3 Basics of Dry or Plasma Etching

Because wet etching etches laterally as fast as vertically (except for the crystallographic etches mentioned above), the mask patterns are not faithfully transferred to the substrate. Instead,

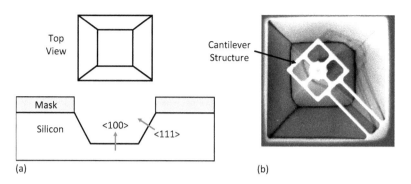

Figure 9.3 (a) Schematic of pyramidal structures formed by KOH etching, stopping on the (111) plane. (b) Suspended cantilever over a pyramidal cavity. (From P. Apte *et al.*, https://www.ee.iitb.ac.in/~apte/PFOLIO .htm). The red arrows indicate crystal directions that are perpendicular to planes in cubic crystals like silicon.

they are enlarged over what is printed in the mask. This led to the development of dry or plasma etching, which is capable of much more directional etching and better transfer of the mask pattern to the underlying layers, as shown in Figure 9.4. We will see that the directionality of the etching process arises because ions are involved in the etching and these ions have a directed velocity perpendicular to the wafer surface due to the presence of electric fields in the plasma system. Neutral species may also be involved in plasma etching, as illustrated in Figure 9.4. Neutrals generally do not provide directional etching. The etching byproducts must be volatile so that they do not coat the etching surface and prevent further etching.

A plasma is a mixture of ions, electrons, neutral molecules or atoms, excited molecules and free radicals (neutral species that are incompletely bonded and very reactive), all existing in an ionized gas mixture. A plasma was first called the fourth state of matter by W. Crookes in 1880, when he described what happens beyond the transformation of matter from solid to liquid to gas as energy (heat) is added to the system. At high enough temperatures, atoms or molecules in a gas eventually ionize, with stars being an example of a high-temperature plasma confined by gravitational fields generating fusion power. Stars are plasmas in thermal equilibrium. Such a thermally generated plasma of ions and electrons, with a common charged particle density $n_i = n_e$ (number of ions equal to number of electrons) and an equilibrium ion and electron temperature $T_i = T_e$, would not be very useful for semiconductor processing because of the very high temperatures involved. Instead, plasmas can easily be generated by electric fields accelerating some free electrons to high enough energy to strip more electrons from atoms or molecules, providing a convenient way to generate a weakly ionized plasma where n_i and n_e may be four to six orders of magnitude lower than the neutral gas density. The most common example of such an electrically generated plasma is a fluorescent light bulb or a neon sign. The electron temperature inside a fluorescent bulb is ~11,000 K, but it does not feel hot because the free electron density is much less than the number of particles in the gas (weakly ionized), so the total amount of heat transferred to the walls by the impact of the electrons is small.

A simple plasma discharge is shown in Figure 9.5(a). A voltage source drives a current through a low-pressure gas between two parallel conducting plates in an evacuated chamber. At

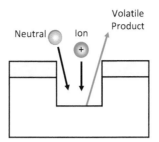

Figure 9.4 Schematic of directional etching in a plasma system.

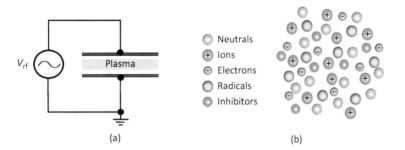

(a)   (b)

Figure 9.5 (a) Simple parallel-plate plasma and (b) schematic of neutrals, ions, electrons and inhibitors in a plasma reaction. Only a few of the much more common neutrals from the source gas are shown in the plasma schematic.

a sufficient voltage the gas "breaks down" and the plasma forms, glowing with a characteristic color depending on the gas mixture. As shown in Figure 9.5(b), a chemically reactive plasma is a complex, weakly ionized gas mixture comprising electrons, ions, neutrals and radicals. The electrons are the light particles that respond and quickly gain energy from the applied fields. The ions, neutrals and radicals are the "heavies" that respond slowly and have energies much lower than the electrons, so that $T_i \ll T_e$. Still, the ionized particles can gain significant energy as they cross high-field sheath regions that surround the plasma and isolate it from the chamber surfaces. We will discuss the formation of the sheath regions in more detail below. The high-energy ions that traverse the sheath region are responsible for the directionality of plasma etching. An excellent introductory reference is the book by Chapman [2], and a much more detailed text by Lieberman and Lichtenberg [3].

   A simple recipe for etching silicon in a plasma discharge indicates some of the issues we will discuss in more detail later. We begin with a simple inert gas, CF_4, a chemically stable tetrahedral molecule, which by itself would not etch silicon. By applying an electric field, an electron can cause dissociation and ionization of the molecule:

$$e^- + CF_4 \rightarrow 2e^- + CF_3^+ + F.$$

$$(9.5)$$

The dissociative ionization creates enough excess electrons to sustain the plasma discharge. Less energetic reactions can also occur to create neutral radicals, or can excite a radical to a higher energy state $F \rightarrow F^*$ that decays and gives out the characteristic glow of the plasma:

$$\begin{aligned} e^- + CF_4 &\rightarrow e^- + CF_3 + F, \\ e^- + CF_3 &\rightarrow e^- + CF_2 + F, \\ e^- + F &\rightarrow e^- + F^*. \end{aligned} \tag{9.6}$$

The free radicals are electrically neutral species that have incomplete bonding, i.e., they have unpaired electrons. Examples are the F radical in these equations and the neutral CF_3 and CF_2 radicals in (9.6). Because of their incomplete bonding, free radicals are highly chemically reactive. They want to bind to other atoms so all electrons are paired, making them highly reactive. The key point is that these neutral, chemically reactive radicals can etch the silicon through reactions like

$$Si + 4F \rightarrow SiF_4, \tag{9.7}$$

where the important point is that the reaction product is volatile. This is why copper cannot be etched in a plasma, as it has no volatile byproducts at normal processing temperatures.

One might wonder, if these are neutrals, how can they provide any directionality in the etch profile? And, indeed, they cannot – these reactions would provide isotropic etching in a plasma system. Thus, the key in plasma systems is to utilize the ion energies which are directed vertically across the plasma sheaths to strike the surface. For a directional or anisotropic etch, there must be high-energy bombardment of the surface by the CF_3^+ ions. The CF_3^+ ions would typically dissociate, producing a C atom and three F atoms that bond to silicon atoms. Once a volatile SiF_2 or SiF_4 molecule is formed, it would be desorbed and pumped away. The ions can also enhance the etching by the CF_2 radicals, which, because of their abundance, provide most of the C on the surface. This fluorocarbon layer can be 0.5 nm thick and needs to be activated or removed by the ions to enable etching. The deposited C is interesting to consider, since SiC is not volatile. Carbon would inhibit the etch rate, unless ejected by ion bombardment, or perhaps by adding some O_2 to the source gas to form CO or CO_2 volatile byproducts. This competition between inhibitors, etchants and ion bombardment is at the heart of plasma etching physics and chemistry.

9.3.1   The Plasma Sheath and Ion Energies

We briefly discussed the formation of a weakly ionized plasma above, where the degree of ionization is approximately 10^{-5} of the neutral gas density. Because the ion energies are so important when discussing plasma etching, we spend a few moments on describing the units used to measure plasma energies.

Plasmas were originally described as the fourth state of matter, as heat (energy) was added to a gas, producing a new ionized state of matter. Thus it seemed natural to describe the characteristic of a plasma in terms of temperature, with ion and electron temperatures T_i and T_e. But these really describe the energy of the plasma:

$$E(J) = kT \quad \text{where } k = 1.38 \times 10^{-23} \text{ J K}^{-1}. \tag{9.8}$$

Because we discuss the plasma potential, sheath fields and ion bombardment energies, another common way to describe the energy of the plasma components is in term of volts, where

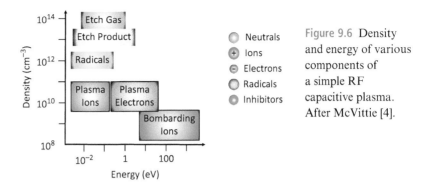

Figure 9.6 Density and energy of various components of a simple RF capacitive plasma. After McVittie [4].

$$E \text{ (J)} = qV \quad \text{where } q = 1.6 \times 10^{-19} \text{ C}. \tag{9.9}$$

Hence, room temperature, 297 K, is equivalent to 0.026 V. Sometimes, the voltage is given in electronvolts to indicate its energy origins, so room temperature would be equivalent to 0.026 eV. Using $kT = 1$ eV $= 1.6 \times 10^{-19}$ J, the conversion factor is $1\,\text{eV} = 11,600$ K. Figure 9.6 plots the densities and energies for the various species in a simple low-pressure capacitive plasma discharge like that in Figure 9.5, using the electronvolt energy scale.

We see in Figure 9.6 that the plasma is weakly ionized (the plasma ions n_i and plasma electrons n_e are orders of magnitude smaller than the neutral gas density) and that the electrons have a much higher temperature (energy) T_e than the ions T_i. This makes sense, as the electrons can quickly respond to the RF energy whereas the heavy ions are much slower to gain energy. However, there is a component of the ions (the bombarding ions) that have surprisingly high energies. It is these ions that impact the substrate and drive the chemical reactions to create the anisotropic etching profiles. This is the key to plasma etching, and to understand it we need to understand how the sheaths form around a plasma and how the sheaths can accelerate ions to such high energies.

9.3.2 Plasma Sheath Formation

Consider a simple plasma configuration such as that shown in Figure 9.7(a). The glowing plasma does not fill the entire space; instead, the glow is isolated from the walls by dark sheaths. These sheaths form naturally because of the basic plasma dynamics.

Consider that while the ion and electron densities are equal in the plasma ($n_i = n_e$), the electron mass m is much lighter than the ion mass M, so that $m/M \ll 1$, and the electron temperature T_e is much higher than the ion temperature T_i, so that $T_i \ll T_e$. This means that the electron thermal velocity is much higher than the ion velocity, so electrons can escape the plasma and recombine at the chamber walls. In the bulk of the plasma itself, there is no electric field because the ion and electron densities are equal. The plasma can be thought of as a conductor, so the $\mathcal{E}$ field is zero and the potential must be constant. The electrons are not confined and the fast-moving electrons can escape the plasma at the edges. Thus, a DC plasma potential must exist to contain the more mobile species and allow the fluxes of positive and

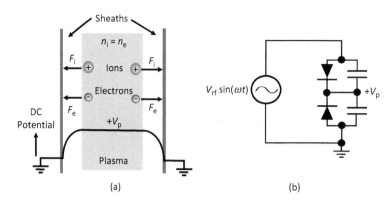

Figure 9.7 (a) Plasma and sheath region, with electric field and ion and electron fluxes indicated. After [3]. (b) A simple circuit model for a typical high-frequency RF plasma.

negative carriers to the walls to be balanced ($F_i = F_e$). It is easy to deduce that, if the walls are grounded, then the plasma must take on a positive potential ($+V_P$) with respect to the walls to retard the electron loss from the plasma and accelerate the ion loss across the sheath to balance the few electrons that do escape the plasma and are responsible for creating the sheath. It is important to realize that this happens naturally without any external attempt to create a plasma potential and is simply driven by the plasma dynamics.

We can make a crude estimate of the plasma potential V_p if we assume that the electrons have a Boltzmann distribution of energies. There is an exponential decline in the number of electrons when there is a barrier of a few eV or equivalently a few T_e (recall that $n \sim e^{-qV/kT}$). Thus we expect that $V_p \sim$ few T_e or its energy equivalent in volts is necessary to create a barrier to confine the electrons. The energy of the bombarding ions is then also $\sim$ few T_e or equivalent. This explains how the bombarding ions gain such high energies, even though their average ion temperature $T_i \ll T_e$, as shown in Figure 9.6, and this leads the ion bombardment to dominate the characteristics of plasma etching systems. (Note that the actual electron distribution has a long high-energy tail rather than the simple Boltzmann distribution mentioned above.)

After this discussion of the plasma dynamics, we can draw a more realistic schematic of the plasma and sheath region, as shown in Figure 9.7. The schematic shows that an electron depletion region forms around the plasma at all surfaces to confine electrons in the plasma. The depletion regions or sheaths are dark because there are few electrons and therefore few electron–neutral collisions giving no excited species that relax to produce light emission. The plasma naturally builds up a positive DC potential relative to the grounded electrode to create the sheath electric field that may be a few kilovolts per centimeter and which confines electrons to the plasma. Most of the electrons get returned to the plasma, while a few make it across, which sustains the sheath potential and ensures that the ion and electron currents are equal at the wall surface. The electric field means the ions gain energy and directionality as they cross the sheath. As the circuit diagram indicates, most of the current across the sheath is electron displacement current across the sheath capacitance as the sheath edges oscillate in response to the applied voltage. The reverse-biased diodes support the field in this equivalent circuit and

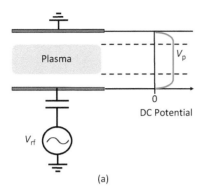
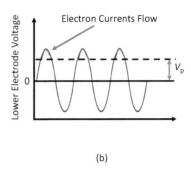

(a)   (b)

Figure 9.8 (a) Development of the plasma potential V_p, and (b) electron currents flowing during a small part of the RF cycle to maintain the sheath voltage.

allow the relatively small ion current to flow and limit the electron current so both match at the wall surface. An alternative way to think about the plasma is to consider the RF current that flows in response to the applied RF field. When the applied field is first turned on, there is no DC plasma voltage. Very quickly, some high-mobility electrons are lost to the walls (this might be considered reverse leakage current in the diodes) and the plasma takes on a positive potential to confine the electrons, shown in Figure 9.8(a). In steady state, the ions can only respond to the DC sheath field and there is a continuous ion bombardment of the electrode. The electrons can follow the RF potential and flow only during a brief period of the cycle, as indicated in Figure 9.8(b). In Figure 9.8(b) we see that the average (DC) voltage on the powered electrode is 0 V but the AC signal swings symmetrically above and below 0 V. It is only during the blue-colored portions of the AC applied voltage that the electrode voltage is more positive than the DC plasma potential. During this portion of the cycle, electrons can easily flow from the plasma to the powered electrode so that the electron and ion currents balance on average. These currents are small and most of the RF current is carried by displacement current in the capacitor, corresponding to sheath oscillations giving a dQ/dt (we do not prove that here).

This model makes it immediately clear that something different must happen if one of the electrodes is smaller than the other (to this point, we have only considered symmetrical electrodes). The RF current must be continuous, so the smaller electrode must have a higher current density I/A, where A is the electrode area, to maintain the same total current as the larger electrode. This requires a higher field at the smaller electrode. Since the plasma is quasi-neutral and cannot sustain a field, the extra field must come from an additional DC field at the smaller electrode, not just V_p. This is illustrated in Figure 9.9, where the additional DC field means very high ion bombardment on the smaller electrode.

In the AC voltage waveform shown in Figure 9.9(b), the electron flow to the powered electrode only occurs during the small blue-colored portion of the AC cycle, but this is sufficient to balance the ion and electron currents to establish a stable DC voltage ($-V_{DC}$) on the powered electrode. In Figure 9.9(b) the average (DC) voltage on the powered electrode is $-V_{DC}$ but the AC signal swings symmetrically above and below this value. The $-V_{DC}$ voltage is a useful way to adjust the ion energies but it is fixed by the initial design of the electrodes. We will discuss

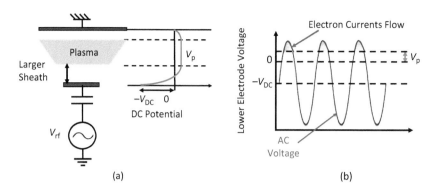

Figure 9.9 Schematic of the higher DC field on the smaller electrode. An additional DC sheath bias ($-V_{DC}$) is created by the system to maintain the same current at the smaller electrode.

better ways to control the ion energies later. In the design shown in Figure 9.9, the only adjustable parameter is the magnitude of the RF voltage source. When this is increased to, for example, increase the plasma density, $-V_{DC}$ will also change because all of the voltages in Figure 9.9(b) are coupled. Decoupling them requires a different system design with additional controls, as we will see in Section 9.6.

9.4   Plasma Chemistry

Now that we have described the physical background for how plasmas and sheaths form, we will briefly discuss the chemistry that occurs in the plasma, which will allow us to understand better the actual etching mechanisms. We show in Figure 9.10 a simple schematic of the main chemistries that occur in a CF_4 plasma due to electron collisions with the molecules, one of the very-well-studied plasma systems. Most of the collisions of relevance are inelastic, where some of the kinetic energy of the electrons is converted into some other form of internal energy after the collision. The inelastic collisions can be divided into three main categories: excitation, dissociation and ionization.

During excitation reactions, a collision with a neutral molecule raises the energy from the ground state to an excited state, which quickly decays with the emission of light. The color of the light emission is characteristic of each particular plasma chemistry. For example, oxygen plasmas glow blue. When a collision leads to dissociation, a gas molecule splits or dissociates into two neutral molecules that now have unsatisfied bonds and are highly reactive, often referred to as "free radicals." An ionization reaction leads to the formation of a positive ion and a secondary electron and is responsible for sustaining a plasma by secondary electron generation. Some more complicated subreactions can also occur, such as dissociative ionization, which both dissociates and ionizes a molecule in a single step. Finally, recombination reactions can occur that tend to offset the plasma generation. The optical spectrograph in Figure 9.10(b) shows some of the complexity that can occur even in a simple CF_4/O_2 discharge, where each peak corresponds to a plasma component molecule.

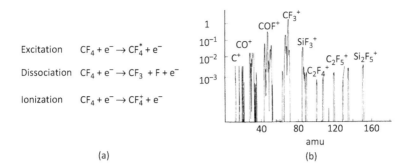

Excitation $CF_4 + e^- \rightarrow CF_4^* + e^-$

Dissociation $CF_4 + e^- \rightarrow CF_3 + F + e^-$

Ionization $CF_4 + e^- \rightarrow CF_4^+ + e^-$

(a)

(b)

Figure 9.10 (a) Simple schematic of the three basic types of reactions that occur in a CF_4 plasma due to electron collisions with the source gas molecules. (b) A typical optical emission spectrum in a CF_4/O_2 discharge. Only some of the spectral peaks are labeled. The horizontal axis is the molecular weight of the various molecules. Reprinted from [5] with the permission of AIP Publishing.

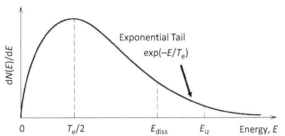

Figure 9.11 Distribution of electron energies with a long, high-energy tail. After [3].

The likelihood of any of these reactions depends on the cross-section for the reaction occurring as well as the densities and energies (temperature) of the species involved. The electron–neutral collisions are very important in that they allow the electrons to be knocked out of phase with the RF field and gain energy over many RF cycles and collisions, giving rise to a high-energy tail of electrons. In general, as indicated in Figure 9.11, the chance of a dissociation reaction occurring (E_{diss}) is much higher than that of an ionization reaction (E_{iz}). This is why electrically generated plasmas are characterized as "weakly ionized." The density of the charged species is much lower than the neutral or free radical density, as shown previously in Figure 9.6.

The situation in a real plasma is much more complicated than that shown in Figure 9.10, as the electrons can continue to interact with the dissociated neutrals, generating further chains of reaction products. In addition to the source gas, the etch byproducts can contribute to the plasma chemistry, as the etch products must be volatile for a useful etching reaction to occur. Indeed, as indicated in Figure 9.10(b), the optical emission spectrum can be much more complicated than a single color, with each wavelength producing a characteristic peak associated with one of the plasma components. Such measurements can monitor plasma conditions and even show when some of the etch byproducts decay away, indicating that the etch process has completed.

Table 9.3 Boiling points for potential etch byproducts from chloride, fluoride and bromide chemistries. Data based on [6].

Elements	Fluorides	Boiling point (°C)	Chlorides	Boiling point (°C)	Bromides	Boiling point (°C)
Si	SiF_4	−86	$SiCl_4$	58	$SiBr_4$	154
Ge	GeF_4	−37	$GeCl_4$	84	$GeBr_4$	186
C	CF_4	−128	CCl_4	77	CBr_4	189
Ga	GaF_3	1000	$GaCl_3$	201	$GaBr_3$	279
As	AsF_3	−63	$AsCl_3$	130	$AsBr_3$	221
In	InF_3	>1200	$InCl_3$	300	—	—
P	PF_3	−101	PCl_3	75	PBr_3	173
Al	AlF_3	1297	$AlCl_3$	178	$AlBr_3$	263
Cu	CuF	1100	$CuCl$	1490	$CuBr$	1345
Ti	TiF_4	284	$TiCl_4$	136	$TiBr_4$	230
W	WF_6	17.5	WCl_5	276	WBr_5	333

9.4.1 Choosing a Plasma Chemistry

Given the enormous complexity of the plasma chemistry, how does one begin to choose reasonable source gases for etching a particular substrate? The main requirement is that the etch byproduct be volatile. The boiling point or vapor pressure of a compound is a good guide to its volatility and can help choose a promising starting gas. The chlorides, fluorides and bromides have the boiling points shown in Table 9.3 and indicate which might be promising for various substrates.

As seen in Table 9.3, SiF_4 is highly volatile, so a fluorine-containing chemistry would be a good choice to etch silicon (e.g., SF_6 or CF_4). It is also clear that copper cannot be easily etched by any of the fluorine, chlorine or bromine chemistries because volatile byproducts do not occur. This kind of information can act as an initial screen for promising source gases for etching different materials. We can now discuss how these various chemistries work in various plasma etching systems.

9.5   Plasma Etching Mechanisms

A general schematic of the plasma etching mechanisms we will consider is shown in Figure 9.12. We will see that plasma etch mechanisms span the whole range from purely chemical, to purely physical, with "mixed-mode" mechanisms being the most important in practice. We discuss each of these briefly in the following sections.

9.5.1 Purely Chemical Plasma Etching (Isotropic Plasma Etching)

It might seem odd to discuss isotropic plasma etching like that illustrated in Figure 9.13, when one of the main reasons for using plasma etching is to have control over the profile

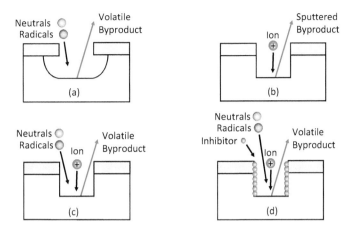

Figure 9.12 Schematic of (a) purely chemical, (b) purely physical, (c) ion-enhanced chemical and (d) ion-enhanced inhibitor etching mechanisms. After Flamm and Donnelly [7].

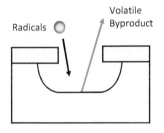

Figure 9.13 Schematic of plasma etching where reactive neutrals (radicals) etch isotropically.

shape. But there are instances, such as photoresist etching, where a plasma process has advantages over a wet etch process. If wet etching is used, toxic acids such as hot sulfuric acid are required to strip the photoresist. Since photoresists are primarily carbon-based polymers, it is reasonable that an oxygen plasma might be a good choice to form volatile carbon compounds such as H_2O and CO or CO_2. An oxygen plasma has further advantages as it can strip "hardened" resist, such as resist that has acted as a mask for high-dose implants. Indeed, this was one of the early uses of plasma etching, and similar systems with higher-density plasmas (which we will discuss later) still dominate the photoresist stripping process. Note that, while both neutrals and free radicals were shown in Figure 9.12(a), we show only the free radicals in Figure 9.13 because they are the principal species involved in most etching systems.

The simplest plasma etching system for stripping photoresist is a barrel etcher, where the wafers are surrounded by a grounded metal grid that captures the bombarding ions and only lets the highly reactive free radicals through to etch the photoresist, as shown in Figure 9.14. The elimination of any bombarding ions hitting the wafers mean there is no risk of ion damage to the substrates.

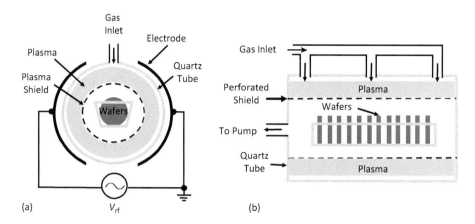

Figure 9.14 A simple barrel etching system using an oxygen plasma to strip photoresist: (a) end view and (b) side view.

Just as wet etching is isotropic and very selective, so too can plasma chemistries be isotropic and very selective. For example, the oxygen plasma will strip photoresist but have no effect on underlying oxide, nitride or silicon layers. Such isotropic plasma etches are even used in the most advanced processes, such as gate-all-around (GAA) field effect transistors (FETs). In that case, a stack of alternating silicon and silicon–germanium layers are grown and an isotropic plasma etch can remove the SiGe layers while leaving Si layers available for surrounding dielectric and gate depositions. Similar etches are often used in forming MEMS membranes by undercutting a lower layer. Plasma gas etching avoids the wettability and surface tension issues that occur with wet etchants, especially at small dimensions.

9.5.2 Purely Physical Plasma Etching (Ion Milling)

At the other extreme end of plasma processing is ion milling, illustrated in Figure 9.15. This process uses a highly directional ion beam to physically sputter or etch substrate atoms. It involves momentum transfer from the incident ion beam to the surface layers and atoms, which gain more than their binding energy so they can leave the surface.

The yield of the process is the number of ejected atoms compared to the number of incident ions and is a maximum for certain intermediate angles of incidence, as shown in Figure 9.16. Note that, if the substrate has topography as shown in Figure 9.16(a), the incident angle can vary with position on the substrate. Typically, an argon-ion plasma is used, as there is no pretense to any chemistry taking place. The process is similar to a low-energy ion implantation process, except that the accelerating fields are the naturally occurring sheath fields in a plasma.

For safety reasons, the grounded electrode is often connected to the chamber walls, effectively increasing the size of that electrode and increasing the sheath fields at the smaller electrode. To maintain very directional etching and to provide high fields, the pressure in the gas is as low as possible (~10^{-3} Torr), consistent with having a plasma. Lower pressure leads to fewer ion collisions with neutrals as they traverse the sheath, giving more anisotropic profiles. But there

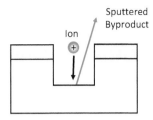

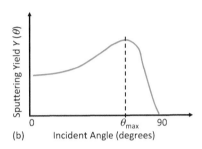

Figure 9.15 Schematic of purely physical plasma etching, producing very anisotropic profiles.

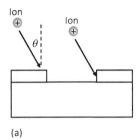

(a)

(b)

Figure 9.16 Sputter yield as a function of incident angle, showing maximum sputter yield (corresponding to maximum sputter etch rate) at an angle of 50°–60°.

is also lower plasma density (fewer gas atoms at lower pressures), so the etch rate decreases. The etch rate or sputter yield depends on the ion and target masses, the ion energy, the binding energy and the angle of incidence. The angle of incidence may limit the kind of structures that can be formed and how structures are designed in the layout. The etch rate, which determines the throughput of the process, depends on the product of the yield and the ion flux. This limits ion milling to research applications, but it is a useful process because it can etch any material even if it does not have volatile byproducts. On the other hand, because it is a purely physical process, it has essentially no selectivity to mask layers or layers underlying the target layer. Perhaps the best description of ion milling is ionic sandblasting, which makes it a general-purpose etch.

9.5.3 Ion-Enhanced Chemical Etching (1 + 1 ≠ 2)

The real key to why plasma etching is the dominant etching technique in semiconductor manufacturing is because there is an unusual synergistic relationship between ion bombardment and chemical etching. The classic experiment that shows the effect is from a 1979 paper by Coburn and Winters [5] and the results are shown in Figure 9.17.

Figure 9.17 shows that either the chemical gas XeF_2 (no plasma) or the Ar^+ ion bombardment alone have very slow silicon etch rates, but when both are used together there is a dramatic order-of-magnitude increase in the etch rates. This synergistic effect is at the heart of anisotropic plasma etching. Because the ions have directionality after crossing the sheath fields, the etching reaction can proceed in an anisotropic fashion, with areas not being bombarded by ions having a very low etch rate. Atomistic-scale simulations suggest that the ions temporarily

Figure 9.17 Plot showing
silicon etch rate in three
regimes: purely chemical
etching, ion-enhanced
chemical etching and purely
physical ion etching.
Data from [8].

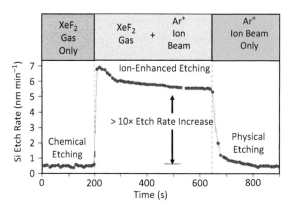

Figure 9.18 Schematic of the ion-
enhanced chemical etching process,
whereby adsorbed reactive species are
activated by ion bombardment to
produce volatile reaction products.

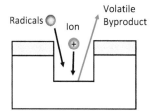

break surface bonds, allowing the reaction species to quickly react and form volatile etch
byproducts.

Previously, the ion-enhanced chemical etching process shown in Figure 9.17 was known as
reactive ion etching, but that is a misnomer. Clearly, there are no reactive ions involved in the
classic example in Figure 9.17, just inert Ar^+ ions. It is the interaction between the ions and the
reactive radicals that drives the etching, so that ion-assisted or ion-enhanced etching is a more
accurate and descriptive name.

Because there is both ion bombardment and chemistry involved, this plasma process can be
both directional and selective. It has higher etch rates than physical etching (Section 9.5.2)
because of the synergistic effects of ion bombardment and chemistry. The etch rate depends on
both the ion and neutral (radical) fluxes (Figure 9.18). Note that, while both neutrals and free
radicals were shown in Figure 9.12(a), we show only the free radicals in Figure 9.18 because they
are the principal species involved in most etching systems.

Controlled experiments such as indicated in Figure 9.19 describe an argon-ion beam directed
at a silicon surface while the flow rate of molecular Cl_2 or atomic Cl etch gas increases. The
results show an initial ion-enhanced etching, then a saturation of the etch rate when the atomic
or molecular flux of chlorine etchant gets much larger than the ion flux (>100). Physically, this
corresponds to maximum adsorption of neutrals on the surface (see Figure 9.19) beyond which
the ions cannot enhance the etch rate any more. More reactive radicals at greater flow rates give
higher etch rates, as expected. These are control experiments, because, in the simple parallel-
plate plasma etch system we have described, turning up the power will increase the ion energies,

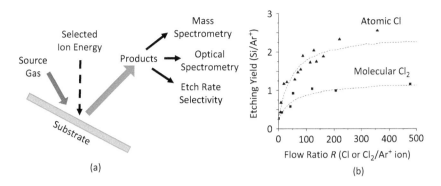

Figure 9.19 (a) Experimental setup for determining plasma etch kinetics. (b) Etch yield versus neutral/ion flow ratio. After Chang *et al.* [9].

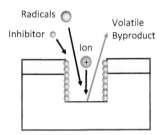

Figure 9.20 Schematic of ion-enhanced inhibitor etching.

but it will also increase ionization and radical generation in the bulk of the plasma. Thus there is no way to independently control both the radicals and the ions in a simple parallel-plate plasma. Later, we will discuss more sophisticated plasma systems that do provide some separate control over the radical density and the ion energies.

Etching systems based on ion-enhanced etching are widely used today in semiconductor fabrication. They can provide a very good combination of selectivity (because of the chemistry driven by the reactive radicals) and anisotropy (because of the ions).

9.5.4 Ion-Enhanced Inhibitor Etching

We consider a special case of ion-enhanced etching, where inhibitors are a major component of the reaction. Ions remove the inhibitor layer to allow chemical etching to occur, as shown in Figure 9.20. Ideally, the ions would remove the inhibitor on the bottom but not on the sidewalls, allowing anisotropic etching to occur. We will examine how one might choose the etchant gas mixtures to take advantage of inhibitors to modify the etch profile and selectivity. Note, again, that, while both neutrals and free radicals were shown in Figure 9.12(a), we show only the free radicals in Figure 9.20 because they are the principal species involved in most etching systems.

For example, the typical chemistries to etch silicon and its dielectrics are fluorine-based. We discussed the canonical CF_4-based etch chemistry in (9.4). But it would be more accurate to

Table 9.4 **Relative etch rates between SiO$_2$ and Si. Data from [10].**

Ion	SiO$_2$	Si
CF$_3$	6.1	0.61
CF$_2$	3.2	0.32
CF	1.8	0.15
C	0.4	—
F	0.4	0.6

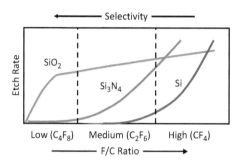

Figure 9.21 Relative etch rates of oxide, nitride and silicon as the F/C ratio increases. After R. J. Shul, Sandia National Laboratory.

describe this fluorine-based chemistry as a CF$_x$-based chemistry, with gases such as CF$_4$, CHF$_3$, C$_2$F$_6$ and C$_4$F$_8$ as the base gases. The process requirements drive the chemistry that is chosen, factors such as selectivity to mask or underlying layers, profile shape and etch rate. The fluorocarbons (C$_x$F$_y$) provide both selectivity and profile control. To improve selectivity, one tries to push the chemistry into fluorine-deficient regimes. This can be done by choosing a lower fluorine/carbon ratio in the source gas (i.e., picking C$_4$F$_8$) or by additive gases like CH$_4$ or H$_2$. As Table 9.4 shows, the F/C ratio determines the relative etch rates between materials like SiO$_2$ and Si.

Table 9.4 shows that the F radical by itself has similar etch rates for oxide and silicon (silicon etches 30% faster), but when a carbon-rich source such as CF is used, then the oxide etches 12 times faster than silicon. The C acts as an inhibitor on the silicon surface, while the oxygen in the SiO$_2$ can desorb the carbon, providing an increase in the etch rate relative to silicon.

A similar effect is seen in Figure 9.21, where the F/C ratio determines the etch rate and selectivity between oxide, nitride and silicon. At low F/C ratios, the oxygen in SiO$_2$ is more effective than the nitrogen in Si$_3$N$_4$ at removing the C inhibitor, so the oxide etches faster than the nitride or silicon. At slightly higher F/C ratios, the silicon nitride begins to etch but the silicon etch rate is still slow. Finally, at high F/C ratios, all three materials etch with similar rates. The process for choosing an etch chemistry then becomes clearer and is depicted in Table 9.5.

It is a combination of the ion bombardment and chemistry that affects the selectivity and etch rates. When deposition is an important part of the chemistry, the ion flux plays an important part in controlling the polymer thickness. As might be imagined, it is not trivial to control the ion flux and the additive additions, especially when the etch byproducts also influence the reactions. This makes plasma etching as much of an art as a science. Figure 9.22 illustrates

Table 9.5 **Process for increasing or decreasing the F/C ratio to change etch selectivity.**

Source gas	CF_4, CHF_3, C_2F_6, C_4F_8
To increase F/C ratio	
Add C scavengers	O_2, CO_2, NO_2
Add F	F_2, SF_6
To decrease F/C ratio	
Add F scavengers	H_2
Add more C	C_xF_y

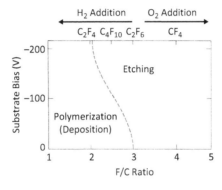

Figure 9.22 Schematic of interacting ion flux, F/C source gas ratio, and O_2 and H_2 additions on etch and deposition landscape. After [5].

the complexity of the ions, F/C ratios, and simple oxygen or hydrogen additives on the etch rates and deposition rates.

The H_2 additions can reduce free F radicals by producing HF, which reduces the amount of atomic fluorine and pushes the system away from etching and towards deposition. On the other hand, oxygen additions react with dissociated CF_4 species (like CF_3 and CF_2), which reduces recombination reactions with F, thereby increasing the amount of free fluorine radicals. The factors that determine whether etching or deposition take place are the stoichiometry of the source gas (F/C ratio in the fluorocarbon systems) and the ion energy or substrate bias. Figure 9.22 makes it clear that, with higher polymer deposition, a higher sheath bias and higher ion flux are needed to enter the etching regime. Consider the case of a C_2F_6 source gas, perhaps with some small H_2 additions. If there is no ion bombardment, such as on the sidewalls, then deposition will take place. At the bottom of the feature, ion bombardment is occurring and the process shifts towards the etch regime. Thus, the sidewalls will be coated with inhibitor and anisotropic etching can be obtained as illustrated in Figure 9.20.

Reducing the F/C ratio greatly improves the etch selectivity of oxide to silicon. This is because the carbon inhibitor formation occurs preferentially on silicon rather than on SiO_2, where it is more likely to form volatile CO compounds. With more inhibitor on the silicon, its etch rate slows relative to that of SiO_2, improving selectivity. Thus, hydrogen additions would improve the selectivity of oxide etching compared to silicon.

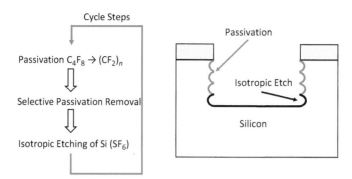

Figure 9.23 Schematic of the Bosch process.

Note that some of the deposition species can come from the mask layers, as these are normally carbon-rich polymers. The deposition species may not be simple C atoms on the surfaces. More often than not, the deposition species have a defined thickness and can sometimes be seen peeling from the sidewalls where they accrue. It is easy to see that CF_2 molecules generated in the plasma can create $(CF_2)_n$ molecular chains, which is a chemical formula similar to Teflon. These passivating layers need to be removed post-etch, often by HF dips or oxygen plasmas.

One particular technique that makes extensive use of inhibitor layers is called the Bosch process, shown in Figure 9.23, which is a patented process that enables very deep silicon etching [11]. This kind of etch is often used in MEMS processes or in through-wafer silicon etches. The process alternates between etching in SF_6 and C_4F_8.

The C_4F_8 step functions in the deposition region indicated in Figure 9.22 and deposits a polymer layer everywhere. The SF_6 plasma step provides the ion bombardment to remove the inhibitor on the bottom surface. At first sight, it is surprising that the third step is an isotropic silicon etch – if one wanted vertical etch profiles, an ion-assisted anisotropic chemistry might seem more reasonable. However, the deliberate polymer deposition on the sidewalls prevents sideways etching and the isotropic etch gives much faster etch rates than an anisotropic etch at the same power levels. Since the goal of the process is to perform deep etches, this unique combination of inhibitor deposition and isotropic etching can provide deep, vertical etch profiles. By alternating many cycles, each of a few seconds in length, a deep etch profile can be generated. The isotropic etch cycle does generate a "scalloped" edge profile, as indicated in Figure 9.23, so the smoothness of the sidewalls depends on the precise switching time between the deposition and etch cycles. The longer the etch cycle, the deeper the scallops. It is a complex process because there are large parameter changes between each step, such as gas flow, power, pressure and bias, that must be controlled. As shown in Figure 9.23, it is a multi-step process involving three key steps:

1. conformal polymer deposition $(CF_2)_n$,
2. polymer removal on the horizontal surface only, and
3. high-etch-rate SF_6 isotropic silicon etch.

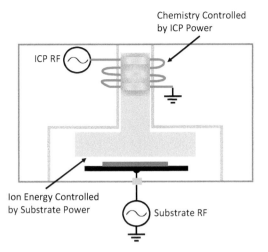

Figure 9.24 Schematic of an inductively coupled plasma system (ICP) that separates bulk plasma generation and the ion energy distribution using different power sources.

9.6 High-Density Plasmas

We have seen that, in a parallel-plate plasma etcher, the plasma chemistry and the sheath bias are intimately coupled. (Recall the discussion in connection with Figure 9.9.) Providing more power affects both the plasma conditions and the ion energies across the sheath. In modern processing, there are extreme requirements for selectivity, high etch rates and low substrate ion damage. However, low ion energies mean that the etch rates are reduced proportionally. In a typical parallel-plate etcher, the etch rate depends on the ion power incident on the substrate. To enable both high etch rates and low ion damage, a new kind of high-density source is required that allows the plasma density to be uncoupled from the ion energy. A solution was to use an inductively coupled plasma (ICP) source, like that shown in Figure 9.24, which allows independent control of the plasma density and the ion energy [12].

Extra power can be inductively coupled into the plasma to generate high-density radicals, while the parallel-plate bias can separately control the ion energy across the sheath. In an ICP system, there is independent control of the ion energy and the radical flux. Typically, the power applied to the source is much larger than the bias power applied to the substrate. In this case, the bias power does not affect the plasma density, providing a "second knob" to optimize wafer damage or selectivity on the wafer. Because of the extra flexibility that this allows, this kind of system is now very widely used in manufacturing. Similar high-density plasma systems where the energy is coupled magnetically (electron cyclotron resonance) or with microwave sources to the plasma also exist but are less common. Because of the stringency of plasma etch requirements, extra power supplies to further fine-tune the voltage or provide pulsed voltages are becoming more common. Of course, while having an "extra knob" provides greater flexibility in an etching system, this also means that optimizing the process for a given application is often more complex.

9.7   Etching with the Chlorides and Bromides

We have concentrated on the fluoride etching reactions above, but, as Table 9.3 shows, there are useful volatile byproducts for the chlorides and bromides. All of the considerations we have discussed above apply equally to the chlorides and bromides. For silicon etching, the fluorides give the highest etch rates, but, as one moves from chloride chemistries (Cl_2, HCl, CCl_4 and the freons $CH_xCl_yF_z$) towards bromide chemistries (HBr and Br_2), the profiles become more anisotropic, as shown in Figure 9.25. In this example, the Br etch chemistry achieves a vertical-to-lateral etch ratio of 100:1 whereas the F-based chemistry is <10:1.

So far, we have generally talked about perfect isotropic or perfect anisotropic etching. In fact, there are intermediate regions between these two extremes. If a perfectly anisotropic profile is to be produced, then the ion-enhanced etch rate in the vertical direction must be very large compared with the spontaneous chemical etch rate in the absence of ion bombardment. For silicon etching, this is demonstrated by the bromine etch profile, as indicated in Figure 9.25. To obtain anisotropic silicon etch profiles, a Cl_2 and HBr mixture is often used to optimize selectivity and anisotropy. The selectivity against oxide improves, especially with small O_2 additions to remove carbon residues. Chloride or bromide chemistries are also essential to etch aluminum, as there are no volatile fluoride byproducts.

On the other hand, the fluorine-based chemistry leads to a tapered profile because the chemical component continues to act to etch laterally as the etching proceeds. Since the top of the profile is exposed longer to the chemical etch component, it undercuts the mask as the etching proceeds, generating a tapered profile, as indicated in Figure 9.25. There may be times when a tapered profile is desired, such as wanting to improve step coverage in a subsequent deposition step or to minimize stress at the corners in a shallow trench profile. Another way to generate a tapered profile is to cause some mask erosion as the etching proceeds. As the mask edge is recessed, a more tapered profile is formed.

For gallium-containing compound semiconductors (GaAs, GaP, GaSb), the chlorine chemistries with argon bombardment (Ar/Cl_2, Ar/BCl_3) provide volatile byproducts. The greater the Ar percentage in the chlorine mixture, the more anisotropic the profile. For aluminum-containing

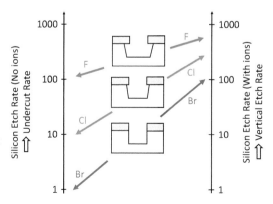

Figure 9.25 Silicon lateral and vertical etch rates in the fluorine, chlorine and bromine chemistries. The relative etch rates are approximate. After Coburn, chapter 1 in Shul and Pearton (eds.), *Handbook of Advanced Plasma Processing Techniques* [13].

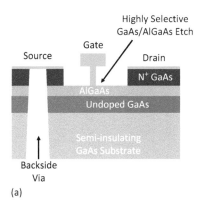
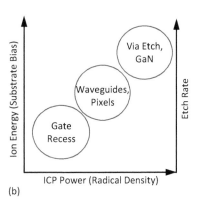

Figure 9.26 (a) GaAs etch profile showing backside via etch to reduce source resistance and separate high-selectivity etch to remove GaAs and selectively stop on AlGaAs. (b) Schematic of the range of etch processes that can be achieved by ICP high-density plasma systems, where the ion energy (*y* axis) and radical density (*x* axis) can be independently controlled.

compound semiconductors (AlGaAs, AlN), chlorine chemistries must be used to generate volatile byproducts.

Compound semiconductor etching has at times been more sophisticated than silicon etching. For example, GaAs-based high-electron-mobility transistors (HEMTs) and heterojunction bipolar transistors (HBTs) have used through-wafer etching to contact the source and minimize source resistance and inductance (Figure 9.26). Etch rates may need to be 10 μm min^{-1} through 100 μm diameter holes with 20:1 selectivity over photoresist masks to be viable in manufacturing.

High-density plasmas are needed to provide such high etch rates and selectivity. The example shown in Figure 9.26 also requires is a very high selectivity etch, which stops on the thin AlGaAs layer. A summary of the different etching regimes that can be achieved by high-density plasmas (such as ICP systems) is shown in Figure 9.26(b). The *x* and *y* axes correspond to the two "knobs" that can be adjusted to change either the substrate bias or the radical density independently, providing a large range of etch processes and selectivity. There are many examples in the literature of etching techniques applied to specific compound semiconductor device structures. Generally, these can be understood using the etching principles discussed in this chapter.

9.8 Plasma Etching Anomalies

There are several anomalous or subtle effects that occur when plasma etching. Most of these can be understood by considering the interplay between the ions and the chemistry. Figure 9.27 shows examples of three common etch anomalies: bowing, microtrenching and aspect-ratio-dependent etching (ARDE).

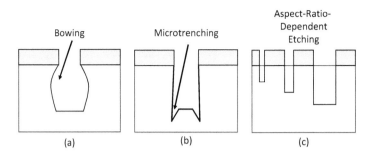

Figure 9.27 (a) Bowing, (b) microtrenching and (c) aspect-ratio-dependent etching (ARDE). After McVittie [4].

Bowing is caused by a broad ion distribution that partially etches the sidewalls. This can be caused by collisions in the sheath if the pressure is not low enough to provide a long mean free path for the ions. The solution is to decrease the pressure and increase the sheath bias. However, we have seen that, in a simple parallel-plate reactor, this will also cause changes in the plasma chemistry that will affect rates. Increased bias may also cause substrate damage. Another solution is to increase the sidewall inhibitor deposition. The ICP high-density plasma equipment described above helps minimize these problems.

Microtrenching is also caused by off-axis ions that scatter off the sidewalls. In other words, the ions do not stick on impact and re-emit uniformly. Instead, they preferentially undergo a low-angle scattering event and preferentially cause enhanced etching at the corners of the trench.

Aspect-ratio-dependent etching has its origins in both off-axis ions and also how the neutrals behave. If some of the ions do stick on impact, then it is possible that a narrow trench receives less ion bombardment at the base of the trench because the ions stick on the sidewalls. This is much less of an issue in a wide trench. The neutrals may also recombine on the sidewalls, reducing the neutral flux density at the bottom of a narrow trench. This will have a significant effect if the neutral transport is a limiting factor in the etch rate. One solution is to try lower pressures to reduce the off-axis ions.

9.9   Etch Simulations

Arguably, simulating etch profiles is not as important as simulating diffusion or oxidation, where the interstitials and vacancies are unseen and the two-dimensional doping profiles are almost impossible to measure, making simulation the only way to capture anomalous process effects. With etching, it is straightforward to cleave a trench and look at the profile either with an optical microscope or with a scanning electron microscope (SEM). Another reason why plasma simulations are less well developed than diffusion simulations is because plasmas are complex and the simulation really needs to extend from the reactor scale (where the plasma is generated) to the feature scale (where the structures are etched). Given this complexity, it is rare that these multi-scale simulations are attempted. Instead, most simulation tools concentrate on feature-scale simulations, assuming that somehow the ion and neutral distributions near the features are generated by optimizing the plasma conditions. Some of these feature-scale simulations still capture the essence of the physics of plasma etching and are described in the next sections.

9.9.1 Moving Boundaries

One of the most challenging mathematical and numerical problems in simulating etching, deposition and oxidation processes is the problem of moving boundaries and interfaces. Points on the surface evolve quickly, with changes in each surface point's position moving with velocities that depend on the normal component of the etch or deposition fluxes. After a given time step Δt, each surface point's new position needs to be calculated.

Although the formulation of a moving boundary problem is simple, the solution is very difficult to obtain. There are two main groups of methods to address this problem, indicated in Figure 9.28, which are called explicit and implicit methods.

The explicit method is the obvious way of moving the interface, consisting of a string of connected points that are updated as the velocity of the points moves in response to the normal component of the etching flux, as in Figure 9.28(a). The point positions are updated according to the ordinary differential equation

$$\frac{\mathrm{d}x}{\mathrm{d}t} = V_\mathrm{n}(x, t), \tag{9.10}$$

where $V_\mathrm{n}(x, t)$ is the velocity normal to the point. This is easily solved for any individual point, but the points then have to be reconnected after updating their positions. Imagine a series of buoys connected by lines or strings that move in response to the current or water velocity. An extra line or buoys might need to be added as they move to provide a good representation of the

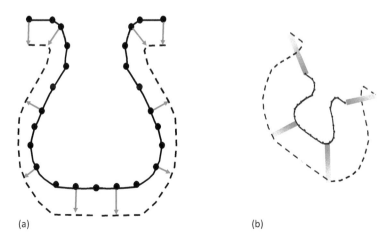

(a)   (b)

Figure 9.28 (a) Explicit evolution of a string of points due to the velocity from the normal component of etching flux. (b) Implicit evolution of a surface cross-section using the level-set method.

boundary shape. A bigger problem comes from more severe topological changes to the boundary. Consider what happens when the etch fronts from two adjacent trenches subject to isotropic etching begin to collide. This corresponds to the buoys getting tangled and creates numerical difficulties, especially when extended to 3D etch profiles.

The implicit method solves for the evolution of a surface that incorporates the boundary, as indicated in Figure 9.28(b). This is known as a level-set method and solves the equation for the evolution of the surface φ according to the equation

$$\frac{d\varphi}{dt} + V_n |\nabla \varphi| = 0 \,. \tag{9.11}$$

A 1D surface is modeled by solving an equation for φ in 2D, and the moving front is defined by the contour line $\varphi = 0$. In 3D the contour line becomes a contour surface and the equation is solved on a 3D grid. It might seem crazy to solve a higher-dimensional surface equation rather than an equation for a string, but it turns out that the level-set approach is much more robust to complex topological changes in the profiles and is easy to generalize to 3D simulations [14]. The differential equation has no complex topological issues to work with, any geometric issues arise only at the stage of defining the surface and, if present, are localized to individual time steps in the evolution. With a tangled string or self-crossing surface in a conventional approach, problems can propagate and multiply. This level-set approach is a late-1980s mathematical invention to moving boundary problems and is now commonly used in simulation tools [15].

There are several additional concepts that are widely used in simulation tools that we now briefly discuss. The first is a simple mathematical means of representing the incoming angular distribution of ions and neutral species as they approach the wafer surface. The second, which we discuss in Section 9.9.3, models whether arriving ions and neutral species "stick" at the location where they originally encounter the wafer surface. This is modeled in terms of a "sticking coefficient." Finally, Section 9.9.4 discusses the concepts of viewing angles and shadowing.

9.9.2 Ion and Neutral Profiles

The arrival angular distribution is illustrated in Figure 9.29. The neutrals (generally free radicals) in a plasma are usually assumed to have a broad angular distribution, since they are unaffected by electric fields and because multiple gas-phase collisions in the plasma will have randomized their directions. The ions, on the other hand, gain energy as they cross the sheaths and can have quite directional distributions. In principle, many system parameters, such as gas pressure, plasma composition and power settings, would affect the specific angular directions of the neutrals and ions in a specific etching system. Monte Carlo simulations have been used to attempt to estimate these distributions in some systems, but, in practice, most etching (and deposition) simulators take a much simpler approach.

The simple approach that is taken is to describe these distributions by a cosine function to a power n. The flux of the species is given by

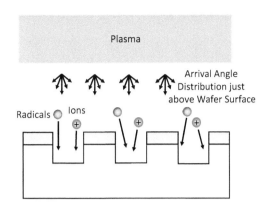

Figure 9.29 Ions and neutral species generated by the plasma arrive at the wafer surface with an angular distribution.

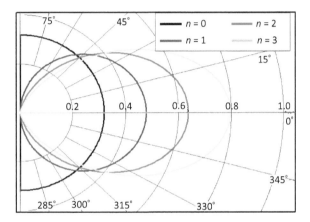

Figure 9.30 Normalized flux distribution for a cosine function to various powers. An ideal cosine function corresponds to $n = 1$. After Silvaco Victory Process User Manual [16].

$$F = F_0 \left(\cos \theta \right)^n , \qquad (9.12)$$

where the isotropic flux F_0 is modified by a cosine function of the emission angle θ to a power n. An ideal cosine function corresponds to $n = 1$, which is often used for neutral species, while higher values of n correspond to more and more directional fluxes. These higher-power cosine functions do a reasonable job of simulating the ion fluxes in plasma systems. The higher the sheath voltages, the more directional the ion flux is and the higher the n value. Lower n values mean that the ion distribution may cause some bowing in the etched profiles, as indicated in Figure 9.30. Values of n as high as 50 or more are often used for highly directed ion fluxes in some simulations. These high n values would produce very anisotropic etching profiles.

9.9.3 Sticking Coefficients

The second concept that is used in most etching and deposition simulations is the "sticking coefficient." Not all incoming etchant molecules necessarily stick on the surface. Many of them do not stick, and bounce off the surface. One might suspect that a glancing angle of incidence would result in a similar bounce at the reciprocal angle from the surface, like a billiard ball collision with the sidewall of the table. Instead, it is as if the incoming molecule loses all memory of its incoming angle and simply emits at an angle consistent with a cosine distribution (with $n = 1$). This process is captured by the sticking coefficient (S_C), where a sticking coefficient of 1 means that the incoming etchant molecule immediately sticks to the surface, and a sticking coefficient of 0.1 means that the incoming molecule bounces around 10 times before attaching to the surface. Simulations of the etching process verify this sticking and re-emission effect. The sticking coefficient S_C is defined as

$$S_C = \frac{F_{reacted}}{F_{incident}},$$ (9.13)

where $F_{incident}$ is the incoming flux and $F_{reacted}$ is the fraction of the incoming flux that "sticks" at the point of incidence. Clearly $0 \leq S_C \leq 1$. Since ions typically arrive at higher velocities because of the electric fields they respond to, they are commonly modeled with an $S_C = 1$. Neutral species generally have $S_C < 1$ and often have very small S_C values (<0.01). The exact values for neutrals in a particular application depend on the type of etching system and the chemistry involved.

Etchant molecules that hit the surface and re-emit in a cosine distribution allow etching in regions that are shadowed by the mask or etch profile. An example of the etching profile with sticking and re-emission is shown in Figure 9.31. With $S_C = 1$, the surface under the mask is not etched at all because the incoming angular distribution does not reach this region. With $S_C < 1$, molecules can bounce or re-emit from other regions and land at the surface under the mask and so etching can occur in this region, as the simulation shows.

9.9.4 Viewing Angles and Shadowing

The implications of the angular distribution of incoming species and their sticking coefficients are illustrated in Figure 9.32, which shows the concept of viewing angle. At each point on the structure being etched (point i, for example), the simulator must geometrically determine what portion of the incoming neutral and ion fluxes that point can "see." Obviously, at point i, only a fraction of the incoming flux vectors would be visible, and so the etch rate at point i would be lower than on a flat surface because the flux density would be lower. Given that ions and neutrals generally have different n values in $(\cos \theta)^n$, point i would also "see" different portions of these fluxes.

If $S_C < 1$, then, in addition to the incoming flux, every other point on the structure will also be emitting molecules. Since generally $S_C < 1$ for neutral species, this re-emission would be dominated by neutral molecules. The viewing angle for point i with respect to point g is again a geometry problem and would be different for each point i and point g on the structure. Simulation of the net etching rate at point i would thus have to sum the incoming fluxes from all possible points g in addition to the plasma-generated incoming flux. This is, in fact, exactly what etch simulators do, and the example in Figure 9.31 showed how a "shadowed" region could get etched in a system in which $S_C < 1$.

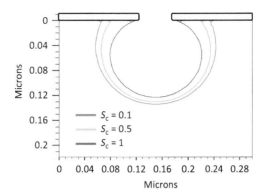

Figure 9.31 Etching profile for an ideal cosine input distribution of neutrals, accounting for shadowing and with various sticking coefficients and also accounting for re-emission with an ideal cosine distribution, allowing for shadowed areas to be etched. From Victory Process [16].

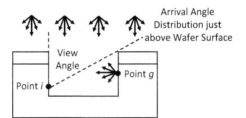

Figure 9.32 Viewing angle concept in calculating the total incoming flux density at a point i.

The final point that should be made regarding etching (and deposition) simulations is that, while, in principle, parameters such as S_C and n could be calculated using methods such as Monte Carlo simulations, these parameters are generally experimentally measured for a given etching or deposition system. We will discuss such experimental measurements in Chapter 10 in Section 10.3.1. For now, the reader should understand qualitatively that low n values usually correspond to more etching on sidewalls such as point i in Figure 9.32, and low S_C values will generally indicate more isotropic etching because molecules will bounce around and reach more parts of the structure being etched. For now, we turn to some simulation examples to illustrate many of the ideas we have discussed in this chapter.

9.9.5 Linear Combination of Isotropic and Directional Etching

The first simulation model we discuss is a simple combination of isotropic and anisotropic etching in various ratios, from completely isotropic to completely directional, as shown in Figure 9.33. Such a simple model, of course, does not capture any of the etching physics. We will discuss some more sophisticated models in subsequent sections in which more of the underlying physics is explicitly incorporated.

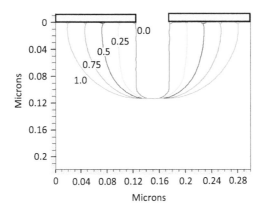

Figure 9.33 Linear combination of isotropic and anisotropic etching, with the $S_C K_I F_I$ term varying from 0% to 100% of the total, showing a progression from completely directional to completely isotropic. From Victory Process [16].

If we assume that the isotropic and anisotropic components of an etching system act completely independently, then the net etching rate at any point on the structure would simply be the linear sum of the two etching rates, i.e., the linear etch model:

$$\text{etch rate} = \frac{1}{N} \left(S_C K_I F_I + K_A F_A \right), \tag{9.14}$$

where F_I is the flux of the isotropic etching species (likely the neutral species), F_A is the flux of the anisotropic etching species (likely the ions), K_I and K_A are the relative etch rates of the two processes and S_C is the sticking coefficient for the isotropic species; $S_C = 1$ is assumed for the anisotropic species (ions). Finally, N is the density of the material being etched (atoms cm^{-3}). A simulator using this model would need to calculate values for F_I and F_A at each point i (Figure 9.32) in the structure for each time step of the simulation.

The simulations in Figure 9.33 illustrate the shapes that are generated as one moves from 100% isotropic to 100% directional etching. These simulations contain no real physics and just present a way to quickly generate a profile that might be imported into a device simulator to test the properties of an isolation region or device structure with varying curvatures.

9.9.6 Ion-Enhanced Chemical Etching (IECE)

The most interesting etching simulations relate to the ion-enhanced chemical etching process, which is at the heart of plasma etching. The simulation model is surprisingly simple (Figure 9.34). First, it is assumed that the neutral etchant molecules (F radicals) chemically attach themselves to the silicon surface and react immediately with Si atoms, forming SiF$_x$ molecules, where x is the number of fluorine atoms contained within one SiF$_x$ molecule. The newly formed SiF$_x$ molecules stay on the surface and cover the Si atoms and prevent further reactions.

The SiF$_x$ molecules can be removed from the surface by one of the following two physical mechanisms:

Example

Silicon is plasma-etched for 5 min and the process follows the linear etch model. The flux of isotropic species, F_I, on an unobstructed flat surface is equal to 2.5×10^{18} atoms cm^{-2} s^{-1}, the sticking coefficient is 0.01 and $K_I = 0.2$. The flux of anisotropic species, F_A, on a flat surface is equal to 1×10^{16} atoms cm^{-2} s^{-1} and $K_A = 1$. The density of Si is 5.0×10^{22} atoms cm^{-3}. A photoresist mask is used so that a wide trench is etched in the Si. (There is no etching of the photoresist.) How far is the Si etched in the vertical direction (away from the mask edge), and how far is the Si etched in the lateral direction right under the mask for these conditions?

Answer

For the linear etch model,

$$\text{etch rate} = \frac{1}{N}(S_C K_I F_I + K_A F_A).$$

In the vertical direction, there would be both isotropic and anisotropic fluxes, and the total etch rate would be the sum of those components. Away from the mask edge, we assume no shadowing effects, so F_I and F_A are the unobstructed chemical and ionic fluxes, respectively, given above. Thus

$$\text{etch rate} = \frac{1}{5 \times 10^{22}}[(0.01)(0.2)(2.5 \times 10^{18}) + (1)(1 \times 10^{16})]$$

$$= 3 \times 10^{-7} \text{cms}^{-1},$$

and therefore

$$\text{etch depth} = (3 \times 10^{-7} \text{cms}^{-1})(300\text{s}) = 9.0 \times 10^{-5} \text{cm}$$

$$= 0.9 \text{ μm}.$$

In the lateral direction, right under the mask edge, there is virtually no anisotropic flux ($F_A \approx 0$). Therefore, only the chemical flux term is used in the etch rate expression and it is assumed that the chemical flux is the same in all directions (due to the isotropic arrival distribution and, especially, the low S_C). Thus

$$\text{etch rate} = \frac{1}{5 \times 10^{22}}[(0.01)(0.2)(2.5 \times 10^{18})]$$

$$= 1 \times 10^{-7} \text{cms}^{-1},$$

and therefore

$$\text{etch depth} = (1 \times 10^{-7} \text{cms}^{-1})(300\text{s}) = 3.0 \times 10^{-5} \text{cm}$$

$$= 0.3 \text{ μm}.$$

There is more etching in the vertical direction than in the lateral direction, and a profile such as that in Figure 9.33 with $S_C K_I F_I \approx 0.33$ occurs in this case.

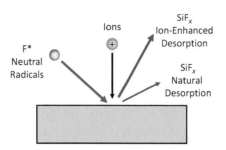

Figure 9.34 Schematic of ion-enhanced and natural desorption of SiF$_x$ byproducts.

1. natural desorption of the reactant molecules or
2. ion-enhanced desorption of the reactants.

It is assumed that the ions can enhance the rate of desorption, which in turn leads to an increase in the etch rate. Once the SiF$_x$ molecule leaves the surface, the Si bonds are exposed again and can react with fluorine radicals from the plasma. The ions are assumed not to sputter the silicon atoms and only create enhanced desorption of the neutral radicals attached to the surface. Thus the ions do not etch by themselves, but promote ion-enhanced etching by enhancing the desorption of the byproduct molecules, as shown in Figure 9.34. The ion energy must exceed some threshold energy (a few eV) for desorption to take place, and the ions get this energy from crossing the plasma sheaths. This leads to two key parameters that determine the etch profile: the surface coverage and the silicon etch rate.

In the absence of ion bombardment, the steady-state surface coverage is given by

$$\theta_{ss} = \frac{J_n}{J_n + d_{by}}, \tag{9.15}$$

where J_n is the neutral flux at a point and d_{by} is the natural desorption rate of the byproducts. If there is no desorption, the surface coverage $\theta_{ss} = 1$ and there is no etching, since the surface is completely covered by reactant molecules. With some natural desorption rate, the surface coverage is somewhat less than 1. With the addition of ion bombardment, the surface coverage is further reduced according to

$$\theta_{ss} = \frac{J_n}{J_n + J_i + d_{by}}, \tag{9.16}$$

where J_i is the ion flux at a point. The etch rate of the silicon is given by

$$R_{Si} = J_n(1 - \theta_{ss}) \tag{9.17}$$

or, more accurately, accounting for the number of incoming radical molecules that stick to the surface (S_C) and the number of incoming fluorine atoms that form the SiF$_x$ byproducts, as

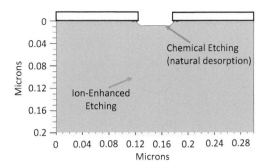

Figure 9.35 Simulation of ion-enhanced chemical etching, showing an approximately 12× enhancement in the chemical etch rate when ion bombardment is present. From Victory Process [16].

$$R_{Si} = J_n \frac{S_C}{x}(1 - \theta_{ss}).$$ (9.18)

This shows that the silicon etch rate is primarily determined by the neutral flux and the surface coverage, where the surface coverage of the reactants is reduced by ion bombardment, allowing more neutrals to react with the surface. A simulation based on this model is shown in Figure 9.35. The simulation shows the chemical etching and the ion-enhanced etching profiles, and the difference in rates is very close to the 12× observed in the classic Coburn and Winters experiment shown in Figure 9.17. Note that this simulation does not account for the additional sputtering of the silicon by the ion beam itself, and relies only on the ion bombardment to reduce the surface coverage of reactants on the surface.

9.9.7 Aspect-Ratio-Dependent Etching (ARDE) Simulation

The ion-enhanced chemical etching model easily accounts for aspect-ratio-dependent etching, as the neutrals or the ions can "stick" to the photoresist mask edges and reduce the flux of the reactants at the surface of a high-aspect-ratio structure, as shown in Figure 9.36. In this example, a series of gaps in the photoresist pattern with increasing size are simulated from left to right. The green etching profiles illustrate that the wider openings show significantly deeper etching.

9.9.8 Simulation Caveats

We have described some of the simple but physically realistic processes involved in simulating plasma etching, but there are certain effects that we did not take into account. We did not include ion sputtering or the deposition of inhibitor layers in any of the simulations, which are important for processes such as Bosch etching. This can lead to strange effects, such as reverse ARDE, where wide features etch slower than narrow features. This is caused by an enhanced inhibitor deposition on wider structures, reducing the etch rate. Some other effects not modeled here are charging effects from the ions and specular reflection (angle of incidence = angle of

Figure 9.36 Simulation of aspect-ratio-dependent etching where ion and neutral fluxes stick on the sidewalls of narrow trenches and reduce the etch rate depending on the aspect ratio. From Victory Process [16].

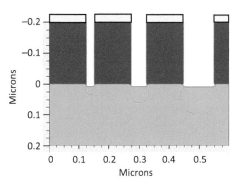

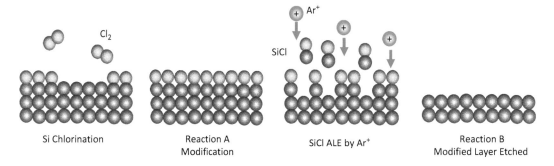

Figure 9.37 Schematic of the two sequential reactions that constitute a cycle of ALE etching. In reaction A, a modified surface layer is formed. In reaction B, a removal step etches the modified layer without disturbing the underlying substrate.

reflection) of the ions. These can cause the bowing and microtrenching, respectively, as indicated in Figure 9.27. However, these effects are relatively easy to add into simulation code if required, and modern simulation codes often make integrating user models into the code relatively easy.

9.10 Atomic Layer Etching (ALE)

Atomic layer etching (ALE) is a technique that removes thin layers of material (often a single monolayer of material) using sequential self-limiting reactions. The tight control of etch variability is a necessity for modern processes, which can tolerate a feature size variability that may only be a few atoms of silicon. ALE separates the etching step into two independent reactions. The first is a surface modification step to form a reactive layer that saturates the surface and is self-limiting. The second step is a removal step to take off only this modified surface layer. We examine the ALE process for silicon etching to provide a specific example. The process is schematically illustrated in Figure 9.37, which shows the process window where ALE etching is viable [17]. The etch step resets the surface to the original pristine condition for

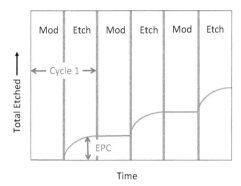

Figure 9.38 Cyclic processing during ALE removes material during each etch per cycle (EPC), with the time needed to switch between the modification and etching reactions shown in gray bars, i.e., to pump out the chlorine species and introduce the argon at the right ion energy. After [17].

the next modification and etching cycle. If too much etching occurs, there is sputtering of the surface layer. The concept is very similar to the better developed process of atomic layer deposition (ALD), discussed in Chapter 10, where sequential reactions allow layer-by-layer deposition versus etching.

Separation of the reactions means that the process runs as repeated cycles, as shown in Figure 9.38. In each cycle, there is a monolayer (approximately) removed in an etch per cycle (EPC). The advantage of the separation is that it decouples the generation, transport and etching by the neutrals, radicals, ions and electrons that occur in normal continuous plasma etching. The self-limiting reactions replace transport-limited reactions in continuous plasma etching. The self-limiting steps in ALE separate the adsorption and etching reactions, allowing for exquisite control of the layer removal [17].

9.10.1 ALE Surface Modification

To discuss the separation of the modification and etching reactions, we focus on the concrete example of etching silicon by a chlorination reaction. Chlorine modifies the silicon surface by chemisorption. Chlorine gas spontaneously adsorbs on a clean silicon surface at room temperature to form $SiCl_x$. The reaction dissociates a chlorine molecule as follows:

$$Cl_2(g) + Si(s) \leftrightarrow SiCl_x(s).$$ (9.19)

One chlorine atom at a time adsorbs onto a silicon dangling bond. The role of the chlorine is to weaken the Si–Si bond underneath the $SiCl_x$ layer and make it easier to remove that layer of silicon atoms in the subsequent etch step. The electronegative Cl atoms have a 4.2 eV binding energy for the Si–Cl bond compared to the 3.4 eV for the Si–Si bond. The electron transfer to the $SiCl_x$ layer weakens the underlying Si–Si bond to an estimated 2.3 eV, making it easier to selectively remove that layer [17]. The kinetics of the chlorine chemisorption follow the Langmuir adsorption model, where the adsorption or condensation rate is limited by the

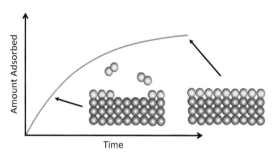

Figure 9.39 Surface coverage by adsorbed molecules versus time, with the kinetics being determined by the number of vacant surface sites according to the Langmuir model [18, 19].

number of available sites. This means adsorption is fast initially and slows as the surface sites get saturated, as shown schematically in Figure 9.39.

Langmuir showed that, in cases of true adsorption, the surface layer is not more than one molecule deep, for, as soon as the surface becomes covered by a single layer, the surface forces are chemically saturated [18]. Although the model is conceptually simple, Langmuir earned a Nobel Prize for his work on surface chemistry [19]. Of key importance to throughput during etching, the kinetics of the surface coverage $\theta(t)$ during adsorption is given by

$$\theta(t) = 1 - e^{-KPt}, \tag{9.20}$$

where θ is the fraction of the surface covered by the adsorbed atoms, K is the Langmuir rate constant, P is the partial pressure of the Cl_2 gas and t is time. Because the adsorption rate is limited by the number of available sites, it is rapid initially and then slowly saturates. The saturation of the surface coverage is desirable, as it provides a self-limiting modification step that can be separated from the subsequent etch step.

The kinetics of adsorption illustrate the major problem with atomic layer etching. If the etch rate is only a monolayer at a time, it is important to quickly cycle through the modification and etching reactions. Saturating the surface with adsorbed chlorine can take 10 s or longer for normal thermal chlorination. To be viable for manufacturing, the adsorption must proceed at a rapid rate. A plasma source can provide high concentrations of reactive molecules and radicals and speed up the kinetics by orders of magnitude, saturating the surface is less than 1 s. Since plasmas are used for the subsequent etch step, relatively conventional advanced plasma sources can be used for ALE [20]. Because the plasma inherently contains ions, neutrals and radicals, advanced plasma sources such as inductively coupled plasmas that separate the plasma generation from the ion energy are required. The ion energies during the surface modifications step must be low enough that simultaneous ion-enhanced etching does not occur, which would negate the principle advantage of ALE, which is to separate the modification and etching steps.

9.10.2 ALE Etch Regimes

The second step in the ALE cycle is to etch the modified surface layer without etching the underlying silicon. The etch can proceed thermally by heating the wafer to above 650 °C to

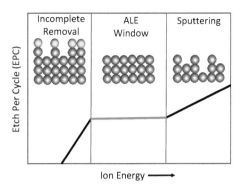

Figure 9.40 Schematic illustration of the etch per cycle versus the ion energy, showing an ALE window with the useful operating regime.

desorb the SiCl$_x$ layer. However, this process is isotropic, because thermal desorption cannot contribute any directionality to the etch process. Because both low temperatures and etch directionality are important, ions from the plasma are used to provide the energy to drive the etch front. Originally, it was thought that low-energy Ar$^+$ ions with approximately 50 eV of energy were required to selectively remove the modified layer. These ion energies are much lower than what is used in continuous ion-enhanced plasma etching. However, they are much higher than the 2.3 eV Si–Si bond under the SiCl$_x$ layer and much higher than the equivalent energy of the 650 °C thermal desorption (≈ 0.1 eV). This indicates that the yield from the ion bombardment is relatively inefficient. A simple explanation is that the SiCl$_x$ molecule must be ejected vertically with sufficient energy to overcome the surface binding, leading to many inefficient collisions. A high flux of ions at the appropriate energy is required, so, once again, advanced plasma systems such as inductively coupled plasma machines are used. There is a window for the ion energy where ALE proceeds, limited on the low-energy end by incomplete removal and on the high-energy end by Ar$^+$ ion sputtering, as shown in Figure 9.40 [17].

Recently, it was discovered that there is an inverse relationship between the ion energy and the etch time that can create a desirable ALE operating window [21]. In other words, it is the power delivered to the modified layer that determines the etch window. To maximize throughput, a high ion energy and a short exposure time can be used in the ALE cycle [21].

9.11 Planarization Methods: Chemical–Mechanical Polishing (CMP)

As the need for planarization became critical in the 1980s, many workers in the field found the idea of polishing the wafer to achieve planarity counterintuitive. Particle control and defects in semiconductor manufacturing are huge yield issues (Chapter 4) and so the idea of introducing particles to "grind" or polish surface topography seemed to be a very bad idea. However, the engineers at IBM who were responsible for Si wafer preparation realized that wafer polishing is actually a critical part of achieving the atomically flat "perfect" Si wafers that are used in Si chip manufacturing (see Figure 3.20 and the associated discussion) [22, 23]. They realized that what appeared at first glance to be a "dirty technology" might, in fact, be useful to planarize the oxide between interconnect levels.

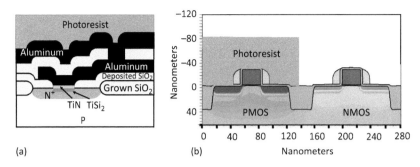

Figure 9.41 Examples of (a) an older multi-level metal technology and (b) a modern CMOS process, both illustrating the need for planarization.

In the early 1980s, IBM was looking to move from three levels of metal to four to support its mainframe logic chips. Adding an additional level of interconnect seems straightforward. The process could be as simple as depositing an inter-metal dielectric layer (typically SiO_2), opening a hole in the dielectric using lithography wherever a contact to the lower level is needed, depositing Al, forming both the contact and the next level interconnect, followed by patterning of the aluminum. Then the next level of inter-metal dielectric material would be deposited. This structure is illustrated in Figure 9.41(a), which is discussed in more detail in Chapter 11. In Figure 9.41(a), the second level of Al has been conformally deposited and photoresist has been deposited in preparation for patterning the second level Al. Because the photoresist is a liquid as applied, it forms a flat upper surface, with the result that its thickness varies considerably over the underlying topology. This, in fact, is how the early multi-level interconnect structures were made.

Obviously, the topography gets rougher and less planar as more levels are added. IBM engineers perfected oxide planarization using CMP, which was a highly confidential process innovation. Slowly, the technology spread outside of IBM and has gradually spread to planarization of many other materials, including Cu metallization.

A more planar topography is desired for two main reasons – lithography limitations and step coverage. The lithography limitations arise partly from the limited depth of focus of submicron lithography instruments. We discussed this issue in Chapter 5 (see Figure 5.22 and the associated derivation and also Figure 5.60). The depth of focus for a modern deep ultraviolet (DUV) 193 nm lithography tool is only about ±250 nm and is much less for extreme ultraviolet (EUV) tools. Nonplanar topography results in non-uniform resist thickness, as shown in Figure 9.41(b), which can result in non-uniform exposure and development because the exposure dose required depends on the thickness of the resist. There are lithographic techniques that address these problems, such as using multilayer resist structures, as discussed in Chapter 5, but these are not always effective or practicable solutions.

It was not at all clear that the CMP industry would be so successful, expanding faster than the overall semiconductor industry. Before CMP was widely available as a planarization technology, several other approaches were developed and used in some manufacturing processes. Perhaps the simplest of these is illustrated in Figure 9.42 and involves an etchback process.

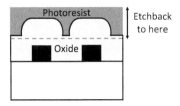

Figure 9.42 Illustration of photoresist etchback process. Photoresist is deposited over rough topography, then the structure is etched back, leaving a smooth top surface of the oxide.

In this technique, the dielectric is deposited, followed by the deposition of a sacrificial layer. The overall structure is then etched back to planarize it. The earliest such process used photoresist as a sacrificial etchback layer on top of oxide. The photoresist was deposited, or spun, over the oxide layer, filling up the spaces and ending up with a nearly flat top surface. After a hard bake, an etch is done that etches both the oxide and photoresist, ideally at the same rate. In other words, an etch chemistry is chosen such that the selectivity $S = 1$. When the bottom of the photoresist is reached, a nearly flat oxide is left behind (dashed red line in Figure 9.42). This results in local planarization, but generally not perfect global planarization because feature sizes vary across typical chips and it is difficult to achieve $S = 1$. This can also be done without a sacrificial layer if the dielectric layer is deposited thick enough. This is because the thicker a material is, the smoother the local topography tends to become. While this process is conceptually simple, it does not achieve the nanometer level of planarization required in modern structures.

The photoresist process described above works because photoresist is a liquid when it is applied and hence forms a flat upper surface. Other types of liquids have also been applied in an effort to planarize back-end structures. Spin-on-glass (SOG) has found application in some manufacturing processes. SOG materials are initially liquids, containing organic siloxanes or inorganic silicates in an alcohol-based solvent. The liquid SOG is spun onto the wafer like photoresist, filling the spaces between features. Then the wafer is baked and cured, driving off the solvents and polymerizing the silicon-containing groups, resulting in the formation of Si–O–Si bonds.

Both photoresist and SOG could be etched back uniformly using ion milling (see Section 9.5.2), which was a well-understood, clean technique using simple vacuum chambers. But manufacturing trials on dynamic random access memory (DRAM) products showed that CMP could outperform other techniques in terms of throughput and yield, cementing its place in the semiconductor manufacturing toolbox.

After the successful demonstration of CMP for planarizing inter-level metal oxides, it was used in the front-end process to planarize shallow trench isolation (STI) trenches. This basic idea of polishing inlaid structures was later transferred to polishing Cu metallization patterns using the so-called damascene patterning method (see Chapter 11). Initially, the desired degree of planarity was approximately 50 nm, but lithography requirements now require close to 0 nm of topography.

Non-planar topography also leads to problems with step coverage and filling of spaces with deposited films. We will see in Chapter 10 how step coverage and filling can be a problem. To make matters worse, step coverage problems compound one another. A large step height on one level can lead not only to thinning of the next level on sidewalls, but also to cusping and overhangs. This, in turn, leads to even worse coverage problems on the next level (see Figure 9.41).

Reducing the step heights and achieving more planar topology through processing techniques allows a degree of planarization (DOP) to be defined as

$$DOP = 1 - \frac{x_{step}^{f}}{x_{step}^{i}}, \qquad (9.21)$$

where x_{step}^{i} is the initial step height in the topography and x_{step}^{f} is the final step height after planarization. These step heights can refer to hole depths as well. A DOP of 0 means that no planarization is achieved, while a DOP of 1 corresponds to complete planarization – the final step height is zero. Figure 9.43 illustrates the two extreme degrees of planarization.

The basic operation of a CMP tool to obtain this planarization is shown in Figure 9.44 [24]. Photographs of modern CMP tools are shown in Figure 9.45 [25]. In a CMP tool, a rotating wafer is pressed against a rotating pad. The polishing slurry consists of chemically active ingredients, such as oxidizers and surfactants, and mechanically active abrasive particles. Pressure, rotation rates, slurry flow rate, and the pH and composition of the slurry are the

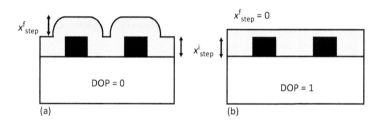

Figure 9.43 Example of degree of planarization (DOP).

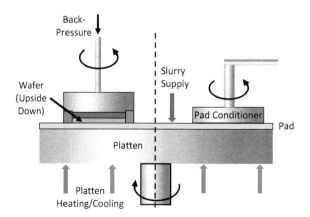

Figure 9.44 Basic CMP planarization tool [24].

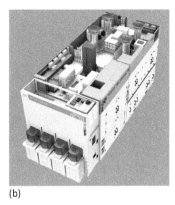

(a)   (b)

Figure 9.45 (a) Applied Materials Reflexion CMP Machine. (b) Applied Opta^{TM} CMP System. Photos courtesy of Applied Materials, Inc.

main process variables. Because chemistry is involved in the polishing process, the polishing rate can be made quite different for different materials.

In general, the chemical reaction at the surface of the wafer makes the surface material susceptible to mechanical abrasion by the silica in the slurry. For silicon dioxide removal, for example, potassium hydroxide in the slurry reacts with the oxide to form a hydrated silicate layer, which is then removed mechanically by abrasion. For metal removal, the slurry solution oxidizes the surface, which is then polished in a similar fashion to the oxide removal. We will consider the details of the chemical and mechanical aspects of the polishing process in the next subsection.

CMP has become, by far, the dominant technology used for planarization. Today, it is used in both front-end and back-end processing. It has allowed back-end structures to incorporate as many as 15 layers of wiring, and it has played a major role in mitigating the very limited depth of focus of modern lithography tools, because it allows a thin and uniform-thickness photoresist layer to be used for imaging. CMP is fundamentally a material removal process and it complements the many etching methods discussed in this chapter. The fundamental difference, of course, is the surface planarization achieved with CMP.

9.11.1 CMP Models and Simulation

We will now discuss in more detail the physical mechanisms involved in CMP. Models and simulation tools have been developed for this complex process. The first models developed were phenomenological in nature. In recent years, however, there has been a significant effort to develop more physically based models. The starting point in all these cases is the Preston equation, which was originally developed to describe glass polishing [26]:

$$RR = K_P PV \, . \tag{9.22}$$

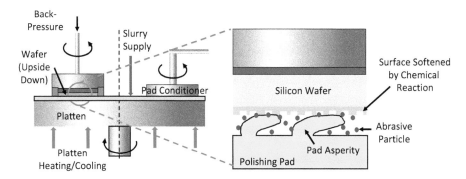

Figure 9.46 Illustration of CMP processes at the macroscopic and microscopic levels. The left side repeats Figure 9.44. The right side shows an expanded, microscopic view of the polishing pad/wafer interface.

Here RR is the material removal rate or etch rate; K_P is Preston's constant (with units of inverse pressure) and is a function of the mechanical and chemical properties of the film and pads, and the composition and properties of the polishing slurry; P is the applied pressure between the wafer and the pad; and V is the relative velocity between the pad and the wafer. Preston's equation should be regarded as empirical because it attempts to connect macroscopic parameters (P and V) to microscopic outputs (RR), as illustrated in Figure 9.46. The applied pressure is typically 2–7 psi, and the platen that holds the polishing pad rotates at 20–80 rpm. Material removal rates are typically hundreds of nanometers per minute.

At a microscopic level, the applied pressure at any point on the wafer depends on the geometry or topography of the surface. Since the pads used are not perfectly rigid and have asperities on their surface, they conform to the surface to a limited degree, but, of course, not exactly. As a result, the pressure at each point on the wafer varies, with protruding points receiving high pressure, and sunken or shadowed points receiving little or no pressure. Because of the semi-rigidity of the pads, the resulting profiles depend on both the stiffness and other properties of the pad, as well as on the density and spacing of elevated features on the wafer surface. It is important to recognize that the CMP process is inherently a nonlocal process. The removal rate at a particular point depends on what the surrounding topography on the wafer looks like. For example, a low-lying point on a wafer surface will see a slower removal rate because the higher topography around it results in lower local pressure at the lower point, i.e., the lower lying point is "sheltered" by the surrounding topography.

The size and spacing of the features on the wafer surface will also play a role in the removal rate because the pad surface is limited in its ability to expand into small depressions on the wafer surface. Finally, the chemistry part of CMP plays the role of "softening" the surface layer on the wafer so that it can be more efficiently polished. This provides material selectivity depending on the choice of the chemistry. Typically, many of these effects are incorporated in experimentally determined values for K_P.

One of the first models of the CMP process was developed by Warnock at IBM [27]. Like many models for CMP, this approach attempts to determine the relative pressure at each point and then calculates the relative removal rate at each point assuming that it is linearly proportional to the local pressure, as given by the Preston equation. In Warnock's model, the local removal rate RR_i at a point i on the surface is given by

$$RR_i \propto \frac{K_i A_i}{S_i}. \tag{9.23}$$

The three terms in this expression take into account the increases or decreases in pressure depending on the local topography as well as the rigidity of the polishing pad and wafer: K_i is called the kinetic factor at point i and is a geometrical factor that takes into account sloped surfaces by calculating an effective vertical component of the horizontal polish rate; A_i is an accelerating factor for points that protrude above neighboring points; while S_i is a shading factor for points that are lower than neighboring points so that the pressure, and hence polishing rate, is decreased. The three factors are calculated at each point for each time step during the simulation, and depend on the geometrical relationship of the point to its neighboring points through various mathematical–geometrical expressions. They are also dependent on the rigidity of the polishing pad. Most of the constants in these expressions are empirically determined, but Figure 9.47 gives an indication of how these parameters would depend on surface topography.

Warnock's model has been implemented in commercially available simulation tools, e.g., Silvaco's Athena [16]. In the Athena implementation of Warnock's model, there are four parameters.

1. KINETIC.FAC – this changes the polishing rate as the surface becomes more vertical. This is K_i in (9.23).
2. LENGTH.FAC – this is the horizontal deformation scale in microns. It is a measure of the polishing pad's flexibility and describes the distance over which shadowing will be felt by a tall feature.
3. HEIGHT.FAC – this is the vertical deformation scale in microns. It measures how much the polishing pad will deform with respect to the height of a wafer feature. The LENGTH.FAC and HEIGHT.FAC parameters are used to calculate A_i and S_i at each point on the structure.
4. KINETIC.FAC.SOFT – this is the polishing rate on a flat surface and is equivalent to the proportionality constant in (9.23).

Figure 9.48 shows an example of a CMP simulation using this model. The initial structure shown in Figure 9.48(a) was chosen to match structures that Warnock used in his original paper [27], so that the simulations could be compared to experimental results. The experimental

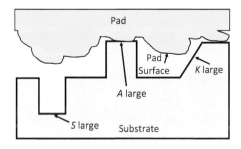

Figure 9.47 Parameters in Warnock's model for CMP. The shape of the pad depends on its physical structure, rigidity and the pad pressure. After [26].

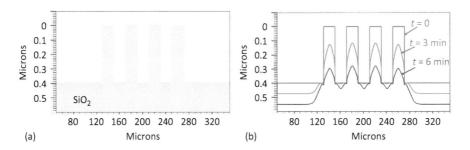

Figure 9.48 Athena simulation of CMP on an SiO$_2$ structure. (a) The starting structure consists of 20 μm SiO$_2$ columns and spaces on an SiO$_2$ substrate. (b) The simulated polished profiles after 3 and 6 min. Nominal simulation parameters were used: KINETIC.FAC.SOFT = 25, HEIGHT.FAC = 0.02, LENGTH. FAC = 4.2 and KINETIC.FAC = 10.

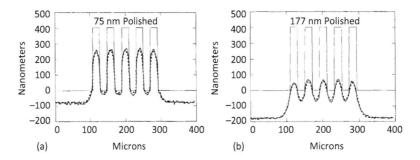

Figure 9.49 Results from Warnock's paper [27], for the same SiO$_2$ structure as simulated in Figure 9.48. The solid lines are experimental results; the dashed lines are simulations. Used with permission of IOP Publishing Ltd., from [27]; permission conveyed through Copyright Clearance Center, Inc.

results from Warnock are shown in Figure 9.49, along with simulations Warnock did using (9.23). The agreement between the simulation and the experimental results in Warnock's paper is quite good, and the Athena results in Figure 9.48(b) reproduce these results quite well. In this example, the only material present is SiO$_2$, so perfect planarization is not achieved, at least for the polishing times simulated.

Figure 9.50(a) shows an Athena simulation of a similar structure but with 5 μm SiO$_2$ columns and spaces. Figure 9.50(b) shows Warnock's experimental results and his simulation results for this structure [27]. The agreement between both simulations and the experimental results is quite good.

Commercially available simulation tools such as Athena often contain a variety of models that the user can select. Athena also contains a "hard polish model" that is less sophisticated than the model illustrated above. The hard polish model is more like a grinding process in which a very simple polishing rate model is used,

$$RR = MAX.HARD(1 - Pf) + MIN.HARD(Pf),\qquad(9.24)$$

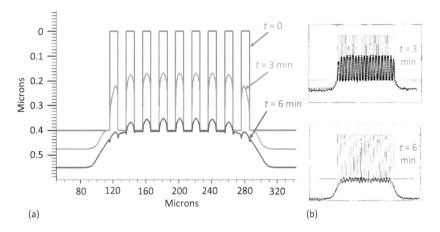

Figure 9.50 (a) Athena simulation of CMP on an SiO_2 structure. The starting structure consists of 5 µm SiO_2 columns and spaces on an SiO_2 substrate. The same simulation parameters as in Figure 9.48 were used. (b) Warnock's experimental and simulation results. Used with permission of IOP Publishing Ltd., from [27]; permission conveyed through Copyright Clearance Center, Inc.

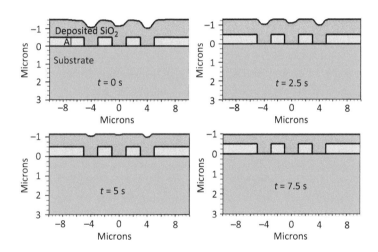

Figure 9.51 Athena simulation of CMP on a simple back-end structure with Al metal lines and chemical vapor deposited (CVD) SiO_2. The hard polish model was used, which results in a perfectly planarized structure after 7.5 s of polishing.

where Pf is a pattern factor that equals 1 on a flat surface and is smaller for regions that stick up above the flat surface. Figure 9.51 shows an example of a CMP simulation using this simple "grinding model" that results in a perfectly planar structure.

A final example using the relatively simple models we have discussed to this point is shown in Figure 9.52. In this example, a tungsten (W) via is polished using CMP to produce a planar structure on which subsequent metal layers could be deposited. When the polishing rate is somewhat different for two different materials and the pad is not perfectly rigid, a phenomenon

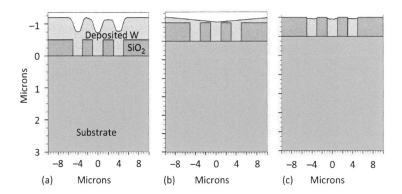

Figure 9.52 Athena simulation of chemical–mechanical polishing of a tungsten via structure: (a) after W deposition by CVD; (b) polishing has almost removed the W outside the vias; and (c) after polishing is complete. Because of the faster polishing of tungsten compared to silicon dioxide and the semi-rigid pad, "dishing" of the tungsten plug can result.

called "dishing" can occur. An over-polish is done to ensure that all the metal is removed from on top of the oxide layer. But metals generally polish faster than oxide in the metal polishing slurry. Because of the semi-rigidity of the pad and the difference in polishing rates, this causes the metal in some places to be removed below the top of the oxide layer. The shape of the resulting metal surface is curved, or "dished." Having less selectivity between materials would reduce this, but would make it harder to stop the polishing without removing too much oxide. Having a more rigid pad would also help, but could cause unequal polishing due to any bowing or thickness non-uniformities in the silicon substrate. Dishing, in general, is worse for wider structures, such as contact pads.

One can also see in Figure 9.52 that the edges of the oxide layer are also slightly sloped. This is called "erosion" and is due to the fact that the presence of a softer material – the metal – allows for more force and polishing on the adjacent harder material – the oxide. This can be important when there are a series of closely spaced metal lines, and the dielectric between them can be over-polished or "eroded."

There has been considerable effort in recent years to develop more sophisticated CMP models, motivated by the desire to be able to accurately simulate the complex back-end and front-end structures that use CMP today. The overall problem involves many levels of modeling, from machine level down to nanometer-scale removal of material. For the interested reader, there is an extensive literature on these topics. Representative papers include [28, 29, 30].

9.12 Summary of Key Ideas

The purpose of etching is to transfer patterns on the wafer into the underlying substrate. Etching often involves selectively removing a layer without etching underlying layers. The degree of selectivity determines the process window available. In general, chemical or wet etching can be extremely selective but usually etches isotropically, in other words, at the same rate in all directions. This means that the etched region is wider than the patterned region. A limited number of wet etchants can take advantage of different etch rates along different crystal planes and produce directional etching in crystalline substrates.

To have more control over the directionality, plasma etching is used extensively in modern processes. Plasmas contain ions, neutrals and reactive radicals. The plasma is contained by sheaths near surfaces, which balance the ions and electrons escaping from the bulk plasma, generally by setting up a positive bias in the plasma, which retards electrons and speeds up ions. The result is that ions are accelerated directionally across the sheaths to the substrate, providing a means for directional etching. Plasma etching can span the range from purely physical to purely chemical etching. A combination of ion and radical etching can act synergistically and provide faster etching than either process in isolation. This ion-enhanced plasma etching is key to improving selectivity and directionality in etch systems. Modern etch machines often separate the bulk plasma generation and the ion energy by using separate power sources for each step. These provide "tuning knobs" to optimize the selectivity and directionality.

Plasmas are often modeled by a combination of an angular energy distribution (a cosine function to a power n, i.e., $(\cos \theta)^n$) for the ions, radicals and neutrals along with a sticking coefficient S_C that determines how well the incoming species stick or bounce off surfaces. These relatively simple models can account for many of the major features of etch profiles, such as the directionality and undercut of the etch profile. Plasma etching has recently been expanded to include atomic layer etching (ALE), capable of etching monolayer by monolayer. ALE achieves this by separating the process into a self-limiting surface modification step driven by the adsorption of molecules that reduce the binding energy of the underlying layer of surface atoms. A separate etching step removes only this modified layer without damaging the underlying substrate. There are limited process windows before the ion-enhanced etching causes sputtering of the substrate, eliminating the usefulness of ALE. In Chapter 10, we will see a companion ALD that enables a monolayer-by-monolayer deposition process. In principle, these processes can provide monolayer control over both the etching and deposition processes.

CMP is a material removal process that combines chemical etching with mechanical polishing to produce a planarized surface. While originally developed for multi-level back-end structures, CMP is now widely also used in front-end processing. This has been driven by the very small depth of focus in modern DUV and EUV lithography systems, which require thin, uniform-thickness layers of photoresist in order to produce high-resolution images. CMP modeling is still fairly empirical, but there are efforts underway to introduce a deeper scientific basis for these models.

9.13 REFERENCES AND NOTES

[1] K. R. Williams, K. Gupta and M. Wasilik, "Etch rates for micromachining processing – Part II," *Journal of Microelectromechanical Systems*, vol. 12, no. 6, pp. 761–778, 2003.

[2] B. N. Chapman, *Glow Discharge Processes: Sputtering and Plasma Etching*, Wiley, 1980.

[3] M. A. Lieberman and A. J. Lichtenberg, *Principles of Plasma Discharges and Materials Processing*, Wiley, 2005.

[4] Jim McVittie, Private communication.

[5]   J. W. Coburn and H. F. Winters, "Plasma etching – a discussion of mechanisms," *Journal of Vacuum Science & Technology*, vol. 16, no. 2, pp. 391–403, 1979.

[6]   C. Cardinaud, M.-C. Peignon and P.-Y. Tessier, "Plasma etching: principles, mechanisms, application to micro- and nano-technologies," *Applied Surface Science*, vol. 164, no. 1–4, pp. 72–83, 2000.

[7]   D. L. Flamm and V. M. Donnelly, "The design of plasma etchants," *Plasma Chemistry and Plasma Processing*, vol. 1, pp. 317–363, 1981.

[8]   J. W. Coburn and H. F. Winters, "Ion- and electron-assisted gas-surface chemistry – an important effect in plasma etching," *Journal of Applied Physics*, vol. 50, no. 5, pp. 3189–3196, 1979.

[9]   J. P. Chang, J. C. Arnold, G. C. Zau, H.-S. Shin and H. H. Sawin, "Kinetic study of low energy argon ion-enhanced plasma etching of polysilicon with atomic/molecular chlorine," *Journal of Vacuum Science & Technology A: Vacuum, Surfaces, and Films*, vol. 15, no. 4, pp. 1853–1863, 1997.

[10] B. A. Heath, "Selective reactive ion beam etching of SiO_2 over polycrystalline Si," *Journal of the Electrochemical Society*, vol. 129, no. 2, p. 396, 1982.

[11] "Method of anisotropically etching silicon," United States Patent 5501893.

[12] J. Hopwood, "Review of inductively coupled plasmas for plasma processing," *Plasma Sources Science and Technology*, vol. 1, no. 2, p. 109, 1992.

[13] R. J. Shul and S. J. Pearton, *Handbook of Advanced Plasma Processing Techniques*, Springer, 2011.

[14] Conor Rafferty, Private communication.

[15] J. A. Sethian, *Level Set Methods and Fast Marching Methods: Evolving Interfaces in Computational Geometry, Fluid Mechanics, Computer Vision, and Materials Science*, Cambridge University Press, 1999.

[16] Victory Process is a Silvaco Process simulation tool. See https://silvaco.com/tcad/victory-process-3d/.

[17] K. J. Kanarik, T. Lill, E. A. Hudson, *et al.*, "Overview of atomic layer etching in the semiconductor industry," *Journal of Vacuum Science & Technology A: Vacuum, Surfaces, and Films*, vol. 33, no. 2, p. 020802, 2015.

[18] I. Langmuir, "The adsorption of gases on plane surfaces of glass, mica and platinum," *Journal of the American Chemical Society*, vol. 40, no. 9, pp. 1361–1403, 1918.

[19] H. Swenson and N. P. Stadie, "Langmuir's theory of adsorption: a centennial review," *Langmuir*, vol. 35, no. 16, pp. 5409–5426, 2019.

[20] A. Goodyear and M. Cooke, "Atomic layer etching in close-to-conventional plasma etch tools," *Journal of Vacuum Science & Technology A: Vacuum, Surfaces, and Films*, vol. 35, no. 1, p. 01A105, 2017.

[21] K. J. Kanarik, S. Tan, W. Yang, I. L. Berry, Y. Pan and R. A. Gottscho, "Universal scaling relationship for atomic layer etching," *Journal of Vacuum Science & Technology A: Vacuum, Surfaces, and Films*, vol. 39, no. 1, p. 010401, 2021.

[22] K. Beyer, "The inception of chemical–mechanical polishing for device applications at IBM," *IBM Micronews*, vol. 5, p. 40, 1999.

[23] K. C. Cadien and L. Nolan, "Chemical mechanical polishing method and practice," in *Handbook of Thin Film Deposition*, 4th edn., eds. K. Seshan and D. Schepis, Elsevier, chap. 10, pp. 317–357, 2018.

[24] "Chemical Mechanical Planarization," MKS Instruments website. See https://www.mksinst.com/n/chemical-mechanical-polishing.

[25] Applied Materials website. See http://www.appliedmaterials.com/products/reflexion-lk-cmp.

[26] F. Preston, "The theory and design of plate glass polishing machines," *Journal of Glass Technology*, vol. 11, no. 44, pp. 214–256, 1927.

[27] J. Warnock, "A two-dimensional process model for chemimechanical polish planarization," *Journal of the Electrochemical Society*, vol. 138, no. 8, p. 2398, 1991.

[28] W. Fan and D. Boning, "Multiscale modeling of chemical mechanical planarization (CMP)," in *Advances in Chemical Mechanical Planarization (CMP)*, Elsevier, pp. 137–167, 2016.

[29] J. Vlassak, "A model for chemical–mechanical polishing of a material surface based on contact mechanics," *Journal of the Mechanics and Physics of Solids*, vol. 52, no. 4, pp. 847–873, 2004.

[30] X. Wang, P. Karra, A. Chandra, *et al.*, "A multi-scale predictive model for wafer surface evolution during a CMP process incorporating slurry evolution," in *Chemical-Mechanical Planarization for ULSI Multilevel Interconnect Conference (CMP-MIC)*, Fremont, CA, March, 2007.

9.14   PROBLEMS

9.1   In a LOCOS oxidation process, Si_3N_4 is used to prevent oxidation in regions where it is present. The nitride layer needs to be removed after LOCOS. Why might pure chemical etching, such as in wet etching, be adequate for removing (etching) the Si_3N_4 layer? See Figure 2.5 in Chapter 2, for example.

9.2   In plasma etching in modern CMOS process flows, generally an etch process is chosen that attempts to achieve good selectivity and good anisotropy. However, there are some situations in which we might choose an etch process that mainly emphasizes selectivity or anisotropy. Give an example from the CMOS process flow we considered in Chapter 2 where we might choose a system that is very selective and not care very much about anisotropy.

9.3   If the etch anisotropy is 0, as in wet etching, what is the undercut or etch bias when etching a 0.5 μm thick film? What is the undercut when the anisotropy is 0.75? Assume no over-etch in each case.

9.4   In a certain process, it is desired to etch the red film in the diagram below. The lithography process (photoresist is the orange layer) can produce 0.25 μm features as shown. The etching process must produce 0.5 μm lines and spaces in the red layer (at the top of the layer), so the pitch of the red features is 1 μm.

(a) What degree of anisotropy is needed in an etch process in order to produce such a structure?

(b) What minimum pitch could be obtained for such a structure with wet etching with the same lithography if the width of the red material must be 0.5 μm at the top?

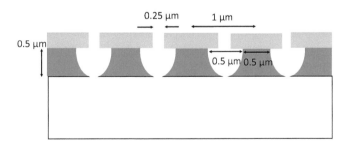

9.5   What are the advantages and disadvantages of ion-enhanced chemical etching versus sputter etching? Cite a hypothetical example of when you might want to use sputter etching rather than the more sophisticated ion-enhanced chemical etching?

9.6   In the CMOS process we discussed in Chapter 2, a polysilicon "dummy" gate was deposited and then etched. Suppose the polysilicon is deposited 50 nm ± 10% thick and the underlying temporary gate oxide is 2 nm thick. The polysilicon etching step needs to over-etch by 10% for manufacturing control. Suppose the selectivity of this etch between polysilicon and SiO_2 is 5:1. If the 2 nm oxide is completely etched, the etch would then start etching the underlying Si (at the same rate as the poly). What is the worst-case depth this process would etch into the silicon?

9.7   One of the most critical etching steps in the CMOS process flow is etching the polysilicon gates, as shown below. Suppose that, in a particular process, the manufacturing control on the polysilicon gate thickness is 3500 Å ± 500 Å. Suppose also that a 10% over-etch has to be done because the etching rate itself can vary by ±10%. The gate oxide in this process is 15 Å thick. The etching process is allowed to etch through the gate oxide and into the silicon source/drain regions, but not deeper than 100 Å into the silicon. Assuming that the poly and the silicon etch at the same rate, what is the minimum required selectivity in the etching process between poly and oxide?

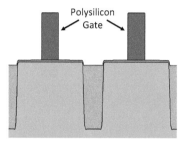

9.8   Copper metallization is usually accomplished with a dual damascene process that we will discuss in detail in Chapter 11. In this process, a layered dielectric is deposited and two successive masks are used to provide inlaid patterns for the Cu vias and the Cu wiring pattern. An Si_3N_4 etch stop is used in the patterning of the oxide layer with each of the two masks, as shown below. If a nominal 1.0 μm SiO_2 layer is to be etched and the etch system has 50:1 selectivity (oxide:nitride), what is the minimum thickness of the Si_3N_4 etch stop required? Assume all layers have a ±10% tolerance on thickness.

9.9   Explain how loading effects can affect endpoint detection.

9.10 It is found that a certain plasma etch chemistry in a certain reactive ion etching (RIE) etch system produces vertical sidewalls with zero etch bias when etching a particular film. Adding chemical A to the etch chemistry results in non-vertical sidewalls, and an etch bias. Adding chemical B to the original etch chemistry results in non-vertical sidewalls, but with zero etch bias. Explain what may be going on.

9.11 (a) In a particular etch process, if selectivity is the biggest concern, which type(s) of etch equipment should be used?

(b) If the biggest concern is ion bombardment damage, which type(s) of etch equipment should be used?

(c) If the biggest concern is obtaining vertical sidewalls, which type(s) of etch equipment should be used?

(d) If the biggest concerns are selectivity *and* vertical sidewalls, which type(s) of etch equipment should be used?

(e) What about selectivity *and* vertical sidewalls *and* damage, while maintaining a reasonable etch rate?

9.12 The figure below was used to describe the ion-enhanced etching process that is critical to achieving both anisotropic and selective etching. Suppose the etching process shown in the middle (XeF_2 + Ar^+) is being used on a structure to etch a silicon film. The material underneath the silicon is not etched by XeF_2 at all. Estimate the selectivity that this etching system would produce in this case. Explain your answer.

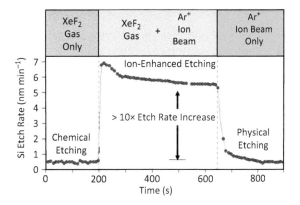

9.13 It is observed that the sidewall slope in an etch process becomes more sloped as the temperature is reduced. Suggest a reason why this might happen.

9.14 Suppose the anisotropy of an etch process is 0.45. What percentage of the etch rate in the vertical direction is due to the chemical/isotropic component and what percentage is ionic/anisotropic, assuming a linear etch mechanism as in Figure 9.30? State all assumptions.

9.15 We wish to etch a WSi_2/poly bilayer structure, as shown below, using one etch process for both the silicide and the poly. The degree of anisotropy of the WSi_2 is 1.0, and the degree of anisotropy of the polysilicon is 0.8. The etch rates of both materials are the same. Assume the selectivities of these layers over the SiO_2 and the photoresist mask are infinite. Also assume no over-etch is done.

(a) Draw the resulting structure after the etch (down to the SiO_2), and indicate important dimensions. Explain your sketch.

(b) Repeat the problem, but now assume that the degree of anisotropy of both the silicide and poly is 0.8, and everything else is the same as in (a). Explain your sketch.

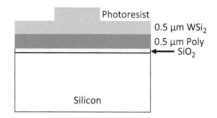

9.16 For a particular plasma etch process in which the linear etch model is applicable, a degree of anisotropy of 0.8 or better is desired. If the unobstructed ionic flux on a flat surface is 3×10^{16} atoms cm^{-2} s^{-1} (with $K_A = 1$), what unobstructed chemical flux would result in an anisotropy of 0.8. For this process, $S_C = 0.01$ and $K_I = 0.1$.

9.17 A particular plasma etch process is used in which the linear etch model applies and both physical/ion and chemical components are present. One can assume that the chemical component is perfectly isotropic, and that the physical or ion component is perfectly anisotropic. It is found that, if the ion flux is tripled, the total etch rate in the vertical direction is doubled.

(a) Based on this, what percentage of the etch in the original case (i.e., before the flux was tripled) in the vertical direction is due to the physical/ion component?

(b) What is the ratio of the lateral to vertical etch rates (again, in the original case)?

9.18 An alternative mathematical formulation to the ion-enhanced chemical etching model discussed in Section 9.9.6 is called the saturation/adsorption model:

$$\text{etch rate} = \frac{1}{N}\left(\frac{1}{1/S_C F_I + 1/K_A F_A}\right).$$

In this expression, the isotropic (chemical) etching term $S_C F_I$ and the anisotropic (ionic) etching term $K_A F_A$ are treated as acting in series rather than in parallel, which is what the linear etch model presumes. Thus, we add reciprocals of the terms, so that, if either term is zero, the total etch rate is zero.

We want to see how the etch rate in the vertical direction might depend on pressure, assuming that the etch follows the above model. Assume that, for a particular etch system, the chemical flux is directly proportional to the pressure, while the ion flux is inversely proportional to the pressure. That is, $F_I = F_I' P$ and $F_A = F_A'/P$ (where P is normalized to 1 atm and unitless). Also assume that the density is 1 atom nm^{-3}, and that $K_A F_A' = S_C F_I' = 1$ atom nm^{-2} s^{-1}.

(a) Plot the vertical etch rate versus pressure P from $P = 0$ to $P = 10$.

(b) Repeat with $K_A F_A' = 40$ atoms nm^{-2} s^{-1} and $S_C F_I' = 1$ atom nm^{-2} s^{-1}.

9.19 In an etch process, there is a finite amount of purely chemical etching without any ion bombardment (i.e., spontaneous chemical etching). In addition, ion bombardment greatly increases the etch rate by facilitating the breaking up of the etch precursor. At high ion flux, the etch rate saturates. No etching occurs when there is only ion bombardment with no chemical component.

(a) Write a generalized etch rate equation that can describe this behavior following the saturation/adsorption model discussed in problem 9.18.

(b) Sketch an etch rate versus ion flux curve for this process for some non-zero chemical flux.

(c) Sketch what the etch profile might look like for this process (i.e., etching through a window in a mask).

9.20 Three different plasma etch processes, (a), (b) and (c) below, each involve ion-enhanced etching but have very different characteristics. The etch rate versus ion flux behavior in each is shown. For each process, sketch what the etch profile would look like if you are using that process to etch the material with a mask, as shown below on the top left (assume zero etch rate for the mask). Explain.

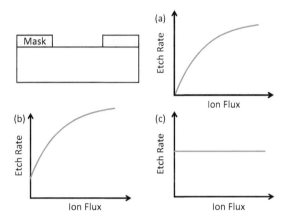

9.21 A perfectly conformal thin film is deposited on an overhang test structure as shown below. If this test structure were etched (unmasked) in three different plasma etching systems, 1, 2 and 3, sketch the resulting thin-film profiles produced in each system. System 1 operates in the chemical etching mode. System 2 operates in the physical etching mode. System 3 operates in the ion-enhanced etching mode. In each case, the etching time is just long enough to completely remove the film on the flat top surface. Explain your sketches!

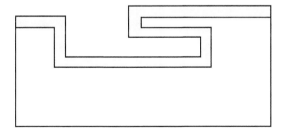

9.22 The test structure in (a) below is used in some etching and deposition experiments. A 1 μm thick thin film is deposited using a perfectly conformal deposition system, as shown in (b) below. The structure is then placed in an ion-enhanced etching system with XeF_2 and Ar^+. This system's etch rate for the deposited film is described by Figure 9.17 in this chapter.

Also presume that the etching system has no selectivity to any of the underlying materials. The etch time is just long enough to etch through the deposited film on the top surface plus a 20% over-etch. Sketch the shape of the resulting structure after etching. Explain the main features in your sketch and label important dimensions quantitatively.

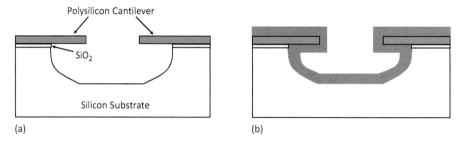

(a)   (b)

9.23 One of the alternatives to CMP planarization is illustrated below. Photoresist is applied on a surface with non-flat topography. Since photoresist is a liquid, it will produce a flat upper surface. If an etch process is then used which has zero selectivity between photoresist and oxide, planarization will result. What type of etching process would you use in this application? Explain why.

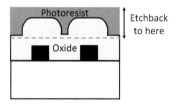

9.24 Aluminum lines 0.5 μm thick are formed on a flat surface. A layer of SiO_2 (at least 1 μm thick) is then deposited over this. A DOP of 1 is desired, but the actual DOP obtained was 0.4. A CMP etchback of the oxide is then done to obtain a net DOP of 1. If the CMP etch rate is 0.12 μm min^{-1}, what is the minimum CMP etch time needed to obtain this net DOP of 1? Assume no variations in thicknesses or CMP etch rates.

10   Deposition Technologies

10.1   Introduction

Multiple deposited layers make up the core of almost all devices, whether micro-electromechanical systems (MEMS) or semiconductor circuits. Successive layers are deposited, patterned and etched to form the complex stacked structures that provide the desired functionality. The range of deposition techniques used varies widely even if we consider a single specific process, such as building a complementary metal-oxide–semiconductor (CMOS) chip. The toolbox of deposition systems is extensive, providing interesting choices for process designers. To provide some structure to this chapter, we divide deposition systems by their thermal profiles, from high-temperature to low-temperature systems, as this often determines their utility at a particular step in a process. It has the advantage of mimicking the historical development, but process engineers use the entire spectrum of systems from the deposition toolbox to develop a novel process.

10.2   Thermal Classification of Deposition Systems

This thermal classification is approximate, with many exceptions. The earliest layers deposited on a silicon substrate were grown oxides at high temperatures (Chapter 6). Today, most thin films are deposited layers of some sort. Table 10.1 shows a simplified thermal classification of layers deposited from high to low temperatures. The top five rows are processes that involve chemistry, and the last three rows are deposition technologies that do not.

10.2.1 Chemical Vapor Deposition (CVD)

We start with chemical vapor deposition (CVD) systems that use reactive gases that undergo thermal decomposition and deposit a layer on the wafer surface. A highly simplified reaction would be the decomposition of silane (SiH_4) to produce a silicon film:

$$SiH_4 \rightarrow Si + 2H_2 . \tag{10.1}$$

Silicon films deposited by CVD may be crystalline if they are deposited on a crystalline substrate with appropriate process conditions. They may also be polycrystalline or amorphous

Table 10.1 **Thermal classification of different deposition processes. Courtesy Ted Kamins [1].**

Process	Abbrev.	Temperature range (°C)	Method
Chemical vapor deposition	CVD	500–1100	Gas-phase, chemical
Low-pressure chemical vapor deposition	LPCVD	500–800	Gas-phase, low-pressure chemistry
Metal–organic chemical vapor deposition	MOCVD	600–1200	Gas-phase chemistry
Plasma-enhanced chemical vapor deposition	PECVD	300–450	Plasma chemistry
Atomic layer deposition	ALD	100–400	Chemistry
Plasma sputtering	Sputtering	100–300	Physical deposition
Evaporation Physical deposition		100–500	Evaporation
Molecular beam evaporation	MBE	100–400	Physical deposition

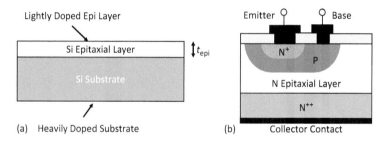

Figure 10.1 (a) Schematic of lightly doped silicon epitaxial layer on a silicon substrate. (b) A power bipolar device that utilizes a lightly doped collector layer on a heavily doped substrate.

if they are deposited on non-crystalline substrates. It may not be obvious why one would want to grow a single-crystal silicon film on a silicon wafer, but, because gases can be ultrapure, this may provide some advantage in yield because of the inherent oxygen and carbon concentrations in Czochralski (CZ) silicon wafers (Chapters 3 and 4). The resistivity of silicon grown by the CZ method is limited to ≈100 ohm cm because of the existence of oxygen donors and other impurities from the quartz crucible. Such impurities can be substantially reduced by gas-phase growth.

One could also grow undoped or lightly doped silicon on a heavily doped substrate, as indicated in Figure 10.1. The only limit is on any dopants that diffuse into the growing layer from the substrate by solid-state diffusion or by gas-phase "autodoping." Such layers can be used to produce highly doped substrate contacts for many types of power transistors, where the active regions are formed in a lightly doped N-type epitaxial layer. The bipolar transistor in Figure 10.1 is one example. For lateral power devices, a thin layer is used to reduce the electric fields. The thickness of the thin drift layer is too small to allow the wafer to be handled, so growing a thin layer on a thick substrate solves these problems.

The same lightly doped structure on a heavily doped substrate or implanted layer can be used to build CMOS devices, with the heavily doped underlying layer helping to avoid a device condition known as "latch-up" in the CMOS devices. As discussed in Chapter 4, the use of epi layers in CMOS technologies can also be used to implement intrinsic gettering. In the most modern 3D fin field effect transistors (FinFETs), the source and drain fins are too thin to be implanted, so they are etched away and a doped region is grown out from the channel using an epitaxial silicon growth process (see Figure 2.30). Another example is where growing a thick silicon layer over a MEMS device can hermetically seal the device.

There are thus many uses for this "epitaxial" growth of silicon on silicon. The term "epitaxy" comes from the Greek roots *epi*, meaning "above," and *taxis*, meaning "an ordered manner." In other words, the added material grows in an ordered manner above the substrate. This means the substrate must be atomically clean, which usually requires a high-temperature anneal in hydrogen to remove any remnants of native oxide on the silicon surface.

If a film of a different crystalline material is grown on a substrate, the term "heteroepitaxy" is used. This might include SiGe on Si, GaN on Si, Si on sapphire (Al_2O_3), AlGaAs on GaAs, etc. The growth of atoms of different size on the substrate causes strain and often causes defects if the layer is too thick. Strain is used to modify the mobility of carriers in the channel of CMOS devices, by growing strained SiGe source/drain regions. For more exotic material combinations, such as GaN on Si, many proprietary tricks are used, which might include intermediate anneals to stabilize the structure. Nevertheless, the same principles of deposition apply in all cases. At an atomic level, depositing atoms find a site on the substrate where they can attach and continue the layer growth.

The growth at an atomistic level occurs by mechanisms like those shown in Figure 10.2. Based on analytical techniques like reflection high-energy electron diffraction (RHEED), the growth is thought to occur layer by layer, perhaps by zipping along kink sites to quickly form a new layer with little stress. A similar mechanism was considered in Chapter 6 during oxidation

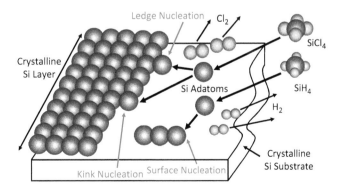

Figure 10.2 Schematic of adsorbed silicon atoms from SiH_4 or $SiCl_4$ on the surface of the wafer, where they migrate to low-energy kink sites or ledge sites. Surface and ledge nucleation can also occur and continue the crystal structure. After Ted Kamins [1].

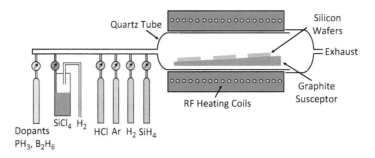

Figure 10.3 Schematic of an atmospheric-pressure chemical vapor deposition (APCVD) system.

Figure 10.4 Schematic of the important gas-phase transport flux (F_1) and the surface reaction flux (F_2) in a CVD deposition system.

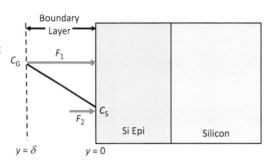

to minimize stress from the growing film. If larger or smaller atoms are grown by epitaxy, processes like surface nucleation can quickly lead to defects when individual nuclei merge. Surface mobility is important for growing defect-free layers.

The equipment for epitaxial growth is relatively simple, as shown in Figure 10.3, though we will discuss some of the complexities that occur a little later. Furnace heating is used mainly for lower temperatures and LPCVD. At normal epitaxial growth temperatures, the hot walls would decompose the gas before it gets to the wafer and create non-uniformity. Most epitaxial systems are "warm wall," with the wall temperature being much cooler than the wafer temperature. The physics of layer growth consists of mass transport of reactants to the surface followed by surface reactions. Similar to the case for oxidation kinetics in Chapter 6, we consider the two most important processes indicated in Figure 10.4 (mass transfer and surface reactions) and equate the fluxes under steady-state conditions. Note that, in the schematic in Figure 10.4, the diffusion flux through the growing layer that plays such an important part in oxidation kinetics does not exist because the reactions occur at the top surface. Instead, the gas-phase transport is important in this case.

The flux F_1 (molecules cm^{-2} s^{-1}) of reactant species from the gas phase to the wafer surface is given by

$$F_1 = h_G(C_G - C_S) \ . \tag{10.2}$$

The $(C_G - C_S)$ term (molecules cm^{-3}) is the concentration gradient of the reactant species across the "boundary layer" between the freely flowing gas region and the substrate surface. This concentration gradient acts as the driving force to move reactants to the surface, with h_G (cm s^{-1}) being the mass transfer coefficient.

The flux F_2 (molecules cm^{-2} s^{-1}) is that consumed by the reaction at the surface and, assuming first-order reaction kinetics, is given as

$$F_2 = k_S C_S,$$ (10.3)

where k_S (cm s^{-1}) is the surface reaction rate and C_S (cm^{-3}) is the concentration of the reacting species at the surface.

Both fluxes describe the flow of molecules per square centimeter per second, and, in steady state, the fluxes must be equal, since they act in series. By equating the fluxes, we can obtain the surface concentration in terms of the known gas concentration:

$$C_S = C_G \left(1 + \frac{k_S}{h_G}\right)^{-1}.$$ (10.4)

The growth rate of the film is now given by

$$v = \frac{F}{N} = \frac{k_S h_G}{(k_S + h_G)} \frac{C_G}{N},$$ (10.5)

where v (cm s^{-1}) is the deposition rate and N (cm^{-3}) is the number of atoms incorporated per unit volume in the film, which, for the case of silicon growth, is 5×10^{22} cm^{-3}. Note that the concentration C_G of reactant molecules may actually be a small fraction of the total gas flow, as usually a carrier gas like H_2 is used to move the reactant gas along the reactor while minimizing the amount of the more expensive reactant gas required for growth. The mole fraction (or yield) of reactant gas that contributes to the deposition, Y, is then a fraction of the total gas concentration, C_T:

$$Y = \frac{C_G}{C_T}.$$ (10.6)

Since the number of molecules is directly related to the pressure, this also corresponds to the partial pressure of the reactant molecules compared to the total pressure. This gives a final equation for the growth velocity of

$$v = \frac{k_S h_G}{(k_S + h_G)} \frac{C_T}{N} Y.$$ (10.7)

From (10.7), we can see that there are two limiting cases, and one of the two fluxes will determine the rate of deposition. If $k_S \ll h_G$, then

$$v \cong k_S \frac{C_T}{N} Y.$$ (10.8)

This is the surface-reaction-controlled case. The mass transfer is relatively fast, while the slower surface reaction determines the deposition rate. The surface concentration C_S approaches the gas concentration C_G and the transport term is relatively flat in Figure 10.5. If $h_G \ll k_S$, then

Example

Calculate the deposition rate for a CVD system in which:

$h_G = 1.0$ cm s^{-1},
$k_S = 10$ cm s^{-1},
partial pressure of incorporating species, $P_G = 1$ Torr,
total pressure, $P_T = 1$ atm $= 760$ Torr,
total concentration in gas phase, $C_T = 1 \times 10^{19}$ cm^{-3},
density of depositing film, $N = 5 \times 10^{22}$ cm^{-3}.

Answer

Using (10.7):

$$v = \frac{k_S h_G}{(k_S + h_G)} \frac{C_T}{N} Y = \frac{1}{(1/k_S + 1/h_G)} \frac{C_T}{N} Y$$

$$= \frac{1}{(1/10 + 1/1)} \left(\frac{1 \times 10^{19}}{5 \times 10^{22}}\right)\left(\frac{1}{760}\right) = 0.14 \ \mu\text{m min}^{-1}.$$

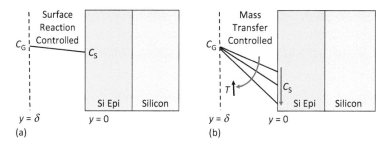

Figure 10.5 (a) Surface-reaction-controlled regime, where the surface reaction is slow (strong function of temperature). (b) Mass-transfer-controlled regime, where the surface reaction is fast (weak function of temperature).

$$v \cong h_G \frac{C_T}{N} Y. \tag{10.9}$$

This is the reactant transfer or gas-phase diffusion-controlled case. Here the surface reaction is fast, and the growth is limited by how fast the reactant molecules can diffuse to the surface. The different regimes are indicated schematically in Figure 10.5.

In contrast to the case of oxidation, where there was a linear–parabolic growth in the limiting regimes, the growth velocity is constant in both regimes for the deposition case. Fundamentally, this is because the gas-phase diffusion process here corresponds to mass transport across a boundary layer of constant thickness. In fact, the mass transport coefficient can be written as

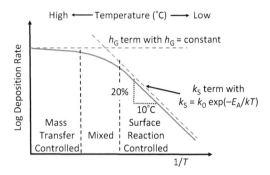

Figure 10.6 Arrhenius plot of the growth velocity (log scale) versus $1/T$, showing the mass-transfer- and surface-reaction-controlled regimes. For a 2 eV surface reaction activation energy, a 10 °C change in temperature causes a 20% change in deposition rate.

$$h_G = \frac{D_G}{\delta}, \tag{10.10}$$

where D_G is the gas-phase diffusion coefficient of the reacting species across a boundary layer of constant thickness δ. Even though there is a diffusion term, the transport coefficient h_G is relatively independent of temperature, because gas-phase diffusion has a much lower temperature dependence than solid-state diffusion. While the growth velocities in both regimes are constant with time, there is a very strong difference in the temperature dependence of the growth kinetics in these two different regimes. The temperature dependence comes about because the surface reactions are exponentially activated with temperature, as shown in Figure 10.6.

The surface reactions are exponentially activated, so that k_S is given by

$$k_S = k_0 \exp\left(-\frac{E_A}{kT}\right), \tag{10.11}$$

where, for silicon deposition, the activation energy $E_A \approx 2$ eV, suggesting the making or breaking of silicon bonds, just as in the case for oxidation in the linear regime. In this temperature regime, a small change in temperature causes a large change in deposition rate.

While these films are being deposited, they can also be doped *in-situ*. The common dopant source gases are arsine (AsH_3), phosphine (PH_3) and diborane (B_2H_6). Since the silicon density is $5 \times 10^{22} \mathrm{cm}^{-3}$ and the doping density might be on the order of 10^{15}–10^{20} cm^{-3} for lightly and highly doped silicon, the amount of dopant source gas is a small fraction of the silicon-containing gas. It can be more problematic to grow a lightly doped epilayer on a heavily doped substrate. The lightly doped layer tends to get extra doping from two sources. The first is simple solid-state diffusion from the doped substrate, which is well modeled by complementary error function diffusion into an infinite medium that we discussed in Chapter 7, given by

$$C(x, t) = \frac{C}{2}\left[\mathrm{erfc}\left(\frac{x}{2\sqrt{Dt}}\right)\right]. \tag{10.12}$$

Here, $x = 0$ is at the substrate–epitaxial interface. We can assume diffusion into an infinite medium because the growth velocity is much faster than the diffusion from the substrate (i.e., $vt \gg \sqrt{Dt}$). The second source of doping is caused by dopant atoms from the front side, backside or sides of the wafer evaporating into the gas phase and redepositing on the growing film. Empirically, this follows an exponential decay and is modeled by

$$C(x) = C_S^* \exp\left(-\frac{x}{L}\right), \qquad (10.13)$$

where C_S^* is an effective surface concentration and L is an experimentally determined decay length. A schematic of this "autodoping" is shown in Figure 10.7.

Some of these anomalous effects can be countered by growing the epitaxial layer faster than the diffusion processes take place. Thus, silane (SiH_4) growth at high temperatures can minimize these effects. Figure 10.8 shows the growth rate in different silicon-containing gases, with silane growth in the mass-flow-limited regime being the fastest. Figure 10.8 shows the growth rate for equal flows of the silicon-containing gases. Silane has the fastest growth rate because it is the least stable of the four precursors, with silicon tetrachloride being the most stable and

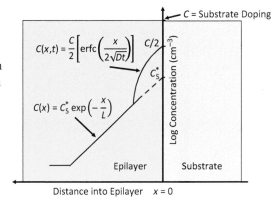

Figure 10.7 Complementary error function doping from substrate solid-state diffusion and exponential doping from evaporating dopant atoms, combining to cause "autodoping" during epitaxial growth.

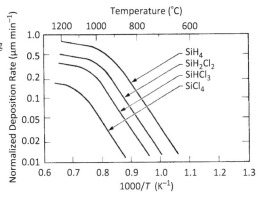

Figure 10.8 Arrhenius plot of normalized deposition rate for equal silicon-containing gas flows using silicon source gases with different molecular weights. After [1, 2].

Table 10.2 **Deposition characteristics of silane, dichlorosilane and silicon tetrachloride typical of the higher-temperature mass-transport-controlled regime. Courtesy of Ted Kamins [1].**

	SiH$_4$	SiH$_2$Cl$_2$	SiCl$_4$
Temperature (°C)	900–1050	950–1100	1100–1200
Reaction	SiH$_4$ → Si + 2H$_2$	SiH$_2$Cl$_2$ → Si + 2HCl	SiCl$_4$ + 2H$_2$ → Si + 4HCl
Reversible	No	Yes	Yes
HCl byproduct	No	Yes	Yes
Gas-phase nucleation	Yes	No	No
Flow control	Easy	Moderate	Moderate
Cost	High	Moderate	Low
Danger	High	Moderate	Moderate
Deposition rate (μm min^{-1})	0.2	<1	>2
Uses	Thin layers	Many layers	Thick layers

consequently providing the lowest growth rates. Silane is a very dangerous, explosive gas because of its instability. Silane is more reactive than the other Si-containing gases, so one would expect the growth rate to be higher. Apart from that, there are other reasons to consider the chlorosilanes indicated in Figure 10.8. For example, the chlorosilanes tend to deposit selectively on silicon rather than on exposed oxides, so growing strained SiGe source/drain regions with Si- and Ge-doped chlorosilanes would form selectively deposited regions in the exposed source/drain region and not on the oxide isolation structures. The chlorine helps remove any silicon that nucleates on the insulation regions. The chlorosilanes are also less likely to decompose in the gas phase, preventing random particulate deposition, as described in Table 10.2.

Note that the deposition rates in Table 10.2 do not match up with the rates in Figure 10.8. The reason is that, because SiH$_4$ is more reactive, the flow of SiH$_4$ must be reduced to avoid unwanted reactions, such as particle formation in the boundary layer and coating of the chamber walls. Therefore, the practically useful growth rate is lower for SiH$_4$ than for some of the other Si-containing gases. This shows that other constraints can limit the behavior in practice.

There is, however, a problem with epitaxial growth in the mass-flow-dominated regime. The uniformity of growth is very sensitive to the gas flow and concentration. Thus, methods to uniformly distribute the gas using showerhead gas distribution systems in single-wafer systems are used. Older methods tried to deposit on several wafers in a chamber and alter the temperature or gas flow along the chamber to account for boundary layer variations and gas depletion effects. These were not too successful in high-volume manufacturing. Other methods had to be developed to allow batch wafer processing, which we discuss in the Section 10.2.4 on low-pressure CVD deposition.

10.2.2 Heteroepitaxy

When different materials are epitaxially grown on top of one another, the term "heteroepitaxy" is used. There are several reasons to grow materials using heteroepitaxy. The first is that wafers

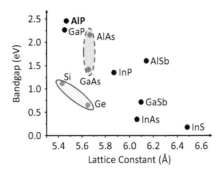

Figure 10.9 Bandgap versus lattice constant for various semiconductors. An example of strained SiGe heteroepitaxy is indicated (yellow region with solid boundary), as well as unstrained heteroepitaxy of GaAs/AlAs (blue region with dashed boundary).

of large size may not be available for materials like GaN, so growing GaN on large silicon wafers is a very interesting prospect. The second is to create bandgaps intermediate between two materials, which can be very important for tuning direct-bandgap semiconductors to the desired operating conditions (for example, the emitted photon wavelength in light-emitting diodes (LEDs)). And, third, by growing layers of different lattice constants on one another, strain can be produced that can provide useful device characteristics.

Figure 10.9 shows a range of semiconductors with their bandgap and lattice constant. Materials with the same lattice constant can be easily grown using lattice-matched unstrained heteroepitaxy without creating defects, such as the GaAs/AlAs materials shown in the blue region. This allows the bandgap of AlGaAs alloys to be tuned in these direct-bandgap materials. Materials with different lattice constants can also be grown using strained heteroepitaxy, such as the SiGe layers indicated in the yellow region. Because both of these applications are very important, we describe them in more detail.

In Chapter 2, Section 2.9, we briefly discussed how strained SiGe heteroepitaxy was used to improve device mobility in P-channel MOS (PMOS) channels starting with the 90 nm node. The gate oxide could no longer be scaled after this node until high-K dielectrics were introduced, so an alternative means of improving device performance was required. The concept is shown in Figure 10.10, where compressive strain in a recessed source/drain is introduced by growing SiGe epitaxy on the Si substrate, enabling increased channel mobility in PMOS devices. The photographs on the left of the figure show two generations of Intel PMOS devices. The upper photo is Intel's 90 nm generation (2003) with a polysilicon gate. The lower photo is Intel's 45 nm High-K metal gate PMOS device (2007). Both have grown SiGe source regions. The right-hand side of Figure 10.10 illustrates the generation of compressive stress when a larger lattice constant Ge or SiGe region is grown on the silicon substrate. The Ge or SiGe grown epitaxial film is forced to adopt the smaller silicon lattice spacing because the thin film is grown on a much thicker substrate. This causes the SiGe regions to push laterally on the PMOS channel region.

The compressive strain introduced in the channel provides a significant boost in channel mobility in PMOS devices, and efforts to increase the Ge concentration were important to continue improving device performance. However, there are limits on the Ge concentration and

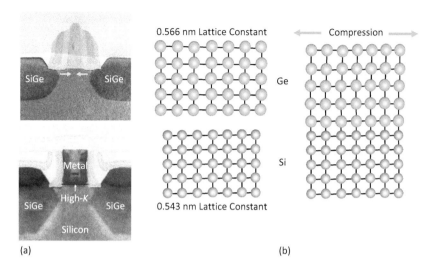

Figure 10.10 (a) Photomicrographs of two generations of recessed SiGe source/drain structures in PMOS devices. Upper photo © IEEE 2004. Reprinted with permission from [3]. Lower photo © IEEE 2007. Reprinted with permission from [4]. (b) A schematic of how the Ge or SiGe lattice constant is compressed during heteroepitaxial growth [1]. This causes the SiGe region in the PMOS source/drain region to push sideways on the silicon channel, creating the compression that increases hole mobility in the channel region.

the thickness of the SiGe layers that can be grown. Beyond a certain threshold or critical thickness, dislocations form in the grown layers. These dislocations cause leakage currents and then device performance is reduced. Transmission electron microscope (TEM) pictures of SiGe layers grown below and above the critical thickness are shown in Figure 10.11(a) and (b), respectively, with many dislocations evident in the thicker SiGe layer [5].

Because of the lattice mismatch, strain increases with layer thickness and with increasing Ge content. The higher the growth temperature and the longer the time, the faster the strain relaxes through dislocation formation. For very high Ge content, an anneal part way through the growth can purposely relax the layer, and a subsequent layer tends to have fewer defects. Similar techniques of growing buffer layers and deliberately annealing them to relax them and confine the defects in the lower layers are used to grow GaN on Si, for example. It is primarily an empirical process to find the optimum growth conditions and achieve usable device layers. A similar concept of growing carbon-doped epitaxial layers on Si can be used to create tensile strain and improve N-channel MOS (NMOS) mobility, but it is more limited because of the low solubility of carbon in silicon [6].

There exists a significant literature describing models to estimate the critical thickness h_c for dislocation formation. For the interested reader, a good summary of these models is contained in [7]. Models for h_c generally rely on mechanical equilibrium or energy balance principles and give expressions for h_c of the form [7]

$$h_c = \frac{r_0}{8\pi f \, \sin\theta \, \cos\phi}\left(\frac{1 - v\cos^2\theta}{1 + v}\right)\ln\left(\frac{h_c}{r_0} - 1\right). \tag{10.14}$$

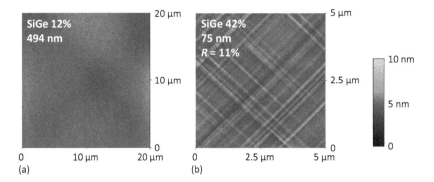

(a)   (b)

Figure 10.11 SiGe layers on Si grown (a) below and (b) above the critical thickness. Above the critical thickness, many dislocations are introduced. Here R is the degree of strain relaxation. The measurements were made with an atomic force microscope (AFM). The vertical scale (out of the page) is shown on the right. Reprinted from [5] with the permission of AIP Publishing.

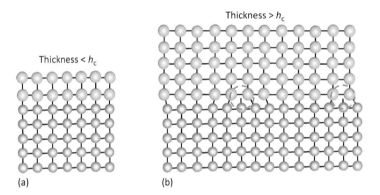

(a)   (b)

Figure 10.12 An illustration of (a) a strained epitaxial layer below h_c in thickness and (b) a relaxed epitaxial layer with dislocations (dashed red circles) forming at a thickness beyond h_c.

Here θ is the angle between the dislocation line and its Burgers vector, ϕ is the angle by which the dislocation glide plane tilts from the film plane, v is Poisson's ratio, r_0 is the core radius of the dislocation and f is the misfit between the substrate and the epitaxial layer, defined by

$$f = \frac{a_f - a_s}{a_s}, \tag{10.15}$$

where a_f and a_s are the film and substrate lattice constants, respectively. For epitaxial film thicknesses below h_c, strained films can be grown without dislocations, as illustrated in Figure 10.12(a). For thicker films, the equilibrium structure is a relaxed film with dislocations, as shown in Figure 10.12(b).

Typical results from these models for h_c are shown in Figure 10.13. It is clear from Figure 10.13(a) that, for lattice mismatches larger than $\approx 1\%$, only very thin strained layers can be grown before h_c is exceeded. The example in Figure 10.13(b) is relevant to PMOS source/

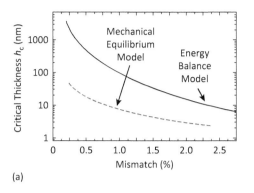
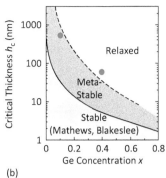

(a)   (b)

Figure 10.13 (a) Typical model predictions for h_c as a function of the mismatch in lattice constant between the epitaxial film and the substrate. (b) Specific example for the $Si_{1-x}Ge_x$ system. After [8]. The red dots correspond to the images in Figure 10.11.

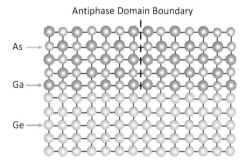

Figure 10.14 Schematic of the formation of antiphase domain boundaries when even a lattice-matched material like GaAs is grown on a Ge substrate.

drain regions such as those shown in Figure 10.10. Higher Ge content is helpful in increasing the hole mobility in the channel region but clearly only quite thin SiGe layers can be grown before dislocations appear. Low growth temperatures can allow metastable layers beyond h_c to be grown. Methods such as grading the Ge composition while the SiGe region is being grown can also help.

Dislocations do not occur when lattice-matched materials are grown by heteroepitaxy, making the layers easy to grow by epitaxy. Of particular interest is GaAs on Ge, which provides a path for high-mobility III–V NMOS devices and, as shown in Figure 10.9, these are lattice-matched materials. However, even these layers are subject to defects because arsenic and gallium layers may not fully cover the Ge surface, leading to regions that intersect to form antiphase domain boundaries, as illustrated in Figure 10.14. Growth conditions and slight variations in the Ge substrate cut orientation can help minimize these defects, which act as scattering centers in devices.

The problems of growing a material with different bandgap and different lattice constant on a substrate, such as GaN on Si, only compound the issues discussed above, as illustrated in Figure 10.15. Presently, GaN on Si is a very interesting combination, as it is enabling for very-high-power electronic systems. GaN's bandgap of 3.4 eV supports blue LEDs and is far above

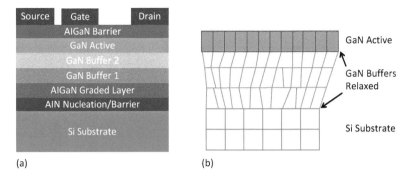

(a)   (b)

Figure 10.15 Schematic illustration of GaN on Si heteroepitaxial growth for a device structure, showing the barrier and buffer layers required to produce device-quality GaN layers.

the materials shown in Figure 10.9. An extended version of Figure 10.9 was shown in Figure 3.4 in Chapter 3. The high bandgap leads to high breakdown fields and provides a path to high-power systems [9].

To grow GaN on Si, a series of barrier layers and buffer layers are required, illustrated in Figure 10.15. The barrier layer is needed to prevent Si interdiffusion into the GaN layer, and the buffer layer is required to reduce the dislocation density caused by the severe lattice mismatch. Several anneals of the buffer layer to cause purposeful relaxation of the strain may be required, along with optimized grading of the buffer layer before a high-quality GaN layer can be grown [10].

Currently, GaN layers can only be grown with high quality on Si (111), whereas most CMOS circuitry is grown on (100) wafers. As discussed in Chapter 3, GaN has a hexagonal crystal structure, while silicon is a cubic crystal. Growing such radically different crystals by hetero-epitaxy is only possible because the surface of the Si (111) crystal has a hexagonal atomic layout that forms a usable seed for the GaN crystal. But the lattice mismatch is huge (17%), with the GaN lattice constant being smaller than the Si (111) "footprint," as illustrated qualitatively in Figure 10.15. With a lattice mismatch this large, h_c would be less than one monolayer by extrapolating the curves in Figure 10.13. To further complicate matters, the thermal expansion coefficients of Si and GaN are very different. This introduces additional strain as the wafer is cooled from growth temperatures. Figure 3.8 showed a TEM cross-section of a GaN on Si epitaxial structure clearly showing the defects illustrated in Figure 10.15. In principle, GaN devices could be integrated with Si devices using the GaN on Si wafer structure, but it is likely that the chiplet packaging solutions described in Chapter 11 will be utilized in the near term.

10.2.3 Epitaxial Lateral Overgrowth (ELO)

Often, different materials are required to be grown in selected areas of a silicon wafer, for example, to integrate light-emitting devices such as InP on Si. This does not require creating an entire InP wafer on silicon, but rather selective heteroepitaxy in masked regions. The technique of epitaxial lateral overgrowth (ELO) through a narrow window can provide better-quality layers than planar heteroepitaxy of lattice-mismatched materials. The sophisticated methods of

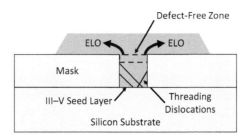

Figure 10.16 Schematic of the epitaxial lateral overgrowth (ELO) method, where masking layers and narrow growth windows inhibit the propagation of threading dislocations into the "wings" or later grown regions over the mask, providing high-quality material on a lattice-mismatched substrate.

buffer engineering mentioned above can significantly reduce the density of dislocations in lattice-mismatched substrates. However, the best modern buffers still contain a density of dislocations of $\approx 10^6$ cm^{-2}, which is too high for many device applications. The ELO method was developed to block threading dislocations from propagating from the substrate to the grown epitaxial layers [11, 12].

A schematic of the ELO process is shown in Figure 10.16. Many dislocations lie at angles to the surface, i.e., on the $\langle 111 \rangle$ planes in silicon, and relax the strain from lattice-mismatched materials grown on the substrate. These threading dislocations propagate through the grown layers and negatively affect device performance. The ELO method aims to trap these threading dislocations under the mask layer and also to terminate the dislocations on the edges of a high-aspect-ratio seed region. The method works best when the opening of the seed region is less than the spacing of the dislocations in the substrate. Even if that is not the case, the dislocations are confined to a narrow region above the seed and do not propagate laterally over the mask. In this way, the mask efficiently blocks dislocations and the density of defects in the laterally grown layers (the wings) is significantly reduced. The defect filtration of substrate dislocations in ELO can be very efficient, leading to orders-of-magnitude lower defects in the wings area than in standard planar heteroepitaxy [13].

Note that the ELO process is really a combination of selective epitaxy with lateral overgrowth. It is important to reduce nucleation on the dielectric while growing material at the epitaxial growth front. This may be achieved by adding HCl to the carrier and reactant gases, using the HCl to reduce the nucleation on the dielectric. Another practical approach is to deposit part of the layer and then "recondition" the dielectric surface by etching nuclei from the dielectric. Because the deposition on the mask is amorphous, the film's etch rate can be much higher than at the crystalline growth front, providing a cyclic deposition/etch process to achieve selective epitaxy.

One of the most spectacular achievements of the ELO method was the breakthrough in the development of long-lifetime GaN/InGaN blue lasers, due to the high efficiency of defect filtration during the lateral growth of GaN on sapphire [14]. Most of the integrated hetero-epitaxy material systems are grown using variations of the metal–organic chemical vapor deposition (MOCVD) or molecular beam epitaxy (MBE) deposition systems described in the following sections. They represent a small but growing and exciting portion of the semiconductor ecosystem based on the fundamentals of silicon processing techniques.

10.2.4 Low-Pressure Chemical Vapor Deposition (LPCVD)

Growth in the mass-transfer-controlled regime (Figure 10.6) provides the fastest growth rates at higher temperatures, compared to growth in the surface-reaction-limited regime. But it is much more difficult to control gas flows and achieve uniformity and reproducibility than it is to control temperature accurately. Thus, growth in the surface-limited regime offers the prospect of uniform, controllable growth rates.

However, throughput is a serious concern – note the scale on the y axis in Figure 10.6 is logarithmic, so the growth rates decrease exponentially in the surface-dominated regime. Luckily, lowering the pressure solves this problem in a somewhat unexpected way. One might think that focusing on the surface reaction processes is the way to solve the problem, but it turns out to be exactly the opposite, and the way to solve the problem is to increase the mass flow transport through the boundary layer. Recall that, in the mass-flow-dominated regime, the growth velocity is

$$v \cong h_G \frac{C_T}{N} Y, \tag{10.16}$$

with

$$h_G = \frac{D_G}{\delta} . \tag{10.17}$$

The key new point is that reducing the pressure increases the mean free path of the reactants, so the diffusivity D_G of the reactants increases. In fact, D_G is inversely proportional to the pressure,

$$D_G = \frac{1}{P_{\text{total}}}. \tag{10.18}$$

Decreasing the total pressure P_{total} from 760 Torr (1 atm) to 1 Torr, a typical LPCVD pressure, increases D_G by 760 times. Decreasing P_{total} also increases the boundary layer thickness δ, but only by a factor of 3–10. The net effect is that the mass transport increases by about 100 times. Since h_G is much larger at lower pressures, the deposition velocity is not limited by the mass transfer of reactants through the boundary layer, and the diffusion-limited growth curve shifts up, as shown in Figure 10.17.

This means that higher growth rates can be achieved at lower pressures, increasing the throughput of the deposition process. The two red dots in Figure 10.17 show possible operating points at atmospheric pressure (lower dot) and at 1 Torr (upper dot). The deposition rate is obviously much higher at lower pressure. But there is a further order of magnitude in productivity possible. Because the deposition rate is no longer sensitive to the gas flows, the wafers can be stacked close together and processed in parallel with no loss of uniformity. This is similar to the oxidation case, where we were able to ignore the gas transport flux and concentrate on only the diffusion flux through the growing layer and the surface reaction rate.

LPCVD systems are not completely insensitive to the mass transport flux, so some care has to be taken in the distributed feed of gases along the tube and wafer spacing. This leads to systems like

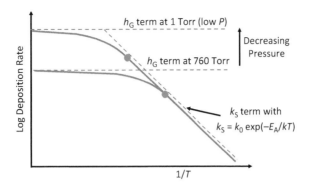

Figure 10.17 Growth velocity versus $1/T$ for APCVD (760 Torr) and LPCVD (1 Torr) systems. The lower total pressure shifts the h_G curve upwards, extending the surface reaction regime to higher temperatures. The red dots show possible operating points.

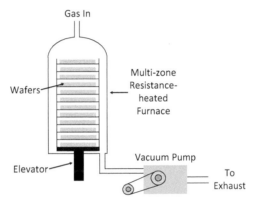

Figure 10.18 Vertical LPCVD reactor with wafers stacked close together, enabled by the deposition reaction operating in the surface-reaction-dominated regime.

that shown in Figure 10.18, where wafers are stacked close together in a vertical LPCVD furnace. Horizontal furnaces are equally viable, but the vertical furnace is more suitable for large-diameter wafers (no cantilever loading system required) and have a smaller footprint in the cleanroom.

10.2.5 Metal–Organic Chemical Vapor Deposition (MOCVD)

There always seems to be an interminable drive towards deposition at lower temperatures. The reason is that it opens up many more applications, allowing depositions on temperature-sensitive materials while not affecting any prior steps, such as doping profiles. To achieve that goal, the solution was to recruit the chemists to develop more thermally unstable gas precursors that decomposed at lower temperatures and allowed deposition of the desired materials. Replacing the silanes or chlorosilanes with more unstable organic–silicon gases such tetraethyl orthosilicate (TEOS) allowed lower-temperature deposition of silicon dioxide via the following reaction:

$$Si(OC_2H_5)_4 \rightarrow SiO_2 + 2(C_2H_5)_2O. \tag{10.19}$$

Note that ethanol is C_2H_5OH, so the "tetra" refers to the four ethanol-like organic groups that attach to the Si, decomposing to give silicon dioxide and a diethyl ether volatile byproduct. We will see a lot more of these clever chemistry solutions later when we discuss atomic layer deposition (ALD).

An even more important application of these metal–organic gas precursors is that they enable other metals that do not have the silane- or chlorosilane-type precursors to be deposited. Examples include the deposition of gallium arsenide and gallium nitride compound semiconductor layers. MOCVD is "just" another chemical vapor deposition method with different precursors, but its key advantage is that it enables a low-temperature deposition reaction, and the enormous diversity of the precursors provides a method for depositing stacks of many different semiconductor compounds. The low-temperature reactions allow abrupt interfaces to be formed between different layers. Much of the credit for developing MOCVD as a viable deposition method goes to Manasevit and Simpson, who first demonstrated that GaAs and Ga–V (V = group V) compounds could be grown by MOCVD [15, 16]. Subsequently, their group demonstrated that many semiconductors could be deposited, such as GaP, GaAsP, GaAsSb, AlAs, AlGaAs, AlP, InP, InAs, InSb, InGaAs, InPAs and InAsSb, along with important wide-bandgap compounds, such as GaN and AlN. The technique is flexible enough to grow essentially all the III–V and II–VI alloys (Table 10.3), including ternary and quaternary compounds. It can grow multilayer heterostructures and superlattices with abrupt interfaces.

Table 10.3 **Summary table of common metals, organics and precursors for semiconductor deposition.**

Metal	+	Organic metal	→	Organic precursor
Common metals		Common organics		Common MOCVD precursors
Gallium				Tri- or tetramethyl (TM)
Aluminum				Triethyl (TE)
Indium		Methyl		Triisopropyl (TiP)
Arsenic				Dimethyl (DM)
Antimony				Diethyl (DE)
Zinc				Trimethylgallium (TMGa)
Cadmium		Ethyl		Triethylgallium (TEGa)
Telluride				Triisopropylgallium (TiPGa)
Germanium				Trimethylaluminum (TMA)
				Trimethylindium (TMIn)
		Isopropyl		Triethylindium (TEIn)
				Trimethylarsenic (TMAs)
				Trimethylantimony (TMSb)
				Dimethylzinc (DMZn)
				Diethylzinc (DEZn)
				Dimethylcadmium (DMCd)
				Dimethyltellurium (DMTe)
				Diethyltellurium (DETe)
				Tetramethylgermanium (TMGe)

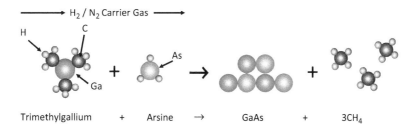

Trimethylgallium   +   Arsine $\rightarrow$ GaAs   + $3CH_4$

Figure 10.19 Schematic (unbalanced) of H_2 or N_2 carrier gas transporting TMGa and AsH$_3$ precursors to a heated substrate to pyrolyze and form GaAs and reactant byproducts. The alternating Ga and As atoms represent the growth of the compound semiconductor layer.

It is simple to synthesize organic precursors like trimethylgallium $Ga(CH_3)_3$, and these precursors can be designed to pyrolyze (decompose due to heat) at low temperatures. This allows reactions to occur between the group III organometallics and the group V hydrides to deposit compound semiconductors, such as

$$Ga(CH_3)_3 + AsH_3 \rightarrow GaAs + 3CH_4 \qquad (10.20)$$

at a temperature around 600 °C. A schematic illustration of the reaction is shown in Figure 10.19.

This reaction can be generalized for the III–V compounds to the following simplified reaction:

$$M(O_P) + VH_3 \rightarrow MV + 3O_PH . \qquad (10.21)$$

Here M is the group III element, such as Al, Ga or In; V is the group V element, such as Sb, P or As; and O_P is an organic precursor subunit, such as CH_3 or C_2H_5. The formation of the semiconductor is basically an acid–base reaction. The group III metal alkyls are strong acids and the group V hydrides are strong bases, making the formation of the binary semiconductor favorable. The reaction is sometimes referred to as a Lewis acid–base reaction, indicating that there is a transfer of two electrons during the product formation.

Although this reaction uses a group V hydride species, MOCVD precursors can be used for both reactants, such as the reaction of trimethylgallium and trimethylantimony to form GaSb via the reaction

$$Ga(CH_3)_3 + Sb(CH_3)_3 \rightarrow GaSb + 3C_2H_6 . \qquad (10.22)$$

However, the hydrides are often preferred, as they are relatively inexpensive, give high-purity layers, and the excess hydrogen can remove C impurities. Common hydrides are PH_3 and AsH_3 for growing the binary, ternary and quaternary compounds (e.g., InP, InGaAs and GaInAsP). NH_3 is used for growing the important wide-bandgap GaN layers. The grown layers are often doped P-type with DMZn or N-type with SiH_4 or SH_2. Zn is a P-type dopant because it has one less electron and replaces the column III element and acts as an acceptor, just like boron in silicon. Sulfur has one more electron than the column V elements and acts as an N-type donor. Silicon as a column IV element can act as either a donor (that is, on Ga sites) or an acceptor (that is, on As sites), but, since arsenic is smaller than gallium and silicon, it tends to occupy gallium sites. Thus, silicon is used as a dopant for the formation of N-type material.

In (10.20), it looks like C could easily be incorporated in the film, leading to impurity doping in the epitaxial layer. To avoid it, excess hydride can be supplied, where the atomic H from the decomposing AsH_3 scavenges the CH_x products from the surface. Often, V/III flow ratios of 100 are used for devices prepared with methyl-based organic precursors and ratios of 10,000 have been used for the NH_3 flow rates when forming gallium or aluminum nitrides. This substantially contributes to the cost of the epitaxial structure.

To produce a metal–organic precursor such as trimethylgallium requires some elementary chemistry. Gallium can be purified by reacting with chlorine, producing gallium trichloride. Standard "methylating agents," like methyl magnesium iodide, are routine chemical reagents that result from the ready insertion of active magnesium or lithium metals into carbon–halogen bonds like methyl iodide. The resultant "organometallic" species will transfer the methyl group to metals that are more noble in the metallic form (as gallium is relative to magnesium). In other words, the formation of trimethylgallium from standard methyl lithium or methyl magnesium halide precursors is thermodynamically favored [17]. A schematic of some of the desirable properties of metal–organic precursors is shown in Figure 10.20.

The cost of the metal–organic precursor at high purity is much more than the cost of a simple metal at similar purity levels. The simple metal can be evaporated in an MBE reactor to grow superlattices of semiconductor layers in a straightforward manner (see Section 10.2.10), but, in spite of enormous effort, the MBE technique was never viable for manufacturing, and MOCVD systems became the method of choice for compound semiconductor production. Because of the competing technology, it was an uphill battle for MOCVD to gain acceptance as a viable deposition method [19]. Early attempts at fabricating compounds with MOCVD incorporated impurities such as carbon and silicon, which degraded the layer properties, such as the device

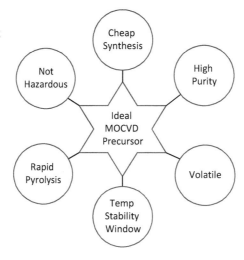

Figure 10.20 Schematic showing some of the desirable properties of an ideal MOCVD precursor. After [18].

mobility. The MOCVD liquid precursor must be volatile, so that it can be transported from a heated bubbler by a carrier gas such as H_2 or N_2 to the heated susceptor where pyrolysis or catalytic decomposition can occur. It must be sufficiently stable during transport in the gas phase so that it does not decompose, and it should not react with the other precursors until it encounters the heated substrate. There, rapid pyrolysis should occur for both components and the semiconductor layer will be formed.

When the organic radical is small, such as trimethyl $(CH_3)_3$, it becomes more stable and the temperature of decomposition is high. When the organic radical is big, such as triethyl $(C_2H_5)_3$, the temperature of decomposition is low and in this case the growth temperature will be low but there is the possibility of unwanted parasitic reactions in the gas phase. The strength of the metal–organic (M–O) bond is a measure of the stability of the organic precursor.

In a short 20 years from the first reports of GaAs growth [13], MOCVD became the system of choice for compound semiconductor production due to improvements in the purity of the precursors and the design of reactors capable of uniform, abrupt layer growth. State-of-the-art performance has been demonstrated for compound semiconductor power and high-frequency devices, lasers, detectors, photonic circuits and LEDs.

For many compounds based on Ga, In, As, Al and P, the growth rates under As-rich conditions have been shown to be linearly dependent on the partial pressure of the metal alkyl species, so that it is limited by the arrival rate of the column III organometallics, indicating a first-order reaction. The growth rate is relatively independent of temperature over a wide range (500–800 °C). The purity of the III–V films does improve monotonically with lower temperatures due to decreased carbon and silicon incorporation in the layers.

Empirically, MOCVD growth is largely mass-transport-limited such that the epilayer growth rate is mostly controlled by diffusion of the group III precursor through a boundary layer of thickness δ. In terms of Figure 10.6, the growth is in the mass-transfer-limited regime, where the mass transfer coefficient $h_G = D_G/\delta$ dominates and D_G has a relatively weak temperature dependence. Control of growth rate and the achievement of uniform film growth depends on achieving a uniform metal alkyl flux over the substrate, illustrating the importance of careful hydrodynamic design. The gas distribution and hydrodynamics in the reactor have a major effect on the epilayer thickness, uniformity and the abruptness of the layer growth. The reactor schematic in Figure 10.21 shows a cold-wall reactor with inductive heating of the substrate with a uniform boundary layer.

After much experimentation and modeling, vertical flow reactors are used in production in preference to horizontal flow reactors. In ideal vertical flow, the gas flows radially outwards in a uniform way and the boundary layer δ is uniform and is proportional to $\sqrt{\eta/\sigma V}$, where η is the gas viscosity, σ is the gas density and V is the gas velocity. A rotating susceptor is used to further improve the gas hydrodynamics and layer uniformity. To prevent premature decomposition of the precursors, the susceptor and substrate must be the hottest part of the growth chamber, so that the gas precursors are cracked near the substrate and the epitaxial film can be grown. Care must be taken to have stable reactants or to minimize contact between the reactants until they reach the hot zone to avoid parasitic reactions. The reactor walls are cooler than the substrate so that decomposition on the walls is minimized.

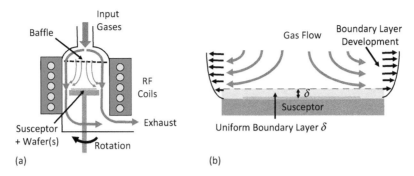

Figure 10.21 (a) Schematic of a vertical flow reactor, with a showerhead baffle to uniformly distribute input gases, which decompose on the RF-heated susceptor. (b) Schematic of the uniform boundary development from a combination of susceptor rotation and careful hydrodynamic system design. After [20].

Figure 10.22 Plasma-enhanced CVD equipment.

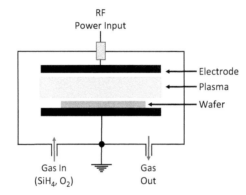

10.2.6 Plasma-Enhanced Chemical Vapor Deposition (PECVD)

In spite of the enhanced deposition rate achieved by low-pressure CVD, it may be still too slow if there is a temperature constraint, since the deposition rate in the surface-controlled regime varies exponentially with temperature. The most obvious example that is widely encountered is when a dielectric layer like silicon dioxide or silicon nitride needs to be deposited on a metal like aluminum, which has a melting point of 660 °C. To improve the deposition rate at very low temperatures, a plasma excitation is used to couple electrical energy instead of thermal energy into the source gas. Plasma-enhanced chemical vapor deposition (PECVD) systems are often parallel-plate plasma systems, where the main purpose of the plasma is to break apart the source gases and enhance the deposition rates, as illustrated in Figure 10.22 and with the plasma physics described in detail in references [21] and [22]. Chapter 9 also provides an introduction to the basic plasma physics (see Section 9.3 in particular).

The plasma provides a very strong source of radicals, so the deposition rate becomes mass-transport-dominated and shows very little temperature dependence. The plasma is composed of ions, electrons and excited neutral species (radicals) in a weakly ionized gas. A large range of films can be deposited (oxides, nitrides, silicon, metals) and the deposited films are usually non-stoichiometric. For example, plasma-deposited silicon dioxide is not SiO_2, but instead SiO_x, where $x < 2$. Silicon nitride is likely to be Si_xN_y:H_z, indicating that the film contains a substantial amount of hydrogen, which may be as much as 10%. The hydrogen comes from the carrier and reactant gases and sometimes provides a useful function, such as when amorphous silicon is deposited and is electrically passivated by the hydrogen additions. Amorphous hydrogenated silicon nitride (SiN:H) layers can provide not just antireflection coatings on silicon solar cells, but also excellent surface and bulk passivation.

The same plasma sheaths that occur in etch plasmas exist to confine the electrons to the plasma because of their higher mobility than the ions. By controlling the ion energy across the sheath, some ion bombardment during deposition can be used to densify the deposited film and improve its properties. For example, some plasma-deposited oxides can approach the properties of thermally grown oxides grown at much higher temperatures [23].

Ions, electrons and radicals are generated in the plasma and, just as in etch plasmas, the radicals are the dominant excited species. A simple example shows the energy needed to create a silane radical versus an ionized silane molecule:

$$SiH_4 + e^- \rightarrow SiH_3^* + H + e^- \quad (\Delta E \sim 3.5 \text{ eV}), \tag{10.23}$$

$$SiH_4 + e^- \rightarrow SiH_4^+ + 2e^- \quad (\Delta E \sim 12.2 \text{ eV}), \tag{10.24}$$

showing that it is much easier to create a radical than an ionized species. It is the combination of radical reactions on the surface along with ion bombardment that determine the film deposition rate and quality.

The same tricks that are used in etching to decouple the plasma parameters from the ion bombardment energies are used in PECVD systems [24]. A high-energy remote source can generate a high density of reactive species, while a tuned plasma sheath voltage can optimize the ion energies. Such systems are called high-density PECVD (HD-PECVD) systems, as illustrated in Figure 10.23.

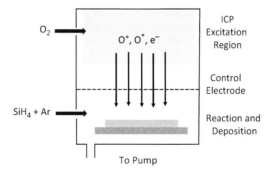

Figure 10.23 A possible configuration for a remote HD-PECVD system for oxide deposition, where one gas is excited far from the wafer in a high-density excitation region (ICP) and the more reactive gas is introduced close to the wafer.

The pressure range has to be such that electrons gain sufficient energy between collisions to excite many – and ionize some – gas molecules, as shown in (10.23) and (10.24). We have seen previously in (10.18) that the diffusivity D_G depends inversely on the pressure. The physical reason for this is that the mean free path has the same pressure dependence. For an electron, an approximation for the mean free path λ is

$$\lambda = \frac{0.005 \ (\text{cm})}{P \ (\text{Torr})}. \tag{10.25}$$

At too high pressures, the neutrals decrease the ability of the electrons to gain sufficient energy, because there are too many collisions. At too low pressures, the electrons do not collide with enough other molecules to sustain the plasma and create species for the reaction. Thus, the pressure range for PECVD is limited to approximately 50 mTorr to 1 Torr, while HD-PECVD can use lower pressures in the main chamber ($\approx$ 5 mTorr) to create longer mean free paths with fewer collisions, giving a more directional ion flux onto the wafers. The high ion and radical densities give a high deposition rate with controlled ion bombardment, causing simultaneous deposition and sputter etching, so that deeper features can be filled with no voids (we discussed sputter etching or ion milling previously in Chapter 9, Section 9.5.2).

There are different ways to control the filling of narrow vias, as shown in Figure 10.24. One way is to recruit the etch engineers to help by tapering the slope of the feature. Another is to use the flexibility of remote high-density PECVD to decouple the plasma generation, chemical reaction and ion bombardment.

This simple example in Figure 10.24 shows how the variation in the angle of the incoming species can cause lips or bulges on corners, and how improving the directionality of the incoming species combined with some simultaneous etching can create conformal layers. Whether the incoming particles stick immediately upon deposition or whether they are re-emitted to deposit elsewhere also affects the profile. We will discuss these angle distributions and sticking coefficients of the depositing species later in this chapter.

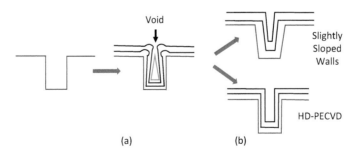

Figure 10.24 (a) Filling of a trench using PECVD (with void), and (b) using etch profiling to taper sidewalls and using high-density PECVD with simultaneous deposition and corner sputter etching to uniformly fill the trench.

10.2.7 Atomic Layer Deposition (ALD)

Atomic layer deposition (ALD) is a newer variant of CVD, developed only in the past couple of decades. But it is not an obvious extension of CVD. Instead, ALD represents a new invention that allows layers with single-atomic-thickness precision to be deposited. The pivotal event that stimulated interest in ALD was when the semiconductor industry realized they needed the exquisite thickness control provided by ALD to deposit high-K gate dielectrics thin enough to replace SiO_2. The chemists came to the rescue and developed highly reactive precursors that deposited a layer by two separate half-reactions on a substrate. In other words, ALD is a two-step process, where, in the first step, one reactant forms a self-limiting monolayer on the surface. Once the first reactant layer has formed, the introduction of the second reactant converts the first reactant layer into the desired solid film, e.g., Al_2O_3. Repeated cycling of the two half-reactions builds up a solid film of the desired thickness [11].

Any solid surface will retain some gas-phase molecules on the surface after excess molecules are removed by a vacuum pump. The forces holding the molecules on the surface can be weak in the case of physical adsorption (physisorption) or strongly bound in the case of chemical adsorption (chemisorption) where chemical bonds form. In the latter case, the chemisorbed molecules tend to form the monolayer of interest for ALD. Additional excess reactants arriving at the surface can only adhere by weak van der Waals forces and can be removed by heating the substrate. Thus, chemisorption tends to produce the desired monolayer of reactant. There is a temperature window where ALD works. If the substrate temperature is too low, more than a monolayer can form or condense on the substrate. If the temperature is too high, the first reactant may desorb from the surface before it can react with the second reactant. But, in general, ALD is a very-low-temperature deposition technique, capable of depositing layers all the way down to room temperature.

As an example, we will examine the deposition of Al_2O_3 in more detail, as it is a generic high-K dielectric that also has applications in optical and coating applications. The two reactants used are trimethylaluminum (TMA) and water (H_2O). Three methyl (CH_3) groups are bonded to the aluminum and form the first gas-phase reactant. One could argue that the first reactant is actually the hydroxyl (OH) group from surface-adsorbed H_2O on the silicon surface. But the starting point of the iterative reactions that comprise the two-step ALD cycle is the $Al(CH_3)_3$ introduction. This reacts with the hydroxyl (OH) group on the Si surface and forms an Al–O bond, releasing a hydrogen with one methyl group, producing a methane molecule (CH_4) as the reaction product. These first steps are shown as (a) and (b) in Figure 10.25, which also shows the remaining steps (c)–(f) in the ALD sequence discussed below.

The first reaction indicated in Figure 10.25(a) and (b) is

$$Al(CH_3)_3(g) + :Si–O–H(s) \rightarrow :Si–O–Al(CH_3)_2(s) + CH_4, \qquad (10.26)$$

where the (g) and (s) refer to gas and surface components. The TMA reacts with the absorbed hydroxyl groups until a monolayer forms and the surface is passivated. Excess TMA and the methane reaction products are pumped away (Figure 10.25(c)).This half-reaction is self-terminating, which leads to the perfect uniformity of ALD. After they are pumped away

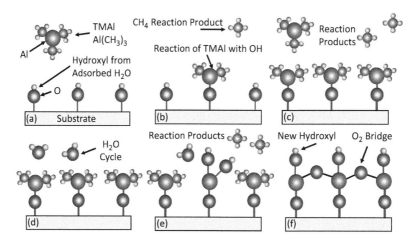

Figure 10.25 The sequence of steps in the ALD process: (a) hydroxylated Si surface; (b) TMA reaction with OH; (c) first monolayer; (d) H₂O pulse; (e) H₂O reaction with methyl groups, releasing methane; and (f) second passivating monolayer. After J. Ruzyllo, Penn State.

(with perhaps an intermediate inert-gas purge cycle), water vapor (H_2O) is pulsed into the reaction chamber, as shown in Figure 10.25(d). The H_2O reacts with the dangling methyl groups on the new surface, forming hydroxyl (OH) surface groups, releasing one methyl group and also forming aluminum–oxygen (Al–O) bridges that release the other methyl group (Figure 10.25(e)). Two methane molecules are released according to the reaction

$$2H_2O + :\text{Si–O–Al(CH}_3)_2(\text{s}) \rightarrow :\text{Si–O–Al(OH)}_2(\text{s}) + 2CH_4. \tag{10.27}$$

The H_2O vapor does not react with the hydroxyl surface groups, again creating a perfect passivation limited to one atomic layer (Figure 10.25(f)).

To summarize, one TMA and one H_2O vapor pulse form one full cycle, which then gets repeated until the desired thickness is deposited. The growth rate is approximately 1 Å (0.1 nm) per cycle. Each cycle of gas pulsing and pumping takes about 3 s. The net result is that atomically controlled layers can be deposited. A picture of two cycles is shown in Figure 10.26 to illustrate the atomic layer control possible with ALD. It should be noted that, while the first monolayer might have some crystalline characteristics, thicker layers would greatly exceed the critical thickness h_c and quickly take on an amorphous character. Thus, while Figure 10.26 pictures a crystalline Al_2O_3 layer, actual deposited layers would be amorphous.

10.2.8 Physical Deposition Systems: Sputtering Systems

We now turn to three deposition methods that do not involve chemistry, that is, they are simple physical vapor deposition (PVD) systems. These are the final three entries in Table 10.1 and we discuss them in the next three sections.

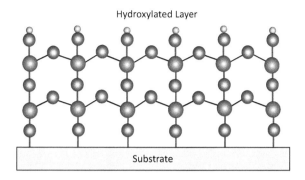

Figure 10.26 Two full cycles deposit two monolayers, with approximately 1 Å per cycle. After J. Ruzyllo, Penn State.

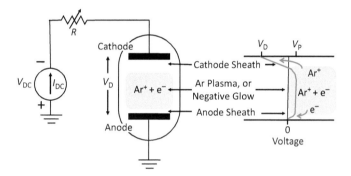

Figure 10.27 A DC plasma created by applying a DC voltage high enough to ionize the low-pressure gas between the cathode and anode. The voltage distribution between the cathode and anode is also shown on the right-hand side.

The first is DC plasma sputter deposition, which is a low-temperature physical technique that can deposit a wide variety of layers. DC plasmas were not discussed in Chapter 9 because they have almost no utility in etching, but they can be useful in deposition systems. They are particularly interesting because a DC plasma is a very simple, basic and reproducible source of plasma. DC plasma deposition is the most basic and inexpensive type of deposition from electrically conductive targets. It simply requires a gas in an enclosure between two electrodes with a high enough DC voltage applied to ionize the gas and generate a plasma. The characteristics of the plasma are controlled by four variables: the voltage applied and the distance between the electrodes, as well as the type of gas used and its pressure. A typical setup is shown in Figure 10.27, where gas in a tube at low pressure is ionized by an applied voltage in series with a resistor to control the current through the plasma [25].

When the voltage is increased sufficiently, avalanche ionization occurs in the gas, generating ions and electrons and producing a luminous glow region in the chamber. The plasma is formed when a small fraction (10^{-6} to 10^{-4}) of the gas atoms or molecules become ionized. As previously seen with RF plasmas in Section 9.3 of Chapter 9, the electrons have energies in the range of electronvolts, which is sufficient to trigger dissociation and ionization and thus sustain the plasma. The ion temperature (the heavy species) is much closer to room temperature. For a fixed distance between the electrodes, the pressure is a very important parameter for the plasma dynamics. The number of gas atoms or molecules is proportional to the pressure. At low pressure, the electron mean free path is large and many electrons reach the anode without colliding with gas atoms. Thus, at low pressure, a high applied voltage is required to give the few colliding electrons enough energy to sustain the plasma. At higher pressures, the electron mean free path is short and the electrons may not gain enough energy from the electric field to ionize the gas atoms due to their frequent collisions. Again, high applied voltages are required to ionize the plasma. Thus, there is an intermediate pressure where the plasma can be initiated with a minimum applied voltage, known as the Paschen curve minimum.

Even though the electrodes are symmetrical, there is a dark sheath near the cathode where the glow is absent and a much smaller anode sheath region. This occurs because of the different mobilities of the ions and electrons. The DC current through the discharge must be continuous, and is carried by ion current at the negative cathode and electron current at the positive anode. To supply equal currents, the electric fields at the anode and cathode must differ by the ratio of the ion and electron mobilities. Recall that drift current is given by

$$J = q\mu n \mathcal{E}. \tag{10.28}$$

The voltages on the anode and cathode are set by the external power supply at $-V_{DC}$ (cathode) and 0 (anode). The plasma self-adjusts to a small positive plasma potential V_P, which retards the electron flow to the grounded anode. The large negative potential at the cathode accelerates the ions from the plasma to the cathode. As the ions bombard the cathode, secondary electrons are emitted and accelerate in the reverse direction towards the plasma, gaining sufficient energy to excite and ionize the atoms in the plasma, producing a self-sustaining discharge. When the electrons gain sufficient energy, they collide with gas atoms, and the glow becomes visible due to excitation of the discharge gas at some distance from the cathode. These simple DC tube plasmas have been studied for more than a hundred years, and the glow can show considerable fine structure, such as striations in the glow region, all dependent on the precise voltage–current characteristics in the plasma [26].

In order for a DC current to flow in the circuit, the cathode must be a conductive target material. Under these conditions, the high ion energies that hit the cathode will cause sputtering of the target material, which is the ejection of target atoms by the impinging ions. The substrate, which is where the sputtered atoms are deposited, is placed on or below the other electrode, which is the anode. A typical system for DC sputtering is shown in Figure 10.28(a). It is a low-aspect-ratio DC glow discharge with a short distance between the anode and cathode. A more practicable implementation for deposition on semiconductor device wafers is also shown in Figure 10.28(b), where a peripheral anode accommodates the DC current flow and minimizes current through the device wafer. Thus, the anode and substrate need not be physically at the same place, but sputter coating still occurs because the emitted atoms can travel to the substrate.

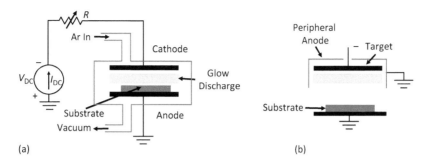

Figure 10.28 (a) A low-aspect-ratio DC sputtering discharge, where the cathode target material is sputtered onto a substrate on the grounded anode. (b) A practical implementation for semiconductor deposition using a peripheral electrode to minimize DC current through the substrate.

Because of the short cathode–anode spacing, the major structural features of the plasma are the cathode sheath, where the ions are accelerated, and the glow region, which extends almost to the anode. A short anode dark space can also be visible where the slightly positive plasma potential returns back to zero at the grounded anode. There is a narrow pressure range where DC sputtering is viable, with the pressure being neither too low nor too high. A typical pressure range is approximately 1–5 Pa ($\approx$ 5–50 mTorr). This pressure range is high enough that many collisions occur in the sheath between the accelerating ions and the working gas neutrals. The consequence is that there are broad energy and directional spectra of ions impinging on the target cathode, resulting in an equally wide range of sputtered atoms, which are often described by a cosine angular distribution, as in (9.12). The angular distribution of incoming ions helps increase the sputtering yield, which is usually a maximum at a 60°–80° angle of incidence (see Figure 9.16). The cosine distribution of sputtered atoms means that the step coverage in sputtering systems is much better than in evaporation systems, which operate at lower pressure and have much higher n values in (9.12). However, the disadvantage of simple DC sputtering is that the deposition rate is relatively low ($\approx$ 10 nm min^{-1}).

In order to improve the deposition rate and expand the useful pressure range, the effective lifetime of electrons in the vicinity of the target must be increased to improve the ionization. This can be achieved by adding permanent magnets behind the cathode to confine the secondary electrons in the $\mathcal{E} \times B$ field. A schematic of such a magnetron sputtering target is shown in Figure 10.29, where the electrons spiral perpendicular to the $\mathcal{E} \times B$ field and generate a dense plasma along the target, often creating a "racetrack" of intense plasma erosion of the target material. The pressure range can be reduced to the range of 0.1–1 Pa ($\approx$ 0.5–5 mTorr) and the deposition rate can improve by an order of magnitude to 100 nm min^{-1} or more.

While Ar atoms are the most common source of working plasma gas, it is sometimes advantageous to add a reactive gas to the sputtered material to improve the deposited film quality and deposition rates. Consider, for example, the deposition of TiN. This is a compound target that could easily be sputtered using an argon plasma, but there are advantages to using an elemental metal target and a reactive nitrogen plasma. First, the purity of elemental metals and gases usually exceeds the purity of fabricated compound targets. Second, the sputtering yield of the compound

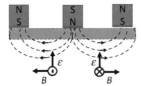

Figure 10.29 A schematic of a DC planar magnetron sputtering target. Permanent magnets behind the cathode target produce a magnetic field that is substantially parallel to the target, creating an $\mathcal{E} \times B$ field that causes electrons to spiral around the cathode surface, increasing the plasma density and thus increasing the sputtering rate.

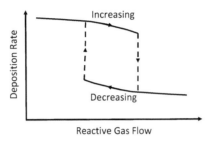

Figure 10.30 Typical experimental hysteresis curve for a reactive sputtering process with increasing and decreasing gas flow rates. The deposition rate does not increase and decrease at the same value with the supply of reactive gas due to target poisoning.

material is substantially lower than the sputtering yield of the elemental target material. Thus, it can be possible to deposit high-purity films at high deposition rates using reactive sputtering.

At first sight, this seems very simple, but the reaction between the elemental target and the reactive gas can cause process stability problems [27]. This occurs because there is a conflict between obtaining high sputtering rates (which requires a clean elemental target) and obtaining a truly stoichiometric compound film (which requires high reactive gas flows). The problem is that high reactive gas flows also cause compound formation at the surface of the sputtering target (target poisoning). Target poisoning causes the deposition rate to decrease as the supply of the reactive gas increases, in a highly nonlinear fashion. A typical experimental processing curve for the deposition rate versus the reactive gas flow for a reactive sputtering process is shown in Figure 10.30. The characteristic feature of this curve is that it exhibits a hysteresis effect.

A simple explanation is that high deposition rates require more reactive gas to form the deposited compound material, but that poisoning of the elemental target can occur suddenly and drop the deposition rate sharply. Less nitrogen is needed, which causes a further increase of the nitride formation on the target. The lower deposition rate needs less reactive gas flow to form the deposited compound material, but finally, at low enough flows, the sputtering action clears the poisoned target of the reactive gas and returns it to its elemental state. This hysteresis effect is one of the key problems in experimental reactive sputtering systems, and it requires very careful process control to maintain reproducible conditions close to unstable operating points.

In summary, DC sputtering is a simple, inexpensive way to deposit material from conducting targets, and is widely used beyond semiconductor manufacturing, having applications in

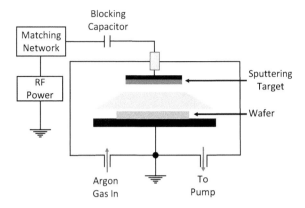

Figure 10.31 Sputtering equipment showing sputtering from the smaller electrode with higher sheath fields.

thin-film coatings for solar cells, hard coatings for industrial applications on machine tools, optical coatings, etc. DC plasmas are the simplest form of plasma system, simply requiring a high voltage between two electrodes in a low-pressure tube of gas to initiate ionization and create a source of plasma. However, the disadvantage is that the target must be conductive to accommodate the DC current flow.

A more flexible deposition system is the RF plasma sputter system. The equipment is identical to a parallel-plate plasma etcher (see Figure 9.9), except that the target is on one electrode and the wafer is on the opposite electrode, as shown in Figure 10.31. Note that the size of the target electrode is smaller than that of the other electrode, because the sheath voltage varies approximately as

$$\frac{V_1}{V_2} \approx \left(\frac{A_2}{A_1}\right)^4, \tag{10.29}$$

where V_1 and V_2 are the voltages at the two sheaths, and A_1 and A_2 are the areas of the corresponding electrodes. The fourth power comes from theory. In practice, the power appears to be between 1 and 2, but the point is that the sheath voltage is inversely proportional to the electrode area to *some* power. This means the ion bombardment is higher on the smaller electrode (the target), allowing atoms to be sputtered off and deposited on the wafer sitting on the larger electrode.

The prototypical sputtering system we show in Figure 10.31 has an RF power supply, which allows the target to be an insulator, along with a matching network to match the output impedance of the RF supply to the plasma impedance. This enables maximum power transfer when the impedances are matched. A blocking capacitor blocks the DC voltage that builds up on the smaller target electrode from the power supply, but couples the RF energy into the plasma. The plasma is composed of argon ions that bombard the target and sputter or knock off target atoms that then deposit on the wafer.

The bombarding ions have to have enough energy (typically 10–20 eV at a minimum) to eject the target atoms. At high enough incident energies, the target atoms get ejected with a cosine

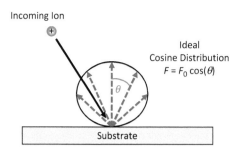

Figure 10.32 Cosine distribution of ejected target atoms from a sputtering target under ion bombardment.

distribution, as indicated in Figure 10.32. The ideal cosine distribution shows that the emitted flux of atoms is reduced by the cosine of the emission angle. But there is a distribution of emission angles, which does help with step coverage compared to more directional deposition techniques (like evaporation).

Because it is a relatively simple system that can sputter many different targets, it finds much use in university research laboratories and similar environments. One typical sputter target company Lesker (www.lesker.com) supplies hundreds of targets from aluminum to zirconium oxide. Even compound targets can be sputter-deposited. If one element of the compound is sputtered more efficiently, the surface concentration of the other element increases. The result is that, in steady state, the deposition reflects the original compound composition of the target, i.e., the sputtering self-corrects to deposit the original target stoichiometry. Multiple materials can be sequentially deposited by opening and closing shutters over different targets in the same vacuum chamber. These very flexible systems are widely used for materials research and exploratory work.

Although we have emphasized the simplicity and flexibility of sputtering systems, there are industrial-scale sputtering machines capable of depositing materials like transparent indium tin oxide (ITO) on large-area thin-film displays (thin-film-transistor liquid-crystal displays, TFT-LCDs). There are also multi-target cluster machines capable of depositing sophisticated layer stacks for magnetic storage applications. This illustrates the usefulness of sputter deposition in various applications.

10.2.9 Evaporation

An even simpler physical deposition method is based on evaporating a target and depositing the material on a distant wafer. The material to be evaporated is heated in a container (a ceramic crucible or tungsten or refractory metal spiral wire) and electrical current or an electron beam is used to heat a small volume of the target in a crucible and evaporate the substance. A schematic of an evaporation system is shown in Figure 10.33. Evaporation systems operate at low pressures ($\sim 10^{-5}$ Torr) to avoid contamination during the deposition. This means the deposition is line-of-sight (a long mean free path) and step coverage is poor. To improve both step coverage and film uniformity, the wafers usually sit on a hemispherical platen that rotates during deposition.

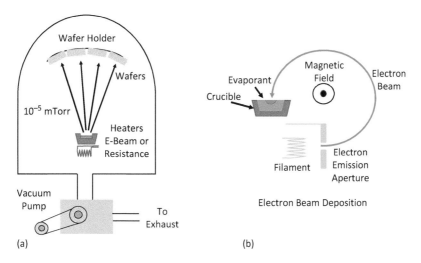

Figure 10.33 (a) Schematic of an evaporation system with an operating pressure of 10^{-5} Torr and (b) electron-beam deposition system.

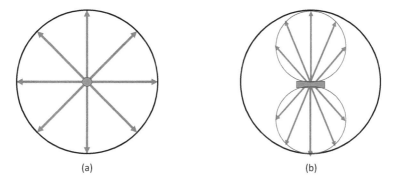

Figure 10.34 (a) Schematic of isotropic deposition from a point source. (b) A typical small planar evaporation source deposits with a cosine angle distribution.

There is another reason to use a hemispherical geometry that relates to the flux emitted from the source. In two dimensions, we can imagine a wire running down the center of a cylinder. Evaporating the wire would cause uniform, isotropic deposition on the inside of the cylinder. If a narrow tape runs down the center of the cylinder and material is evaporated from it, the deposition characteristics are quite different and are represented by the cosine distribution we discussed earlier. The same effects occur in a spherical geometry. A purely point source evaporates isotropically and uniformly coats the exposed sphere, as shown in Figure 10.34. In practice, evaporation sources act more like a small planar source and the flux is a cosine distribution, so, to obtain a uniform distribution, the planar source should be at the base of the sphere, as shown in Figure 10.35.

Figure 10.35 also illustrates why the cosine flux from a source at the base of the sphere provides a uniform coating on the sphere. If we consider sections with the same solid angle, then

Table 10.4 **Typical evaporation sources available in a university research laboratory. Taken from the Stanford Nanofabrication Facility cleanroom.**

Evaporation sources
Ag, Al, Au, Co, Cr, Cu, Fe, Ge, In, Mo, Ni, Pd, Pt, Si, Sn, Ti, Ta, W, Er, Hf, Ir, Ru, Tb, Y, Nb, V, Zr, NiO, SiO$_2$

Figure 10.35 A planar source placed at the base of a sphere deposits the same thickness on wafers mounted around the sphere, because the reduced flux at an angle θ from the vertical is exactly compensated for by a shorter distance to the source, if the emission is an ideal cosine distribution.

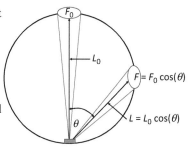

in a cone directly above the source the flux is F_0, while in a cone at an angle θ from the vertical the flux is reduced to $F = F_0 \cos(\theta)$ from the cosine flux distribution. But the distance of that cone from the source (the length of the vector F) is reduced by the same $\cos(\theta)$ factor. The reduction in flux is exactly compensated for by the closer distance to the source, so the deposition rate is the same all around the sphere when a planar source is placed at the base of the sphere (Figure 10.35).

In practice, equipment manufacturers move the source somewhere between the center and the edge of the deposition chamber to achieve uniform deposition. The ideal $\cos(\theta)$ flux profile may also be more directional in practice, leading to a $(\cos\theta)^n$ distribution, where $n > 1$ creates a narrower and more directional flux. We will discuss the flux distributions more carefully in the simulation section (Section 10.3).

Almost any element can be evaporated, though it is difficult to evaporate alloys or compounds because the constituents evaporate at different rates, unless separate sources are used. These are flexible systems and are used in many university research laboratories. A sample of the evaporation sources available in the Stanford Nanofabrication Facility is listed in Table 10.4, though there are often different evaporation machines for the "clean" and "contaminated" sources, referenced to CMOS process requirements, for mostly historical reasons.

10.2.10 Molecular Beam Epitaxy (MBE)

The final physical deposition method is molecular beam epitaxy (MBE), which in many ways is the ultimate physical deposition system based on evaporation. It operates at extremely low base pressures (10^{-10} Torr), so it is an ultra-clean system capable of growing atomic layer thicknesses of crystalline material on crystalline starting substrates (such as GaAs wafers). The chamber is

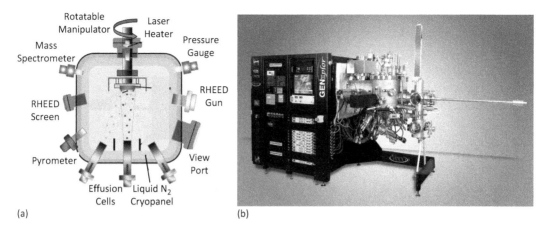

(a)

(b)

Figure 10.36 (a) Schematic of an MBE chamber, showing the source effusion cells, RHEED analytical monitoring of layer thickness and other ports. (b) A commercially available Veeco MBE system, showing various stainless-steel ports and chambers. Photo courtesy Veeco Corp., used with permission.

usually stainless steel that may be baked using external heaters to remove any surface contaminants. The main chamber is surrounded by other bolt-on stainless-steel modules using soft copper gaskets – these modules contain analytical instruments to monitor the deposition rate, substrate temperature, source temperature, etc. A schematic of an MBE chamber is shown in Figure 10.36(a).

The growth rate is typically one monolayer per second and it is low enough that surface migration of the impinging species on a heated substrate can ensure layer-by-layer epitaxial growth. Simple mechanical shutters in front of the beam sources can turn on and off the deposition, leading to changes in composition and doping that are abrupt at the atomic scale. Because it is performed in an ultra-high-vacuum environment, it can be controlled *in-situ* by surface diagnostic methods such as reflection high-energy electron diffraction (RHEED). A low-angle electron beam glances off the surface, and the periodicity of characteristic oscillations in the intensity of the detected beam correspond exactly to the growth of a single monolayer, enabling real-time control of the layer-by-layer growth. This eliminates much of the guesswork in layer growth, and allows very sophisticated structures to be fabricated.

Figure 10.36 shows that the constituent elements are evaporated from special sources known as effusion sources or Knudsen cells under the control of individual shutters. The vaporized elements travel by line of sight through the vacuum chamber before settling and crystallizing on a heated substrate. In a Knudsen cell, evaporation occurs as effusion or emission of atoms from an isothermal enclosure with a small orifice. With the cell maintained at a uniform, constant temperature, the evaporant's equilibrium vapor pressure is known. The aperture is small, so the vapor lost through it does not significantly affect the cell's internal pressure, causing the evaporant's emission rate to be constant. The evaporating surface within the enclosure is large compared with the orifice and maintains an equilibrium pressure P_{eq} inside the effusion cell. Under these conditions, the orifice constitutes an evaporating surface with an effusion pressure equal to that in the main cell. The small aperture means that the vapor emerges in a cosine distribution (the flux emitted at any given angle to the cell's normal is proportional to the cosine of that angle).

The deposition of semiconductors from the III–V group, such as gallium arsenide and aluminum/indium gallium arsenide, has been particularly well studied, because of their superior high-frequency properties and their unique optical properties among the direct-bandgap III–V semiconductors. These MBE systems are primarily used for research applications because of the very low growth rates. As a result, these machines are often found in university and corporate research labs.

10.3 Deposition Models and Simulations

Collectively, the deposition methods discussed above provide a very powerful set of deposition options. Just as in the case of etching, deposition simulations can provide insight into the physics of deposition, but are not critical, like diffusion or oxidation simulations, because it is easy to obtain a cross-section of the material deposited in a trench or via and see the results in an electron microscope. However, simulations have played an important role in understanding some of the physical mechanisms of deposition systems, and it is for that reason we include them here.

We begin with some simple simulations of deposited layers that span the range from conformal to directional depositions. As shown in Figure 10.37, changing the ratio that describes the coverage on a sidewall versus a planar surface can emulate many different types of deposition systems. These simulations contain no physics, but are able to generate profiles that approximate many deposition systems and can be useful to feed into device simulation programs to understand how devices operate. For example, an accurate description of a spacer layer in a transistor is critical to the operation of the device. The simulations might correspond to directional incoming distributions due to ion-assisted deposition (such as PECVD), or directional atoms or molecules from the thermal energy and low pressure in evaporation systems. A ratio of 1 corresponds to perfectly conformal deposition like CVD. A ratio of 0.5

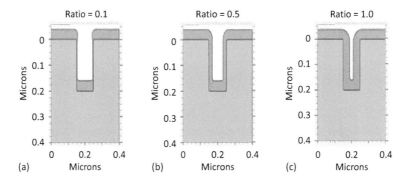

Figure 10.37 Simulations of deposited layers with different ratios of sidewall to planar coverage.

may be typical of a sputtering system at higher pressures. A ratio of 0.1 might correspond to poor sidewall coverage from an evaporation system.

10.3.1 Physically Based Deposition Simulations

Physically based deposition simulations begin with the two key parameters: the cosine angle distribution of the incoming particles or ions, and their sticking coefficient when they hit the surface [28, 29]. Once these key parameters are known, the feature-scale simulations can accurately account for shading effects and redeposition from molecules that do not immediately stick on the surface.

A very clever test structure to elucidate these key parameters is the overhang test structure developed by McVittie [30]. The structure is shown in Figure 10.38, along with an indication of how it enables the sticking coefficient and input angle distribution to be determined. Simulations of a range of cosine input angle exponents (the n in $(\cos \theta)^n$) and sticking coefficients S_C are shown in Figure 10.39.

The simulations in Figure 10.39 show that the higher n values produce a more directional incoming beam that is able to fill directly underneath openings. A high sticking coefficient causes the incoming atoms or ions to stick where they hit and gives rise to poor coverage on the sidewalls or under the overhang. A low sticking coefficient allows the particles to bounce around.

The experimental results in Figure 10.40 show a series of depositions on an overhang test structure as the sticking coefficient increases from 0.01, for tungsten CVD (Figure 10.40(a)), to 1.0, for aluminum sputtering (Figure 10.40(c)). The TEOS layer in Figure 10.40(b) has an intermediate sticking coefficient. Note that the step coverage decreases as the sticking coefficient increases. Fitting simulations with various n and S_C values to experimental results like these allows effective n and S_C values to be determined for a particular deposition system.

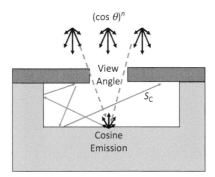

Figure 10.38 Overhang test structure to determine the input angle distribution and sticking coefficient. Shadowing is accounted for through the view angle of the incoming atoms. Note that the incoming atoms emit with an ideal cosine emission during each bounce after they strike the surface.

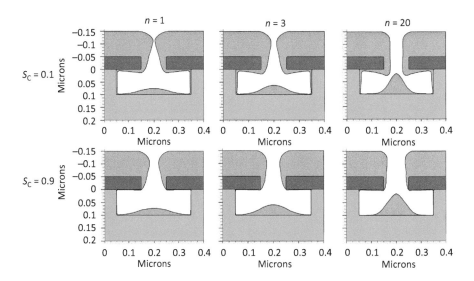

Figure 10.39 Simulations of deposition on an overhang test structure. The columns are for $n = 1, 3$ and 20 in the $(\cos\theta)^n$ exponent. The rows are for sticking coefficients $S_C = 0.1$ and 0.9. Simulated with Victory Process [31].

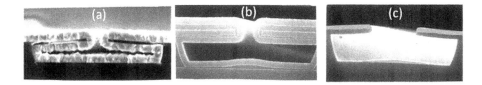

Figure 10.40 Experimental results for: (a) conformal tungsten CVD deposition ($WF_6 + H_2$); (b) TEOS oxide deposition, where the darker oxide layer is surrounded by a conformal polysilicon layer for image contrast; and (c) a sputtered aluminum layer, showing high sticking on landing. Courtesy of Jim McVittie.

10.3.2 Two- versus Three-Dimensional Deposition Simulations

There are subtle effects that occur when 3D structures are subject to depositions, due to the viewing angle and the shading of the incoming particles. An example is shown in Figure 10.41 for a trench and a via structure with the same depth. Because the trench can "see" incoming particles along its length, it has a wider viewing angle for the incoming species than a 3D via.

The results of these depositions are shown in Figure 10.42 and show that the thickness on the trench sidewalls is larger than in the via due to these view and shadowing effects, which can only be captured by accurate 3D simulations. These 3D simulations use the level-set numerical methods discussed in Chapter 9. These methods are particularly important in cases when the profiles on the trench edges begin to merge, creating keyhole-type structures, as shown in Figure 10.43. To avoid these, a more directional fill is needed with a lower sticking coefficient to produce a more conformal fill. However, structures similar to those shown in Figure 10.43

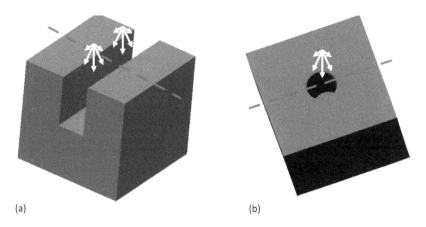

(a)   (b)

Figure 10.41 Viewing angle for (a) a trench structure and (b) a via structure with the same aspect ratio or etch depth. The trench is effectively a 2D structure while the via is a 3D structure. The dashed red lines show cross-sections that are considered in Figure 10.42. Simulated with Victory Process [31].

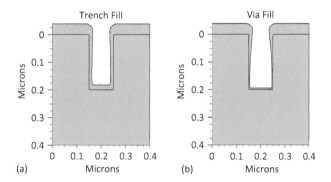

Figure 10.42 Depositions in (a) a trench versus (b) a via differ because of 2D versus 3D view angles and shadowing effects. The cosine exponent is $n = 3$ for the incoming particles and the sticking coefficient is $S_C = 0.2$. These represent cross-sections from the structures in Figure 10.41.

were used by Intel in high-volume manufacturing, because the air gap lowers the dielectric constant between adjacent conducting lines in the interconnect layers (see Figure 11.24).

10.4 Summary of Key Ideas

We divided deposition systems into those that are primarily chemical (variations of chemical vapor deposition) and those that are primarily physical (sputtering or evaporation). In general, the physical deposition systems are capable of operating at very low temperatures down to room temperature, because no thermal energy is needed. CVD systems require thermal energy to decompose the gas species, but there is a continued

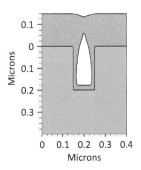

Figure 10.43 Creation of a "keyhole" due to incomplete filling in a trench. An ideal cosine distribution ($n = 1$) of incoming particles and an $S_C = 0.2$ were used in the simulation.

drive to develop more unstable gas species (like the metal–organic precursors) that allow operation at lower temperatures. Chemists have made very significant contributions to deposition systems, including the latest development of two separate, self-limiting half-reactions that enable atomic layer by ALD of a wide variety of materials.

Deposition systems can deposit dielectrics such as oxides and nitrides at low temperatures, enabling the layering of metal interconnects separated by dielectric layers in back-end-of-line (BEOL) processing (discussed in Chapter 11). Polysilicon, silicon and alternative semiconductors can also be deposited by variations of CVD and epitaxy. The possibility of integrating optical components onto silicon wafers is enabled by heteroepitaxy, where buffer layers or epitaxial lateral overgrowth enable the growth of lattice-mismatched semiconductors on silicon wafers.

The kinetics of deposition are controlled by the mass transport flux of reactants to the surface and the subsequent chemical reaction flux at the interface. The balance between these two fluxes determines whether the deposition kinetics is controlled by surface reactions or mass transport. Growth in the surface-limited regime offers the prospect of uniform, controllable growth rates provided the temperature is controlled accurately, as the surface reactions are exponentially activated. But the growth rates are much lower than in the mass-flow-controlled regime, so low-pressure CVD systems were developed to enable operation in that regime. By lowering the pressure, the mass transport increased dramatically and enabled stable operation at higher temperatures in the surface-limited regime with high deposition rates. Further improvements in CVD systems, such as metal–organic precursors (MOCVD) and plasma-enhanced (PECVD) deposition, enabled operation at even lower pressures. Physical deposition systems like sputtering and evaporation are very flexible, and capable of depositing a wide variety of materials at low temperatures, but are primarily used in research and niche applications because of low throughput.

All of these systems can be physically modeled by a combination of an angular energy distribution (a cosine function to a power n, i.e., $(\cos\theta)^n$) for the incoming reactants along with a sticking coefficient S_C that determines how well the incoming species stick or bounce off surfaces. These relatively simple models can account for many of the major features of deposition profiles, such as how conformable the layers are or how well they coat sidewalls in trenches, etc.

It is likely that increasingly interdisciplinary teams of electrical engineers, materials scientists and chemists will contribute to further advances in both atomic layer by ALD and etching, enabling very sophisticated structures with atomic precision to be constructed in the future.

10.5 REFERENCES AND NOTES

[1] Dr. Ted Kamins, private communication.

[2] F. C. Eversteyn, "Chemical reaction engineering in the semiconductor industry," *Philips Research Reports*, vol. 29, pp. 45–66, 1974.

[3] K. Mistry, M. Armstrong, C. Auth, *et al.*, "Delaying forever: uniaxial strained silicon transistors in a 90nm CMOS technology," in *Digest of Technical Papers. 2004 Symposium on VLSI Technology*, IEEE, pp. 50–51, 2004.

[4] K. Mistry, C. Allen, C. Auth, *et al.*, "A 45nm logic technology with high-k + metal gate transistors, strained silicon, 9 Cu interconnect layers, 193nm dry patterning, and 100% Pb-free packaging," in *2007 IEEE International Electron Devices Meeting*, IEEE, pp. 247–250, 2007.

[5] J. Hartmann, A. Abbadie and S. Favier, "Critical thickness for plastic relaxation of SiGe on Si(001) revisited," *Journal of Applied Physics*, vol. 110, p. 083529, 2011.

[6] S. G. Thomas, M. Bauer, M. Stephens and J. Kouvetakis, "Precursors for group IV epitaxy for micro/opto-electronic applications," *Solid State Technology*, vol. 52, no. 4, pp. 12–16, 2009.

[7] S. M. Hu, "Misfit dislocations and critical thickness of heteroepitaxy," *Journal of Applied Physics*, vol. 69, pp. 7901–7903, 1991.

[8] R. People and J. C. Bean, "Calculation of critical layer thickness versus lattice mismatch for Ge_xSi_{1-x}/Si strained-layer heterostructures," *Applied Physics Letters*, vol. 47, pp. 322–324, 1985.

[9] K. J. Chen, O. Häberlen, A. Lidow, *et al.*, "GaN-on-Si power technology: devices and applications," *IEEE Transactions on Electron Devices*, vol. 64, no. 3, pp. 779–795, 2017.

[10] A. Dadgar, "Sixteen years GaN on Si," *Physica Status Solidi (b)*, vol. 252, no. 5, pp. 1063–1068, 2015.

[11] R. W. Johnson, A. Hultqvist and S. F. Bent, "A brief review of atomic layer deposition: from fundamentals to applications," *Materials Today*, vol. 17, no. 5, pp. 236–246, 2014.

[12] Z. R. Zytkiewicz, "Epitaxial lateral overgrowth of semiconductors," in *Springer Handbook of Crystal Growth*, eds. G. Dhanaraj, K. Byrappa, V. Prasad and M. Dudley, Springer, pp. 999–1039, 2010.

[13] B.-Y. Tsaur, R. W. McClelland, J. C. C. Fan, *et al.*, "Low-dislocation-density GaAs epilayers grown on Ge-coated Si substrates by means of lateral epitaxial overgrowth," *Applied Physics Letters*, vol. 41, no. 4, pp. 347–349, 1982.

[14] S. Nakamura, M. Senoh, S. Nagahama, *et al.*, "InGaN/GaN/AlGaN-based laser diodes with modulation-doped strained-layer superlattices grown on an epitaxially laterally overgrown GaN substrate," *Applied Physics Letters*, vol. 72, no. 2, pp. 211–213, 1998.

[15] H. M. Manasevit, "Single-crystal gallium arsenide on insulating substrates," *Applied Physics Letters*, vol. 12, no. 4, pp. 156–159, 1968.

[16] H. Manasevit and W. Simpson, "The use of metal-organics in the preparation of semiconductor materials: I. Epitaxial gallium–V compounds," *Journal of the Electrochemical Society*, vol. 116, no. 12, p. 1725, 1969.

[17] Professor Chris Chidsey, private communication.

[18] A. Devi, "'Old chemistries' for new applications: perspectives for development of precursors for MOCVD and ALD applications," *Coordination Chemistry Reviews*, vol. 257, no. 23–24, pp. 3332–3384, 2013.

[19] H. Manasevit, "Recollections and reflections of MO-CVD," *Journal of Crystal Growth*, vol. 55, no. 1, pp. 1–9, 1981.

[20] C. A. Wang, "Early history of MOVPE reactor development," *Journal of Crystal Growth*, vol. 506, pp. 190–200, 2019.

[21] B. N. Chapman, *Glow Discharge Processes: Sputtering and Plasma Etching*, Wiley, 1980.

[22] M. A. Lieberman and A. J. Lichtenberg, *Principles of Plasma Discharges and Materials Processing*, Wiley, 2005.

[23] J. Batey and E. Tierney, "Low-temperature deposition of high-quality silicon dioxide by plasma-enhanced chemical vapor deposition," *Journal of Applied Physics*, vol. 60, no. 9, pp. 3136–3145, 1986.

[24] S. Sivaram, *Chemical Vapor Deposition: Thermal and Plasma Deposition of Electronic Materials*, Springer, 2013.

[25] J. T. Gudmundsson and A. Hecimovic, "Foundations of DC plasma sources," *Plasma Sources Science and Technology*, vol. 26, no. 12, p. 123001, 2017.

[26] J. E. Greene, "Tracing the recorded history of thin-film sputter deposition: from the 1800s to 2017," *Journal of Vacuum Science & Technology A: Vacuum, Surfaces, and Films*, vol. 35, no. 5, p. 05C204, 2017.

[27] K. Strijckmans, R. Schelfhout and D. Depla, "Tutorial: hysteresis during the reactive magnetron sputtering process," *Journal of Applied Physics*, vol. 124, no. 24, p. 241101, 2018.

[28] W. G. Oldham, A. Neureuther, C. Sung, J. Reynolds and S. Nandgaonkar, "A general simulator for VLSI lithography and etching processes: Part II – Application to deposition and etching," *IEEE Transactions on Electron Devices*, vol. 27, no. 8, pp. 1455–1459, 1980.

[29] J. C. Rey, L. Y. Cheng, J. P. McVittie and K. C. Saraswat, "Monte Carlo low pressure deposition profile simulations," *Journal of Vacuum Science & Technology A: Vacuum, Surfaces, and Films*, vol. 9, no. 3, pp. 1083–1087, 1991.

[30] L. Y. Cheng, J. P. McVittie and K. C. Saraswat, "New test structure to identify step coverage mechanisms in chemical vapor deposition of silicon dioxide," *Applied Physics Letters*, vol. 58, no. 19, pp. 2147–2149, 1991.

[31] Victory Process is a Silvaco Process simulation tool. See https://silvaco.com/tcad/victory-process-3d/.

10.6   PROBLEMS

10.1 What are the two commonly observed rate-limiting steps in silicon epitaxial growth? Under what conditions do they normally dominate the overall deposition rate?

10.2 In an epitaxial deposition, under mass-transfer-limited conditions, is it more important to control the reactor temperature or the source gas composition in the gas stream to obtain reproducible results? Why?

10.3 In a reactor used for epitaxial growth, the wafers are normally placed flat on the susceptor, and epi grows on the top side only. If the same reactor were used to oxidize wafers, by introducing O_2 rather than SiH_4 (or another Si gas source), SiO_2 would grow on both sides of the wafer. Explain why SiO_2 grows on both sides and epi grows only on the top side.

10.4 For CVD deposition of a film, it is found that the mass transfer coefficient $h_G = 10.0 \text{ cm s}^{-1}$ and the surface reaction rate coefficient $k_S = 1 \times 10^7 \exp(-1.9 \text{ eV}/kT) \text{ cm s}^{-1}$. For a deposition at 900 °C, which CVD system would you recommend using: a cold-walled, horizontal graphite susceptor type, or a hot-walled, stacked wafer type? Explain your answer.

10.5 Calculate the deposition rate for an LPCVD system with the same parameter values as given in the example following (10.7), but at a reduced total pressure, so that h_G is increased by 100×. Assume that the partial pressure of the incorporating species, P_G, remains the same, and that C_T decreases by the same factor as the total pressure.

10.6 (a) Plot the deposition rate (on a log scale) versus $1/T$ (kelvin) in the range 600–1200 °C for a CVD system with the following parameter values: $h_G = 0.5$ cm s^{-1}, $k_S = 4 \times 10^6 \exp(-1.45 \text{ eV}/kT)$ cm s^{-1}, partial pressure of incorporating species = 1 Torr, total pressure = 1 atm and $C_T/N = 1/10{,}000$. Identify the reaction- and mass-transfer-limited regimes.

(b) Redo the problem when the total pressure is decreased to 1 Torr, so that h_G increases by 100 times. Assume that the partial pressure of the incorporating species remains the same, and that C_T decreases by the same factor as the total pressure. (Since both the total pressure and the partial pressure of the incorporating species equal 1 Torr, this means that the gas is made up of only the incorporating species in this case.)

10.7 In a particular LPCVD deposition system, the deposition rate is experimentally measured to be 0.1 μm h^{-1} at 800 °C and 0.02 μm h^{-1} at 700 °C. At much higher temperatures, the deposition rate is found to experimentally saturate at 1 μm h^{-1}. What is the highest temperature at which this system should be used to deposit films in a system configured like a typical LPCVD system with closely spaced wafers? Explain.

10.8 A silicon epitaxial layer is deposited nominally at 700 °C. The system being used operates in the surface-reaction-controlled regime in which temperature control is very important ($E_A = 1.6$ eV). If the deposition rate needs to be controlled to ±5%, what temperature control (± how many degrees Celsius) is needed in the epitaxial reactor? State any assumptions.

10.9 We used the analogy of the Deal–Grove oxidation model when we discussed growth laws for deposited CVD films. The Deal–Grove model predicts linear–parabolic growth versus time for oxidation, whereas the CVD model we developed predicts only linear growth. Explain physically why these models are so different in their thickness versus time predictions.

10.10 The figure below is for a CVD deposition system. In a particular application, this system has to be operated at the temperature indicated by the dot on the growth velocity curve. Given this constraint, how would you arrange the wafers in this system, laying flat horizontally or stacked side-by-side vertically? Explain your choice.

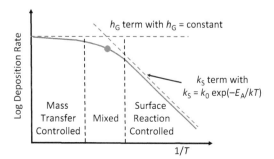

10.11 Gas-phase collisions play a role in many deposition systems. From kinetic theory, the mean free path of a gas particle is

$$\lambda = \frac{kT}{\sqrt{2}\pi d^2 P},$$

where $k = 1.36 \times 10^{-22}\,cm^3$ atm K^{-1}, T (K) is the temperature, d (cm) is the collision diameter of the molecule (approximately 4×10^{-8} cm, or about 0.4 nm, for most molecules of interest) and P (atm) is the pressure. Calculate the mean free path of a particle in the gas phase of a deposition system, and estimate the number of collisions it experiences in traveling from the source to the substrate in each of the cases below. Assume that, in each case, the molecular collisional diameter is 0.4 nm, the source-to-substrate distance is 5 cm, and the number of collisions is approximately equal to the source-to-substrate distance divided by the mean free path.
   (a) An evaporation system in which the pressure is 10^{-5} Torr and the temperature is 25 °C.
   (b) A sputter deposition system in which the pressure is 3 mTorr and the temperature is 25 °C.
   (c) An LPCVD system in which the pressure is 1 Torr and the temperature is 600 °C.
   (d) An APCVD system in which the pressure is 1 atm and the temperature is 600 °C.

10.12 Using sketches, show qualitatively how changing the gas pressure in a standard PVD system ($S_C = 1$ for this type of system) can affect the deposition profile when depositing a material on a narrow trench structure.

10.13 How does the ability to fill the bottom of a narrow trench using sputter deposition change as the target is moved further away from the wafer? Neglect any gas-phase collision effects.

10.14 The figure below shows a simulation of an LPCVD deposition over a narrow trench. The various contours are for equal time steps during the deposition. Explain why the simulation shows a time-independent deposition rate on the top surfaces (the lines are all equally spaced), but the deposition rate decreases with time on the sidewalls and the bottom of the trench (the lines get closer together as the deposition proceeds.

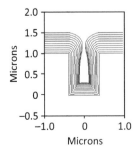

10.15 Explain how asymmetric depositions can occur on a wafer in a sputter deposition system. Asymmetric deposition means that thicker deposition occurs on one side of a feature (a step, for example) than the other. Also suggest a way to decrease the asymmetry. Use schematic diagrams in your explanations.

10.16 What value of n in the arrival angle distribution is desired for good step coverage over a step in topography? For good bottom filling of a via? Explain with schematic diagrams. What value of S_C, the sticking coefficient, do you want in each case? Explain with diagrams.

10.17 The figure below is a deposition simulation with $n = 1$ and $S_C = 1$. Explain physically why the deposition thickness varies across the bottom of the trench.

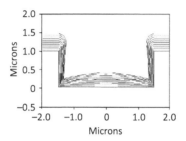

10.18 We have seen how conformal coverage in low-pressure systems results from a low sticking coefficient, which causes multiple re-emission and redeposition processes within the feature topography. Describe how, in atmospheric CVD systems, gas-phase collisions by the reactants within the topographical features can lead to good coverage and filling.

10.19 An SiO_2 layer is deposited by LPCVD over a 1 μm wide, 0.5 μm high metal line. The sticking coefficient for this process is 0.01 and the arrival angle distribution parameter n is equal to 1. Very conformal deposition is obtained, but the degree of planarization (DOP) is 0. Would changing n or S_C improve (increase) the DOP? Explain.

10.20 If you wanted to simulate thermal oxidation using a deposition simulator like Athena or Victory Process, what values of n and S_C would you choose? Why? Would such a simulation accurately predict the oxide shape on a trench structure for example? Why or why not?

10.21 The pictures below show simulations of thin-film deposition on a trench structure. In the three left-hand pictures, S_C is held constant and n is varied. In the three right-hand pictures, n is held constant and S_C is varied. In each case, describe what approximate values of n and S_C must have been used in the simulations. Explain your answers.

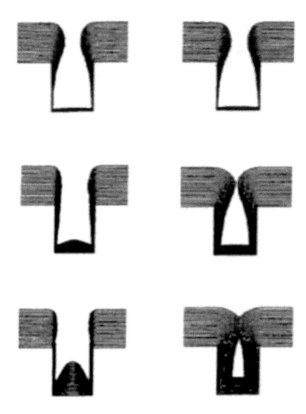

10.22 A polysilicon overhang test structure is sketched below. It is to be used in some deposition experiments. Sketch a simple process flow that would construct such a test structure starting with a silicon substrate. Explain the steps you have chosen.

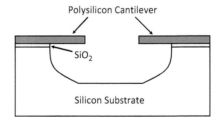

10.23 The pictures below (courtesy J. McVittie) show experimental results in some deposition experiments. In both cases, the outermost film was deposited just to provide contrast. The experiment is about the layer labeled "Deposited film." One of these experiments was done in an LPCVD system, the other in a PECVD system. Which is which? Explain your answer.

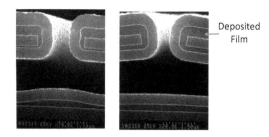

Deposited Film

10.24 The test structure shown below is used in a series of deposition experiments. Three different types of deposition systems are used. System 1 is an LPCVD system that operates with $n = 1$ and $S_C = 0.1$. System 2 is a sputter deposition system that operates with $n = 2$ and $S_C = 1$. System 3 is an HD-PECVD deposition system that operates with a strong directed ion flux and $n = 10$ and $S_C = 1$. Sketch the shapes of each of the thin-film layers deposited by these systems on three separate sketches. Explain your sketches.

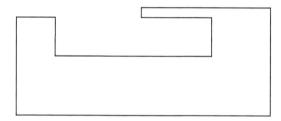

10.25 Air gaps (white in the diagrams below) between metal lines have been used in some manufacturing processes to reduce capacitances. A company is studying this possibility using a deposition simulator. In the simulations below, metal 1 (M1) and metal 2 (M2) layers are separated by a deposited dielectric layer (the middle layer). In the simulations of this layer, $S_C = 1$ in each case and n is varied from low (≈ 1) to medium ($\approx$ single digits) to high (>10). Which simulation corresponds to which of these values? Explain your reasoning.

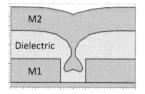 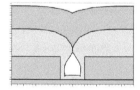 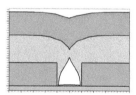

10.26 Two new experimental deposition/etching systems are being evaluated. The first system is characterized by $n = 1$ and $S_C = 0.01$. The second system is an HD-PECVD system characterized by $n = 10$ and $S_C = 1$. Each system can be used for either deposition or etching. Suppose the standard test structure shown below with a polysilicon cantilever beam is used and two experiments are done. In each case, the etching is just long enough to remove the deposited film on the top surface of the cantilever. Sketch the deposit and etch results in each case and explain your sketches.

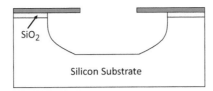

SiO$_2$

Silicon Substrate

(a) Deposit with $n = 1$ and $S_C = 0.01$; and etch with HD-PECVD system with $n = 10$ and $S_C = 1$.

(b) Deposit with HD-PECVD system with $n = 10$ and $S_C = 1$; and etch with $n = 1$ and $S_C = 0.01$.

10.27 The test structure shown below is used in two deposition system experiments. The two systems are:

(a) an Al evaporation system ($n \approx 50$, $S_C = 1$); and

(b) an LPCVD SiO$_2$ system using silane as the Si source ($n = 1$, $S_C \approx 0.3$).

Sketch the shape of the deposited films in each case. Explain your sketches.

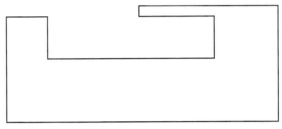

10.28 ALD deposition systems are often referred to as "self-limiting" in terms of the deposited film thickness. Explain physically, in your own words, why this happens.

10.29 In an argon DC plasma sputtering system, a large negative voltage $(-V_C)$ is applied to the cathode electrode. This results in the rather unusual voltage profile inside the plasma system. Explain, in your own words, why the voltage profile looks the way that it does by discussing the physical processes present inside the plasma system.

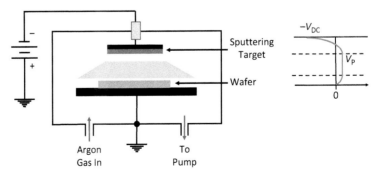

11   Back-End (Interconnect) Processing Technologies

11.1   Introduction

In the silicon complementary metal-oxide–semiconductor (CMOS) process discussed in Chapter 2, the "back-end" (wiring) portion of the process flow was described (see Figure 11.1(a)) with tungsten (W) vias, two layers of Cu wiring and two layers of deposited dielectric. In the discussion in Chapter 2, perhaps an inkling of the actual complexity of this part of the process was given through the brief discussion of TiN or TaN barrier/adhesion layers, chemical vapor deposition (CVD) of tungsten, chemical–mechanical polishing (CMP) to planarize the W layer, deposition of a copper (Cu) seed layer, followed by a thicker electroplated Cu layer, and a "dual damascene" lithography and etching process to pattern the Cu layers. An actual image of a similar back-end silicon CMOS structure is also shown in Figure 11.1(b).

The typical hierarchy of interconnects in a modern silicon CMOS chip is shown in Figure 11.2 [1]. Interconnects can either be local or global depending on the length of the wire. In general, local interconnects are the first, or lowest, level of interconnects. They usually connect gates, sources and drains in MOS technology and connect nearby transistors to realize circuit functions. Local interconnects can tolerate higher metal resistance than global interconnects, since they are not very long. In today's technologies, local interconnects are made from Cu, but the metal lines are smaller in cross-section because metal resistance is not as important as it is in longer-distance wires. Intermediate length and global interconnects, also mostly made of Cu today, are generally all of the interconnect levels above the local interconnect level. They often cover large distances between different circuit blocks and different parts of the chip. Note in Figure 11.2 that the thickness and lateral dimension of the metal layers generally increase towards the top of the structure in order to decrease interconnect resistance (indicated by the sloping green lines).

Ohmic contacts connect an interconnect to active regions or devices in the silicon substrate. A high-resistivity dielectric layer (pre-metal dielectric in Figure 11.2), which is often silicon dioxide, separates the active regions from the first metal layer, and electrical contact is made between the interconnect and the active regions in the silicon through openings in that dielectric layer. Connections between two levels of interconnects are usually called vias. They are made through openings in the different levels of the inter-metal dielectrics (IMDs), which separate interconnect levels from each other. Today, IMD layers are usually made from materials with lower dielectric constants than SiO_2. Examples of these structures are shown in Figure 11.1.

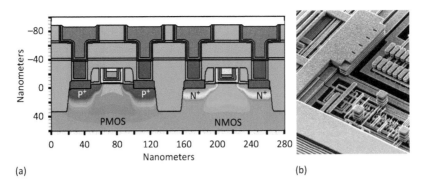

(a)   (b)

Figure 11.1 (a) A 2D cross-section of the final CMOS integrated circuit discussed in Chapter 2, with a P-channel MOS device on the left and an N-channel MOS device on the right. Only the first level of Cu metallization is shown. (b) A 3D scanning electron micrograph (SEM) of a portion of the interconnect structure in an actual silicon chip with the dielectric layers etched away so the Cu metal layers and vias are clearly seen. IBM was the first to introduce Cu interconnects in 1997. © Photo reprint courtesy of IBM Corporation.

Figure 11.2 Hierarchy of interconnects in a modern silicon CMOS chip [1]. This figure is based on a figure in the last version (2013) of the ITRS.

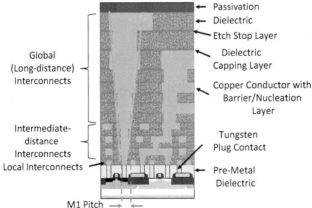

In this chapter, we discuss in much more detail the technologies needed to fabricate such structures. It will become apparent that the complexity of these processes is comparable to front-end (transistor) technology today. The innovations using new materials, high-resolution lithography and detailed electrical modeling characteristic of modern transistor structures are also critical for modern back-end technologies. In the early decades of the semiconductor industry, back-end technology was simply about wires to connect transistors to make functional circuits; today, the back-end structure is an equal partner with the front-end transistors in determining the performance of integrated circuits (ICs).

11.2 Interconnect Speed Limitations: the Need for Low *R* and *C*

We will see in this chapter that much of the innovation in back-end materials and process technology that has occurred over the past 50 years has been aimed at three issues [2, 3]:

1. Increasing the number of interconnect levels so that increasing numbers of active components can be successfully wired into complex circuits and systems.
2. Reducing the resistance *R* of the metal wires.
3. Decreasing the parasitic capacitance *C* that exists between wires in a complex back-end structure.

Reducing *R* and *C* is driven by the need to make interconnects "faster" so that higher-performance circuits can be built. In this section, we will analyze the limitations of interconnects and see directly why lower *R* and lower *C* are so important. Much of the rest of this chapter is aimed at material and process innovations to accomplish this.

We can understand the main issues using the very simple structure shown in Figure 11.3. Obviously, this is greatly simplified from real structures, such as in Figure 11.1(b). The line resistance *R*, in ohms, of one of the interconnects is given by

$$R = \rho \frac{L}{WH},\tag{11.1}$$

where ρ is the interconnect resistivity, and *L*, *W* and *H* are the interconnect length, width and height, respectively. Treating the capacitors simply as parallel-plate capacitors (neglecting fringing fields), the total capacitance associated with each line is

$$C = K_D \varepsilon_0 \frac{WL}{x_D} + K_D \varepsilon_0 \frac{HL}{L_S},\tag{11.2}$$

where x_D and K_D are the oxide thickness and dielectric constant, respectively, and ε_0 is the permittivity of free space. The first term represents the line-to-substrate capacitance, C_S, and

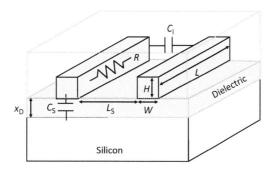

Figure 11.3 Simple interconnect structure for *RC* analysis. The two lines on top are metal interconnects of dimensions *W*, *L* and *H*, sitting on a dielectric layer such as SiO$_2$. The dielectric material is also between and above the metal lines (dotted blue lines). After [3].

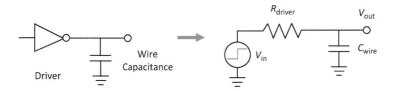

Figure 11.4 Electrical equivalent circuit for very short wires for which $R_{\text{wire}} \ll R_{\text{driver}}$.

the second term the coupling capacitance between adjacent lines, C_I. It is assumed that the lines are surrounded by dielectric on all sides.

To calculate how fast a signal can propagate along an interconnect, we need to use a simple electrical circuit model. The appropriate model depends on the length of the interconnect. For very short interconnect lines, for which $R_{\text{wire}} \ll R_{\text{driver}}$, the simplest electrical representation is shown in Figure 11.4. The distributed wire capacitance is lumped into a single C_{wire} and the resistance in the circuit comes from the CMOS driver circuit.

Using Kirchhoff's current law, the response of this circuit is described by

$$C_{\text{wire}} \frac{dV_{\text{out}}}{dt} + \frac{V_{\text{out}} - V_{\text{in}}}{R_{\text{driver}}} = 0, \tag{11.3}$$

which has a solution

$$V_{\text{out}}(t) = V_{\text{in}}\left(1 - e^{-t/\tau}\right), \tag{11.4}$$

where $\tau = R_{\text{driver}}C_{\text{wire}}$ and is the circuit time constant. The output voltage rises (or falls) exponentially and the time required to go from 10% to 90% of its final value is a time given by 2.2τ, a number often used to describe the speed of the circuit. Minimizing R_{driver} and C_{wire} clearly speeds up the circuit.

For longer wires in which $R_{\text{wire}} \gg R_{\text{driver}}$, the simplest circuit replaces R_{driver} with R_{wire} in Figure 11.4 and the circuit response is similar. For longer lines, of course, the wire really should be treated as a distributed RC network. Modern electronic design automation (EDA) tools from companies like Cadence, Synopsys and Silvaco can extract more realistic capacitance and resistance values from actual chip layouts and then use these values in circuit simulation tools based on SPICE to predict circuit performance. These more complex models provide some additional insight, but they do not change the principal conclusions we will reach in this section, namely that R and C must be minimized in back-end technologies, so we will use the very simple model in Figure 11.4.

Using (11.1) and (11.2), we have

$$\tau = R_{\text{wire}}C_{\text{wire}} = \rho L^2 K_D \varepsilon_0 \left(\frac{1}{Hx_D} + \frac{1}{WL_S}\right). \tag{11.5}$$

This simple result allows us to draw the main conclusions we will need in this chapter as we consider back-end technologies. We see directly that τ and hence interconnect delay are both directly proportional to ρ, K_D and L^2. Physically, the delay is proportional to L^2 because both R and C are proportional to L, so the RC product goes as L^2. Thus, we need to minimize interconnect resistance and parasitic capacitance, and we also note that because of the L^2 dependence, the longest wires in a particular circuit (global interconnects) are the most critical in determining circuit speed.

Example

Calculate the interconnect delay time (2.2τ) according to (11.5) for the following conditions: Al interconnect ($\rho = 3.0 \times 10^{-6}$ Ω cm), SiO$_2$ dielectric ($K_D = 3.9$), $L = 1$ cm (global interconnect), $H = W = L_S = 0.1$ μm and $x_D = 1$ μm.

Answer

From the equation we have

$$t_D = 2.2\tau = 2.2\rho L^2 K_D \varepsilon_0 \left(\frac{1}{H x_D} + \frac{1}{W L_S} \right)$$

$$= 2.2(3 \times 10^{-6} \ \Omega \ \text{cm})(1 \ \text{cm})^2 (3.9)(8.86 \times 10^{-14} \text{F cm}^{-1})$$

$$\times \left(\frac{1}{(10^{-5}\text{cm})(10^{-4}\text{cm})} + \frac{1}{(10^{-5}\text{cm})(10^{-5}\text{cm})} \right)$$

$$= 25.1 \ \text{ns}.$$

This simple example illustrates that modern chips that operate at gigahertz frequencies cannot use simple interconnects like this 1 cm long wire with delays that would be 25 times the clock period.

It is also interesting to observe in (11.5) the impact of technology scaling on interconnect delay. In Chapter 1 we discussed "ideal" or Dennard scaling (Figure 1.16). This type of device scaling dominated silicon technology until about the year 2000. Ideal scaling implies that all lateral and vertical dimensions scale by the same factor k, where k was typically 1.4 for each new generation in that time period (giving a device density increase of $k^2 = 2$, see Table 1.1). If L, H, W, L_S and x_D all scale by k, then τ would not change, implying that the interconnect delays would remain relatively constant as technology scaled. There are two problems with this. First, since scaling does improve transistor speed, even if interconnects are unchanged with scaling, their relative contribution to overall circuit performance will increase. Second, ideal scaling only deals with individual transistors and ignores the fact that chip area increases with time, which means that worst-case global interconnects get longer, not shorter, since these interconnects may have to extend from one side of the chip to the opposite side in the worst case. Thus L in (11.5) actually does not scale for the longest interconnects, again implying that the

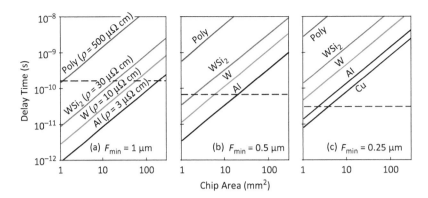

Figure 11.5 Global interconnect and gate time delay versus chip area A for various minimum feature sizes F_{min}, for various interconnect materials. The diagonal lines are the interconnect delays for various materials. The horizontal dashed line is the gate delay. The interconnect delay corresponds to the longest global interconnect in the circuit. After [3].

interconnect contribution to delay increases with technology generations. These observations about long global interconnects also imply that chip architecture and design are very important elements in chip performance, since minimizing long-distance interconnects obviously can help.

One of the early studies of the impact of interconnect delays on chip performance from the 1980s [3] illustrated the problem, showing results like those in Figure 11.5. This analysis used relatively simple models like those shown in Figure 11.3. In this analysis, three generations of then-current technology were compared, from 1 μm to 0.25 μm minimum feature sizes. The horizontal dashed line on each plot is the gate delay (transistor speed), which scales as $1/k$ and gets faster due to scaling with each generation. The solid diagonal lines show the worst-case interconnect delay for different interconnect materials as a function of the chip size. The worst-case interconnect delay is assumed to be for a wire that connects devices across the full dimension of the chip. In this simple analysis, the longest global interconnect line length was assumed to scale as $\sqrt{A}$, which means that the longest L^2 increases as A, and this explains the unity slope of the lines in Figure 11.5.

While the models used in this analysis were greatly simplified, there are general observations that can be made. First, the lower resistivity of Cu wires produces lower interconnect delays, as (11.5) would suggest. Second, it is obvious why it would be poor design practice to use materials like polysilicon or even WSi_2 or W for global interconnects. Finally, for even modest chip sizes, interconnect delays dominate gate delays even in these 30-year-old technologies.

These simple calculations suggested even 30 years ago that back-end interconnect technology needed careful attention in high-performance circuits. A number of strategies have evolved to mitigate these issues. First, obviously R and C need to be minimized, and we will discuss materials and technology approaches to these goals throughout this chapter. Second, the interconnect hierarchy illustrated in Figure 11.2 has evolved. The lower levels of wiring that use smaller, higher-resistance wires are used only for short-range interconnect, whereas the much thicker and wider upper levels are used for global interconnects, reducing R in these longer wires. Note also in Figure 11.2 that the dielectric layers are thicker in the upper metal lines, which helps to reduce R.

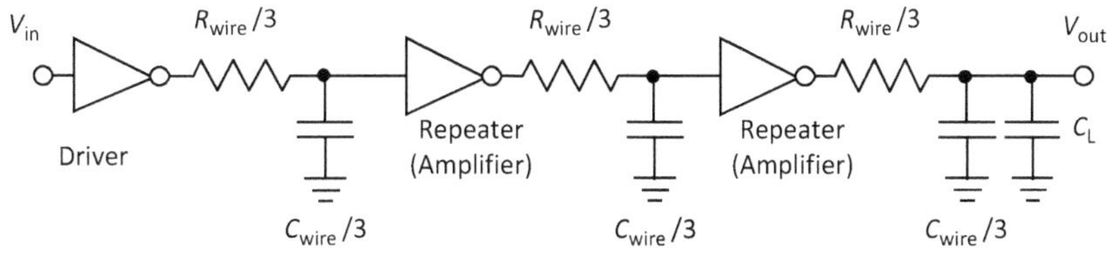

Figure 11.6 Example of breaking up a long interconnect line into sections (three in this case) and inserting repeaters (amplifiers) between each section.

There are also a number of circuit and architecture options that can be used to minimize interconnect delays. Blocks in a large chip that need to communicate with each other often should be physically placed near each other on the chip to minimize L. Repeaters are another powerful idea. Equation (11.5) shows that the delay is proportional to L^2. Suppose a long line were cut into three pieces and a repeater (amplifier) placed between each section, as in Figure 11.6. Each section would have an RC delay of 1/9 of the original line, because L, and hence R and C, are reduced by 1/3. The repeater would introduce some additional delay, but the line itself would now have an overall delay proportional to L rather than L^2. In practice, very long lines can be split into multiple sections with multiple repeaters. There is a considerable literature on how to design such interconnects, e.g., [4, 5, 6], but the basic conclusion is that wire delays can be made proportional to wire length rather than L^2, which is a significant improvement. Today, sophisticated EDA tools are used to optimize the design of back-end structures and chip layout, to achieve the highest performance possible. Finally, there is significant research currently underway to explore optical interconnect methods and perhaps free-space broadcast connections, but neither of these is likely to impact chip interconnects in the near future.

11.3 Historical Development and Basic Concepts

The earliest silicon ICs used aluminum lines connecting active regions – transistors and resistors. A combination of thermally grown silicon dioxide passivated the silicon surface and isolated the aluminum interconnects from those regions in the substrate where electrical contact was not wanted. Where contact was desired, openings in the oxide were made so that the aluminum could come into contact with the silicon. Figure 11.7 shows an early structure.

Pure aluminum was thus used for both the contact material and the interconnect. Why aluminum? First, it has a low electrical resistivity at room temperature. (See Table 11.1 for a listing and properties of commonly used interconnect materials.) Second, it adheres well to silicon and silicon dioxide. Third, it makes a low-resistance electrical contact to heavily doped Si. The latter two characteristics are related to the fact that the oxide of Al is very stable, more stable than SiO_2. This means that if Al is deposited on top of SiO_2, it will react with it even at low temperatures, forming a thin layer of Al_2O_3 at the interface. This forms a glue layer, binding the Al to the SiO_2. In the case of Al-to-Si contacts, the Al will reduce any native Si oxide on top of

Table 11.1 **Properties of interconnect materials [7, 8].**

Material	Thin-film resistivity ($\mu\Omega$ cm)	Melting point (°C)
Al	2.7–3.0	660
Cu	1.7–2.0	1084
Ag	≈1.6	961
Au	≈2.2	1064
W	8–15	3410
Ti	40–70	1670
PtSi	28–35	1229
$TiSi_2$	13–16	1540
WSi_2	30–70	2165
$CoSi_2$	15–20	1326
NiSi	14–20	992
TiN	50–150	~2950
$Ti_{0.3}W_{0.7}$	75–200	~2200
Polysilicon (heavily doped)	450–1000	1417

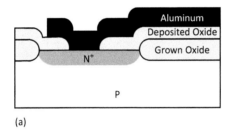
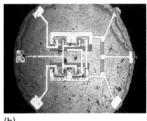

(a)   (b)

Figure 11.7 (a) Schematic of early back-end structure with Al interconnect making contact with a highly doped Si active region. (b) The chip is a photograph of the first integrated circuit from Fairchild Semiconductor in 1959, which used a simple Al wiring technology. (b) Photo reprinted with permission from onsemi.

the Si. Without this reaction, the ubiquitous native oxide on the silicon could prevent ohmic contact.

Fortunately, the Al_2O_3 that is formed allows Al to diffuse through it into the Si, so that good ohmic contact between the Al and Si is obtained. A fourth benefit of using Al is that its presence assists the annealing out of interface traps at the Si/SiO_2 interface, presumably by converting H_2O to free H, which passivates the traps. (See Chapter 6 where this process is described in more detail.) For all these reasons, contacts and interconnects made of aluminum were instrumental in the development of the first ICs.

As the complexity of silicon ICs increased over time, junction depths became shallower and more layers of interconnect were needed to wire the thousands and later millions and billions of transistors into circuits. The simple Al/SiO_2 structure shown in Figure 11.7 simply was not capable of realizing such chips, and, over time, new materials and new back-end innovations were gradually introduced. Four major changes gradually happened in the 50-plus years between 1970 and today:

1. Contacts evolved to become multilayer structures with adhesion layers and barrier layers.
2. Planarization methods became necessary as the number of wiring levels gradually increased from one or two to as many as 15 today.
3. Cu replaced Al as the primary metal used for interconnects.
4. "Low-K" dielectric materials gradually replaced SiO_2 in back-end structures.

We will consider each of these changes in the sections below.

11.3.1 Evolution of Contact Structures

The first innovations that were introduced in back-end processing had to do with contact structures, and were driven by contact reliability and contact resistance. As device structures were scaled down in size beginning in the 1960s and 1970s, junction depths were correspondingly reduced (see Figure 1.16 for a discussion of ideal scaling). One of the first major changes in back-end structures was a result of the fact that Al contacts to shallow junctions suffer from reliability issues. As discussed earlier, Al makes a good contact to silicon by reducing the native oxide on the surface. To ensure that the Al reduces the native Si oxide and that the Al is in good physical contact to the Si, a sintering anneal at about 450 °C is generally performed after Al deposition and patterning. This anneal, in a hydrogen ambient, is also done to anneal out interface traps at the Si/SiO_2 interface, as discussed in Chapter 6. Subsequent processing at about this same temperature can also occur, to deposit dielectric films used in the back-end structure.

The equilibrium solubility of Si in Al is significant. At 450 °C, the solubility is about 0.5 at.% (atomic percent), and at 500 °C it is about 1 at.%. This means that pure Al in contact with Si will want to absorb Si from the substrate up to its solubility level for that temperature. The kinetics of the process involve the diffusion of Si in the Al, so the time to reach equilibrium depends on the amount of Al acting as a sink for the Si. Unfortunately, the diffusivity of Si in polycrystalline Al is quite high, and the amount of Al nearby can be large (wires connecting to other devices). So a significant amount of Si gets drawn up into the Al. This creates voids in the remaining Si, which are quickly filled by the overlying Al. If the Al penetrated uniformly into the substrate, the situation might be manageable, at least for relatively deep junctions.

However, the situation is actually worse than that. This is because, when the Al reduces the native oxide, it does so non-uniformly. Therefore, the Al penetrates in localized spots. Al "spikes" occur in the Si substrate, typically penetrating in excess of 1 μm deep. Figure 11.8 illustrates this, showing the Al penetrating in some areas deeper than the diffused junction, which would short the junction. Recall that Al is a column III element and hence is a P-type dopant in silicon, so this problem is particularly acute for contact to

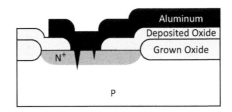

Figure 11.8 Schematic of Si–Al contact region showing spiking of Al into the Si active region. This is due to Si diffusing into the Al to satisfy the solubility requirement, with Al filling the resulting voids in the Si substrate.

N^+ regions. As a practical matter, pure Al contacts to Si cannot be used for junction depths of less than 2–3 μm.

One solution to the Al spiking problem is to use Al films that already have Si in them, so that the solubility requirement is already fulfilled. Sputtering as a deposition method allows a Si/Al alloy to be easily deposited, so Al films with approximately 1 at.% Si have been commonly used. This is the solubility of Si in Al at about 500 °C, the highest temperature the Al sees during processing. The spiking problem is reduced, but another problem arises. When the contact structure is cooled down below 500 °C, or subsequent thermal steps are done at less than 500 °C, the solubility of Si in Al is lower. The now-excess Si precipitates out, usually at the Al/Si interface, where heterogeneous nucleation can occur, leaving Si nodules. In fact, there is enough Al dissolved in these nodules to make them P-type, causing higher contact resistance to N^+ regions.

The use of Al(Si) to solve the spiking problem was adequate as long as the contact resistance was not limiting circuit performance. But, as scaling continued in the 1970s and 1980s, contact areas were reduced in size and the contact resistance did become an issue, so another solution to Al–Si interaction and spiking was needed. The solution was the use of barrier layers. Barrier layers are placed between the Al and Si to prevent, or at least slow down, any interaction. Such barriers need to be a barrier for chemical interdiffusion between Si and Al at processing temperatures (up to 450–500 °C). They also need to be thermally stable. They should have low stress. This usually means that their coefficients of thermal expansion need to be close to that of Si (2.6×10^{-6} °C^{-1}). They need to adhere well to Si and Al, as well as to SiO$_2$. This usually means that there is some interfacial reaction between the barrier and Si and Al. Finally, they should have good electrical conductivity and a low contact resistance to both Si and Al.

Barrier layers have been classified into three types: passive barriers, stuffed barriers and sacrificial barriers [9]. Passive barriers are chemically inert to both Si and Al, and are good diffusion barriers as well. TiN is an example of this type. Some materials are chemically inert, but diffusion along their grain boundaries can be quite significant. However, it has been found that by "stuffing" the grain boundaries with other atoms or molecules, the diffusion of other species can be greatly reduced, making them diffusion barriers as well. The reduction of diffusion by stuffing the grain boundaries is a combination of a physical effect (physically blocking the diffusing species) and a chemical effect (the stuffed species, such as nitrogen or oxygen, chemically bond with the diffusing species, such as Al, stopping their diffusion). A Ti–W alloy with nitrogen stuffing the grain boundaries (achieved by sputtering the Ti–W in an N_2 ambient) is an example of a stuffed barrier.

Sacrificial barriers are those in which the barrier material is "sacrificed" in order to prevent reaction between the Al and Si. The material, such as Ti, reacts with either the Al or Si or both, forming aluminides or silicides. But, by doing so, it prevents the Al and Si from interacting, since the reactions with the barrier material are favored both kinetically and thermodynamically over the reaction between the Al and Si themselves. However, once the sacrificial barrier material is consumed, the barrier is gone, and so these types of barriers are only effective for limited processing temperatures and times.

There are many choices for barrier layers, some of which are listed in Table 11.1. One of the first barrier layers for contacts was platinum silicide. It is formed by depositing Pt on the wafer, and then annealing at low temperatures (250–350 °C). Because the surface Si is consumed during the formation reaction, a new, clean, interface is obtained, and very good electrical contact is made. It is also "self-aligned" because the silicide is only formed where the Si is exposed through the oxide contact opening (and the unreacted Pt is chemically removed from other areas). Therefore, an extra mask is not needed and no alignment overlap is required. PtSi acts satisfactorily as a barrier between Al and Si, but only if subsequent processing is done at about 350 °C or below, and junction depths underneath are greater than about 0.3 μm. That is because PtSi will react with Al, forming Pt–Si–Al intermetallic compounds, eventually allowing the Al to spike into the Si substrate. Once junction depths got shallower than 0.3 μm and processing temperatures got above 350 °C, another barrier was necessary.

Refractory metals by themselves have had limited use as barriers. Ti, in particular, is attractive, since it reduces and breaks through any native oxide, and adheres well to both Si and SiO_2. It forms very-low-resistance contacts, silicon diffusion through it is slow and therefore it prevents the Si dissolving in the Al. However, it reacts with Al to form $TiAl_3$ at low temperatures. Above 500 °C, Ti will also react with the Si to form $TiSi_2$. It thus acts as a sacrificial barrier, acting as a barrier only until it is consumed, which happens rather quickly above 400 °C. So while Ti is a good contact/adhesion layer, it is not a very good barrier layer. Ti is thus often used in conjunction with a barrier layer on top.

Refractory metal silicides, such as $TiSi_2$, have also been used as barrier layers in contacts. They will also react with Al, but at a somewhat higher temperature (about 500 °C for $TiSi_2$) than the noble metal silicides, such as PtSi. However, diffusion along their grain boundaries limits their effectiveness as barriers to below 400 °C. Therefore, like Ti, refractory metal silicides are usually used in contact structures as the adhesion/contact layer, with a better barrier layer on top. An advantage in using refractory metal silicides in contacts is the ability to form self-aligned contact structures. Ti can be deposited over both an oxide layer and into the openings in the contacts without a mask. But during an anneal it will only form a silicide in the contacts where Si is exposed. The unreacted Ti over the oxide can be etched off, leaving the silicide self-aligned to the contact.

As the processing temperatures were raised, the next commonly used barrier after PtSi was titanium–tungsten. This alloy, generally about 30 at.% or 10 wt.% (weight percent) Ti, acts as a diffusion barrier between the Si and Al, especially when stuffed with nitrogen. While pure W is a good barrier itself, adding the Ti to it helps it adhere to underlying layers and increases its corrosion resistance.

Refractory metal compounds, notably TiN, are very popular barrier layers. These are commonly fine-grained structures, with grain sizes below 10 nm. Diffusion through them is very slow, making them impermeable to silicon and most other species. Incorporation of oxygen or nitrogen in these films makes them even better barriers. They are also chemically stable and chemically inert with most other layers. There may be some reaction with Al, forming some AlN and $TiAl_3$ at temperatures greater than 550 °C. (The AlN, in fact, may form a thin barrier layer itself, improving the barrier properties of the structure even more.)

The electrical resistivity of TiN is low enough for use as both a contact material and a local interconnect. The contact resistance of TiN to Si is somewhat higher than those of PtSi, Ti or $TiSi_2$. Therefore, TiN is usually used in a bilayer structure with TiN on top of Ti, with the Ti either remaining unreacted or forming $TiSi_2$. The best properties of both layers are utilized: the good adhesion and low contact resistance of the Ti or $TiSi_2$, and the good barrier properties of TiN. This is accomplished without much more processing, since the TiN can easily be deposited immediately after the Ti, just by adding nitrogen to the input in the deposition chamber. Or TiN can be formed on top of $TiSi_2$ by depositing Ti on top of the Si substrate in the contact opening and annealing in a nitrogen ambient. The Ti reacts with the Si below it to form the silicide while simultaneously reacting with the nitrogen in the ambient to form TiN on top. We will consider this process in detail later in this chapter (Figure 11.31).

Figure 11.9 shows a multilayer contact structure using $TiSi_2$/TiN contact/barrier layers, which is a typical structure used with Al interconnects. The $TiSi_2$ makes good electrical contact to the Si, and provides good adhesion. The TiN serves as a barrier to interaction between the Al and both the $TiSi_2$ and Si. There are, of course, other material options that can also provide the combination of low contact resistance, adhesion and diffusion barrier. We will see additional examples throughout this chapter.

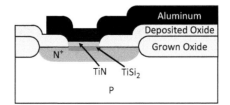

Figure 11.9 Typical multilayer contact structure for Al interconnects with $TiSi_2$/TiN barrier layers between the Si and Al.

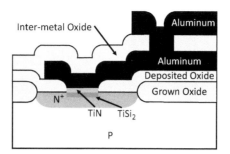

Figure 11.10 Illustration of early two-level metal structure.

11.3.2 Planarization Methods to Permit Multilayer Interconnects

As circuits became larger and more complex in the 1970s and 1980s, additional interconnect levels became necessary in order to be able to wire the increasing transistor numbers into complex circuits. Adding an additional level of interconnect seems straightforward. The process could be as simple as depositing an inter-metal dielectric layer (typically SiO_2), opening a hole in the dielectric using lithography wherever a contact to the lower level is needed, depositing Al, forming both the contact and the next level interconnect, followed by patterning of the aluminum. Then the next level of inter-metal dielectric material would be deposited. This structure is illustrated in Figure 11.10. This, in fact, is how the early multi-level interconnect structures were made. Looking at the contact, or via in this case, Al is now making contact to Al, so spiking into Si is not a problem. Good electrical contact between two layers of Al can be achieved if the interface is clean.

Obviously the topography gets rougher and less planar as more levels are added, as Figure 11.10 illustrates. A more planar topography is desired for two main reasons – lithography limitations and step coverage. The lithography limitations arise partly from the limited depth of focus of submicron lithography instruments. We discussed this issue in Chapter 5 (see Figure 5.22 and the associated derivation and also Figure 5.60). The depth of focus for a modern deep ultraviolet (DUV) 193 nm lithography tool is only about ±250 nm and is much less for extreme ultraviolet (EUV) tools. In addition, non-planar topography results in non-uniform resist thickness, which can result in non-uniform exposure and development. There are lithographic techniques that address these problems, such as using multilayer resist structures, as discussed in Chapter 5, but these are not always effective or practicable solutions.

Non-planar topography also leads to problems with step coverage and filling of spaces with deposited films. We saw in Chapter 10 how step coverage and filling can be a problem. Such problems lead to higher electrical resistance in the line, as well as potential reliability problems. To make matters worse, step coverage problems compound one another. A large step height on one level can lead not only to thinning of the next level on sidewalls, but also to cusping and overhangs. This in turn leads to even worse coverage problems on the next level.

Reducing the step heights and achieving more planar topology through processing techniques is called "planarization." We defined the degree of planarization (DOP) in Chapter 9 (see (9.21)) as

$$DOP = 1 - \frac{x^f_{step}}{x^i_{step}}, \tag{11.6}$$

where x^i_{step} is the initial step height in the topography and x^f_{step} is the final step height after planarization (see Figure 9.43). These step heights can refer to hole depths as well. A DOP of 0 means that no planarization is achieved, while a DOP of 1 corresponds to complete planarization – the final step height is zero. Several methods have been developed to improve the planarity of multi-level interconnect structures. Some involve the metals, while others involve the dielectric materials, both of which are discussed below.

One way to help achieve planar metal layers is by using a W contact or via fill, also called a W plug. Figure 11.11(a) shows the topography above a contact or via that uses Al as both the contact material and the next level of interconnect. The topography over the contact is not planar. Achieving a planar or smooth topography over the contact can be achieved by using a via fill or plug structure, resulting in the structure in Figure 11.11(b).

The structure in Figure 11.11(b) can be fabricated in two ways. One method involves a selective W deposition into contact or via holes cut in the dielectric layer, which fills the contacts or vias without depositing any W over the top of the dielectric. The other, much more common, method uses a blanket W deposition, which fills the contact or via as well as depositing on the dielectric. The W deposition thickness is greater than the depth of the via so that the via is overfilled. The W on the dielectric is then removed through CMP, a process we described in Chapter 9. This is the process that was used in the CMOS process in Chapter 2 (see Figure 2.23). As was done in that process, usually a thin TiN barrier layer is deposited first by sputtering and then the W is deposited by CVD. This process of filling holes or trenches in a dielectric level with the metal, followed by CMP, rather than depositing, patterning and etching the metal on top of a dielectric layer, is called the "damascene" method. Its name comes from the ancient practice in the Middle East of inlaying metal in wood or ceramics for decoration. It can be used for vias or interconnect layers, or both, and we will see later that it is essential for Cu interconnects, since Cu cannot be easily etched.

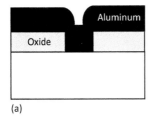
(a)

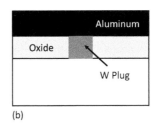
(b)

Figure 11.11 (a) Using Al as both the contact and the next level interconnect layer results in non-planar topography. (b) Using a tungsten plug as the contact along with CMP results in planar topography.

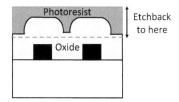

Figure 11.12 Illustration of photoresist etchback process. Photoresist is deposited over rough topography, then the structure is etched back, leaving a smooth top surface of the oxide.

Before CMP was widely available as a planarization technology, several other approaches were developed and used in some manufacturing processes. The simplest of these was described in Chapter 9 (Figure 9.42), which is repeated in Figure 11.12. As we saw in Chapter 9, in this technique the dielectric is deposited, followed by the deposition of a sacrificial layer. The overall structure is then etched back to planarize it. The earliest such processes used photoresist as a sacrificial etchback layer on top of oxide. The photoresist was deposited, or spun, over the oxide layer, filling up the spaces and ending up with a nearly flat top surface. After a hard bake, an etch is done, which etches both the oxide and photoresist, ideally at the same rate. In other words, an etch chemistry is chosen such that the etch selectivity $S = 1$. When the bottom of the photoresist is reached, a nearly flat oxide is left behind (dashed red line in Figure 11.12). This results in local planarization, but generally not perfect global planarization. This can also be done without a sacrificial layer if the dielectric layer is deposited thick enough. This is because the thicker a material is, the smoother the local topography tends to become. While this process is conceptually simple, it does not achieve the nanometer level of planarization required in modern structures.

The photoresist process described above works because photoresist is a liquid when it is applied and hence forms a flat upper surface. Other types of liquids have also been applied in an effort to planarize back-end structures. Spin-on-glass (SOG) has found application in some manufacturing processes. This is similar to the photoresist etchback technique, but here the spun-on material is left behind and serves as an inter-metal dielectric. SOG materials are initially liquids, containing organic siloxanes or inorganic silicates in an alcohol-based solvent. The liquid SOG is spun onto the wafer like photoresist, filling the spaces between features. Then the wafer is baked and cured, driving off the solvents and polymerizing the silicon-containing groups, resulting in the formation of Si–O–Si bonds. SOG is often sandwiched between layers of CVD SiO_2 in an inter-metal dielectric structure, as shown in Figure 11.13.

The SOG process is often combined with an etchback. After deposition and curing, the SOG is etched back so that no SOG remains over any metal region that will have a via made through it. This is so that, when via holes are cut through the inter-metal dielectric, they do not pass through any SOG, as also illustrated in Figure 11.13(b). If the vias touch the SOG (Figure 11.13(a)), residual gases and moisture from the SOG can contaminate the contacts or vias and cause large increases in contact or via resistance, so-called "poisoned vias." SOG films made from organic-based solutions, containing carbon, are more susceptible to this than those

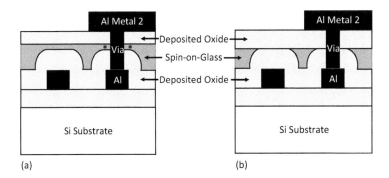

Figure 11.13 (a) Back-end structure without an etchback step, showing poisoned via (region labeled with *) where the via metal is in direct contact with the spin-on-glass (SOG) layer. (b) The back-end structure with an etchback step does not have poisoned vias since the via metal is not in direct contact with the SOG layer.

made from inorganic solutions. With SOG, local planarization and limited global planarization can be achieved.

As the need for planarization became critical in the 1980s, many workers in the field found the idea of polishing the wafer to achieve planarity counterintuitive. Particle control and defects in semiconductor manufacturing are huge yield issues (Chapter 4) and so the idea of introducing particles to "grind" or polish surface topography seemed to be a very bad idea. However, the engineers at IBM who were responsible for Si wafer preparation realized that wafer polishing is actually a critical part of achieving the atomically flat "perfect" Si wafers that are used in Si chip manufacturing (see Figure 3.20 and the associated discussion) [10, 11]. The IBM engineers perfected oxide planarization using CMP first but the technology gradually spread to planarization of many other materials, including Cu metallization.

The basic operation of a CMP tool was described in Chapter 9 (Figures 9.44 and 9.45). In a CMP tool, a rotating wafer is pressed against a rotating pad. The polishing slurry consists of chemically active ingredients, such as oxidizers and surfactants, and mechanically active abrasive particles. Pressure, rotation rates, slurry flow rate, and the pH and composition of the slurry are the main process variables. Because chemistry is involved in the polishing process, the polishing rate can be made quite different for different materials. In the W plug polishing example in Figure 11.11(b), this means that the polishing rate can be designed to slow down significantly once the W is removed from the oxide surface and SiO_2 is exposed to the CMP tool, leaving a planarized surface with the W located only in the via holes. An example of this process was discussed in Chapter 9 using CMP simulations (Figure 9.52). Chapter 9 also discusses in some detail physical models and simulation tools for the CMP process.

CMP has become, by far, the dominant technology used for planarization. Today, it is used in both front-end and back-end processing. It has allowed back-end structures to incorporate as many as 15 layers of wiring, and it has played a major role in mitigating the very limited depth of focus of modern lithography tools because it allows a thin and uniform-thickness photoresist layer to be used for imaging. An example of a 1998 (before Cu) back-end metallization structure is shown in Figure 11.14, showing many of the concepts discussed to this point in this chapter.

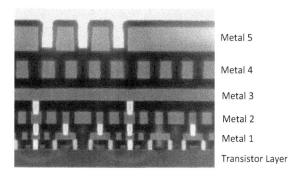

Metal 5

Metal 4

Metal 3

Metal 2

Metal 1

Transistor Layer

Figure 11.14 Five-layer interconnect structure used in Intel's 0.25 μm technology in 1998 [12]. The metal layers are Ti/Al–Cu/Ti/TiN. The Ti and TiN layers are adhesion and barrier layers. The small percentage of Cu was added to the Al for electromigration reasons (discussed later). White W plugs are visible in via holes connecting the Al wiring levels. CMP was used to planarize the structure at each wiring level. © Intel, reprinted with permission.

11.3.3 The Transition from Al to Cu

From the beginnings of the silicon IC industry in the 1960s until the late 1990s, Al was exclusively used for interconnects. While Al is not the lowest-resistivity metal (Au, Ag and Cu are lower in Table 11.1), Al has other properties that made it simple to use and, for a long time, its resistivity was adequate for the chips that were being designed and built. Al is straightforward to deposit and etch, it adheres well to Si and SiO_2, and it produces low-resistance contacts to Si because it reduces thin SiO_2 layers, as we have discussed.

The first modification to pure Al interconnects in the 1970s and 1980s was the addition of Si at about 1 at.%. This prevents spiking of Al into the silicon substrate caused by silicon's relatively high solubility and diffusivity in Al (Figure 11.8). Even with the use of barrier layers such as TiN, Si is often still used in Al interconnects. This is because the presence of the Si reduces the driving force for interdiffusion and hence decreases the chance of barrier failure.

Almost all of the rest of the modifications made to the Al interconnect structure in the 1980s were a result of aluminum's low melting point and ease of deformation. Two phenomena that occur with aluminum, which have caused significant problems and have required modifications in the Al interconnect, are hillock growth and electromigration.

Hillock growth occurs when the Al thin film is subjected to high compressive stresses. Because of the polycrystalline structure of Al, coupled with its low melting point and resulting susceptibility to plastic deformation, the Al will deform under such stresses. High stresses can easily be generated due to Al's relatively high coefficient of thermal expansion ($23 \times 10^{-6}\,^{\circ}C^{-1}$, while Si is $2.6 \times 10^{-6}\,^{\circ}C^{-1}$). If Al is deposited on a Si substrate (or even on another film, such as SiO_2, on a Si substrate) at low temperatures and the structure is heated, high compressive stresses in the Al can be generated because the Al, which is tightly constrained to the wafer, will want to expand much more than the Si. To relieve the stress, portions of the Al can squeeze up, forming small hills or "hillocks," as illustrated in Figure 11.15. The movement of the Al occurs

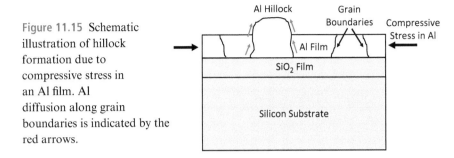

Figure 11.15 Schematic illustration of hillock formation due to compressive stress in an Al film. Al diffusion along grain boundaries is indicated by the red arrows.

primarily along grain boundaries in the Al, since diffusion is usually much faster along grain boundaries than through the interior of the grains. Whole grains of Al can be pushed or grow upwards, forming hillocks about the size of the grains.

Hillocks can result in electrical shorts between interconnect levels, as well as cause the surface topography to be rough, making lithography and etching more difficult. While the overlying films can help suppress hillock formation by acting as a mechanical barrier, this does not completely eliminate the problem. In fact, so much stress can build up that an overlying SiO_2 layer can crack. When this happens, the Al can protrude from the crack in long whiskers, increasing the chances of a short.

The addition of elements that have limited solubility in Al, such as Cu, has been found to suppress hillock formation. The excess atoms segregate and precipitate preferentially at the Al grain boundaries. This reduces the Al diffusion along the grain boundaries and suppresses hillock formation.

In a similar manner, voids can form when the Al is subjected to tensile stresses. If the Al is deposited at a higher temperature, then, upon cooling, the Al will want to shrink more than the Si below it. Since the Si is more mechanically rigid than the Al, the stress is relieved by vacancy movement and agglomeration in the Al, with voids forming. This can greatly increase the interconnect resistance and even cause open circuits. The addition of Cu to the Al can suppress void formation in Al as well as suppressing hillock formation.

Perhaps the biggest problem encountered with Al interconnects is that of electromigration, which we will discuss in more detail later in this chapter (Section 11.4.4). One of the requirements of an interconnect is that it is stable not only during processing, but also during circuit operation. Electromigration is a phenomenon that occurs during circuit operation and has very adverse effects. When an electric current flows through aluminum (on the order of 0.1–$0.5\,MA\,cm^{-2}$, which is commonly achieved in integrated circuits), the electrons can actually transfer enough momentum to the Al atoms to cause them to diffuse. This diffusion is faster along grain boundaries, and it can cause a build-up of Al in some regions, resulting in hillocks, and can cause a depletion in other regions, resulting in voids. As a result, shorts and open circuits in the interconnects can occur during the normal operation of the circuit. Figure 11.16 shows an SEM of an actual interconnect where this has occurred. Because most of the Al atoms and vacancies diffuse along the grain boundaries, electromigration is very dependent on the grain structure, including the size and crystallographic orientation of the grains. It also depends on over- and underlying layers, and on the processing history of the interconnects.

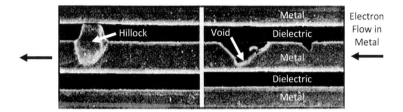

Figure 11.16 SEM top view of hillock and voids that have formed due to electromigration in an Al(Cu, Si) line. From [13] © 1996. Reprinted by permission from Springer Nature.

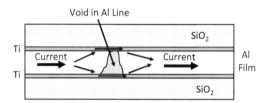

Figure 11.17 Illustration of shunt layers to minimize electromigration problems. Thin layers of metals, such as Ti, are used above and below the Al interconnect line.

Substantial effort has been made to study and remedy this phenomenon [14]. One of the early remedies for electromigration was to add Cu to the Al. Similar to the mechanism for suppressing hillocks, adding Cu suppresses grain boundary diffusion, and hence slows down electromigration. Adding too much Cu, however, causes problems with etching of the interconnect as well as corrosion, and 4 wt.% Cu is the most that is usually added.

Thus, the first modifications to Al interconnects involved switching from pure Al to Al with about 1–2 wt.% Si and 0.5–4 wt.% Cu. Adding these elements to the Al increases the sheet resistivity of the interconnect by as much as 35%, but this was a necessary trade-off in order to prevent the failure of circuits due to electromigration, and hillock and void formation.

Other methods have been used to inhibit electromigration, or, in some cases, to negate its effect on the integrity of the interconnect lines. One method uses shunt lines. Another interconnect material is layered below, above or within the Al layer, as shown in Figure 11.17. If electromigration does cause an opening in the Al line, the current can be shunted through the other material, maintaining the electrical integrity of that interconnect. The material does not have to have as low a resistivity as Al, since current will only flow through it for a short distance. But it should have good resistance to electromigration, and good adhesion or barrier properties. Layers of Ti, Ti–W, TiN or TiSi$_2$ have been commonly used for this purpose.

These additional layers in interconnects help in electromigration protection for two other reasons. If they have high melting points and strong mechanical properties, they can be more rigid than the Al and act as physical or mechanical barriers to hillock and void formation. If Ti is used, it will actually react with Al to form TiAl$_3$, which is even harder mechanically than Ti,

Figure 11.18 A typical metallization scheme used in Al interconnect IC technology circa 2000, showing the multilayer structures that evolved.

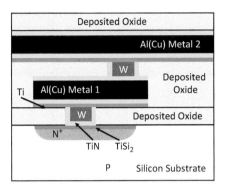

and will reduce the effects of electromigration. However, if too much TiAl$_3$ forms, it can increase the resistivity of the interconnect. If TiN is used, it is best to have excess Ti in it, which will react with the Al to form TiAl$_3$. The multilayer interconnect structures also help with barrier and adhesion issues. Ti can help with adhesion and electrical contact. TiN can serve not only as a good barrier, but also as an adhesion layer, particularly in conjunction with W plugs. TiN provides for good contact between Al and W and prevents any reaction between them. A TiN layer can also be used in the interconnect stack to serve as a barrier between the Ti and the Al.

Putting all of these pieces together led to Al interconnects in the 1990s with structures like those shown in Figure 11.18. A W plug is used in both the contact and the vias, resulting in good planarity. At the bottom of the contact, TiSi$_2$ is used to make good electrical contact to the Si, and also serves to strap the sources and drains, providing for lower sheet resistance along those paths. In both the contacts and the vias, TiN surrounds the W, promoting adhesion to the oxide. The global interconnect structure in this example is Ti/TiN/Al(Cu)/TiN. Cu in the Al suppresses electromigration and hillock formation. Ti helps to improve electromigration resistance and promotes adhesion. TiN is a barrier between the Al and Ti to limit TiAl$_3$ formation. The Ti/TiN and TiN layers on the bottom and top of each interconnect provide an electrical shunt in case of void formation in the Al. They also suppress hillock and void formation by providing mechanical barriers to plastic deformation. Revisiting Figure 11.14, we see an actual chip, from 1998, that incorporates many of these features.

By 1990 it was clear that the resistance associated with Al interconnects was going to be an issue as scaling continued and chips grew larger (longer interconnects). *RC* delays in interconnects were discussed earlier in this chapter, where we saw the importance of low resistance in interconnect materials. It is apparent from Table 11.1 that Cu, Ag and Au have lower resistivities than Al, and hence in principle could reduce the resistance of interconnect lines in ICs.

Silver has corrosion problems and poor resistance to electromigration, and gold has just marginally lower resistivity than Al along with device contamination problems. Copper has a lower resistivity, 1.72 Ω cm versus 2.7 Ω cm for Al. Perhaps more importantly, Cu has much better resistance to electromigration than Al. Higher resistance to electromigration also allows for circuit operation at higher current densities in the interconnects, resulting in potentially faster circuit speed.

However, there are problems with using copper interconnects, which led to Al interconnects being dominant until the late 1990s. Copper is difficult to plasma-etch since the byproducts, copper halides, are not volatile at low temperatures. Corrosion of copper is a problem, especially if the halides are not all removed. A damascene process, single or dual, is the obvious solution to this problem, since the copper in this process does not have to be plasma-etched, just polished back using CMP. Another problem with copper interconnects is that Cu is a harmful contaminant in Si devices, and Cu atoms can move quickly through SiO_2 and organic dielectrics. Cladding layers, of TiN, Si_3N_4 or Ta-based alloys, for example, must be used as barriers around the copper to prevent this. But this can reduce the effectiveness of using the lower-resistivity material if cladding layers of higher-resistivity material must be used on all sides of the interconnect, thus reducing the cross-sectional area of the Cu.

A representative dual damascene process flow used today for Cu interconnects is shown in Figure 11.19. This figure provides some of the details that were not shown in the CMOS overview in Chapter 2, specifically in Figure 2.24.

There are many possible variations on this process flow. In Figure 11.19(a), a multilayer stack of materials is deposited on top of the metal 1, Cu. In the earliest Cu technologies, the dielectric was often SiO_2 and the etch stop was Si_3N_4. Today, lower-dielectric-constant materials are used, as discussed in the next subsection. In Figure 11.19(b), photolithography is used to define the pattern that will become the metal 2 pattern, and this pattern is etched in a hard mask material, which could be Si_3N_4 or another material that is not attacked by the subsequent etching processes. In Figure 11.19(c), a second lithography step is used to define the via pattern that connects metal 2 to metal 1. The via pattern is etched down to the etch stop in the middle of the deposited stack. This particular process flow allows both lithography steps to be performed

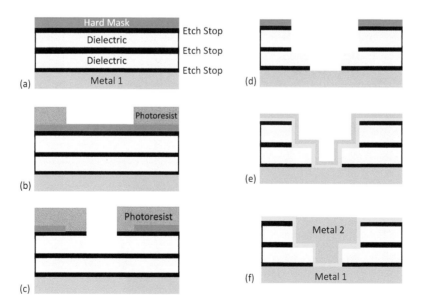

Figure 11.19 Example of dual damascene process used today for Cu interconnects.

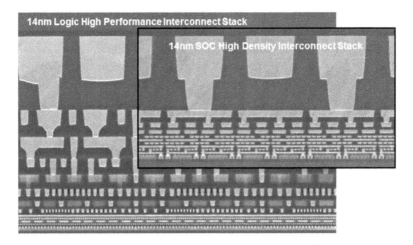

Figure 11.20 Examples of modern silicon CMOS chips with many levels of Cu wiring. These images are from Intel's 14 nm technologies. © 2016 IEEE. Reprinted with permission from [15].

on a relatively flat surface, which minimizes depth of focus issues in high-resolution lithography.

In Figure 11.19(d), the hard mask is used to define the metal 2 pattern. Etching through the top etch stop layer and top dielectric layer stops on the middle etch stop layer for the metal 2 pattern. At the same time, the via hole etching is completed by etching through the bottom dielectric. The etching is finished by removing the etch stop layers at the bottoms of the two patterns. In Figure 11.19(e), a barrier/adhesion layer, which could be TaN/Ta, is first deposited to keep the Cu from diffusing into the surrounding layers, and then a thin seed Cu layer is deposited to provide a base for the electroplating of the main Cu metal 2 layer. In Figure 11.19(f), the metal 2 Cu layer has been electroplated to completely fill the metal 2 pattern, and then CMP has been used to polish the Cu back to create the planar structure shown.

In modern silicon CMOS technologies, this dual damascene process is repeated as many as 15 times to produce a back-end structure with up to 15 levels of wiring. Figure 11.20 shows examples of such chips. Comparing Figure 11.7 with Figure 11.20, it is obvious that 50 years of back-end technology development has resulted in a much more complicated, and much higher-performance, interconnect technology.

11.3.4 The Transition to Low-K Dielectrics

Switching to Cu approximately 20 years ago reduced the R in wiring RC delays to a practical minimum in terms of metal resistivities. Only Ag has a slightly lower resistivity and, as discussed earlier, the processing difficulties with Ag imply that it will likely not be used in integrated circuits. The other opportunity is to reduce C.

The capacitance of a parallel-plate capacitor is given by

$$C = \frac{K\varepsilon_0}{d}A \,, \tag{11.7}$$

where ε_0 is the permittivity of free space, d is the distance between the plates of the capacitor, A is the area and K is the dielectric constant of the material between the capacitor plates (the ratio of this material's permittivity to ε_0). In interconnect structures, d and A are geometric factors that depend on the size and spacing between the Al or Cu lines. In terms of process technology, K is the parameter we can adjust. In air or vacuum, $K = 1$, so this is the minimum value. For SiO_2, $K = 3.9$. Reducing C thus implies finding materials with $K < 3.9$.

The dielectric constant K of a material can typically be described by the Clausius–Mossotti equation [16]

$$\frac{K-1}{K+2} = \frac{N\alpha}{3\varepsilon_0} \,, \tag{11.8}$$

where N is the number of molecules per unit volume (density) and α is the total polarizability. A material containing polar chemical bonds can be represented as having electric dipoles due to the charge separation in the bonds. This increases the dielectric constant because such dipoles can align with an external electric field, adding the field of the dipole to the external field. A capacitor made with such a material stores more charge for a given applied voltage and therefore has a higher capacitance. Dipole formation can be the result of electronic polarization (electron displacement), distortion polarization (ion displacement) or orientation polarization (molecule displacement). There are thus two approaches to reducing K. Either α can be reduced by finding materials with chemical bonds that have lower polarizability than Si–O; or N can be reduced by finding materials with lower density than SiO_2.

Figure 11.21 illustrates these two general approaches, which can be combined to achieve even lower K values [17, 18]. Table 11.2 classifies the various approaches to achieving low-K materials [19].

The first new materials that were introduced into silicon manufacturing were silica-based and replaced some Si–O bonds with less polar Si–F or Si–C bonds. Fluorinated silicon glass (SiOF) was used in 0.18 μm technology (late 1990s) and realized K values of 3.5–3.8. The next generation of low-K materials were organosilicate glasses (SiCOH) in which the Si–O bond was replaced by the less polarizable Si–CH$_3$ bond. The K value of SiCOH materials is 2.6–3.0 depending on the

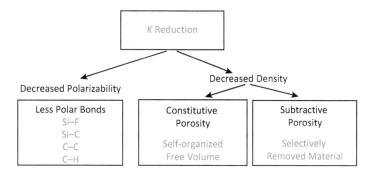

Figure 11.21 Approaches to materials with lower K. After [18].

Table 11.2 **Low-K materials classification. After [19].**

Classification	Material	Dielectric constant, K
Silica based	SiOF (fluorinated silicon glass, FSG)	≈ 3.5
	SiCOH (organosilicate glasses)	≈ 2.8
Silsesquioxane (SSQ) based	Hydrogen silsesquioxane (HSQ)	≈ 3.0
	Methyl silsesquioxane (MSQ)	≈ 2.5
Polymer	Poly(arylene ether) (PAE)	≈ 2.6
	Polyimide	≈ 2.3
	Parylene-N (parylene-F)	≈ 2.4
	Teflon (PTFE)	≈ 2.0
Amorphous carbon	C–C (C–F)	≈ 2.0
Porous	Porous SiCOH, MSQ, PAE	<2.0
Air gaps	Air	1.0

number of CH_3 groups built into the material. These materials were used in some 130 nm and 90 nm technologies (from 2000 to 2010). Both F and C increase the interatomic distance or volume of silica films, and hence part of the reduction in K is due to a density decrease (self-organized free volume in Figure 11.21). The CH_3 molecule also has a larger volume, but it is hydrophilic, which is generally a bad thing because this means that these films can absorb water vapor. Water has extremely polar O–H bonds and a K value close to 80, so even a small amount of absorbed water significantly increases the material's K value.

SiCOH materials have a lower limit on K of about 2.6. To achieve even lower K values, porosity is required. This can be achieved by the subtractive porosity approach shown in Figure 11.22 (discussed below). The deposited material is a mixture of SiCOH and an organic additive such as alpha-terpinene (ATRP), bicycloheptadiene (BCHD) or cyclooctane (C_8H_{16}) [19]. A thermal or UV curing process after the deposition removes the additive, leaving a porous film. With processes like this, K values as low as 2 can be achieved. The materials in Table 11.2 are generally deposited using plasma-enhanced chemical vapor deposition (PECVD) for higher-K materials, or by liquid spin-on methods for lower-K materials. SiCOH materials with organic additives can be deposited by PECVD methods with K values as low as 2.1 [20]. These porous films are often called pSiCOH films. An excellent review of progress in these areas is [21].

Back-end dielectrics obviously need to provide good electrical isolation between metal lines. This implies a high breakdown field strength (>5 MV cm^{-1}) and low leakage (bulk and surface resistivity should exceed 10^{15} Ω cm). However, some of the most difficult challenges associated with low-K materials arise from process integration issues. Compared to SiO_2, low-K materials tend to be mechanically weak, thermally unstable and often poorly compatible with other materials (able to absorb chemicals, etc.). Ref. [18] suggests five requirements for successful integration of low-K materials:

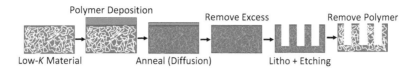

Figure 11.22 Example of one approach to mitigate the reduced mechanical strength of low-K dielectric materials. In this example, a low-K dielectric is given temporary mechanical strength by filling its pores with a polymer, which is later removed. After [22].

1. *Hydrophobicity.* As pointed out previously, water has extremely ionic O–H bonds, and materials that absorb water quickly increase their K value.
2. *Mechanical stability.* With Cu wiring, damascene technology and CMP are necessarily part of the back-end process. Low-K materials must be able to withstand the mechanical stresses inherent in CMP. Mechanical properties generally deteriorate as porosity increases.
3. *Thermal stability.* Back-end processing temperatures can be as high as 400–450 °C. Some low-K organic polymers begin to decompose at such temperatures.
4. *Chemical and physical stability.* Back-end processing involves plasma etching, cleaning and lithography, all of which must be tolerated by the low-K materials.
5. *Compatibility with other materials.* Issues such as coefficient of thermal expansion, barrier material deposition and adhesion may all be important, depending on the particular processes used.

The mechanical strength of low-*K* materials (item 2 above) is a significant challenge for integrating these materials into back-end structures. Two examples of approaches to help mitigate these issues are shown in Figures 11.22 and 11.23 [22]. In Figure 11.22, the pores in a low-*K* material are "stuffed" with a polymer material. Proper choice of the polymer minimizes the degradation (increase in *K*) that results from the "stuffing." The polymer stuffing provides mechanical strength for processes like lithography and etching. Depending on the desired final *K* value, the polymer can be removed after processing. In Figure 11.23, a replacement low-*K* dielectric process is illustrated that uses a temporary "stronger" dielectric while the metal CMP process is being performed. That temporary dielectric is then replaced with the low-*K* material.

Not surprisingly, low-*K* materials remains an active area of research and development. The "ultimate" technology, an air dielectric, has even been incorporated selectively in some silicon CMOS processes, as shown in Figure 11.24. Intel claimed that the incorporation of these air gaps provided a 17% improvement in interconnect *RC* performance.

11.4 Key Process Technologies and Back-End Issues

The evolution of back-end technologies described in the previous sections required the development of a number of key process technologies, which we have not discussed in detail yet in this text. One of these key technologies, CMP, which was critical for back-end structures, has

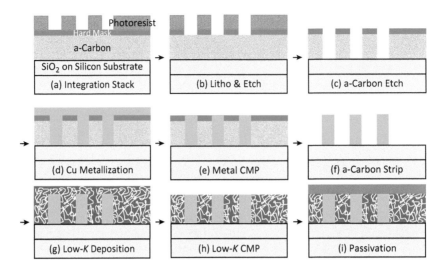

Figure 11.23 Example of another approach to mitigate the reduced mechanical strength of low-K dielectric materials. In this example, a stronger dielectric material is used during processing and replaced with a low-K material at the end of the process. After [22].

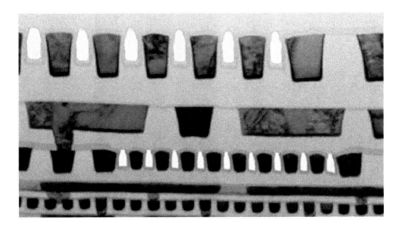

Figure 11.24 Back-end wiring in one version of Intel's 14 nm technology. The white areas are air gaps used in two of the metal levels (M4 and M6); the black areas are the Cu metal wires; and the gray areas are the back-end dielectric layers. © 2015 IEEE. Reprinted with permission from [23].

also become widely used in front-end processes, and hence is a general material removal technology. As such, we discussed the CMP process in detail in Chapter 9. The sections below discuss other key technologies, their underlying physical principles and, where appropriate, introduce simulation tools to model these processes. We also discuss electromigration, a significant reliability concern in interconnects.

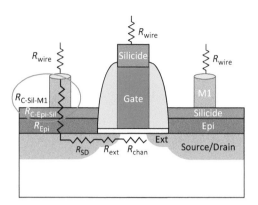

Figure 11.25 Schematic of a MOS device cross-section showing the device resistance components along with contact resistances and metal interconnect resistance.

11.4.1 Metal–Semiconductor Contacts

Figure 11.25 is based on Figure 7.1 where we discussed the resistances associated with a typical device structure such as an MOS transistor. We have added the wire resistances to Figure 11.25, since they are basically the interconnects that are the subject of this chapter. In Chapter 7 we discussed many of the components of the total resistance in a device like this, relating them to dopant profiles. However, the connection point between the front-end transistor structure and the back-end interconnects is through the metal–semiconductor contact, the topic of this section. In Figure 11.25, this is shown in two pieces (circled in red), $R_{\text{C-Epi-Sil}}$, which is the contact resistance between the silicide and the silicon, and $R_{\text{C-Sil-M1}}$, which is the contact resistance between the silicide and metal 1 (M1). Electrical contact resistances between two metals or between a silicide and a metal ($R_{\text{C-Sil-M1}}$) are generally small, so in this section we will deal with a general contact resistance between a metal and a semiconductor ($R_{\text{C-Epi-Sil}}$), which generally dominates and needs to be made as small as possible in order to minimize the total resistance in the circuit.

Metal–semiconductor (Schottky) junctions are discussed in detail in all basic semiconductor device textbooks, e.g., [24], so we will not derive the results describing the contact resistance of such structures here. Rather, we will present and discuss the results in the context of process technology, and refer the interested reader to references such as [24] for further discussion of the physics and mathematics behind these results.

Metal–semiconductor junctions can be rectifying (Schottky) or ohmic. Figure 11.26 shows the band diagrams for these modes of current transfer through such a junction. Because of the work function difference between the metal and semiconductor, a barrier, ϕ_{B}, exists at the interface. Nature provides a wide variety of metal work functions, so, in principle, it should be possible to choose a metal that produces a small ϕ_{B} contact to silicon device regions. In practice, this is not generally possible: first, because there are other requirements on which metals we may choose (resistivity, for example), and, second, because surface states at the metal–semiconductor junction tend to pin the Fermi level at the interface deep in the silicon bandgap.

As we have seen, metal silicide to silicon is a more common contact structure in silicon technology. The same band diagrams as in Figure 11.26 hold for this contact as well, and the same Fermi-level pinning issues hold for metal silicide/silicon contacts. Table 11.3 shows

Table 11.3 Experimentally measured barrier heights to N- and P-type silicon for common silicides. From [24].

Silicide	MoSi$_2$	ZrSi$_2$	TiSi$_2$	CoSi$_2$	WSi$_2$	NiSi$_2$	Pd$_2$Si	PtSi$_2$
ϕ_{BN} (V)	0.55	0.55	0.61	0.65	0.67	0.67	0.75	0.87
ϕ_{BP} (V)	0.55	0.55	0.49	0.45	0.43	0.43	0.35	0.23

Figure 11.26 Band diagrams and *I–V* characteristics for metal–semiconductor junctions or contacts, illustrating different modes of current transfer: (a) Schottky or rectifying contact; and (b) tunneling or ohmic contact.

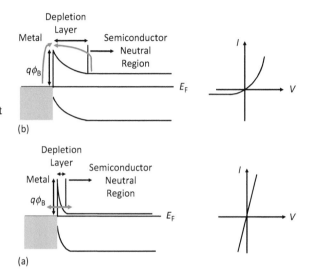

experimentally measured values for ϕ_B to N- and P-type silicon for several commonly used silicides. It is obvious from the experimentally measured values for ϕ_{BN} and ϕ_{BP} that Fermi-level pinning at the interface plays a dominant role, since the numbers are generally similar for all the silicides in spite of the very different metal work functions.

Thermionic emission, shown in Figure 11.26(a), is a process in which thermal energy allows some carriers to surmount the barrier and does allow some current to flow over a Schottky barrier when a voltage is applied. The population of carriers in the semiconductor decreases exponentially with energy above the conduction band edge, so, as voltage is applied to the junction to decrease the potential barrier on the semiconductor side (positive on the metal, negative on the semiconductor), exponentially more carriers are able to surmount the barrier.

From standard texts [24], the current density versus voltage (*J–V*) relationship of such a structure is given by

$$J_S = RT^2 \exp\left(-\frac{q\phi_B}{kT}\right)\left[\exp\left(\frac{qV_{MS}}{kT}\right) - 1\right],\tag{11.9}$$

where J_S is the current density (A cm^{-2}),

$$R = \frac{4\pi q m_{\mathrm{e}} k^2}{h^3} \qquad (11.10)$$

is the Richardson constant (120 A cm^{-2} K^{-2} in silicon and 400 A cm^{-2} K^{-2} in SiC) and V_{MS} is the applied voltage across the Schottky junction. The result is a current that increases exponentially with voltage, as shown in Figure 11.26(a). In the reverse direction, with the metal negative and the semiconductor positive, almost no current flows because ϕ_{B} is not reduced by reverse bias (the metal is a conductor and does not support applied voltage), thus very few carriers can surmount the barrier from the metal to the semiconductor.

The specific contact resistivity, ρ_{c}, using (11.9) is

$$\rho_{\mathrm{c}} = \left[\frac{\partial V_{\mathrm{MS}}}{\partial J_{\mathrm{S}}}\right]_{V_{\mathrm{MS}}=0} = \frac{kT}{qJ_{\mathrm{S}}} \ \Omega \ \mathrm{cm}^2 . \qquad (11.11)$$

Rectifying contacts are not ohmic contacts, as is obvious from the I–V characteristic in Figure 11.26(a). If we pick a modest current density of 1 A cm^{-2} and substitute numbers into (11.11) we find that $\rho_{\mathrm{c}} = 2.5 \times 10^{-3}$ Ω cm^2, a number that is far too large for contacts in modern devices (see Figure 11.27). As a result, we need a different approach to achieve a contact that can carry significant currents in either direction with a low voltage drop.

If the semiconductor is very heavily doped, as in Figure 11.26(b), the depletion region is very narrow (a few nanometers) and significant quantum mechanical tunneling of electrons through this barrier becomes possible. The depletion region thickness is given by

$$x_{\mathrm{D}} = \sqrt{\frac{2\varepsilon_{\mathrm{S}}\phi_{\mathrm{B}}}{qN_{\mathrm{D}}}}, \qquad (11.12)$$

where N_{D} is the doping concentration in the semiconductor. In fact, the tunneling probability increases exponentially with decreasing barrier thickness. The tunneling current that flows in such a barrier is given by [24]

$$J_{\mathrm{S}} = \frac{1}{2} q N_{\mathrm{D}} v_{\mathrm{th}} \exp\left[-\frac{H(\phi_{\mathrm{B}} - V_{\mathrm{MS}})}{\sqrt{N_{\mathrm{D}}}}\right], \qquad (11.13)$$

where

$$H = \frac{4\pi}{h} \sqrt{\frac{\varepsilon_{\mathrm{S}} m_{\mathrm{n}}}{q}} . \qquad (11.14)$$

The specific contact resistivity for a tunneling contact using (11.13) is [24]

$$\rho_{\mathrm{c}} = \left[\frac{\partial V_{\mathrm{MS}}}{\partial J_{\mathrm{S}}}\right]_{V_{\mathrm{MS}}=0} = \frac{2\exp(H\phi_{\mathrm{B}}/\sqrt{N_{\mathrm{D}}})}{qv_{\mathrm{th}}H\sqrt{N_{\mathrm{D}}}} \ \Omega \ \mathrm{cm}^2$$

$$= \rho_{\mathrm{c0}} \exp(H\phi_{\mathrm{B}}/\sqrt{N_{\mathrm{D}}}) \ \Omega \ \mathrm{cm}^2. \qquad (11.15)$$

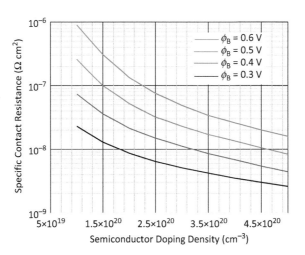

Figure 11.27 Calculated
specific contact resistance
for various values of ϕ_B
and doping concentration
in the silicon. After [24, 25].

The contact resistivity decreases as N_D increases. Generally, in order for x_D to be less than ≈ 2.5 nm, and thus for efficient tunneling to occur, $N_D > 5 \times 10^{19}$ cm^{-3} is required. Equation (11.15) is also applicable to P-type semiconductors if ϕ_{BP}, m_p and N_A are used. Figure 11.27 shows calculated values of ρ_c using a slightly more sophisticated model than (11.15).

In a modern silicon very-large-scale integration (VLSI) device, if we had a contact area of 25 nm $\times$ 25 nm, and used a representative value of ρ_c, shown in Figure 11.27 (1 $\times$ 10^{-8} Ω cm^2), this would produce a contact resistance of about 1.3 kΩ. This might be too large a value to be acceptable, compared to other resistances in the device and in the circuit surrounding it. Reducing the contact resistance would require either a smaller value of ϕ_B and/or a significantly higher doping concentration in the silicon. Each of these is a difficult technological challenge, doping because of dopant solubility issues and ϕ_B because of Fermi-level pinning issues. This is why research continues today on methods to lower ρ_c values, e.g., [26, 27].

Ohmic contacts to compound semiconductor devices follow the same physics as described above for silicon devices. The specific metals that are commonly used vary with the specific compound material. For the interested reader, several representative references on this topic are [28, 29, 30, 31, 32, 33]. Two examples are shown in Figure 11.28.

In GaAs MESFET devices (Figure 11.28(a)), in which the gate is a rectifying Schottky contact, a refractory gate metal (tungsten nitride in this example) is used. The ohmic S/D contacts use an alloyed Au–Ge metal. Ge is an N-type dopant in GaAs and it outdiffuses from the metal system, producing a highly doped N$^+$ region with a tunneling ohmic contact. The upper layers of the metallurgy (Ni/Au) help with adhesion and thicken the layer to reduce the overall resistivity. Contacts to P-type GaAs are often made with alloys of Au–Zn or Au–In, with the Zn or In providing the highly doped P$^+$ region. There are many other choices available for both ohmic contacts and rectifying contacts in GaAs [33]. The deeper N$^+$ regions forming the S/D regions are generally made using ion implantation.

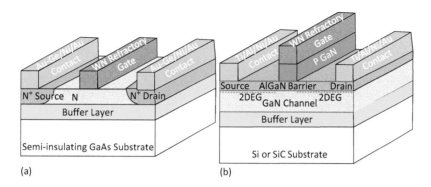

Figure 11.28 Examples of compound semiconductor contact technologies. (a) GaAs metal–semiconductor field effect transistor (MESFET) with both ohmic (Au–Ge/Ni/Au) and Schottky contacts (WN). (b) GaN high-electron-mobility transistor (HEMT) device, which also uses both ohmic and rectifying (Schottky) contacts.

The structure in Figure 11.28(b) is a GaN HEMT, a device that is finding increasing use in power switching applications today. The ohmic contacts over the S/D use a somewhat similar metallurgy to the GaAs device in Figure 11.28(a), but the mechanism is different. The Ti reacts with the AlGaN layer during a high-temperature anneal (typically 850 °C for 30 s), forming an interfacial TiN region and creating N vacancies. These vacancies act as donors in AlGaN, producing an N^+ layer under the contact. This creates an ohmic tunneling contact to the AlGaN [30, 31, 34]. As in the GaAs example, the gate contact in the GaN example is WN, a refractory metal that forms a Schottky or rectifying contact. There are many other choices for both ohmic and rectifying contacts to GaN devices. However, the basic principles for contacts to compound semiconductor devices remain those shown in Figure 11.26 and described above for silicon devices. Back-end technologies for compound semiconductor devices are discussed in more detail in Section 11.7.

Finally, it should be pointed out that extracting experimental data for contact resistance is not as straightforward as it might seem. In principle, one could simply build a structure that forced current through a contact and then measure the voltage drop caused by the contact. The contact resistance would then simply be the ratio of the voltage drop divided by the current from Ohm's law ($R = V/I$). The specific contact resistivity (Ω cm^2) would then be obtained by multiplying by the contact area. This simple approach makes many assumptions, however. The most important of these is that the current flow through the contact is uniform, and so the effective area of the contact is its physical area. This is rarely true because current flow is highly two- or even three-dimensional in small structures, so that the effective area of the contact can be quite different from its physical area. There is an extensive literature on the design of test structures to correctly extract ρ_c. See [35], for example.

11.4.2 Silicide Structures and Technology

As we have seen, silicide structures are very common in silicon back-end technologies. In this section, we will consider these technologies more deeply and also use simulation tools to illustrate and understand silicide formation processes. Table 11.4 lists some of the properties of common silicides.

Table 11.4 **Properties of silicides. After [36].**

Silicide	Thin-film resistivity ($\mu\Omega$ cm)	Sintering temp. (°C)	Stable on Si up to (°C)	Reaction with Al at (°C)	Thickness of Si consumed per nm of metal (nm)	Thickness of resulting silicide per nm of metal (nm)	Barrier height to n-Si (eV)
PtSi	28–35	250–400	~750	250	1.12	1.97	0.84
$TiSi_2$(C54)	13–16	700–900	~900	450	2.27	2.51	0.58
$TiSi_2$(C49)	60–70	500–700			2.27	2.51	
WSi_2	30–70	1000	~1000	500	2.53	2.58	0.67
Co_2Si	~70	300–500			0.91	1.47	
CoSi	100–150	400–600			1.82	2.02	
$CoSi_2$	14–20	600–800	~950	400	3.64	3.52	0.65
NiSi	14–20	400–600	~650		1.83	2.34	
$NiSi_2$	40–50	600–800			3.65	3.63	0.66
$MoSi_2$	40–100	800–1000	~1000	500	2.56	2.59	0.64
$TaSi_2$	35–55	800–1000	~1000	500	2.21	2.41	0.59

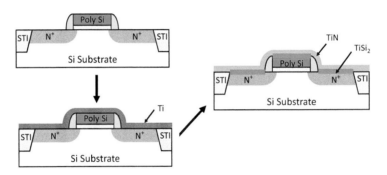

Figure 11.29 Generic process flow for $TiSi_2$ formation.

Silicides can be deposited directly as a compound using sputtering, CVD or other methods, but it is also common for the metal to be deposited, and then a chemical reaction with Si from the substrate can be used to form the silicide. This has the advantage of only forming the silicide in regions where there is a silicon source. A generic process flow is illustrated in Figure 11. 29 for the case of $TiSi_2$.

The process is shown for a polysilicon gate NMOS device. After stripping any oxide on the silicon and polysilicon surfaces, a thin layer of Ti is deposited, often by sputtering. Typically 50–100 nm of Ti might be deposited, as shown lower left. The structure is then reacted at ≈ 600 °C for a time sufficient to allow $TiSi_2$ and TiN formation, usually less than a minute. During this time, two chemical reactions take place. On the top surface of the Ti, reaction with the ambient N_2 produces TiN. On the bottom surface of the Ti in the regions where a Si source is present, $TiSi_2$ is formed. If an inert ambient were used (Ar, for example), no TiN would be formed and unreacted Ti would be left in regions in which there is no Si source.

What happens next depends on the particular technology being implemented. In some cases, a lithography step would be used to pattern the TiN layer, leaving it in places where it can be used as a local interconnect [37]. From Table 11.1, the resistivity of TiN is $\approx$ 50–100 $\mu\Omega$ cm, much higher than that of Cu, which would be used in the metal layers in most of the interconnect structure. Thus the TiN layer could only be used to interconnect structures relatively close to each other, and hence is called a "local" interconnect. In other technologies in which the TiN is not used as a local interconnect, no lithography would be needed and the TiN, and any unreacted Ti, would then be selectively etched off using a solution of $NH_4OH:H_2O_2:H_2O$ (1:1:5) or using a plasma etching process. Finally, a second anneal is generally used to lower the resistivity of the silicide and the TiN if it is present. This is because, below about 750 °C, a metastable phase of $TiSi_2$ forms, called the "C49" phase, which has a higher resistivity (60–70 $\mu\Omega$ cm). An 800 °C anneal for 30–60 s is required to transform it to the equilibrium "C54" phase, with its lower resistivity of 13–20 $\mu\Omega$ cm. Both of these phases are listed in Table 11.4.

The thickness of the $TiSi_2$ is dependent on how much Ti is deposited and the time and temperature of the anneals. Using the information in Table 11.4, for a 100 nm layer of Ti, and reaction of all 100 nm of the Ti with the underlying Si, 227 nm of the Si will be consumed, and 250 nm of $TiSi_2$ will be formed. This would give a film sheet resistance of approximately $(15 \times 10^{-6} \ \Omega \text{ cm})/(2500 \times 10^{-8} \text{ cm}) = 0.6 \ \Omega/\text{square}$. But less silicide is produced if the anneal is done in N_2, since the top part of the Ti reacts with the nitrogen to form TiN instead of the silicide. The actual silicide thickness in this case depends on the relative kinetics of the silicide and nitride formation reactions, and will depend on anneal time and temperature. We will consider simulation models for these processes shortly.

Several things are accomplished by this silicide formation process. First, the doped regions in the silicon (N^+ and P^+) are coated by the surface silicide, significantly reducing their resistivity and effectively increasing the area of contacts on these regions to the full diffused area, which reduces contact resistance. Second, the resistivity of the polysilicon gate is also reduced by the silicide layer on its surface. Finally, if the TiN is used for local interconnects, an extra layer of wiring is available below all of the upper Cu wire layers. In the CMOS process in Chapter 2, $TiSi_2$ was used on the doped regions in the silicon, although no TiN was used for local interconnects (see Figure 2.16). Since that process used a high-K metal gate structure, the $TiSi_2$ layer on top of the gate was also unnecessary.

One additional point about these silicide processes is illustrated in Figure 11.30. The formation of MSi_2, where M is the metal, requires either the metal M or the Si to diffuse through the forming MSi_2 film so that the reaction can continue. In the first case (Figure 11.30(a)) the new MSi_2 forms on the bottom surface of the MSi_2 layer. In the latter case (Figure 11.30(b)), the MSi_2 forms on the top surface. The difference can be important. Note in Figure 11.30(b) that the MSi_2 layer will grow up the side of the sidewall spacer. This potentially could short the silicon N^+ region to the gate, causing the transistor to be non-functional after the excess unreacted metal is removed.

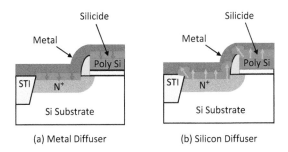

(a) Metal Diffuser   (b) Silicon Diffuser

Figure 11.30 Schematic illustration of silicide formation for a silicide that forms (a) by diffusion of metal species and (b) by diffusion of silicon. In the latter case, lateral encroachment of the silicide over the spacer can occur.

In general, then, we would likely prefer a silicide in which the metal was the diffusing species. If Si is the diffuser, careful control over the time and temperature of the formation anneal would be needed. For metal-rich silicides such as M_2Si, the dominant diffusing species are usually metal atoms. On the other hand, in the formation of monosilicide MSi and disilicide MSi_2, silicon atoms are generally the dominant diffusing species [38]. However, there are exceptions. The dominant diffusing species in the growth of $TiSi_2$, $CoSi_2$, WSi_2 and NiSi are Si, Co, Si and Ni, respectively. Of course, this issue is only one factor that a chip manufacturer would have to consider in picking a particular silicide.

In Chapter 6 the thermal oxidation of silicon was modeled using the linear–parabolic Deal–Grove model. Oxygen diffuses through the growing SiO_2 layer and reacts with Si at the Si–SiO_2 interface. This is also often called a diffusion–reaction model. For short times and thin oxides, the reaction between the Si and O rate-limits the oxidation process, and the oxidation rate is linear with time. For long times and thick oxides, the diffusion of oxygen or H_2O through the oxide limits the oxidation process, and the overall oxidation rate is parabolic in time ($\propto \sqrt{t}$) since diffusion slows down as the oxide thickens. The model derived in Chapter 6 is repeated in (11.16):

$$\frac{x_s^2}{B} + \frac{x_s}{B/A} = t + \tau \quad \text{or} \quad x_s = \frac{A}{2}\left\{\sqrt{1 + \frac{t + \tau}{A^2/4B}} - 1\right\}. \tag{11.16}$$

Here, B and B/A, respectively, are the parabolic and linear rate constants, x_s is the silicide thickness and τ is an initial time used when the silicide thickness is not zero to start, as discussed in Chapter 6. Physically, B models the diffusion process and B/A models the chemical reaction process. When applied to the silicide growth process, obviously different values for B and B/A would be used compared to the oxidation case.

When the parabolic term in (11.16) dominates, these expressions reduce simply to

$$x_s^2 \cong B(t + \tau). \tag{11.17}$$

Physically, this corresponds to the diffusion process (Si in the case of $TiSi_2$) being the rate-limiting step. The B value is primarily determined by the Si diffusivity in $TiSi_2$.

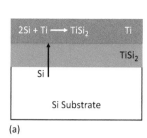
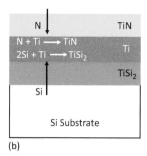

(a)   (b)

Figure 11.31 Illustration of TiSi$_2$ silicide formation. (a) TiSi$_2$ formation only, showing Si diffusion through the silicide forming new silicide at the TiSi$_2$/Ti interface. (b) TiSi$_2$ formation in a nitrogen ambient with simultaneous TiN formation on the top. Nitrogen diffuses through the TiN layer, forming new TiN at the TiN/Ti interface.

When the linear term in (11.16) dominates, the growth model reduces to

$$x_s \cong \frac{B}{A}(t + \tau).$$

(11.18)

Physically, this corresponds to the interface reaction forming new TiSi$_2$ being the rate-limiting step. The value of B/A is largely determined by the rate of this reaction.

Figure 11.31 shows the 1D physical structure for the case of TiSi$_2$. Figure 11.31(a) illustrates silicide formation in an inert ambient (Ar), so that only TiSi$_2$ is formed. This is often called the salicide (self-aligned silicide) process. This looks much like the Deal–Grove oxidation model, except that the reaction is occurring at the top surface of the TiSi$_2$, not the bottom surface as it does in oxidation. In Figure 11.31(b), the reaction takes place in an N$_2$ atmosphere, so two reactions occur simultaneously, forming TiN on the top of the Ti layer and TiSi$_2$ on the bottom. In the case of Ti as the reacting metal, the diffusing species are N and Si, as illustrated.

In the simpler case in Figure 11.31(a), either the diffusion of the Si or the reaction at the interface can dominate the overall formation kinetics, depending on the conditions, leading to linear–parabolic behavior. For oxidation of silicon, there is a net increase in material volume, leading to a compressive stress. For TiSi$_2$ formation, one might expect the opposite to be true. This is because, from Table 11.4, for every 1 nm of Ti consumed, 2.27 nm of Si are consumed, producing 2.51 nm of silicide. Therefore, for every $1 + 2.27 = 3.27$ nm of reactants, only 2.51 nm of product is formed, leading to a net volume reduction of about 23%. Hence, tensile stresses should result in the TiSi$_2$ film. However, experimentally, it is found that either compressive or tensile stresses can develop in silicide films depending on the conditions of formation.

Experimental silicide thickness versus time data of Pico and Lagally [39] have been used by Cea and Law [40] to extract values for B and B/A for TiSi$_2$ formation. Plotting x_s versus $(t + \tau)/x_s$ gives $-A$ as the y intercept and B as the slope. By this method, B, the parabolic rate constant, and B/A, the linear rate constant, can be easily extracted from experimental thickness–time data. Cea and Law's values for B and B/A are given in Table 11.5. For Si (100), B/A is "large," meaning that the Si–Ti reaction is very fast and the formation is limited by the diffusion

Table 11.5 The *B* and *B/A* values for TiSi$_2$ formation [40].

	B ($\mu m^2\ min^{-1}$)	B/A ($\mu m\ min^{-1}$)
Si (100)	$1.85 \times 10^{11} \exp(-2.5\ eV/kT)$	"large"
Si (111) Si	"large"	$8.9 \times 10^{11} \exp(-2.5\ eV/kT)$

Example

TiSi$_2$ is formed by depositing 55 nm of Ti on a Si (100) substrate and annealing in an argon atmosphere. If the anneal is done long enough so that all the Ti is reacted, how thick is the TiSi$_2$ that is formed, and how much Si is consumed? Repeat the problem if the anneal is done at 700 °C for 30 s using Cea and Law's values in Table 11.5 for *B* and *B/A*.

Answer

When the annealing is long enough so that all the Ti is reacted, 2.51 nm of TiSi$_2$ is formed for each 1.00 nm of Ti, consuming 2.27 nm of Si (Table 11.4). Since the anneal is done in an inert, argon ambient, no TiN is formed. Therefore, we find that

$$\text{amount of TiSi}_2 \text{ formed} = 55 \text{ nm} \times 2.51 = 138 \text{ nm},$$

$$\text{amount of Si consumed} = 55 \text{ nm} \times 2.27 = 125 \text{ nm}.$$

For a 700 °C, 30 s anneal, not all of the Ti may have reacted. We must therefore use the Deal–Grove-like equations. For Si (100), Table 11.5 shows that B/A = "large." The "large" B/A value means that the growth is not limited by the Si–Ti reaction. Therefore, using (11.17), and letting τ be zero, the silicide thickness x_s is given by

$$x_s \cong \sqrt{Bt}.$$

At 700 °C, we have

$$B = 1.85 \times 10^{11} \exp\left(-\frac{2.5 \text{ eV}}{(8.63 \times 10^{-5} eV\ K^{-1})(973 \text{ K})}\right) \mu m^2\ min^{-1}$$

$$= 0.022\ \mu m^2\ min^{-1}.$$

Therefore, x_s, the amount of TiSi$_2$ formed, is

$$x_s \cong \sqrt{(0.022\ \mu m^2 min^{-1})(0.5 \text{ min})} = 105 \text{ nm}.$$

The amount of Ti needed to form this = 105 nm/2.51 = 42 nm. The amount of Si consumed = 42 nm × 2.27 = 95.3 nm.

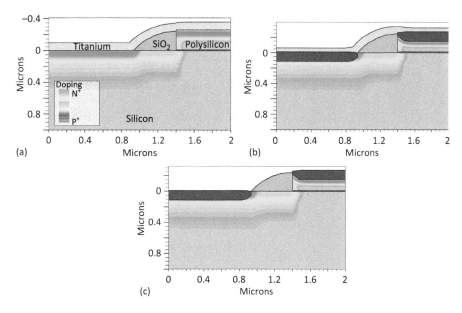

Figure 11.32 Athena simulation of TiSi$_2$ salicide process: (a) 0.1 µm of Ti is conformally deposited on the N$^+$ source and poly gate regions of an NMOS transistor; (b) a 2.5 min, 675 °C anneal forms ≈ 0.1 µm TiSi$_2$; and (c) the excess Ti has been etched off.

of Si, as governed by the parabolic rate constant, B. Therefore, any sufficiently large value of B/A will work (and will have no effect on the outcome). Alternatively, for Si (111), the formation is limited by the Si–Ti reaction, and the parabolic rate constant is "large."

In small structures where 2D and 3D effects are important, it is necessary to include stress effects in the oxidation process due to volume expansion in constrained regions. Stress-dependent oxidation parameters and appropriate viscoelastic flow models were discussed in Chapter 6. Silicide formation can be modeled in a similar fashion by incorporating stress-dependent B and B/A values, as was done in the oxidation case, when these parameters are known for specific silicide processes. These types of models are incorporated in simulation tools like Silvaco's Athena [41]. However, it should be kept in mind that physically correct simulation of the silicidation process is complicated. Silicides are generally polycrystalline materials that have fast diffusion paths along grain boundaries. Diffusion along these grain boundaries can be affected by stresses in the films. Many silicides have multiple phases that can coexist. So the simulation tools that have been developed should be regarded as more qualitatively correct than quantitatively correct in most cases.

Figure 11.32 shows an example of an Athena simulation of the salicide process (a self-aligned silicide process). A 0.1 µm layer of Ti is conformally deposited on a structure that is the left half of an NMOS transistor with a 1.2 µm long gate (e.g., 1980s technology). The CMOS example in Chapter 2 (Figure 2.16) would use a thinner Ti layer and a much thinner TiSi$_2$ layer. The anneal in Figure 11.32 is 675 °C for 2.5 min in an inert Ar atmosphere. This forms ≈ 0.1 µm of TiSi$_2$, after which the excess Ti is etched off. The color contours in the silicon are the doping contours. Note

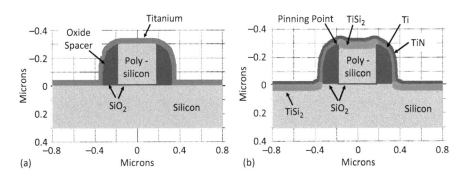

Figure 11.33 Simulation of TiSi$_2$ formation using FLOOPS [44] on a 0.35 μm wide gate structure. (a) Before formation anneal step. (b) After formation anneal step: 30 s at 650 °C in a nitrogen atmosphere. TiSi$_2$ is formed everywhere the Ti is in contact with or close to Si. TiN is formed on top of the Ti layer.

that, during the formation of the silicide, significant amounts of dopant are incorporated into the silicide. To model this process, the simulation needs to account for dopant segregation at the TiSi$_2$/Si interface as well as dopant diffusion during the anneal. The amount of silicide formed in this example is more than would be the case in smaller device structures with shallower junctions.

A second example of silicide simulation is shown in Figure 11.33. This simulation uses the University of Florida developed FLOOPS process simulator that uses models similar to those described above for silicide formation [42]. In addition to the linear–parabolic growth model for silicides, Cea and Law [40] also included some stress-dependent parameters in FLOOPS to account for the volume changes that occur during silicide formation.

As with the oxidation models presented in Chapter 6, the B and B/A parameters in silicide formation may be affected by stress. There is some experimental evidence that Si diffusion in TiSi$_2$ is, in fact, retarded by high compressive stress levels, and this is incorporated in Cea's simulations [43] by including a value of 1.3 nm^3 for V_D, the activation volume for the stress-dependent Si diffusivity in TiSi$_2$. This is utilized in (6.24), repeated here

$$D(\text{stress}) = D \exp\left(-\frac{PV_D}{kT}\right), \tag{11.19}$$

where P is the hydrostatic pressure (positive for compression) in the growing silicide. This will locally affect the value of B since $B = 2DC^*/N_1$ in the Deal–Grove model. In regions of high stress, or hydrostatic pressure, such as at corners, the diffusivity of the silicon through the silicide will be reduced and the silicide growth rate will be reduced.

The formation of TiSi$_2$ in Figure 11.33 is done in a nitrogen ambient. As discussed previously, this causes TiN to form on top of the silicide and on top of the oxide spacer, inhibiting lateral encroachment of the TiSi$_2$, and perhaps serving as a local interconnect. The model for TiN formation is similar to that for TiSi$_2$ formation. In this case, nitrogen diffuses through the forming TiN, reacting with the Ti at the Ti/TiN interface, as illustrated in Figure 11.31. This reaction can also be modeled with the Deal–Grove linear–parabolic model. This requires simulating two reacting and moving interfaces simultaneously.

In Figure 11.33, 40 nm of Ti is deposited on a 0.35 µm polysilicon gate structure, with adjoining source/drain regions. The structure is annealed for 30 s at 650 °C in a nitrogen ambient. The model parameters are the same as in the previous simulation, except that parameters for TiN have been added. Based on limited experimental data, the linear and parabolic rate constants for TiN growth were set to be 20% of the values for $TiSi_2$.

The simulated structure after the rapid thermal silicidation anneal in a nitrogen ambient is shown in Figure 11.33(b). $TiSi_2$ forms as before over the polysilicon gate and Si source/drain regions, but, in addition, TiN forms over the entire structure. Not all of the Ti over the silicon is reacted, with about 7.5 nm remaining. Approximately 32.5 nm of Ti and 44 nm of Si are consumed, producing approximately 50 nm of $TiSi_2$ and 27 nm of TiN over the silicide. In the simulated structure, some lateral encroachment of the silicide over the oxide spacers is evident, as well as downward bowing of the silicide over the gate. Both of these features are characteristic of experimental $TiSi_2$ formation and are more pronounced in small-scale structures. The lateral encroachment is due to the Si diffusion through the silicide and over the oxide to react with the Ti there. The continued formation of TiN over the oxide spacer would block the lateral growth of $TiSi_2$ and prevent it from encroaching further. In addition, nitrogen is believed to stuff the grain boundaries of any unreacted Ti, slowing the Si diffusion and helping retard the lateral growth – however, this is not explicitly modeled here.

The bowing is believed to be due to the mechanical pinning of the bottom of the silicide at each end on the top of each oxide spacer [44, 45]. This pinning point is shown in Figure 11.33. As the $TiSi_2$ is formed, Si leaves the polysilicon at the $TiSi_2$/polysilicon interface, diffusing up to the $TiSi_2$/Ti interface, where it forms more $TiSi_2$. The $TiSi_2$ then moves down to fill the space left from the vacant Si. If pinning of the silicide occurs at the top of the oxide spacers, consumption of Si to form the silicide can only occur in the middle of the gate and not the ends. All this leads to a bowed downward structure.

11.4.3 Copper Electroplating Technology

In this section, we will discuss a deposition technology that is unique to back-end structures, Cu electroplating. This was not discussed in Chapter 10 as a general material deposition method because the use of electroplating is highly specialized and primarily restricted to back-end structures. For the most part, back-end materials are deposited using methods described in Chapter 10. Sputtering is very common for Al and Al/Cu when they are used for the interconnects. Sputtering is also commonly used for metals used in silicide formation and often for barrier and adhesion layers. CVD methods are often used for W when it is used for vias. CVD is also commonly used for SiO_2 and Si_3N_4 back-end dielectrics. Finally, both PECVD and liquid spin-on technologies are used for low-K dielectrics, as we discussed earlier.

As we saw previously in this chapter, the transition from Al-based interconnects to Cu-based interconnects required major changes in the processes used for depositing and defining the metal structures, largely because Cu cannot be easily plasma-etched. Because of this, the dual damascene process illustrated in Figure 11.19 was widely adopted for Cu. While the barrier

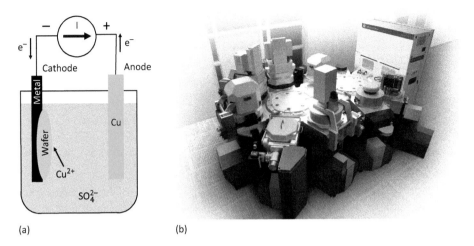

(a)   (b)

Figure 11.34 (a) Cu electroplating basics. (b) Applied Endura® copper barrier/seed PVD system: a high-vacuum integrated metal deposition system, including surface preparation, PVD TaN barrier, CVD Co liners and PVD copper seed films. Electroplating on the seed Cu layer is done in a separate machine. Photo courtesy Applied Materials Inc.

layer (often TiN or a Ta-based material) and the seed Cu layer in this process are usually deposited by sputtering, most of the copper is deposited by electroplating.

Cu electroplating is, in principle, a very simple process, as illustrated in Figure 11.34 [46]. The solution in its most basic form is a mixture of $CuSO_4$, H_2SO_4 (sulfuric acid) and H_2O. The $CuSO_4$ is the plating source, and the sulfuric acid provides conductivity in the solution. The Cu diffusion barrier (often Ta/TaN) and the Cu seed layer deposited everywhere on the patterned dielectric as part of the dual damascene process (Figure 11.19) provide a conductive path across the wafer necessary for the electroplating process. An external power supply, usually operated as a current source, drives the deposition reaction. At the anode and cathode, the basic reactions are

$$\text{anode} \qquad Cu \rightarrow Cu^{2+} + 2e^- , \tag{11.20}$$

$$\text{cathode (wafer)} \quad Cu^{2+} + 2e^- \rightarrow Cu . \tag{11.21}$$

The current delivered to a conductive surface during electroplating is directly proportional to the amount of metal deposited, from Faraday's law of electrolysis,

$$W = ItA_W/nF , \tag{11.22}$$

where W is the weight deposited in grams, t is time in seconds, I is current in amps, F is the Faraday constant (96,500 $C\ mol^{-1}$), n is the number of electrons transferred per atom deposited ($n = 2$ in the case of Cu) and A_W is the atomic weight. Controlling the current and the time in electroplating thus allows a known thickness of Cu to be deposited on the wafer. Manufacturing electroplating machines normally operate at current densities of 3–15 $mA\ cm^{-2}$, which deposits Cu at ≈0.1–0.4 $\mu m\ min^{-1}$.

When no external potential is applied to a wafer sitting in a plating solution, and no current is forced through the circuit, an equilibrium potential exists between the wafer and the solution, the rest potential. When an external power source is applied overcoming this potential, current will begin to flow. The current–voltage relationship that holds under typical plating conditions is known as the Tafel equation:

$$I = I_0 \, e^{-\alpha n F V / RT}, \tag{11.23}$$

where I_0 is the exchange current density, α is the charge transfer coefficient, n is the number of electrons transferred per atom deposited ($n = 2$ in the case of Cu), V is the applied voltage, F is the Faraday constant and R is the gas constant [47]. In the exponential current region (Tafel region), the deposition rate is largely determined by reaction rate kinetics at the cathode (wafer), and this is normally the regime in which Cu plating is done. At higher current densities, mass transport in the solution can become rate-limiting, and this is generally not desirable because it is less reproducible and uniform.

The standard electromotive force (EMF) for an electrochemical reaction, such as (11.20) and (11.21), is expressed as a voltage relative to some reference. Normally, in electrochemistry, that reference is the reaction of hydrogen gas on a platinum electrode to form hydrogen ions and electrons. In Table 11.6, this reaction is given the reference potential of 0.00 V. The standard reduction potentials for a number of other metals are given in Table 11.6. Metals that are thermodynamically unstable, such as Na, have negative reduction potentials, which means that ions form in a reaction with a positive free energy. Stable metals, such as Au, have positive reduction potentials, which means that the reaction creating the ions has a negative free energy. Cu has a positive but fairly small reduction potential, which means that, when a voltage more negative than E_0 is applied to the wafer in Figure 11.34, plating will begin. Applying a potential more positive than E_0 to the cathode favors dissolution of the Cu into ions in the solution. Typical electroplating systems operate with applied voltages a few tenths of a volt away from E_0.

Actual manufacturing processes for Cu plating in chip manufacturing are, of course, much more complex than the simple beaker shown in Figure 11.34. A typical machine used in manufacturing is shown in Figure 11.34(b). Part of the reason for the complex machine is that an integrated approach involving wafer cleaning, barrier metal deposition, Cu seed

Table 11.6 **Electrochemical series of standard E_0 values for reduction of metal ions [47].**

Reaction	Reduction potential (V)
$Au^+ + e^- = Au$	1.88
$Ag^+ + e^- = Ag$	0.799
$Cu^{2+} + 2e^- = Cu$	0.345
$2H^+ + 2e^- = H_2$	0.00
$Fe^{2+} + 2e^- = Fe$	−0.409
$Al^{3+} + 3e^- = Al$	−1.76
$Na^+ + e^- = Na$	−2.71

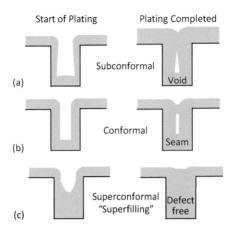

Figure 11.35 Challenges in plating Cu to completely fill via holes in modern structures.

deposition and finally Cu plating is usually used in order to avoid exposing the wafers to ambient conditions between steps. This prevents Cu oxidation and contamination issues.

But even the simple electrochemistry illustrated in the beaker is actually not adequate in modern devices. Figure 11.35 illustrates some of the challenges in designing a plating process. The dual damascene process shown in Figure 11.19 requires that the Cu plate small via holes, filling them completely so there are no voids. Even perfectly conformal plating does not guarantee that this will consistently happen, as the example in Figure 11.35(b) shows. Seams or voids can cause problems with interconnect reliability due to electromigration, as we will discuss in Section 11.4.4. As a result, most Cu plating systems today are designed to operate in the "super-conformal" or "superfilling" mode, as shown in Figure 11.35(c). How is this achieved?

What is desired is a deposition or plating rate that is higher at the bottom of the via or trench than it is on the sides or top. The plating rate at any point on the surface is proportional to the local current density, since this is an electrochemical process. The current associated with the process will take the path of least resistance. Since the current flows through a fluid between anode and cathode, fluid flow dynamics could well play a role in the uniformity of current density on the wafer surface, so the first order of business is to design the plating chamber so that very uniform flow occurs. This generally involves diffusers and baffles and careful chamber design using computational fluid dynamics tools. However, this by itself is not sufficient because, in principle, perfect current density uniformity would lead to conformal deposition in which the deposition rate is the same at each point on the wafer surface, not "superfilling."

The region very near the wafer surface is generally regarded as a boundary layer or stagnant region in which there is no fluid flow [48]. This region is typically tens of microns thick, and transport in this region is governed by diffusion. Modeling of this transport process generally assumes that the concentrations of Cu^{2+} ions and any other "additives" in the solution are uniform at the edge of the stagnant layer. When the current reaches the wafer surface, it encounters a voltage barrier because of the electrochemical reaction that takes place there to plate the Cu. Chemistry tells us that this potential barrier gets larger as the current density increases (the Tafel kinetic expression [46]), so there is no reason that some point A on the surface should receive a higher current density than another point B, unless one of the following applies [48]:

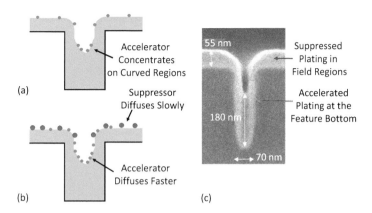

Figure 11.36 (a, b) Physical models explaining the "superfilling" concept in Cu electroplating. (c) Experimental Cu plating profile partway through the deposition, showing "superfilling." Photo reprinted from [49] with permission from Elsevier.

1. The ohmic pathway to point A is significantly more favorable than to point B.
2. The Cu^{2+} has been significantly depleted at point B compared to point A.
3. The rate constant for electrodeposition is larger at point A than at point B.

Practical Cu plating systems are designed so that neither 1 nor 2 is true. This leaves us with 3 as a pathway to achieving our goal. Cu plating solutions include small concentrations of additives, "accelerators," "suppressors" and "levelers." These are surface-active molecules in parts per million (ppm) concentrations, such as polyethylene glycol and bis(3-sulfopropyl) disulfide. There are a number of physical explanations and models for how these additives work. Two models are shown in Figure 11.36.

In Figure 11.36(a), additive molecules that preferentially adsorb in locations of high surface curvature, and which participate in the charge transfer reaction, speed up the deposition rate on the bottom of vias, achieving the "superfilling" concept we desire. In Figure 11.36(b), two types of molecules are shown. The accelerators are smaller, faster-diffusing molecules that reach the via bottoms before the larger "suppressor" molecules, again achieving the "superfilling" goal. "Leveller" molecules are also usually added that reduce the Cu growth rate on protrusions and edges to yield a final surface that is smoother than it otherwise would be. Commercially available plating mixtures typically contain three or more organic additives, and also chloride ions from HCl, which are adsorbed by both the anode and cathode and help with dissolution kinetics. All of these additives are in the few ppm to several hundred ppm concentration range, and the specific molecules are generally proprietary to the manufacturer. A significant literature on mechanisms and models for the additive molecules exists, e.g., [46, 47, 48, 49, 50, 51].

11.4.4 Electromigration

The phenomenon of electromigration is a significant problem in interconnect technology. We consider this topic in this book because electromigration is significantly impacted by material

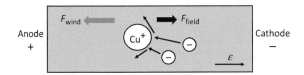

Figure 11.37 Illustration of the forces acting on lattice atoms (Cu in this case) in an interconnect wire carrying current.

properties and by processing choices in back-end structures. As we discussed earlier in connection with Figure 11.16, when high currents pass through relatively soft polycrystalline metals with high atomic diffusivities, such as Al or Cu, the electrons can transfer some of their momentum to the metal atoms, causing them to move, as illustrated in Figure 11.37. These mechanisms become important when current densities reach $\approx 10^6$–10^7 A cm^{-2}, values that are quite possible in chip interconnects. For example, 1 mA (a typical transistor current) flowing through a conductor with a cross-sectional area of 100 nm × 100 nm (10^{-10} cm^2, a typical M1 metal cross-section) is a current density of 10^7 A cm^{-2}.

There are actually two forces acting on the lattice Cu atoms, as illustrated in Figure 11.37. Because of the resistance associated with the interconnect, an $\mathcal{E}$ field exists that would tend to drive the Cu$^+$ atoms to the right. Since the metal ions are shielded to some extent by the sea of free electrons, this force is normally very small and can be neglected. The second, and dominant, force is the "wind" due to colliding electrons that transfers momentum and drives the Cu$^+$ atoms to the left. This force is given by [52]

$$F_{\text{wind}} = Z^* q \mathcal{E} = Z^* q \rho J .$$

(11.24)

Here Z^* is the effective ion valence or charge, a negative number for electron conductors on the order of -4 to -8 for Al, and usually empirically determined for polycrystalline films; $\mathcal{E}$ is the electric field accelerating the electrons; ρ is the electrical resistivity of the metal; and J is the current density (A cm^{-2}).

The atomic flux F induced by F_{wind} (in the direction of the electron flow, or opposite to the current) equals the force times the mobility times the concentration, and is given by

$$F = \frac{DC}{kT} Z^* q \mathcal{E} = \frac{DC}{kT} Z^* q \rho J .$$

(11.25)

Here C is the atomic concentration and we have used the Einstein relation $\mu = D/kT$, where D is the diffusivity of the atoms. The latter is thermally activated and of the form

$$D = D_0 \exp\left(-\frac{E_A}{kT}\right),$$

(11.26)

so that E_A is the activation energy for diffusion. Interconnect materials have multiple possible diffusion paths, including within the lattice, usually via vacancy mechanisms, along grain boundaries and along surfaces, each with its own E_A value. Generally, in Al interconnects, grain boundary diffusion dominates; and in Cu interconnects, surface diffusion often dominates. One would expect that grain structure, including grain size and the location of grain boundaries, would be important in electromigration behavior.

This current-induced diffusion can lead to void and hillock formation, as illustrated earlier in Figure 11.16. Void formation can lead to an open circuit, while a hillock that extends to make contact with another interconnect line can lead to a short circuit. Either of these will lead to circuit failure. The failure rate is often modeled empirically, with the following equation for the mean time to failure (MTTF),

$$\text{MTTF} = AJ^{-n} \exp\left(\frac{E_A}{kT}\right), \tag{11.27}$$

where J is the current density, n is typically close to 2, A is a constant that depends on film structure (grain size, etc.) and processing, and E_A is the activation energy for electromigration and is often associated with the diffusion processes described by (11.24).

For Al interconnects, E_A ranges from 0.5 to 0.8 eV (the activation energy of lattice diffusion in bulk Al films is about 1.4 eV). The lower value for E_A in (11.27) implies that grain boundary diffusion is the likely mechanism. For Cu interconnects, $E_A \approx 0.9$ eV, which is believed to be due to a surface diffusion mechanism. The parameters for this equation are usually determined under accelerated testing conditions – higher than usual current and temperature. Testing of new interconnect structures or materials is also usually performed at elevated temperatures and at very high J values so that failures occur in reasonable test times. Equation (11.27) allows such data to be extrapolated to MTTF values that are more typical of normal operating temperatures and currents.

The original theoretical basis for (11.27), especially with respect to the n dependence of the current density, was provided by Black [53], and the equation is often referred to as Black's equation. He argued that n should equal 2 based on the momentum exchange between the moving electrons and the metal atoms. Other justifications for n being equal to 2, or at least being in the range of 1–3 for Al interconnects, have been based on Joule heating considerations, critical vacancy [54] or stress [55] levels, and by void nucleation and growth models [56]. Often a value of 2 is used in modeling for Al interconnects. For Cu interconnects, n values between 1.1 and 1.3 are often used depending on the dominant failure mode [57].

The E_A value in (11.27) is a critical factor in determining the resistance of metal interconnects to electromigration. Generally, both Al and Cu films are polycrystalline, although the grain size and grain orientation can depend on the deposition and processing conditions. As illustrated in Figure 11.38, there are three possible diffusion processes for the metal ions, each with a different E_A. The E_A value is primarily determined by the binding energy of the metal atom in the metal lattice. Bulk diffusion refers to atoms fully bonded in the crystal and has the highest E_A. Atoms in these positions require vacancies nearby in order to diffuse (recall the discussion of dopant diffusion mechanisms in Chapter 7). Grain boundary diffusion has a lower E_A because atoms in those positions have strained and often asymmetric bonds. Surface diffusion is also characterized by a lower E_A value, which also depends on adjacent materials.

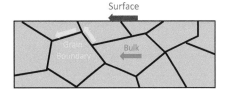

Diffusion process	Activation energy E_A	
	Al	Cu
Bulk diffusion	1.2	2.3
Grain boundary	0.7	1.2
Surface	0.8	0.8

Figure 11.38 Diffusion mechanisms in polycrystalline metal thin films. The E_A values are taken from [57].

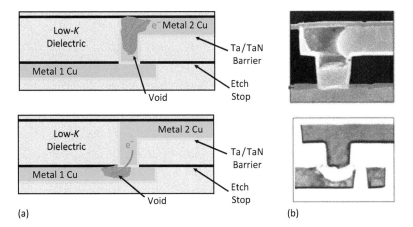

Figure 11.39 (a) Illustration of void formation processes in modern multi-level Cu interconnects. After [57]. (b) Experimental results similar to the drawings on the left. Photos © 2002 IEEE. Reprinted with permission from [58].

In the case of Al interconnects, grain boundary diffusion normally dominates since this process has the smallest E_A. As a result, electromigration robustness can be improved by adding a small percentage of Cu to the Al films, since the Cu preferentially segregates to the grain boundaries and slows down Al diffusion along this path. This mitigation was discussed briefly in Section 11.3.3. In the case of Cu interconnects, surface diffusion normally dominates, since that process has the smallest E_A value. Not surprisingly, there is considerable effort to understand the impact of barrier layers and barrier materials on this surface diffusion process, since, if it could be eliminated, Cu has much higher E_A values for grain boundary diffusion, which would make it much more resistant to electromigration. Changing the barrier materials and/or process deposition steps can have a significant effect on E_A [58].

Geometry and multi-material issues often make the situation in real interconnects more complex than simple 1D models might suggest. Figure 11.39 illustrates a couple of examples.

In these structures, the current flows through barrier layers and also turns corners. Also, the cross-sectional area of the via may be different than that of the metal lines, resulting in a higher or lower current density. Recall that, in the dual damascene process, used for Cu interconnects (Figure 11.22), the Cu is polished using CMP to produce a planarized structure. This means that the top surface of the Cu can have polishing defects. Typically, this surface would be coated

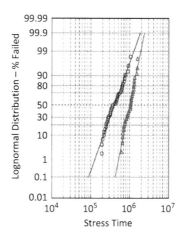

Figure 11.40 Experimental data on electromigration failures using a lognormal plot. The stress time would typically be in seconds, but the units depend on the stress, current density and temperature. The blue data show longer lifetimes and would result from either lower-temperature or lower-current-density stress conditions. After [59].

with an etch stop material to set up the next level of Cu metal. Defects at this interface can reduce the E_A value, making surface diffusion more likely. Experimentally, voids due to electromigration are often found along the top surface of Cu interconnects (see Figure 11.39). There is a very extensive literature on experimental results for electromigration in both Al and Cu interconnects, e.g., [59].

Experiments involving electromigration are always part of developing any new technology. Typically, test structures involving all metal levels and vias are designed and fabricated, and these are then tested at higher-than-normal operating temperatures and currents to accelerate their failure. The times when individual test structures fail are then plotted on a lognormal plot, an example of which is shown in Figure 11.40. Assuming the data are well fitted by an expression like (11.27), values for E_A, n, etc., can be experimentally determined. In Figure 11.40, the mean time to failure is indicated by the horizontal dashed pink line at the 50% point on the curve. General goals include a steep distribution (well-controlled manufacturing process) and a long MTTF.

11.5 Future Evolution of Back-End Structures

In many ways, the future directions of back-end technology, at least for silicon VLSI chips, are clear – reduce R and reduce C. Reducing C means introducing dielectric materials that have lower K (dielectric constant) values. We discussed many of the issues associated with lower-K materials in Section 11.3.4. There continues to be a lot of research into such materials, and undoubtedly there will be progress. The fundamental challenge is that lower K implies lower mechanical strength, which complicates integration issues, particularly when processes like CMP are used. The ultimate limit is $K = 1$ and, as we saw earlier, some current technologies have introduced local air gaps, where $K = 1$.

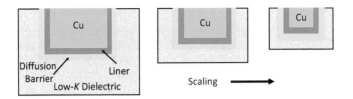

Figure 11.41 Scaling the size of Cu interconnects. The diffusion barrier and liner do not scale.

Reducing R superficially implies finding metals with lower resistivity than that of Cu. But, as we saw earlier, there really are no viable metals that have lower bulk resistivity than Cu. However, the situation is actually more complicated than just considering bulk resistivity. Figure 11.41 illustrates what happens when Cu interconnects are scaled down in size.

Copper is a fast diffuser in dielectrics and in silicon, and it creates deep-level traps if it is allowed to migrate into the silicon where sensitive devices are located. Thus, a diffusion barrier is required, and the thickness of this barrier does not scale as the overall dimensions of the wire decrease. In addition, we saw in our discussion of electromigration that surface diffusion is the dominant mechanism for Cu atom diffusion under high current densities. Thus, a liner, optimized to control this surface diffusion, is used with Cu interconnects. Typically a TaN/Ta structure is used with an overall thickness of 2–4 nm. The liner and barrier materials have much lower conductivities than Cu and so, as the overall size of the wire is reduced, the effective resistivity of the wire increases. As a specific example, the conductivity of a 10 nm metal line that does not require a barrier/liner would be expected to outperform the conductivity of a metal that has 2× lower resistivity but requires a 2 nm barrier/liner.

There is one additional factor that also needs to be considered. As we discussed in connection with Figure 11.37, electromigration is caused by momentum transfer between electrons that are moving in the Cu conductor and the Cu atoms. The mean free path is a measure of the average distance that electrons travel before scattering off Cu lattice atoms. Intuitively, the longer this distance is, the more energy the electrons will gain and the more energy they will transfer to the Cu lattice atoms. Thus a metal with a longer mean free path would be expected to be more susceptible to electromigration issues. Metals that have shorter mean free paths should be more resistant to electromigration. Another way of thinking about this issue is that metals that have higher melting temperatures than Cu (or Al) might be less susceptible to electromigration issues, since a higher melting temperature generally implies stronger atomic bonds and therefore less susceptibility to atomic diffusion.

The above arguments suggest that metals other than Cu might be considered for interconnects even if they have higher bulk resistivities, if they do not require diffusion barriers and liners, and if they have shorter mean free paths for electrons than does Cu. Given these issues, metals like Ru and Co have received a lot of current interest. These metals do not require barrier layers because of their low diffusivities, they have shorter mean free paths for electrons and they have higher melting temperatures than Cu, suggesting that they will have better resistance to electromigration. Specific data for back-end metals currently used or being considered for use in back-end technologies are listed in Table 11.7.

Table 11.7 **Resistivity, electron mean free path and activation energy associated with electromigration for metals used or being considered for back-end interconnects. Data with * are estimated from the melting temperature. After [60].**

Material	Symbol	Bulk resistivity, ρ_0 ($\mu\Omega$ cm)	Electron mean free path, λ (nm)	Activation energy for electromigration, E_A (eV)
Aluminum	Al	2.73	14.9	0.6–0.7
Copper	Cu	1.67	39.3	1.35
Tungsten	W	5.49	14.2	2.47*
Titanium	Ti	55.6	0.83	1.42*
Tantalum	Ta	12.4	3.81	2.22*
Cobalt	Co	6.25	11.8/7.77	2.06*
Ruthenium	Ru	7.59	10.2	2.96*

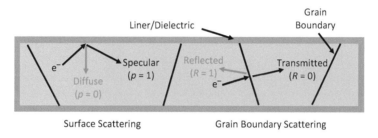

Figure 11.42 Scattering mechanisms for electrons travelling through small metal lines. After [61].

Because the bulk resistivities of these metals are higher than that of Cu, they will make the most difference in circuit performance if they replace Cu in the lowest levels of the metal hierarchy (M1, M2, etc.), where the metal wire dimensions are the smallest and the scaling issues illustrated in Figure 11.41 are most critical. The activation energies associated with electromigration listed in Table 11.7 are much higher for materials listed below Al and Cu, suggesting (see (11.27)) that these materials should be much more resistant to electromigration issues.

In very small interconnect lines, two scattering mechanisms dominate, as illustrated in Figure 11.42, i.e., surface scattering and grain boundary scattering [61]. To first order, the effects of scattering can be incorporated into an effective resistivity for the metal line as

$$\rho_{\text{eff}} = \rho_0 + \rho_0\lambda \frac{3(1-p)}{4d} + \rho_0\lambda \frac{3R}{2D(1-R)}, \qquad (11.28)$$

where ρ_0 is the metal bulk resistivity, λ is the electron mean free path, d is the metal linewidth, D is the grain size, p is the surface scattering specularity and R is the grain boundary reflection coefficient. Both p and R range between 0 and 1, and are phenomenological parameters that depend strongly on material properties. The effective resistivity ρ_{eff} is higher than the metal bulk resistivity because of the contributions of these scattering mechanisms.

Surface scattering is more important as the wire size d decreases. Specular scattering is analogous to light reflecting off a mirror and, if $p = 1$, there is no loss of electron momentum parallel to the surface and hence no resistivity increase due to this scattering mechanism. Diffuse scattering causes a randomization of the electron's momentum and therefore does increase ρ_{eff}. Experimental and theoretical results have shown that achieving specular scattering is highly dependent on an atomically smooth interface between the metal and the liner material. Specular scattering also critically depends on the liner material having few available states into which the electron could scatter, so insulators which have few states at the Fermi level generally produce higher p values. Clearly, material choices for the liner and processing conditions that maintain an abrupt atomically flat interface are key.

The grain size D in small interconnects typically scales with the wire dimension, so smaller wires tend to have smaller grain sizes and hence more scattering due to grain boundaries. Processing methods that increase grain size clearly help. If the grain size can be made roughly 10λ, then grain boundary scattering generally is negligible. If the grain size is small enough to affect scattering, then the reflection coefficient R becomes key. Many different types of grain boundaries are found in metal wires, ranging from twin boundaries with an $R \approx 0.01$, to random boundaries with $R \approx 0.6$–0.7 [61]. Grain boundaries can sometime be "passivated" with dopants to reduce R. We saw an example of this in the electromigration discussion earlier in which Cu was added to Al to passivate its grain boundaries. Processing conditions also play a role, with the goal being to produce grain boundaries that have low energy barriers and hence higher R values.

Both of the scattering terms in (11.28) are proportional to $\rho_0 \lambda$. As wires become small, and surface and grain boundary scattering become dominant, it is reasonable to expect that metals with the lowest $\rho_0 \lambda$ product would be the best candidates for interconnects. Table 11.8 shows theoretical values for $\rho_0 \lambda$ for a variety of metals [61]. There are obviously several metals that potentially offer better performance than Cu when wire geometries are very small.

Many current researchers are investigating these possibilities [62, 63, 64, 65, 66]. An example of recent experimental data is shown in Figure 11.43 for small-geometry interconnects made from some of these new materials [60]. The first commercial introductions of these new metals are now being reported [62, 63, 64, 65, 66]. Even more "exotic" materials, such as metal alloys like AlNi or

Table 11.8 Theoretical values for $\rho_0 \lambda$ for various metals [61].

Material	Symbol	Theoretical $\rho_0 \lambda$ ($\times 10^{-16}$ Ω m^2)
Copper	Cu	6.7
Molybdenum	Mo	6.0
Aluminum	Al	5.0
Tantalum	Ta	4.2
Ruthenium	Ru	5.1/3.8
Iridium	Ir	3.7
Platinum	Pt	3.4
Rhodium	Rh	3.2

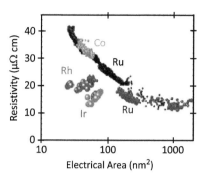

Figure 11.43 Experimental data from imec on small-geometry interconnects made from proposed new single-metal interconnect materials. © 2018 IEEE. Reprinted with permission from [60].

even three-metal alloys, and completely new materials, like carbon nanotubes, are being investigated as possible interconnect materials, although at this point it is not clear that they would offer major advantages over the more traditional single-metal materials discussed here.

Beyond the incremental improvements that may be achieved by introducing new metals like Ru, Rh, Ir and Co, there are not many options for major leaps in back-end RC performance, at least within the current technology options. Of course "incremental progress" has been the history of the semiconductor industry for many decades, and there is no reason to believe that this will end any time soon.

In addition to steady progress in lowering R and C, there are certainly ideas for radically different back-end technologies. Interconnects are basically communication links, carrying information from one part of a chip to another part. Longer-distance communication links today are dominated by RF links or by optical links, not by wires. So a long-standing question is whether it makes sense, and whether it is technically possible, to use such technologies on a chip. Either approach would seem to make possible communicating across a chip at the speed of light. Optical links between parts of a chip would require conversion of electrical signals to photonic signals, waveguides to carry the photons across a chip and finally conversion back to an electronic signal at the receiving end. While technologies exist to do all these things, integration on a silicon chip is a formidable challenge, and to this point such an approach has not been feasible. Similar arguments apply to the possibility of broadcasting information from one part of a chip to another. So, while these are interesting and perhaps long-term possibilities, for now they remain research topics, and would seem to be many years away from practical implementation for on-chip communications.

11.6 Si CMOS Back-End Technology Revisited

We can summarize many of the elements of back-end technology we have been discussing to this point, by revisiting the CMOS process flow introduced in Chapter 2. The detailed steps are shown sequentially in Figure 11.44. These simulations use the modeling capabilities of Silvaco's Athena and Victory Process [41]. Note that the layer thicknesses used in these simulations are meant to be illustrative only. Specific values used in a particular process flow would vary from one manufacturer to another, and also depend on the specific chip design. The following list describes each of the corresponding panels in Figure 11.44.

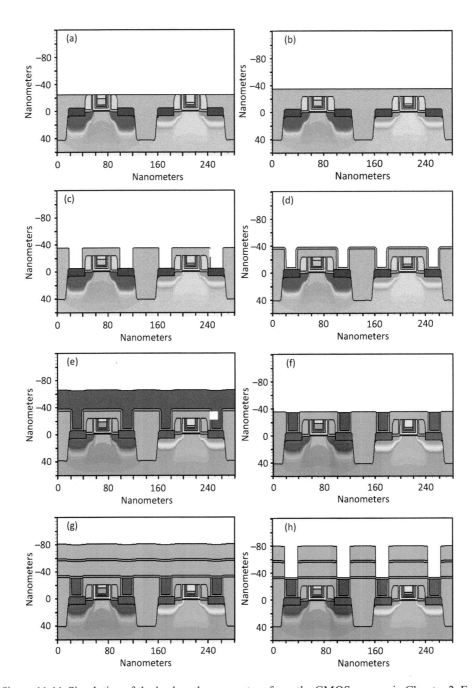

Figure 11.44 Simulation of the back-end process steps from the CMOS process in Chapter 2. Each panel is described in the corresponding item in the list in the text.

(a) This is the structure after the completion of the front-end (transistor) fabrication steps (Figure 2.18 in Chapter 2). The PMOS device is on the left, and the NMOS on the right. TiSi$_2$ layers cover the source/drain regions of the transistors to provide good ohmic

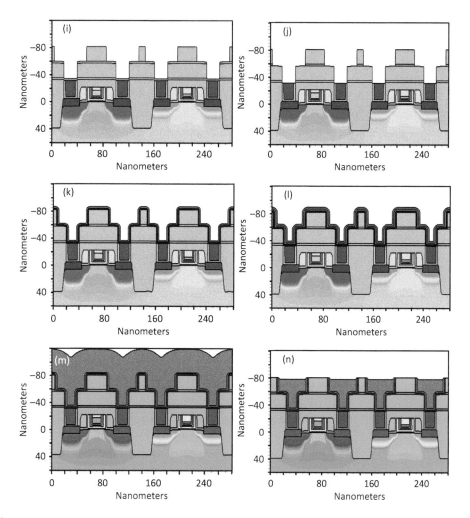

Figure 11.44 (cont.)

contacts to the silicon and to reduce the contact resistance. The high-K metal gate structure is apparent in the gate regions.

(b) A 20 nm SiO_2 layer is uniformly deposited on the wafer surface. This covers the gate structures and electrically isolates them from the metal layers to be deposited. In some process flows, this step might not be necessary if an existing SiO_2 layer is present.

(c) Photolithography is used to define the contact holes allowing the metal lines to contact the device source/drain regions. Contacts to the gate regions would be made outside the plane of these simulations (into or out of the plane of the diagram). Photoresist would be deposited and then exposed in a lithography tool. After developing, plasma etching would be used to etch the SiO_2 to form the contact holes, using the resist as a mask. The resist would then be stripped, resulting in the structure in (c).

(d) A 2–3 nm TiN adhesion layer is conformally deposited by sputtering or another convenient method. This layer helps the W, to be deposited in the next step, adhere well to the SiO_2.

(e) Tungsten is deposited using sputtering or another convenient method. This deposition needs to fill the contact holes, without leaving voids, and overfill the contact holes as shown.

(f) CMP is used to planarize the structure. The planarization removes both the W and the TiN from the upper SiO_2 surface. The hard SiO_2 serves as polishing stop. There is a slight amount of dishing in the W vias, which is shown in more detail in the simulation in Figure 9.52.

(g) The dielectric stack that will be used for the Cu dual damascene process flow is deposited using LPCVD or another convenient method. The stack consists of four sequential layers: a 2–3 nm Si_3N_4 etch stop, a 20 nm SiO_2 layer, a second 2–3 nm Si_3N_4 etch stop and finally a second 20 nm SiO_2 layer. In some process flows, a hard mask would be deposited on top of the upper SiO_2 layer if the process required a hard mask rather than a photoresist mask during the etching to follow. This was illustrated in Figure 11.19.

(h) The first of two sequential lithography steps is performed, defining the vias that will allow the Cu to connect to the W in the contact holes. Photoresist would be deposited and then exposed in a lithography tool. After developing, plasma etching would be used to sequentially etch the top SiO_2, then the top Si_3N_4 etch stop and finally the lower SiO_2 layer to form the via pattern, using the resist as a mask. Note that the bottom Si_3N_4 etch stop is not etched at this point because it will be needed in the second lithography step that follows. The resist would then be stripped, resulting in the structure in (h).

(i) The second lithography step is performed, defining the interconnect pattern for the Cu in metal level 1. Photoresist would be deposited and then exposed in a lithography tool. After developing, plasma etching would be used to etch the top SiO_2, stopping on the upper Si_3N_4 etch stop. The resist would then be stripped, resulting in the structure in (i).

(j) The Si_3N_4 etch stop layers remaining at the bottom of each of the lithography patterns are now removed using unmasked plasma etching. This etch would need to be selective to the etch stop material (Si_3N_4 in this case) so that the exposed SiO_2 layers were not significantly etched. Note that the dual damascene process described in Figure 11.19, which used a hard mask, avoids the problem of doing the second lithography over significant topography, which the process shown here requires.

(k) An adhesion/barrier layer such as TaN/Ta is deposited conformally with an overall thickness of 2–4 nm (see Figure 11.41 for more detail). These layers must cover all surfaces, and so a deposition method producing a conformal layer is critical.

(l) A seed Cu layer is next deposited to enable the Cu electroplating step that comes next. This layer must also be conformal.

(m) Copper electroplating overfills the via and M1 layer pattern. This would be done using the methods described in detail in Section 11.4.3.

(n) CMP is used to planarize the structure using the SiO_2 as a polishing stop, resulting in the final structure shown in (n).

The dual damascene process shown in Figure 11.44 and described above (or the process shown in Figure 11.19) would be repeated as many as 15 times to produce the complex interconnect structures illustrated in Figure 11.20. As Figure 11.20 illustrates, the layer thicknesses generally increase in the upper metal layers, providing lower resistance and lower capacitance for longer-distance interconnects.

11.7 Back-End Technologies for Compound Semiconductors

Compound semiconductors have found many applications, particularly in areas in which their superior electronic properties compared to silicon are critical. Many of them are direct-bandgap materials, which means that photon emission is efficient, and so all light-emitting diodes (LEDs), lasers and other such devices are based on compound semiconductors. In addition, many of these materials have higher electron mobilities than silicon, so, in applications that require very high frequencies, materials like GaAs are commonly used.

In all of these applications, the circuit complexity is far less than in today's silicon chips, so that the back-end metallization tends to be much simpler than in silicon. Typically, only a few levels of metal are required. Nevertheless, there are some features of back-end processes in compound materials that are both different than in silicon and interesting from a process technology perspective. We will briefly discuss a few of these features in this section.

The first compound semiconductor devices to be widely used were GaAs MESFETs, shown in Figure 11.45(a). These are FETs, but they use a Schottky diode gate to control the current flow between source and drain. The electron mobility in GaAs is ≈ 8500 cm^2 V^{-1} s^{-1}, approximately 5–6 times larger than in Si. GaAs MESFETs have been used since the 1970s for high-frequency amplification. The pseudomorphic high-electron-mobility transistor (pHEMT) device in Figure 11.45(b) is a more recent GaAs-based device structure that makes use of the band structure in an epitaxially grown AlGaAs/GaAs or AlGaAs/InGaAs structure to produce a two-dimensional electron gas (2DEG) physically separated from the doping atoms that provide the electrons. This results in even higher mobilities because scattering off dopant atoms is eliminated. Today, these devices are used in high-frequency communication systems (cell/mobile phones, base stations, etc.) and in many other high-frequency applications. Figure 11.46 shows an example of such a chip. The component count is these chips is obviously orders of magnitude smaller than in today's multi-billion-transistor silicon chips.

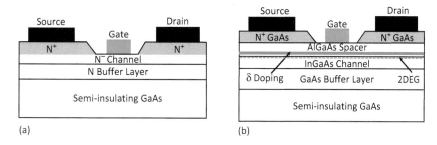

Figure 11.45 (a) GaAs MESFET device structure. (b) GaAs pHEMT device structure.

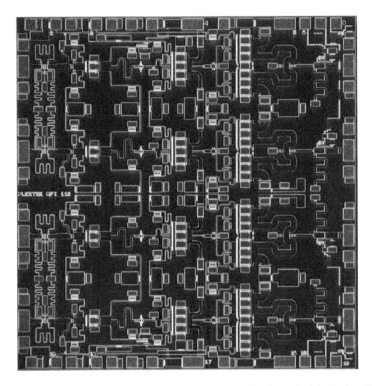

Figure 11.46 Photomicrograph of a 28 GHz power amplifier fabricated with GaAs pHEMT devices. Reprinted from [67] with permission from *Compound Semiconductor*.

There are many different options for fabricating structures like those shown in Figure 11.45. The starting point is a semi-insulating GaAs substrate, which uses Cr doping (a deep-level impurity) to pin the Fermi level towards the middle of the bandgap and hence makes the material semi-insulating. This prevents the absorption of high-frequency energy by free carriers in the substrate. The entire structure can be grown using epitaxy. Alternatively, the N^+ source/ drain regions can be ion-implanted using Si as a dopant. The gate region is typically etched as shown in order to control the threshold voltage of the device.

The metallization needs to provide low-resistance ohmic contacts to the source and drain and a rectifying Schottky contact to the gate. There are many choices for metals. A common choice for the ohmic contacts is a Ni/AuGe sandwich that is sintered at $\approx 450\,°C$ to reduce the contact resistance. During the alloying step, Ge (an N-type dopant) outdiffuses from the metal, producing an N^{++} surface layer and a low-resistance tunneling contact, as discussed in Figure 11.27. Rectifying gate contacts are typically made with a Ti/Pd/Au stack or a Ti/Pt/Au stack, although many other choices exist, such as the refractory metal WN gates shown in Figure 11.28.

Many of these metal stacks are complicated to etch, especially using plasma etching methods, because the chemistry to achieve volatile byproducts for each of the metal constituents is very complicated. We saw earlier in this chapter that etching complications with Cu (no volatile

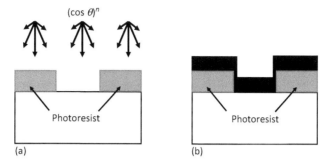

Figure 11.47 Lift-off principle shown for a photoresist profile that does not achieve clean lift-off.

byproducts) in silicon technologies led to the use of damascene and dual damascene process flows so that the Cu metal did not have to be etched. In GaAs technologies, a common solution is to use lift-off, a process we have not yet described, which is discussed next.

The lift-off principle is illustrated in Figure 11.47. Photoresist is deposited and then defined using standard lithography. In this example, the lithography produces vertical edges, something we generally try to achieve in high-resolution printing, as discussed in Chapter 5. Next, a metal or a metal stack is deposited. As we saw in Chapter 10, deposition systems generally have an angular distribution associated with the incoming metal atoms. We modeled this in Chapter 10 with a $(\cos \theta)^n$ angular distribution. The net result is that any practical deposition system will deposit some metal on the sides of the photoresist steps. Lift-off then involves dissolving the photoresist, which should lift off the metal on top of the resist, leaving only the metal in the middle. Clearly, this is not going to work in this example because the metal is continuous. So we need a change in geometry to make this work. Two solutions are illustrated in Figure 11.48.

The process in Figure 11.48(a) uses a negative resist. These resists are removed during developing in regions that are not exposed. Since the exposure process, as we saw in Chapter 5, starts from the top of the resist and progresses to the bottom, if the exposure time is controlled, the bottom part of the resist will develop faster than the top part because it is less exposed. This can result in the re-entrant profile shown, which then allows the lift-off process to work reliably. In Figure 11.48(b), an underlayer material, often a different resist not sensitive to the exposing wavelength, is used. After developing the top photoresist, the lower resist is over-developed, which results in undercutting. This also allows the lift-off process to work reliably.

The lift-off process is thus really an application of some of the lithography, deposition and etching principles we discussed in earlier chapters of this book. It has been widely used in compound semiconductor chip fabrication for decades.

11.8 Multi-Chip Modules, Chiplets, Interposers and Three-Dimensional Integrated Circuits

In recent years, as scaling of lateral dimensions on silicon chips has become increasingly difficult, other approaches to realizing complex systems on a chip have been aggressively

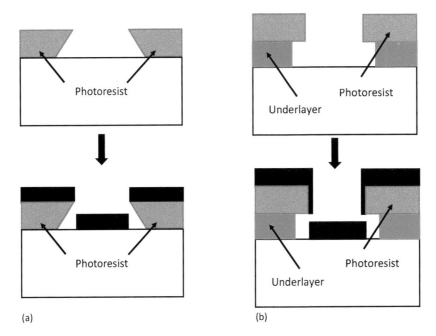

Figure 11.48 Two solutions to the lift-off problem: (a) use of a negative resist, which is exposed more at the top than the bottom; (b) use of an underlayer that is undercut after photoresist patterning.

pursued. Most of these approaches involve back-end technologies and interconnects. We will briefly discuss some of these methods in this section, because it is clear that they are going to be very important in future systems, and they will have an impact on how chips and systems are architected and built.

Printed circuit (PC) boards have been used for decades to assemble chips and other electronic components into complex systems. Figure 11.49 shows a simple example. PC boards consist of alternating layers of Cu and a non-conductive substrate material, often phenolic paper impregnated with phenol formaldehyde resin or woven fiberglass impregnated with an epoxy resin. The Cu layers are patterned typically using lithography and drilled; plated via holes between the Cu layers provide interconnects between the Cu layers.

The Cu layers and patterns can be produced either by subtractive (etching) methods or by additive processes like electroplating. When etching methods are used, wet chemical etching using ammonium persulfate or ferric chloride is used. The isotropic etching typical of these processes is not an issue because of the large dimensions used on PC boards. The technology is similar in concept to the multilayer metal back-end technologies discussed in this chapter, although much simpler to implement because the dimensions are much larger. Chips historically were typically soldered onto the board using through holes in the boards. Today, surface-mounted components are more common using small metal tabs on the components or solder bumps on the surface of the components which are flip-chip mounted on the board. These methods provide much higher component densities on the PC board.

Figure 11.49 Printed circuit board containing silicon chips and other electronic components.

Continued scaling of silicon VLSI technology has made possible very complex "system-on-chip" (SOC) ICs. Today, chips that contain more than ten billion (10B) transistors on a single chip are being manufactured, and chips with >100B transistors will clearly be possible in the near future. One recent example is the Apple M2 chip shown in Figure 11.50(a), which has 20B transistors and is built in 5 nm technology at TSMC. Chips with this level of complexity are very expensive to design. These costs can only be recovered when the market for the chip is very large, as is the case for the computer market for which the M2 was designed.

An increasing number of system designers prefer an approach such as that shown in Figure 11.50(b), which is sometimes called 2.5D integration, since it begins to use the third dimension to build complex systems. In this approach, multiple more specialized chips ("chiplets") are packaged together to realize a complex system. One can think of this approach as taking the PC board in Figure 11.49 and scaling its feature sizes (wire dimensions) to much smaller sizes. The "PC board" in this case is often called an "interposer" and it is often made from silicon, using silicon chip manufacturing technologies. The advantages of this approach include the following.

1. The most advanced (and most expensive) chip technology only needs to be used for chiplets for which it is needed. Much cheaper (older) technology nodes can be used for chiplets that do not require the fastest and densest transistors.
2. Components from a variety of suppliers can be combined together to implement a system. For example, a microprocessor from one vendor can be combined with memory chips from other manufacturers. Thus "best-in-class" components can be used, as opposed to all subsystem components coming from the same supplier in the SOC approach.
3. In general, "chiplets" will be much smaller chips than a complete SOC. This means manufacturing yields will be higher and therefore costs much lower than a large SOC.
4. Some of the chiplets may be non-silicon chips, making possible interesting new systems. For example, photonic chips made from compound semiconductor materials or power

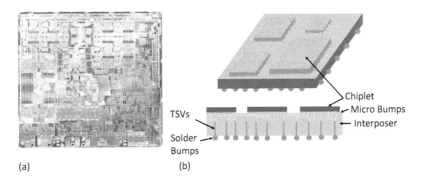

Figure 11.50 (a) System-on-chip (SOC), an Apple M2. Copyright © Apple Corporation, reprinted with permission. (b) Complex system realized by packaging multiple chips, each more specialized than the SOC, in a package. This figure was briefly discussed in Chapter 1 (Figure 1.25).

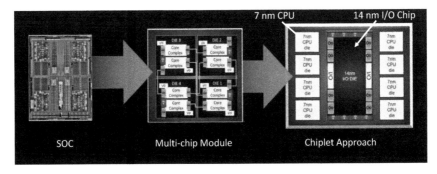

Figure 11.51 Partitioning of a multi-core microprocessor SOC into chiplets (AMD EPYC^{TM} processor architecture).

management chips made from wide-bandgap materials like GaN or SiC can be easily combined with silicon CMOS chips.

5. More rapid innovation, since individual chiplets can be redesigned to upgrade system performance without building a completely new SOC.

In order to implement high-performance systems with this approach, the packaging methods must be high performance. Generally, this means focusing on the same issues that we described in this chapter for high-performance back-end technologies – low R and low C. There is a lot of innovation currently occurring in this area. System partitioning also becomes a very important design issue. How should one "deconstruct" a complex SOC into smaller specialized parts that could be used to realize a system with performance equivalent to or better than the SOC? Figure 11.51 shows one example of this, from AMD, that shows a multi-core processor SOC being deconstructed initially into four identical cores and then into simpler cores with the common functions like input/output (I/O) implemented on an older-generation technology chip.

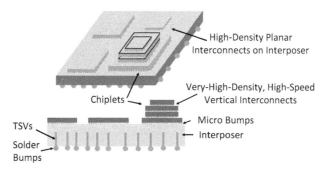

Figure 11.52 Complex system assembly using chiplets, stacked chips, multiple technologies and both lateral and vertical inter-chip connections.

Figure 11.52 expands on the concepts in Figure 11.50 to include stacked chiplets for complex system packaging. Lateral interconnects between chips are implemented with the interposer. Vertical interconnects between stacked chips are implemented by wafer-to-wafer bonding. All of the technologies illustrated in Figure 11.52 are active areas of current research and development in both industrial and academic laboratories around the world.

The final very active R&D area that we will mention is 3D monolithic integration. As discussed in the introduction to this chapter, silicon ICs have historically contained a single layer of active devices along with multiple layers of interconnects above those devices. As lateral scaling of feature sizes becomes more difficult, stacking multiple devices vertically would provide a way to continue to increase device density beyond simple lateral scaling. Thus, "3D" integration is a very active area of research today. The stacked NMOS and PMOS transistors briefly described in Chapter 2 (Figures 2.32 and 2.33) are one example of this. The stacked chips interconnected by wafer-to-wafer bonding in Figure 11.52 are another example, although this example is not 3D integration on a single wafer. Single-wafer, monolithic, 3D integration is being widely explored, and we briefly discuss one commercially important example below.

Memory arrays, especially NAND chips, have a circuit arrangement that stacks devices in series, as shown in Figure 11.53. Technologies have been developed to physically stack these memory cells vertically in a monolithic structure, greatly increasing the effective transistor density.

NAND memory arrays were built using side-by-side 2D memory transistors until about 2013, when the limits of further lateral scaling of these devices were reached. In 3D NAND technologies, dozens or hundreds of layers are vertically deposited on top of each other to form the stacked memory cells. Highly anisotropic etching is then used to etch vertical columns through the deposited layers. The etched trenches are then refilled, as shown in Figure 11.53. The control gates are contacted on the sides of the memory array (not shown, but in the dimension into the plane of the figure). These technologies provide very dense memory arrays but require very sophisticated etching and deposition technologies. The techniques discussed in Chapters 9 and 10 for etching and depositing films with very anisotropic structures are very important in these technologies. As traditional (lateral) scaling approaches increasingly approach fundamental limits, 3D approaches to density increases will become much more common.

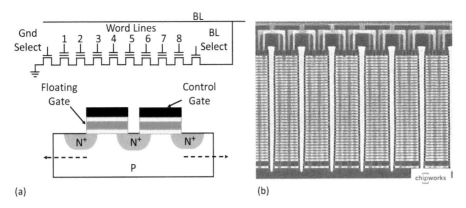

(a)   (b)

Figure 11.53 (a) Schematic of a 3D NAND structure showing the circuit arrangement and vertical stacking of devices. (b) A Samsung 32-layer NAND chip cross-section. Photo from Chipworks (now TechInsights).

11.9 Summary of Key Ideas

In this chapter, we have discussed back-end technology – the structures and fabrication techniques to electrically connect devices to each other and to the outside world. As structures and devices get smaller and chips larger, the importance of interconnects and back-end structures increases – even dominates – over the active devices in terms of circuit performance. In silicon technologies, the back-end structures include a variety of components: contacts to active areas, local interconnects, global interconnects, vias between interconnect levels, first-level dielectrics and inter-metal dielectrics to physically and electrically separate the different interconnects, and finally a passivation layer to protect the entire circuit. We saw how Si back-end structures have evolved considerably from the simple $Si/SiO_2/Al$ structures in the first integrated circuits. Today's technology is characterized by multi-level Cu-based interconnects, each level itself made up of multilayer metal structures with multilayer, usually low-K, dielectrics separating them. Adhesion, contact, barrier and shunt layers are used in the interconnects, while barrier and planarization layers are utilized in the dielectric structures. There have been many driving forces at work in this development, some of the most important of which include the need for additional interconnect levels, the need for planarized structures, lower processing temperatures and the minimization of reliability problems. In addition, there has always been an effort to make the process as simple and cost-effective as possible.

We described the typical processing steps that are commonly used in back-end technology. These include: the formation of local interconnects, such as silicide structures; the fabrication of the first-level dielectrics; contact formation often using tungsten plugs and CMP; global interconnects with stacked multilayer structures; inter-metal dielectric deposition and planarization, using different combinations of PECVD dielectrics and spin-on dielectrics; and final passivation with PECVD oxide/nitride layers. We also briefly considered back-end technologies in compound semiconductor chips. In many of these technologies, lift-off is used to form the metal patterns, primarily to avoid the need to etch complex metal structures.

Back-end structures and technology will continue to evolve as devices get smaller, processing temperatures decrease, and even higher performance is required. Copper and low-K dielectrics were first introduced 20 years ago, and today new metals like Co and perhaps Ru are beginning to be used. While it used to be the case that back-end technology was simply wires to connect active devices together, today the complexity of back-end technology rivals front-end processes, and the R&D resources being focused on back-end technology also rivals front-end investments. Pathways for continued incremental progress exist for many years into the future.

11.10 REFERENCES AND NOTES

[1] "International Technology Roadmap for Semiconductors," (ITRS), Interconnect chapter, 2013. See http://www.itrs2.net/itrs-reports.html.

[2] For example, B. Nikolic, "The Wire," UC Berkeley Lecture Notes. See http://bwrcs.eecs.berkeley.edu/Classes/icdesign/ee141_f01/Notes/chapter4.pdf.

[3] K. C. Saraswat and F. Mohammadi, "Effect of scaling of interconnections on the time delay of VLSI circuits," *IEEE Transactions on Electron Devices*, vol. 29, no. 4, pp. 645–650, 1982.

[4] G. Chen and E. G. Friedman, "Low-power repeaters driving *RC* and *RLC* interconnects with delay and bandwidth constraints," *IEEE Transactions on Very Large Scale Integration (VLSI) Systems*, vol. 14, no. 2, pp. 161–172, 2006.

[5] M. A. El-Moursy and E. G. Friedman, "Optimum wire sizing of *RLC* interconnect with repeaters," *Integration*, vol. 38, no. 2, pp. 205–225, 2004.

[6] S. Rajendar, P. Chandrasekhar, M. A. Rani and R. Naresh, "Performance analysis of alternate repeaters for on-chip interconnections in nanometer technologies," *Procedia Materials Science*, vol. 10, pp. 344–352, 2015.

[7] Y.-L. Cheng, C.-Y. Lee and Y.-L. Huang, "Copper metal for semiconductor interconnects," in *Noble and Precious Metals – Properties, Nanoscale Effects and Applications*, eds. M. S. Seehra and A. D. Bristow, IntechOpen, chap. 10, 2018.

[8] L. Taylor and E. Howard, *Metals Handbook*, Vol. 1, *Properties and Selection of Metals*, American Society for Metals, 1961.

[9] M.-A. Nicolet, "Diffusion barriers in thin films," *Thin Solid Films*, vol. 52, no. 3, pp. 415–443, 1978.

[10] K. Beyer, "The inception of chemical–mechanical polishing for device applications at IBM," *IBM Micronews*, vol. 5, p. 40, 1999.

[11] K. C. Cadien and L. Nolan, "Chemical mechanical polishing method and practice," in *Handbook of Thin Film Deposition*, 4th edn., eds. K. Seshan and D. Schepis, Elsevier, chap. 10, pp. 317–357, 2018.

[12] A. Brand, A. Haranahalli, N. Hsieh, *et al.*, "Intel's 0.25 micron, 2.0 volts logic process technology," *Intel Technology Journal*, vol. 3, no. Q3, pp. 1–8, 1998.

[13] S. Bauguess, L. Liu, M. Dreyer, M. Griswold and E. Hurley, "The effects of accelerated stress conditions on electromigration failure kinetics and void morphology," *MRS Online Proceedings Library*, vol. 428, pp. 93–99, 1996.

[14] C. Volkert, "Electromigration in interconnects," in *Encyclopedia of Materials: Science and Technology*, Elsevier Science, pp. 2550–2557, 2001.

[15] K. Fischer, H. K. Chang, D. Ingerly, *et al.*, "Performance enhancement for 14nm high volume manufacturing microprocessor and system on a chip processes," in *2016 IEEE International Interconnect Technology Conference/Advanced Metallization Conference (IITC/AMC)*, IEEE, pp. 5–7, 2016.

[16] K. Maex, M. Baklanov, D. Shamiryan, F. Lacopi, S. Brongersma and Z. S. Yanovitskaya, "Low dielectric constant materials for microelectronics," *Journal of Applied Physics*, vol. 93, no. 11, pp. 8793–8841, 2003.

[17] Y.-L. Cheng and C.-Y. Lee, "Porous low-dielectric-constant material for semiconductor microelectronics," in *Nanofluid Flow in Porous Media*, eds. M. S. Kandelousi, S. Ameen, M. S. Akhtar and H.-S. Shin, IntechOpen, chap. 11, 2018.

[18] D. Shamiryan, T. Abell, F. Iacopi and K. Maex, "Low-*k* dielectric materials," *Materials Today*, vol. 7, no. 1, pp. 34–39, 2004.

[19] Y.-L. Cheng, C.-Y. Lee and C.-W. Haung, "Plasma damage on low-*k* dielectric materials," in *Plasma Science and Technology – Basic Fundamentals and Modern Applications*, eds. H. Jelassi and D. Benredjem, IntechOpen, chap. 15, 2018.

[20] A. Grill and V. Patel, "Ultralow-*k* dielectrics prepared by plasma-enhanced chemical vapor deposition," *Applied Physics Letters*, vol. 79, no. 6, pp. 803–805, 2001.

[21] A. Grill, S. M. Gates, T. E. Ryan, S. V. Nguyen and D. Priyadarshini, "Progress in the development and understanding of advanced low *k* and ultralow *k* dielectrics for very large-scale integrated interconnects – state of the art," *Applied Physics Reviews*, vol. 1, no. 1, p. 011306, 2014.

[22] M. R. Baklanov, "Challenges in implementation of low-k dielectrics in advanced ULSI interconnects," Presentation at *AVS 62th Symposium*, October 2015. See https://nanoandgiga.com/ngc2017/documents/presentation_baklanov.pdf.

[23] K. Fischer, M. Agostinelli, C. Allen, *et al.*, "Low-*k* interconnect stack with multi-layer air gap and tri-metal-insulator-metal capacitors for 14nm high volume manufacturing," in *2015 IEEE International Interconnect Technology Conference and 2015 IEEE Materials for Advanced Metallization Conference (IITC/MAM)*, IEEE, pp. 5–8, 2015.

[24] C. Hu, *Modern Semiconductor Devices for Integrated Circuits*, Prentice Hall, 2010.

[25] M. C. Ozturk, Private communication. (Originally in "Advanced Contact Formation," Review of SRC Center for Front End Processes, 1999.)

[26] A. Agrawal, J. Lin, M. Barth, *et al.*, "Fermi level depinning and contact resistivity reduction using a reduced titania interlayer in n-silicon metal–insulator–semiconductor ohmic contacts," *Applied Physics Letters*, vol. 104, no. 11, p. 112101, 2014.

[27] S.-M. Koh, E. Y.-J. Kong, B. Liu, C.-M. Ng, G. S. Samudra and Y.-C. Yeo, "Contact-resistance reduction for strained n-FinFETs with silicon–carbon source/drain and platinum-based silicide contacts featuring tellurium implantation and segregation," *IEEE Transactions on Electron Devices*, vol. 58, no. 11, pp. 3852–3862, 2011.

[28] A. Baca, F. Ren, J. Zolper, R. Briggs and S. Pearton, "A survey of ohmic contacts to III–V compound semiconductors," *Thin Solid Films*, vol. 308, pp. 599–606, 1997.

[29] L. J. Brillson, "Contacts for compound semiconductors: Schottky barrier type," in *Reference Module in Materials Science and Materials Engineering*, Elsevier, 2016. [Originally in *Encyclopedia of Materials: Science and Technology*, 2nd edn., Elsevier, pp. 1587–1595, 2001.]

[30] E. Y. Chang and Y.-K. Lin, "Ohmic contacts with low contact resistance for GaN HEMTs," in *2019 19th International Workshop on Junction Technology (IWJT)*, IEEE, pp. 1–2, 2019.

[31] Y. Lu, X. Ma, L. Yang, *et al.*, "High RF performance AlGaN/GaN HEMT fabricated by recess-arrayed ohmic contact technology," *IEEE Electron Device Letters*, vol. 39, no. 6, pp. 811–814, 2018.

[32] M. Murakami and T. Oku, "Development of ohmic contacts for compound semiconductors," in *Proceedings of 4th International Conference on Solid-State and IC Technology*, IEEE, pp. 374–378, 1995.

[33] B. Welch, D. Nelson, Y. Shen and R. Venkataraman, "Metallization technology for GaAs integrated circuits," in *VLSI Metallization*, eds. N. G. Einspruch, S. S. Cohen and G. S. Gildenblat (VLSI Electronics Microstructure Science, vol. 15), Elsevier, pp. 393–450, 1987.

[34] S. Bouzid-Driad, H. Maher, M. Renvoise, *et al.*, "Optimization of AlGaN/GaN HEMT Schottky contact for microwave applications," in *2012 7th European Microwave Integrated Circuit Conference*, IEEE, pp. 119–122, 2012.

[35] D. K. Schroder, *Semiconductor Material and Device Characterization*, Wiley, 2015.

[36] S. P. Murarka, *Silicides for VLSI Applications*, Academic Press, 2012.

[37] R. W. Mann, L. A. Clevenger, P. D. Agnello and F. R. White, "Silicides and local interconnections for high-performance VLSI applications," *IBM Journal of Research and Development*, vol. 39, no. 4, pp. 403–417, 1995.

[38] L. Chen, "Metal silicides: an integral part of microelectronics," *JOM (The Journal of the Minerals, Metals & Materials Society)*, vol. 57, no. 9, pp. 24–30, 2005.

[39] C. A. Pico and M. G. Lagally, "Kinetics of titanium silicide formation on single-crystal Si: experiment and modeling," *Journal of Applied Physics*, vol. 64, no. 10, pp. 4957–4967, 1988.

[40] S. Cea and M. Law, "Two dimensional simulation of silicide growth and flow," in *Proceedings of International Workshop on Numerical Modeling of Processes and Devices for Integrated Circuits (NUPAD V)*, IEEE, pp. 113–116, 1994.

[41] Athena and Victory Process are Silvaco process simulation tools. See https://silvaco.com/tcad/.

[42] FLOOPS is a process simulator developed by Professor Mark Law at the University of Florida.

[43] S. M. Cea, "Multidimensional viscoelastic modeling of silicon oxidation and titanium silicidation," Ph.D. Thesis, University of Florida, 1996.

[44] H. Norström, K. Maex and A. Romano-Rodriguez, "Formation of $CoSi_2$ and $TiSi_2$ on narrow poly-Si lines," *Microelectronic Engineering*, vol. 14, no. 3–4, pp. 327–339, 1991.

[45] H. Norstrom, K. Maex and P. Vandenabeele, "Thermal-stability and interface bowing of submicron $TiSi_2$/polycrystalline silicon," *Thin Solid Films*, vol. 198, no. 1–2, pp. 53–66, 1991.

[46] J. K. Jhothiraman and R. Balachandran, "Electroplating: applications in the semiconductor industry," *Advances in Chemical Engineering and Science*, vol. 9, no. 2, pp. 239–261, 2019.

[47] J. Reid, "Damascene copper electroplating," in *Handbook of Semiconductor Manufacturing Technology*, CRC Press, pp. 16-1–16-47, 2017.

[48] P. C. Andricacos, C. Uzoh, J. O. Dukovic, J. Horkans and H. Deligianni, "Damascene copper electroplating for chip interconnections," *IBM Journal of Research and Development*, vol. 42, no. 5, pp. 567–574, 1998.

[49] R. Akolkar, "Current status and advances in damascene electrodeposition," in *Encyclopedia of Interfacial Chemistry: Surface Science and Electrochemistry*, ed. K. Wandelt, Elsevier, pp. 24–31, 2018.

[50] C. L. Beaudry and J. O. Dukovic, "Faraday in the fab: a look at copper plating equipment for on-chip wiring," *Electrochemical Society Interface*, vol. 13, no. 4, p. 40, 2004.

[51] J. Newman and K. E. Thomas-Alyea, *Electrochemical Systems*, Wiley, 2012.

[52] E. Arzt and W. D. Nix, "A model for the effect of line width and mechanical strength on electromigration failure of interconnects with 'near-bamboo' grain structures," *Journal of Materials Research*, vol. 6, no. 4, pp. 731–736, 1991.

[53] J. R. Black, "Electromigration – a brief survey and some recent results," *IEEE Transactions on Electron Devices*, vol. 16, no. 4, pp. 338–347, 1969.

[54] M. Shatzkes and J. Lloyd, "A model for conductor failure considering diffusion concurrently with electromigration resulting in a current exponent of 2," *Journal of Applied Physics*, vol. 59, no. 11, pp. 3890–3893, 1986.

[55] M. Korhonen, P. Børgesen, K.-N. Tu, and C. Y. Li, "Stress evolution due to electromigration in confined metal lines," *Journal of Applied Physics*, vol. 73, no. 8, pp. 3790–3799, 1993.

[56] J. Lloyd, "Electromigration failure," *Journal of Applied Physics*, vol. 69, no. 11, pp. 7601–7604, 1991.

[57] J. Lienig and M. Thiele, *Fundamentals of Electromigration-Aware Integrated Circuit Design*, Springer, 2018.

[58] A. H. Fischer and A. von Glasow, "Electromigration and stressmigration failure mechanism studies in copper interconnects," in *SEMICON 2002*, 2002. See https://citeseerx.ist.psu.edu/document?repid=rep1&type=pdf&doi=404eaa0597f8fe2baae457bd2e03d1764f8bb7e8.

[59] M. Lin, Y. Lin, K. Chang, K. Su and T. Wang, "Copper interconnect electromigration behaviors in various structures and lifetime improvement by cap/dielectric interface treatment," *Microelectronics Reliability*, vol. 45, no. 7–8, pp. 1061–1078, 2005.

[60] K. Croes, C. Adelmann, C. J. Wilson, *et al.*, "Interconnect metals beyond copper: reliability challenges and opportunities," in *2018 IEEE International Electron Devices Meeting (IEDM)*, IEEE, pp. 5.3.1–5.3.4, 2018.

[61] D. Gall, "The search for the most conductive metal for narrow interconnect lines," *Journal of Applied Physics*, vol. 127, p. 050901, 2020.

[62] N. Bekiaris, Z. Wu, H. Ren, *et al.*, "Cobalt fill for advanced interconnects," in *2017 IEEE International Interconnect Technology Conference (IITC)*, IEEE, pp. 1–3, 2017.

[63] S. Dutta, S. Kundu, A. Gupta, *et al.*, "Highly scaled ruthenium interconnects," *IEEE Electron Device Letters*, vol. 38, no. 7, pp. 949–951, 2017.

[64] S. Dutta, K. Moors, M. Vandemaele and C. Adelmann, "Finite size effects in highly scaled ruthenium interconnects," *IEEE Electron Device Letters*, vol. 39, no. 2, pp. 268–271, 2018.

[65] X. Zhang, H. Huang, R. Patlolla, *et al.*, "Ruthenium interconnect resistivity and reliability at 48 nm pitch," in *2016 IEEE International Interconnect Technology Conference/Advanced Metallization Conference (IITC/AMC)*, IEEE, pp. 31–33, 2016.

[66] C. Auth, A. Aliyarukunju, M. Asoro, *et al.*, "A 10nm high performance and low-power CMOS technology featuring 3rd generation FinFET transistors, self-aligned quad patterning, contact over active gate and cobalt local interconnects," in *2017 IEEE International Electron Devices Meeting (IEDM)*, IEEE, pp. 29.1.1–29.1.4, 2017.

[67] L. Devlin, A. Dearn, S. Glynn, G. Pearson, R. Smith and M. Tahir, "Designing MMICs for millimetre-wave 5G," *Compound Semiconductor*, May 18, 2020. See https://compoundsemiconductor.net/article/111294/Designing_MMICs_for_millimetre-wave_5G/feature.

11.11 PROBLEMS

11.1 Calculate the percentage increase in the interconnect *RC* delay according to (11.5) if the thicknesses H and x_D, remain constant while the lateral dimensions W and L_S scale with (and equal) F_{min}. Assume that F_{min} is decreased from 0.5 to 0.35 μm, and H and x_D equal 0.5 μm. Also assume that the interconnect length L remains constant.

11.2 In a simple back-end structure such as in Figure 11.3, normally the wiring would be covered with a layer of SiO_2 as part of the fabrication process. A designer is interested to analyze how much faster her circuit might work if the upper layer of SiO_2 were not used and the chip were left with no upper passivation layer (just air). Assume in this technology that $H = 1$ μm, $W = 1$ μm, $L_S = 1$ μm and $x_{ox} = 1$ μm. What is the percentage improvement in interconnect delay with air instead of SiO_2 between the metal lines? You may use the simple *RC* model for delay represented by (11.5).

11.3 As illustrated in Figure 11.24, in 14 nm technology, an air gap is sometimes intentionally introduced between metal wires to reduce the *RC* delay. For the simple back-end structure shown below, $H = 1$ μm, $W = 1$ μm, $W' = 0.4$ μm, $L_S = 1$ μm and $x_{ox} = 1$ μm. Oxide partially fills the area between the metal wires with an air gap in the middle. Assume that the cross-section of the air gap is also rectangular. What is the percentage improvement in interconnect delay with this design instead of filling all the space between metal wires with oxide? You may use the simple *RC* model for delay represented by (11.5).

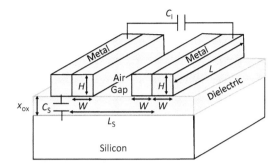

11.4   A new generation of technology is introduced by a silicon chip manufacturer that reduces all vertical and lateral dimensions by a factor of 2. Given the simple back-end structure shown in Figure 11.3, how much faster should the RC delay due to the interconnects be in the new technology compared to the previous technology? Calculate a quantitative answer.

11.5   Aluminum spiking in Si (Figure 11.8) would not be a problem if the Si diffusion in Al were small and little Si diffused into the Al, leaving voids in the Si for the Al to fill. Just how much does the Si diffuse into the Al? Assume that the back-end thermal processing is 60 min at 450 °C. If the diffusivity of Si in Al at 450 °C is about 40 $\mu m^2 \, min^{-1}$, calculate how far the Si will diffuse into the Al. (Calculate the distance in which the Si concentration falls to 50% of the surface concentration.)

11.6   In 0.35 μm technology, the junction depth (before silicidation) is about 100 nm. If you want to leave 50 nm of Si after silicidation to ensure low leakage current, how much $TiSi_2$ is formed and how much Ti is needed if all of the deposited Ti film is consumed?

11.7   A source/drain implant is done with arsenic with a dose of $1 \times 10^{15} \, cm^{-2}$ and an energy of 40 keV. A titanium silicide layer is then formed on top of the source/drain regions to reduce the sheet resistance of those regions. This is done by depositing Ti on the surface and annealing. Some 55 nm of silicide is formed, which consumes the top surface of the Si in the source/drain regions and, as a result, reduces the amount of As dopant in the Si. What is the peak concentration of As in the Si in the source/drain regions after the silicidation process? (Assume that the implant is done directly into the Si with no oxide, and that no dopant diffusion or segregation occurs during the silicidation.)

11.8   What are two reasons why the damascene process might be used instead of depositing a metal layer and then etching it away where it is not desired?

11.9   In the dual damascene metallization process described in Figure 11.19, a Si_3N_4 etch stop is used in the patterning of the oxide layer. If a nominal 1.0 μm SiO_2 layer is to be etched and the etch system has 50:1 selectivity (oxide:nitride), what is the minimum thickness of the Si_3N_4 etch stop required? Assume all layers have a ±10% tolerance on thickness.

11.10   The tungsten planarization process shown below was an early method of achieving planar structures without using CMP. In this process, a thick layer of W is deposited that overfills the via hole and produces an approximately flat surface. The W is then etched back until just the via hole is filled with W. Explain why designers might need to be constrained in terms of the via hole sizes they can use when this technology is being used.

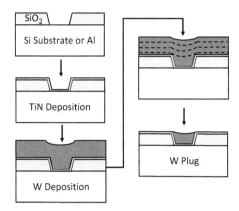

11.11 A damascene process is used to fabricate a tungsten via through an SiO_2 dielectric layer. First, the SiO_2 dielectric layer is deposited, with a thickness of 1 μm. A via hole is etched in the SiO_2, and W is blanket deposited by CVD. Right after the W deposition, the DOP equals 1.0, and the thickness of the W directly above the dielectric layer is equal to 0.8 μm. A plasma etchback of the W layer is now done to remove the W that covers the top of the dielectric layer, and leaving the W only in the via hole. If the etch rate of the W in the etchback process is 5.0 nm s^{-1}, and the etch selectivity of W with respect to SiO_2 is 4:1, what is the profile of the structure after 180 s of etching? Specify the heights of the W and SiO_2 layers in nanometers. (Neglect any variations in thicknesses or etch rate.)

11.12 An SiO_2 layer is deposited by LPCVD over a single 1 μm wide, 0.5 μm high metal line. The deposition flux and time are such that 0.25 μm of SiO_2 would be deposited on a flat surface. The sticking coefficient for this process is 0.01 and the arrival angle distribution parameter n is equal to 1. Very conformal deposition is obtained, but the DOP is 0. Would changing n or S_C improve (increase) the DOP? Explain.

11.13 A new metal M is being evaluated to replace the Ti we used in the silicide process in the CMOS process (Chapter 2). Metal M is reacted in an N_2 ambient and it forms MN and MSi_2, as did the Ti in our CMOS process. It is found experimentally that the MN nitride layer has linear growth kinetics and the MSi_2 silicide layer has parabolic growth kinetics. What can you conclude about the physical mechanisms responsible for the formation of the MN and MSi_2 layers? Explain.

11.14 Back-end structures (wires) are increasingly limiting silicon chip performance because of R and C. Research is currently underway to evaluate metals like Ru, Ir and Co as possible replacements for Cu, at least in some wiring levels. These metals have significantly higher bulk resistivities than Cu. Discuss in your own words why they would be even considered as wiring metals in silicon chips.

Appendices

A.1   Basics of Semiconductor Materials in Equilibrium

This appendix provides a brief review of the basic properties of semiconductor materials in equilibrium. The ideas discussed here are used throughout this book. For readers already familiar with this material, this appendix can be skimmed or skipped, or simply used as a reference when reading the book.

Semiconductors are a class of materials that have the unique property that their electrical conductivity can be controlled over a very wide range by the introduction of dopants. While this property can easily be observed in crystalline, polycrystalline or amorphous semiconductor materials, crystalline materials provide the most reproducible properties and the highest-performance devices and are almost always used in integrated circuits (ICs). Dopants are atoms that generally contain either one more or one fewer electrons in their outermost shell than the host semiconductor. They provide one extra electron or one missing electron (a "hole") compared to the host atoms. These excess electrons and holes are mobile and carry the current in semiconductor devices. The key to building semiconductor devices and ICs lies in the ability to control the local doping and hence the local electronic properties of a semiconductor crystal.

Consider silicon as a representative semiconductor. As illustrated in Figure A1.1(a), silicon has four electrons in its outermost shell. These are known as the valence electrons, since they are the participants in chemical reactions and chemical bonding. When silicon atoms combine to form a solid crystal, a particularly stable electronic arrangement can be formed if the silicon atoms form covalent or shared electron bonds with their four nearest neighbors, as illustrated in Figure A1.1 in a two-dimensional (2D) representation of the crystal. Only the valence shell of electrons is shown in Figure A1.1(b), and the covalent bonds are represented by the gray ovals. This arrangement is favored because each silicon atom can then "fill up" its outermost shell to a total of eight shared electrons. (In the periodic table, elements with full outer electron shells are the inert gases, like Ar and Ne, which are chemically rather inert and very stable.)

We will use the simple 2D representation in Figure A1.1 to discuss the basic properties of semiconductors, but, of course, real crystals are three-dimensional (3D) structures. The 3D unit cell of silicon (the diamond structure) is shown in Figure A1.2 (see Figure 3.3 in the main text). Crystal structure is discussed in much more detail in Chapter 3, but the silicon unit cell is cubic, as Figure A1.2 shows. The elemental semiconductors Si and Ge both have this basic crystal structure. The unit cell shown in Figure A1.2 is repeated in all three dimensions to form the crystalline substrates that are used to fabricate ICs. Other semiconductors may have more

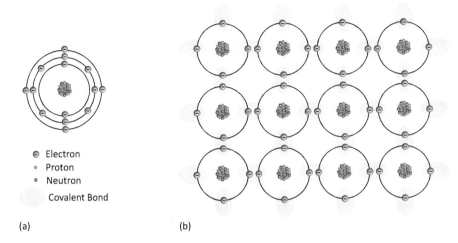

(a)   (b)

Figure A1.1 (a) Simple representation of a Si atom with 14 electrons, four of which are in the outermost or valence shell. (b) Bonded in a crystal, only the valence electrons are shown and these are shared (gray regions) resulting in a stable "full" outer shell.

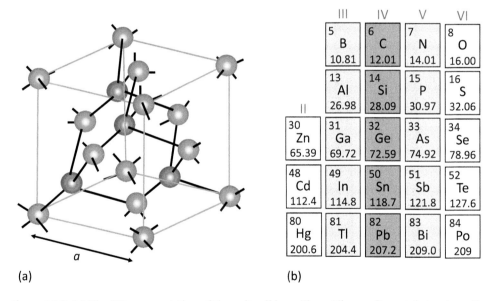

(a)   (b)

Figure A1.2 (a) The 3D representation of the unit cell in a silicon (diamond) crystal structure. For Si, $a = 0.543$ nm. The unit cell is two interleaved FCC (face-centered cubic) cells. The solid black bars represent the covalent bonds between the atoms in the crystal. Each Si atom is bonded to four surrounding atoms. This is discussed in more detail in Chapter 3. (b) The portion of the periodic table relevant to semiconductor materials and doping. Elemental semiconductors are in column IV. Compound semiconductors are combinations of elements from columns III and V (or II and VI).

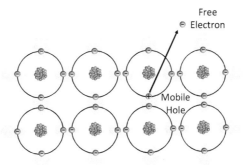

Figure A1.3 Electron (−) and hole (+) pair generation represented by a broken bond in the crystal. Both carriers are mobile and can carry currents in devices.

complex unit cells (Chapter 3), but in all cases the unit cell repeats in 3D to form the bulk material used for device fabrication.

The covalent bonding arrangement shown in Figure A1.1 occurs in all elemental semiconductors that lie in column IV of the periodic table (Figure A1.2(b)). This same type of bonding arrangement can be produced using mixtures of elements from other columns of the periodic table. For example, GaAs consists of alternating Ga (column III) and As (column V) atoms, which have an average of four electrons per atom, and so the same covalent bonding arrangement works. More complex examples like $Hg_xCd_{1-x}Te$ are also possible. Thus, nature provides many possible materials that can act as semiconductors.

At temperatures above absolute zero, thermal energy can break some of the Si–Si bonds, as illustrated in Figure A1.3. This creates both a free or mobile electron and a mobile hole (or missing electron).

The concentrations of electrons and holes are exactly equal in pure semiconductors, and are referred to as the intrinsic carrier concentration n_i. Figure A1.4 plots this concentration as a function of temperature. At room temperature (RT in Figure A1.4), n_i has a value of about $1.4 \times 10^{10}\,\text{cm}^{-3}$ in silicon. Since there are about $5 \times 10^{22}\,\text{cm}^{-3}$ atoms in silicon, less than 1 in $10^{12}\,\text{cm}^{-3}$ bonds are broken at room temperature. As a result, pure silicon is a very poor conductor. The mobile carriers carry electrical currents in devices. With so few of these present in pure silicon, the currents would be far too small to be useful in devices.

In addition to Si and Ge, n_i is also plotted in Figure A1.4 for several other semiconductors that are commercially interesting today. GaAs is used in a number of high-frequency applications. SiC and GaN are providing new opportunities in power semiconductor devices. The lower n_i values, especially in SiC and GaN, reflect the fact that the covalent bonds are stronger in these materials, which implies that fewer will be broken at any given temperature. These materials are known as "wide-bandgap" semiconductors, for reasons that we will shortly discuss.

Fortunately, semiconductors have the property that they can be doped. This process is illustrated in Figure A1.5. Doping could be accomplished by the gas-phase diffusion process we discussed in connection with Figures 1.5 and 1.6. However, today, it is generally accomplished by a process called ion implantation, which is discussed in detail in Chapter 8. Doping

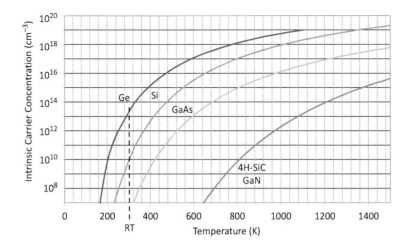

Figure A1.4 Intrinsic carrier concentration in several semiconductor materials. The vertical dashed line (RT) indicates room temperature.

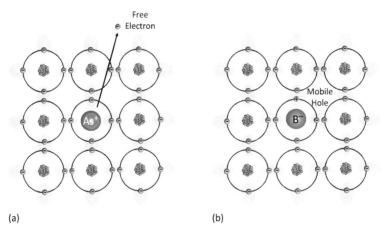

Figure A1.5 Doping of column IV semiconductors using elements (a) from column V (As) or (b) from column III (B) of the periodic table.

results in a column V (As, for example) or a column III (B, for example) atom replacing a silicon atom in the crystal. Such dopants either contribute an extra electron (column V) to the crystal and become N-type dopants, or they contribute a hole (column III) and become P-type dopants. The electrons or holes are introduced on a one-for-one basis by the dopants. As a result, to the extent that we can control the doping concentration accurately, we can precisely control the free electron and hole concentrations and therefore the conductivity of the silicon. The symbols N_D and N_A are used to refer to the N-type (donor) and P-type (acceptor) concentrations, respectively. In semiconductor devices, this doping is done locally, as illustrated in Figure 1.7, using photolithography and masking.

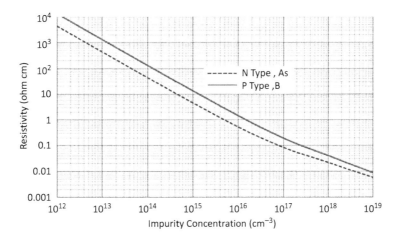

Figure A1.6 Resistivity (ρ) versus doping for N- and P-type silicon. In this plot, it is assumed that $n = N_D$ in N-type silicon, and that $p = N_A$ in P-type silicon, i.e., that all donors and acceptors are ionized. After aplusphysics.com. See https://www.aplusphysics.com/courses/honors/microe/silicon.html.

It is important to note that, in Figure A1.5, the As atom is labeled As^+ and the B atom is labeled B^-. This is because, when these atoms act as donors or acceptors and contribute mobile carriers, they are left with a net charge of the opposite sign to the mobile carrier they provide. The fact that the donor and acceptor atoms are charged in the crystal has important consequences, which we discuss in Chapter 7 on dopant diffusion. The charged dopant atoms are also critical in normal room-temperature device operation because they set up internal electric fields in devices.

Ion implantation permits the controlled introduction of parts per million to parts per hundred of dopant atoms. As a result, the conductivity of silicon can be controlled over a very wide range, permitting many types of semiconductor devices to be fabricated. Figure A1.6 illustrates the range of conductivity it is possible to achieve in silicon by doping. The vertical axis in this plot, resistivity, is a term we will define shortly. The terms N^+, N^-, P^{++}, P^{--} (or $P^=$), etc. are often used in semiconductor devices to describe relative levels of doping in a particular region. While there are no formal definitions of the ranges of doping that each of these terms represents, the following approximate definitions are often used.

$$
\begin{aligned}
N^{--} \text{ or } P^{--}: & \quad N_D \text{ or } N_A < 10^{14} \text{ cm}^{-3}; \\
N^- \text{ or } P^-: & \quad 10^{14} \text{ cm}^{-3} < N_D \text{ or } N_A < 10^{16} \text{ cm}^{-3}; \\
N \text{ or } P: & \quad 10^{16} \text{ cm}^{-3} < N_D \text{ or } N_A < 10^{18} \text{ cm}^{-3}; \\
N^+ \text{ or } P^+: & \quad 10^{18} \text{ cm}^{-3} < N_D \text{ or } N_A < 10^{20} \text{ cm}^{-3}; \\
N^{++} \text{ or } P^{++}: & \quad N_D \text{ or } N_A > 10^{20} \text{ cm}^{-3}.
\end{aligned}
$$

Since the lattice density in silicon is 5×10^{22} cm^{-3}, even the heaviest doped regions, N^{++} and P^{++}, normally use doping concentrations less than 1%.

In compound semiconductors whose elements come from columns II, III, V or VI of the periodic table, additional doping options are possible. For example, in GaAs, if a Zn atom

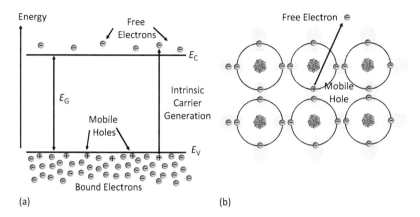

Figure A1.7 (a) Band and (b) bond model representations of an undoped (intrinsic) semiconductor. Bonded electrons are at energies below E_V. Free electrons are at energies above E_C. The process of intrinsic carrier generation is represented in each model.

replaced a Ga atom, it should act as a P-type dopant because Zn has one fewer valence electrons than Ga. If a Se atom replaced an As atom, it should act as an N-type dopant because Se has one more valence electron than As. Interestingly, Si could be either a P-type dopant (if it replaces an As atom) or an N-type dopant (if it replaces a Ga atom). Not all of the many possible combinations actually are useful in practice.

When semiconductors are doped, they become more conductive. A very useful measure of the resistance of a semiconductor material is called the resistivity (the vertical axis in Figure A1.7), which is defined as

$$\rho = \frac{1}{qn\mu_n + qp\mu_p}, \tag{A1.1}$$

where n and p are the free electron and hole concentrations, respectively, μ_n and μ_p are the mobilities of the two carriers and q is the charge on an electron. In lightly doped silicon, $\mu_n \approx 1400$ cm^2 V^{-1} s^{-1} and $\mu_p \approx 470$ cm^2 V^{-1} s^{-1}. The values of μ_n and μ_p vary with temperature, doping and electric field in semiconductors. The concentrations n and p are determined by the dopant concentrations N_D and N_A through relationships we will describe shortly. In general, only one of the two terms in (A1.1) is significant since $n \gg p$ in N doped material and $p \gg n$ in P doped material. The resistivity plotted in Figure A1.7 is for silicon. Similar curves apply to other semiconductor materials, although μ_n and μ_p are different in other materials.

The electrical properties of semiconductors are often described through the use of two types of models. The first of these is the "bond" model that we used beginning with Figure A1.1. The second model is the "band" model. Both of these approaches model the same physical phenomena. They are simply different ways of explaining the same underlying mechanisms. We make use of both approaches in this book because some concepts are easier to understand with the bond model, others with the band model.

Figure A1.7 illustrates both models for the case of an undoped "intrinsic" semiconductor such as silicon. In the band model representation, the vertical direction is the energy of electrons in the crystal. From the bond model, we can see that, in a perfect crystal, electrons are either bound to host silicon atoms or they are free. A finite amount of energy is required to free the electron (that is, to break a Si–Si bond). In the band diagram, we represent this with the bound electron energy levels below the valence band (E_V) and the free electron energy levels above the conduction band (E_C). The energy gap between E_V and E_C represents the energy needed to free the electron. In a perfect crystal, there are no allowed energy levels between E_V and E_C (an electron is either free or it is bound). Thus, the breaking of a Si–Si bond, which creates one free electron and one free hole, is illustrated in the band model by elevating an electron up to the conduction band.

In Figure A1.8, the introduction of dopants is represented in the bond model and in the band diagram by the E_D and E_A energy levels. Consider the E_D level first, which represents a donor atom (As in Si, for example). When an As atom is introduced into a Si crystal, it brings with it an extra electron in its outermost shell, beyond the four needed for covalent bonding in the silicon lattice. This electron will be bound weakly to the As atom because the As atom also brings along with it a positive charge in its nucleus (proton) to balance each electron. The binding energy of this fifth valence electron is usually rather small, since it is not needed for the four covalent bonds the As atom forms in the silicon lattice. As a result, the fifth electron can be quite easily freed. We represent this in the band diagram with an energy level E_D that is close to E_C. The small difference in energy between E_C and E_D represents the energy that must be supplied to free the fifth electron. When $E_C - E_D$ is small, we refer to the dopant as a "shallow" donor. For such donors, except at very low temperatures, this energy is always available due to the thermal energy of the crystal, and so each donor provides one free electron in the crystal. As discussed earlier, once the As atom has donated its extra electron to the crystal, the As will have a net positive charge, As$^+$:

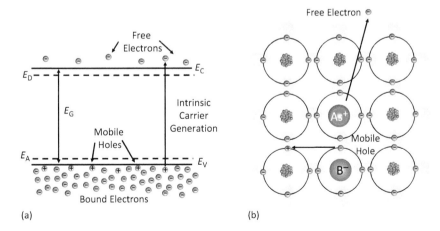

(a)   (b)

Figure A1.8 (a) Band and (b) bond model representations of a doped semiconductor, silicon in this example. Here E_D and E_A represent the donor and acceptor energy levels, respectively. Both P- and N-type doping are represented in each model.

$$\mathrm{As} \rightarrow \mathrm{As}^+ + \mathrm{e}^- \quad \text{or} \quad \mathrm{As} \rightarrow \mathrm{As}^+ + \mathrm{n}. \tag{A1.2}$$

P-type dopants are represented in an analogous manner, as also shown in Figure A1.8 for both the bond and band models. There, E_A represents the energy level introduced into the crystal by the P-type dopant. Such dopants have one fewer valence electron per atom than the host crystal. The hole or missing electron is weakly bound to the B atom. An electron from a neighboring Si atom can exchange positions with the hole, effectively allowing the hole to move through the crystal as a positively charged carrier. The energy difference between E_V and E_A represents this binding energy of the hole to the B atom and is small in the case of a "shallow" acceptor. As is the case with N-type dopants, thermal energy supplies enough energy to free the hole at all but the very lowest temperatures, assuming that $E_A - E_V$ is small.

An analogy to bubbles in a liquid is often used to describe hole behavior in the band model. Bubbles tend to rise to the surface. Thus energy needs to be provided to these bubbles or holes to push them down from the E_A level into the valence band, where they become mobile positively charged carriers. Once the boron gives up its hole to the crystal, it has a net negative charge, B^-:

$$\mathrm{B} \rightarrow \mathrm{B}^- + \mathrm{h}^+ \quad \text{or} \quad \mathrm{B} \rightarrow \mathrm{B}^- + \mathrm{p}. \tag{A1.3}$$

The concepts that we have discussed to this point regarding semiconductors can be quantified mathematically, which we will now proceed to do. The result will be a series of mathematical expressions that will allow us to calculate important properties of these materials (intrinsic carrier concentration, n, p, etc.).

We begin by considering the band model of semiconductors a little more deeply. The electrons around the nucleus in Si atoms, for example, occupy discrete energy levels, and those levels can be the same in different atoms when they are physically distant from each other. When the atoms are brought close together in a crystal structure, the Pauli exclusion principle states that each quantum state can be occupied by no more than one electron in an electron system such as a crystal. Thus the allowed energy levels split into a band of closely spaced allowed levels, as shown in Figure A1.9.

Mathematically, the density of states can be derived from quantum mechanics, as follows:

$$D_C(E) = \frac{4\pi}{h^3}(m_e^*)^{3/2}(E - E_C)^{1/2}, \quad E \geq E_C, \tag{A1.4}$$

$$D_V(E) = \frac{4\pi}{h^3}(m_h^*)^{3/2}(E_V - E)^{1/2}, \quad E \leq E_V. \tag{A1.5}$$

Here, $D_C(E)$ and $D_V(E)$ have dimensions of number per cm³ per electronvolt ($\mathrm{cm}^{-3}\,\mathrm{eV}^{-1}$); m_e and m_h are the electron and hole density of states effective mass; and h is Planck's constant. In general, electrons in the conduction band will tend to occupy the lowest energy levels available (close to E_C), and holes in the valence band will similarly bubble up to occupy the lowest energy levels available to them (near E_V).

In addition to knowing what the allowed energy levels are within the crystal, we also need to know what the probability is that a given level is occupied. Mathematically, this probability is

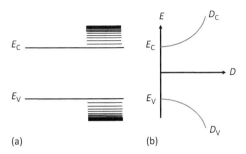

Figure A1.9 (a) Energy bands shown as a collection of discrete states above E_C and below E_V. (b) The density of states D increases with energies further from the band edges.

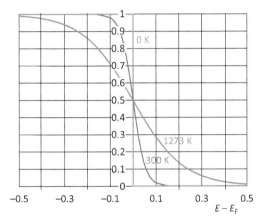

Figure A1.10 Probability $f(E)$ of an electron occupying an energy level E at three different temperatures.

described by Fermi–Dirac statistics, which comes from statistical mechanics, and is plotted in Figure A1.10:

$$f(E) = \frac{1}{1 + \exp[(E - E_F)/kT]}. \qquad (A1.6)$$

Here, $f(E)$ is the probability that an electron occupies a level at energy E, and is plotted in Figure A1.10 at three different temperatures of interest; E_F is called the Fermi level and is by definition the energy level at which $f(E) = 0.5$. At 0 K, $f(E)$ is a step function; at higher temperatures, there is a finite probability that energy levels above E_F are occupied; 300 K corresponds to room temperature and is relevant to the normal temperatures at which semiconductor devices operate; a temperature of 1273 K (1000 °C) is a typical processing temperature used in device fabrication, and so that curve is relevant to topics we cover in this book.

For energy levels well above E_F, the exponential dominates in the denominator of (A1.6) and the Fermi–Dirac probability function reduces to the Boltzmann distribution given by

$$f(E) \cong \exp\left(-\frac{E - E_F}{kT}\right). \qquad (A1.7)$$

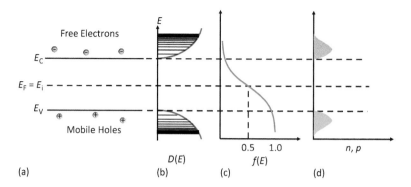

Figure A1.11 Density of allowed states, probability function, and resulting electron and hole populations in a semiconductor crystal. In an undoped semiconductor, $n = p = n_i$, and E_F is equal to E_i at approximately the middle of the bandgap.

Using the density of states (allowed energy levels) and the probability $f(E)$, we can now calculate the total number of electrons and holes present in the conduction and valence bands, respectively. Figure A1.11 illustrates graphically how we do this.

In Figure A1.11(c), $f(E)$ is the probability distribution for electrons, which, for an undoped crystal, has a value of 0.5 at the middle of the forbidden band. The density of allowed states, $D(E)$, is plotted from (A1.7) and (A1.8). Note that there are no allowed energy levels in the forbidden band. Above E_C or below E_V, the densities increase as the square root of the energy. The multiple lines in this part of Figure A1.11(b) are meant to represent the discrete allowed states that exist at any energy E. For a crystal with an appreciable number of atoms, the discrete energy levels are so close together and so numerous that they appear to be a continuous distribution, as shown in (A1.7) and (A1.8). The product of the $f(E)$ and $D(E)$ curves in Figure A1.11(d) represents the electron population at any energy. Since we are interested only in the free electrons, we have shown this population only in the conduction band. In an analogous way, the product of $D(E)$ and $1 - f(E)$ represents the hole population, which we have shown only in the valence band, since this is where holes are mobile. Note that most of the free electrons are located fairly close to E_C and most of the free holes close to E_V, because the respective probability functions rapidly fall towards zero further away from the band edges. This is in spite of the fact that the number of allowed energy levels increases away from E_C or E_V.

Usually, in device physics or in process physics, we are interested in the total number of electrons in the conduction band and the total number of holes in the valence band. These quantities are the areas under the curves in Figure A1.11(d) and can be calculated as follows:

$$n = \int_{E_C}^{\infty} f(E) D(E) \, dE \cong N_C \exp\left(-\frac{E_C - E_F}{kT}\right),
\tag{A1.8}$$

$$p = \int_{-\infty}^{E_V} [1 - f(E)] D(E) \, dE \cong N_V \exp\left(-\frac{E_F - E_V}{kT}\right),
\tag{A1.9}$$

where

$$N_C = 2\left(\frac{2\pi m_e^* kT}{h^2}\right)^{3/2} \quad \text{and} \quad N_V = 2\left(\frac{2\pi m_h^* kT}{h^2}\right)^{3/2}, \tag{A1.10}$$

Here, N_C and N_V are often called the effective densities of states in the conduction and valence bands. They have values of $3.2 \times 10^{19}\,\mathrm{cm}^{-3}$ and $1.8 \times 10^{19}\,\mathrm{cm}^{-3}$, respectively, in silicon at room temperature.

In integrating (A1.8) and (A1.9), we have made use of the Boltzmann distribution (A1.7), which is valid when the Fermi level is at least a few kT away from the conduction and valence bands and from any other allowed energy levels in the bandgap. Equations (A1.8) and (A1.9) only apply as long as this is true. When E_F approaches E_C, E_V or other allowed energy levels in the bandgap, the full Fermi–Dirac distribution function must be used to describe the electrons populating the various energy levels. This may be required at low temperatures when not all donors or acceptors are ionized and E_F can approach E_D or E_A. Fermi–Dirac statistics is also generally required when doping levels exceed $10^{19}\,\mathrm{cm}^{-3}$ in silicon, since then E_F moves up into the conduction band or down into the valence band, and allowed energy levels exist near E_F. Such heavily doped semiconductors are often called degenerate and act more like metals than semiconductors. We will, in general, use the simple results in (A1.8) and (A1.9) in this text. However, there will be cases where this is not valid. We will consider some of those cases a little later in this appendix.

In an undoped semiconductor, $n = p = n_i$, and so

$$np = n_i^2. \tag{A1.11}$$

Although we have not proven it, (A1.11) holds in doped as well as in undoped semiconductors. This is often called the law of mass action, and is an important relationship in describing carrier concentrations in semiconductors. An important implication of this equation is that, if we increase the electron population in the crystal by introducing N-type dopants, n will increase and p will decrease. Physically, the reason for this is that the processes of generation of hole–electron pairs by Si–Si bond breaking and the reverse recombination process are constantly occurring in a crystal to produce an equilibrium number of carriers. If we introduce more electrons by doping, we will increase the probability of electrons and holes encountering each other in the crystal and hence the probability of recombination. An increase in n will therefore drive the hole population down, as described by (A1.11).

By combining (A1.11) with (A1.8) and (A1.9), we arrive at the result that

$$np = n_i^2 = N_C N_V \exp\left(-\frac{E_G}{kT}\right) \cong 1.52 \times 10^{33} T^3 \exp\left(-\frac{1.21\ \mathrm{eV}}{kT}\right), \tag{A1.12}$$

where $E_G = E_C - E_V$. The approximate expression on the right uses numbers appropriate for silicon. Values for the various parameters in (A1.12) are given in Table A1.1 for the semiconductors of current interest. These values were used in plotting Figure A1.4. Note that the

Table A1.1 **Expressions for bandgap, density of states and intrinsic carrier concentration for various semiconductors of interest. Temperature T is in kelvins (K).**

Material	E_G	N_C	N_V	n_i at RT
Si	$1.17 - 4.73 \times 10^{-4}\left(\dfrac{T^2}{T+636}\right)$	$6.2 \times 10^{15} T^{1.5}$	$3.5 \times 10^{15} T^{1.5}$	1.45×10^{10} cm^{-3}
Ge	$0.742 - 4.8 \times 10^{-4}\left(\dfrac{T^2}{T+235}\right)$	$1.98 \times 10^{15} T^{1.5}$	$9.6 \times 10^{14} T^{1.5}$	2.1×10^{13} cm^{-3}
GaAs	$1.52 - 5.41 \times 10^{-4}\left(\dfrac{T^2}{T+204}\right)$	$8.63 \times 10^{13} T^{1.5}$	$1.83 \times 10^{15} T^{1.5}$	1.0×10^{6} cm^{-3}
4H-SiC	$3.63 - 6.5 \times 10^{-4}\left(\dfrac{T^2}{T+1300}\right)$	$1.83 \times 10^{15} T^{1.5}$	$1.83 \times 10^{15} T^{1.5}$	1.6×10^{-8} cm^{-3}
GaN	$3.47 - 7.7 \times 10^{-4}\left(\dfrac{T^2}{T+600}\right)$	$4.3 \times 10^{14} T^{1.5}$	$8.9 \times 10^{15} T^{1.5}$	3.3×10^{-10} cm^{-3}

expressions in Table A1.1 are designed to approximately fit over a very wide temperature range. As a result, the numbers calculated from these expressions are approximate. Note in Table A1.1 that the bandgap shrinks with temperature. Physically, this occurs because, as a material is heated, thermal expansion causes the atoms to physically move further apart. This weakens the covalent bonds and increases n_i.

At very high doping concentrations, the n_i value calculated from (A1.12) must be changed to account for what are called "heavy doping effects." Fundamentally, such effects occur because the simple picture of a Si crystal with widely separated doping atoms is not a correct picture at high doping concentrations. At n or p values greater than about 10^{19} cm^{-3}, the doping atoms are on average close enough to each other that they begin to interact and to affect the Si crystal structure. Generally, this results in a reduction in the Si bandgap and an increase in n_i.

By combining (A1.8), (A1.9) and (A1.12), we obtain the following expressions:

$$n = n_i \exp[(E_F - E_i)/kT], \tag{A1.13}$$

$$p = n_i \exp[(E_i - E_F)/kT], \tag{A1.14}$$

where E_i is the Fermi-level position in undoped or intrinsic material in which $n = p = n_i$. Mathematically, these equations say that, as n increases, E_F moves above E_i, and, when p increases, E_F moves below E_i. Figure A1.12 illustrates these ideas. In the N-doped example in Figure A1.12(b), the donor atoms represented by E_D contribute free electrons. Since E_F is the energy level at which the probability of finding an electron is 0.5, it moves up in the bandgap, reflecting the higher electron concentration compared to the intrinsic case in Figure A1.12(a). Since $np = n_i^2$, the hole concentration decreases. Exactly the opposite happens in the P-doped

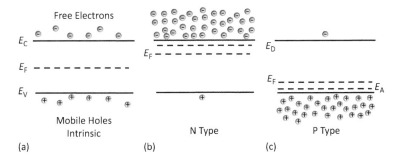

Figure A1.12 In N-type material (b), E_F is above the middle of the bandgap. In P-type material (c), E_F lies below the middle.

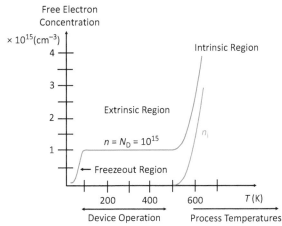

Figure A1.13 Behavior of free carrier concentration versus temperature. Arsenic in silicon is qualitatively illustrated as a specific example $(N_D = 10^{15}\ \text{cm}^{-3})$. Note that at high temperatures n_i becomes larger than the 10^{15} doping and $n \approx n_i$.

case in Figure A1.12(c). In general, the closer E_F is to E_C, the higher the N-type doping is, and, similarly, the closer E_F is to E_V, the higher the P-type doping is. Thus, these diagrams provide a visual representation of the doping level in the semiconductor.

Figure A1.13 illustrates the general behavior of the free carrier concentration versus temperature, for a specific case of a shallow donor at a concentration of $1 \times 10^{15}\ \text{cm}^{-3}$ in silicon. At temperatures below about 100 K, freezeout of the donor is observed and the free electron concentration drops. By "freezeout" we mean that there is not sufficient thermal energy for all the donors to donate their fifth electron ($E_C - E_D$ energy is needed). At temperatures above about 600 K, n_i begins to dominate the doping and n increases. We refer to this n_i-dominated regime as the intrinsic region. For lower temperatures, the material is extrinsic or dopant-dominated.

Devices are almost always operated in the temperature regime in which the material is extrinsic, i.e., the doping controls the behavior. In the processing world, however, much higher temperatures are involved and the material is often intrinsic. This difference has significant impact throughout this book. Note also that the value of n_i that is appropriate for device operation is the room-temperature value of $1.45 \times 10^{10}\ \text{cm}^{-3}$ in silicon, whereas much higher

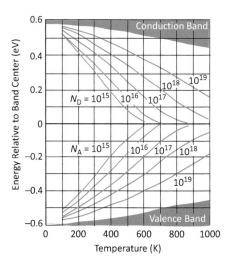

Figure A1.14 Fermi-level position (red curves) in the forbidden band for a given doping level, as a function of temperature for Si. An energy of 0 corresponds to the middle of the bandgap (E_i).

values (10^{16}–10^{18} cm^{-3}) are appropriate in the processing world because of the higher temperatures involved.

Figure A1.14 illustrates some of these concepts. The dark bands at the top and bottom represent the conduction and valence band edges. Note that the bandgap of the silicon decreases as temperature increases, as we saw in Table A1.1. The series of solid curves plot the position of the Fermi level for a specific doping concentration, as temperature changes. Each of these curves moves closer to E_i as T increases, reflecting the fact that the material becomes more intrinsic at higher temperatures. Normally, in studying semiconductor device physics, we are concerned with temperatures around 300 K. In the processing world, however, temperatures can be much higher. This means that, in modeling process phenomena, we will use values of n_i which are much larger, and values for E_G which are significantly smaller, and we will analyze materials which are more often intrinsic. Throughout this book, we will see the implications of these observations.

In our discussion to this point, we have used the Boltzmann distribution (A1.7) rather than the full Fermi–Dirac probability function (A1.6). There are many situations in which this approximation is not valid. Some semiconductor dopants are not "shallow," that is, the E_D or E_A levels are not very close to the respective band edges. This results in incomplete ionization even at room temperature. Wide-bandgap materials like SiC and GaN are good examples. All of the practical P-type dopants in those materials do not have shallow levels. In addition, at high doping concentrations, E_F moves close to the band edges, and hence, even if the donor or acceptor levels are shallow, E_F will get close to these levels, and again the Boltzmann approximation is not valid. Finally, at low temperatures, even shallow acceptors or donors will not be fully ionized.

In each of these instances, a more accurate approach uses the full Fermi–Dirac distribution. In those situations,

$$N_D^+ = \frac{N_D}{1 + g_D \exp[(E_F - E_D)/kT]},$$

$$N_A^- = \frac{N_A}{1 + g_A \exp[(E_A - E_F)/kT]},$$

(A1.15)

where N_D^+ and N_A^- are the ionized donor and acceptor concentrations, and g_D and g_A are the degeneracy factors for the acceptor and donor levels. The latter account for the fact that electrons with different spins can occupy the same energy level; typically $g_D = 2$ and $g_A = 4$. Therefore, we have

$$n = N_D^+ = \frac{N_D}{1 + g_D \exp[(E_F - E_D)/kT]}$$

$$= \frac{N_D}{1 + g_D \exp[(E_F - E_D + E_C - E_C)/kT]} \qquad (A1.16)$$

$$= \frac{N_D}{1 + g_D(n/N_C) \exp[(E_C - E_D)/kT]}.$$

Letting $\gamma_n = (N_C/g_D)\exp[-(E_C - E_D)/kT]$, we have $n^2 + \gamma_n n - \gamma_n N_D = 0$, which has the solution

$$n = N_D^+ = \frac{\gamma_n}{2}\left(\sqrt{\left(1 + \frac{4N_D}{\gamma_n}\right)} - 1\right). \qquad (A1.17)$$

Similarly, letting $\gamma_p = (N_V/g_A)\exp[-(E_A - E_V)/kT]$, we obtain

$$p = N_A^- = \frac{\gamma_p}{2}\left(\sqrt{\left(1 + \frac{4N_A}{\gamma_p}\right)} - 1\right). \qquad (A1.18)$$

Thus, given a donor or acceptor level ($E_C - E_D$ or $E_A - E_V$) and a temperature, we can calculate the fraction of donors or acceptors that are ionized, which will give us n and p. Table A1.2 lists common dopants and their energy levels in several semiconductor materials used today in fabricating devices and ICs.

Table A1.2 Typical donors and acceptors, with corresponding energy levels for semiconductors.

Semiconductor	Donors	$E_C - E_D$ (meV)	Acceptors	$E_A - E_V$ (meV)
Si	P, As, Sb	44–49	B, Al, Ga	45–65
			In	160
Ge	P, As, Sb	10–12	B, Al, Ga	10
GaAs	Si, Ge, S, Sn	6	Mg, Si, Be, C	25–35
4H-SiC	N on C site	60–140	Al on Si site	200–240
	P on Si site	60–130	B on Si site	280–350
GaN	Si on Ga site	15	Mg on Ga site	160
			Zn	340

Example

A sample of silicon that is doped with both N- and P-type dopants is shown in Figure A1.15. Calculate n, p and E_F for: (a) room temperature, a typical device operating condition; (b) 1000 °C, a typical high processing temperature, which might be used during device fabrication; and (c) 50 K, a low temperature where incomplete ionization could be an issue, even though the donors and acceptors are shallow.

Answer

In general, in problems of this type, there are three unknowns, n, p and E_F, assuming that the dopants are fully ionized and that $N_D = N_D^+$ and $N_A = N_A^-$. There are three independent equations taken from those discussed above, which must be solved simultaneously. (There are actually four unknowns including n_i, but that may be calculated directly from (A1.12) since the temperature is known.) Generally, we know N_D, N_A and T, as in this problem. The three equations with three unknowns are as follows. Note that equations (A1.19) and (A1.20) and equations (A1.21) and (A1.22) are pairs of equations, which are not independent but are related through (A1.11):

$$\left.\begin{array}{l} n = \dfrac{1}{2}\left[(N_D^+ - N_A^-) + \sqrt{(N_D^+ - N_A^-)^2 + 4n_i^2}\right], \\[2ex] p = \dfrac{1}{2}\left[(N_A^- - N_D^+) + \sqrt{(N_A^- - N_D^+)^2 + 4n_i^2}\right], \end{array}\right\} \qquad \text{(A1.19), (A1.20)}$$

$$\left.\begin{array}{l} n = n_i \exp[(E_F - E_i)/kT], \\[2ex] p = n_i \exp[(E_i - E_F)/kT], \end{array}\right\} \qquad \text{(A1.21), (A1.22)}$$

$$np = n_i^2. \qquad \text{(A1.11)}$$

These equations also explicitly illustrate the idea of compensation. When both donors and acceptors are present, they compensate each other, and only the net doping is generally important. This concept is easy to understand in connection with Figure A1.15. If both E_D and E_A are present, then the free electron from a donor atom will simply "fall" to the lower energy level available at E_A. Thus one donor and one acceptor are effectively eliminated or cancel each other.

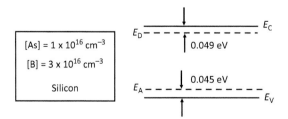

Figure A1.15   Silicon doped with both N- and P-type dopants.

In virtually every case of interest, physical insight can greatly simplify this situation, usually to the point where simultaneous equations do not have to be solved. In those cases in which simplifications cannot be made, there are computer programs available which efficiently calculate a solution. In solving this example, we will take the approach of first using physical insight to simplify the problem.

(a) *Room-temperature solution.* At room temperature, it is reasonable to assume that all of the dopants are ionized, since both As ($E_D = 0.049$ eV) and B ($E_A = 0.045$ eV) are shallow dopants (close to their respective band edges). Furthermore, the net doping will be P type and $N_A - N_D$ is much greater than n_i, so we have directly from (A1.20) and (A1.11) that

$$p = N_A - N_D = 2 \times 10^{16} \text{cm}^{-3},$$

$$n = \frac{n_i^2}{p} = \frac{2.1 \times 10^{20}}{2 \times 10^{16}} = 1.05 \times 10^4 \text{cm}^{-3}.$$

The Fermi-level position can now be calculated directly from (A1.22),

$$p = n_i \exp[(E_i - E_F)/kT].$$

Therefore

$$E_i - E_F = kT\ln\left(\frac{p}{n_i}\right) = (8.62 \times 10^{-5}\text{eV K}^{-1})(298 \text{ K})\ln\left(\frac{2 \times 10^{16}}{1.45 \times 10^{10}}\right)$$
$$= 0.36 \text{ eV}.$$

Another observation at this point may be helpful. To first order, it is the case that energy levels above E_F are not occupied by electrons, while levels below E_F are occupied. (Recall that E_F by definition is the level at which the probability of finding an electron is exactly 0.5. The probability drops rapidly towards 0 above E_F and approaches 1 below E_F as shown in Figure A1.11.) The result in Figure A1.16(a) shows E_F to be well below E_D, which means that the donor levels are not occupied or have donated their electrons. Also, E_F is well above E_A, which means that the acceptor levels are occupied or have accepted electrons. This is consistent with our assumption of complete ionization.

(b) *High-temperature solution.* At 1000 °C, we can calculate directly from (A1.12) that $n_i = 7.14 \times 10^{18}\text{cm}^{-3}$. This is much larger than the donor or acceptor concentrations, so that the material will be intrinsic. Therefore, $n = p = n_i$ and the Fermi level will be essentially in the middle of the bandgap. Note that E_G is significantly smaller at this temperature than it is at room temperature (0.77 eV from the expression in Table A1.1). This solution is shown in Figure A1.16(b). The solution is consistent with all the donors and acceptors being ionized, since E_F is well below E_D and well above E_A.

(c) *Low-temperature solution.* At 50 K, it is likely that not all of the acceptors will be ionized. However, it should still be the case that the donors are all ionized because the material should still be net P type, E_F will be in the lower half of the bandgap and therefore well below E_D. Thus the net N_A should still be $2 \times 10^{16}\text{cm}^{-3}$. We need to find how many of those acceptors are actually ionized. We can use (A1.18),

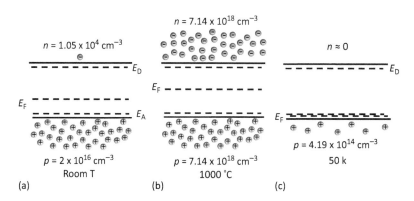

Figure A1.16 Solutions for the example at room temperature, 1000 °C and 50 K.

$$p = N_A^- = \frac{\gamma_p}{2}\left(\sqrt{\left(1 + \frac{4N_A}{\gamma_p}\right)} - 1\right) \quad \text{with} \quad \gamma_p = \frac{N_V}{g_A}\exp\left[-\frac{(E_A - E_V)}{kT}\right].$$

From Table A1.1, $N_V = 3.5 \times 10^{15} T^{1.5} = 1.23 \times 10^{18} \text{cm}^{-3}$. Thus,

$$\gamma_p = \frac{N_V}{g_A}\exp\left[-\frac{(E_A - E_V)}{kT}\right]$$

$$= \frac{1.23 \times 10^{18}}{4}\exp\left(-\frac{(0.045 \text{ eV})}{(8.62 \times 10^{-5})(50)}\right)$$

$$= 8.98 \times 10^{12},$$

$$p = N_A^- = \frac{\gamma_p}{2}\left(\sqrt{\left(1 + \frac{4N_A}{\gamma_p}\right)} - 1\right)$$

$$= \frac{8.98 \times 10^{12}}{2}\left(\sqrt{\left(1 + \frac{(4)(2 \times 10^{16})}{8.98 \times 10^{12}}\right)} - 1\right)$$

$$= 4.19 \times 10^{14} \text{cm}^{-3},$$

$$n_i^2 = N_C N_V \exp\left(-\frac{E_G}{kT}\right)$$

$$= (2.19 \times 10^{18})(1.23 \times 10^{18})\exp\left(-\frac{1.17 \text{ eV}}{(8.62 \times 10^{-5})(50)}\right)$$

$$\cong 0,$$

so $n \approx 0$. The position of the Fermi level can also now be calculated using (A1.9):

$$
\begin{aligned}
E_F - E_i &= kT \ln\left(\frac{N_V}{p}\right) \\
&= (8.62 \times 10^{-5})(50) \ln\left(\frac{1.23 \times 10^{18}}{4.19 \times 10^{14}}\right) \\
&= 0.0344 \text{ eV}.
\end{aligned}
$$

The Fermi level is actually below E_A, reflecting the fact that many of the acceptors are not ionized at this low temperature. The solution at 50 K is also shown in Figure A1.16(c).

The concepts we have been using to describe semiconductors and their band structure are not sufficient to describe many advanced aspects of device performance. For example, in all of our discussion of band diagrams, we have used one-dimensional pictures to represent the material properties. In actual fact, such pictures should be two-dimensional, with the vertical axis being electron energy (as in our figures) and the horizontal axis being the crystal momentum (k). These more complete representations show that the maximum of the valence band and the minimum of the conduction band in general do not occur at the same value of momentum and are discussed in most device textbooks. This means that, in any electron transition from the valence to the conduction band, we must account not only for the energy change of the electron, but also for its momentum change.

This becomes important in some aspects of device operation. For example, efficient light emission, such as in light-emitting diodes (LEDs), is generally only possible in semiconductors in which the conduction band minimum and the valence band maximum occur at the same k value in momentum space. This is referred to as a direct-bandgap semiconductor. In this case, when an electron falls from the conduction band back to the valence band and recombines there with a hole, the resulting energy can be transferred to a photon (which has essentially no momentum). This occurs in many compound semiconductors, such as GaAs, and importantly in GaN, which is also a direct-bandgap semiconductor and is today the material used for LEDs in lighting applications.

Silicon, germanium and SiC are not direct-bandgap semiconductors since the maximum of the valence band and the minimum of the conduction band occur at different positions on the k axis. As a result, in these materials, the recombination process generally occurs through an intermediate energy level in the bandgap (such as a gold atom), which captures the electron and hole necessary for recombination and transfers the excess momentum to a phonon (lattice vibration). This sequence of events conserves both energy and momentum, and is referred to as Shockley–Read–Hall (SRH) recombination after the scientists who first developed a theoretical description of it.

We will need to understand the indirect recombination process through SRH traps a little more carefully in order to understand the role that impurities like Au, Cu, Fe and other heavy metals play in silicon technology. These elements act as catalysts for recombination or "traps"

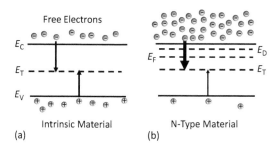

Figure A1.17 Shockley–Read–Hall recombination process. The trap level captures each carrier separately, effecting recombination. In N-type material, the relative sizes of the arrows represent the easy capture of the majority carrier (electron) and the more difficult capture of the minority carrier (hole).

in silicon. Consider the deep energy level in Figure A1.17(a) and assume that it represents an allowed energy level in the semiconductor bandgap.

Elements like Au, Cu and Fe introduce such levels in silicon, as do lattice defects like vacancies and interstitials, which we discuss in more detail in Chapter 3. In indirect-bandgap materials like silicon, the direct recombination of an electron and a hole is not a very probable process because of the need to conserve momentum. What is much more likely, and what generally occurs in silicon, is recombination through an intermediate trap level. The trap level first captures either a hole or an electron and then captures the other carrier, completing the recombination event. The excess momentum is given to the crystal through the trapping atom, which generally sits on a lattice site.

These steps are illustrated in Figure A1.17(b) for the specific case of an N-type semiconductor. In N-type material, there are many electrons present, so the likelihood of one of them encountering the trap is much higher than for holes, which are few in number. The typical condition for the trap is therefore to have captured an electron and be waiting for a hole to come into close enough proximity to be captured. It is thus the minority carrier capture process that limits the overall recombination rate, and we usually speak of the minority carrier lifetime as representing the time required for recombination to occur.

This bulk lifetime to first order is given by

$$\tau_R = \frac{1}{\sigma v_{th} N_t},$$ (A1.23)

where τ_R is the recombination lifetime, σ is the capture cross-section of the trap (on the order of the atom cross-sectional area or about 10^{-16} cm^{-2}), v_{th} is the minority carrier thermal velocity (about 10^7 cm s^{-1}) and N_t is the density of traps per cm^3. We use this equation in Chapter 4 to estimate the allowable contamination levels in silicon crystals (Section 4.2).

The band diagram in Figure A1.17(b) illustrates another way of visualizing the recombination process. In N-type material, E_F will be in the upper half of the bandgap, above the trap level. Energy levels below E_F are normally occupied by electrons, since the probability of finding an electron at a level below E_F is close to one. Thus, we again conclude that the trap level will normally be occupied by an electron and that the hole (minority carrier) capture will be the rate-limiting step. A similar argument in P-type material would show that the capture of the minority carrier electron is the rate-limiting step. This process of recombination is important in many types of devices, but especially in devices that depend on the behavior of minority carriers. This includes bipolar transistors but also many of the devices that are interesting in power applications such as thyristors and insulated gate bipolar transistors (IGBTs).

The inverse process – generation – through SRH traps is also possible and is often the dominant carrier generation process in indirect-bandgap semiconductors like silicon. In this process, the trap level first generates an electron or hole, and then generates the other species. These events can then be repeated to supply carriers to the semiconductor when they are needed. In this case, the material is characterized by a generation rather than a recombination lifetime. Depletion regions are a very common example in which such generation takes place. In general, the two lifetimes τ_R (recombination) and τ_G (generation) are not the same, since the processes involved in each are somewhat different (capture of a free carrier versus emission of a carrier from a trap). In silicon, τ_G is normally longer than τ_R. Equation (A1.23) provides a simple estimate of carrier lifetime, and we will often not have to distinguish between τ_G and τ_R in this text.

SRH theory gives the rate of recombination (or generation) as

$$U = \frac{np - n_i^2}{\tau\left(p + n + 2n_i \cosh[(E_T - E_i)/kT]\right)}, \tag{A1.24}$$

where τ is the carrier lifetime. In equilibrium, $np = n_i^2$ and U goes to zero. In this case, recombination and generation exactly balance each other. When excess carriers are present, $np > n_i^2$ and U is positive (net recombination). When $np < n_i^2$, U is negative (net generation). Note that the rate of recombination (or generation) is maximized when the trap level is near the middle of the bandgap.

At the surface of the semiconductor, the crystal structure terminates, usually at a thermally grown SiO_2 layer in silicon devices. Traps at such an interface can also act as SRH centers and can both generate and recombine carriers. The surface equivalent of (A1.23) is

$$s = \sigma_S v_{th} N_{it}, \tag{A1.25}$$

where σ_S is the capture cross-section of the surface trap (cm^{-2}) and N_{it} is the density of traps per cm^2 at the surface. In this equation, s has units of cm s^{-1} and is called the surface recombination velocity. Thus s related to the rate at which carriers recombine at the surface; $1/\tau_R$ is the equivalent bulk recombination rate. When surface effects dominate, an analogous expression to (A1.24) gives the net rate of recombination or generation:

$$U = \frac{s(np - n_i^2)}{p + n + 2n_i \cosh[(E_T - E_i)/kT]}. \tag{A1.26}$$

The fact that the minority carrier lifetime τ_R is inversely proportional to N_t implies that, if we can control N_t, then the lifetime is a design parameter in indirect-bandgap semiconductors. This is exactly what is done in these materials. In some cases, it is desirable to have as long a lifetime as possible, and minimizing N_t through processes like "gettering" is used to achieve this goal. We discuss gettering in Chapter 4. In other applications, a specific value of lifetime is desired and a controlled number of traps are introduced during the fabrication. One example of this is in power switching devices in which lifetime affects the trade-off between device on-resistance and switching speed. We also consider the controlled introduction of traps in Chapter 4 for these kinds of applications. Generally, these traps are introduced through controlled concentrations of transition metals like Au, or by slightly damaging the crystal through electron irradiation or other means. In direct-bandgap materials like GaAs, in which carriers can recombine without an intermediate trap level, minority carrier lifetimes are generally very short (on the order of nanoseconds). In addition, these lifetimes are fundamental properties of the material and are not generally available to the process or device designer as design variables.

A general reference for much of the material discussed in this appendix is C. Hu, *Modern Semiconductor Devices for Integrated Circuits*, Pearson, 2010. Most other texts on semiconductor devices also cover these topics.

Problems

A set of homework problems covering the topics in this appendix is included at the end of Chapter 1.

A.2 Standard Prefixes

Prefix	Symbol	Value
tera	T	10^{12}
giga	G	10^{9}
mega	M	10^{6}
kilo	k	10^{3}
milli	m	10^{-3}
micro	μ	10^{-6}
nano	n	10^{-9}
pico	p	10^{-12}
femto	f	10^{-15}

A.3 Useful Conversions

---

1 atm = 760 Torr (exactly)

1 atm = 1.013×10^5 Pa

1 Pa = $1\,J\,m^{-3}$ = 10 dyne cm^{-2}, 1 MPa = 10^7 dyne cm^{-2}

1 eV = 1.60219×10^{-19} J

1 cal = 4.184 J

1 J = 10^7 erg

1 erg = 2.39×10^{-8} cal

1 N = $1\,J\,m^{-1}$ = 10^5 dyne

1 = 1 ohm amp = 1 Ω A

1 J = 1 volt coulomb = 1 V C = 1 coulomb amp ohm = 1 C A Ω

1 J = 10^7 dyne cm

1 coulomb = 1 C = 1 amp second = 1 A s = 1 farad volt = 1 F V

1 amp = 1 A = 1 farad volt/second = $1\,F\,V\,s^{-1}$

1 nm = 10^{-3} μm = 10^{-7} cm = 10^{-9} m = 10 Å

$T\,(K) = T\,(°C) + 273.15$

A.4 Physical Constants

Quantity	Symbol	Value
Avogadro constant	N_{AV}	6.022×10^{23} mole^{-1}
Boltzmann constant	k	1.381×10^{-23} J K$^{-1} = 8.617 \times 10^{-5}$ eV K^{-1}
Electron charge	q	1.602×10^{-19} C
Electron rest mass	m_e	9.1×10^{-28} g
Gas constant	R	1.987 cal mole^{-1} K^{-1}
Permittivity in vacuum	ε_0	8.854×10^{-14} F cm^{-1}
Planck constant	h	6.626×10^{-34} J s $= 4.136 \times 10^{-15}$ eV s
Speed of light in vacuum	c	2.998×10^{10} cm s^{-1}
Thermal voltage at 300 K	kT/q	0.0259 V

A.5　Physical Properties of Silicon

Property	Value
Bandgap at 300 K	1.107 eV
Electron majority carrier mobility at 300 K	1400 cm^2 V^{-1}s^{-1}
Hole majority carrier mobility at 300 K	470 cm^2 V^{-1}s^{-1}
Crystal structure	Cubic diamond
Melting point	1417 °C
Density	2.328 g cm^{-3}
Thermal conductivity k_S at 300 K	0.358 cal (s cm °C)$^{-1}$
Latent heat of fusion	340 cal g^{-1}
Linear thermal expansion coefficient $\Delta L/L\Delta T$	2.6 × 10^{-6} °C^{-1}
Young's modulus	$Y_{100} = 1.3 \times 10^{12}$ dyne cm^{-2}
	$Y_{111} = 1.9 \times 10^{12}$ dyne cm^{-2}
Lattice constant	0.5431 nm
Specific heat	0.7 J g^{-1} °C^{-1}
Thermal diffusivity	0.92 cm^{2}s^{-1}
Atomic density	5 × 10^{22} cm^{-3}
Atomic weight	28.09
Breakdown field	≈ 3 × 10^5 Vcm^{-1}
Effective density of states (valence band) N_V	1.8 × 10^{19} cm^{-3}
Effective density of states (conduction band) N_C	3.2 × 10^{19} cm^{-3}
Density of states effective mass of electrons m_e^*	1.08m_e　($m_e = 9.1 \times 10^{-28}$ g)
Density of states effective mass of holes m_p^*	0.81m_e
Peak electron velocity v_{SAT}	1 × 10^7 cms^{-1}
Peak hole velocity v_{SAT}	1 × 10^7 cms^{-1}
Electron affinity χ	4.05 eV
Intrinsic carrier concentration n_i at 25 °C	1.45 × 10^{10} cm^{-3}
Intrinsic resistivity	2.3 × 10^5 Ω cm
Relative permittivity	11.9

A.6 Properties of Insulators Used in Semiconductor Technology

	SiO_2	Si_3N_4	SiO_xN_y
Structure	Amorphous	Amorphous	Amorphous
Resistivity (Ω cm)	10^{14}–10^{16}	$\approx 10^{14}$	
Density (g cm^{-3})	2.27	3.1	
Dielectric constant, K	3.8–3.9	7.5	5–6
Dielectric strength (V cm^{-1})	$(5$–$10) \times 10^6$	$\approx 10^7$	$\approx 5 \times 10^6$
Energy gap (eV)	≈ 9	≈ 5	
Expansion coefficient (°C^{-1})	5×10^{-7}		
Refractive index	1.46	2.05	1.6–1.9
Thin-film stress (dyne cm^{-2})	$(2$–$4) \times 10^9$	$(9$–$10) \times 10^9$	$(1$–$6) \times 10^9$
Thermal conductivity (W cm^{-1} °C^{-1})	0.014		
Infrared absorption peak (μm)	9.3	11.5–12	9–12
Etch rate in buffered HF (nm min^{-1})	100	0.5–1	2–40
Melting point (°C)	≈ 1700		
Electron mobility (cm^2 V^{-1} s^{-1})	20–40		
Hole mobility (cm^2 V^{-1} s^{-1})	$\approx 2 \times 10^{-5}$		
Surface resistivity (Ω/sq)	10^{16}–10^{19}		

A.7 Color Chart for Thermally Grown SiO₂ Films Observed Perpendicularly under Daylight Fluorescent Lighting

Film thickness (nm)	Color	Film thickness (nm)	Color
50	Tan	630	Violet-red
70	Brown	680	"Bluish"
100	Dark violet to red-violet	720	Blue-green to green
120	Royal blue	770	"Yellowish"
150	Light blue to metallic blue	800	Orange
170	Metallic to light yellow-green	820	Salmon
200	Light gold or yellow	850	Light red-violet
220	Gold with slight yellow-orange	860	Violet
250	Orange to melon	890	Blue
270	Red-violet	920	Blue-green
300	Blue to blue-violet	950	Dull yellow-green
310	Blue	970	Yellow
320	Blue to blue-green	990	Orange
340	Light green	1000	Carnation pink
350	Green to yellow-green	1020	Violet-red
360	Yellow-green	1050	Red-violet
370	Green-yellow	1060	Violet
390	Yellow	1070	Blue-violet
410	Light orange	1100	Green
420	Carnation pink	1110	Yellow-green
440	Violet-red	1120	Green
460	Red-violet	1180	Violet
470	Violet	1190	Red-violet
480	Blue-violet	1210	Violet-red
490	Blue	1240	Carnation pink to salmon
500	Blue-green	1250	Orange
520	Green	1280	"Yellowish"
540	Yellow-green	1320	Sky blue to green-blue
560	Green-yellow	1400	Orange
570	Yellow to "yellowish"	1450	Violet
580	Light orange or yellow to pink	1460	Blue-violet
600	Carnation pink	1500	Blue

A.8   Irwin Curves

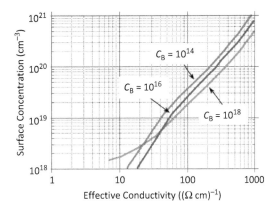

Figure A8.1 Irvin's curves for N-type erfc diffusion profiles.

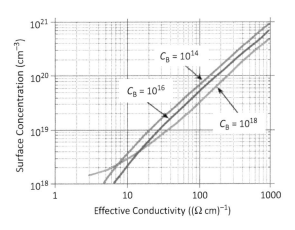

Figure A8.2 Irvin's curves for P-type erfc diffusion profiles.

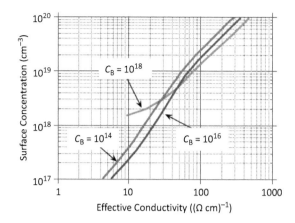

Figure A8.3 Irvin's curves for N-type Gaussian diffusion profiles.

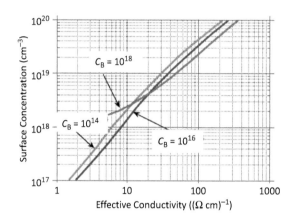

Figure A8.4 Irvin's curves for P-type Gaussian diffusion profiles.

A.9 **Error Function**

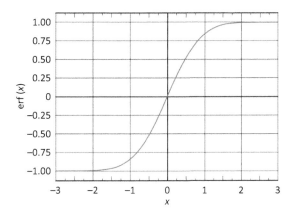

Figure A9.1 Erf plot.

Properties

The error function is defined by

$$\mathrm{erf}(x) = \frac{2}{\sqrt{\pi}} \int_0^x \exp(-\eta^2)\, d\eta,$$

so that

$$\mathrm{erfc}(x) = 1 - \mathrm{erf}(x) = \frac{2}{\sqrt{\pi}} \int_x^\infty \exp(-\eta^2)\, d\eta.$$

The error function is an odd function: $\mathrm{erf}(-x) = -\mathrm{erf}(x)$,
which also means that $\mathrm{erf}(0) = 0$.
The limit as $x \to \infty$ is $\mathrm{erf}(\infty) = 1$,
which also means $\mathrm{erf}(-\infty) = -1$.
The derivative of the error function is

$$\frac{d}{dx}\mathrm{erf}(x) = \frac{2}{\sqrt{\pi}} \exp(-x^2).$$

A useful approximation with better than 4% accuracy is

$$\mathrm{erfc}(x) = \frac{1.1 \exp(-x^2)}{|x| + \sqrt{x^2 + 1.21}},$$

which is taken from M. Abramowitz and I. A. Stegun, *Handbook of Mathematical Functions: with Formulas, Graphs and Mathematical Tables* (NBS Applied Mathematics Series 55), National Bureau of Standards, 1964 (and since reprinted elsewhere).

Asymptotic approximations:

$$\mathrm{erf}(x) \to \frac{\exp(-x^2)}{2\sqrt{\pi}} \ \text{ for } x \gg 1,$$

$$\mathrm{erf}(x) \to \frac{2x}{\sqrt{\pi}} \ \ \text{ for } x \ll 1.$$

x	$\mathrm{erf}(x)$	x	$\mathrm{erf}(x)$	x	$\mathrm{erf}(x)$
0.00	0.00	0.34	0.369365	0.68	0.663782
0.01	0.011283	0.35	0.379382	0.69	0.670840
0.02	0.022565	0.36	0.389330	0.70	0.677801
0.03	0.033841	0.37	0.399206	0.71	0.684666
0.04	0.045111	0.38	0.409009	0.72	0.691433
0.05	0.056372	0.39	0.418739	0.73	0.698104
0.06	0.067622	0.40	0.428392	0.74	0.704678
0.07	0.078858	0.41	0.437969	0.75	0.711156
0.08	0.090078	0.42	0.447468	0.76	0.717537
0.09	0.101281	0.43	0.456887	0.77	0.723822
0.10	0.112463	0.44	0.466225	0.78	0.730010
0.11	0.123623	0.45	0.475482	0.79	0.736103
0.12	0.134758	0.46	0.484655	0.80	0.742101
0.13	0.145867	0.47	0.493745	0.81	0.748003
0.14	0.156947	0.48	0.502750	0.82	0.753811
0.15	0.167996	0.49	0.511668	0.83	0.759524
0.16	0.179012	0.50	0.520500	0.84	0.765143
0.17	0.189992	0.51	0.529244	0.85	0.770668
0.18	0.200936	0.52	0.537899	0.86	0.776100
0.19	0.211840	0.53	0.546464	0.87	0.781440
0.20	0.222703	0.54	0.554939	0.88	0.786687
0.21	0.233522	0.55	0.563323	0.89	0.791843
0.22	0.244296	0.56	0.571616	0.90	0.796908
0.23	0.255023	0.57	0.579816	0.91	0.801883
0.24	0.265700	0.58	0.587923	0.92	0.806768
0.25	0.276326	0.59	0.595936	0.93	0.811564
0.26	0.286900	0.60	0.603856	0.94	0.816271
0.27	0.297418	0.61	0.611681	0.95	0.820891
0.28	0.307880	0.62	0.619411	0.96	0.825424
0.29	0.318283	0.63	0.627046	0.97	0.829870
0.30	0.328627	0.64	0.634586	0.98	0.834231
0.31	0.338908	0.65	0.642029	0.99	0.838508
0.32	0.349126	0.66	0.649377		
0.33	0.359279	0.67	0.656628		

x	erf(x)	x	erf(x)	x	erf(x)
1.00	0.842701	1.34	0.941914	1.68	0.982493
1.01	0.846811	1.35	0.943762	1.69	0.983153
1.02	0.850838	1.36	0.945561	1.70	0.983790
1.03	0.854784	1.37	0.947312	1.71	0.984407
1.04	0.858650	1.38	0.949016	1.72	0.985003
1.05	0.862436	1.39	0.950673	1.73	0.985578
1.06	0.866144	1.40	0.952285	1.74	0.986135
1.07	0.869773	1.41	0.953852	1.75	0.986672
1.08	0.873326	1.42	0.955376	1.76	0.987190
1.09	0.876803	1.43	0.956857	1.77	0.987691
1.10	0.880205	1.44	0.958297	1.78	0.988174
1.11	0.883533	1.45	0.959695	1.79	0.988641
1.12	0.886788	1.46	0.961053	1.80	0.989091
1.13	0.889971	1.47	0.962373	1.81	0.989525
1.14	0.893082	1.48	0.963654	1.82	0.989943
1.15	0.896124	1.49	0.964898	1.83	0.990347
1.16	0.899096	1.50	0.966105	1.84	0.990736
1.17	0.902000	1.51	0.967277	1.85	0.991111
1.18	0.904837	1.52	0.968413	1.86	0.991472
1.19	0.907608	1.53	0.969516	1.87	0.991821
1.20	0.910314	1.54	0.970586	1.88	0.992156
1.21	0.912956	1.55	0.971623	1.89	0.992479
1.22	0.915534	1.56	0.972628	1.90	0.992790
1.23	0.918050	1.57	0.973603	1.91	0.993090
1.24	0.920505	1.58	0.974547	1.92	0.993378
1.25	0.922900	1.59	0.975462	1.93	0.993656
1.26	0.925236	1.60	0.976348	1.94	0.993923
1.27	0.927514	1.61	0.977207	1.95	0.994179
1.28	0.929734	1.62	0.978038	1.96	0.994426
1.29	0.931899	1.63	0.978843	1.97	0.994664
1.30	0.934008	1.64	0.979622	1.98	0.994892
1.31	0.936063	1.65	0.980376	1.99	0.995111
1.32	0.938065	1.66	0.981105		
1.33	0.940015	1.67	0.981810		

x	erf(x)	x	erf(x)	x	erf(x)
2.00	0.995322	2.34	0.999065	2.68	0.999849
2.01	0.995525	2.35	0.999111	2.69	0.999858
2.02	0.995719	2.36	0.999155	2.70	0.999866
2.03	0.995906	2.37	0.999197	2.71	0.999873
2.04	0.996086	2.38	0.999237	2.72	0.999880
2.05	0.996258	2.39	0.999275	2.73	0.999887
2.06	0.996423	2.40	0.999312	2.74	0.999893
2.07	0.996582	2.41	0.999346	2.75	0.999899
2.08	0.996734	2.42	0.999379	2.76	0.999905
2.09	0.996880	2.43	0.999411	2.77	0.999910
2.10	0.997021	2.44	0.999441	2.78	0.999916
2.11	0.997155	2.45	0.999469	2.79	0.999920
2.12	0.997284	2.46	0.999497	2.80	0.999925
2.13	0.997407	2.47	0.999523	2.81	0.999929
2.14	0.997525	2.48	0.999547	2.82	0.999933
2.15	0.997639	2.49	0.999571	2.83	0.999937
2.16	0.997747	2.50	0.999593	2.84	0.999941
2.17	0.997851	2.51	0.999614	2.85	0.999944
2.18	0.997951	2.52	0.999635	2.86	0.999948
2.19	0.998046	2.53	0.999654	2.87	0.999951
2.20	0.998137	2.54	0.999672	2.88	0.999954
2.21	0.998224	2.55	0.999689	2.89	0.999956
2.22	0.998308	2.56	0.999706	2.90	0.999959
2.23	0.998388	2.57	0.999722	2.91	0.999961
2.24	0.998464	2.58	0.999736	2.92	0.999964
2.25	0.998537	2.59	0.999751	2.93	0.999966
2.26	0.998607	2.60	0.999764	2.94	0.999968
2.27	0.998674	2.61	0.999777	2.95	0.999970
2.28	0.998738	2.62	0.999789	2.96	0.999972
2.29	0.998799	2.63	0.999800	2.97	0.999973
2.30	0.998857	2.64	0.999811	2.98	0.999975
2.31	0.998912	2.65	0.999821	2.99	0.999976
2.32	0.998966	2.66	0.999831	3.00	0.999978
2.33	0.999016	2.67	0.999841		

A.10 List of Important Symbols

A – area (cm^2)

C – capacitance (F)

C – concentration (cm^{-3})

D – diffusivity, experimentally measured ($\text{cm}^2\,\text{s}^{-1}$)

d – diffusivity of mobile species ($\text{cm}^2\,\text{s}^{-1}$)

E – energy (eV)

F – flux ($\text{cm}^{-2}\,\text{s}^{-1}$)

$F_{\min}$ – minimum feature size (nm)

I – current (A)

I – optical intensity ($\text{J}\,\text{cm}^{-2}\,\text{s}^{-1}$)

I – interstitial (Si interstitial in Si crystal, unless otherwise designated)

J – current density ($\text{A}\,\text{cm}^{-2}$)

K – relative permittivity or dielectric constant

m – mass (g)

n – electron concentration (cm^{-3})

n – index of refraction (—)

N – density (atoms cm^{-3})

P – pressure (atm) or (Torr)

Q – dose (cm^{-2})

R – resistance (Ω)

t – time (s)

T – temperature (K) (unless denoted by °C)

v – velocity ($\text{cm}\,\text{s}^{-1}$)

v – deposition or etch rate (velocity of surface) ($\text{cm}\,\text{s}^{-1}$)

V – voltage or potential difference (V)

V – volume (cm^3)

V – vacancy (in Si unless otherwise designated)

$\mathcal{E}$ – electric field ($\text{V}\,\text{cm}^{-1}$)

ε – permittivity ($\text{F}\,\text{cm}^{-1}$)

μ – carrier mobility ($\text{cm}^2\,\text{V}^{-1}\,\text{s}^{-1}$)

σ – conductivity ($\text{S}\,\text{m}^{-1}$) or (($\Omega\,\text{m})^{-1}$)

σ – stress (dyne cm^{-2}) or (Pa)

τ – lifetime (s)

ρ – electrical resistivity ($\Omega\,\text{cm}$)

ρ_S – sheet resistance (Ω/sq)

ρ_C – specific contact resistivity ($\Omega\,\text{cm}^2$)

A.11 List of Common Acronyms

AC	– alternating current
A/C	– amorphous/crystalline interface in Si
AFM	– atomic force microscopy
APCVD	– atmospheric-pressure chemical vapor deposition
ARC	– antireflection coating
ARDE	– aspect-ratio-dependent etching
Athena	– Silvaco Inc. process simulator
BCC	– body-centered cubic unit cell
BJT	– bipolar junction transistor
BOE	– buffered oxide etch (same as BHF, buffered HF)
BPSG	– borophosphosilicate glass
CA(R)	– chemically amplified (resist)
CVD	– chemical vapor deposition
CZ	– Czochralski crystal growth method
DC	– direct current
DI	– deionized, or a very pure, water
DOF	– depth of focus
DOP	– degree of planarization
DRAM	– dynamic random access memory
DQN	– diazonaphthoquinone, a resist material
DUV	– deep ultraviolet
e-beam	– electron-beam lithography
EGS	– electronic-grade silicon
EMP	– electron microprobe
EOR	– end-of-range defects
erf	– error function
erfc	– complementary error function
EUV	– extreme ultraviolet
FCC	– face-centered cubic unit cell
FET	– field effect transistor
FZ	– float zone crystal growth method
HDP	– high-density plasma
HEPA	– high-efficiency particulate air (filter)
HF	– high-frequency
HMDS	– hexamethyldisilane, an adhesion promoter
IC	– integrated circuit
ICP	– inductively coupled plasma
IMD	– inter-metal dielectric

IMP	– ionized metal deposition (same as IPVD)
INSOL	– insoluble portion of polymer base in chemically amplified resist
ITRS	– International Technology Roadmap for Semiconductors
LDD	– lightly doped drain
LF	– low-frequency
LOCOS	– local oxidation of silicon
LPCVD	– low-pressure chemical vapor deposition
LSS	– implant range theory developed by Lindhard, Scharff and Schiøtt
MBE	– molecular beam epitaxy
MGS	– metallurgical grade silicon
MOS	– metal-oxide–semiconductor
MTF	– modulation transfer function
MTTF	– mean time to failure
NMOS	– N-channel MOS transistor
NPN	– bipolar transistor with N-type emitter and collector
NTD	– neutron transmutation doping
NTRS	– National Technology Roadmap for Semiconductors
OISF	– oxidation-induced stacking fault
OPC	– optical proximity correction
OPD	– optical path difference
PAC	– photoactive compound
PAG	– photo-acid generator
PEB	– post-exposure bake
PECVD	– plasma-enhanced chemical vapor deposition
PMOS	– P-channel MOS transistor
PSG	– phosphosilicate glass
PSM	– phase shift mask
PVD	– physical vapor deposition (e.g., evaporation, sputter deposition)
RCA	– a wafer cleaning process
RF	– radio frequency, usually 13.56 MHz
RIE	– reactive ion etching
RTA	– rapid thermal annealing
RTO	– rapid thermal oxidation
SC-1, SC-2	– wafer cleaning solutions, part of RCA cleaning process
SEM	– scanning electron microscopy
SIA	– Semiconductor Industry Association
SIMOX	– process for SOI, named for "separation by implanted oxygen"
SIMS	– secondary ion mass spectrometry
SMIF	– standard mechanical interface (box)
SOG	– spin-on-glass
SOI	– silicon on insulator
SOL	– soluble portion of polymer base in CA resist
SRH	– Shockley–Read–Hall recombination

STM   – scanning tunneling microscope
SUPREM   – Stanford University Process Engineering Models
TCAD   – technology computer-aided design
TED   – transient enhanced diffusion
TEM   – transmission electron microscope
TEOS   – tetraethyl orthosilicate, tetraethoxysilane, $Si(OC_2H_5)_4$
UV   – ultraviolet light, light with a wavelength of about 0.2–0.4 μm
VLSI   – very-large-scale integration, usually $>10^5$ components

Index

Numbers in *italics* indicate index terms that occur in figures and/or in tables